INTRODUCTION TO HEAT TRANSFER

INTRODUCTION TO HEAT TRANSFER

Frank P. Incropera
David P. DeWitt

School of Mechanical Engineering
Purdue University

John Wiley & Sons

New York · Chichester · Brisbane · Toronto · Singapore

Library of Congress Cataloging in Publication Data:

Incropera, Frank P.

Introduction to heat transfer.

Includes bibliographies and index.
1. Heat–Transmission. I. DeWitt, David P., 1934–
II. Title.
QC320.I46 1985 536′.2 84-19691
ISBN 0-471-80126-7

Printed in the United States of America

10 9 8 7 6 5 4 3 2 1

PREFACE

In preparing to revise this text, we decided to solicit feedback from colleagues who have used or were otherwise familiar with the first edition. Our solicitation was formalized by the publisher, who sent questionnaires to 119 faculty and received 82 responses. This questionnaire addressed key features of the first edition. Matters of particular interest included:

- The extent to which exact solutions, numerical methods, and computer usage should be considered in the treatment of conduction.
- The extent to which boundary layer solutions should be developed.
- Whether the treatments of heat transfer with phase change and heat exchangers were adequate.
- Whether the treatment of radiation was too detailed.
- Whether the treatments of heat and mass transfer should be integrated.

Many of the responses revealed considerable insights into (and convictions concerning) heat transfer pedagogy. Collectively, they provided an invaluable resource for planning the second edition.

Although most respondents felt that the previous edition's coverage of conduction analysis was satisfactory for a first course, 18 percent felt that more could be done to illustrate exact methods of solving multidimensional and transient forms of the heat equation. Fifty-four percent of the respondents felt that the treatment of numerical methods could be improved, and 48 percent did not object to inclusion of stock codes in the text to facilitate computer usage. However, many of the 52 percent who did object to the use of such codes voiced their opposition strongly. One group felt that stock codes fostered a superficial understanding of heat transfer, while another group held that students should (and can) write their own codes.

Although most respondents thought that the treatment of boundary layer behavior was adequate (or excessive), 18 percent felt that more should be done to obtain boundary layer solutions. Small numbers (less than 15 percent) suggested that the treatment of boiling and condensation be expanded and that compact heat exchangers be included in the discussion of heat transfer equipment. Although 71 percent of the respondents were pleased with the material on radiation, many of the others felt that the material on surface spectral and directional effects was too detailed and called for simplification. A slight majority (56 percent) favored an integrated treatment of convection heat and mass transfer, but much of the minority opinion that opposed the inclusion of mass transfer was strongly worded.

All of this input stimulated change; however, it did not cause us to alter our original objectives. In fact, these objectives were reaffirmed. Simply stated, they are:

1. To convey an appreciation for the principles of heat transfer.
2. To establish the relationship of these principles to thermal system behavior and to develop methodologies for predicting this behavior.

Implicit in the first objective is our belief that students should be comfortable with the physical mechanisms and conditions of heat transfer. What, for example, are the mechanisms associated with transfer by conduction, convection, and radiation? What is thermal energy storage? Thermal energy generation? A buoyancy force? A blackbody? When is heat transfer one dimensional or time invariant? Such concepts are introduced throughout the text, and comprehension is essential if the student is to be able to identify heat transfer modes and conditions pertaining to a physical system.

Implicit in the second objective is our belief that students should learn to deal rationally with problems involving thermal system design and performance. They should be able to discern relevant transport processes and simplifying assumptions, identify key dependent and independent variables, and choose or develop appropriate expressions for evaluating the dependent variables. In short, students must be able to effect the kind of engineering analysis that, although not always exact, often provides useful information concerning the behavior of a particular system or process.

Preparation of this edition has been strongly motivated by the foregoing objectives, as well as by the survey of opinion on the first edition. The responses to this survey have led us to make many small changes that have improved the book's overall quality. We have also made larger changes that have enhanced the role of exact solutions in the heat and boundary layer equations, improved the discussion of numerical solution methods, expanded the treatment of heat transfer with phase change (to include more recent results), and expanded the treatment of heat exchangers (to include compact geometries). *Perhaps the most significant change has been the preparation of two versions of the text, one with and a second without mass transfer.*

To enhance the student's appreciation of exact methods of heat transfer analysis, we have illustrated use of the separation of variables method for two-dimensional conduction in Chapter 4, presented (without derivation) series solutions for the three Heisler chart geometries in Chapter 5, and illustrated use of the similarity method of forced-convection boundary layer analysis in Chapter 7. In all three cases, our intent is to have students recognize the existence of such solutions, the form which they take, and their advantages and limitations. In the third case our aim is to have students appreciate the manner in which convection coefficients may be determined theoretically. However, it is not our objective to foster competence in the application of these methods. Although the equations of heat transfer provide fertile material for enhancing student skills in applied mathematics, we do not believe that significant effort should be devoted to this purpose in a first course in heat transfer.

We have improved our treatment of numerical methods in Chapters 4 and 5. We eliminated the relaxation method and placed more emphasis on other iterative methods. We have also increased our emphasis on implicit methods, while expanding our discussion of stability criteria associated with explicit methods. We have not, however, included any stock computer programs or subroutines that could be used to effect these solutions, since such software is accessible in most computer libraries and/or is easily written by the student.

Two versions of the book have been prepared. One version, *Fundamentals of Heat and Mass Transfer*, is the second edition of the previously entitled *Fundamentals of Heat Transfer*. Although it is concerned primarily with heat transfer, it includes mass transfer effects *to whatever extent they can be inferred by analogy to heat transfer*. The manner in which this is done varies with the mass transfer mode. Because engineering applications of mass transfer by diffusion are limited, all topics related to this mode are treated separately in a single chapter at the end of the text. Therefore, any instructor wishing to ignore the subject may readily do so. In contrast, because engineering applications of convection mass transfer are extensive and because the process can make a significant contribution to total energy exchange at a vapor–liquid interface, its treatment is integrated with that of convection heat transfer. This integration permits a *unified* treatment of heat, mass, and momentum transfer in the study of boundary layer behavior and provides excellent opportunities for developing important similarity concepts.

The second version of the book, *Introduction to Heat Transfer*, is intended for users who do not wish to consider mass transfer. It is simply the first version, without mass transfer.

Chapters 1 through 5 are equivalent in both versions. Chapter 1 develops the physical basis of heat transfer by conduction, convection, and radiation and emphasizes the important role played by the requirement of energy conservation. Through carefully selected examples and problems, the student is encouraged to develop the facility to identify the relevant heat transfer processes for a variety of practical situations. Chapter 2 provides a detailed introduction to the conduction process. The nature and origin of Fourier's law are discussed, and general forms of the conduction heat equation are developed. The physical significance of each of the terms in the equation is emphasized, along with the nature of the simplifications that often may be made. Chapters 3, 4, and 5 are structured along traditional lines, with methods of solution developed for one-dimensional, two-dimensional, and transient conduction problems.

Chapters 6, 7, and 8 of *Fundamentals of Heat and Mass Transfer* provide an integrated treatment of heat and species transfer by forced convection. In Chapter 6 the concepts of the velocity, thermal, and concentration boundary layers are developed, and the related conservation equations are derived. The physical significance of the various terms in the equations and their relation to the wall shear stress and to convection heat and mass transfer are emphasized. Similarity parameters relevant to convection transfer are obtained by non-dimensionalizing the conservation equations, and the important role that they play in the generalization of heat and mass transfer results is developed. The dimensionless forms of the conservation equations are also used to infer important boundary layer analogies.

In many respects Chapter 6 is a cornerstone to the treatment of convection transfer. It sets forth important physical principles before providing the student with the tools to perform engineering calculations. In Chapters 7 and 8 these tools are presented for external and internal flows, respectively. Chapters 6, 7, and 8 of *Introduction to Heat Transfer* are organized in precisely the same manner, except that the mass transfer content has been deleted.

Chapters 9 (free convection), 10 (boiling and condensation), 11 (heat exchangers), 12 (surface radiation), and 13 (enclosure radiation exchange) are equivalent in both versions of the book. Mass transfer effects have been deleted from Chapter 14 (multimode heat transfer) in *Introduction to Heat Transfer*. In both versions, however, the material of Chapter 14 can contribute greatly to developing student confidence in using the tools of heat transfer analysis, and the instructor is urged to include it in the course requirements. Examples related to industrial and environmental applications illustrate the manner in which previously developed tools may be systematically integrated for effective problem solving. Emphasis is placed on engineering problems for which the student must establish objectives, make appropriate assumptions, and develop solutions that best respond to the objectives. Chapter 15 (mass diffusion) appears only in *Fundamentals of Heat and Mass Transfer*.

A significant number of respondents expressed concern about the coverage of surface radiation in Chapter 12 and noted that this material is difficult for students to assimilate. We are in sympathy with this concern and are cognizant of our own students' difficulties in grasping the subject. However, Chapter 12 was written with the conviction that nature is spectral and directional and that too many important engineering applications involve nondiffuse, nongray behavior. Hence, in an effort to systematically deal with directional and spectral effects, we have had to include details that require careful reading by even the best of students. We attempt to help our students by pointing out that their difficulties may be verbal, rather than physical or mathematical, in nature. We urge them to carefully consider each new term and to fully digest the contents of Table 12.3. Because we devote almost four weeks to radiation (Chapters 12 and

13), the conscientious student has time to achieve a good level of understanding and to develop confidence in using the material.

Although we have opted to retain our approach to radiation, we recognize that some instructors may be unable to commit sufficient time for satisfactory coverage of the material. In such cases we recommend that directional effects be de-emphasized. This may be done by having the students read Sections 12.2 and 12.7 for background and not holding them responsible for application of the contents. In other sections, expressions involving hemispherical (total and spectral) properties could be emphasized, and those involving directional properties could be treated lightly. Problems that deal exclusively with spectral, total, and hemispherical properties could be assigned. To paraphrase several of the respondents who treat the subject in this manner, "Although coverage is limited, students are at least aware of the importance of other effects and know where to go if they had to deal with them."

Other key features of the first edition have been retained and expanded. For example, we continue to stress the importance of conservation principles, particularly the first law of thermodynamics, in heat transfer analysis. General formulations of the energy conservation requirement are provided early in the text and are used as a basis for systematically treating all subsequent applications, whether to a differential or to a macroscopic control volume. The importance of the first law to heat transfer analysis is firmly established through numerous examples and problems.

We have also expanded our emphasis on the aspect of heat transfer that seems to hold the greatest appeal to students, namely, the large number of applications. Many of these applications deal with familiar topics such as energy conversion and conservation, space heating and cooling, propulsion, material and chemical processing, and thermal protection. To whatever extent possible, the relationship of heat transfer to the *behavior* of related *thermal systems* is stressed. Reference to applications is interwoven with text and example material, and many assigned problems are couched in terms of actual thermal systems. Such problems reinforce the relationship of the course to the *design process* by having the student make decisions concerning the configuration and/or operation of the system.

A goal of all engineering courses is to condition the student to perform engineering analyses that provide useful information concerning system design or performance. We have attempted to foster this goal by advocating a systematic approach to problem solving. The approach involves delineating all relevant processes on a schematic diagram of the physical system, making appropriate assumptions, identifying relevant variables, and introducing appropriate rate and conservation equations to predict thermal system behavior. We feel that students adopting this methodology are better able to apply heat transfer to industrial problems.

This text has evolved from many years of teaching heat transfer and from experiences gained through use of the first edition. In its preparation we have tried to remain conscious of student needs and difficulties, to be true to our objectives, and to incorporate many of the constructive suggestions made by our colleagues. We thank all those who participated in the survey and/or took the time to personally communicate their views. We would be remiss if we did not offer special thanks to Professor J. T. Pearson, who has been a constant source of suggestions and encouragement.

West Lafayette, Indiana

Frank P. Incropera
David P. DeWitt

CONTENTS

SYMBOLS

A	area, m^2
A_c	cross-sectional area, m^2
A_{ff}	free-flow area in compact heat exchanger core (minimum cross-sectional area available for flow through the core), m^2
A_{fr}	heat exchanger frontal area, m^2
A_p	area of prime (unfinned) surface, m^2
A_s	surface area, m^2
a	acceleration, m/s^2
Bi	Biot number
Bo	Bond number
C	heat capacity rate, W/K
C_D	drag coefficient
C_f	friction coefficient
C_t	thermal capacitance, J/kg
c	specific heat, J/kg · K; speed of light, m/s
c_p	specific heat at constant pressure, J/kg · K
c_v	specific heat at constant volume, J/kg · K
D	diameter, m
D_h	hydraulic diameter, m
E	thermal (sensible) internal energy, J; electric potential, V; emissive power, W/m^2
Ec	Eckert number
$\dot{E}_g$	rate of energy generation, W
$\dot{E}_{in}$	rate of energy transfer into a control volume, W
$\dot{E}_{out}$	rate of energy transfer out of control volume, W
$\dot{E}_{st}$	rate of increase of energy stored within a control volume, W
e	thermal internal energy per unit mass, J/kg; surface roughness, m
F	force, N; heat exchanger correction factor; fraction of blackbody radiation in a wavelength band; view factor
Fo	Fourier number
f	friction factor; similarity variable
G	irradiation, W/m^2; mass velocity, $kg/s \cdot m^2$

Gr	Grashof number
Gz	Graetz number
g	gravitational acceleration, m/s^2
g_c	gravitational constant, 1 kg · m/N · s^2 or 32.17 ft · lb$_m$/lb$_f$ · s^2
h	convection heat transfer coefficient, W/m^2 · K; Planck's constant
h_{fg}	latent heat of vaporization, J/kg
h_{rad}	radiation heat transfer coefficient, W/m^2 · K
I	electric current, A; radiation intensity, W/m^2 · sr
i	electric current density, A/m^2; enthalpy per unit mass, J/kg
J	radiosity, W/m^2
Ja	Jakob number
j_H	Colburn j factor for heat transfer
k	thermal conductivity, W/m · K; Boltzmann's constant
L	characteristic length, m
M	mass, kg; number of heat transfer lanes in a flux plot; reciprocal of the Fourier number for finite-difference solutions
$\mathcal{M}$	molecular weight, kg/kmol
m	mass, kg
$\dot{m}$	mass flow rate, kg/s
N	number of temperature increments in a flux plot; total number of tubes in a tube bank; number of surfaces in an enclosure
N_L, N_T	number of tubes in longitudinal and transverse directions
Nu	Nusselt number
NTU	number of transfer units
P	perimeter, m; general fluid property designation
P_L, P_T	dimensionless longitudinal and transverse pitch of a tube bank
Pe	Peclet number ($RePr$)
Pr	Prandtl number
p	pressure, N/m^2
Q	heat transfer, J
q	heat transfer rate, W
$\dot{q}$	rate of energy generation per unit volume, W/m^3
q'	heat transfer rate per unit length, W/m
q''	heat flux, W/m^2
R	cylinder radius, m
$\mathcal{R}$	universal gas constant
Ra	Rayleigh number
Re	Reynolds number
$R_{t,c}$	thermal contact resistance, K/W
R_e	electric resistance, Ω
R_f	fouling factor, m^2 · K/W
$R_{m,n}$	residual for the m,n nodal point
R_t	thermal resistance, K/W
r_o	cylinder or sphere radius, m
r, ϕ, z	cylindrical coordinates
r, θ, ϕ	spherical coordinates
S	shape factor for two-dimensional conduction, m
S_c	solar constant
S_D, S_L, S_T	diagonal, longitudinal, and transverse pitch of a tube bank, m

St	Stanton number
T	temperature, K
t	time, s
U	overall heat transfer coefficient, $W/m^2 \cdot K$
u, v, w	mass average fluid velocity components, m/s
V	volume, m^3; fluid velocity, m/s
v	specific volume, m^3/kg
$\dot{W}$	rate at which work is performed, W
We	Weber number
X, Y, Z	components of the body force per unit volume, N/m^3
x, y, z	rectangular coordinates, m
x_c	critical location for transition to turbulence, m
$x_{fd,h}$	hydrodynamic entry length, m
$x_{fd,t}$	thermal entry length, m

Greek Letters

α	thermal diffusivity, m^2/s; heat exchanger surface area per unit volume, m^2/m^3; absorptivity
β	volumetric thermal expansion coefficient, K^{-1}
Γ	mass flow rate per unit width in film condensation, $kg/s \cdot m$
δ	hydrodynamic boundary layer thickness, m
δ_t	thermal boundary layer thickness, m
ε	emissivity; porosity of a packed bed; heat exchanger effectiveness
ε_f	fin effectiveness
ε_H	turbulent diffusivity for heat transfer, m^2/s
ε_M	turbulent diffusivity for momentum transfer, m^2/s
η	similarity variable
η_f	fin efficiency
η_o	fin temperature effectiveness
θ	zenith angle, rad; temperature difference, K
κ	absorption coefficient, m^{-1}
λ	wavelength, μm
μ	viscosity, $kg/s \cdot m$
ν	kinematic viscosity, m^2/s; frequency of radiation, s^{-1}
ρ	mass density, kg/m^3; reflectivity
σ	Stefan-Boltzmann constant; electrical conductivity, $1/\Omega \cdot m$; normal viscous stress, N/m^2; surface tension, N/m; ratio of heat exchanger minimum cross-sectional area to frontal area
Φ	viscous dissipation function, s^{-2}
ϕ	azimuthal angle, rad
ψ	stream function, m^2/s
τ	shear stress, N/m^2; transmissivity
ω	solid angle, sr

Subscripts

abs	absorbed
adv	advection
am	arithmetic mean

b	base of an extended surface; blackbody
c	cross-sectional; cold fluid
cr	critical insulation thickness
cond	conduction
conv	convection
CF	counterflow
D	diameter; drag
e	excess; emission
evap	evaporation
f	fluid properties; fin conditions; saturated liquid conditions
fd	fully developed conditions
g	saturated vapor conditions
H	heat transfer conditions
h	hydrodynamic; hot fluid
i	inner surface of an annulus; initial condition; tube inlet condition; incident radiation
L	based on characteristic length
l	saturated liquid conditions
lm	log mean condition
M	momentum transfer condition
m	mean value over a tube cross section
max	maximum fluid velocity
o	center or midplane condition; tube outlet condition; outer
R	reradiating surface
r, ref	reflected radiation
rad	radiation
S	solar conditions
s	surface conditions; solid properties
sat	saturated conditions
sky	sky conditions
sur	surroundings
t	thermal
tr	transmitted
v	saturated vapor conditions
x	local conditions on a surface
λ	spectral
∞	freestream conditions

Superscripts

$'$	fluctuating quantity
$*$	dimensionless quantity

Overbar

$\bar{}$	surface average conditions; time mean

One
INTRODUCTION

From the study of thermodynamics, you have learned that energy can be transferred by interactions of a system with its surroundings. These interactions are called work and heat. However, thermodynamics deals with the end states of the process during which an interaction occurs and provides no information concerning the nature of the interaction or the time rate at which it occurs. The objective of this text is to extend thermodynamic analysis through study of the *modes* of heat transfer and through development of relations to calculate heat transfer *rates*. In this chapter we lay the foundation for much of the material treated in the text. We do so by raising several questions. *What is heat transfer? How is heat transferred? Why is it important to study it?* In answering these questions, we will begin to appreciate the physical mechanisms that underlie heat transfer processes and the relevance of these processes to our industrial and environmental problems.

1.1 WHAT AND HOW?

A simple, yet general, definition provides sufficient response to the question: What is heat transfer?

Heat transfer (or heat) is energy in transit due to a temperature difference.

Whenever there exists a temperature difference in a medium or between media, heat transfer must occur.

As shown in Figure 1.1, we refer to different types of heat transfer processes as *modes*. When a temperature gradient exists in a stationary medium, which may be a solid or a fluid, we use the term *conduction* to refer to the heat transfer that will occur across the medium. In contrast, the term *convection* refers to heat transfer that will occur between a surface and a moving fluid when they are at different temperatures. The third mode of heat transfer is termed *thermal radiation*. All surfaces of finite temperature emit energy in the form of

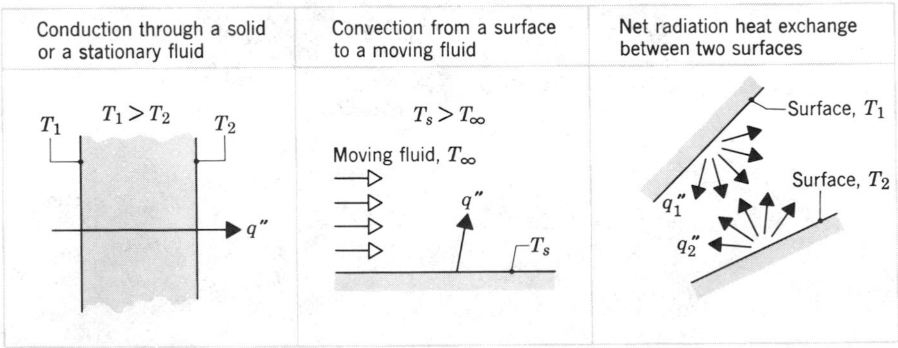

Figure 1.1 Conduction, convection, and radiation heat transfer modes.

electromagnetic waves. Hence, in the absence of an intervening medium, there is net heat transfer by radiation between two surfaces at different temperatures.

1.2 PHYSICAL ORIGINS AND RATE EQUATIONS

As engineers it is important that we understand the *physical mechanisms* that underlie the heat transfer modes and that we be able to use the rate equations that quantify the amount of energy being transferred per unit time.

1.2.1 Conduction

At mention of the word "conduction," we should immediately conjure up concepts of *atomic* and *molecular activity*, for it is processes at these levels that sustain this mode of heat transfer. Conduction may be viewed as the transfer of energy from the more energetic to the less energetic particles of a substance due to interactions between the particles.

The physical mechanism of conduction is most easily explained by considering a gas and using ideas familiar from your thermodynamics background. Consider a gas in which there exists a temperature gradient and assume that there is *no* bulk motion. The gas may occupy the space between two surfaces that are maintained at different temperatures, as shown in Figure 1.2. We associate the temperature at any point with the energy of the gas molecules in the vicinity of the point. This energy is related to the random translational motion, as well as to the internal rotational and vibrational motions, of the molecules. Moreover, higher temperatures are associated with higher molecular energies, and when neighboring molecules collide, as they are constantly doing, a transfer of energy from the more energetic to the less energetic molecules must occur. In the presence of a temperature gradient, energy transfer by conduction must then occur in the direction of decreasing temperature. This transfer is

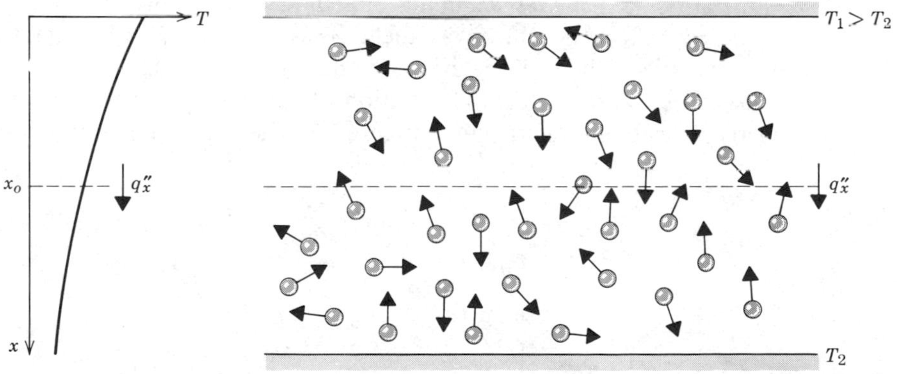

Figure 1.2 Association of conduction heat transfer with diffusion of energy due to molecular activity.

evident from Figure 1.2. The hypothetical plane at x_o is constantly being crossed by molecules from above and below due to their *random* motion. However, molecules from above are associated with a larger temperature than those from below, in which case there must be a *net* transfer of energy in the positive x direction. We may speak of the net transfer of energy by random molecular motion as a *diffusion* of energy.

The situation is much the same in liquids, although the molecules are more closely spaced and the molecular interactions are stronger and more frequent. Similarly, in a solid, conduction may be attributed to atomic activity in the form of lattice vibrations. The modern view is to ascribe the energy transfer to *lattice waves* induced by atomic motion. In a nonconductor, the energy transfer is exclusively via these lattice waves; in a conductor it is also due to the translational motion of the free electrons. We treat the important properties associated with conduction phenomena in Chapter 2 and in Appendix A.

Examples of conduction heat transfer are legion. The exposed end of a metal spoon suddenly immersed in a cup of hot coffee will eventually be warmed due to the conduction of energy through the spoon. On a winter day there is significant energy loss from a heated room to the outside air. This loss is principally due to conduction heat transfer through the wall that separates the room air from the outside air.

It is possible to quantify heat transfer processes in terms of appropriate *rate equations.* These equations may be used to compute the amount of energy being transferred per unit time. For heat conduction, the rate equation is known as *Fourier's law.* For the one-dimensional plane wall shown in Figure 1.3, having a temperature distribution $T(x)$, the rate equation is expressed as

$$q''_x = -k\frac{dT}{dx} \tag{1.1}$$

The *heat flux* q''_x (W/m^2) is the heat transfer rate in the x direction *per* unit area perpendicular to the direction of transfer, and it is proportional to the temperature gradient, dT/dx, in this direction. The proportionality constant k is a *transport* property known as the *thermal conductivity* (W/m·K) and is a characteristic of the wall material. The minus sign is a consequence of the fact that heat is transferred in the direction of decreasing temperature. Under the steady-state conditions shown in Figure 1.3, where the temperature distribution

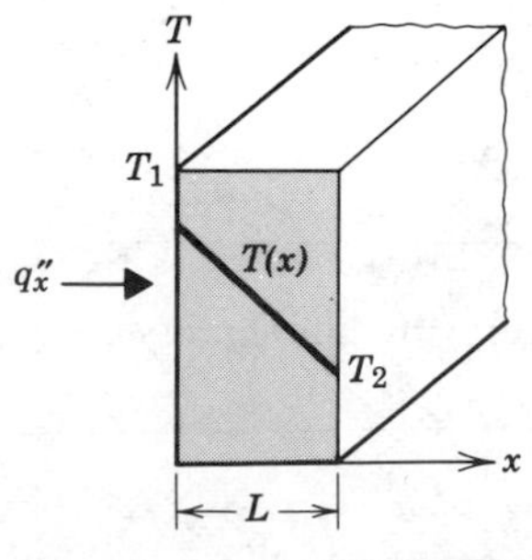

Figure 1.3 One-dimensional heat transfer by conduction (diffusion of energy).

is linear, the temperature gradient may be expressed as

$$\frac{dT}{dx} = \frac{T_2 - T_1}{L}$$

and the heat flux is then

$$q''_x = -k\frac{T_2 - T_1}{L}$$

or

$$q''_x = k\frac{T_1 - T_2}{L} = k\frac{\Delta T}{L} \tag{1.2}$$

Note that this equation provides a *heat flux*, that is, the rate of heat transfer per *unit area*. The *heat rate* by conduction, q_x (W), through a plane wall of area A is then the product of the flux and the area, $q_x = q''_x \cdot A$.

EXAMPLE 1.1

The wall of an industrial furnace is constructed from 0.15 m thick fireclay brick having a thermal conductivity of 1.7 W/m · K. Measurements made during steady-state operation reveal temperatures of 1400 and 1150 K at the inner and outer surfaces, respectively. What is the rate of heat loss through a wall which is 0.5 m by 3 m on a side?

SOLUTION

Known:

Steady-state conditions with prescribed wall thickness, area, thermal conductivity, and surface temperatures.

Find: Wall heat loss.

Schematic:

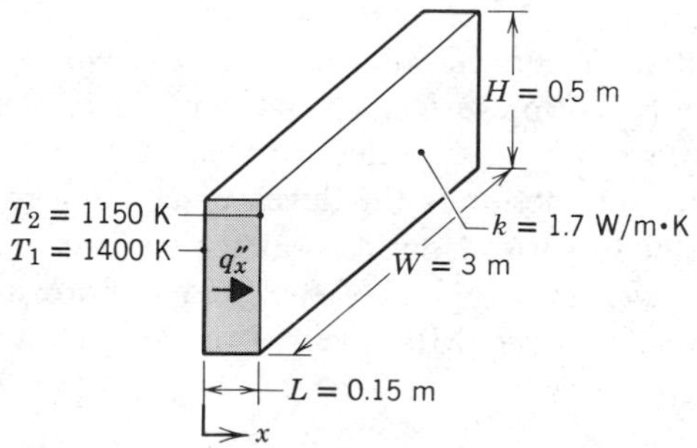

Assumptions:

1. Steady-state conditions.
2. One-dimensional conduction through the wall.
3. Constant properties.

Analysis:

Since heat transfer through the wall is by conduction, the heat flux may be determined from Fourier's law. Using Equation 1.2

$$q_x'' = k\frac{\Delta T}{L} = 1.7\ \text{W/m}\cdot\text{K} \times \frac{250\ \text{K}}{0.15\ \text{m}} = 2833\ \text{W/m}^2$$

The heat flux represents the rate of heat transfer through a section of unit area. The wall heat loss is then

$$q_x = (HW)q_x'' = (0.5\ \text{m} \times 3.0\ \text{m})\ 2833\ \text{W/m}^2 = 4250\ \text{W} \qquad \triangleleft$$

Comments:

1. Note direction of heat flow.
2. Note distinction between heat flux and heat rate.

1.2.2 Convection

The convection heat transfer *mode* is comprised of *two mechanisms.* In addition to energy transfer due to *random molecular motion* (diffusion), there is also energy being transferred by the *bulk*, or *macroscopic*, *motion* of the fluid. This fluid motion is associated with the fact that, at any instant, large numbers of molecules are moving collectively or as aggregates. Such motion, in the presence of a temperature gradient, will give rise to heat transfer. Because the molecules in the aggregate retain their random motion, the total heat transfer is then due to a superposition of energy transport by the random motion of the molecules and by the bulk motion of the fluid. It is customary to use the term *convection* when referring to this cumulative transport and the term *advection* when referring to transport due to bulk fluid motion.

We are especially interested in convection heat transfer, which occurs between a fluid in motion and a bounding surface when the two are at different temperatures. Consider fluid flow over the heated surface of Figure 1.4. A consequence of the fluid–surface interaction is the development of a region in the fluid through which the velocity varies from zero at the surface to a finite value u_∞ associated with the flow. This region of the fluid is known as the *hydrodynamic*, or *velocity*, *boundary layer*. Moreover, if the surface and flow

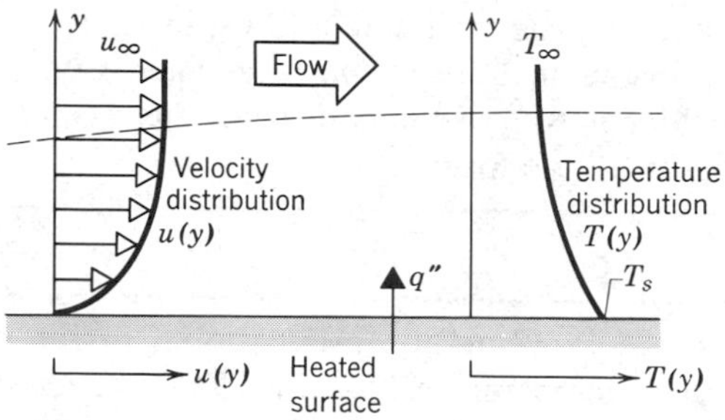

Figure 1.4 Boundary layer development in convection heat transfer.

temperatures differ, there will be a region of the fluid through which the temperature varies from T_s at $y = 0$ to T_∞ in the outer flow. This region, called the *thermal boundary layer*, may be smaller, larger, or the same size as that through which the velocity varies. In any case, if $T_s > T_\infty$, convection heat transfer will occur between the surface and the outer flow.

The convection heat transfer mode is sustained both by random molecular motion and by the bulk motion of the fluid within the boundary layer. The contribution due to random molecular motion (diffusion) generally dominates near the surface where the fluid velocity is low. In fact, at the interface between the surface and the fluid ($y = 0$), heat is transferred by this mechanism only. The contribution due to bulk fluid motion originates from the fact that the boundary layers *grow* as the flow progresses in the x direction. Appreciation of boundary layer phenomena is essential to understanding convection heat transfer. It is for this reason that the discipline of fluid mechanics plays a vital role in our later analysis of convection.

Convection heat transfer may be categorized according to the nature of the flow. We speak of *forced convection* when the flow is caused by some external means, such as by a fan, a pump, or atmospheric winds. In contrast, for *free* (or *natural*) *convection* the flow is induced by buoyancy forces in the fluid. These forces arise from density variations caused by temperature variations in the fluid. An example is the free convection heat transfer that occurs from a hot pavement to the atmosphere on a still day. Air that is in contact with the hot pavement has a lower density than that of the cooler air above the pavement. Hence, a circulation pattern exists in which the warm air moves up from the pavement and the cooler air moves downward. However, in the presence of atmospheric winds, heat transfer from the pavement to the air is likely to be dominated by forced convection, even though the free convection mode still exists.

We have described the convection heat transfer mode as energy transfer occurring within a fluid due to the combined effects of conduction and bulk fluid motion. In general, the energy that is being transferred is the *sensible*, or internal thermal, energy of the fluid. However, there are convection processes for which there is, in addition, *latent* heat exchange. This latent heat exchange is generally associated with a phase change between the liquid and vapor states of the fluid. Two special cases of interest in this text are *boiling* and *condensation.*

Regardless of the particular nature of the convection heat transfer mode, the appropriate rate equation is of the form

$$q'' = h(T_s - T_\infty) \tag{1.3}$$

where q'', the convective *heat flux* (W/m^2), is proportional to the difference between the surface and fluid temperatures, T_s and T_∞, respectively. This expression is known as *Newton's law of cooling*, and the proportionality constant h ($W/m^2 \cdot K$) is referred to as the *convection heat transfer coefficient*, the *film conductance*, or the *film coefficient*. It encompasses all the effects that influence the convection mode. It depends on conditions in the boundary layer, which are influenced by surface geometry, the nature of the fluid motion, and a number of the fluid thermodynamic and transport properties. Moreover, any study of convection ultimately reduces to a study of the means by which h may be determined. Although consideration of these means is deferred to Chapter 6, it is important to note that convection heat transfer will frequently appear as a boundary condition in the solution of conduction problems (Chapters 2 to 5). In the solution of such problems we presume h to be known, using typical values given in Table 1.1. Note the range of values associated with the various convection processes.

1.2.3 Radiation

Thermal radiation is energy emitted by matter that is at a finite temperature. Although we focus primarily on radiation from solid surfaces, emission may also occur from liquids and gases. Regardless of the form of matter, the emission may be attributed to changes in the electron configurations of the constituent atoms or molecules. The energy of the radiation field is transported by electromagnetic waves (or alternatively, photons). While the transfer of energy by conduction or

Table 1.1 Typical values of the convection heat transfer coefficient

PROCESS	h ($W/m^2 \cdot K$)
Free convection	5–25
Forced convection	
Gases	25–250
Liquids	50–20,000
Convection with phase change	
Boiling or condensation	2500–100,000

convection requires the presence of a material medium, radiation does not. In fact, radiation transfer occurs most efficiently in a vacuum.

The *maximum* flux (W/m²) at which radiation may be emitted from a surface is given by the *Stefan-Boltzmann law*

$$q'' = \sigma T_s^4 \tag{1.4}$$

where T_s is the *absolute* temperature (K) of the surface and σ is the *Stefan-Boltzmann constant* ($\sigma = 5.67 \times 10^{-8}\ \mathrm{W/m^2 \cdot K^4}$). Such a surface is called an ideal radiator or *blackbody*. The heat flux emitted by a real surface is less than that of the ideal radiator and is given by

$$q'' = \varepsilon \sigma T_s^4 \tag{1.5}$$

where ε is a radiative property of the surface called the *emissivity*. This property indicates how efficiently the surface emits compared to an ideal radiator.

Equation 1.5 determines the rate at which energy is *emitted* by a surface. Determination of the *net* rate at which *radiation* is *exchanged between surfaces* is generally a good deal more complicated. However, a special case that occurs very frequently in practice involves the net exchange between a small surface and a much larger surface that completely surrounds the smaller one (Figure 1.5). The surface and the surroundings are separated by a gas that has no effect on the radiation transfer. The net rate of radiation heat exchange between the surface and its *surroundings*, expressed per unit area of the surface, is

$$q'' = \frac{q}{A} = \varepsilon \sigma (T_s^4 - T_{sur}^4) \tag{1.6}$$

In this expression, A is the surface area and ε is its emissivity, while T_{sur} is the temperature of the surroundings. For this special case, the area and emissivity of the surroundings do not influence the net heat exchange rate. Additional restrictions on this equation are discussed in Chapter 13.

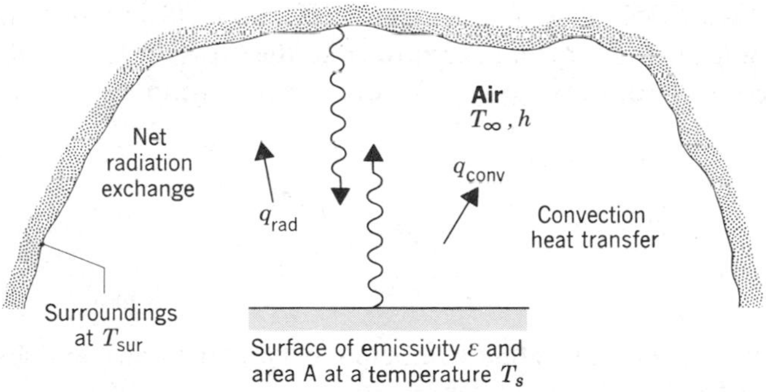

Figure 1.5 Radiation exchange between a surface and its surroundings.

There are many applications for which it is convenient to express the net radiation heat exchange in the form

$$q_{\text{rad}} = h_r A(T_s - T_{\text{sur}}) \tag{1.7}$$

where, from Equation 1.6, the *radiation heat transfer coefficient* h_r is

$$h_r \equiv \varepsilon\sigma(T_s + T_{\text{sur}})(T_s^2 + T_{\text{sur}}^2) \tag{1.8}$$

Here we have modeled the radiation mode in a manner similar to convection. In this sense we have *linearized* the radiation rate equation, making the heat rate proportional to a temperature difference rather than to the difference between two temperatures to the fourth power. Note, however, that h_r depends strongly on temperature, while the temperature dependence of the convection heat transfer coefficient h is generally weak.

The surface within the surroundings may also simultaneously transfer heat by convection to the adjoining gas (Figure 1.5). The total rate of heat transfer *from* the surface is then the sum of the heat rates due to the two modes. That is,

$$q = q_{\text{conv}} + q_{\text{rad}}$$

or,

$$q = hA(T_s - T_\infty) + \varepsilon A\sigma(T_s^4 - T_{\text{sur}}^4) \tag{1.9}$$

Note that the convection heat transfer rate q_{conv} is simply the product of the flux given by Equation 1.3 and the surface area.

EXAMPLE 1.2

An uninsulated steam pipe passes through a room in which the air and walls are at 25°C. The outside diameter of the pipe is 70 mm, and its surface temperature and emissivity are 200°C and 0.8, respectively. If the coefficient associated with free convection heat transfer from the surface to the air is 15 W/m$^2 \cdot$K, what is the rate of heat loss from the surface per unit length of pipe?

SOLUTION

Known:

Uninsulated pipe of prescribed diameter, emissivity, and surface temperature in a room with fixed wall and air temperatures.

Find: Pipe heat loss per unit length, q' (W/m).

Schematic:

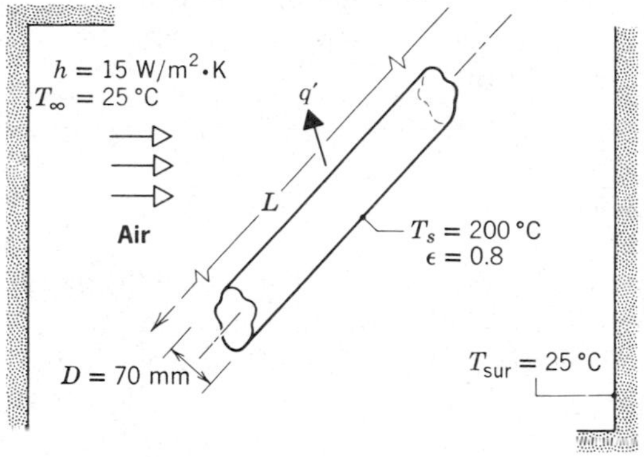

Assumptions:

1. Steady-state conditions exist.
2. Radiation exchange between the pipe and thc room is between a small surface enclosed within a much larger surface.

Analysis:

Heat loss from the pipe is by convection to the room air and by radiation exchange with the walls. Hence, from Equation 1.9, with $A = \pi DL$,

$$q = h(\pi DL)(T_s - T_\infty) + \varepsilon(\pi DL)\sigma(T_s^4 - T_{sur}^4)$$

The heat loss per unit length of pipe is then

$$q' = \frac{q}{L} = 15\ \text{W/m}^2 \cdot \text{K}\ (\pi \times 0.07\ \text{m})(200 - 25)^\circ\text{C}$$

$$+\,0.8(\pi \times 0.07\ \text{m})\ 5.67 \times 10^{-8}\ \text{W/m}^2 \cdot \text{K}^4\ (473^4 - 298^4)\ \text{K}^4$$

$$q' = 577\ \text{W/m} + 421\ \text{W/m} = 998\ \text{W/m} \qquad \triangleleft$$

Comments:

1. Note that temperature may be expressed in two ways (i.e., in °C or K units) when evaluating the temperature difference for a convection (or conduction) heat transfer rate. However, temperature must be expressed in kelvins (K) when evaluating a radiation transfer rate.
2. In this situation the radiation and convection heat transfer rates are comparable because T_s is large compared to T_{sur} and the coefficient associated with free convection is small. For more moderate values of T_s

and the larger values of h associated with forced convection, the effect of radiation may often be neglected. The radiation heat transfer coefficient may be computed from Equation 1.8, and for the conditions of this problem its value is $h_r = 11\ \text{W/m}^2 \cdot \text{K}$.

1.2.4 Relationship to Thermodynamics

At this point it is important to take note of the fundamental differences that exist between heat transfer and thermodynamics. Although thermodynamics is concerned with the heat interaction and the vital role it plays in the first and second laws, it considers neither the mechanisms that provide for heat exchange nor the methods that exist for computing the *rate* of heat exchange. Thermodynamics is concerned with *equilibrium* states of matter, where an equilibrium state necessarily precludes the existence of a temperature gradient. Although thermodynamics may be used to determine the amount of energy required in the form of heat for a system to pass from one equilibrium state to another, it does not acknowledge that *heat transfer is inherently a nonequilibrium process*. For heat transfer to occur, there must be a temperature gradient, hence thermodynamic nonequilibrium. The discipline of heat transfer therefore seeks to do what thermodynamics is inherently unable to do. It seeks to quantify the *rate* at which heat transfer occurs in terms of the degree of thermal nonequilibrium. This is done through the rate equations for the three modes, expressed by Equations 1.1, 1.3, and 1.6.

1.3 THE CONSERVATION OF ENERGY REQUIREMENT

The subjects of thermodynamics and heat transfer are highly complementary. For example, heat transfer is an extension of thermodynamics in that it considers the *rate* at which energy is transported. Moreover, in many heat transfer analyses the first law of thermodynamics (the *law of conservation of energy*) plays an important role. It is therefore useful to consider general statements for this law.

1.3.1 Conservation of Energy for a Control Volume

In our application of the conservation laws, we first need to identify the control volume, a fixed region of space bounded by a control surface through which energy and matter may pass. With respect to a control volume, a form of the energy conservation requirement that is most useful for heat transfer analyses may be stated as follows.

The rate at which thermal and mechanical energy enters a control volume minus the rate at which this energy leaves the control volume must equal the rate at which this energy is stored in the control volume.

If the inflow of energy exceeds the outflow, there will be an increase in the amount of energy stored (accumulated) in the control volume; if the converse is true, there will be a decrease in energy storage. If the inflow of energy equals the outflow, a *steady-state* condition must prevail in which there will be no change in the amount of energy stored in the control volume.

Consider applying energy conservation to the control volume shown in Figure 1.6. The first step is to identify the control surface by drawing a dashed line. The next step is to identify the energy terms. These include the rate of energy entering and leaving *through* the control surface, $\dot{E}_{in}$ and $\dot{E}_{out}$. Also, thermal energy may enter the control volume due to conversion from other energy forms. We refer to this process as *energy generation*, and the rate at which it occurs is designated as $\dot{E}_g$. The rate of change of energy stored within the control volume, dE_{st}/dt, is designated as $\dot{E}_{st}$. A general form of the energy conservation requirement may then be expressed on a *rate* basis as

$$\dot{E}_{in} + \dot{E}_g - \dot{E}_{out} = \dot{E}_{st} \tag{1.10a}$$

Equation 1.10a may be applied at any *instant of time*. An alternative form may be expressed for a *time interval*, Δt, and is obtained by integrating Equation 1.10a over time:

$$E_{in} + E_g - E_{out} = \Delta E_{st} \tag{1.10b}$$

In words this relation says that the amounts of energy inflow and generation act to increase the amount of energy stored within the control volume, whereas outflow acts to decrease the stored energy.

The inflow and outflow rate terms $\dot{E}_{in}$ and $\dot{E}_{out}$ are *surface phenomena*. That is, they are associated exclusively with processes occurring at the control surface, and the rate at which they occur is proportional to surface area. The most common situation will involve energy inflow and outflow due to heat transfer by the conduction, convection, and/or radiation modes. In a situation involving fluid flow across the control surface, $\dot{E}_{in}$ and $\dot{E}_{out}$ also include energy transported by the fluid into and out of the control volume. This energy may be composed of potential, kinetic, and thermal forms. However, for most of the heat transfer problems encountered in this text, the potential and kinetic energy forms are negligible. The inflow and outflow terms may also include work interactions.

The *thermal energy generation rate term* $\dot{E}_g$ is associated with the rate of

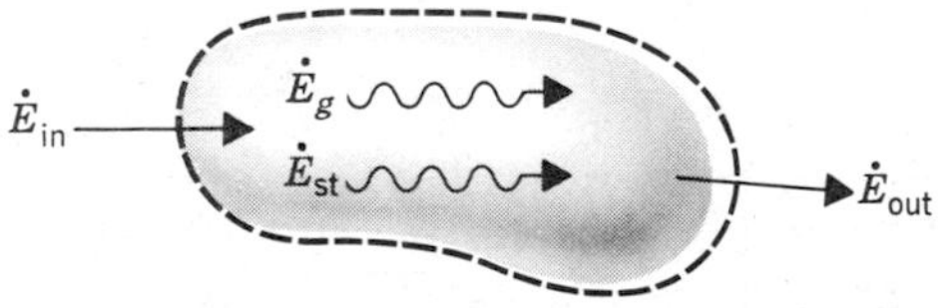

Figure 1.6 Conservation of energy for a control volume.

conversion from some other energy form (chemical, electrical, electromagnetic, or nuclear) to thermal energy. It is a *volumetric phenomenon.* That is, it occurs within the control volume, and its rate is proportional to the volume. For example, an exothermic chemical reaction may be occurring, converting chemical to thermal energy. The net effect is an increase in the thermal energy of the control volume. Another source of thermal energy is the conversion from electrical energy that occurs due to resistance heating when an electric current is passed through a conductor.

It is important not to confuse the physical process of *energy storage* with that of *energy generation.* Although energy generation may certainly contribute to energy storage, the two processes are fundamentally different. Energy storage is also a *volumetric phenomenon*, but it is simply associated with an increase ($\dot{E}_{st} > 0$) or decrease ($\dot{E}_{st} < 0$) in the energy of the matter occupying the control volume. Under steady-state conditions there is, of course, no change in energy storage ($\dot{E}_{st} = 0$).

EXAMPLE 1.3

A long conducting rod of diameter D and electrical resistance per unit length R'_e is initially in thermal equilibrium with the ambient air and its surroundings. This equilibrium is disturbed when an electrical current I is passed through the rod. Develop an equation that could be used to compute the variation of the rod temperature with time during passage of the current.

SOLUTION

Known:

Temperature of a rod of prescribed diameter and electrical resistance changes with time due to passage of an electrical current.

Find: Equation that governs temperature change with time for the rod.

Schematic:

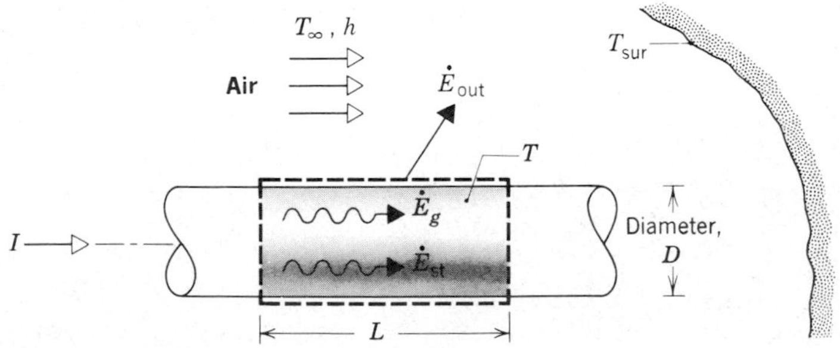

Assumptions:

1. At any time t the temperature of the rod is uniform.
2. Constant properties exist.
3. Radiation exchange between the outer surface of the rod and the surroundings is between a small surface and a large enclosure.

Analysis:

Applying Equation 1.10a to a control volume of length L about the rod, it follows that

$$\dot{E}_g - \dot{E}_{out} = \dot{E}_{st}$$

where energy generation is due to the electric resistance heating,

$$\dot{E}_g = I^2 R'_e L$$

Energy outflow is due to convection and net radiation from the surface, Equations 1.3 and 1.6, respectively,

$$\dot{E}_{out} = h(\pi DL)(T - T_\infty) + \varepsilon\sigma(\pi DL)(T^4 - T^4_{sur})$$

and the change in energy storage is due to the temperature change,

$$\dot{E}_{st} = \frac{d}{dt}(\rho VcT)$$

The term $\dot{E}_{st}$ is associated with the rate of change in the internal energy of the rod, where ρ and c are the mass density and the specific heat, respectively, of the rod material, and V is the volume of the rod, $V = (\pi D^2/4)L$. Substituting the rate equations into the energy balance, it follows that

$$I^2 R'_e L - h(\pi DL)(T - T_\infty) - \varepsilon\sigma(\pi DL)(T^4 - T^4_{sur}) = \rho c\left(\frac{\pi D^2}{4}\right)L\frac{dT}{dt}$$

Hence

$$\frac{dT}{dt} = \frac{I^2 R'_e - \pi Dh(T - T_\infty) - \pi D\varepsilon\sigma(T^4 - T^4_{sur})}{\rho c(\pi D^2/4)} \qquad \triangleleft$$

Comments:

The above equation could be solved for the time dependence of the rod temperature by integrating numerically. A steady-state condition would eventually be reached for which $dT/dt = 0$. The rod temperature is then determined by an algebraic equation of the form

$$\pi Dh(T - T_\infty) + \pi D\varepsilon\sigma(T^4 - T^4_{sur}) = I^2 R'_e$$

1.3.2 The Surface Energy Balance

We will frequently have occasion to apply the conservation of energy requirement at the surface of a medium. In this special case the control surface includes no mass or volume and appears as shown in Figure 1.7. Accordingly, the generation and storage terms of the conservation expression, Equation 1.10a, are no longer relevant and it is only necessary to deal with surface phenomena. For this case the conservation requirement then becomes

$$\dot{E}_{in} - \dot{E}_{out} = 0 \tag{1.11}$$

Even though thermal energy generation may be occurring in the medium, the process would not affect the energy balance at the control surface. This conservation requirement holds for both *steady-state* and *transient* conditions.

In Figure 1.7 three heat transfer terms are shown for the control surface. On a unit area basis they are conduction from the medium *to* the control surface (q''_{cond}), convection *from* the surface to a fluid (q''_{conv}), and net radiation exchange from the surface to the surroundings (q''_{rad}). The energy balance then takes the form

$$q''_{cond} - q''_{conv} - q''_{rad} = 0 \tag{1.12}$$

and we can express each of the terms according to the appropriate rate equations, Equations 1.2, 1.3, and 1.6.

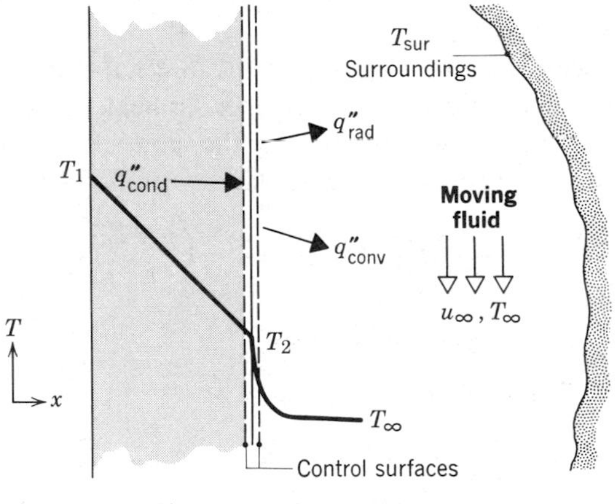

Figure 1.7 The energy balance for conservation of energy at the surface of a medium.

EXAMPLE 1.4

The hot combustion gases of a furnace are separated from the ambient air and its surroundings, which are at 25°C, by a brick wall 0.15 m thick. The brick has a thermal conductivity of 1.2 W/m·K and a surface emissivity of 0.8. Under steady-state conditions an outer surface temperature of 100°C is measured. Free convection heat transfer to the air adjoining this surface is characterized by a convection coefficient $h = 20\ \mathrm{W/m^2 \cdot K}$. What is the brick inner surface temperature?

SOLUTION

Known:

Outer surface temperature of a furnace wall of prescribed thickness, thermal conductivity, and emissivity.

Find: Wall inner surface temperature.

Schematic:

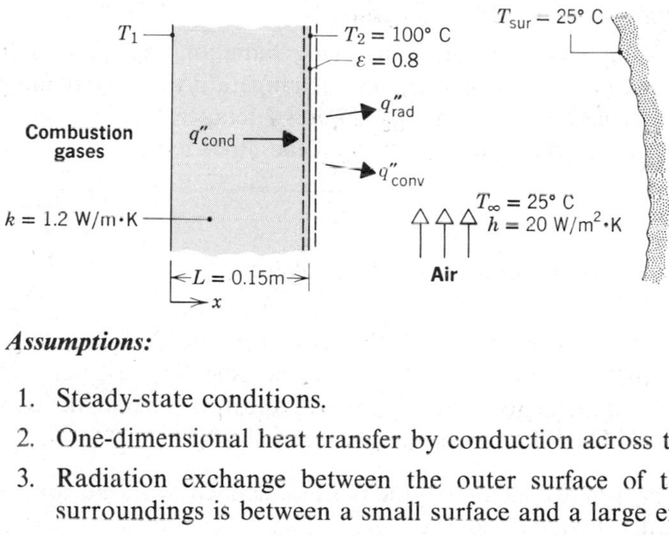

Assumptions:

1. Steady-state conditions.
2. One-dimensional heat transfer by conduction across the wall.
3. Radiation exchange between the outer surface of the wall and the surroundings is between a small surface and a large enclosure.

Analysis:

The inside surface temperature may be obtained by performing an energy balance at the outer surface. From Equation 1.11

$$\dot{E}_{in} - \dot{E}_{out} = 0$$

it follows that, on a unit area basis,

$$q''_{cond} - q''_{conv} - q''_{rad} = 0$$

or, rearranging and substituting from Equations 1.2, 1.3, and 1.6,

$$k\frac{T_1 - T_2}{L} = h(T_2 - T_\infty) + \varepsilon\sigma(T_2^4 - T_{sur}^4)$$

Therefore, substituting the appropriate numerical values

$$1.2\ \mathrm{W/m\cdot K}\,\frac{(T_1 - 373)\ \mathrm{K}}{0.15\ \mathrm{m}} = 20\ \mathrm{W/m^2\cdot K}\ (373 - 298)\ \mathrm{K}$$

$$+\,0.8(5.67 \times 10^{-8}\ \mathrm{W/m^2\cdot K^4})(373^4 - 298^4)\ \mathrm{K^4}$$

$$= 1500\ \mathrm{W/m^2} + 520\ \mathrm{W/m^2} = 2020\ \mathrm{W/m^2}$$

Solving for T_1,

$$T_1 = 373\ \mathrm{K} + \frac{0.15\ \mathrm{m}}{1.2\ \mathrm{W/m\cdot K}}\,(2020\ \mathrm{W/m^2}) = 625\ \mathrm{K} = 352°\mathrm{C} \qquad \triangleleft$$

Comments:

1. Note that the contribution of radiation to heat transfer from the outer surface is significant. The relative contribution would diminish, however, with increasing h and/or decreasing T_2.
2. When using energy balances involving radiation exchange and other modes, it is good practice to express all temperatures in kelvin units. This practice is *necessary* when the unknown temperature appears in the radiation term and in one or more of the other terms.

1.3.3 Application of the Conservation Laws: Methodology

In addition to being familiar with the transport rate equations described in Section 1.2, the heat transfer analyst must be able to work with the energy conservation requirements of Equations 1.10 and 1.11. The application of these balances is simplified if a few basic rules are followed.

1. The appropriate control volume must be defined, with the control surface represented by a dashed line.
2. The relevant energy processes should be identified. Each process should be shown on the control volume by an appropriately labeled arrow.
3. The conservation equation should then be written, and appropriate rate expressions should be substituted for the terms in the equation.

In connection with applying the energy conservation equation (1.10) note that this application may be in terms of a *finite* control volume or a *differential* (infinitesimal) control volume. In the first case an equation would be obtained that governs the gross behavior of the system. In contrast, for the second case a differential equation would be obtained that could be solved for conditions at each point in the system. Both types of control volume are used extensively in this text.

1.4 ANALYSIS OF HEAT TRANSFER PROBLEMS: METHODOLOGY

A major objective of this text is to prepare you to solve engineering problems that involve heat transfer processes. To this end numerous problems are provided at the end of each chapter. In working these problems you will gain a deeper appreciation for the fundamentals of the subject, and you will gain confidence in your ability to apply these fundamentals to the solution of engineering problems.

It is our conviction that there exists a preferred approach to problem solving. In contrast to a somewhat random, "plug-and-chug" approach, we advocate the use of a systematic procedure characterized by a prescribed format. We consistently employ the procedure in all the examples of this text, and we require our students to use it in their solution of problems. It consists of the following steps.

1. KNOWN: After carefully reading the problem, state briefly and concisely what is known about the problem. Do not repeat the problem statement.
2. FIND: State briefly and concisely what must be found.
3. SCHEMATIC: Draw a schematic of the physical system. If application of the conservation laws is anticipated, represent the required control surface by dashed lines on the schematic. Identify relevant heat transfer processes by appropriately labeled arrows on the schematic.
4. ASSUMPTIONS: List all pertinent simplifying assumptions.
5. PROPERTIES: Compile property values needed for subsequent calculations and identify the table from which they are obtained.
6. ANALYSIS: Begin your analysis by applying appropriate conservation laws, and introduce rate equations as needed. Develop the analysis as completely as possible before substituting numerical values. Perform the calculations needed to obtain the desired results.
7. COMMENTS: Discuss your results. Such a discussion may include a summary of key conclusions, an inference of trends, and a critique of the original assumptions.

The importance of following steps 1 through 4 should not be underestimated. These steps provide the opportunity to systematically think about a problem before effecting its solution.

1.5 RELEVANCE OF HEAT TRANSFER

In the vernacular of the time, heat transfer is indeed a relevant subject, not to mention an inherently fascinating part of the engineering sciences. We will devote much time to acquiring an understanding of heat transfer effects and to developing the skills needed to predict heat transfer rates. What is the value of this knowledge, and to what kinds of problems may it be applied?

Heat transfer phenomena play an important role in many industrial and environmental problems. As an example, consider the vital area of energy production and conversion. There is not a single application in this area that does not involve heat transfer effects in some way. In the generation of electrical power, whether it be through nuclear fission or fusion, the combustion of fossil fuels, magnetohydrodynamic processes, or the use of geothermal energy sources, there are numerous heat transfer problems that must be solved. These problems involve conduction, convection, and radiation processes and relate to the design of systems such as boilers, condensers, and turbines. One is often confronted with the need to maximize heat transfer rates and to maintain the integrity of materials in high-temperature environments. One is also confronted with the thermal pollution problem associated with the discharge of large amounts of waste heat from a power plant to the environment. Numerous heat transfer considerations relate to the design of cooling towers to alleviate the environmental problems associated with this discharge.

On a smaller scale there are many heat transfer problems related to the development of solar energy conversion systems for space heating and air conditioning, as well as for electric power production. Heat transfer processes also affect the performance of propulsion systems, such as the internal combustion and rocket engines. Heat transfer problems arise in the design of conventional space and water heating systems, in the design of incinerators and cryogenic storage equipment, in the cooling of electronic equipment, in the design of refrigeration and air conditioning systems, and in many other industrial problems. Heat transfer processes are also relevant to air and water pollution and strongly influence local and global climate.

1.6 UNITS AND DIMENSIONS

The physical quantities of heat transfer are specified in terms of *dimensions*, which are measured in terms of units. Four *basic* dimensions are required for the development of heat transfer; they are length (L), mass (M), time (t), and temperature (T). All other physical quantities of interest may be related to these four basic dimensions.

In the United States it has been customary to measure dimensions in terms of an *English system of units*, for which the *base units* are

DIMENSION		UNIT
Length (L)	→	foot (ft)
Mass (M)	→	pound mass (lb_m)
Time (t)	→	second (s)
Temerature (T)	→	degree Fahrenheit (°F)

The units required to specify other physical quantities may then be inferred from this group. For example, the dimension of force is related to mass through

Newton's second law of motion

$$F = \frac{1}{g_c} Ma \tag{1.13}$$

where the acceleration a has units of feet per square second and g_c is a proportionality constant. If this constant is arbitrarily set equal to unity and made *dimensionless*, the dimensions of force are $(F) = (M) \cdot (L)/(t)^2$ and the unit of force is

$$1 \text{ poundal} = 1 \text{ lb}_m \cdot \text{ft/s}^2$$

Alternatively, one could work with a system of basic dimensions that includes both mass and force. However, in this case the proportionality constant must have the dimensions $(M) \cdot (L)/(F) \cdot (t)^2$. Moreover, if one defines the pound force (lb_f) as a unit of force that will accelerate one pound mass by 32.17 ft/s^2, the proportionality constant must be of the form

$$g_c = 32.17 \text{ lb}_m \cdot \text{ft/lb}_f \cdot \text{s}^2$$

The units of work may be inferred from its definition as the product of a force times a distance, in which case the units are $\text{ft} \cdot \text{lb}_f$. The units of work and energy are, of course, equivalent, although it is customary to use the British thermal unit (Btu) as the unit of thermal energy. One British thermal unit will raise the temperature of 1 lb_m of water at 68°F by 1°F. It is equivalent to 778.16 $\text{ft} \cdot \text{lb}_f$, which is termed the *mechanical equivalent of heat*.

In recent years there has been a strong trend toward worldwide usage of a standard set of units. In 1960 the SI (Système International d'Unités) system of units was defined by the Eleventh General Conference on Weights and Measures and recommended as a worldwide standard. In response to this trend, the American Society of Mechanical Engineers (ASME) has required the use of SI units in all of its publications since July 1, 1974. For this reason and because it is operationally more convenient than the English system, the SI system is used for the calculations of this text. However, because for some time to come, engineers will also have to work with results expressed in the English system, you should be able to convert from one system to the other. For your convenience conversion factors are provided on the inside back cover of the text.

The SI *base* units required for this text are summarized in Table 1.2. With regard to these units note that 1 mol is the amount of substance that has as many atoms or molecules as there are atoms in 12 g of carbon-12 (^{12}C); this is the gram-mole (mol). Although the mole has been recommended as the unit quantity of matter for the SI system, it is more consistent to work with the kilogram-mole (kmol, kg-mole). One kmol is simply the amount of substance that has as many atoms or molecules as there are atoms in 12 kg of ^{12}C. So long as the use is consistent within a given problem, no difficulties arise in using either mol or kmol. The molecular weight of a substance is the mass associated with a mole or kilogram-mole. For oxygen, as an example, the molecular weight $\mathscr{M}$ is 16 g/mol or 16 kg/kmol.

Table 1.2 SI base and supplementary units

QUANTITY AND SYMBOL	UNIT AND SYMBOL
Length (L)	meter (m)
Mass (M)	kilogram (kg)
Concentration (C)	mole (mol)
Time (t)	second (s)
Electric current (I)	ampere (A)
Thermodynamic temperature (T)	kelvin (K)
Plane angle[a] (θ)	radian (rad)
Solid angle[a] (ω)	steradian (sr)

[a] Supplementary unit.

Although the SI unit of temperature is the kelvin, use of the Celsius temperature scale remains widespread. Zero on the Celsius scale (0°C) is equivalent to 273.15 K on the thermodynamic scale,[1] in which case

$$T\,(\mathrm{K}) = T\,(^\circ\mathrm{C}) + 273.15$$

However, temperature *differences* are equivalent for the two scales and may be denoted as °C or K. Also, although the SI unit of time is the second, other units of time (minute, hour, and day) are so common that their use with the SI system is generally accepted.

The SI units comprise a coherent form of the metric system. That is, all remaining units may be derived from the base units using formulas that do not involve any numerical factors. *Derived* units for selected quantities are listed in Table 1.3. Note that force is measured in newtons, where a 1-N force will accelerate a 1-kg mass at 1 $\mathrm{m/s^2}$. Hence 1 N = 1 $\mathrm{kg \cdot m/s^2}$. The unit of pressure ($\mathrm{N/m^2}$) is often referred to as the pascal. In the SI system there is one unit of energy (thermal, mechanical, or electrical), called the joule (J), and 1 J = 1 N · m. The unit for energy rate, or power, is then J/s, where one joule per second is equivalent to one watt (1 J/s = 1 W). Since it is frequently necessary to work with extremely large or small numbers, a set of standard prefixes has been introduced to simplify matters (Table 1.4). For example, 1 megawatt = 10^6 W, and 1 micrometer (μm) = 10^{-6} m.

Table 1.3 SI derived units for selected quantities

QUANTITY	NAME AND SYMBOL	FORMULA	EXPRESSION IN SI BASE UNITS
Force	newton (N)	$\mathrm{m \cdot kg/s^2}$	$\mathrm{m \cdot kg/s^2}$
Pressure and stress	pascal (Pa)	$\mathrm{N/m^2}$	$\mathrm{kg/m \cdot s^2}$
Energy	joule (J)	$\mathrm{N \cdot m}$	$\mathrm{m^2 \cdot kg/s^2}$
Power	watt (W)	J/s	$\mathrm{m^2 \cdot kg/s^3}$

[1] The degree symbol is retained for designation of the Celsius temperature (°C) to avoid confusion with use of C for the unit of electrical charge (coulomb).

Table 1.4 Multiplying prefixes

PREFIX	ABBREVIATION	MULTIPLIER
pico	p	10^{-12}
nano	n	10^{-9}
micro	μ	10^{-6}
milli	m	10^{-3}
centi	c	10^{-2}
hecto	h	10^{2}
kilo	k	10^{3}
mega	M	10^{6}
giga	G	10^{9}
tera	T	10^{12}

1.7 SUMMARY

Although much of the material of this chapter will be discussed in greater detail, you should now have a reasonable overview of heat transfer. You should be aware of the several modes of transfer and their physical origins. Moreover, given a physical situation, you should be able to perceive the relevant transport phenomena. The importance of developing this facility must not be underestimated. You will be devoting much time to acquiring the tools needed to calculate heat transfer phenomena. However, before you can begin to use these tools to solve practical problems, you must have the intuition to determine what is happening physically. In short, you must be able to look at a problem and identify the pertinent transport phenomena. The examples and problems at the end of this chapter should help you to begin developing this intuition.

You should also appreciate the significance of the rate equations and feel comfortable in using them to compute transport rates. These equations, summarized in Table 1.5, *should be committed to memory*. You must also

Table 1.5 Summary of heat transfer rate processes

MODE	MECHANISM	RATE EQUATION	EQUATION NUMBER	TRANSPORT PROPERTY OR COEFFICIENT
Conduction	Diffusion of energy due to random molecular motion	$q''_x\ (\text{W/m}^2) = -k\dfrac{dT}{dx}$	(1.1)	k (W/m · K)
Convection	Diffusion of energy due to random molecular motion plus energy transfer due to bulk motion (advection)	$q''\ (\text{W/m}^2) = h(T_s - T_\infty)$	(1.3)	h (W/m² · K)
Radiation	Energy transfer by electromagnetic waves	$q''\ (\text{W/m}^2) = \varepsilon\sigma(T_s^4 - T_{sur}^4)$ or $q\ (\text{W}) = h_r A(T_s - T_{sur})$	(1.6) (1.7)	ε h_r (W/m² · K)

recognize the importance of the conservation laws and the need to carefully identify control volumes. With the rate equations, the conservation laws may be used to solve numerous problems involving heat transfer.

EXAMPLE 1.5

A closed container filled with hot coffee is in a room whose air and walls are at a fixed temperature. Identify all heat transfer processes that contribute to cooling of the coffee. Comment on features that would contribute to a superior container design.

SOLUTION

Known:

Hot coffee is separated from its cooler surroundings by a plastic flask, an air space, and a plastic cover.

Find: Relevant heat transfer processes.

Schematic:

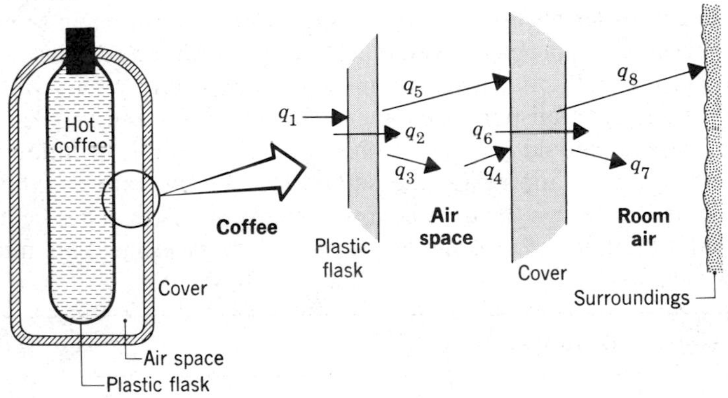

Pathways for energy transfer from the coffee are as follows.

q_1: free convection from the coffee to the flask

q_2: conduction through the flask

q_3: free convection from the flask to the air

q_4: free convection from the air to the cover

q_5: net radiation exchange between the outer surface of the flask and the inner surface of the cover

q_6: conduction through the cover

q_7: free convection from the cover to the room air

q_8: net radiation exchange between the outer surface of the cover and the surroundings

Comments:

Design improvements are associated with (1) use of aluminized (low emissivity) surfaces for the flask and cover to reduce net radiation, and (2) evacuating the air space or using a filler material to retard free convection.

PROBLEMS

1.1 A heat rate of 3 kW is conducted through a section of an insulating material of cross-sectional area 10 m^2 and of thickness 2.5 cm. If the inner (hot) surface temperature is at 415°C and the thermal conductivity of the material is 0.2 W/m · K, what is the outer surface temperature?

1.2 A concrete wall, which has a surface area of 30 m^2 and is 0.30 m thick, separates warm room air from cold ambient air. The temperature of the inner surface of the wall is known to be 25°C, while the outer surface is at −15°C. The thermal conductivity of the concrete is 1 W/m · K. What is the heat loss through the wall?

1.3 The heat flux through a wood slab 50 mm thick, whose inner and outer surface temperatures are 40 and 20°C, respectively, has been determined to be 40 W/m^2. What is the thermal conductivity of the wood?

1.4 The inner and outer surface temperatures of a glass window 5 mm thick are 15 and 5°C. What is the heat loss through a window that is 1 m by 3 m on a side? The thermal conductivity of glass is 1.4 W/m · K.

1.5 A freezer compartment consists of a cubical cavity that is 2 m on a side. Assume the bottom to be perfectly insulated. What is the minimum thickness of styrofoam insulation ($k = 0.030$ W/m · K) which must be applied to the top and side walls to ensure a heat load of less than 500 W, when the inner and outer surfaces are −10 and 35°C?

1.6 What is the thickness required of a masonry wall having thermal conductivity 0.75 W/m · K if the heat rate is to be 80% of the heat rate through a composite structural wall having a thermal conductivity 0.25 W/m · K and a thickness of 100 mm? Both walls are subjected to the same surface temperature difference.

1.7 The convection heat transfer coefficient between a surface at 40°C and ambient air at 20°C is $20 \text{ W/m}^2 \cdot \text{K}$. Calculate the heat flux leaving the plate by convection.

1.8 Air at 300°C flows over a flat plate of dimensions 0.50 m by 0.25 m. If the convection heat transfer coefficient is $250 \text{ W/m}^2 \cdot \text{K}$, determine the heat transfer rate from the air to one side of the plate when the plate is maintained at 40°C.

1.9 An electric resistance heater is embedded in a long cylinder of diameter 30 mm. When water with a temperature of 25°C and velocity of 1 m/s flows crosswise over the cylinder, the power per unit length required to maintain the surface at a uniform temperature of 90°C is 28 kW/m. When air, also at 25°C, but with a velocity of

10 m/s is flowing, the power per unit length required to maintain the same surface temperature is 400 W/m. Calculate and compare the convection coefficients for the flows of water and air.

1.10 A water-cooled, spherical object of diameter 10 mm and emissivity 0.9 is maintained at 80°C when placed in a large vacuum oven whose walls are maintained at 400°C. What is the net heat transfer rate from the oven walls to the object?

1.11 A surface of area 0.5 m^2, emissivity 0.8, and temperature 150°C is placed in a large, evacuated chamber whose walls are maintained at 25°C. What is the rate at which radiation is emitted by the surface? What is the net rate at which radiation is exchanged between the surface and the chamber walls?

1.12 The variation of temperature with position in a wall is shown below for a specific time, t_1, during a transient (time-varying) process.

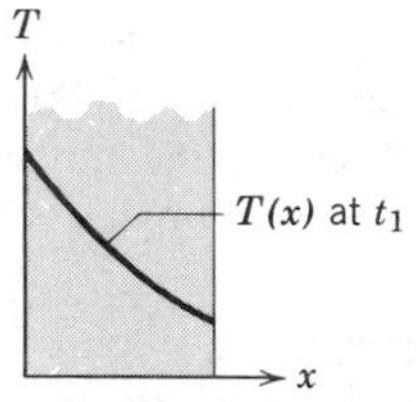

Is the wall being heated or cooled?

1.13 A solid aluminum sphere of emissivity ε is initially at an elevated temperature and is cooled by placing it in a chamber. The walls of the chamber are maintained at a lower temperature, and a cold gas is circulated through the chamber. Obtain an equation that could be used to predict the variation of the aluminum temperature with time during the cooling process. Do not attempt to solve.

1.14 An aluminum plate 4 mm thick is mounted in a horizontal position, and its bottom surface is well insulated. A special, thin coating is applied to the top surface such that it absorbs 80% of any incident solar radiation, while having an emissivity of 0.25. The density ρ and specific heat c of aluminum are known to be 2700 kg/m^3 and 900 J/kg · K, respectively.

(a) Consider conditions for which the plate is at a temperature of 25°C and its top surface is suddenly exposed to ambient air at $T_\infty = 20°C$ and to solar radiation that provides an incident flux of 900 W/m^2. The convection heat transfer coefficient between the surface and the air is $h = 20$ W/m^2 · K. What is the initial rate of change of the plate temperature?

(b) What will be the equilibrium temperature of the plate when steady-state conditions are reached?

1.15 A spherical, stainless steel (AISI 302) canister is used to store reacting chemicals that provide for a uniform heat flux q_i'' to its inner surface. The canister is suddenly

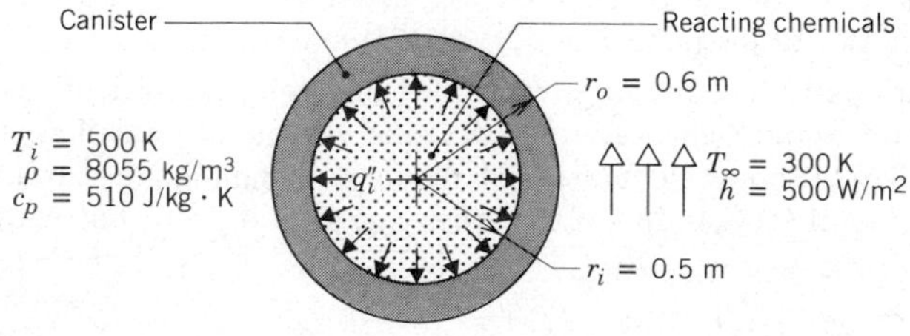

submerged in a liquid bath of temperature $T_\infty < T_i$, where T_i is the initial temperature of the canister wall.

(a) Assuming negligible temperature gradients in the canister wall and a constant heat flux q_i'', develop an equation that governs the variation of the wall temperature with time during the transient process. What is the initial rate of change of the wall temperature if $q_i'' = 10^5$ W/m^2?

(b) What is the steady-state temperature of the wall?

1.16 A slab of ice in a thin-walled container 10 mm thick and 300 mm on each side is placed on a well-insulated pad. At its top surface, the ice is exposed to ambient air for which $T_\infty = 25$°C and the convection coefficient is 25 W/m$^2 \cdot$ K. Neglecting heat transfer from the sides and assuming the ice-water mixture remains at 0°C, how long will it take to completely melt the ice? The density and latent heat of fusion of ice are 920 kg/m^3 and 334 kJ/kg, respectively.

1.17 The coating on a plate is to be cured by exposure to an infrared lamp providing a radiant flux of 2000 W/m^2 incident upon the coating. The coating absorbs 80% of the incident radiant flux and has an emissivity of 0.50. The surroundings are large compared to the plate and at a temperature of 30°C. The convection heat transfer coefficient between the plate surface and the ambient air at 20°C is 15 W/m$^2 \cdot$ K. Determine the temperature of the plate.

1.18 The back side of a metallic plate is perfectly insulated while the front side absorbs a solar radiant flux of 800 W/m^2. The convection coefficient between the plate and the ambient air is 12 W/m$^2 \cdot$ K.

(a) Neglecting radiation exchange with the surroundings, calculate the temperature of the plate under steady-state conditions if the ambient air temperature is 20°C.

(b) For the same ambient air temperature, calculate the temperature of the plate if its surface emissivity is 0.8 and the temperature of the surroundings is also 20°C.

1.19 The heat flux by diffusion through a plane wall to a surface is known to be 400 W/m^2. Determine the surface temperature for each of the following conditions.

(a) Convection between the surface and an airstream at 20°C with a heat transfer coefficient $h = 10$ W/m$^2 \cdot$ K.

(b) The same convection process occurs along with radiative heat transfer between the surface and cold surroundings at -150°C, with a radiative heat transfer coefficient of $h_r = 5$ W/m$^2 \cdot$ K.

1.20 A surface whose temperature is maintained at 400°C is separated from an airflow by a layer of insulation 25 mm thick for which the thermal conductivity is 0.1 W/m $\cdot$ K. If the air temperature is 35°C and the convection coefficient between the air and the outer surface of the insulation is 500 W/m$^2 \cdot$ K, what is the temperature of this outer surface?

1.21 A large wall is of thickness $L = 0.05$ m and thermal conductivity $k = 0.7$ W/m $\cdot$ K. At the front surface, which has an emissivity of 0.8, there is radiation exchange with large surroundings and heat transfer by convection to air. The air and surroundings are at 300 K, and the convection coefficient is 20 W/m$^2 \cdot$ K. If the front surface has a temperature of 400 K, what is the temperature of the back surface?

1.22 An experiment to determine the convection coefficient associated with airflow over the surface of a thick steel casting involves insertion of thermocouples in the casting at distances of 10 and 20 mm from the surface along a hypothetical line normal to the surface. The steel has a thermal conductivity of 15 W/m$\cdot$K. If the thermocouples measure temperatures of 50 and 40°C in the steel when the air temperature is 100°C, what is the convection coefficient?

1.23 Plate glass at 600°C is cooled by passing air over its surfaces such that the convection heat transfer coefficient is $h = 5\ \text{W/m}^2 \cdot \text{K}$. To prevent cracking, it is known that the temperature gradient must not exceed 15°C/mm at any point in the glass during the cooling process. If the thermal conductivity of the glass is 1.4 W/m · K and its surface emissivity is 0.8, what is the lowest temperature of the air that can be initially used for the cooling? Assume that the temperature of the air equals that of the surroundings.

1.24 A solar flux of 700 W/m² is incident upon a flat-plate solar collector used to heat water. The area of the collector is 3 m², and 90% of the solar radiation passes through the cover glass and is absorbed by the absorber plate. The remaining 10% is reflected away from the collector. Water flows through the tube passages on the back side of the absorber plate and is heated from an inlet temperature T_i to an outlet temperature T_o. The cover glass operating at a temperature of 30°C, has an emissivity of 0.94 and experiences radiation exchange with the sky at −10°C. The convection coefficient between the cover glass and the ambient air at 25°C is 10 W/m²·K.

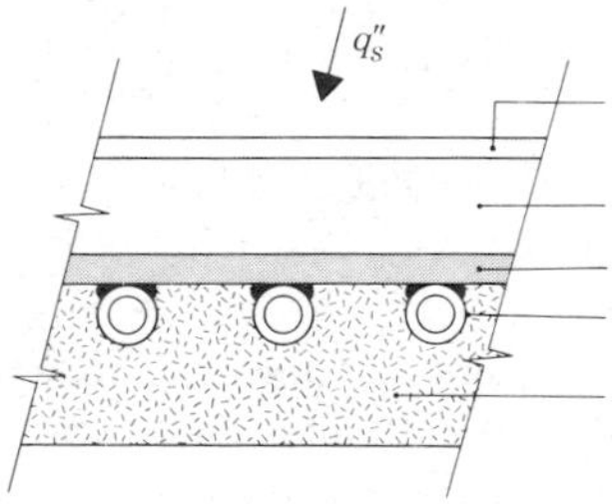

(a) Perform an overall energy balance on the collector to obtain an expression for the rate at which useful heat is collected per unit area of the collector, q''_u. Determine the value of q''_u.

(b) Calculate the temperature rise of the water, $T_o - T_i$, if the flow rate is 0.01 kg/s. Assume the specific heat of the water to be 4179 J/kg · K.

(c) The collector efficiency η is defined as the ratio of the useful heat collected to the rate at which solar energy is incident on the collector. What is the value of η?

1.25 Consider a flat plate solar collector operating under steady-state conditions. Solar radiation is incident, per unit surface area of the collector, at a rate q''_S (W/m²). The cover glass is completely transparent to this radiation, and the fraction of the radiation that is absorbed by the blackened absorber plate is designated as α (the absorptivity). The fraction of the radiation not absorbed by the absorber plate $(1 - \alpha)$ may be assumed to be reflected through the cover plate and back to the atmosphere and space.

Useful energy is obtained from the collector by passing a working fluid through copper tubing that is brazed to the underside of the absorber plate. The tubing forms a serpentine arrangement for which fluid, at a constant flow rate $\dot{m}$ and specific heat c_p, is heated from an inlet temperature T_i to an outlet temperature T_o. Although the bottom of the collector may be assumed to be perfectly insulated (no heat loss), there will be heat loss from the absorber plate due to convection across the air space and radiation exchange with the cover plate. Assuming that the

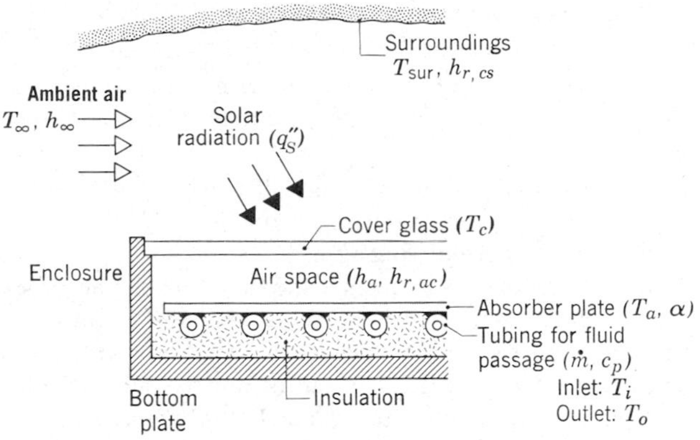

absorber and cover plates have uniform temperatures T_a and T_c, respectively, the parallel convection and radiation heat fluxes may be expressed as $h_a(T_a - T_c)$ and $h_{r,ac}(T_a - T_c)$. The quantity h_a is the convection heat transfer coefficient associated with the air space, while $h_{r,ac}$ is the radiation heat transfer coefficient associated with the absorber plate–cover plate combination. The cover glass also transfers heat by convection to the ambient air, $h_\infty(T_c - T_\infty)$, and exchanges energy in the form of radiation with its surroundings, $h_{r,cs}(T_c - T_{sur})$. The effective temperature T_{sur} of the sky and surrounding surfaces seen by the cover glass is generally less than the ambient air temperature.

(a) Write an equation for the rate at which useful energy $q_u(W)$ is collected by the working fluid, expressing your result in terms of $\dot{m}$, c_p, T_i, and T_o.

(b) Perform an energy balance on the absorber plate. Use this balance to obtain an expression for q_u in terms of q''_S, α, T_a, T_c, h_a, $h_{r,ac}$, and A (the surface area of the absorber and cover plates).

(c) Perform an energy balance on the cover plate.

(d) Perform an overall energy balance on the entire collector, working wih a control volume about the collector. Compare your result with that obtained in parts b and c.

(e) The collector efficiency η is defined as the ratio of the useful heat collected to the rate at which solar energy is incident on the collector. Obtain an expression for η.

(f) Comment on what effect the value of $\dot{m}$ will have on T_a, T_o, and η. What would happen to T_a if the cover plate were removed?

1.26 In analyzing the performance of a thermal system, the engineer must be able to identify the relevant heat transfer processes. Only then can the system behavior be properly quantified. For the following systems identify the pertinent processes, designating them by an appropriately labeled arrow on a sketch of the system. Answer additional questions that appear in the problem statement. *Note:* In considering problems involving heat transfer in the natural environment (outdoors), recognize that solar radiation is comprised of long and short wavelength components. If this radiation is incident on a *semitransparent medium,* such as water or glass, two things will happen to the nonreflected portion of the radiation. The long wavelength component will be absorbed at the surface of the medium, whereas the short wavelength component will be transmitted by the surface.

(a) Identify the heat transfer processes that determine the temperature of an asphalt pavement on a summer day. Write an energy balance for the surface of the pavement.

(b) Consider an exposed portion of your body (e.g., your forearm with a short-sleeved shirt) while you are sitting in a room. Identify all heat transfer processes that occur at the surface of your skin.

In the interest of conserving fuel and funds, an engineer's spouse insists on keeping the thermostat of their home at 15°C (59°F) throughout the winter months. The engineer is able to tolerate this condition if the outside ambient air temperature exceeds −10°C (14°F) but complains of being cold if the ambient temperature falls much below this value. Is the engineer imagining things?

(c) Consider an incandescent light source that consists of a tungsten filament enclosed in an air-filled glass bulb. Assuming steady-state operation with the filament at a temperature of approximately 2900 K, list all the pertinent heat transfer processes for (i) the filament and (ii) the glass bulb.

(d) In recent years there has been considerable interest in developing building materials that have improved insulating qualities. The development of such materials would do much to enhance energy conservation by reducing space heating requirements.

It has been suggested that superior structural and insulating qualities could be obtained by using the composite shown. The material consists of a honeycomb, with cells of square cross section, sandwiched between solid slabs. The cells are filled with air, and the slabs, as well as the honeycomb matrix, are fabricated from plastics of low thermal conductivity. Identify all heat transfer processes pertinent to the performance of the composite. Suggest ways in which this performance could be enhanced.

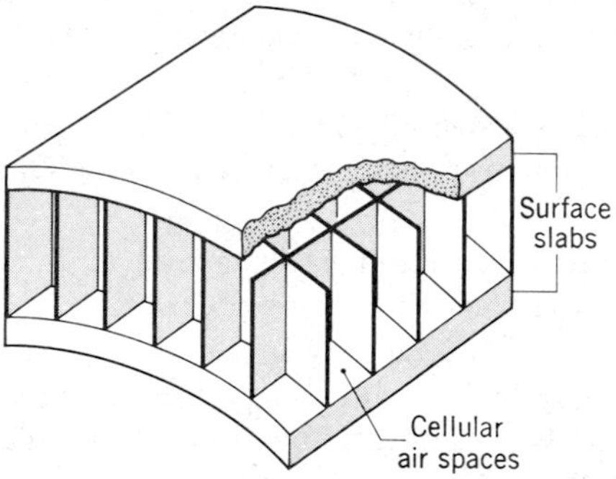

(e) The number of panes in a window can strongly influence the heat loss from a heated room to the outside ambient air. Compare the single- and double-paned units shown by identifying relevant heat transfer processes for each case.

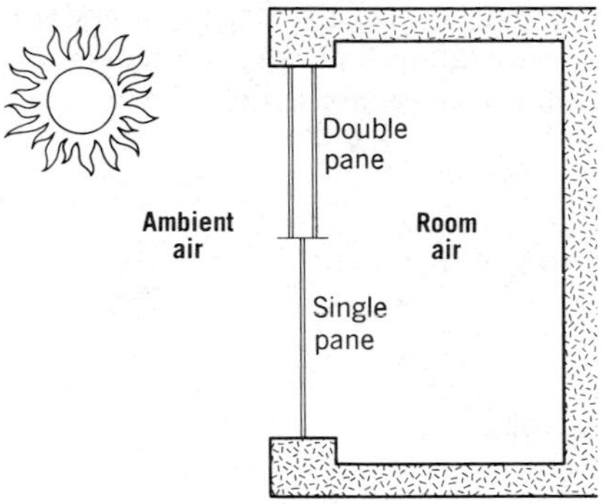

(f) In a typical flat plate solar collector, energy is collected by a working fluid that is circulated through tubes that are in good contact with the back face of an absorber plate. The back face is insulated from the surroundings, and the absorber plate receives solar radiation on its front face, which is typically covered by one or more transparent plates. Identify the relevant heat transfer processes, first for the absorber plate with no cover plate and then for the absorber plate with a single cover plate.

(g) Evacuated-tube solar collectors are capable of improved performance relative to flat plate collectors. The design consists of an inner tube enclosed in an outer tube that is transparent to solar radiation. The annular space between the tubes is evacuated. The outer, opaque surface of the inner tube absorbs solar radiation, and a working fluid is passed through the tube to collect the solar energy. The collector design generally consists of a row of such tubes arranged in front of a reflecting panel.

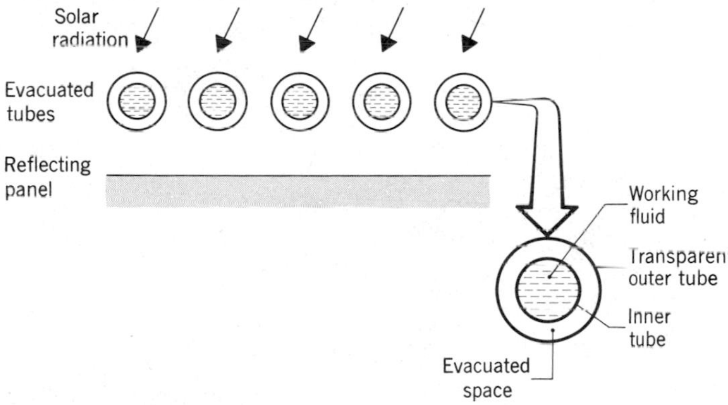

Identify all heat transfer processes relevant to the performance of this device.

(h) The solar energy collector design shown on the next page has been used for agricultural applications. Air is blown through a long duct whose cross section is in the form of an equilateral triangle. One side of the triangle is comprised of a double-paned, semitransparent cover, while the other two sides are constructed from aluminum sheets painted flat black on the inside and covered on the outside with a layer of styrofoam insulation. During sunny periods air entering the system is heated for delivery to either a greenhouse, grain drying unit, or a storage system.

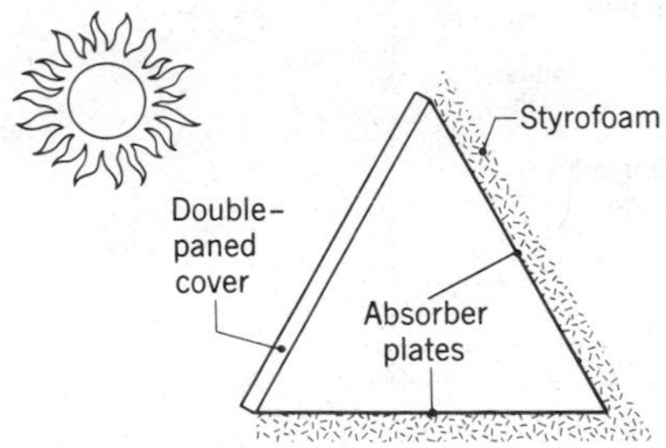

Identify all heat transfer processes associated with the cover plates, the absorber plate(s), and the air.

(i) A thermocouple junction is used to measure the temperature of a hot gas stream flowing through a channel by inserting the juntion into the mainstream of the gas. The surface of the channel is cooled such that its temperature is well below that of the gas. Identify the heat transfer processes associated with the junction surface. Will the junction sense a temperature that is less than, equal to, or greater than the gas temperature? A radiation shield is a small, open-ended tube that encloses the thermocouple junction, yet allows for passage of the gas through the tube. How does use of such a shield improve the accuracy of the temperature measurement?

(j) A double-glazed, glass fire screen is inserted between a woodburning fireplace and the interior of a room. The screen consists of two vertical glass plates that are separated by a space through which room air may flow (the space is open at the top and bottom). Identify the heat transfer processes associated with the fire screen.

Two

INTRODUCTION TO CONDUCTION

Recall that conduction (heat transfer by diffusion) refers to the transport of energy in a medium due to a temperature gradient, and the physical mechanism is that of random atomic or molecular activity. In this chapter we consider in greater detail the conduction rate equation and the relationship of energy conservation to the conduction process. In Chapters 3 to 5 we consider applications involving various geometrical and time-dependent conditions.

2.1 THE CONDUCTION RATE EQUATION

Although the conduction (energy diffusion) rate equation, Fourier's law, was introduced in Section 1.2, it is now appropriate to consider its origin. Fourier's law is phenomenological; that is, it is developed from observed phenomena rather than being derived from first principles. Hence, we view the rate equation as a generalization based on much experimental evidence. For example, consider the steady-state conduction experiment of Figure 2.1. A cylindrical rod of known material is insulated on its lateral surface, while its end faces are maintained at different temperatures with $T_1 > T_2$. The temperature difference causes conduction heat transfer in the positive x direction. We are able to measure the heat transfer rate q_x, and we seek to determine how q_x depends on the following variables: ΔT, the temperature difference; Δx, the rod length; and A, the cross-sectional area.

We might imagine first holding ΔT and Δx constant and varying A. If we do so, we find that q_x is directly proportional to A. Similarly, holding ΔT and A constant, we observe that q_x varies inversely with Δx. Finally, holding A and Δx constant, we find that q_x is directly proportional to ΔT. The collective effect is then

$$q_x \propto A\frac{\Delta T}{\Delta x}$$

In changing the material (e.g., from a metal to a plastic), we would find that the above proportionality remains valid. However, we would also find that, for equal values of A, Δx, and ΔT, the value of q_x would be smaller for the plastic than for the metal. This suggests that the proportionality may be converted to an equality by introducing a coefficient that is a measure of the material behavior. Hence, we write

$$q_x = kA\frac{\Delta T}{\Delta x}$$

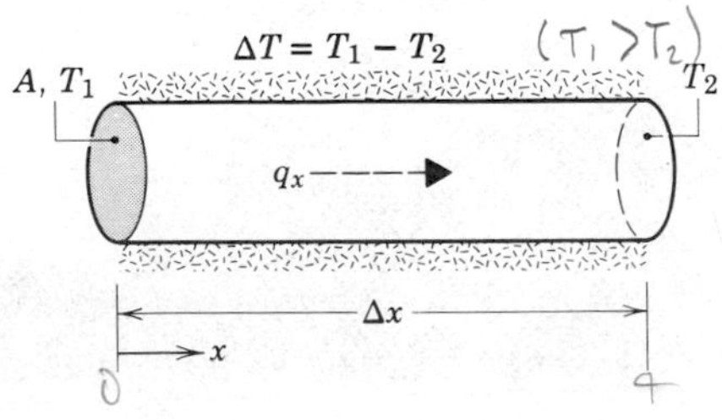

Figure 2.1 Steady-state heat conduction experiment.

where k, the *thermal conductivity* (W/m · K), is an important *property* of the material. Evaluating this expression in the limit as $\Delta x \rightarrow 0$, we obtain for the heat *rate*

$$q_x = -kA\frac{dT}{dx} \tag{2.1}$$

Fourier's Law

or for the heat *flux*

$$q''_x = \frac{q_x}{A} = -k\frac{dT}{dx} \tag{2.2}$$

Recall that the minus sign is necessary because heat is always transferred in the direction of decreasing temperature.

Fourier's law, as written in Equation 2.2, implies that the heat flux is a directional quantity. In particular the direction of q''_x is *normal* to the cross-sectional area A. Or, more generally, the direction of heat flow will always be normal to a surface of constant temperature, called an *isothermal* surface. Figure 2.2 illustrates the direction of heat flow q''_x in a one-dimensional coordinate system when the *temperature gradient* dT/dx is negative. From Equation 2.2, it follows that q''_x is positive. Note that the isothermal surfaces are planes normal to the x direction.

Recognizing that the heat flux is a vector quantity, we can write a more general statement of the conduction rate equation (*Fourier's law*) as follows:

$$\mathbf{q}'' = -k\nabla T = -k\left(\mathbf{i}\frac{\partial T}{\partial x} + \mathbf{j}\frac{\partial T}{\partial y} + \mathbf{k}\frac{\partial T}{\partial z}\right) \tag{2.3}$$

where ∇ is the three-dimensional del operator and $T(x, y, z)$ is the scalar temperature field. It is implicit in Equation 2.3 that the heat flux vector is in a direction perpendicular to the isothermal surfaces. An alternative form of Fourier's law is therefore

$$q''_n = -k\frac{\partial T}{\partial n} \tag{2.4}$$

where q''_n is the heat flux in a direction n, which is normal to an *isotherm*, as shown for the two-dimensional case in Figure 2.3. The heat transfer is sustained by a temperature gradient along $\mathbf{n}$. Note also that the heat flux vector can be resolved into components such that, in Cartesian coordinates, the general

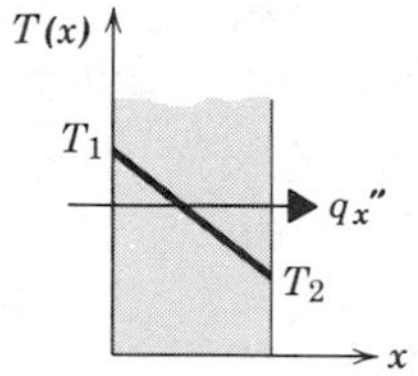

Figure 2.2 The relationship between coordinate system, heat flow direction, and temperature gradient in one dimension.

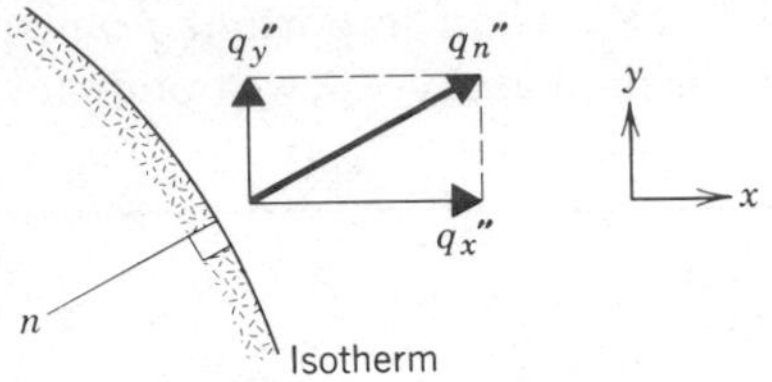

Figure 2.3 The heat flux vector normal to an isotherm in a two-dimensional coordinate system.

expression for $\boldsymbol{q}''$ is

$$\boldsymbol{q}'' = \boldsymbol{i}q_x'' + \boldsymbol{j}q_y'' + \boldsymbol{k}q_z'' \tag{2.5}$$

where, from Equation 2.3, it follws that

$$q_x'' = -k\frac{\partial T}{\partial x} \qquad q_y'' = -k\frac{\partial T}{\partial y} \qquad q_z'' = -k\frac{\partial T}{\partial z} \tag{2.6}$$

Fourier's Law

Each of these expressions relates the heat flux *across a surface* to the temperature gradient in a direction perpendicular to the surface. It is also implicit in Equation 2.3 that the medium in which the conduction occurs is *isotropic*. For such a medium the value of the thermal conductivity is independent of the coordinate direction.

Since Fourier's law provides the basis of conduction heat transfer, its key features are summarized as follows. It is *not* an expression that may be derived from first principles; it is instead a generalization based on experimental evidence. It is also an expression that *defines* an important material property, the thermal conductivity. In addition, Fourier's law is a vector expression indicating that the heat flux is normal to an isotherm and in the direction of decreasing temperature. Finally, note that Fourier's law applies for all matter regardless of its state: solid, liquid, or gas.

2.2 THE THERMAL PROPERTIES OF MATTER

Using Fourier's law necessitates knowledge of the thermal conductivity. This property, which is referred to as a *transport property*, provides an indication of the rate at which energy is transferred by the diffusion process. It depends on the physical structure of matter, atomic and molecular, which is related to the state of the matter. In this section we consider various forms of matter, identifying important aspects of their behavior and presenting typical property values.

2.2.1 Thermal Conductivity

From Fourier's law, Equation 2.6, the thermal conductivity is defined as

$$k \equiv -\frac{q_x''}{(\partial T/\partial x)}$$

It follows that, for a prescribed temperature gradient, the conduction heat flux increases with increasing thermal conductivity. Recalling the physical mechanism associated with conduction (Section 1.2.1), it also follows that, in general, the thermal conductivity of a solid is larger than that of a liquid, which is larger than that of a gas. As illustrated in Figure 2.4, the thermal conductivity of a solid may be more than four orders of magnitude larger than that of a gas. This trend is due largely to the difference in the intermolecular spacing for the two states.

The Solid State In the modern view of materials, a solid may be comprised of free electrons and of atoms bound in a periodic arrangement called the lattice. Accordingly, transport of thermal energy is due to two effects: the migration of free electrons and lattice vibrational waves. These effects are additive, such that the thermal conductivity k is the sum of the electronic component k_e and the lattice component k_l

$$k = k_e + k_l$$

To a first approximation, k_e is inversely proportional to the electrical resistivity ρ_e. For pure metals, which are of low ρ_e, it is therefore not surprising that k_e is much larger than k_l. In contrast, for alloys, which are of substantially larger ρ_e, the contribution of k_l to k is no longer negligible. For nonmetallic solids, k is determined primarily by k_l, which depends on the frequency of interactions between the atoms of the lattice. The regularity of the lattice arrangement has an important effect on k_l, with crystalline (well-ordered) materials like quartz having a higher thermal conductivity than amorphous materials like glass. In fact, for crystalline, nonmetallic solids such as diamond and beryllium oxide, k_l can be quite large, exceeding values of k associated with good conductors, such as aluminum.

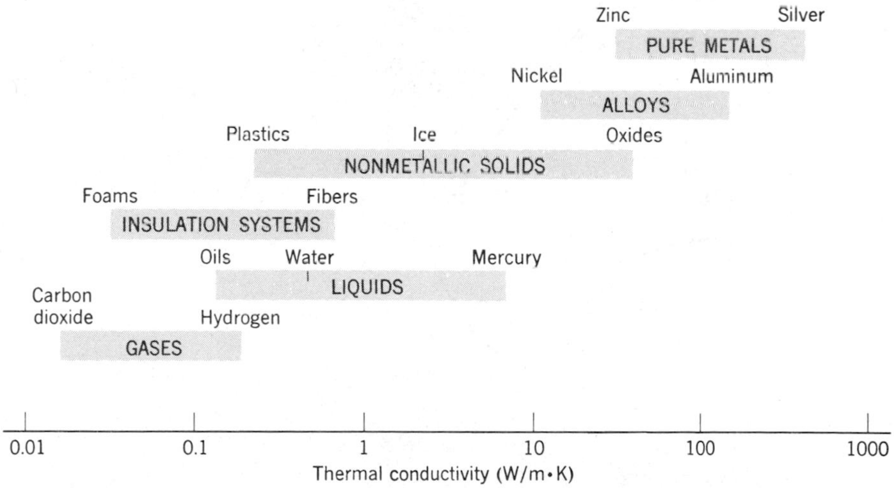

Figure 2.4 Range of thermal conductivity for various states of matter at normal temperatures and pressure.

The temperature dependence of k is shown in Figure 2.5 for representative metallic and nonmetallic solids. Values for selected materials of technical importance are also provided in Table A.1 (metallic solids) and Tables A.2 and A.3 (nonmetallic solids). More detailed treatments of thermal conductivity are available in the literature [1].

Insulation Systems Thermal insulations are comprised of low thermal conductivity materials combined to achieve an even lower system thermal conductivity. In *fiber-*, *powder-*, and *flake*-type insulations, the solid material is finely dispersed throughout an air space. The effective conductivity of the system depends on several characteristics, which include the thermal conductivity and surface radiative properties of the solid material and the nature and volumetric fraction of the air or void space. A special parameter of the system is its bulk density (solid mass/total volume), which depends strongly on the manner in which the solid material is interconnected.

If small voids or hollow spaces are formed by bonding or fusing portions of

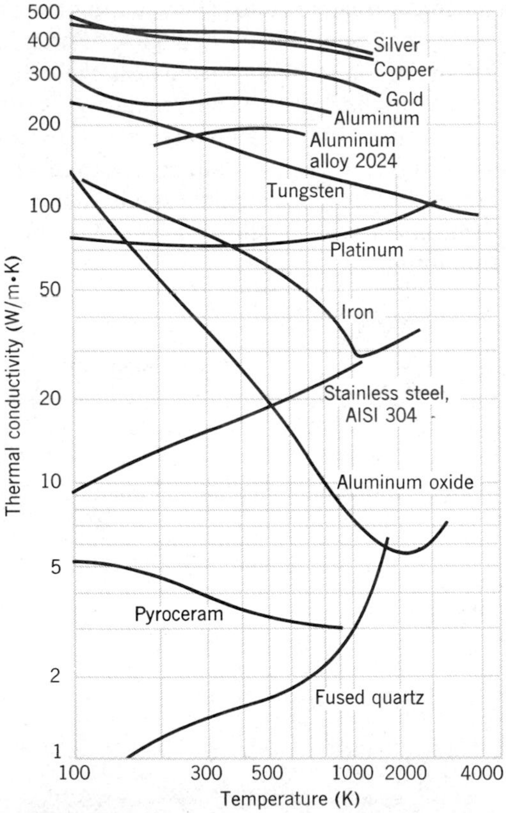

Figure 2.5 The temperature dependence of the thermal conductivity of selected solids.

the solid material, a rigid matrix is created. When these spaces are sealed from each other, the system is referred to as a cellular insulation. Examples of such rigid insulations are *foamed* systems, particularly those made from plastic and glass materials. *Reflective* insulations are composed of multilayered, parallel, thin sheets or foils of high reflectivity, which are spaced to reflect radiant heat back to its source. The spacing between the foils is designed to restrict the motion of air, and in high performance insulations, the space is even evacuated. In all types of insulation, evacuation of the air in the void space will reduce the effective thermal conductivity of the system.

It is important to recognize that heat transfer through any of these insulation systems may include several modes: conduction through the solid materials; conduction or convection through the air in the void spaces; and, if the temperature is sufficiently high, radiation exchange between the surfaces of the solid matrix. The *effective* thermal conductivity accounts for all of these effects, and values for selected insulation systems are summarized in Table A.3. Additional background information and data are available in the literature [2, 3].

The Fluid State Since the intermolecular spacing is much larger and the motion of the molecules is more random for the fluid state than for the solid state, thermal energy transport is less effective. The thermal conductivity of gases and liquids is therefore generally smaller than that of solids.

The effect of temperature, pressure and the nature of the species on the thermal conductivity of a gas may be explained in terms of the kinetic theory of gases [4]. From this theory it is known that the thermal conductivity is directly proportional to the number of particles per unit volume n, the mean molecular speed $\bar{c}$, and the mean free path λ, which is the average distance traveled by a molecule before experiencing a collision. Hence

$$k \propto n\bar{c}\lambda$$

Because $\bar{c}$ increases with increasing temperature and decreasing molecular mass, the thermal conductivity of a gas increases with increasing temperature and decreasing molecular weight. These trends are shown in Figure 2.6. However, because n and λ are directly and indirectly proportional to the gas pressure, respectively, the thermal conductivity is independent of pressure. This assumption is appropriate for the gas pressures of interest in this text. Accordingly, although the values of k presented in Table A.4 pertain to atmospheric pressure or the saturation pressure corresponding to the prescribed temperature, they may be used over a much wider range.

Molecular conditions associated with the liquid state are more difficult to describe, and physical mechanisms for explaining the thermal conductivity are not well understood [5]. As shown in Figure 2.7, the thermal conductivity of nonmetallic liquids generally decreases with increasing temperature, the notable exceptions being glycerine and water. This property is insensitive to pressure except near the critical point. Also, it generally follows that thermal conductivity

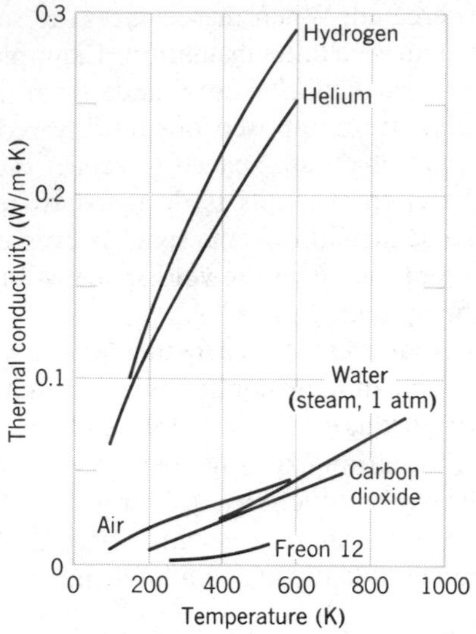

Figure 2.6 The temperature dependence of the thermal conductivity of selected gases at normal pressures.

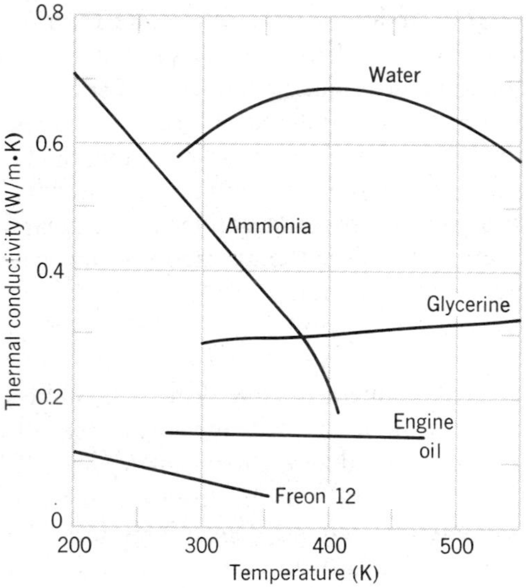

Figure 2.7 The temperature dependence of the thermal conductivity of selected nonmetallic liquids under saturated conditions.

decreases with increasing molecular weight. Values of the thermal conductivity are generally tabulated as a function of temperature for the saturated state of the liquid. Tables A.5 and A.6 present such data for several common liquids.

Metallic liquids are useful as heat transfer media in high heat flux situations, such as occur in nuclear power plants. The thermal conductivity of such liquids is given in Table A.7. Note that the values are much larger than for the nonmetallic liquids [6].

2.2.2 Other Relevant Properties

In our analysis of heat transfer problems, it will be necessary to use many properties of matter. These properties are generally referred to as *thermophysical* properties and include two distinct categories, *transport* and *thermodynamic* properties. The transport properties include the diffusion rate coefficients such as k, the thermal conductivity (for heat transfer) and ν, the kinematic viscosity (for momentum transfer). Thermodynamic properties, on the other hand, define the equilibrium state of a system. Density and specific heat are two such properties used extensively in thermodynamic analysis. Such properties are provided in the tables of Appendix A. At this point, it would be useful for you to become acquainted with the information contained in these tables.

The accuracy of engineering calculations depends on the accuracy with which the thermophysical properties are known [7–9]. Numerous examples could be cited of flaws in equipment and process design or failure to meet performance specifications that were attributable to misinformation associated with the selection of key property values used in the initial system analysis. Selection of reliable property data is an integral part of any careful engineering analysis. The casual use of data from the literature or handbooks, which have not been well characterized or evaluated, is to be avoided. Recommended data values for many thermophysical properties can be obtained from Reference 10. This reference, available in most institutional libraries, was prepared by the Thermophysical Properties Research Center (TPRC) at Purdue University. The TPRC maintains a continuing program to provide current, comprehensive coverage on thermophysical properties [11].

EXAMPLE 2.1

The thermal diffusivity α is the controlling transport property for transient conduction and is equal to $k/\rho c_p$. Using the appropriate values of k, ρ, and c_p from Appendix A, calculate α for the following materials at the prescribed temperatures:

Pure aluminum, 300 and 700 K

Silicon carbide, 1000 K

Paraffin, 300 K

SOLUTION

Known: Definition of the thermal diffusivity α.

Find: Numerical values of α for selected materials and temperatures.

Properties:

Table A.1, pure aluminum (300 K):

$$\left.\begin{aligned} \rho &= 2702 \text{ kg/m}^3 \\ c_p &= 903 \text{ J/kg}\cdot\text{K} \\ k &= 237 \text{ W/m}\cdot\text{K} \end{aligned}\right\} \quad \begin{aligned} \alpha &= \frac{k}{\rho c_p} = \frac{237 \text{ W/m}\cdot\text{K}}{2702 \text{ kg/m}^3 \times 903 \text{ J/kg}\cdot\text{K}} \\ &= 97.1 \times 10^{-6} \text{ m}^2\text{/s} \end{aligned} \quad \triangleleft$$

Table A.1, pure aluminum (700 K):

$\rho = 2702 \text{ kg/m}^3$ at 300 K

$c_p = 1090 \text{ J/kg}\cdot\text{K}$ at 700 K (by linear interpolation)

$k = 225 \text{ W/m}\cdot\text{K}$ at 700 K (by linear interpolation)

Hence

$$\alpha = \frac{k}{\rho c_p} = \frac{225 \text{ W/m}\cdot\text{K}}{2702 \text{ kg/m}^3 \times 1090 \text{ J/kg}\cdot\text{K}} = 76 \times 10^{-6} \text{ m}^2\text{/s} \quad \triangleleft$$

Table A.2, silicon carbide (1000 K):

$$\left.\begin{aligned} \rho &= 3160 \text{ kg/m}^3 && \text{at 300 K} \\ c_p &= 1195 \text{ J/kg}\cdot\text{K} && \text{at 1000 K} \\ k &= 87 \text{ W/m}\cdot\text{K} && \text{at 1000 K} \end{aligned}\right\} \quad \begin{aligned} \alpha &= \frac{87 \text{ W/m}\cdot\text{K}}{3160 \text{ kg/m}^3 \times 1195 \text{ J/kg}\cdot\text{K}} \\ &= 23 \times 10^{-6} \text{ m}^2\text{/s} \end{aligned} \quad \triangleleft$$

Table A.3, paraffin (300 K):

$$\left.\begin{aligned} \rho &= 900 \text{ kg/m}^3 \\ c_p &= 2890 \text{ J/kg}\cdot\text{K} \\ k &= 0.020 \text{ W/m}\cdot\text{K} \end{aligned}\right\} \quad \begin{aligned} \alpha &= \frac{k}{\rho c_p} = \frac{0.020 \text{ W/m}\cdot\text{K}}{900 \text{ kg/m}^3 \times 2890 \text{ J/kg}\cdot\text{K}} \\ &= 7.7 \times 10^{-9} \text{ m}^2\text{/s} \end{aligned} \quad \triangleleft$$

Comments:

1. Note temperature dependence of the thermophysical properties of aluminum and silicon carbide. For example, for silicon carbide, $\alpha(1000\text{ K}) \approx 0.1 \times \alpha(300\text{ K})$; hence properties of this material have a strong temperature dependence.
2. The physical interpretation of α is that it provides a measure of heat transport (k) relative to energy storage (ρc_p). In general, metallic solids have higher α, while nonmetallics (e.g., paraffin) have lower values of α.
3. Linear interpolation of property values is generally acceptable for engineering purposes.

4. Use of the low temperature (300 K) density at higher temperatures ignores thermal expansion effects but is also acceptable for engineering purposes.

2.3 THE HEAT DIFFUSION EQUATION

A major objective in a conduction analysis is to determine the *temperature field* in a medium resulting from conditions that are imposed on its boundaries. That is, we wish to know the *temperature distribution*, which represents how temperature varies with position in the medium. Once this distribution is known, the heat flux at any point in the medium or on its surface may be computed from Fourier's law. Other important quantities of interest may also be determined. For the solid medium, knowledge of the temperature distribution could be used to ascertain structural integrity through determination of thermal stresses, expansions, and deflections. The temperature distribution could also be used to optimize the thickness of an insulating material or to determine the compatibility of special coatings or adhesives used with the material.

We now consider the manner in which the temperature distribution can be determined. The approach follows the methodology described in Section 1.3.3 of applying the energy conservation requirement. That is, we define a differential control volume, identify the relevant energy transfer processes, and introduce the appropriate rate equations. The result is a differential equation whose solution, for prescribed boundary conditions, provides the temperature distribution in the medium.

Consider a homogeneous medium in which temperature gradients exist and the temperature distribution $T(x, y, z)$ is expressed in Cartesian coordinates. Following the methodology of applying conservation of energy (Section 1.3.3), we first define an infinitesimally small (differential) control volume, $dx \cdot dy \cdot dz$, as shown in Figure 2.8. The second step is to consider the energy processes that are relevant to this control volume. If there are temperature gradients, conduction heat transfer will occur across each of the control surfaces. The conduction heat rates perpendicular to each of the control surfaces at the x, y, and z coordinate locations are indicated by the terms q_x, q_y, and q_z, respectively. The conduction heat rates at the opposite surfaces can then be expressed as a Taylor series expansion where, neglecting higher order terms,

$$q_{x+dx} = q_x + \frac{\partial q_x}{\partial x} dx \tag{2.7a}$$

$$q_{y+dy} = q_y + \frac{\partial q_y}{\partial y} dy \tag{2.7b}$$

$$q_{z+dz} = q_z + \frac{\partial q_z}{\partial z} dz \tag{2.7c}$$

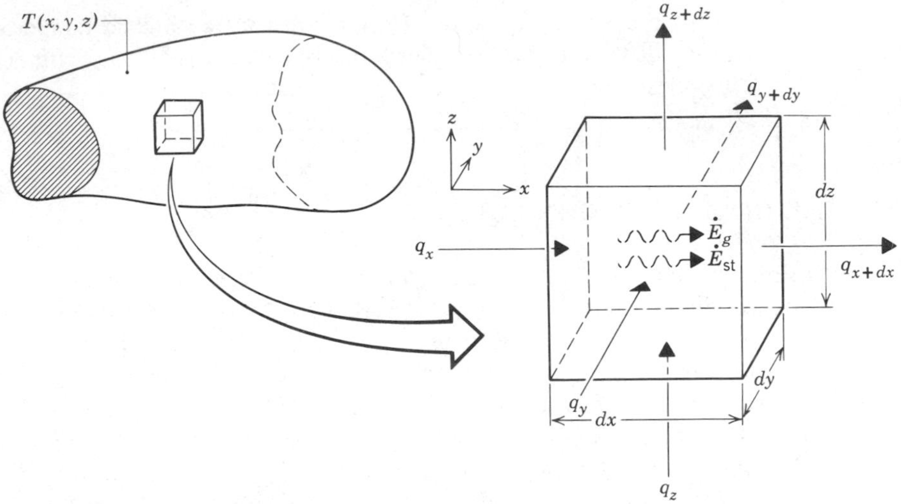

Figure 2.8 Differential control volume, $dx\,dy\,dz$, for conduction analysis in Cartesian coordinates.

In words, Equation 2.7a simply states that the x component of the heat transfer rate at $x + dx$ is equal to the value of this component at x plus the amount by which it changes with respect to x times dx.

Within the medium there may also be an *energy source* term associated with the rate of thermal energy generation. This term is represented as

$$\dot{E}_g = \dot{q}\,dx\,dy\,dz \tag{2.8}$$

where $\dot{q}$ is the rate at which energy is generated per unit volume of the medium (W/m^3). In addition there may occur changes in the amount of the internal thermal energy stored by the material in the control volume. On a rate basis, we express this *energy storage* term as

$$\dot{E}_{st} = \rho c_p \frac{\partial T}{\partial t}\,dx\,dy\,dz \tag{2.9}$$

where $\rho c_p\,\partial T/\partial t$ is the time rate of change of the internal energy of the medium per unit volume.

Once again it is important to note that the terms $\dot{E}_g$ and $\dot{E}_{st}$ represent different physical processes. The energy generation term $\dot{E}_g$ is a manifestation of some energy conversion process involving thermal energy on one hand and chemical, electrical, or nuclear energy on the other. The term is positive (a *source*) if thermal energy is being generated in the material at the expense of some other energy form; it is negative (a *sink*) if thermal energy is being consumed. In contrast the energy storage term $\dot{E}_{st}$ refers specifically to the rate of change of internal energy stored by the matter.

The third step in the methodology outlined in Section 1.3.3 is to express conservation of energy using the foregoing rate equations. On a *rate* basis, the general form of the conservation of energy requirement is

$$\dot{E}_{\text{in}} + \dot{E}_g - \dot{E}_{\text{out}} = \dot{E}_{\text{st}} \tag{1.10a}$$

Hence, recognizing that the conduction rates constitute the energy inflow, $\dot{E}_{\text{in}}$, and outflow, $\dot{E}_{\text{out}}$, and substituting Equations 2.8 and 2.9, we obtain

$$q_x + q_y + q_z + \dot{q}\,dx\,dy\,dz - q_{x+dx} - q_{y+dy} - q_{z+dz} = \rho c_p \frac{\partial T}{\partial t}\,dx\,dy\,dz \tag{2.10}$$

Substituting from Equations 2.7 we then obtain

$$-\frac{\partial q_x}{\partial x}\,dx - \frac{\partial q_y}{\partial y}\,dy - \frac{\partial q_z}{\partial z}\,dz + \dot{q}\,dx\,dy\,dz = \rho c_p \frac{\partial T}{\partial t}\,dx\,dy\,dz \tag{2.11}$$

The conduction heat rates may be evaluated from Fourier's law,

$$q_x = -k\,dy\,dz \frac{\partial T}{\partial x} \tag{2.12a}$$

$$q_y = -k\,dx\,dz \frac{\partial T}{\partial y} \tag{2.12b}$$

$$q_z = -k\,dx\,dy \frac{\partial T}{\partial z} \tag{2.12c}$$

where each heat flux component of Equation 2.6 has been multiplied by the appropriate control surface (differential) area to obtain the heat transfer rate. Substituting Equations 2.12 into Equation 2.11 and dividing out the dimensions of the control volume ($dx\,dy\,dz$), we obtain

$$\frac{\partial}{\partial x}\left(k\frac{\partial T}{\partial x}\right) + \frac{\partial}{\partial y}\left(k\frac{\partial T}{\partial y}\right) + \frac{\partial}{\partial z}\left(k\frac{\partial T}{\partial z}\right) + \dot{q} = \rho c_p \frac{\partial T}{\partial t} \tag{2.13}$$

Equation 2.13 is the general form, in Cartesian coordinates, of the *heat diffusion equation.* This equation, usually known as the *heat equation,* provides the basic tool for heat conduction analysis. From its solution, we can obtain the temperature distribution $T(x, y, z)$ as a function of time. The apparent complexity of this expression should not obscure the fact that it describes an important physical condition, that is, conservation of energy. You should have a clear understanding of the physical significance of each term appearing in the equation. For example, the term $\partial(k\partial T/\partial x)/\partial x$ describes the *net* conduction heat flux *into* the control volume for the x-coordinate direction. That is,

$$\frac{\partial}{\partial x}\left(k\frac{\partial T}{\partial x}\right) = q''_x - q''_{x+dx} \tag{2.14}$$

with similar expressions applying for the fluxes in the y and z directions. In words, the heat equation, Equation 2.13, therefore states that *at any point in a medium the net conduction heat rate into a unit volume plus the volumetric rate of thermal energy generation must equal the rate of change of thermal energy stored within the volume.*

It is often possible to work with simplified versions of Equation 2.13. For example, if the thermal conductivity is a constant, the heat equation is

$$\frac{\partial^2 T}{\partial x^2} + \frac{\partial^2 T}{\partial y^2} + \frac{\partial^2 T}{\partial z^2} + \frac{\dot{q}}{k} = \frac{1}{\alpha}\frac{\partial T}{\partial t} \tag{2.15}$$

const. thermal conductivity

where $\alpha = k/\rho c_p$ is the *thermal diffusivity*. This important thermophysical property is the ratio of the thermal conductivity k of the medium to the thermal capacitance ρc_p. A large value of α (large k and/or low ρc_p) implies that a medium is more effective in transferring energy by conduction than it is in storing energy.

Additional simplifications of the general form of the heat equation are often possible. For example, under *steady-state* conditions, there can be no change in the amount of energy storage; hence Equation 2.13 reduces to

$$\frac{\partial}{\partial x}\left(k\frac{\partial T}{\partial x}\right) + \frac{\partial}{\partial y}\left(k\frac{\partial T}{\partial y}\right) + \frac{\partial}{\partial z}\left(k\frac{\partial T}{\partial z}\right) + \dot{q} = 0 \tag{2.16}$$

S-S

Moreover, if the heat transfer is *one-dimensional* (e.g., in the x direction) and there is *no energy generation*, Equation 2.16 reduces to

$$\frac{d}{dx}\left(k\frac{dT}{dx}\right) = 0 \tag{2.17}$$

The important implication of this result is that *under steady-state, one-dimensional conditions with no energy generation*, the heat flux is a constant in the direction of transfer ($dq''_x/dx = 0$).

The heat equation may also be expressed in cylindrical and spherical coordinates. The differential control volumes for these two coordinate systems are shown in Figures 2.9 and 2.10. Applying the methodology used for the Cartesian coordinates, the general forms of the heat equation are,

Cylindrical Coordinates:

$$\frac{1}{r}\frac{\partial}{\partial r}\left(kr\frac{\partial T}{\partial r}\right) + \frac{1}{r^2}\frac{\partial}{\partial \phi}\left(k\frac{\partial T}{\partial \phi}\right) + \frac{\partial}{\partial z}\left(k\frac{\partial T}{\partial z}\right) + \dot{q} = \rho c_p \frac{\partial T}{\partial t} \tag{2.18}$$

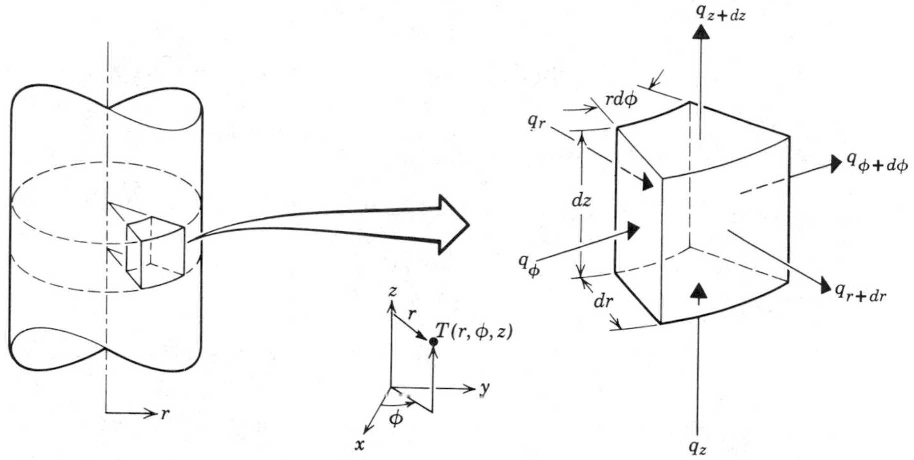

Figure 2.9 Differential control volume, $dr \cdot r\,d\phi \cdot dz$, for conduction analysis in cylindrical coordinates (r, ϕ, z).

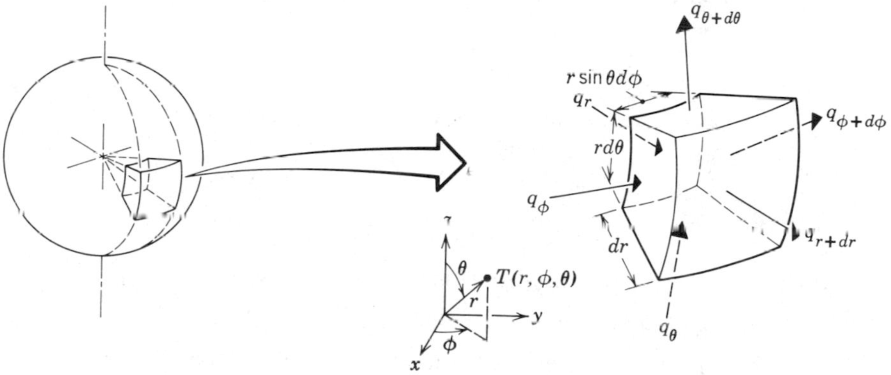

Figure 2.10 Differential control volume, $dr \cdot r \sin\theta\, d\phi \cdot r\,d\theta$, for conduction analysis in spherical coordinates (r, ϕ, θ).

Spherical Coordinates:

$$\frac{1}{r^2}\frac{\partial}{\partial r}\left(kr^2\frac{\partial T}{\partial r}\right) + \frac{1}{r^2 \sin^2\theta}\frac{\partial}{\partial \phi}\left(k\frac{\partial T}{\partial \phi}\right)$$

$$+\frac{1}{r^2 \sin\theta}\frac{\partial}{\partial \theta}\left(k \sin\theta\frac{\partial T}{\partial \theta}\right) + \dot{q} = \rho c_p \frac{\partial T}{\partial t} \quad (2.19)$$

Since it is important that you be able to apply conservation principles to differential control volumes, you should attempt to derive one of these expressions (see Problems 2.23 and 2.24).

EXAMPLE 2.2

The temperature distribution across a wall 1 m thick at a certain instant of time is given as

$$T(x) = a + bx + cx^2$$

where T is in degrees Celsius and x is in meters, while $a = 900°\text{C}$, $b = -300°\text{C/m}$, and $c = -50°\text{C/m}^2$. A uniform heat generation, $\dot{q} = 1000\ \text{W/m}^3$, is present in the wall of area $10\ \text{m}^2$ having the properties $\rho = 1600\ \text{kg/m}^3$, $k = 40\ \text{W/m}\cdot\text{K}$, and $c_p = 4\ \text{kJ/kg}\cdot\text{K}$.

1. Determine the rate of heat transfer entering the wall ($x = 0$) and leaving the wall ($x = 1$ m).
2. Determine the rate of change of energy storage in the wall.
3. Determine the time rate of temperature change at $x = 0$, 0.25, and 0.5 m.

SOLUTION

Known:

Temperature distribution $T(x)$ at a given instant of time t within a one-dimensional wall with uniform heat generation.

Find:

1. Heat rates entering, q_{in} ($x = 0$), and leaving, q_{out} ($x = 1$), the wall.
2. Rate of change of energy storage in the wall, $\dot{E}_{st}$.
3. Time rate of temperature change at $x = 0$, 0.25, and 0.5 m.

Schematic:

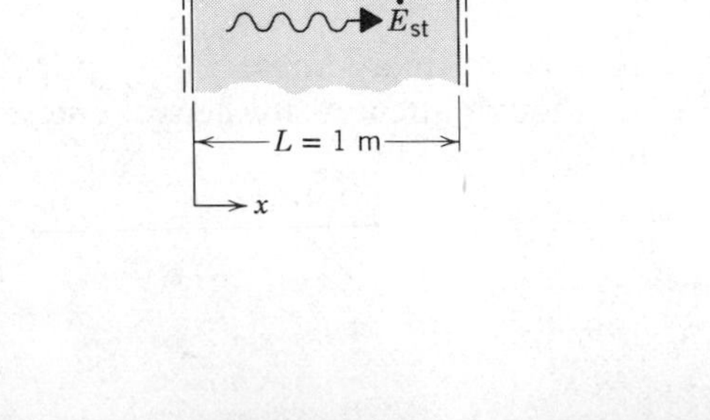

Assumptions:

1. One-dimensional conduction in the x direction.
2. Homogeneous medium with constant properties.
3. Uniform internal heat generation, $\dot{q}$ (W/m^3).

Analysis:

1. Recall that once the temperature distribution is known for a medium, it is a simple matter to determine the conduction heat transfer rate at any point in the medium, or at its surfaces, by using Fourier's law. Hence, the desired heat rates may be determined by using the prescribed temperature distribution with Equation 2.1. Accordingly,

$$q_{in} = q_x(0) = -kA \left.\frac{\partial T}{\partial x}\right|_{x=0} = -kA(b+2cx)_{x=0}$$

$$q_{in} = -bkA = 300°\text{C/m} \times 40\ \text{W/m} \cdot \text{K} \times 10\ \text{m}^2 = 120\ \text{kW} \qquad \triangleleft$$

Similarly,

$$q_{out} = q_x(L) = -kA \left.\frac{\partial T}{\partial x}\right|_{x=L} = -kA(b+2cx)_{x=L}$$

$$q_{out} = -(b+2cL)kA = -[-300°\text{C/m}$$
$$+ 2(-50°\text{C/m}^2) \times 1\ \text{m}] \times 40\ \text{W/m} \cdot \text{K} \times 10\ \text{m}^2 = 160\ \text{kW} \qquad \triangleleft$$

2. The rate of change of energy storage in the wall $\dot{E}_{st}$ may be determined by applying an overall energy balance to the wall. Using Equation 1.10a for a control volume about the wall,

$$\dot{E}_{in} + \dot{E}_g - \dot{E}_{out} = \dot{E}_{st}$$

where $\dot{E}_g = \dot{q}AL$, it follows that

$$\dot{E}_{st} = \dot{E}_{in} + \dot{E}_g - \dot{E}_{out} = q_{in} + \dot{q}AL - q_{out}$$

$$\dot{E}_{st} = 120\ \text{kW} + 1000\ \text{W/m}^3 \times 10\ \text{m}^2 \times 1\ \text{m} - 160\ \text{kW} = -30\ \text{kW} \qquad \triangleleft$$

3. The time rate of change of the temperature at any point in the medium may be determined from the heat equation, Equation 2.15, rewritten as

$$\frac{\partial T}{\partial t} = \frac{k}{\rho c_p} \frac{\partial^2 T}{\partial x^2} + \frac{\dot{q}}{\rho c_p}$$

From the prescribed temperature distribution, it follows that

$$\frac{\partial^2 T}{\partial x^2} = \frac{\partial}{\partial x}\left(\frac{\partial T}{\partial x}\right) = \frac{\partial}{\partial x}(b+2cx) = 2c = 2(-50°\text{C/m}^2) = -100°\text{C/m}^2$$

Note that this derivative is independent of position in the medium.

Hence the time rate of temperature change is also independent of position and is given by

$$\frac{\partial T}{\partial t} = \frac{40\ \text{W/m}\cdot\text{K}}{1600\ \text{kg/m}^3 \times 4\ \text{kJ/kg}\cdot\text{K}} \times (-100^\circ\text{C/m}^2)$$

$$+ \frac{1000\ \text{W/m}^3}{1600\ \text{kg/m}^3 \times 4\ \text{kJ/kg}\cdot\text{K}}$$

$$\frac{\partial T}{\partial t} = -6.25 \times 10^{-4}\,^\circ\text{C/s} + 1.56 \times 10^{-4}\,^\circ\text{C/s} = -4.69 \times 10^{-4}\,^\circ\text{C/s} \qquad \triangleleft$$

Comments:

1. From the above result it is evident that the temperature at every point within the wall is decreasing with time.
2. Fourier's law can always be used to compute the conduction heat rate from knowledge of the temperature distribution, even for unsteady conditions with internal heat generation.

2.4 BOUNDARY AND INITIAL CONDITIONS

To determine the temperature distribution in a medium, it is necessary to solve the appropriate form of the heat equation. However, such a solution depends on the physical conditions existing at the *boundaries* of the medium and, if the situation is time dependent, on conditions existing in the medium at some *initial time*. With regard to the *boundary conditions*, there exist several common physical effects, which are simply expressed in mathematical form. Because the heat equation is second order in the spatial coordinates, two boundary conditions must be expressed for each coordinate needed to describe the system. Because the equation is first order in time, however, only one condition, referred to as the *initial condition*, must be specified.

The three kinds of boundary conditions commonly encountered in heat transfer are summarized in Table 2.1. The conditions are specified at the surface $x = 0$ for a one-dimensional system. The heat transfer is in the positive x direction with the temperature distribution, which may be time dependent, designated as $T(x,t)$. The first condition corresponds to a situation for which the surface is maintained at a fixed or constant temperature T_s. The second condition corresponds to the existence of a fixed or constant heat flux q_s'' at the surface. Note that this heat flux is related to the temperature gradient at the surface by Fourier's law, Equation 2.6, which may be expressed as

$$q_x''(0) = -k\frac{\partial T}{\partial x}\bigg|_{x=0}$$

Table 2.1 Boundary conditions for the heat diffusion equation at the surface ($x = 0$)

T (x,t)

1. Constant surface temperature

$$T(0,t) = T_s \tag{2.20}$$

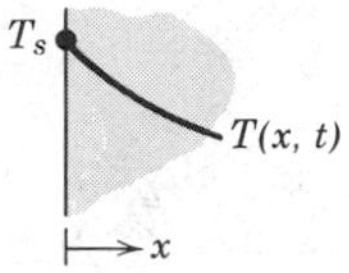

2. Constant surface heat flux

(a) Finite heat flux

$$-k\left.\frac{\partial T}{\partial x}\right|_{x=0} = q_s'' \tag{2.21}$$

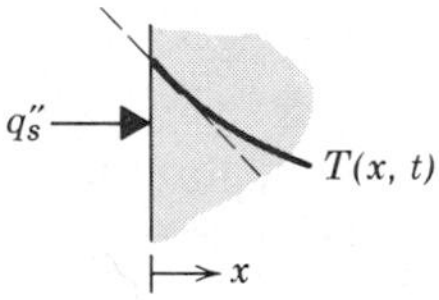

(b) Adiabatic or insulated surface

$$\left.\frac{\partial T}{\partial x}\right|_{x=0} = 0 \tag{2.22}$$

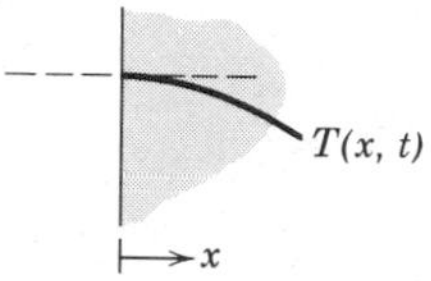

3. Convection surface condition

$$-k\left.\frac{\partial T}{\partial x}\right|_{x=0} = h[T_\infty - T(0,t)] \tag{2.23}$$

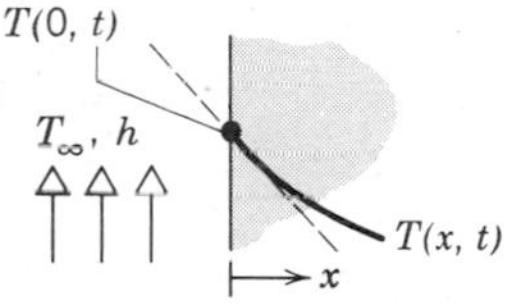

A special case of this condition corresponds to the *perfectly insulated*, or *adiabatic*, surface for which $\partial T/\partial x|_{x=0} = 0$. The third condition corresponds to the existence of convection heating (or cooling) at the surface. Note the relation of this last condition to the surface energy balance discussed in Section 1.3.2. Physical situations in which the foregoing boundary conditions are realized will be illustrated by the examples and problems of this text.

EXAMPLE 2.3

A long copper bar of rectangular cross section, whose width w is much greater than its thickness L, is maintained in contact with a heat sink at its lower surface, and the temperature throughout the bar is approximately equal to that of the sink, T_o. Suddenly, an electric current is passed through the bar and an airstream of temperature T_∞ is passed over the top surface, while the bottom surface continues to be maintained at T_o. Obtain the differential equation and the boundary and initial conditions that could be solved to determine the temperature as a function of position and time in the bar.

SOLUTION

Known:

Copper bar initially in thermal equilibrium with a heat sink is suddenly heated by passage of an electric current.

Find:

Differential equation and boundary and initial conditions needed to determine temperature as a function of position and time within the bar.

Schematic:

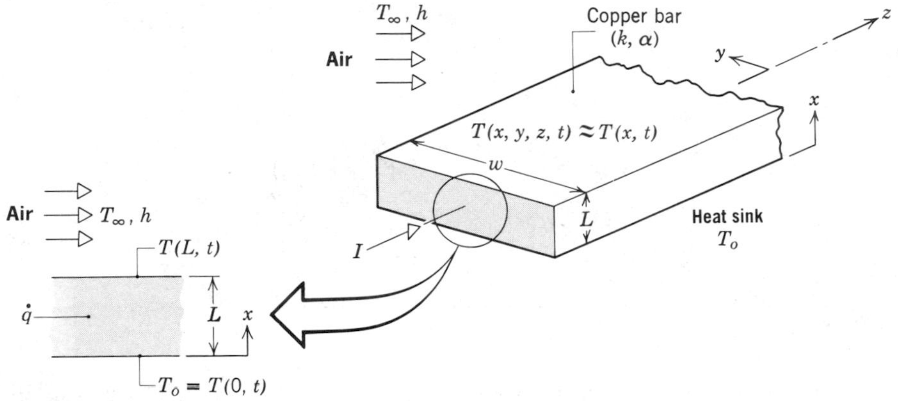

Assumptions:

1. Since $w \gg L$, side effects are negligible and heat transfer within the bar is primarily one dimensional in the x direction.
2. Uniform volumetric heat generation, $\dot{q}$.
3. Constant properties.

Analysis:

The temperature distribution is governed by the heat equation (2.13), which, for the one-dimensional and constant property conditions of the present problem, reduces to

$$\frac{\partial^2 T}{\partial x^2} + \frac{\dot{q}}{k} = \frac{1}{\alpha}\frac{\partial T}{\partial t} \qquad (1) \ \triangleleft$$

where the temperature is a function of position and time, $T(x,t)$. Recognize that this differential equation is second order in the spatial coordinate x, and first order in time t. Hence there must be two boundary conditions for the x direction and one condition, termed the initial condition, for time. The boundary condition appropriate for the bottom surface corresponds to case 1 of Table 2.1. In particular, since the temperature of this surface is maintained at a value, T_o, which is fixed with time, it follows that

$$T(0,t) = T_o \qquad (2) \quad \triangleleft$$

In contrast the convection surface condition, case 3 of Table 2.1, is appropriate for the top surface. Hence

$$-k\frac{\partial T}{\partial x}\bigg|_{x=L} = h[T(L,t) - T_\infty] \qquad (3) \quad \triangleleft$$

The initial condition is inferred from recognition that, before the change in conditions, the bar is at a uniform temperature T_o. Hence

$$T(x,0) = T_o \qquad (4) \quad \triangleleft$$

If T_o, T_∞, $\dot{q}$, and h are known, Equations 1 to 4 may be solved to obtain the temperature distribution $T(x,t)$ after the change in conditions.

Comments:

1. The temperature of the top surface $T(L, t)$ will change with time. This temperature is an unknown and may be obtained after finding $T(x,t)$.
2. How do you expect the temperature to vary with x at different times after the change in conditions? See Problem 2.25.

2.5 SUMMARY

The primary purposes of this chapter were to improve your understanding of the conduction rate equation and to make you familiar with the heat diffusion equation. You should know the origin and implications of Fourier's law, and you should understand the key thermal properties and how they vary for different substances. You should also know the physical meaning of each term appearing in the heat diffusion equation. To what form does this equation reduce for simplified conditions, and what kinds of boundary conditions may be used for its solution? In short, you must now have a fundamental understanding of the conduction process and its mathematical description. In the three chapters that follow, we undertake conduction analyses for numerous systems and conditions.

REFERENCES

1. Klemens, P. G., "Theory of the Thermal Conductivity of Solids," in R. P. Tye, Ed., *Thermal Conductivity*, Vol. 1, Academic Press, London, 1969.
2. Mallory, John F., *Thermal Insulation*, Reinhold Book Corp., New York, 1969.
3. American Society of Heating, Refrigeration and Air Conditioning Engineers, *Handbook of Fundamentals*, Chapters 17 and 31, ASHRAE, New York, 1972.

4. Vincenti, W. G., and C. H. Kruger, Jr., *Introduction to Physical Gas Dynamics*, Wiley, New York, 1965.
5. McLaughlin, E., "Theory of the Thermal Conductivity of Fluids," in R. P. Tye, Ed., *Thermal Conductivity*, Vol. 2, Academic Press, London, 1969.
6. Foust, O. J., Ed., "Sodium Chemistry and Physical Properties," in *Sodium-NaK Engineering Handbook*, Vol. 1, Gordon & Breach, New York, 1972.
7. Sengers, J. V., and M. Klein, Eds., "The Technical Importance of Accurate Thermophysical Property Information," National Bureau of Standards Technical Note No. 590, 1980.
8. Najjar, M. S., K. J. Bell, and R. N. Maddox, *Heat Transfer Eng.*, **2**, 27, 1981.
9. Hanley, H. J. M., and M. E. Baltatu, *Mech. Eng.*, **105**, 68, 1983.
10. Touloukian, Y. S., and C. Y. Ho, Eds., *Thermophysical Properties of Matter, The TPRC Data Series* (13 volumes on thermophysical properties: thermal conductivity, specific heat, thermal radiative, thermal diffusivity, and thermal linear expansion), Plenum Press, New York, 1970 through 1977.
11. Center for Information and Numerical Data Analysis and Synthesis (CINDAS), Purdue University, 2595 Yeager Road, West Lafayette, IN 47906.

PROBLEMS

2.1 Assume steady-state, one-dimensional heat conduction through the axisymmetric shape shown below.

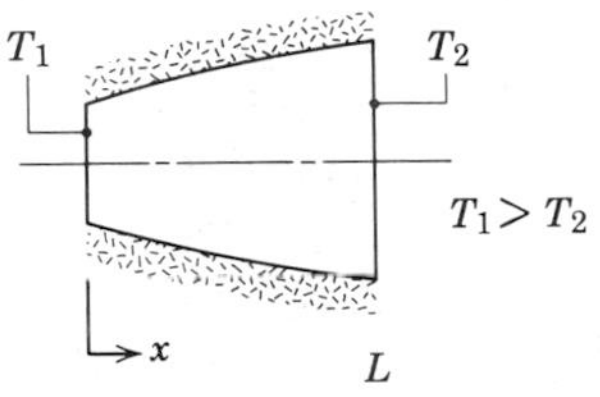

Assuming constant properties and no internal heat generation, sketch the temperature distribution on T–x coordinates. Briefly explain the shape of the curve shown.

2.2 A hot water pipe with outside radius r_1 has a temperature T_1. A thick insulation applied to reduce the heat loss has an outer radius r_2 and temperature T_2. On T–r coordinates, sketch the temperature distribution in the insulation for one-dimensional, steady-state heat transfer with constant properties. Give a brief explanation, justifying the shape of the curve shown.

2.3 A spherical shell with inner radius r_1 and outer radius r_2 has surface temperatures T_1 and T_2, respectively, where $T_1 > T_2$. Sketch the temperature distribution on T–r coordinates assuming steady-state, one-dimensional conduction with constant properties. Briefly justify the shape of the curve shown.

2.4 Assume steady-state, one-dimensional heat conduction through the symmetric shape shown.

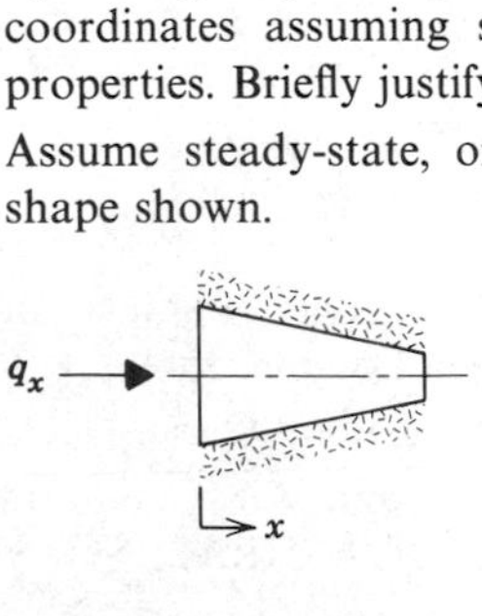

Assuming that there is no internal heat generation, derive an expression for the thermal conductivity $k(x)$ for these conditions: $A(x)=(1-x)$, $T(x)=300(1-2x-x^3)$, and $q=6000$ W where A is in square meters, T in kelvins, and x in meters.

2.5 A solid, truncated cone serves as a support for a system that maintains the top (truncated) face of the cone at a temperature T_1, while the base of the cone is at a temperature $T_2 < T_1$.

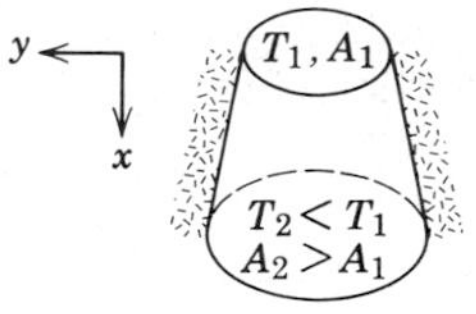

The thermal conductivity of the solid depends on temperature according to the relation $k = k_o - aT$, where a is a positive constant, and the sides of the cone are well insulated. Do the following quantities increase, decrease, or remain the same with increasing x: the heat transfer rate q_x, the heat flux q''_x, the thermal conductivity k, and the temperature gradient dT/dx?

2.6 To determine the effect of the temperature dependence of the thermal conductivity on the temperature distribution in a solid, consider a material for which this dependence may be represented as

$$k = k_o + aT$$

where k_o is a positive constant and a is a coefficient that may be positive or negative. Sketch the steady-state temperature distribution associated with heat transfer in a plane wall for three cases corresponding to $a > 0$, $a = 0$, and $a < 0$.

2.7 One-dimensional, steady-state conduction without heat generation occurs in the system shown. The thermal conductivity is 25 W/m · K and the thickness L is 0.5 m.

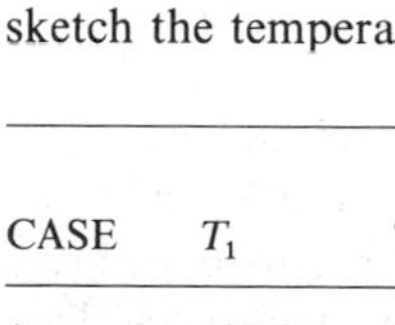

Determine the unknown quantities for each case in the accompanying table and sketch the temperature distribution, indicating the direction of the heat flux.

CASE	T_1	T_2	dT/dx (K/m)	q''_x (W/m^2)
1	400 K	300 K		
2	100°C		−250	
3	80°C		+200	
4		−5°C		4000
5	30°C			−3000

2.8 Consider steady-state conditions for one-dimensional conduction in the system below having a thermal conductivity $k = 50$ W/m · K and a thickness $L = 0.25$ m, with no internal heat generation.

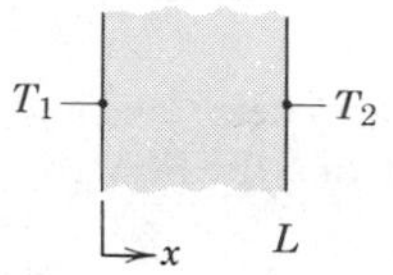

Determine the heat flux and the unknown quantity for each case and sketch the temperature distribution, indicating the direction of the heat flux.

CASE	T_1 (°C)	T_2 (°C)	dT/dx (K/m)
1	50	−20	
2	−30	−10	
3	70		160
4		40	−80
5		30	200

2.9 Consider a plane wall 100 mm thick and of thermal conductivity 100 W/m · K. Steady-state conditions are known to exist with $T_1 = 400$ K and $T_2 = 600$ K. Determine the heat flux q''_x and the temperature gradient dT/dx for the coordinate systems shown.

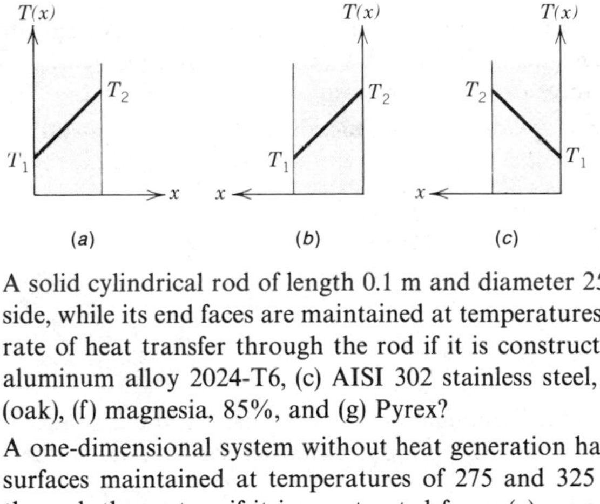

2.10 A solid cylindrical rod of length 0.1 m and diameter 25 mm is well insulated on its side, while its end faces are maintained at temperatures of 100 and 0°C. What is the rate of heat transfer through the rod if it is constructed from (a) pure copper, (b) aluminum alloy 2024-T6, (c) AISI 302 stainless steel, (d) silicon nitride, (e) wood (oak), (f) magnesia, 85%, and (g) Pyrex?

2.11 A one-dimensional system without heat generation has a thickness of 20 mm with surfaces maintained at temperatures of 275 and 325 K. Determine the heat flux through the system if it is constructed from: (a) pure aluminum, (b) plain carbon steel, (c) AISI 316 stainless steel, (d) Pyroceram, (e) Teflon, and (f) concrete.

2.12 A TV advertisement by a well-known insulation manufacturer states: it isn't the thickness of the insulating material that counts, it's the R-value. The ad shows that to obtain an R-value of 19, you need 18 ft of rock, 15 in. of wood, or just 6 in. of the manufacturer's insulation. Is this advertisement technically reasonable? If you are like most TV viewers, you don't know the R-value is defined as L/k, where L (in.) is the thickness of the insulation and k (Btu · in./hr · ft^2 · °F) is the thermal conductivity of the material.

2.13 In the two-dimensional body illustrated, the gradient at surface A is found to be $\partial T/\partial y = 30$ K/m. What are $\partial T/\partial y$ and $\partial T/\partial x$ at the surface B?

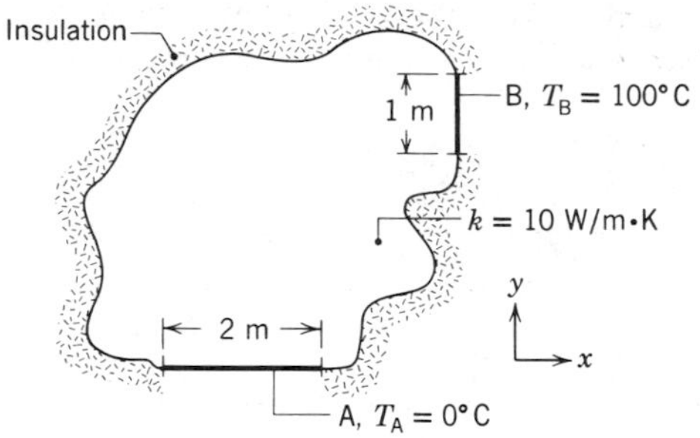

2.14 At a given instant of time the temperature distribution within an infinite homogeneous body is given by the function

$$T(x, y, z) = x^2 - 2y^2 + z^2 - xy + 2yz$$

Assuming constant properties and no internal heat generation, determine the regions where the temperature changes with time.

2.15 A cylinder of radius r_o, length L, and thermal conductivity k is immersed in a fluid of convection coefficient h and unknown temperature T_∞. At a certain instant the temperature distribution in the cylinder is $T(r) = a + br^2$, where a and b are constants. Obtain expressions for the heat transfer rate at r_o and the fluid temperature.

2.16 Uniform internal heat generation at $\dot{q}_1 = 5 \times 10^7$ W/m^3 is occurring in a cylindrical nuclear reactor fuel rod of 50 mm diameter, and under steady-state conditions the temperature distribution is of the form $T(r) = a + br^2$, where T is in degrees Celsius and r is in meters, while $a = 800$°C and $b = -4.167 \times 10^5$°C/m^2. The fuel rod properties are $k = 30$ W/m · K, $\rho = 1100$ kg/m^3, and $c_p = 800$ J/kg · K.

(a) What is the rate of heat transfer per unit length of the rod at $r = 0$ (the centerline) and at $r = 25$ mm (the surface)?

(b) If the reactor power level is suddenly increased to $\dot{q}_2 = 10^8$ W/m^3, what is the initial time rate of temperature change at $r = 0$ and $r = 25$ mm?

2.17 The steady-state temperature distribution in a one-dimensional wall of thermal conductivity 50 W/m · K and thickness 50 mm is observed to be T(°C)$= a + bx^2$, where $a = 200$°C, $b = -2000$°C/m^2, and x is in meters.

(a) What is the heat generation rate $\dot{q}$ in the wall?

(b) Determine the heat fluxes at the two wall faces. In what manner are these heat fluxes related to the heat generating rate?

2.18 The temperature distribution across a wall 0.3 m thick at a certain instant of time is $T(x) = a + bx + cx^2$ where T is in degrees Celsius and x is in meters, $a = 200$°C, $b = -200$°C/m and $c = 30$°C/m^2. The wall has a thermal conductivity of 1 W/m · K.

(a) On a unit surface area basis, determine the rate of heat transfer into and out of the wall and the rate of change of energy stored by the wall.

(b) If the cold surface is exposed to a fluid at 100°C, what is the convection coefficient?

2.19 A salt-gradient solar pond is a shallow body of water that consists of three distinct fluid layers and is used to collect solar energy. The upper- and lowermost layers are well mixed and serve to maintain the upper and lower surfaces of the central layer at uniform temperatures T_1 and T_2, where $T_2 > T_1$. Although there is bulk fluid motion in the mixed layers, there is no such motion in the central layer.

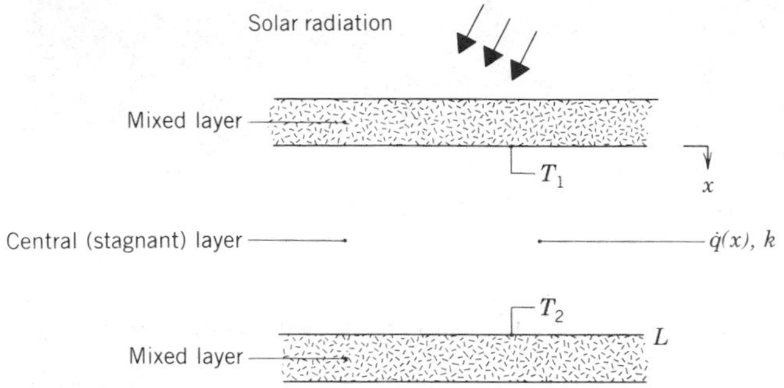

Consider conditions for which solar radiation absorption in the central layer provides nonuniform heat generation of the form $\dot{q} = Ae^{-ax}$, and the temperature distribution in the central layer is

$$T(x) = -\frac{A}{ka^2}e^{-ax} + Bx + C$$

The quantities A (W/m^3), a (1/m), B (K/m), and C (K) are known constants having the prescribed units, and k is the thermal conductivity, which is also constant.

(a) Obtain expressions for the rate at which heat is transferred per unit area from the lower mixed layer to the central layer and from the central layer to the upper mixed layer.

(b) Determine whether conditions are steady or transient.

(c) Obtain an expression for the rate at which thermal energy is generated in the entire central layer, per unit surface area.

2.20 The cylindrical system illustrated has negligible variation of temperature in the r and z directions. Assume that $\Delta r = r_o - r_i$ is small compared to r_i and denote the length in the z direction, normal to the page, as L.

(a) Beginning with a properly defined control volume and considering energy generation and storage effects, derive the differential equation that prescribes the variation in temperature with the angular coordinate ϕ. Compare your result with Equation 2.18.

(b) For steady-state conditions with no internal heat generation and constant properties, determine the temperature distribution $T(\phi)$ in terms of the constants T_1, T_2, r_i, and r_o. Is this distribution linear in ϕ?

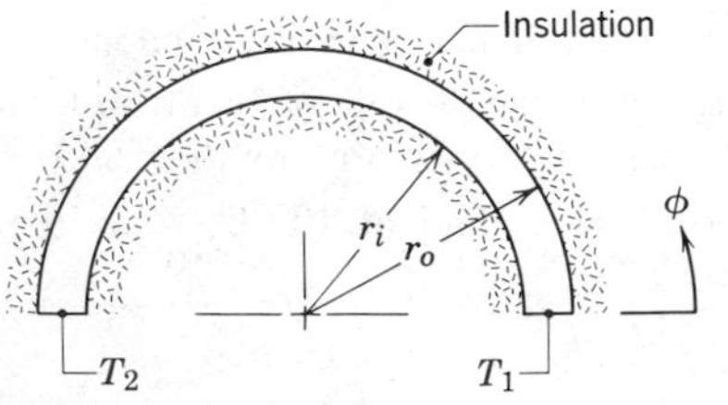

(c) For the conditions of part b write the expression for the heat rate q_ϕ.

2.21 Beginning with a differential control volume in the form of a cylindrical shell, derive the heat diffusion equation for a one-dimensional, cylindrical, radial coordinate system with internal heat generation. Compare your result with Equation 2.18.

2.22 Beginning with a differential control volume in the form of a spherical shell, derive the heat diffusion equation for a one-dimensional, spherical, radial coordinate system with internal heat generation. Compare your result with Equation 2.19.

2.23 Derive the heat diffusion equation, Equation 2.18, for cylindrical coordinates beginning with the differential control volume shown in Figure 2.9.

2.24 Derive the heat diffusion equation, Equation 2.19, for spherical coordinates beginning with the differential control volume shown in Figure 2.10.

2.25 In Example 2.3, we considered a copper bar that was initially at a uniform temperature and suddenly was heated by the passage of an electric current. Recall that the emphasis of the problem was to identify boundary and initial conditions that must be used in the conduction analysis. Assume $T_\infty > T_o$.

(a) On T–x coordinates, sketch the temperature distributions for the following conditions: initial condition ($t \leq 0$), steady-state condition ($t \rightarrow \infty$), and for two intermediate time conditions. Assume that the electric current is sufficiently large that the bar outer surface ($x = L$) will eventually be hotter than the air.

(b) On q''_x–t coordinates, sketch the heat flux at the bar faces. That is, show qualitatively how $q''_x(0,t)$ and $q''_x(L,t)$ vary with time.

2.26 The one-dimensional system of mass M with constant properties and no internal heat generation shown in the figure is initially at a uniform temperature T_i. The electrical heater is suddenly energized providing a uniform heat flux q''_o at the surface $x = 0$. The boundaries at $x = L$ and elsewhere are perfectly insulated.

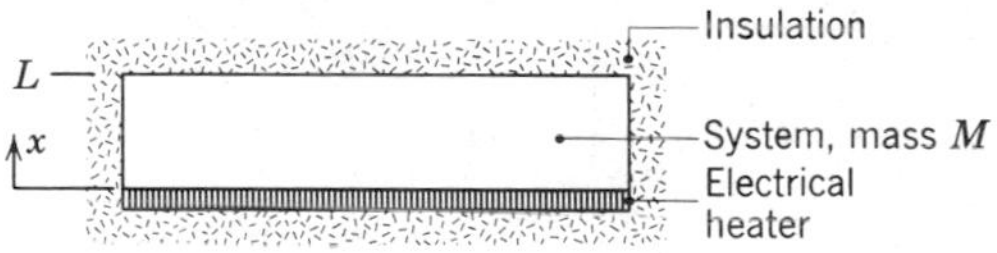

(a) Write the differential equation and identify the boundary and initial conditions that could be used to determine the temperature as a function of position and time in the system.

(b) On T–x coordinates, sketch the temperature distributions for the initial condition ($t \leq 0$) and for several times after the heater is energized. Will a steady-state temperature distribution ever be reached?

(c) On q''_x–t coordinates, sketch the heat flux $q''_x(x,t)$ at the planes $x = 0$, $x = L/2$, and $x = L$ as a function of time.

(d) After a period of time designated t_e has elapsed, the heater power is switched off. Assuming that the insulation is perfect, the system will eventually reach a final uniform temperature T_f. Derive an expression that can be used to determine T_f as a function of the parameters q''_o, t_e, T_i, and the system characteristics M, c_p, and A_s (the heater surface area).

2.27 The plane wall with constant properties and no internal heat generation shown in the figure is initially at a uniform temperature T_i. Suddenly the surface at $x = L$ is exposed to a heating process with a fluid at T_∞ having a convection heat transfer coefficient h. The boundary at $x = 0$ is perfectly insulated.

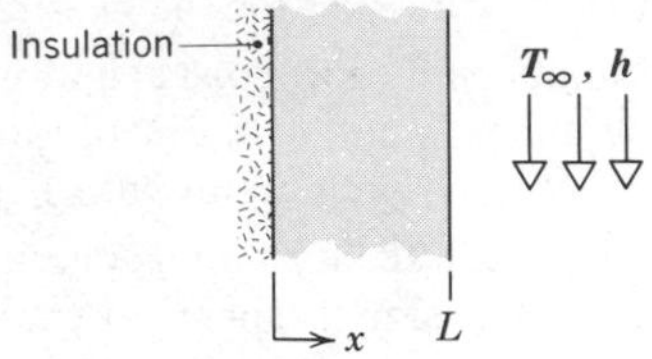

(a) Write the differential equation and identify the boundary and initial conditions that could be used to determine the temperature as a function of position and time in the wall.

(b) On T–x coordinates, sketch the temperature distributions for the following conditions: initial condition ($t \leq 0$), steady-state condition ($t \to \infty$), and two intermediate times.

(c) On q_x''–t coordinates, sketch the heat flux at the locations $x = 0$ and $x = L$. That is, show qualitatively how $q_x''(0,t)$ and $q_x''(L,t)$ vary with time.

(d) Write an expression for the total energy transferred to the wall per unit volume of the wall (J/m^3).

2.28 The plane wall illustrated has constant properties and no internal generation and is initially at a uniform temperature T_i. Suddenly, the surface at $x = L$ is exposed to a heating process with a fluid at T_∞ having a convection coefficient h. At the same instant, the electrical heater is energized providing a constant heat flux q_o'' at $x = 0$.

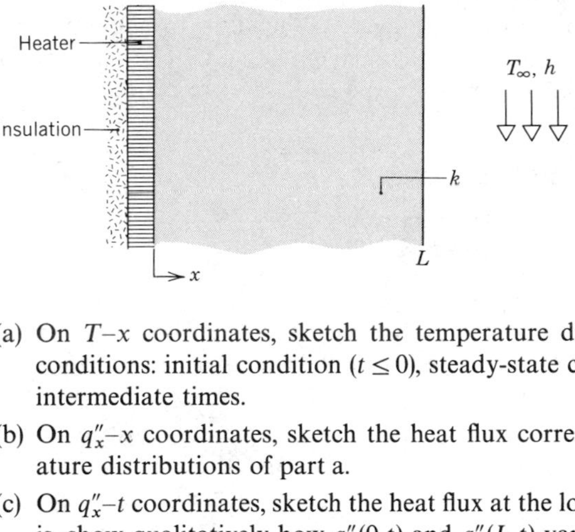

(a) On T–x coordinates, sketch the temperature distributions for the following conditions: initial condition ($t \leq 0$), steady-state condition ($t \to \infty$), and for two intermediate times.

(b) On q_x''–x coordinates, sketch the heat flux corresponding to the four temperature distributions of part a.

(c) On q_x''–t coordinates, sketch the heat flux at the locations $x = 0$ and $x = L$. That is, show qualitatively how $q_x''(0,t)$ and $q_x''(L,t)$ vary with time.

(d) Derive an expression for the steady-state temperature at the heater surface, $T(0, \infty)$, in terms of q_o'', T_∞, k, h, and L

Three

ONE-DIMENSIONAL, STEADY-STATE CONDUCTION

In this chapter we treat situations for which heat is transferred by diffusion under *one-dimensional, steady-state* conditions. The term "one-dimensional" refers to the fact that only one coordinate is needed to describe the spatial variation of the dependent variables. Hence in a *one-dimensional* system, temperature gradients exist along only a single coordinate direction, and heat transfer occurs exclusively in that direction. The system is characterized by *steady-state* conditions if the temperature at each point is independent of time. Despite their inherent simplicity, one-dimensional, steady-state models may be used to accurately represent numerous engineering systems.

We begin our consideration of one-dimensional, steady-state conduction by discussing heat transfer with no internal generation (Sections 3.1 to 3.3). The objective is to determine the temperature distribution and heat transfer rate in common geometries. The concept of thermal resistance (analogous to electrical resistance) is introduced as an aid to solving conduction heat transfer problems. The effect of internal heat generation on the temperature distribution and heat rate is then treated (Section 3.4). Finally, conduction analysis is used to describe the performance of extended surfaces or fins, wherein the role of convection at the external boundary must be considered (Section 3.5).

3.1 THE PLANE WALL

For one-dimensional conduction in a plane wall, temperature is a function of the x coordinate only and heat is transferred exclusively in this direction. In Figure 3.1a, a plane wall separates two fluids of different temperatures. Heat transfer occurs by convection from the hot fluid at $T_{\infty,1}$ to one surface of the wall at $T_{s,1}$, by conduction through the wall, and by convection from the other surface of the wall at $T_{s,2}$ to the cold fluid at $T_{\infty,2}$.

We begin by considering conditions *within* the wall. We first determine the temperature distribution, from which we can obtain the conduction heat transfer rate.

3.1.1 Temperature Distribution

The temperature distribution in the wall can be determined by solving the heat diffusion equation with the proper boundary conditions. For steady-state conditions with no distributed source or sink of energy within the wall, the appropriate form of the heat equation, Equation 2.17, is

$$\frac{d}{dx}\left(k\frac{dT}{dx}\right)=0 \tag{3.1}$$

If the thermal conductivity of the wall material is assumed to be constant, the equation may be integrated twice to obtain the *general solution*

$$T(x)=C_1x+C_2 \tag{3.2}$$

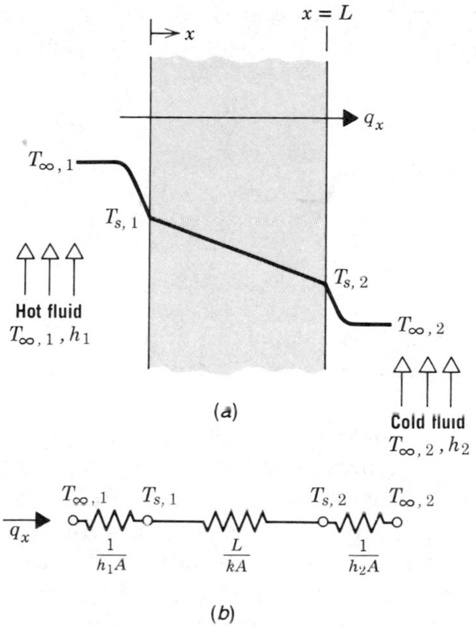

Figure 3.1 Heat transfer through a plane wall.
(*a*) Temperature distribution.
(*b*) Equivalent thermal circuit.

To obtain the constants of integration, C_1 and C_2, boundary conditions must be introduced. We choose to work with the constant surface conditions

$$T(0) = T_{s,1} \qquad \text{and} \qquad T(L) = T_{s,2}$$

Applying the condition at $x = 0$ to the general solution, it follows that

$$T_{s,1} = C_2$$

Similarly, at $x = L$

$$T_{s,2} = C_1 L + C_2 = C_1 L + T_{s,1}$$

in which case

$$\frac{T_{s,2} - T_{s,1}}{L} = C_1$$

Substituting into the general solution, the temperature distribution is then

$$T(x) = (T_{s,2} - T_{s,1})\frac{x}{L} + T_{s,1} \tag{3.3}$$

From this result it is evident that, *for one-dimensional, steady-state conduction in a plane wall with no heat generation and constant thermal conductivity, the temperature varies linearly with x.*

Now that we have the temperature distribution, we may use Fourier's law, Equation 2.1, to determine the conduction heat transfer rate. That is,

$$q_x = -kA\frac{dT}{dx} = \frac{kA}{L}(T_{s,1} - T_{s,2}) \tag{3.4}$$

Note that A is the area of the wall *normal* to the direction of heat transfer and that the heat flux is

$$q_x'' = \frac{q_x}{A} = \frac{k}{L}(T_{s,1} - T_{s,2}) \tag{3.5}$$

Note also that the heat flux is a constant independent of x. This result is consistent with Equation 3.1, which implies that there is no change in the heat flux for the x direction.

In the foregoing paragraphs we have used the *standard approach* to solving conduction problems. That is, the general solution for the temperature distribution is obtained from the appropriate form of the heat equation. The boundary conditions are then applied to obtain the particular solution, which is used with Fourier's law to determine the heat transfer rate. Note that our method has been to prescribe surface temperatures at $x = 0$ and $x = L$ as boundary conditions, even though it is the fluid temperatures, and not the surface temperatures, that are typically known. However, since adjoining fluid and surface temperatures are easily related through a surface energy balance, it is a simple matter to express Equations 3.3 to 3.5 in terms of fluid, rather than surface, temperatures. Equivalent results could be obtained directly by using the surface energy balances as boundary conditions in evaluating the constants of Equation 3.2 (see Problem 3.1).

3.1.2 Thermal Resistance

At this point we note that a very important concept is suggested by Equation 3.4. In particular, there exists an analogy between the diffusion of heat and electrical charge. Just as an electrical resistance may be associated with the conduction of electricity, a thermal resistance may be associated with the conduction of heat. Defining resistance as the ratio of a driving potential to the corresponding transfer rate, it follows from Equation 3.4 that the *thermal resistance for conduction* is

$$R_{t,\text{cond}} \equiv \frac{T_{s,1} - T_{s,2}}{q_x} = \frac{L}{kA} \tag{3.6}$$

Similarly, for electrical conduction in the same system, Ohm's law provides an electrical resistance of the form

$$R_e = \frac{E_{s,1} - E_{s,2}}{I} = \frac{L}{\sigma A} \tag{3.7}$$

The analogy between Equations 3.6 and 3.7 is obvious. A thermal resistance may also be associated with heat transfer by convection at a surface. From Newton's law of cooling,

$$q = hA(T_s - T_\infty) \tag{3.8}$$

the *thermal resistance for convection* is then

$$R_{t,\text{conv}} \equiv \frac{T_s - T_\infty}{q} = \frac{1}{hA} \tag{3.9}$$

Circuit representations provide a useful tool for both conceptualizing and quantifying heat transfer problems. The *equivalent thermal circuit* for the plane wall with convection surface conditions is shown in Figure 3.1*b*. The heat transfer rate may be determined from separate consideration of each element in the network. That is,

$$q_x = \frac{T_{\infty,1} - T_{s,1}}{1/h_1 A} = \frac{T_{s,1} - T_{s,2}}{L/kA} = \frac{T_{s,2} - T_{\infty,2}}{1/h_2 A} \tag{3.10}$$

In terms of the *overall temperature difference* $T_{\infty,1} - T_{\infty,2}$ and the *total thermal resistance* R_{tot}, the heat transfer rate may also be expressed as

$$q_x = \frac{T_{\infty,1} - T_{\infty,2}}{R_{\text{tot}}} \tag{3.11}$$

Because the conduction and convection resistances are in series and may be summed, it follows that

$$R_{\text{tot}} = \frac{1}{h_1 A} + \frac{L}{kA} + \frac{1}{h_2 A} \tag{3.12}$$

3.1.3 The Composite Wall

Equivalent thermal circuits may also be used for more complex systems, such as *composite walls.* Such walls may involve any number of series and parallel thermal resistances due to layers of different materials. Consider the series composite wall of Figure 3.2. The one-dimensional heat transfer rate for this system may be expressed as

$$q_x = \frac{T_{\infty,1} - T_{\infty,4}}{\Sigma R_t} \tag{3.13}$$

where $T_{\infty,1} - T_{\infty,4}$ is the *overall* temperature difference and the summation includes all thermal resistances. Hence,

$$q_x = \frac{T_{\infty,1} - T_{\infty,4}}{[(1/h_1 A) + (L_A/k_A A) + (L_B/k_B A) + (L_C/k_C A) + (1/h_4 A)]} \tag{3.14}$$

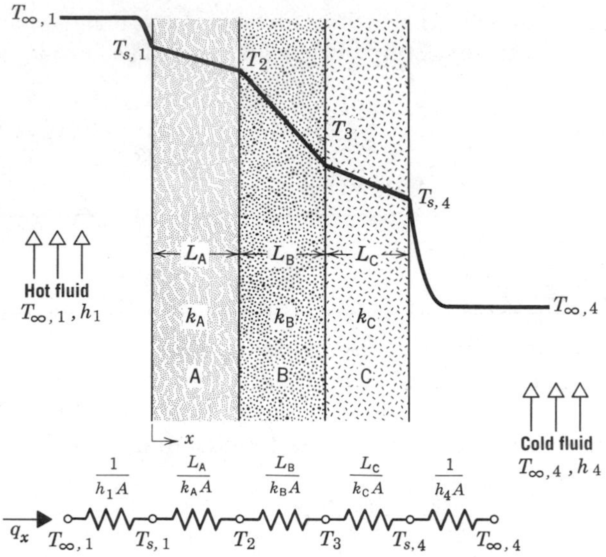

Figure 3.2 Equivalent thermal circuit for a series composite wall.

Alternatively, the heat transfer rate can be related to the temperature difference and resistance associated with each element. For example,

$$q_x = \frac{T_{\infty,1} - T_{s,1}}{(1/h_1 A)} = \frac{T_{s,1} - T_2}{(L_A/k_A A)} = \frac{T_2 - T_3}{(L_B/k_B A)} = \cdots \tag{3.15}$$

With composite systems it is often convenient to work with an *overall heat transfer coefficient*, U, which is defined by an expression analogous to Newton's law of cooling. Accordingly,

$$q_x \equiv UA\Delta T \tag{3.16}$$

where ΔT is the overall temperature difference. Note that the overall heat transfer coefficient is related to the total thermal resistance. From Equations 3.13 and 3.16 we see that $UA = 1/R_{tot}$, in which case, for the composite wall of Figure 3.2,

$$U = \frac{1}{R_{tot} A} = \frac{1}{[(1/h_1) + (L_A/k_A) + (L_B/k_B) + (L_C/k_C) + (1/h_4)]} \tag{3.17}$$

In general, we may write

$$R_{tot} = \Sigma R_t = \frac{\Delta T}{q} = \frac{1}{UA} \tag{3.18}$$

Composite walls may also be characterized by series-parallel configurations, such as that shown in Figure 3.3. Although the heat flow is now

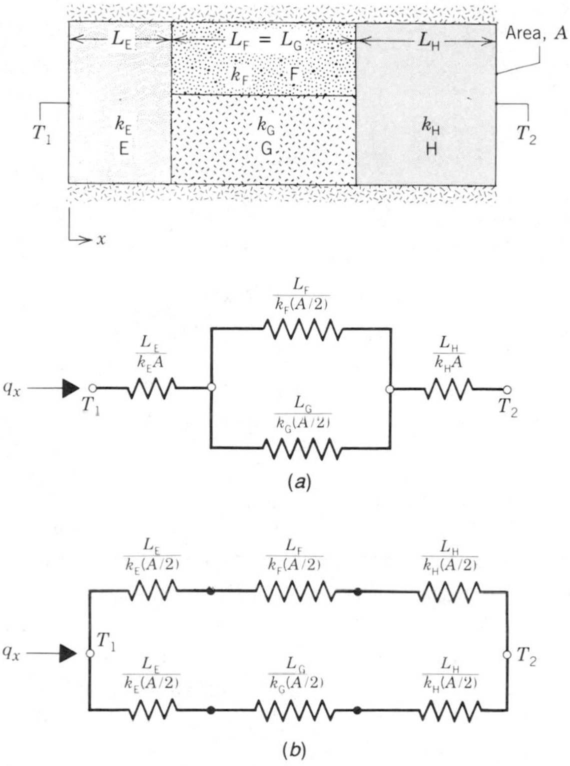

Figure 3.3 Equivalent thermal circuit for a series-parallel composite wall.

two-dimensional, it is often reasonable to assume one-dimensional conditions. Subject to this assumption, two different thermal circuits may be used. For case (*a*) it is presumed that surfaces normal to the x direction are isothermal, while for case (*b*) it is assumed that surfaces parallel to the x direction are adiabatic. Different results are obtained for R_{tot}, and the corresponding values of q bracket the actual heat transfer rate. These differences increase with increasing $|k_F - k_G|$, as two-dimensional effects become more significant.

3.1.4 Contact Resistance

Although neglected until now, it is important to recognize that, in composite systems, the temperature drop across the interface between materials may be appreciable. This temperature change may be attributed to what is known as the *thermal contact resistance*, $R_{t,c}$. The effect is shown in Figure 3.4, and for a unit area the resistance is defined as

$$R''_{t,c} = \frac{T_A - T_B}{q''_x} \tag{3.19}$$

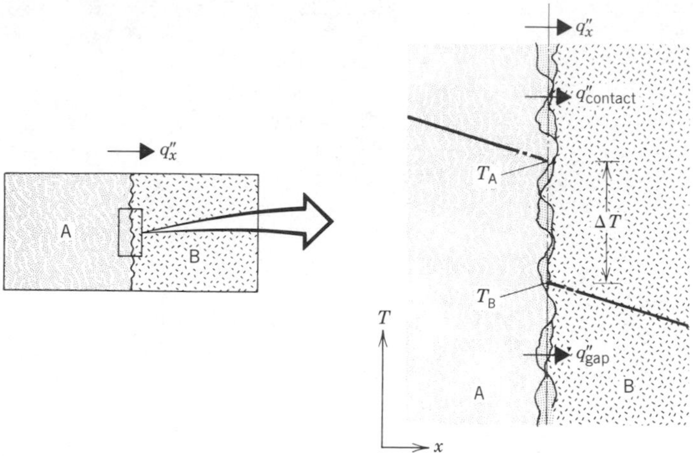

Figure 3.4 Temperature drop due to thermal contact resistance.

The existence of a finite contact resistance is due principally to surface roughness effects. Contact spots are interspersed with gaps that are, in most instances, air filled. Heat transfer is therefore due to conduction across the actual contact area and to conduction (or natural convection) and radiation across the gaps. The contact resistance may therefore be viewed as two parallel resistances: that due to the contact spots and that due to the gaps. If the contact area is small, as it is for rough surfaces, the major contribution to the resistance is made by the gaps. The resistance decreases with decreasing surface roughness and increasing joint pressure.

Although theories have been developed for the prediction of $R''_{t,c}$, the most reliable results are those which have been obtained experimentally. The effect of loading on metallic interfaces can be seen in Table 3.1*a*, which presents the approximate range of thermal resistance values under vacuum conditions. The variation of thermal resistance for an aluminum interface due to interfacial fluids of different thermal conductivity is shown in Table 3.1*b*. More detailed discussions of thermal contact resistance are given by Fried [1] and Schneider [2].

Table 3.1 Thermal contact resistance for (*a*) metallic interfaces under vacuum conditions and (*b*) aluminum interface (10-μm surface roughness, 10^5 N/m^2) with different interfacial fluids [1].

THERMAL RESISTANCE, $R''_{t,c} \times 10^4$ (m$^2 \cdot$K/W)

(a) Vacuum Interface			**(b) Interfacial Fluid**	
Contact pressure	100 kN/m^2	10,000 kN/m^2	Air	2.75
Stainless steel	6–25	0.7–4.0	Helium	1.05
Copper	1–10	0.1–0.5	Hydrogen	0.720
Magnesium	1.5–3.5	0.2–0.4	Silicone oil	0.525
Aluminum	1.5–5.0	0.2–0.4	Glycerine	0.265

EXAMPLE 3.1

A leading manufacturer of household appliances is proposing a self-cleaning oven design that involves use of a composite window separating the oven cavity from the room air. The composite is to consist of two high temperature plastics (A and B) of thicknesses $L_A = 2L_B$ and thermal conductivities $k_A = 0.15$ W/m·K and $k_B = 0.08$ W/m·K. During the self-cleaning process, the oven wall and air temperatures, T_w and T_a, are 400°C, while the room air temperature T_∞ is 25°C. The inside convection and radiation heat transfer coefficients h_i and h_r, as well as the outside convection coefficient h_o, are each approximately 25 W/m²·K. What is the minimum window thickness, $L = L_A + L_B$, needed to ensure a temperature that is 50°C or less at the outer surface of the window? This temperature must not be exceeded for safety reasons.

SOLUTION

Known:

The properties and relative dimensions of plastic materials used for a composite oven window, and conditions associated with self-cleaning operation.

Find: Composite thickness $L_A + L_B$ needed to ensure safe operation.

Schematic:

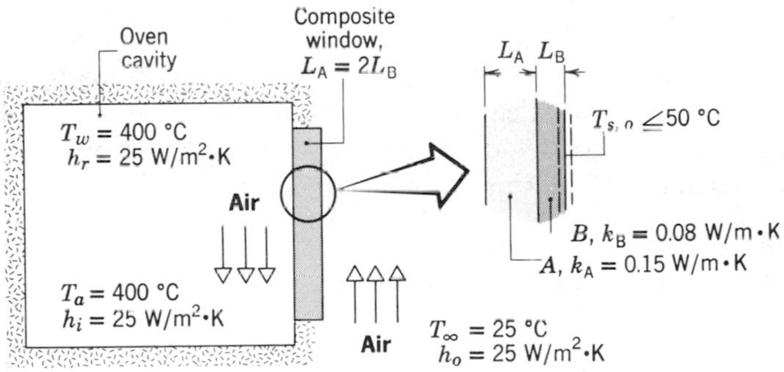

Assumptions:

1. Steady-state conditions exist.
2. Conduction through the window is one-dimensional.
3. Contact resistance is negligible.
4. Radiation absorption *within* the window is negligible; hence no internal heat generation (radiation exchange between window and oven walls occurs at the window inner surface).

5. Radiation exchange between window outer surface and surroundings is negligible.
6. Each plastic is homogeneous with constant properties.

Analysis:

The thermal circuit can be constructed by recognizing that there is resistance to heat flow associated with convection at the outer surface, conduction in the plastics, and convection and radiation at the inner surface. Accordingly, the circuit and the resistances are of the following form.

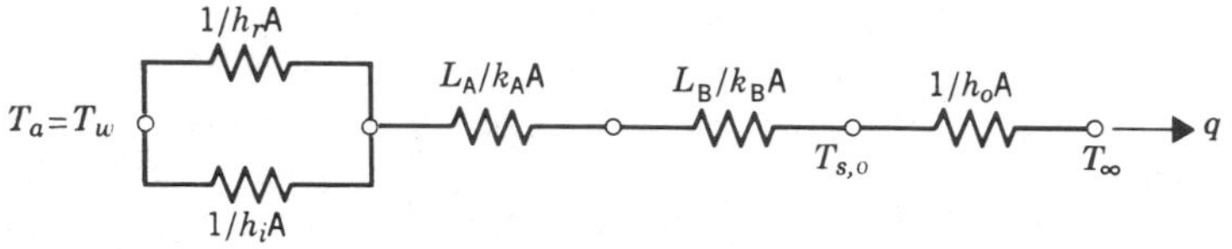

Since the outer surface temperature of the window, $T_{s,o}$, is prescribed, the required window thickness may be obtained by applying an energy balance at this surface. That is, from Equation 1.11

$$\dot{E}_{\text{in}} = \dot{E}_{\text{out}}$$

where, from Equation 3.18, with $T_w = T_a$,

$$\dot{E}_{\text{in}} = q = \frac{T_a - T_{s,o}}{\Sigma R_t}$$

and from Equation 3.8

$$\dot{E}_{\text{out}} = q = h_o A(T_{s,o} - T_\infty)$$

The total thermal resistance between the oven cavity and the outer surface of the window includes an equivalent resistance associated with convection and radiation, which act in parallel at the inner surface of the window, and the conduction resistances of the window materials. Hence

$$\Sigma R_t = \left(\frac{1}{1/h_i A} + \frac{1}{1/h_r A}\right)^{-1} + \frac{L_{\text{A}}}{k_{\text{A}} A} + \frac{L_{\text{B}}}{k_{\text{B}} A}$$

or

$$\Sigma R_t = \frac{1}{A}\left(\frac{1}{h_i + h_r} + \frac{L_{\text{A}}}{k_{\text{A}}} + \frac{L_{\text{A}}}{2k_{\text{B}}}\right)$$

Substituting into the energy balance, it follows that

$$\frac{T_a - T_{s,o}}{(h_i + h_r)^{-1} + (L_{\text{A}}/k_{\text{A}}) + (L_{\text{A}}/2k_{\text{B}})} = h_o(T_{s,o} - T_\infty)$$

Hence, solving for L_{A},

$$L_A = \frac{(1/h_o)(T_a - T_{s,o})/(T_{s,o} - T_\infty) - (h_i + h_r)^{-1}}{(1/k_A + 1/2k_B)}$$

$$L_A = \frac{0.04 \text{ m}^2 \cdot \text{K/W} \left(\dfrac{400 - 50}{50 - 25}\right) - 0.02 \text{ m}^2 \cdot \text{K/W}}{(1/0.15 + 1/0.16) \text{ m} \cdot \text{K/W}} = 0.0418 \text{ m}$$

Since $L_B = L_A/2 = 0.0209$ m,

$$L = L_A + L_B = 0.0627 \text{ m} = 62.7 \text{ mm} \quad \triangleleft$$

Comments:

The self-cleaning operation is a transient process, as far as the thermal response of the window is concerned, and steady-state conditions may not be reached in the time required for cleaning. However, the steady-state condition provides the maximum possible value of $T_{s,o}$ and hence is well suited for the design calculation.

3.2 AN ALTERNATIVE CONDUCTION ANALYSIS

The conduction analysis of Section 3.1 was performed using the *standard approach*. That is, the heat equation was solved to obtain the temperature distribution, Equation 3.3, and Fourier's law was then applied to obtain the heat transfer rate, Equation 3.4. However, an alternative approach may be used for the conditions presently of interest. Considering conduction in the system of Figure 3.5, we recognize that, for *steady-state conditions* with *no heat generation* and *no heat loss from the sides*, the heat transfer rate q_x must be a constant independent of x. That is, for any differential element dx, $q_x = q_{x+dx}$. This condition is, of course, a consequence of the energy conservation requirement, and it must apply even if the area varies with position $A(x)$ and the thermal conductivity varies with temperature $k(T)$. Moreover, even though the temper-

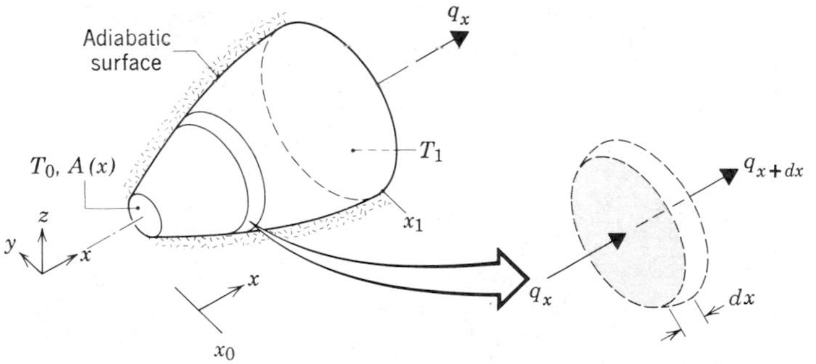

Figure 3.5 System with a constant conduction heat transfer rate.

ature distribution may be two dimensional, varying with x and y, it is often reasonable to neglect the y variation and to assume a *one-dimensional* distribution in x.

For the above conditions it is possible to work exclusively with Fourier's law when performing a conduction analysis. In particular, since the conduction rate is a *constant*, the rate equation may be *integrated*, even though neither the rate nor the temperature distribution is known. Consider Fourier's law, Equation 2.1, which may be applied to the system of Figure 3.5. Although we may have no knowledge of the value of q_x or the form of $T(x)$, we do know that q_x is a constant. Hence we may express Fourier's law in the integral form

$$q_x \int_{x_0}^{x} \frac{dx}{A(x)} = -\int_{T_0}^{T} k(T)\, dT \tag{3.20}$$

The cross-sectional area may be a known function of x, and the material thermal conductivity may vary with temperature in a known manner. If the integration is performed from a point x_0 at which the temperature T_0 is known, the resulting equation provides the functional form of $T(x)$. Moreover, if the temperature $T = T_1$ at some $x = x_1$ is also known, integration between x_0 and x_1 provides an expression from which q_x may be computed. Note that, if the area A is uniform and k is independent of temperature, Equation 3.20 reduces to

$$\frac{q_x \Delta x}{A} = -k\Delta T \tag{3.21}$$

where $\Delta x = x_1 - x_0$ and $\Delta T = T_1 - T_0$.

We frequently elect to solve diffusion problems by working with integrated forms of the diffusion rate equations. However, the limiting conditions for which this may be done should be firmly fixed in our minds: *steady-state* and *one-dimensional* transfer with *no heat generation.*

EXAMPLE 3.2

The diagram shows a conical section fabricated from pyroceram. It is of circular cross section with the diameter $D = ax$, where $a = 0.25$. The small end is at $x_1 = 50$ mm and the large end at $x_2 = 250$ mm. The end temperatures are $T_1 = 400$ K and $T_2 = 600$ K, while the lateral surface is well insulated.

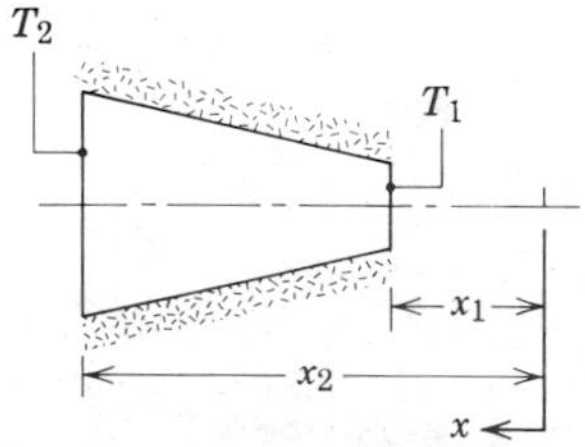

1. Derive an expression for the temperature distribution $T(x)$ in symbolic form, assuming one-dimensional conditions. Sketch the temperature distribution.
2. Calculate the heat rate q_x through the cone.

SOLUTION

Known:

Conduction in a circular conical section having a diameter $D = ax$, where $a = 0.25$.

Find:

1. Temperature distribution $T(x)$.
2. Heat transfer rate q_x.

Schematic:

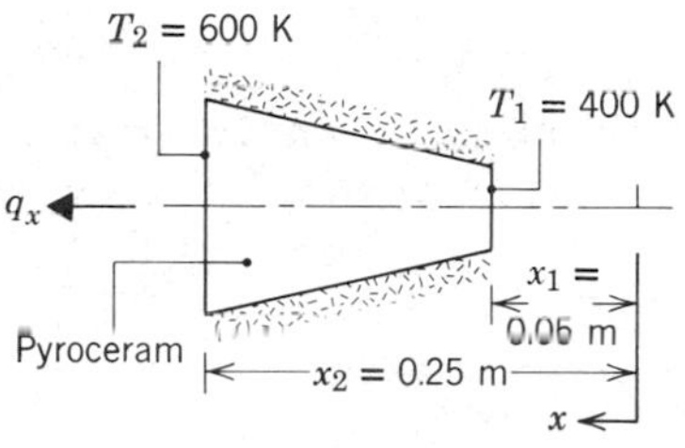

Assumptions:

1. Steady-state conditions.
2. One-dimensional conduction in the x direction.
3. No internal heat generation.
4. Constant properties.

Properties:

Table A.2, pyroceram (500 K): $k = 3.46$ W/m · K.

Analysis:

1. Since heat conduction occurs under steady-state, one-dimensional conditions with no internal heat generation, the heat transfer rate q_x is a constant independent of x. Accordingly, Fourier's law, Equation 2.1,

may be used to determine the temperature distribution

$$q_x = -kA\frac{dT}{dx}$$

With $A = \pi D^2/4 = \pi a^2 x^2/4$ and separating variables

$$\frac{4q_x dx}{\pi a^2 x^2} = -k\,dT$$

Integrating from x_1 to any x within the cone, and recalling that q_x and k are constants, it follows that

$$\frac{4q_x}{\pi a^2}\int_{x_1}^{x}\frac{dx}{x^2} = -k\int_{T_1}^{T} dT$$

with the result

$$\frac{4q_x}{\pi a^2}\left(-\frac{1}{x}+\frac{1}{x_1}\right) = -k(T - T_1)$$

or solving for T

$$T(x) = T_1 - \frac{4q_x}{\pi a^2 k}\left(\frac{1}{x_1}-\frac{1}{x}\right)$$

Although q_x is a constant, it is as yet an unknown. However, it may be determined by evaluating the above expression at $x = x_2$ where $T(x_2) = T_2$. Hence

$$T_2 = T_1 - \frac{4q_x}{\pi a^2 k}\left(\frac{1}{x_1}-\frac{1}{x_2}\right)$$

and solving for q_x

$$q_x = \frac{\pi a^2 k(T_1 - T_2)}{4[(1/x_1)-(1/x_2)]}$$

Substituting for q_x into the expression for $T(x)$, the temperature distribution becomes

$$T(x) = T_1 + (T_1 - T_2)\left[\frac{(1/x)-(1/x_1)}{(1/x_1)-(1/x_2)}\right] \qquad \triangleleft$$

From this result, temperature may be calculated as a function of x and the distribution is as shown.

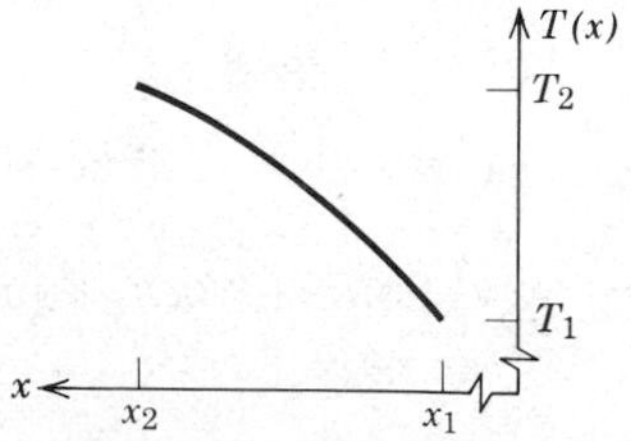

Note that, since $dT/dx = -4q_x/k\pi a^2 x^2$ from Fourier's law, it follows that the temperature gradient and heat flux decrease with increasing x.

2. Substituting numerical values into the foregoing result for the heat transfer rate, it follows that

$$q_x = \frac{\pi(0.25)^2 \times 3.46\ \text{W/m}\cdot\text{K}\ (400-600)\ \text{K}}{4\left(\dfrac{1}{0.05\ \text{m}} - \dfrac{1}{0.25\ \text{m}}\right)} = -2.12\ \text{W} \quad \triangleleft$$

Comments:

When the parameter a increases, the one-dimensional assumption becomes less appropriate. That is, the assumption worsens when the cross-sectional area change with distance is more pronounced.

3.3 RADIAL SYSTEMS

Cylindrical and spherical systems often experience temperature gradients in the radial direction only and may therefore be treated as one dimensional. Moreover, under steady-state conditions with no heat generation, such systems may be analyzed by using the *standard* method, which begins with the appropriate form of the heat equation, or the *alternative* method, which begins with the appropriate form of Fourier's law. In this section, the cylindrical system is analyzed by means of the standard method and the spherical system by means of the alternative method.

3.3.1 The Cylinder

A common example is the hollow cylinder, whose inner and outer surfaces are exposed to fluids at different temperatures (Figure 3.6). For steady-state conditions, with no heat generation, the appropriate form of the heat equation, Equation 2.18, is

$$\frac{1}{r}\frac{d}{dr}\left(kr\frac{dT}{dr}\right) = 0 \tag{3.22}$$

where, for the moment, k is treated as a variable. The physical significance of this result becomes evident if we also consider the appropriate form of Fourier's law. The rate at which energy is conducted across any cylindrical surface in the solid may be expressed as

$$q_r = -kA\frac{dT}{dr} = -k(2\pi rL)\frac{dT}{dr} \tag{3.23}$$

where $A = 2\pi rL$ is the area normal to the direction of heat transfer. From Equations 3.22 and 3.23 it follows that the conduction *heat transfer rate* q_r (*not* the heat flux q_r'') is a *constant in the radial direction.*

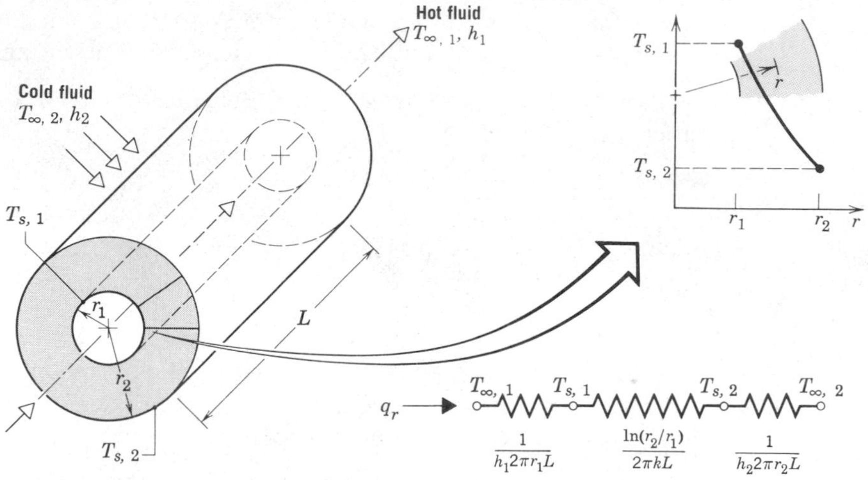

Figure 3.6 Hollow cylinder with convective surface conditions.

We may determine the temperature distribution in the cylinder by solving Equation 3.22 and applying appropriate boundary conditions. Assuming the value of k to be constant, Equation 3.22 may be integrated twice to obtain the general solution

$$T(r) = C_1 \ln r + C_2 \tag{3.24}$$

To obtain the constants of integration C_1 and C_2, we introduce the following boundary conditions.

$$T(r_1) = T_{s,1} \qquad \text{and} \qquad T(r_2) = T_{s,2}$$

Applying these conditions to the general solution, we then obtain

$$T_{s,1} = C_1 \ln r_1 + C_2 \qquad \text{and} \qquad T_{s,2} = C_1 \ln r_2 + C_2$$

Solving for C_1 and C_2 and substituting back into the general solution, we then obtain

$$T(r) = \frac{T_{s,1} - T_{s,2}}{\ln (r_1/r_2)} \ln \left(\frac{r}{r_2}\right) + T_{s,2} \tag{3.25}$$

Note that the temperature distribution associated with radial conduction through a cylinder is logarithmic, not linear, as it is for the plane wall under the same conditions.

If the temperature distribution, Equation 3.25, is now used with Fourier's law, Equation 3.23, we obtain the following expression for the heat transfer rate

$$q_r = \frac{2\pi L k (T_{s,1} - T_{s,2})}{\ln (r_2/r_1)} \tag{3.26}$$

From this result it is evident that, for conduction in hollow cylinders, the thermal resistance is of the form

$R_{t,conv} = \frac{1}{2\pi r h L}$

$$R_{t,\text{cond}} = \frac{\ln (r_2/r_1)}{2\pi L k} \qquad (3.27)$$

hollow cylinders

Note that since the value of q_r is independent of r, the foregoing result could have been obtained by using the alternative method, that is, by integrating Equation 3.23.

Consider now the composite system of Figure 3.7. Recalling how we treated the composite plane wall and neglecting the interfacial contact resistances, the heat transfer rate may be expressed as

$$q_r = \frac{T_{\infty,1} - T_{\infty,4}}{\dfrac{1}{2\pi r_1 L h_1} + \dfrac{\ln (r_2/r_1)}{2\pi k_A L} + \dfrac{\ln (r_3/r_2)}{2\pi k_B L} + \dfrac{\ln (r_4/r_3)}{2\pi k_C L} + \dfrac{1}{2\pi r_4 L h_4}} \qquad (3.28)$$

The foregoing result may also be expressed in terms of an overall heat transfer coefficient. That is,

$$q_r = \frac{T_{\infty,1} - T_{\infty,4}}{R_{\text{tot}}} = U_1 A_1 (T_{\infty,1} - T_{\infty,4}) \qquad (3.29)$$

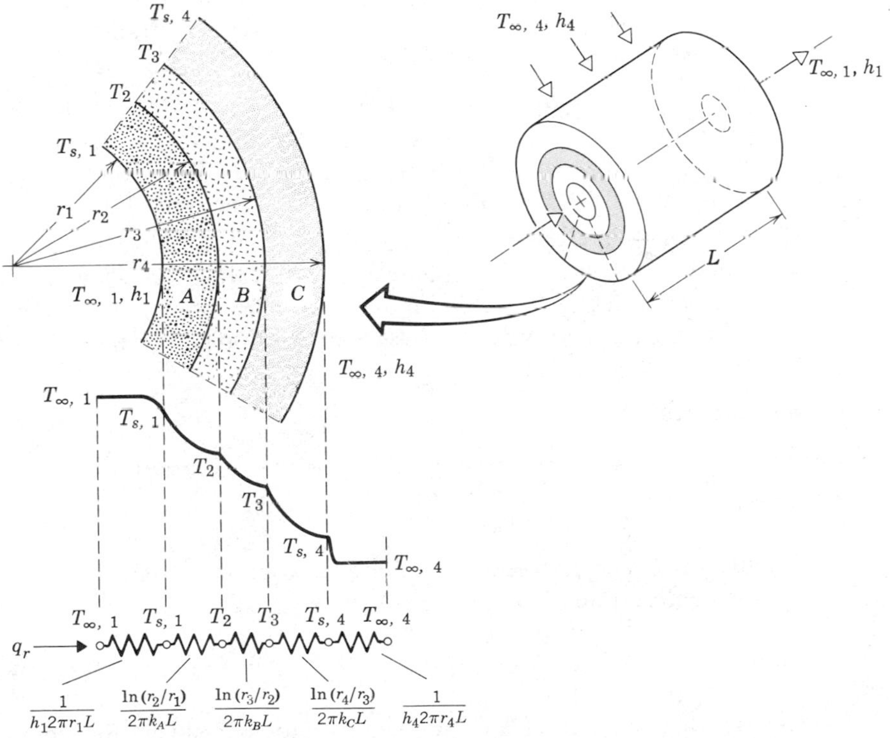

Figure 3.7 Temperature distribution for a composite cylindrical wall.

where $A_1 = 2\pi r_1 L$ and

$$U_1 = \frac{1}{\dfrac{1}{h_1} + \dfrac{r_1}{k_A}\ln\dfrac{r_2}{r_1} + \dfrac{r_1}{k_B}\ln\dfrac{r_3}{r_2} + \dfrac{r_1}{k_C}\ln\dfrac{r_4}{r_3} + \dfrac{r_1}{r_4}\dfrac{1}{h_4}} \tag{3.30}$$

Equation 3.29 defines U in terms of the inside surface area A_1 of the composite cylinder. This definition is *arbitrary*, and the overall coefficient may also be defined in terms of A_4 or any of the intermediate areas. Note that

$$U_1 A_1 = U_2 A_2 = U_3 A_3 = U_4 A_4 = (\Sigma R_t)^{-1} \tag{3.31}$$

and the specific forms of U_2, U_3, and U_4 may be inferred from Equation 3.28.

EXAMPLE 3.3

The possible existence of an optimum insulation thickness for radial systems is suggested by the presence of competing effects associated with an increase in this thickness. In particular, although the conduction resistance increases with the addition of insulation, the convection resistance decreases due to increasing outer surface area. Hence there may exist an insulation thickness that minimizes heat loss by maximizing the total resistance to heat transfer. Resolve this question by considering the following system.

1. A thin-walled copper tube of radius r_i is used to transport a low temperature refrigerant and is at a temperature T_i that is less than that of the ambient air at T_∞ around the tube. Is there an optimum thickness associated with application of insulation to the tube?
2. Confirm the above result by computing the total thermal resistance per unit length of tube for a 10-mm-diameter tube having the following insulation thicknesses: 0, 2, 5, 10, 20, and 40 mm. The insulation is composed of cellular glass, and the outer surface convection coefficient is $5\ \mathrm{W/m^2 \cdot K}$.

SOLUTION

Known:

Radius r_i and temperature T_i of a thin-walled copper tube to be insulated from the ambient air.

Find:

1. Whether there exists an optimum insulation thickness that minimizes the heat transfer rate.

2. Thermal resistance associated with using cellular glass insulation of varying thickness.

Schematic:

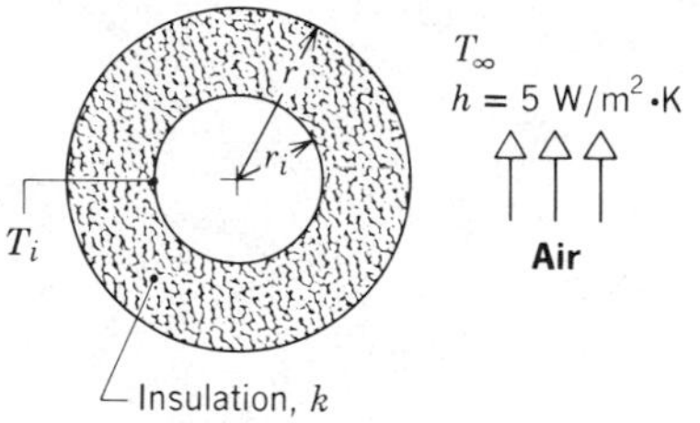

Assumptions:

1. Steady-state conditions.
2. One-dimensional heat transfer in the radial (cylindrical) direction.
3. Negligible tube wall thermal resistance.
4. Constant properties for insulation.
5. Negligible radiation exchange between insulation outer surface and surroundings.

Properties:

Table A.3, cellular glass (285 K, assumed): $k = 0.055\ \text{W/m}\cdot\text{K}$.

Analysis:

1. The resistance to heat transfer between the refrigerant and the air is dominated by conduction in the insulation and convection in the air. The thermal circuit is therefore

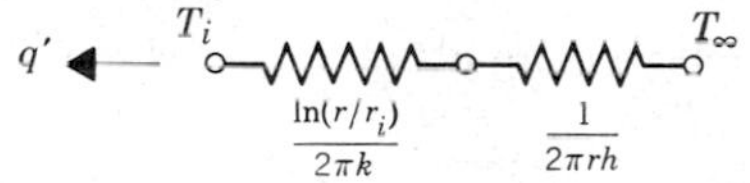

where the conduction and convection resistances per unit length follow from Equations 3.27 and 3.9, respectively. The total thermal resistance per unit length of tube is then

$$R'_{\text{tot}} = \frac{\ln(r/r_i)}{2\pi k} + \frac{1}{2\pi rh}$$

where the rate of heat transfer per unit length of tube is

$$q' = \frac{T_\infty - T_i}{R'_{\text{tot}}}$$

An optimum insulation thickness would be associated with the value of r that minimized q' or maximized R'_{tot}. Such a value could be obtained from the requirement that

$$\frac{dR'_{tot}}{dr} = 0$$

Hence

$$\frac{1}{2\pi kr} - \frac{1}{2\pi r^2 h} = 0$$

or

$$r = \frac{k}{h}$$

To determine whether the foregoing result maximizes or minimizes the total resistance, the second derivative must be evaluated. Hence

$$\frac{d^2R'_{tot}}{dr^2} = -\frac{1}{2\pi kr^2} + \frac{1}{\pi r^3 h}$$

or, at $r = k/h$

$$\frac{d^2R'_{tot}}{dr^2} = \frac{1}{\pi(k/h)^2}\left(\frac{1}{k} - \frac{1}{2k}\right) = \frac{1}{2\pi k^3/h^2} > 0$$

Since this result is always positive, it follows that $r = k/h$ is the insulation radius for which the total resistance is a minimum, not a maximum. Hence an *optimum* insulation thickness *does not exist.*

From the above result it makes more sense to think in terms of a *critical insulation radius*

$$r_{cr} \equiv \frac{k}{h}$$

below which q' increases with increasing r and above which q' decreases with increasing r.

2. With $h = 5\ \text{W/m}^2 \cdot \text{K}$ and $k = 0.055\ \text{W/m} \cdot \text{K}$, the critical radius is

$$r_{cr} = \frac{0.055\ \text{W/m} \cdot \text{K}}{5\ \text{W/m}^2 \cdot \text{K}} = 0.011\ \text{m}$$

Hence $r_{cr} > r_i$ and heat transfer will increase with the addition of insulation up to a thickness of

$$r_{cr} - r_i = (0.011 - 0.005)\ \text{m} = 0.006\ \text{m}$$

The thermal resistances corresponding to the prescribed insulation thicknesses may be calculated and are summarized as follows.

INSULATION THICKNESS $(r - r_i)$ (mm)	INSULATION RADIUS r (m)	THERMAL RESISTANCES (m · K/W) R'_{cond}	R'_{conv}	R'_{tot}
0	0.005	0	6.37	6.37
2	0.007	0.97	4.55	5.52
5	0.010	2.00	3.18	5.18
6	$r_{cr} = 0.011$	2.28	2.89	5.17
10	0.015	3.18	2.12	5.30
20	0.025	4.66	1.27	5.93
40	0.045	6.35	0.71	7.06

Comments:

1. The effect of the critical radius is revealed by the fact that even for 20 mm of insulation, the total resistance is not as large as the value for no insulation.
2. The problem of lowering the total resistance through the application of insulation exists only for small tubes and low convection coefficients, such that $r_{cr} > r_i$. For moderate to large values of r_i and/or h, $r_i > r_{cr}$ and the addition of any amount of insulation will increase the total resistance.

3.3.2 The Sphere

Now consider applying the alternative method to analyzing conduction in the hollow sphere of Figure 3.8. For the differential control volume of the figure, energy conservation requires that $q_r = q_{r+dr}$ for steady-state, one-dimensional conditions with no heat generation. The appropriate form of Fourier's law is

$$q_r = -kA\frac{dT}{dr} = -k(4\pi r^2)\frac{dT}{dr} \tag{3.32}$$

where $A = 4\pi r^2$ is the area normal to the direction of heat transfer.

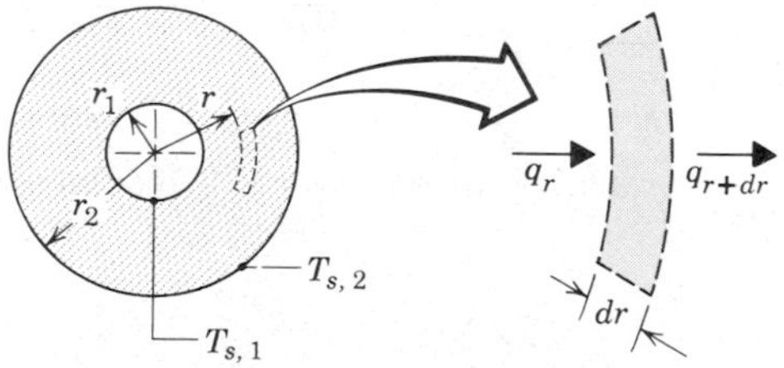

Figure 3.8 Conduction in a spherical shell.

Acknowledging that q_r is a constant, independent of r, Equation 3.32 may be expressed in the integral form

$$\frac{q_r}{4\pi}\int_{r_1}^{r_2}\frac{dr}{r^2} = -\int_{T_{s,1}}^{T_{s,2}} k(T)\,dT \tag{3.33}$$

Assuming constant k, we then obtain

$$q_r = \frac{4\pi k(T_{s,1} - T_{s,2})}{(1/r_1) - (1/r_2)} \tag{3.34}$$

Remembering that the thermal resistance is defined as the temperature difference divided by the heat transfer rate, we obtain

$$R_{t,\text{cond}} = \frac{1}{4\pi k}\left(\frac{1}{r_1} - \frac{1}{r_2}\right) \tag{3.35}$$

hollow sphere

Note that the temperature distribution and Equations 3.34 and 3.35 could have been obtained by using the standard approach, which would begin with the appropriate form of the heat equation.

Spherical composites may be treated in much the same way as composite walls and cylinders, where appropriate forms of the total resistance and overall heat transfer coefficient may be determined.

EXAMPLE 3.4

A spherical, thin-walled metallic container is used to store liquid nitrogen at 77 K. The container has a diameter of 0.5 m and is covered with an evacuated, reflective insulation system composed of silica powder. The insulation is 25 mm thick, and its outer surface is exposed to ambient air at 300 K. The convection coefficient is known to be 20 $W/m^2 \cdot K$. The latent heat of vaporization and the density of liquid nitrogen are 2×10^5 J/kg and 804 kg/m^3, respectively.

1. What is the rate of heat transfer to the liquid nitrogen?
2. What is the rate of liquid boil-off?

SOLUTION

Known:

Liquid nitrogen is stored in a spherical container that is insulated and exposed to ambient air.

Find:

1. The rate of heat transfer to the nitrogen.
2. The mass rate of nitrogen boil-off.

Schematic:

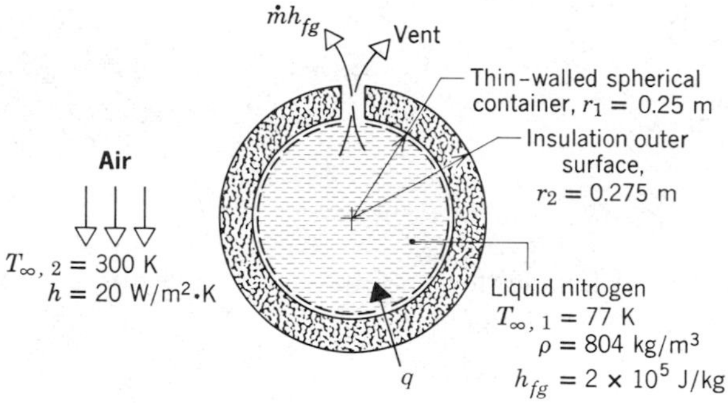

Assumptions:

1. Steady-state conditions.
2. One-dimensional transfer in the radial direction.
3. Negligible resistance to heat transfer through the container wall and from the container to the nitrogen.
4. Constant properties.
5. Negligible radiation exchange between outer surface of insulation and surroundings.

Properties:

Table A.3, evacuated silica powder (300 K): $k = 0.0017\ \mathrm{W/m \cdot K}$.

Analysis:

1. The thermal circuit involves a conduction and convection resistance in series and is of the form

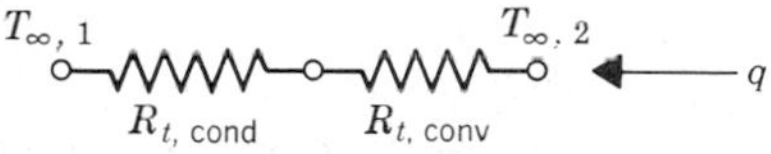

where, from Equation 3.35,

$$R_{t,\text{cond}} = \frac{1}{4\pi k}\left(\frac{1}{r_1} - \frac{1}{r_2}\right)$$

and from Equation 3.9

$$R_{t,\text{conv}} = \frac{1}{h 4\pi r_2^2}$$

The rate of heat transfer to the liquid nitrogen is then

$$q = \frac{T_{\infty,2} - T_{\infty,1}}{(1/4\pi k)[(1/r_1) - (1/r_2)] + (1/h4\pi r_2^2)}$$

Hence

$$q = [(300 - 77)\ \text{K}] \div$$

$$\left[\frac{1}{4\pi(0.0017\ \text{W/m}\cdot\text{K})}\left(\frac{1}{0.25\ \text{m}} - \frac{1}{0.275\ \text{m}}\right) + \frac{1}{(20\ \text{W/m}^2\cdot\text{K})4\pi(0.275\ \text{m})^2}\right]$$

$$q = \frac{223}{17.02 + 0.05}\ \text{W} = 13.06\ \text{W} \qquad \triangleleft$$

2. Performing an energy balance for a control surface about the nitrogen, it follows from Equation 1.11 that

   $$\dot{E}_{\text{in}} - \dot{E}_{\text{out}} = 0$$

   where $\dot{E}_{\text{in}} = q$ and $\dot{E}_{\text{out}} = \dot{m}h_{fg}$ is associated with the loss of latent energy due to boiling. Hence

   $$q - \dot{m}h_{fg} = 0$$

   and the boil-off $\dot{m}$ is

   $$\dot{m} = \frac{q}{h_{fg}}$$

   $$\dot{m} = \frac{13.06\ \text{J/s}}{2 \times 10^5\ \text{J/kg}} = 6.53 \times 10^{-5}\ \text{kg/s}$$

   The loss per day is

   $$\dot{m} = 6.53 \times 10^{-5}\ \text{kg/s} \times 3600\ \text{s/h} \times 24\ \text{h/day} = 5.64\ \text{kg/day} \qquad \triangleleft$$

   or on a volumetric basis

   $$\dot{V} = \frac{\dot{m}}{\rho} = \frac{5.64\ \text{kg/day}}{804\ \text{kg/m}^3} = 0.007\ \text{m}^3/\text{day} = 7\ \text{liters/day}$$

Comments:

1. $R_{t,\text{conv}} \ll R_{t,\text{cond}}$.
2. With a container volume of $(4/3)(\pi r_1^3) = 0.065\ \text{m}^3 = 65$ liters, the daily loss amounts to (7 liters/65 liters) 100% = 10.8% of capacity.

3.4 CONDUCTION WITH THERMAL ENERGY GENERATION

In the preceding section we considered conduction problems for which the temperature distribution in a medium was determined solely by conditions at the boundaries of the medium. We now want to consider the additional effect on the temperature distribution of processes that may be occurring *within* the medium. In particular, we wish to consider situations for which thermal energy is being *generated* due to *conversion* from some other energy form.

A common thermal energy generation process involves the conversion from *electrical to thermal energy* in a current-carrying medium (*ohmic* or *resistance heating*). The rate at which energy is generated by passing a current I through a medium of electrical resistance R_e is

$$\dot{E}_g = I^2 R_e \tag{3.36}$$

If this power generation (W) occurs uniformly throughout the medium of volume V, the volumetric generation rate (W/m^3) is then

$$\dot{q} \equiv \frac{\dot{E}_g}{V} = \frac{I^2 R_e}{V} \tag{3.37}$$

Energy generation may also occur as a result of the deceleration and absorption of neutrons in the fuel element of a nuclear reactor or exothermic chemical reactions occurring within a medium. Endothermic reactions would, of course, have the inverse effect (a thermal energy sink) of converting thermal energy to chemical bonding energy. Finally, a conversion from electromagnetic to thermal energy may occur due to the absorption of radiation within the medium. The process may occur, for example, because gamma rays are absorbed in external nuclear reactor components (cladding, thermal shields, pressure vessels, etc.) or because visible radiation is absorbed in a semitransparent medium. Remember not to confuse energy generation with energy storage (Section 1.3.1).

3.4.1 The Plane Wall

Consider the plane wall of Figure 3.9*a*, in which there is uniform energy generation per unit volume ($\dot{q}$ is constant) and the surfaces are maintained at $T_{s,1}$ and $T_{s,2}$. For constant thermal conductivity k, the appropriate form of the heat equation, Equation 2.16, is

$$\frac{d^2T}{dx^2} + \frac{\dot{q}}{k} = 0 \tag{3.38}$$

The general solution is

$$T = -\frac{\dot{q}}{2k}x^2 + C_1 x + C_2 \tag{3.39}$$

where C_1 and C_2 are the constants of integration. For the prescribed boundary conditions,

$$T(-L) = T_{s,1} \quad \text{and} \quad T(L) = T_{s,2}$$

the constants may be evaluated and are of the form

$$C_1 = \frac{T_{s,2} - T_{s,1}}{2L} \quad \text{and} \quad C_2 = \frac{\dot{q}}{2k}L^2 + \frac{T_{s,1} + T_{s,2}}{2}$$

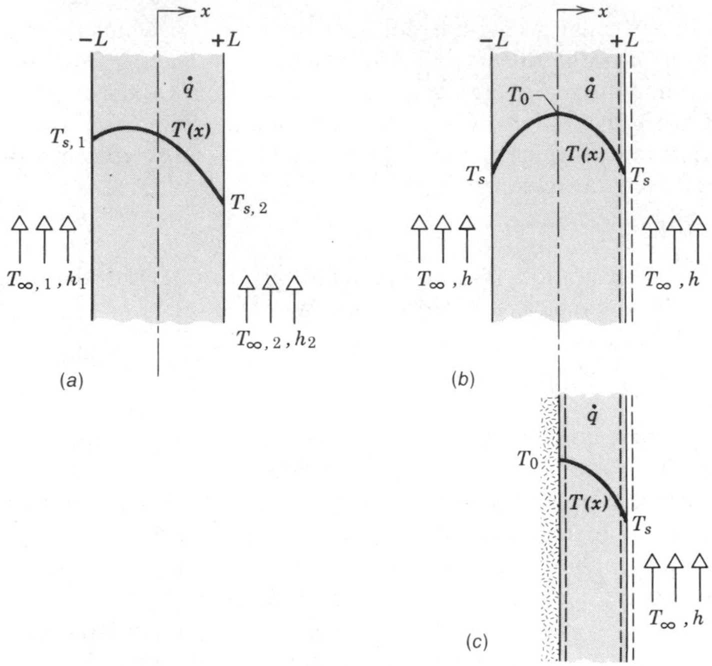

Figure 3.9 Conduction in a plane wall with uniform heat generation. (*a*) Asymmetrical boundary conditions. (*b*) Symmetrical boundary conditions. (*c*) Adiabatic surface at midplane.

in which case the temperature distribution is

$$T(x) = \frac{\dot{q}L^2}{2k}\left(1 - \frac{x^2}{L^2}\right) + \frac{T_{s,2} - T_{s,1}}{2}\frac{x}{L} + \frac{T_{s,1} + T_{s,2}}{2} \tag{3.40}$$

The heat flux at any point in the wall may, of course, be determined by using Equation 3.40 with Fourier's law. Note, however, that with generation the heat flux is no longer independent of x.

The preceding result simplifies when both surfaces are maintained at a common temperature, $T_{s,1} = T_{s,2} \equiv T_s$. The temperature distribution is then *symmetrical* about the midplane, Figure 3.9*b*, and is given by

$$T(x) = \frac{\dot{q}L^2}{2k}\left(1 - \frac{x^2}{L^2}\right) + T_s \tag{3.41}$$

The maximum temperature exists at the midplane

$$T(0) \equiv T_0 = \frac{\dot{q}L^2}{2k} + T_s \tag{3.42}$$

in which case the temperature distribution, Equation 3.41, may be expressed as

$$\frac{T(x) - T_0}{T_s - T_0} = \left(\frac{x}{L}\right)^2 \tag{3.43}$$

It is important to note that at the plane of symmetry in Figure 3.9*b*, the temperature gradient is zero, $(dT/dx)_{x=0} = 0$. Accordingly, this plane may be represented by the *adiabatic* surface shown in Figure 3.9*c*. One implication of this result is that Equation 3.41 also applies to plane walls that are perfectly insulated on one side ($x = 0$) and maintained at a fixed temperature T_s on the other side ($x = L$).

To use the foregoing results the surface temperature(s) T_s must be known. However, a common situation is one for which it is the temperature of an adjoining fluid, T_∞, and not T_s, which is known. It then becomes necessary to relate T_s to T_∞. This relation may be developed by applying a surface energy balance. Consider the surface at $x = L$ for the symmetrical plane wall (Figure 3.9*b*) or the insulated plane wall (Figure 3.9*c*). Neglecting radiation and substituting the appropriate rate equations, the energy balance given by Equation 1.11 reduces to

$$-k\left.\frac{dT}{dx}\right|_{x=L} = h(T_s - T_\infty) \tag{3.44}$$

Substituting from Equation 3.41 to obtain the temperature gradient at $x = L$, it follows that

$$T_s = T_\infty + \frac{\dot{q}L}{h} \tag{3.45}$$

Hence T_s may be computed from knowledge of T_∞, $\dot{q}$, L, and h.

Equation 3.45 may also be obtained by applying an *overall* energy balance to the plane wall of Figure 3.9*b* or 3.9*c*. For example, relative to a control volume about Figure 3.9*c*, the rate at which energy is generated within the wall must be balanced by the rate at which energy leaves via convection at the boundary. Equation 1.10a reduces to

$$\dot{E}_g = \dot{E}_{out} \tag{3.46}$$

or, for a unit surface area,

$$\dot{q}L = h(T_s - T_\infty) \tag{3.47}$$

Solving for T_s, Equation 3.45 is obtained.

Equation 3.45 may be combined with Equation 3.41 to eliminate T_s from the temperature distribution, which is then expressed in terms of the known quantities $\dot{q}$, L, k, h and T_∞. The same result may be obtained directly by using Equation 3.44 as a boundary condition to evaluate the constants of integration appearing in Equation 3.39 (see Problem 3.48).

EXAMPLE 3.5

A plane wall is a composite of two materials, A and B. The wall of material A has heat generation $\dot{q} = 1.5 \times 10^6\ \mathrm{W/m^3}$, $k_A = 75\ \mathrm{W/m \cdot K}$, and thickness $L_A = 50$ mm. The wall material B has no generation, with $k_B = 150\ \mathrm{W/m \cdot K}$

and thickness $L_B = 20$ mm. The inner surface of material A is well insulated, while the outer surface of material B is cooled by a water stream with $T_\infty = 30°C$ and $h = 1000\ W/m^2 \cdot K$.

1. Sketch the temperature distribution that exists in the composite under steady-state conditions.
2. Determine the temperature T_0 of the insulated surface and the temperature T_2 of the cooled surface.

SOLUTION

Known:

Plane wall of material A with internal heat generation is insulated on one side and bounded by a second wall of material B, which is without heat generation and is subjected to convection cooling.

Find:

1. Sketch of steady-state temperature distribution in the composite.
2. Inner and outer surface temperatures of the composite.

Schematic:

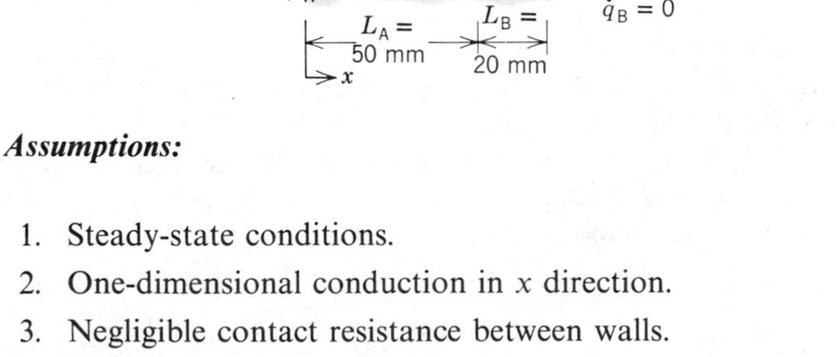

Assumptions:

1. Steady-state conditions.
2. One-dimensional conduction in x direction.
3. Negligible contact resistance between walls.
4. Inner surface of A adiabatic.
5. Constant properties for materials A and B.

Analysis:

1. From the prescribed physical conditions, the temperature distribution in the composite is known to have the following features, as shown.
   (a) Parabolic in material A.
   (b) Zero slope at insulated boundary.
   (c) Linear in material B.
   (d) Slope change $= k_B/k_A = 2$ at interface.
   The temperature distribution in the water is characterized by
   (e) Large gradients near the surface.

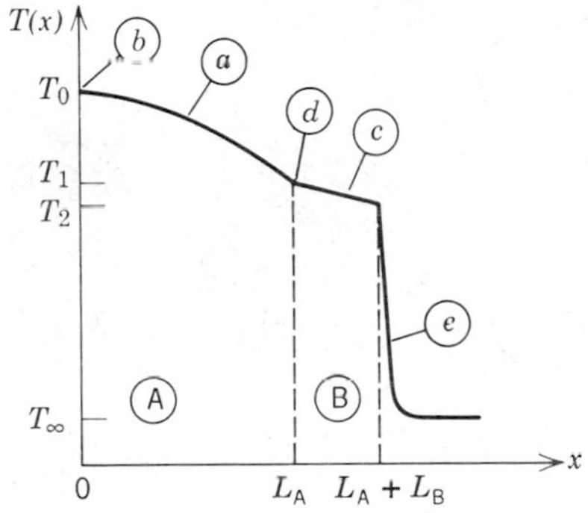

2. The outer surface temperature T_2 may be obtained by performing an energy balance on a control volume about material B. Since there is no generation in this material, it follows that, for steady-state conditions and a unit surface area, the heat flux into the material at $x = L_A$ must equal the heat flux from the material due to convection at $x = L_A + L_B$. Hence

$$q'' = h(T_2 - T_\infty) \tag{1}$$

The heat flux q'' may be determined by performing a second energy balance on a control volume about material A. In particular, since the surface at $x = 0$ is adiabatic, there is no inflow and the rate at which energy is generated must equal the outflow. Accordingly, for a unit surface area,

$$\dot{q}L_A = q'' \tag{2}$$

Combining Equations 1 and 2, the outer surface temperature is

$$T_2 = T_\infty + \frac{\dot{q}L_A}{h}$$

$$T_2 = 30°\text{C} + \frac{1.5 \times 10^6\ \text{W/m}^3 \times 0.05\ \text{m}}{1000\ \text{W/m}^2 \cdot \text{K}} = 105°\text{C} \qquad \triangleleft$$

From Equation 3.42 the temperature at the insulated surface is

$$T_0 = \frac{\dot{q}L_A^2}{2k_A} + T_1 \tag{3}$$

where T_1 may be obtained from the following thermal circuit.

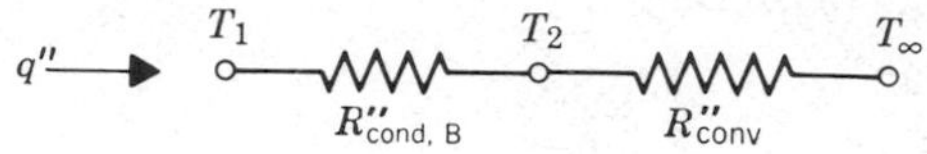

That is,

$$T_1 = T_\infty + (R''_{cond,B} + R''_{conv})q''$$

where the resistances for a unit surface area are

$$R''_{cond,B} = \frac{L_B}{k_B} \qquad R''_{conv} = \frac{1}{h}$$

Hence

$$T_1 = 30°C + \left(\frac{0.02\,m}{150\,W/m\cdot K} + \frac{1}{1000\,W/m^2\cdot K}\right)1.5 \times 10^6\,W/m^3 \times 0.05\,m$$

$$T_1 = 30°C + 85°C = 115°C$$

Substituting into Equation 3,

$$T_0 = \frac{1.5 \times 10^6\,W/m^3\,(0.05\,m)^2}{2 \times 75\,W/m\cdot K} + 115°C$$

$$T_0 = 25°C + 115°C = 140°C$$ ◁

Comments:

1. Material A, having heat generation, cannot be represented by a thermal circuit element.
2. $(T_2 - T_\infty)/(T_1 - T_2) = R''_{conv}/R''_{cond,B} = 7.5$.

3.4.2 Radial Systems

Heat generation may occur in a variety of radial geometries. Consider the long, solid cylinder of Figure 3.10, which could represent a current-carrying wire or a fuel element in a nuclear reactor. For steady-state conditions the rate at which heat is generated within the cylinder must equal the rate at which heat is convected from the surface of the cylinder to a moving fluid. This condition allows the surface temperature to be maintained at a fixed value of T_s.

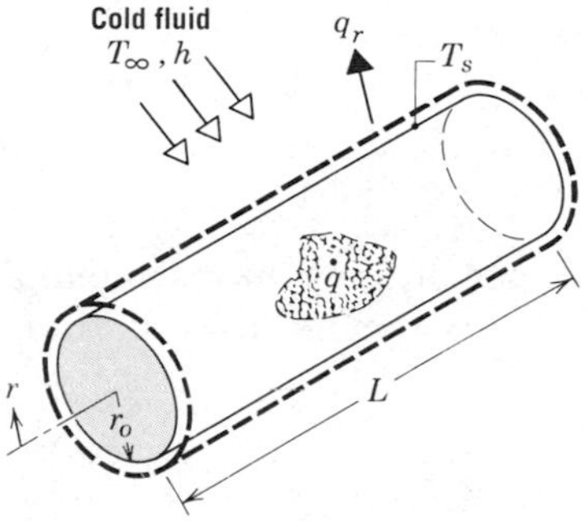

Figure 3.10 Conduction in a solid cylinder with uniform heat generation.

To determine the temperature distribution in the cylinder, we begin with the appropriate form of the heat equation. For constant thermal conductivity k, Equation 2.18 reduces to

$$\frac{1}{r}\frac{d}{dr}\left(r\frac{dT}{dr}\right)+\frac{\dot{q}}{k}=0 \tag{3.48}$$

Separating variables and assuming uniform generation, this expression may be integrated to obtain

$$r\frac{dT}{dr}=-\frac{\dot{q}}{2k}r^2+C_1 \tag{3.49}$$

Repeating the procedure, the general solution for the temperature distribution becomes

$$T(r)=-\frac{\dot{q}}{4k}r^2+C_1\ln r+C_2 \tag{3.50}$$

To obtain the constants of integration C_1 and C_2, we apply the boundary conditions

$$\left.\frac{dT}{dr}\right|_{r=0}=0 \qquad \text{and} \qquad T(r_o)=T_s$$

The first condition results from the symmetry of the situation. That is, for the solid cylinder the centerline is a line of symmetry for the temperature distribution and the temperature gradient must be zero. Recall that similar conditions existed at the midplane of a wall having symmetrical boundary conditions (Figure 3.9*b*). From the foregoing symmetry condition at $r=0$ and Equation 3.49, it is evident that $C_1=0$. Using the surface boundary condition at $r=r_o$ with Equation 3.50, we then obtain

$$C_2=T_s+\frac{\dot{q}}{4k}r_o^2 \tag{3.51}$$

The temperature distribution is therefore

$$T(r) = \frac{\dot{q}r_o^2}{4k}\left(1 - \frac{r^2}{r_o^2}\right) + T_s \tag{3.52}$$

Evaluating Equation 3.52 at the centerline and dividing the result into Equation 3.52, we obtain the temperature distribution in nondimensional form

$$\frac{T(r) - T_s}{T_o - T_s} = 1 - \left(\frac{r}{r_o}\right)^2 \tag{3.53}$$

where T_o is the centerline temperature. The heat rate at any radius in the cylinder may, of course, be evaluated by using Equation 3.52 with Fourier's law.

To relate the surface temperature, T_s, to the temperature, T_∞, of the cold fluid, either a surface energy balance or an overall energy balance may be used. Choosing the second approach, we obtain

$$\dot{q}(\pi r_o^2 L) = h(2\pi r_o L)(T_s - T_\infty)$$

or

$$T_s = T_\infty + \frac{\dot{q}r}{2h} \tag{3.54}$$

The foregoing approach may also be used to obtain the temperature distribution in solid spheres and in cylindrical and spherical shells for a variety of boundary conditions.

EXAMPLE 3.6

Consider a long solid tube, insulated at the outer radius r_o and cooled at the inner radius r_i with uniform heat generation $\dot{q}$ (W/m^3) within the solid.

1. Obtain the general solution for the temperature distribution in the tube.
2. In a practical application a limit would be placed on the maximum temperature that is permissible at the insulated surface ($r = r_o$). Specifying this limit as T_o, identify appropriate boundary conditions that could be used to determine the arbitrary constants appearing in the general solution. Determine these constants and the corresponding form of the temperature distribution.
3. Determine the heat removal rate per unit length of tube.
4. If the coolant is available at a temperature T_∞, obtain an expression for the convection coefficient that would have to be maintained at the inner surface to allow for operation at prescribed values of T_o and $\dot{q}$.

SOLUTION

Known:

Solid tube with uniform heat generation is insulated at the outer surface and cooled at the inner surface.

Find:

1. General solution for the temperature distribution $T(r)$.
2. Appropriate boundary conditions and the corresponding form of the temperature distribution.
3. Heat removal rate.
4. Convection coefficient at the inner surface.

Schematic:

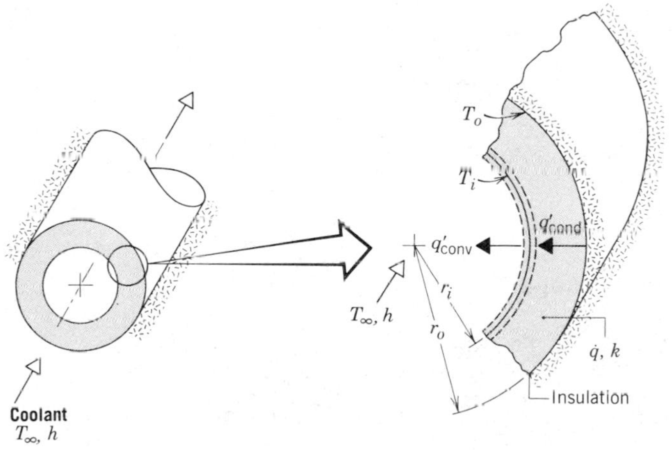

Assumptions:

1. Steady-state conditions.
2. One-dimensional radial conduction.
3. Constant properties.
4. Uniform volumetric heat generation.
5. Outer surface adiabatic.

Analysis:

1. To determine $T(r)$, the appropriate form of the heat equation, Equation 2.18, must be solved. For the prescribed conditions, this expression reduces to Equation 3.48, and the general solution is given by Equation 3.50. Hence this solution applies in a cylindrical shell, as well as in a solid cylinder (Figure 3.10). $T(r) = -\frac{\dot{q}}{4k} r^2 + C_1 \ln r + C_2$
2. Two boundary conditions are needed to evaluate C_1 and C_2, and in this problem it is appropriate to specify both conditions at r_o. Invoking the prescribed temperature limit,

$$T(r_o) = T_o \tag{1}$$

and applying Fourier's law, Equation 3.23, at the adiabatic outer surface

$$\left.\frac{dT}{dr}\right|_{r_o} = 0 \tag{2}$$

Using Equations 3.50 and 1, it follows that

$$T_o = -\frac{\dot{q}}{4k} r_o^2 + C_1 \ln r_o + C_2 \tag{3}$$

Similarly, from Equations 3.49 and 2

$$0 = -\frac{\dot{q}}{2k} r_o^2 + C_1 \tag{4}$$

Hence, from Equation 4,

$$C_1 = \frac{\dot{q}}{2k} r_o^2 \tag{5}$$

and from Equation 3

$$C_2 = T_o + \frac{\dot{q}}{4k} r_o^2 - \frac{\dot{q}}{2k} r_o^2 \ln r_o \tag{6}$$

Substituting Equations 5 and 6 into the general solution, Equation 3.50, it follows that

$$T(r) = T_o + \frac{\dot{q}}{4k}(r_o^2 - r^2) - \frac{\dot{q}}{2k} r_o^2 \ln \frac{r_o}{r} \tag{7}$$

3. The heat removal rate may be determined by obtaining the conduction rate at r_i or by evaluating the total generation rate for the tube. From Fourier's law

$$q'_r = -k 2\pi r \frac{dT}{dr}$$

Hence substituting from Equation 7 and evaluating the result at r_i,

$$q'_r(r_i) = -k2\pi r_i\left(-\frac{\dot{q}}{2k}r_i + \frac{\dot{q}}{2k}\frac{r_o^2}{r_i}\right) = -\pi\dot{q}(r_o^2 - r_i^2) \tag{8}$$

Alternatively, because the tube is insulated at r_o, the rate at which heat is generated in the tube must equal the rate of removal at r_i. That is, for a control volume about the tube, the energy conservation requirement, Equation 1.10a, reduces to $\dot{E}_g - \dot{E}_{out} = 0$, where $\dot{E}_g = \dot{q}\pi(r_o^2 - r_i^2)L$ and $\dot{E}_{out} = q'_{cond}L = -q'_r(r_i)L$. Hence

$$q'_r(r_i) = -\pi\dot{q}(r_o^2 - r_i^2) \tag{9}$$

4. Applying the energy conservation requirement, Equation 1.11, to the inner surface, it follows that

$$q'_{cond} = q'_{conv}$$

or

$$\pi\dot{q}(r_o^2 - r_i^2) = h2\pi r_i(T_i - T_\infty)$$

Hence

$$h = \frac{\dot{q}(r_o^2 - r_i^2)}{2r_i(T_i - T_\infty)} \tag{10}$$

where T_i may be obtained by evaluating Equation 7 at $r = r_i$.

Comments:

Note that, through application of Fourier's law in part 3, the sign on $q'_r(r_i)$ was found to be negative, Equation 8, implying that heat flow is in the negative r direction. However, in applying the energy balance, we acknowledged that heat flow was *out of* the wall. Hence we expressed q'_{cond} as $-q'_r(r_i)$ and we expressed q'_{conv} in terms of $(T_i - T_\infty)$, rather than $(T_\infty - T_i)$.

3.4.3 Application of Resistance Concepts

We conclude our discussion of heat generation effects with a word of caution. In particular, when such effects are present, the heat transfer rate is not a constant, independent of the spatial coordinate. Accordingly, it would be *incorrect* to use the thermal resistance concepts and the related heat rate equations developed in Sections 3.1 and 3.3.

3.5 HEAT TRANSFER FROM EXTENDED SURFACES

The term *extended surface* is commonly used in reference to a solid that experiences energy transfer by conduction within its boundaries, as well as energy transfer by convection (and/or radiation) between its boundaries and the surroundings. Such a system is shown schematically in Figure 3.11. A strut is used to provide mechanical support to two walls that are at different temperatures. A temperature gradient in the x direction sustains heat transfer by conduction internally, at the same time that there is energy transfer by convection from the surface.

Although there are many different situations that involve combined conduction-convection effects, the most frequent application is one in which an extended surface is used specifically to *enhance* the heat transfer rate between a solid and an adjoining fluid. Such an extended surface is termed a *fin*.

Consider the plane wall of Figure 3.12*a*. If T_s is fixed, there are two ways in which the heat transfer rate may be increased. The convection coefficient h could be increased by increasing the fluid velocity, and/or the fluid temperature T_∞ could be reduced. However, many situations would be encountered in which increasing h to the maximum possible value is either insufficient to obtain the desired heat transfer rate or the associated costs are prohibitive. Such costs are related to the blower or pump power requirements needed to increase h through increased fluid motion. Moreover, the second option of reducing T_∞ is often

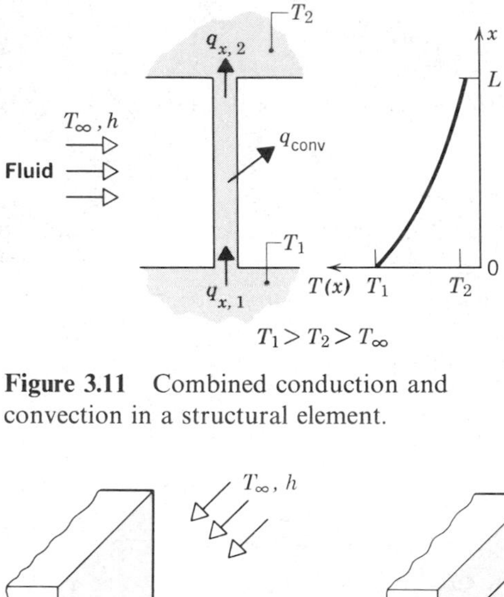

Figure 3.11 Combined conduction and convection in a structural element.

Figure 3.12 Use of fins to enhance heat transfer from a plane wall. (*a*) Bare surface. (*b*) Finned surface.

impractical. Examining Figure 3.12*b*, however, we see that there exists a third option. That is, the heat transfer rate may be increased by increasing the surface area across which the convection occurs. This may be done by employing *fins* that *extend* from the wall into the surrounding fluid. The thermal conductivity of the fin material has a strong effect on the temperature distribution along the fin and therefore influences the degree to which the heat transfer rate is enhanced.

You are already familiar with several fin applications. Consider the arrangement for cooling engine heads on motorcycles and lawnmowers or for cooling electric power transformers. Consider also the tubes with attached fins used to promote heat exchange between air and the working fluid of an air conditioner. Two common finned-tube arrangements are shown in Figure 3.13.

Different fin configurations are illustrated in Figure 3.14. A *straight fin* is any extended surface that is attached to a *plane wall*. It may be of uniform cross-sectional area, or its cross-sectional area may vary with the distance x from the wall. An *annular fin* is one that is circumferentially attached to a cylinder, and its cross section varies with radius from the centerline of the cylinder. The foregoing fin types have rectangular cross sections, whose area may be expressed as a product of the fin thickness t and the width w for straight fins or the circumference $2\pi r$ for annular fins. In contrast a *pin fin*, or *spine*, is an extended surface of circular cross section. Pin fins may also be of uniform or nonuniform cross section.

3.5.1 A General Conduction Analysis

As engineers we are primarily interested in knowing the extent to which particular extended surfaces or fin arrangements could improve heat dissipation from a surface to the surrounding fluid. To determine the heat transfer rate

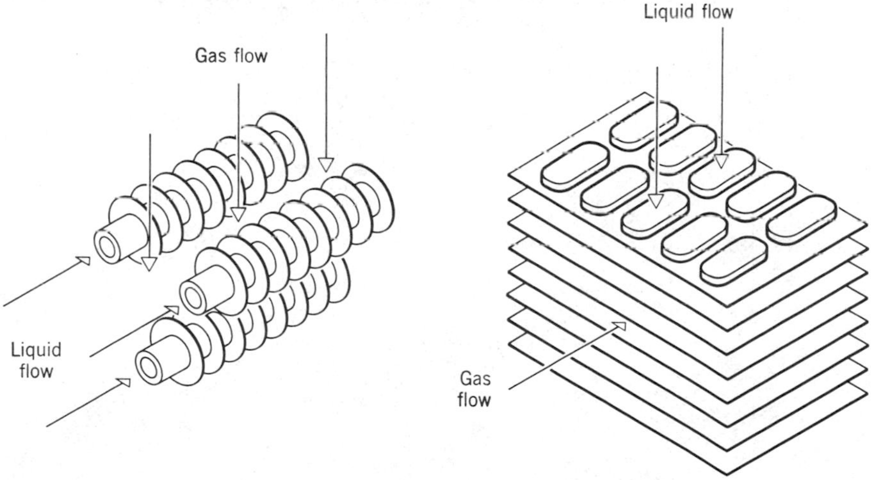

Figure 3.13 Schematic of typical finned-tube heat exchangers.

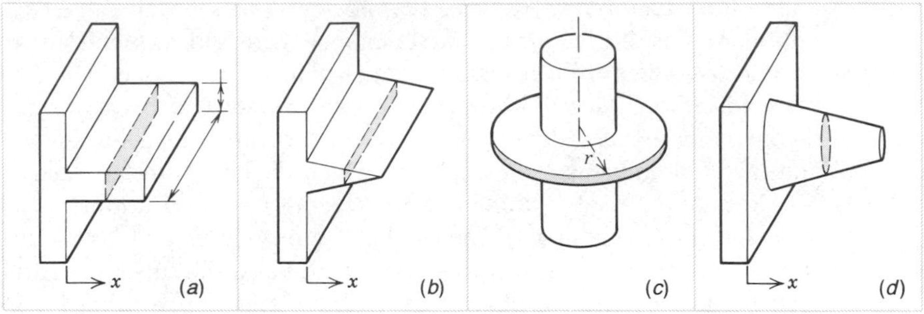

Figure 3.14 Fin configurations. (*a*) Straight fin of uniform cross section. (*b*) Straight fin of nonuniform cross section. (*c*) Annular fin. (*d*) Pin fin.

associated with a fin, we must first obtain the temperature distribution along the fin. As we have done for previous systems, we begin by performing an energy balance on an appropriate differential element. Consider the extended surface of Figure 3.15. The analysis is simplified if certain assumptions are made. We choose to assume one-dimensional conditions in the longitudinal (x) direction, even though conduction within the fin is actually two dimensional. The rate at which energy is convected to the fluid from any point on the fin surface must be balanced by the rate at which energy reaches that point due to conduction in the transverse (y,z) direction. However, in practice the fin is thin, in which case temperature changes in the longitudinal direction are much larger than those in the transverse direction. Hence we may assume one-dimensional conduction in the x direction. We will consider steady-state conditions and also assume that the thermal conductivity is constant, that radiation from the surface is negligible, that heat generation effects are absent, and that the convection heat transfer coefficient h is uniform over the surface.

Applying the conservation of energy requirement, Equation 1.10a, to the

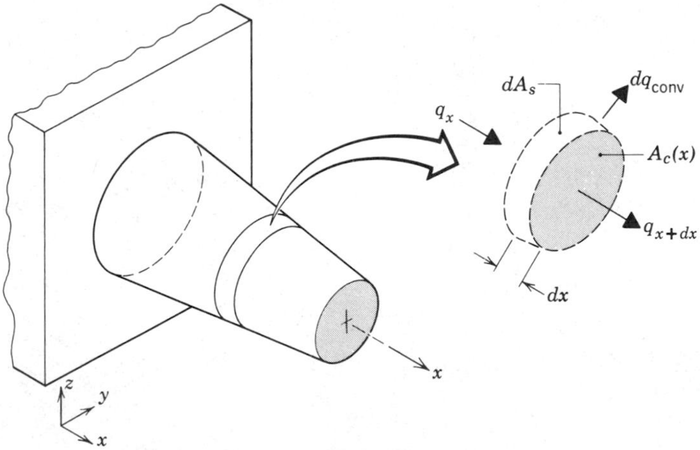

Figure 3.15 Energy balance for an extended surface.

differential element of Figure 3.15, we obtain

$$q_x = q_{x+dx} + dq_{\text{conv}} \tag{3.55}$$

From Fourier's law we know that

$$q_x = -kA_c \frac{dT}{dx} \tag{3.56}$$

where A_c is the *cross-sectional* area, which may vary with x. Since the conduction heat rate at $x + dx$ may be expressed as

$$q_{x+dx} = q_x + \frac{dq_x}{dx} dx \tag{3.57}$$

it follows that

$$q_{x+dx} = -kA_c \frac{dT}{dx} - k \frac{d}{dx}\left(A_c \frac{dT}{dx}\right) dx \tag{3.58}$$

The convection heat transfer rate may be expressed as

$$dq_{\text{conv}} = h\,dA_s(T - T_\infty) \tag{3.59}$$

where dA_s is the *surface* area of the differential element. Substituting the foregoing rate equations into the energy balance, Equation 3.55, we obtain

$$\frac{d}{dx}\left(A_c \frac{dT}{dx}\right) - \frac{h}{k}\frac{dA_s}{dx}(T - T_\infty) = 0$$

or

$$\frac{d^2T}{dx^2} + \left(\frac{1}{A_c}\frac{dA_c}{dx}\right)\frac{dT}{dx} - \left(\frac{1}{A_c}\frac{h}{k}\frac{dA_s}{dx}\right)(T - T_\infty) = 0 \tag{3.60}$$

This result provides a general form of the energy equation for one-dimensional conditions in an extended surface. Its solution for appropriate boundary conditions would provide the temperature distribution, which could then be used with Equation 3.56 to calculate the conduction rate at any x.

3.5.2 Fins of Uniform Cross-Sectional Area

To solve Equation 3.60 it is necessary to be more specific about the geometry. We begin with the simplest case of straight and pin fins of uniform cross section (Figure 3.16). Each fin is attached to a base surface of temperature $T(0) = T_b$ and extends into a fluid of temperature T_∞.

For the prescribed fins, A_c is a constant and $A_s = Px$, where A_s is the surface area measured from the base to x and P is the fin perimeter. Accordingly, $dA_c/dx = 0$ and $dA_s/dx = P$, and Equation 3.60 reduces to

$$\frac{d^2T}{dx^2} - \frac{hP}{kA_c}(T - T_\infty) = 0 \tag{3.61}$$

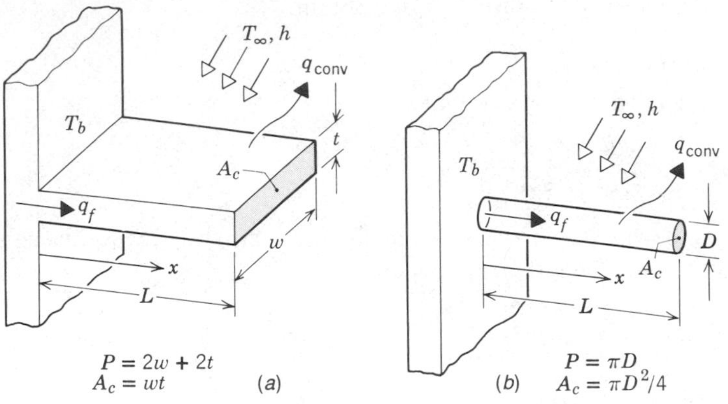

Figure 3.16 Fins of uniform cross section. (*a*) Straight fin. (*b*) Pin fin.

To simplify the form of this equation, we transform the dependent variable by defining an *excess temperature* θ as

$$\theta(x) \equiv T(x) - T_\infty \tag{3.62}$$

where, since T_∞ is a constant, $d\theta/dx = dT/dx$. Substituting Equation 3.62 into Equation 3.61, we then obtain

$$\frac{d^2\theta}{dx^2} - m^2\theta = 0 \tag{3.63}$$

where

$$m^2 \equiv \frac{hP}{kA_c} \tag{3.64}$$

Equation 3.63 is a linear, homogeneous, second-order differential equation with constant coefficients. Its general solution is of the form

$$\theta(x) = C_1 e^{mx} + C_2 e^{-mx} \tag{3.65}$$

By substitution it may be readily verified that Equation 3.65 is indeed a solution to Equation 3.63.

To evaluate the constants C_1 and C_2 of Equation 3.65, it is necessary to specify appropriate boundary conditions. One such condition may be specified in terms of the temperature at the *base* of the fin ($x = 0$).

$$\theta(0) = T_b - T_\infty \equiv \theta_b \tag{3.66}$$

The second condition, specified at the fin tip ($x = L$), may correspond to any one of four different physical situations.

The first condition, case A, considers convection heat transfer from the fin tip. Applying an energy balance to a control surface about this tip (Figure 3.17) we obtain

$$hA_c[T(L) - T_\infty] = -kA_c \left.\frac{dT}{dx}\right|_{x=L}$$

or

$$h\theta(L) = -k \left.\frac{d\theta}{dx}\right|_{x=L} \tag{3.67}$$

That is, the rate at which energy is transferred to the fluid by convection from the tip must equal the rate at which energy reaches the tip by conduction through the fin. Substituting Equation 3.65 into Equations 3.66 and 3.67, we obtain, respectively,

$$\theta_b = C_1 + C_2 \tag{3.68}$$

and

$$h(C_1 e^{mL} + C_2 e^{-mL}) = km(C_2 e^{-mL} - C_1 e^{mL})$$

Solving for C_1 and C_2, it may be shown, after some manipulation, that

$$\frac{\theta}{\theta_b} = \frac{\cosh m(L - x) + (h/mk)\sinh m(L - x)}{\cosh mL + (h/mk)\sinh mL} \tag{3.69}$$

The form of this temperature distribution is shown schematically in Figure 3.17. Note that the magnitude of the temperature gradient decreases with increasing x. This trend is a consequence of the reduction in the conduction heat transfer $q_x(x)$ with increasing x due to continuous convection losses from the fin surface.

We are also interested in the total heat transferred by the fin. From Figure 3.17 it is evident that the fin heat transfer rate q_f may be evaluated in two alternative ways, both of which involve use of the temperature distribution. The

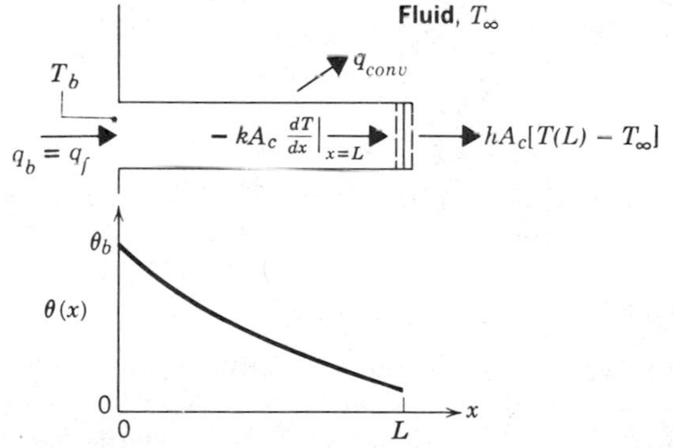

Figure 3.17 Conduction and convection in a fin of uniform cross section.

simpler procedure, and the one which we will use, involves applying Fourier's law at the fin base. That is,

$$q_f = q_b = -kA_c \frac{dT}{dx}\bigg|_{x=0} = -kA_c \frac{d\theta}{dx}\bigg|_{x=0} \tag{3.70}$$

Hence, knowing the temperature distribution, $\theta(x)$, q_f may be evaluated, giving

$$q_f = \sqrt{hPkA_c}\,\theta_b \frac{\sinh mL + (h/mk)\cosh mL}{\cosh mL + (h/mk)\sinh mL} \tag{3.71}$$

Note that conservation of energy dictates that the rate at which heat is transferred by convection from the fin must equal the rate at which it is conducted through the base of the fin. Accordingly, the alternative formulation for q_f is

$$q_f = \int_{A_f} h[T(x) - T_\infty]\, dA_s$$

$$q_f = \int_{A_f} h\theta(x)\, dA_s \tag{3.72}$$

where A_f is the *total*, including the tip, *fin surface area.*

The second tip condition, case B, corresponds to the assumption that the convective heat loss from the fin tip is negligible, in which case the tip may be treated as adiabatic and

$$\frac{d\theta}{dx}\bigg|_{x=L} = 0 \tag{3.73}$$

Substituting from Equation 3.65 and dividing by m, we then obtain

$$C_1 e^{mL} - C_2 e^{-mL} = 0$$

Using this expression with Equation 3.68 to solve for C_1 and C_2 and substituting the results into Equation 3.65, we obtain

$$\frac{\theta}{\theta_b} = \frac{\cosh m(L-x)}{\cosh mL} \tag{3.74}$$

Using this temperature distribution with Equation 3.70, the fin heat transfer rate is then

$$q_f = \sqrt{hPkA_c}\,\theta_b \tanh mL \tag{3.75}$$

q_{conv}, (at tip) = 0

In the same manner, we can obtain the fin temperature distribution and heat transfer rate for case C, where the temperature is prescribed at the fin tip. That is, the second boundary condition is $\theta(L) = \theta_L$, and the resulting expressions are of the form

$$\frac{\theta}{\theta_b} = \frac{(\theta_L/\theta_b)\sinh mx + \sinh m(L-x)}{\sinh mL} \tag{3.76}$$

$$q_f = \sqrt{hPkA_c}\,\theta_b \frac{\cosh mL - \theta_L/\theta_b}{\sinh mL} \tag{3.77}$$

at fin tip

The *very long fin*, case D, is an interesting extension of these results. In particular, as $L \to \infty$, $\theta_L \to 0$ and it is easily verified that

$$\frac{\theta}{\theta_b} = e^{-mx} \tag{3.78}$$

$$q_f = \sqrt{hPkA_c}\,\theta_b \tag{3.79}$$

very long fin

The foregoing results are summarized in Table 3.2. A table of hyperbolic functions is provided in Appendix B.1.

Analysis of fin thermal behavior is, of course, a good deal more complex if the fin is of nonuniform cross section. For such cases the second term of Equation 3.60 must be retained, and the solutions are no longer in the form of simple exponential or hyperbolic functions. The development of such solutions is beyond the scope of this text (although working results are presented in the following section), and the interested student is referred to Schneider [3].

Table 3.2 Temperature distribution and heat loss for fins of uniform cross section

CASE	TIP CONDITION ($x = L$)	TEMPERATURE DISTRIBUTION θ/θ_b	FIN HEAT TRANSFER RATE q_f
A	Convection heat transfer: $h\theta(L) = -kd\theta/dx\vert_{x=L}$	$\dfrac{\cosh m(L-x) + (h/mk)\sinh m(L-x)}{\cosh mL + (h/mk)\sinh mL}$ (3.69)	$M\dfrac{\sinh mL + (h/mk)\cosh mL}{\cosh mL + (h/mk)\sinh mL}$ (3.71)
B	Adiabatic: $d\theta/dx\vert_{x=L} = 0$	$\dfrac{\cosh m(L-x)}{\cosh mL}$ (3.74)	$M \tanh mL$ (3.75)
C	Prescribed temperature: $\theta(L) = \theta_L$	$\dfrac{(\theta_L/\theta_b)\sinh mx + \sinh m(L-x)}{\sinh mL}$ (3.76)	$M\dfrac{(\cosh mL - \theta_L/\theta_b)}{\sinh mL}$ (3.77)
D	Infinite fin ($L \to \infty$): $\theta(L) = 0$	e^{-mx} (3.78)	M (3.79)

$\theta \equiv T - T_\infty$ $\quad m^2 \equiv hP/kA_c$

$\theta_b = \theta(0) = T_b - T_\infty$ $\quad M \equiv \sqrt{hPkA_c}\,\theta_b$

EXAMPLE 3.7

A very long rod 25 mm in diameter has one end maintained at 100°C. The surface of the rod is exposed to ambient air at 25°C with a convection heat transfer coefficient of 10 $W/m^2 \cdot K$.

1. What are the heat losses from rods constructed of pure copper and type AISI 316 stainless steel?
2. Estimate how long the rods must be to be considered infinite.

SOLUTION

Known: A long circular fin exposed to ambient air.

Find:

1. Heat loss q, when rod is fabricated from copper or stainless steel.
2. How long rods must be to assume infinite length.

Schematic:

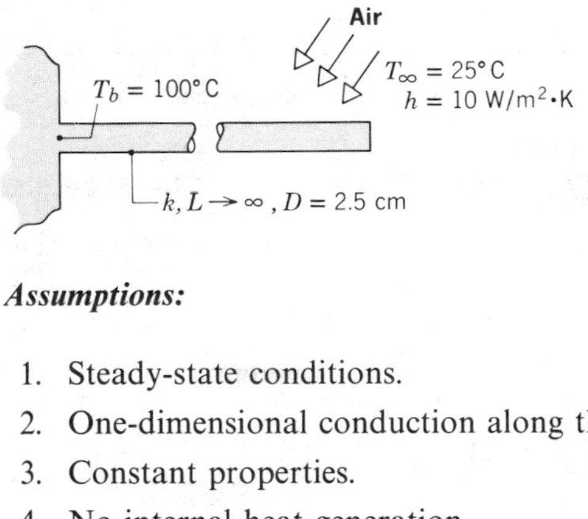

Assumptions:

1. Steady-state conditions.
2. One-dimensional conduction along the rod.
3. Constant properties.
4. No internal heat generation.
5. Negligible radiation exchange with surroundings.
6. Uniform heat transfer coefficient.
7. Infinitely long rod.

Properties:

Table A.1, copper $[T = (T_b + T_\infty)/2 = 62.5°C \approx 335 \text{ K}]$: $k = 398$ $W/m \cdot K$.
Table A.1, stainless steel, AISI 316 (335 K): $k = 14$ $W/m \cdot K$.

Analysis:

1. From Equation 3.79 the heat loss is

   $$q_f = \sqrt{hPkA_c}\,\theta_b$$

   Hence for copper,

   $$q_f = [10\ \text{W/m}^2\cdot\text{K} \times \pi \times 0.025\ \text{m} \times 398\ \text{W/m}\cdot\text{K}$$

   $$\times \frac{\pi}{4}(0.025\ \text{m})^2]^{1/2}(100-25)^\circ\text{C} = 29.4\ \text{W} \qquad \triangleleft$$

   and for stainless steel,

   $$q_f = [10\ \text{W/m}^2\cdot\text{K} \times \pi \times 0.025\ \text{m} \times 14\ \text{W/m}\cdot\text{K}$$

   $$\times \frac{\pi}{4}(0.025\ \text{m})^2]^{1/2}(100-25)^\circ\text{C} = 5.5\ \text{W} \qquad \triangleleft$$

2. Since there is no heat loss from the tip of an infinitely long rod, an estimate of the validity of this approximation may be made by comparing Equations 3.75 and 3.79. To a satisfactory approximation, the expressions provide equivalent results if $\tanh mL \geq 0.99$ or $mL \geq 2.65$. Hence a rod may be assumed to be infinitely long if

   $$L \geq L_\infty \equiv \frac{2.65}{m} = 2.65\left(\frac{kA_c}{hP}\right)^{1/2}$$

   For copper,

   $$L_\infty = 2.65\left[\frac{398\ \text{W/m}\cdot\text{K} \times (\pi/4)(0.025\ \text{m})^2}{10\ \text{W/m}^2\cdot\text{K} \times \pi(0.025\ \text{m})}\right]^{1/2} = 1.32\ \text{m} \qquad \triangleleft$$

   For stainless steel,

   $$L_\infty = 2.65\left[\frac{14\ \text{W/m}\cdot\text{K} \times (\pi/4)(0.025\ \text{m})^2}{10\ \text{W/m}^2\cdot\text{K} \times \pi(0.025\ \text{m})}\right]^{1/2} = 0.25\ \text{m} \qquad \triangleleft$$

Comments:

The foregoing results suggest that the fin heat transfer rate may be accurately predicted from the infinite fin approximation if $mL \gtrsim 2.65$. However, if the infinite fin approximation is to accurately predict the temperature distribution $T(x)$, a larger value of mL would be required. Note the effect of thermal conductivity on q_f and L_∞.

3.5.3 Fin Performance

Recall that fins are used to increase the heat transfer from a surface by increasing the effective surface area. However, the fin itself represents a conduction resistance to heat transfer from the original surface. For this reason, there is no assurance that the heat transfer rate will be increased through the use of fins. An assessment of this matter may be made by evaluating the *fin effectiveness* ε_f. It is defined as the *ratio of the fin heat transfer rate to the heat transfer rate that would exist without the fin.* Therefore

$$\varepsilon_f = \frac{q_f}{hA_{c,b}\theta_b} \tag{3.80}$$

where $A_{c,b}$ is the fin cross-sectional area at the base. In any rational design the value of ε_f should be as large as possible, and in general, the use of fins may rarely be justified unless $\varepsilon_f \gtrsim 2$.

Subject to any one of the four tip conditions that have been considered, the effectiveness for a fin of uniform cross section may be obtained by dividing the appropriate expression for q_f in Table 3.2 by $hA_{c,b}\theta_b$. For the infinite fin approximation (case D), the result is

$$\varepsilon_f = \left(\frac{kP}{hA_c}\right)^{1/2} \tag{3.81}$$

and several important trends may be inferred. Obviously, fin effectiveness is enhanced by the choice of a material of high thermal conductivity. Aluminum alloys and copper come to mind. However, although copper is superior from the standpoint of thermal conductivity, aluminum alloys are the more common choice because of additional benefits related to lower cost and weight. Fin effectiveness is also enhanced by increasing the ratio of the perimeter to the cross-sectional area. For this reason the use of *thin*, but closely spaced, fins is preferred in most engineering applications. Equation 3.81 also suggests that the use of fins can better be justified under conditions for which the convection coefficient h is small. Hence from Table 1.1 it is evident that the need for fins may be stronger when the fluid is a gas rather than a liquid and particularly when the surface heat transfer is by *free* convection. If fins are to be used on a surface separating a gas and a liquid, they are generally placed on the gas side, which is the side of lower convection coefficient.

Another measure of fin thermal performance is provided by the *fin efficiency* η_f. The maximum driving potential for convection is the temperature difference between the base ($x = 0$) and the fluid, $\theta_b = T_b - T_\infty$. Hence, the maximum rate at which a fin could dissipate energy is the rate that would exist *if* the entire fin surface were at the base temperature. However, since any fin is characterized by a finite conduction resistance, a temperature gradient must exist along the fin and the above condition is an idealization. A logical definition of fin efficiency is therefore

$$\eta_f \equiv \frac{q_f}{q_{max}} = \frac{q_f}{hA_f\theta_b} \tag{3.82}$$

where A_f is the total surface area of the fin. From its definition, it is evident that knowledge of η_f may be used as a basis for computing the actual fin heat transfer rate.

Equation 3.82 has proved to be particularly useful for treating fins of nonuniform cross-sectional area. Since none of the results presented in Table 3.2 pertain to such fins, it is necessary to return to Equation 3.60 to determine the solution for the particular form of $A_c(x)$ and $A_s(x)$. For numerous fin configurations, such solutions are conveniently presented in graphical form [3]. In Figure 3.18 the fin efficiency is plotted for three common *straight fin* profiles: the *rectangular fin* of uniform cross-sectional area and the *triangular* ($y \sim x$) and *parabolic* ($y \sim x^2$) *fins* of nonuniform cross-sectional area. Figure 3.19 presents results for the *circular* (or *annular*) *fin* of *rectangular* profile. The foregoing results are presented in terms of a corrected length L_c. To use these results with Equation 3.82, it is necessary to calculate the maximum heat transfer rate in the following manner. For the rectangular, triangular and parabolic fins,

$$q_{max} \equiv hPL_c\theta_b \tag{3.83a}$$

and for the annular fin,

$$q_{max} \equiv 2\pi h(r_{2c}^2 - r_1^2)\theta_b \tag{3.83b}$$

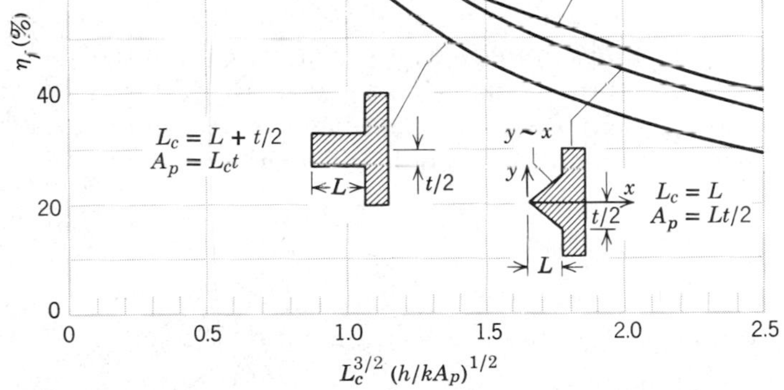

Figure 3.18 Efficiency of straight fins (rectangular, triangular, and parabolic profiles).

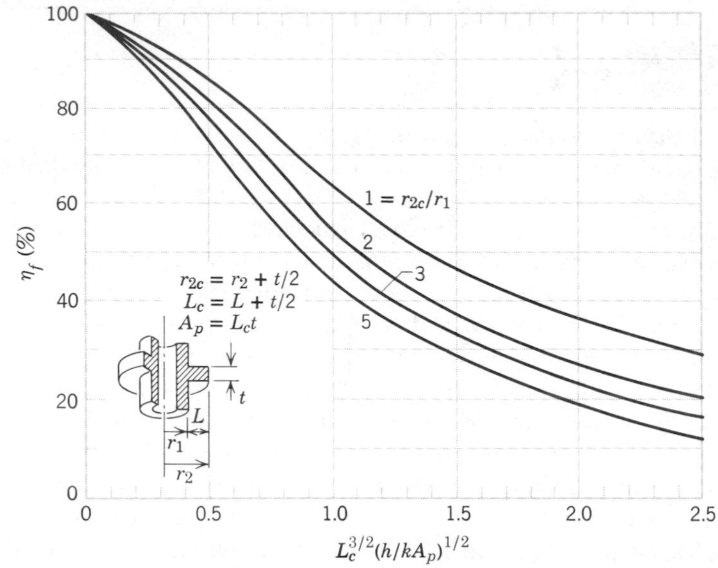

Figure 3.19 Efficiency of annular fins of rectangular profile.

Extended surfaces find use in numerous engineering applications. For comprehensive discussions of thermal effects, the treatments by Schneider [3] and by Kern and Kraus [4] should be consulted.

EXAMPLE 3.8

The cylinder barrel of a motorcycle is constructed of 2024-T6 aluminum alloy and is of height $H = 0.15$ m and outside diameter $D = 50$ mm. Under typical operating conditions the outer surface of the cylinder is at a temperature of 500 K and is exposed to ambient air at 300 K, with a convection coefficient of 50 $W/m^2 \cdot K$. Annular fins of rectangular profile are typically added to increase heat transfer to the surroundings. Assume that five such fins, which are of thickness $t = 6$ mm, length $L = 20$ mm and equally spaced, are added. What is the increase in heat transfer due to addition of the fins?

SOLUTION

Known: Operating conditions of a finned motorcycle cylinder barrel.

Find: Increase in heat transfer associated with using fins.

Schematic:

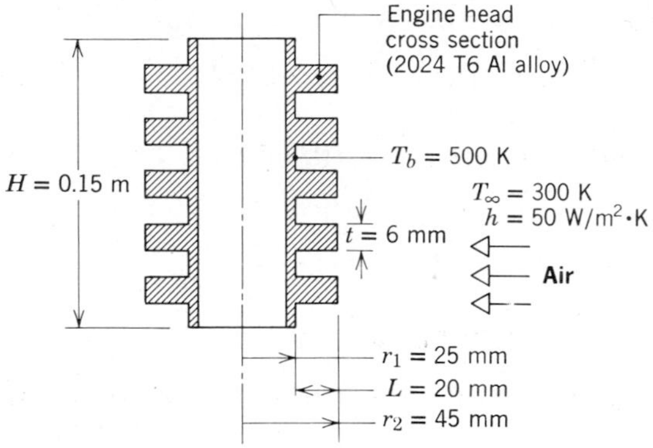

Assumptions:

1. Steady-state conditions.
2. One-dimensional radial conduction in fins.
3. Constant properties.
4. No internal heat generation.
5. Negligible radiation exchange with surroundings.
6. Uniform convection coefficient over outer surface (with or without fins).

Properties:

Table A.1, 2024-T6 aluminum ($T = 400$ K): $k = 186$ W/m · K.

Analysis:

With the fins in place, the heat transfer rate is

$$q = q_f + q_o$$

From Equation 3.82, the fin heat transfer rate is

$$q_f = N\eta_f q_{\max}$$

where N is the number of fins and from Equation 3.83b

$$q_{\max} = 2\pi h(r_{2c}^2 - r_1^2)(T_b - T_\infty)$$

The heat transfer from the exposed (*prime*) cylinder surface is

$$q_o = hA_o(T_b - T_\infty)$$

where

$$A_o = (H - Nt)(2\pi r_1)$$

Hence

$$q = N\eta_f h 2\pi(r_{2c}^2 - r_1^2)(T_b - T_\infty) + h(H - Nt)(2\pi r_1)(T_b - T_\infty)$$

The fin efficiency may be obtained from Figure 3.19 with

$$r_{2c} = r_2 + \frac{t}{2} = 0.048 \text{ m}, \qquad L_c = L + \frac{t}{2} = 0.023 \text{ m}$$

$$\frac{r_{2c}}{r_1} = 1.92, \qquad A_p = L_c t = 1.38 \times 10^{-4} \text{ m}^2, \qquad L_c^{3/2}\left(\frac{h}{kA_p}\right)^{1/2} = 0.15$$

Hence from Figure 3.19, $\eta_f \approx 0.95$. It follows that

$$q = 5\{0.95 \times 50 \text{ W/m}^2\cdot\text{K} \times 2\pi[(0.048^2 - 0.025^2)\text{ m}^2] \times (500 - 300)\text{ K}\}$$

$$+ 50 \text{ W/m}^2\cdot\text{K } (0.15 - 5 \times 0.006)(2\pi \times 0.025)\text{ m}^2 \times (500 - 300)\text{ K}$$

Hence

$$q = 5\{100.22\}\text{ W} + 188.50\text{ W} = 690\text{ W} \qquad \triangleleft$$

Without the fins, the heat transfer rate is

$$q_{wo} = hA_{wo}(T_b - T_\infty)$$

where

$$A_{wo} = H \times 2\pi r_1$$

Hence

$$q_{wo} = 50\text{ W/m}^2\cdot\text{K } (0.15 \times 2\pi \times 0.025)\text{ m}^2\,(200\text{ K}) = 236\text{ W} \qquad \triangleleft$$

Comments:

Although the fins significantly increase heat dissipation from the cylinder, considerable improvement could still be obtained by increasing the number of fins (by reducing both t and the separation between fins).

3.6 SUMMARY

Despite the inherent mathematical simplicity, one-dimensional, steady-state heat transfer occurs in numerous engineering applications. Although one-dimensional, steady-state conditions may not apply exactly, the assumptions may often be made to obtain results of reasonable accuracy. You should therefore be thoroughly familiar with the means by which such problems are treated. In particular you should be comfortable with the use of equivalent thermal circuits and with the expressions for the conduction resistances that pertain to each of the three common geometries. You should also be familiar with how the heat equation and Fourier's law may be used to obtain temperature distributions and the corresponding fluxes. The implications of an internally distributed source of energy should also be clearly understood. Finally, you should appreciate the important role that extended surfaces can play in the design of thermal systems and should have the facility to effect design and performance calculations for such surfaces.

REFERENCES

1. Fried, E., "Thermal Conduction Contribution to Heat Transfer at Contacts," in R. P. Tye, Ed., *Thermal Conductivity*, Vol. 2, Academic Press, London, 1969.
2. Schneider, P. J., "Conduction, Interface Resistance," in W. M. Rohsenow and J. P. Hartnett, Eds., *Handbook of Heat Transfer*, McGraw-Hill, New York, 1973.
3. Schneider, P. J., *Conduction Heat Transfer*, Addison-Wesley, Reading, MA, 1955.
4. Kern, D. Q., and A. D. Kraus, *Extended Surface Heat Transfer*, McGraw-Hill, New York, 1972.

PROBLEMS

3.1 Consider the plane wall of Figure 3.1 separating hot and cold fluids at temperatures $T_{\infty,1}$ and $T_{\infty,2}$, respectively. Using surface energy balances as boundary conditions at $x = 0$ and $x = L$ (see Equation 2.23), obtain the temperature distribution within the wall and the heat flux in terms of $T_{\infty,1}$, $T_{\infty,2}$, h_1, h_2, k, and L.

3.2 The rear window of an automobile is defogged by passing warm air at 40°C over its inner surface, and the associated convection coefficient is $30\ \text{W/m}^2 \cdot \text{K}$. Under conditions for which the outside ambient air temperature is -10°C and the associated convection coefficient is $65\ \text{W/m}^2 \cdot \text{K}$, what are the inner and outer surface temperatures of the window? The window glass is 4 mm thick.

3.3 The rear window of an automobile is defogged by attaching a thin, transparent, film-type heating element to its inner surface. By electrically heating this element, a uniform heat flux may be established at the inner surface. What is the electrical power that must be provided per unit window area to maintain an inner surface temperature of 15°C when the interior air temperature and convection coefficient are 25°C and $10\ \text{W/m}^2 \cdot \text{K}$ and the exterior (ambient) air temperature and convection coefficient are -10°C and $65\ \text{W/m}^2 \cdot \text{K}$? The window glass is 4 mm thick.

3.4 In a manufacturing process, a transparent film is being bonded to a substrate as shown in the sketch. To cure the bond at a temperature T_0, a radiant source is used to provide a heat flux q''_0 (W/m^2), all of which is absorbed at the bonded surface. The back of the substrate is maintained at T_1 while the free surface of the film is exposed to air at T_∞ and a convection heat transfer coefficient h.

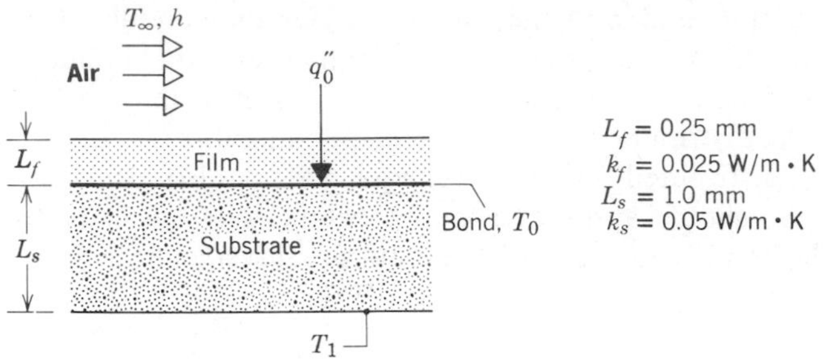

(a) Show the thermal circuit representing the steady-state heat transfer situation. Be sure to label *all* elements, nodes, and heat rates. Leave in symbolic form.

(b) Assume the following conditions: $T_\infty = 20°C$, $h = 50\ W/m^2 \cdot K$, and $T_1 = 30°C$. Calculate the heat flux q_0'' that is required to maintain the bonded surface at $T_0 = 60°C$.

3.5 Both copper and stainless steel (AISI 304) are being considered as wall material for a liquid-cooled rocket nozzle. The cooled exterior of the wall is maintained at 150°C, while combustion gases within the nozzle are at 2750°C. The gas-side heat transfer coefficient is $h_i = 2 \times 10^4\ W/m^2 \cdot K$, and the radius of the nozzle is much larger than the wall thickness. Thermal limitations dictate that the temperature of the copper not exceed 540°C; that of the steel must not exceed 980°C. What is the maximum wall thickness that could be employed for each of the two materials? If the nozzle is constructed with the maximum wall thickness, which material would be preferred?

3.6 The *wind chill*, which is experienced on a cold, windy day, is related to increased heat transfer from exposed human skin to the surrounding atmosphere. Consider a layer of fatty tissue that is 3 mm thick and whose interior surface is maintained at a temperature of 36°C. On a calm day the convection heat transfer coefficient at the outer surface is 25 $W/m^2 \cdot K$, but with 30 km/h winds it reaches 65 $W/m^2 \cdot K$. In both cases the ambient air temperature is −15°C.

(a) What is the ratio of the heat loss per unit area from the skin for the calm day to that for the windy day?

(b) What will be the skin outer surface temperature for the calm day? For the windy day?

(c) What temperature would the air have to assume on the calm day to produce the same heat loss occurring with the air temperature at −15°C on the windy day?

3.7 A thermopane window consists of two pieces of glass 7 mm thick that enclose an air space 7 mm thick. The window separates room air at 20°C from outside ambient air at −10°C. The convection coefficient associated with the inner (room-side) surface is 10 $W/m^2 \cdot K$, while that associated with the outer (ambient air) side is 80 $W/m^2 \cdot K$. What is the heat loss through a window which is 0.8 m long by 0.5 m wide? The air enclosed between the panes may be assumed to be stagnant.

3.8 The wall of a building is a composite consisting of a 100-mm layer of common brick, a 100-mm layer of glass fiber (paper faced, 28 kg/m^3), a 10-mm layer of gypsum plaster (vermiculite), and a 6-mm layer of pine panel. If the inside convection coefficient is 10 $W/m^2 \cdot K$ and the outside convection coefficient is 70 $W/m^2 \cdot K$, what are the total resistance and the overall coefficient for heat transfer?

3.9 The composite wall of an oven consists of three materials, two of which are of known thermal conductivity, $k_A = 20$ W/m · K and $k_C = 50$ W/m · K, and known thickness, $L_A = 0.30$ m and $L_C = 0.15$ m. The third material, B, which is sandwiched between materials A and C, is of known thickness, $L_B = 0.15$ m, but unknown thermal conductivity k_B.

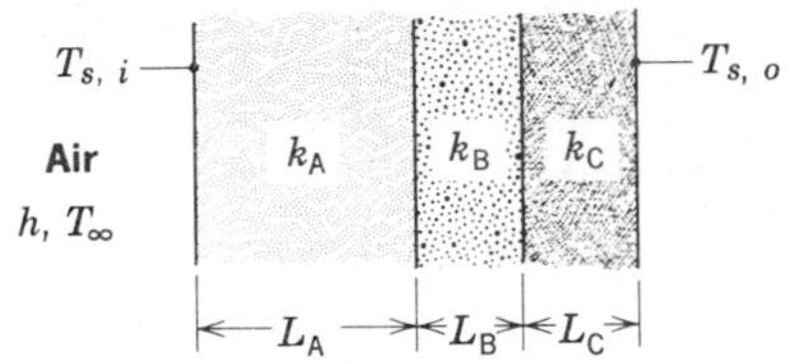

Under steady-state operating conditions, measurements reveal an outer surface temperature of $T_{s,o} = 20°C$, an inner surface temperature of $T_{s,i} = 600°C$, and an oven air temperature of $T_\infty = 800°C$. The inside convection coefficient h is known to be 25 W/m² · K. What is the value of k_B?

3.10 A house has a composite wall of wood, fiberglass insulation, and plaster board, as indicated in the sketch. On a cold winter day the convection heat transfer coefficients are $h_o = 60$ W/m² · K and $h_i = 30$ W/m² · K. The total wall surface area is 350 m².

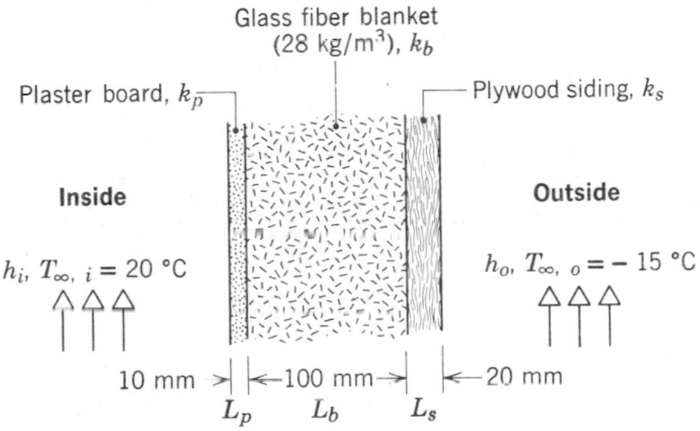

(a) Determine a symbolic expression for the total thermal resistance of the wall, including inside and outside convection effects for the prescribed conditions.

(b) Determine the total heat loss through the wall.

(c) If the wind were blowing violently, raising h_o to 300 W/m² · K, determine the percentage increase in the heat loss.

(d) What is the controlling resistance that determines the amount of heat flow through the wall?

3.11 Consider the composite wall of problem 3.10 under conditions for which the inside air is still characterized by $T_{\infty,i} = 20°C$ and $h_i = 30$ W/m² · K. However, use the more realistic conditions for which the outside air is characterized by a diurnal (time) varying temperature of the form

$$T_{\infty,o}(\mathrm{K}) = 273 + 5\sin\left(\frac{2\pi}{24}t\right) \qquad 0 \le t \le 12h$$

$$T_{\infty,o}(\mathrm{K}) = 273 + 11 \sin\left(\frac{2\pi}{24}t\right) \qquad 12 \le t \le 24h$$

with $h_o = 60\ \mathrm{W/m^2 \cdot K}$. Estimate the daily heat loss through the wall if its total surface area is $200\ \mathrm{m^2}$.

3.12 A composite wall of height H and of unit length normal to the page is insulated at its ends and is comprised of four different materials, arranged as shown.

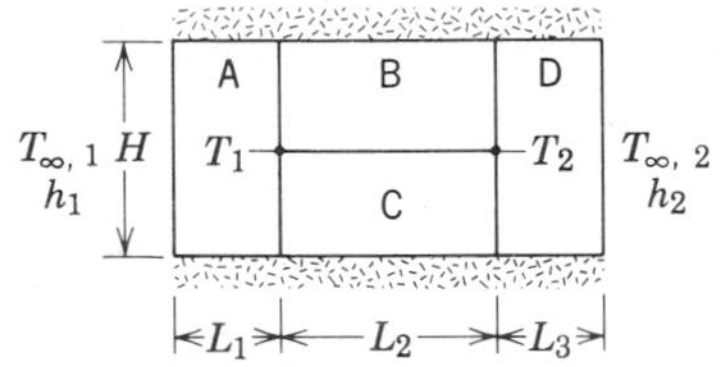

(a) Sketch the thermal circuit of the system.

(b) Consider a wall for which $H = 3$ m, $H_B = H_C = 1.5$ m, $L_1 = L_3 = 0.05$ m, $L_2 = 0.10$ m, $k_A = k_D = 50\ \mathrm{W/m \cdot K}$, $k_B = 10\ \mathrm{W/m \cdot K}$, and $k_C = 1\ \mathrm{W/m \cdot K}$. Under conditions for which $T_{\infty,1} = 200°\mathrm{C}$, $h_1 = 50\ \mathrm{W/m^2 \cdot K}$, $T_{\infty,2} = 25°\mathrm{C}$, and $h_2 = 10\ \mathrm{W/m^2 \cdot K}$, what is the rate of heat transfer through the wall? What are the interface temperatures, T_1 and T_2?

3.13 Consider a composite wall that includes an 8-mm-thick hardwood siding, 40 mm by 130 mm hardwood studs on 0.65-m centers with glass fiber insulation (paper faced, $28\ \mathrm{kg/m^3}$), and a 12-mm layer of gypsum (vermiculite) wall board.

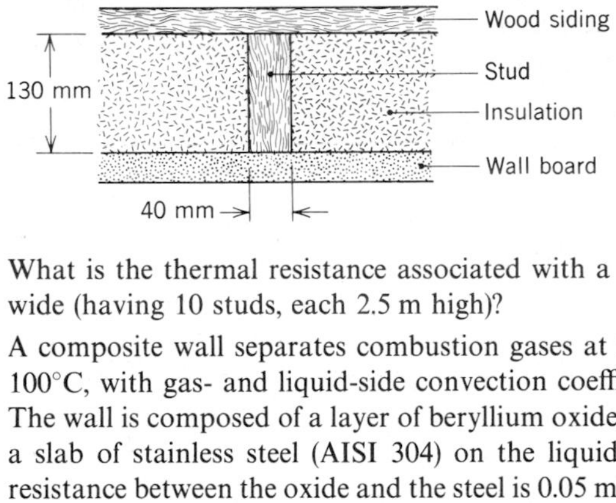

What is the thermal resistance associated with a wall that is 2.5 m high by 6.5 m wide (having 10 studs, each 2.5 m high)?

3.14 A composite wall separates combustion gases at 2600°C from a liquid coolant at 100°C, with gas- and liquid-side convection coefficients of 50 and $1000\ \mathrm{W/m^2 \cdot K}$. The wall is composed of a layer of beryllium oxide on the gas side 10 mm thick and a slab of stainless steel (AISI 304) on the liquid side 20 mm thick. The contact resistance between the oxide and the steel is $0.05\ \mathrm{m^2 \cdot K/W}$. What is the heat loss per unit surface area of the composite? Sketch the temperature distribution from the gas to the liquid.

3.15 Two stainless steel plates 10 mm thick are subjected to a contact pressure of 1 bar under vacuum conditions for which there is an overall temperature drop of 100°C across the plates. What is the heat flux through the plates? What is the temperature drop across the contact plane?

3.16 Consider a plane composite wall that is composed of two materials of thermal conductivities $k_A = 0.1\ \mathrm{W/m \cdot K}$ and $k_B = 0.04\ \mathrm{W/m \cdot K}$ and thicknesses $L_A = 10$ mm and $L_B = 20$ mm. The contact resistance at the interface between the two materials is known to be 0.30 $m^2 \cdot K/W$. Material A adjoins a fluid at 200°C for which $h = 10$ $W/m^2 \cdot K$, and material B adjoins a fluid at 40°C for which $h = 20$ $W/m^2 \cdot K$.

(a) What is the rate of heat transfer through a wall that is 2 m high by 2.5 m wide?

(b) Sketch the temperature distribution.

3.17 The diagram shows a conical section fabricated from pure aluminum. It is of circular cross section having diameter $D = ax^{1/2}$, where $a = 0.5\ m^{1/2}$. The small end is located at $x_1 = 25$ mm and the large end at $x_2 = 125$ mm. The end temperatures are $T_1 = 600$ K and $T_2 = 400$ K, while the lateral surface is well insulated.

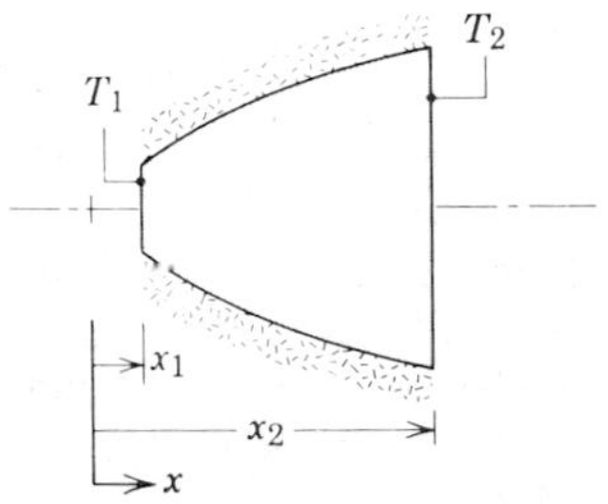

(a) Derive an expression for the temperature distribution $T(x)$ in symbolic form, assuming one-dimensional conditions. Sketch the temperature distribution.

(b) Calculate the heat rate q_x.

3.18 A truncated solid cone is of circular cross section, and its diameter is related to the axial coordinate by an expression of the form $D = ax^{3/2}$, where $a = 1.0\ m^{-1/2}$.

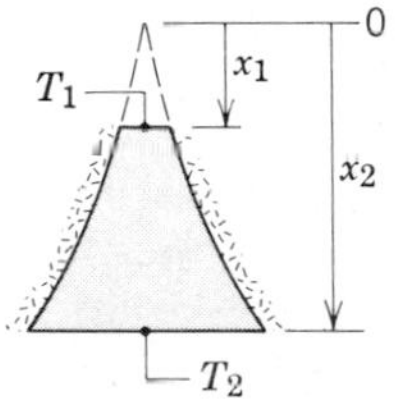

The sides are well insulated, while the top surface of the cone at x_1 is maintained at T_1 and the bottom surface at x_2 is maintained at T_2.

(a) Obtain an expression for the temperature distribution $T(x)$.

(b) What is the rate of heat transfer across the cone if it is constructed of pure aluminum with $x_1 = 0.075$ m, $T_1 = 100$°C, $x_2 = 0.225$ m, and $T_2 = 20$°C.

3.19 From Figure 2.5 it is evident that, over a wide temperature range, the temperature dependence of the thermal conductivity of many solids may be approximated by a linear expression of the form $k = k_o + aT$, where k_o is a positive constant and a is a coefficient that may be positive or negative. Obtain an expression for the heat flux across a plane wall whose inner and outer surfaces are maintained at T_0 and T_1, respectively. Sketch the forms of the temperature distribution corresponding to $a > 0$, $a = 0$, and $a < 0$.

3.20 Consider a tube wall of inner and outer radii r_i and r_o, whose temperatures are maintained at T_i and T_o, respectively. The thermal conductivity of the cylinder is temperature dependent and may be represented by an expression of the form

$k = k_o(1 + aT)$, where k_o and a are constants. Obtain an expression for the heat transfer per unit length of the tube. What is the thermal resistance of the tube wall?

3.21 The thermal conductivity of a material varies with temperature according to the relation $k = k_o(1 + aT^2)$, where k_o and a are constants. Obtain an expression for the thermal resistance per unit surface area of a plane wall constructed from this material.

3.22 A steam pipe of 0.12 m outside diameter is insulated with a 20-mm-thick layer of calcium silicate. If the inner and outer surfaces of the insulation are at temperatures of $T_{s,1} = 800$ K and $T_{s,2} = 490$ K, respectively, what is the heat loss per unit length of the pipe?

3.23 A thin electrical heater is wrapped around the outer surface of a long cylindrical tube whose inner surface is maintained at a temperature of 5°C. The tube wall has inner and outer radii of 25 and 75 mm, respectively, and a thermal conductivity of 10 W/m · K. The thermal contact resistance between the heater and the outer surface of the tube (per unit length of the tube) is $R'_{t,c} = 0.01$ m · K/W. The outer surface of the heater is exposed to a fluid with $T_\infty = -10$°C and a convection coefficient of $h = 100$ W/m^2 · K. Determine the heater power per unit length of tube required to maintain the heater at $T_o = 25$°C.

3.24 A thin electrical heater is fitted between two long, concentric cylinders of inner and outer radii of 20 and 40 mm, respectively. The inner solid cylinder (A) has a thermal conductivity of $k_A = 0.15$ W/m · K. The outer cylinder (B) has a thermal conductivity of $k_B = 1.5$ W/m · K and its surface is subjected to convection with a fluid of temperature $T_\infty = -15$°C and heat transfer coefficient 50 W/m^2·K. There is negligible thermal contact resistance between the cylinder surfaces and the heater.

(a) Determine the electrical power per unit length of the cylinders (W/m) that is required to maintain the outer surface of cylinder B at 5°C.

(b) What is the temperature at the center of cylinder A?

3.25 A long cylindrical rod of radius 10 cm consists of a nuclear reacting material ($k = 0.5$ W/m · K) generating 24,000 W/m^3 uniformly throughout its volume. This rod is encapsulated within another cylinder having an outer radius of 20 cm and a thermal conductivity of 4 W/m · K. The outer surface is surrounded by a fluid at 100°C, and the convection coefficient between the surface and the fluid is 20 W/m^2 · K. Find the temperatures at the interface between the two cylinders and at the outer surface.

3.26 A special coating, which is applied to the inner surface of a plastic tube, is cured by placing a cylindrical radiation heat source within the tube. The space between the tube and the source is evacuated, and the source delivers a uniform heat flux q''_1, which is absorbed at the inner surface of the tube. The outer surface of the tube is maintained at a uniform temperature, $T_{s,2}$.

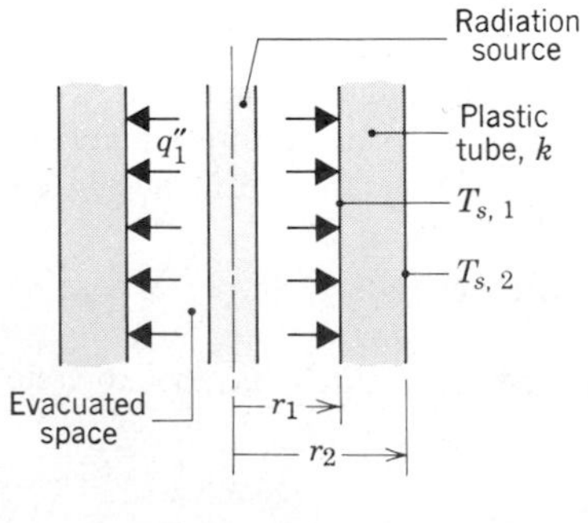

Develop an expression for the temperature distribution $T(r)$ in the tube wall in terms of q_1'', $T_{s,2}$, r_1, r_2, and k. If the inner and outer tube radii are $r_1 = 25$ mm and $r_2 = 38$ mm, what is the power required per unit length of the radiation source to maintain the inner surface at $T_{s,1} = 150°C$, while the outer surface is fixed at $T_{s,2} = 25°C$? The tube wall thermal conductivity is $k = 10$ W/m·K.

3.27 An electrical current of 700 A flows through a stainless steel cable having a diameter of 5 mm and an electrical resistance of 6×10^{-4} Ω/m (i.e., per meter of cable length). The cable is in an environment having a temperature of 30°C, and the total coefficient associated with convection and radiation between the cable and the environment is approximately 25 W/m^2·K.

(a) If the cable is bare, what is its surface temperature?

(b) If a very thin coating of electrical insulation is applied to the cable, with a contact resistance of 0.02 m^2·K/W, what are the insulation and cable surface temperatures?

(c) There is some concern about the ability of the insulation to withstand elevated temperatures. What thickness of this insulation ($k = 0.5$ W/m·K) will yield the lowest value of the maximum insulation temperature? What is the value of the maximum temperature when this thickness is used?

3.28 The radiation heat gage shown in the diagram is made from constantan metal foil, which is coated black and is in the form of a circular disk of radius R and thickness t. The gage is located in an evacuated enclosure. The incident radiation flux absorbed by the foil, q_i'', diffuses toward the outer circumference and into the larger copper ring, which acts as a heat sink at the constant temperature $T(R)$. Two copper lead wires are attached to the center of the foil and to the ring to complete a thermocouple circuit that allows for measurement of the temperature difference between the foil center and the foil edge, $\Delta T = T(0) - T(R)$.

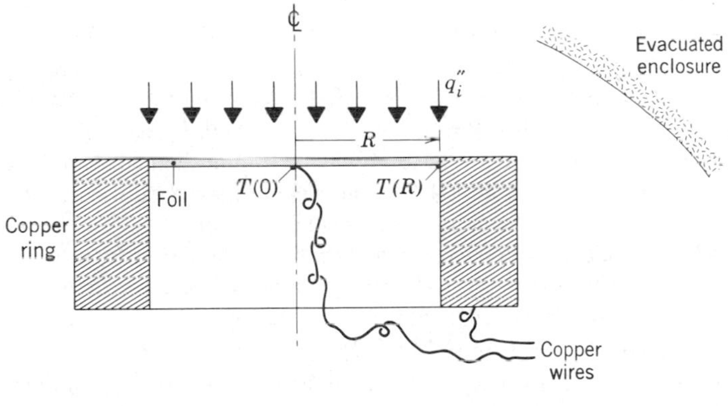

Obtain the differential equation that determines $T(r)$, the temperature distribution in the foil, under steady-state conditions. Solve this equation to obtain an expression relating ΔT to q_i''. You may neglect radiation exchange between the foil and its surroundings.

3.29 A 0.20-m diameter, thin-walled steel pipe is used to transport saturated steam at a pressure of 20 bars in a room for which the air temperature is 25°C and the convection heat transfer coefficient at the outer surface of the pipe is 20 W/m^2·K.

(a) What is the heat loss per unit length from the bare pipe (no insulation)? What is the heat loss per unit length if a 50-mm-thick layer of insulation (magnesia, 85%)

is added? The steel and magnesia may each be assumed to have an emissivity of 0.8.

(b) The costs associated with generating the steam and installing the insulation are known to be $4/10^9 J and $100/m of pipe length, respectively. If the steam line is to operate 7500 h/yr, how many years are needed to pay back the initial investment in insulation?

3.30 Steam at a temperature of 250°C flows through a steel pipe (AISI 1010) of 60 mm inside diameter and 75 mm outside diameter. The convection coefficient between the steam and the inner surface of the pipe is 500 $W/m^2 \cdot K$, while that between the outer surface of the pipe and the surroundings is 25 $W/m^2 \cdot K$. The pipe emissivity is 0.8, and the temperature of the air and the surroundings is 20°C. What is the heat loss per unit length of pipe?

3.31 Steam flowing through a long, thin-walled pipe maintains the pipe wall at a uniform temperature of 500 K. The pipe is covered with an insulation blanket comprised of two different materials, A and B.

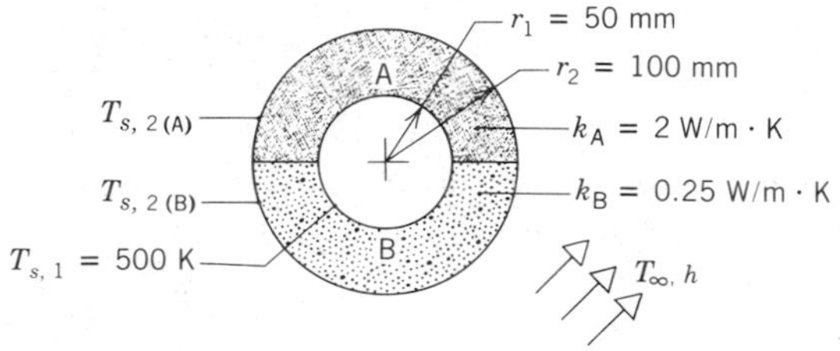

The interface between the two materials may be assumed to have an infinite contact resistance, and the entire outer surface is exposed to air for which $T_\infty = 300$ K and $h = 25$ $W/m^2 \cdot K$.

(a) Sketch the thermal circuit of the system. Label (using the above symbols) all pertinent nodes and resistances.

(b) For the prescribed conditions, what is the total heat loss from the pipe? What are the outer surface temperatures $T_{s,2(A)}$ and $T_{s,2(B)}$?

3.32 A Bakelite coating is to be used with a 10-mm-diameter conducting rod, whose surface is maintained at 200°C by passage of an electrical current. The rod is in a fluid at 25°C, and the convection coefficient is 140 $W/m^2 \cdot K$. What is the critical radius associated with the coating? What is the heat transfer rate per unit length for the bare rod and for the rod with a coating of Bakelite that corresponds to the critical radius? How much Bakelite should be added to reduce the heat transfer associated with the bare rod by 25%?

3.33 A storage tank consists of a cylindrical section that has length and inner diameter of $L = 2$ m and $D_i = 1$ m, respectively, and two hemispherical end sections. The tank is constructed from 20-mm-thick glass (Pyrex) and is exposed to ambient air for which the temperature is 300 K and the convection coefficient is 10 $W/m^2 \cdot K$. The tank is used to store heated oil, which maintains the inner surface at a temperature of 400 K. Determine the electrical power that must be supplied to a heater submerged in the oil if the prescribed conditions are to be maintained. Radiation effects may be neglected, and the Pyrex may be assumed to have a thermal conductivity of 1.4 $W/m \cdot K$.

3.34 In Example 3.3, an expression was derived for the critical insulation radius for an insulated, cylindrical tube. Derive the expression that would be appropriate for an insulated sphere.

3.35 A hollow aluminum sphere, with an electrical heater in the center, is used in tests to determine the thermal conductivity of insulating materials. The inner and outer radii of the sphere are 0.15 and 0.18 m, respectively, and testing is done under steady-state conditions with the inner surface of the aluminum maintained at 250°C. In a particular test, a spherical shell of insulation is cast on the outer surface of the sphere to a thickness of 0.12 m. The system is in a room for which the air temperature is 20°C and the convection coefficient at the outer surface of the insulation is 30 $W/m^2 \cdot K$. If 80 W is dissipated by the heater under steady-state conditions, what is the thermal conductivity of the insulation?

3.36 A composite spherical shell of inner radius $r_1 = 0.25$ m is constructed from lead of outer radius $r_2 = 0.30$ m and AISI 302 stainless steel of outer radius $r_3 = 0.31$ m. The cavity is filled with radioactive wastes that generate heat at a rate of $q = 5 \times 10^5$ W/m^3. It is proposed to submerge the container in oceanic waters that are at a temperature of $T_\infty = 10°C$ and provide a uniform convection coefficient of $h = 500$ $W/m^2 \cdot K$ at the outer surface of the container. Are any problems associated with this proposal?

3.37 The energy transferred from the anterior chamber of the eye through the cornea varies considerably depending on whether a contact lens is worn. Treat the eye as a spherical system and assume the system to be at steady state. The convection coefficient h_o is unchanged with and without the contact lens in place. The cornea and the lens cover one-third of the spherical surface area.

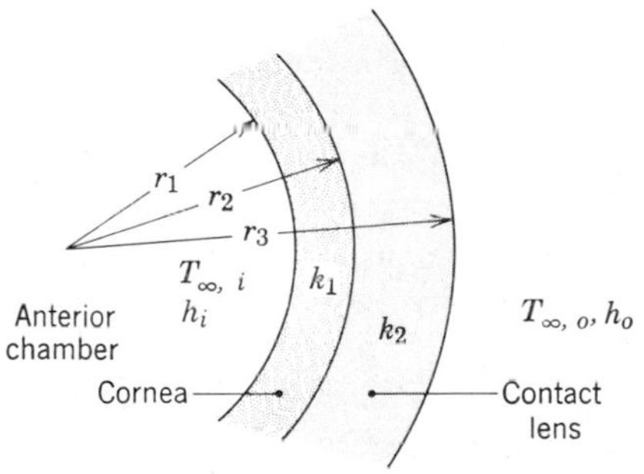

Values of the parameters representing this situation are as follows:

$r_1 = 10.2$ mm	$r_2 = 12.7$ mm	$r_3 = 16.5$ mm
$T_{\infty,i} = 37°C$	$T_{\infty,o} = 21°C$	
$k_1 = 0.35$ $W/m \cdot K$	$k_2 = 0.80$ $W/m \cdot K$	
$h_i = 12$ $W/m^2 \cdot K$	$h_o = 6$ $W/m^2 \cdot K$	

(a) Construct the thermal circuits, labeling all potentials and flows for the systems excluding the contact lens and including the contact lens. Write resistance elements in terms of appropriate parameters.

(b) Determine the heat loss from the anterior chamber with and without the contact lens in place.

(c) Discuss the implication of your results.

3.38 The outer surface of a hollow sphere of radius r_2 is subjected to a uniform heat flux q_2''. The inner surface at r_1 is held at a constant temperature $T_{s,1}$.

(a) Develop an expression for the temperature distribution $T(r)$ in the sphere wall in terms of q_2'', $T_{s,1}$, r_1, r_2, and the thermal conductivity of the wall material k.

(b) If the inner and outer tube radii are $r_1 = 50$ mm and $r_2 = 100$ mm, what heat flux q_2'' is required to maintain the outer surface at $T_{s,2} = 50°C$, while the inner surface is at $T_{s,1} = 20°C$? The thermal conductivity of the wall material is $k = 10$ W/m · K.

3.39 A spherical shell of inner and outer radii r_i and r_o, respectively, is filled with a heat-generating material that provides for a uniform volumetric generation rate (W/m^3) of $\dot{q}$. The outer surface of the shell is exposed to a fluid having a temperature T_∞ and a convection coefficient h. Obtain an expression for the steady-state temperature distribution $T(r)$ in the shell, expressing your result in terms of r_i, r_o, $\dot{q}$, h, T_∞, and the thermal conductivity k of the shell material.

3.40 The steady-state temperature distribution in a composite plane wall of three different materials, each of constant thermal conductivity, is shown below.

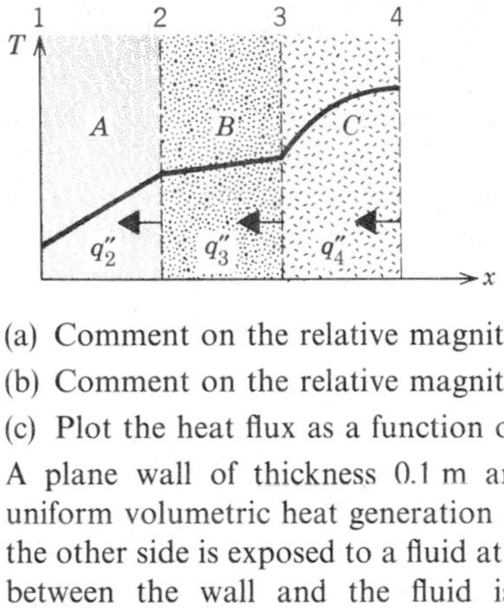

(a) Comment on the relative magnitudes of q_2'' and q_3'' and of q_3'' and q_4''.

(b) Comment on the relative magnitudes of k_A and k_B and of k_B and k_C.

(c) Plot the heat flux as a function of x.

3.41 A plane wall of thickness 0.1 m and thermal conductivity 25 W/m · K having uniform volumetric heat generation of 0.3 MW/m^3 is insulated on one side, while the other side is exposed to a fluid at 92°C. The convection heat transfer coefficient between the wall and the fluid is 500 W/m^2 · K. Determine the maximum temperature in the wall.

3.42 Consider the one-dimensional wall shown. The outer surfaces are exposed to a fluid at 25°C and a convection heat transfer coefficient of 1000 W/m^2 · K. The middle wall B experiences uniform heat generation $\dot{q}_B$, while there is no generation in walls A and C. The temperatures at the interfaces are $T_1 = 260°C$ and $T_2 = 210°C$.

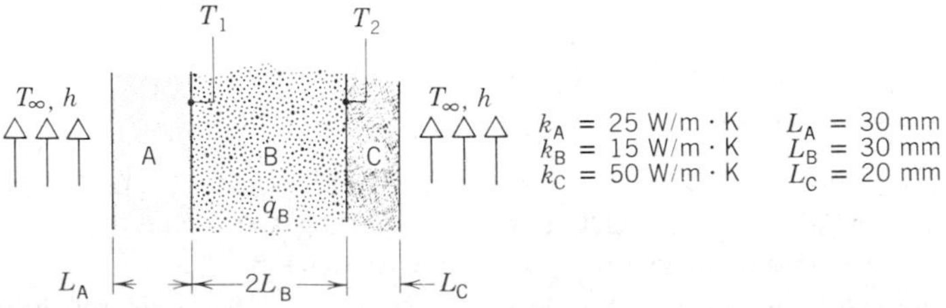

(a) Assuming negligible contact resistance at the interfaces, determine the volumetric heat generation $\dot{q}_B$.

(b) Sketch the temperature distribution, showing its important features.

3.43 Consider the composite plane wall of Problem 3.42 subjected to the same convection conditions. The middle wall experiences a uniform heat generation of $\dot{q}_B = 4 \times 10^6$ W/m³, while the outer walls have no generation. Determine the temperatures of the interfaces T_1 and T_2 and the heat fluxes through the walls A and B.

3.44 When passing an electrical current I, a copper bus bar of rectangular cross section (6 mm × 150 mm) experiences uniform heat generation at a rate $\dot{q}$ (W/m³) given by $\dot{q} = aI^2$, where $a = 0.015$ W/m³ · A². If the bar is in ambient air with $h = 5$ W/m² · K and its maximum temperature must not exceed that of the air by more than 30°C, what is the allowable current capacity for the bus bar?

3.45 A semiconductor material of thermal conductivity $k = 2$ W/m·K and electrical resistivity $\rho = 2 \times 10^{-5}$ Ω·m is used to fabricate a cylindrical rod 10 mm in diameter and 40 mm long. The longitudinal surface of the rod is well insulated, while the ends are maintained at temperatures of 100 and 0°C. If the rod carries a current of 10 A, what is its midpoint temperature? What is the heat rate at each of the ends?

3.46 A rear window defroster in an automobile consists of uniformly distributed high resistance wires molded into the glass. When power is applied to the wires, uniform heat generation may be assumed to occur within the window. During operation, heat that is generated is transferred by convection from both the interior and exterior surfaces of the window. However, the convection coefficient on the warmer, interior side h_i is less than that on the exterior side h_o, due to the effects of vehicle speed and atmospheric winds. On the same coordinate system, sketch the steady-state temperature distributions that would exist in the glass *before* the defroster has been turned on and *after* the defroster has been on for some time.

3.47 A nuclear fuel element of thickness $2L$ is covered with a steel cladding of thickness b. Heat generated within the nuclear fuel at a rate $\dot{q}$ is removed by a fluid at T_∞, which adjoins one surface and is characterized by a convection coefficient h. The other surface is well insulated, and the fuel and steel have thermal conductivities of k_f and k_s, respectively.

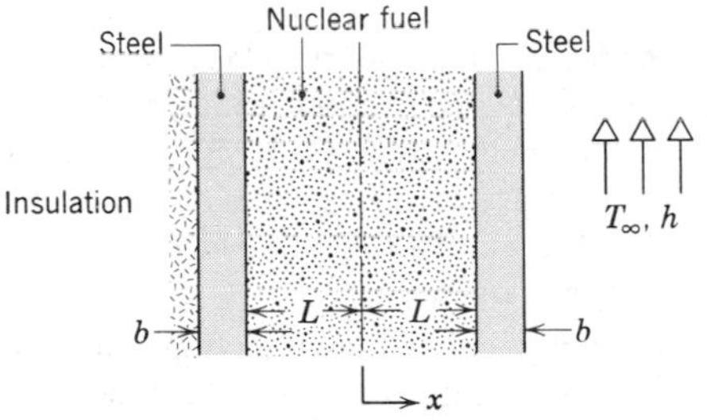

(a) Obtain an equation for the temperature distribution $T(x)$ in the nuclear fuel. Express your results in terms of $\dot{q}$, k_f, L, b, k_s, h, and T_∞.

(b) Sketch the temperature distribution $T(x)$ for the entire system.

3.48 Consider the plane wall of Figure 3.9*c*, which experiences uniform generation with one boundary insulated and the other subjected to convection cooling. Using a surface energy balance as a boundary condition at $x = L$, Equation 3.44, obtain the temperature distribution within the wall in terms of $\dot{q}$, L, k, h, and T_∞. Using Fourier's law, find the heat flux to the cold fluid.

3.49 The exposed surface ($x = 0$) of a plane wall of thermal conductivity k is subjected to microwave radiation that causes volumetric heating to vary as

$$\dot{q}(x) = \dot{q}_o\left(1 - \frac{x}{L}\right)$$

where $\dot{q}_o$ (W/m^3) is a constant. The boundary at $x = L$ is perfectly insulated, while the exposed surface is maintained at a constant temperature T_o. Determine the temperature distribution $T(x)$ in terms of x, L, k, $\dot{q}_o$, and T_o.

3.50 Consider a plane wall of thickness L, which is to act as shielding for a nuclear reactor. The inner surface ($x = 0$) receives gamma radiation that is partially absorbed within the shielding and has the effect of an internally distributed heat source. In particular heat is generated per unit volume within the shielding according to the relation

$$\dot{q}(x) = q''_o \alpha e^{-\alpha x}$$

where q''_o is the incident radiation flux and α is a property (the absorption coefficient) of the shielding material.

(a) If the inner ($x = 0$) and outer ($x = L$) surfaces of the shielding are maintained at temperatures of T_1 and T_2, respectively, what is the form of the temperature distribution in the shielding?

(b) Obtain an expression that could be used to determine the x location in the shield at which the temperature is a maximum.

3.51 A quartz window of thickness L serves as a viewing port in a furnace used for annealing steel. The inner surface ($x = 0$) of the window is irradiated with a uniform heat flux q''_o due to emission from hot gases in the furnace. A fraction, β, of this radiation may be assumed to be absorbed at the inner surface, while the remaining radiation is partially absorbed as it passes through the quartz. The volumetric heat generation due to this absorption may be described by an expression of the form

$$\dot{q}(x) = (1 - \beta)q''_o \alpha e^{-\alpha x}$$

where α is the absorption coefficient of the quartz. Convection heat transfer occurs from the outer surface ($x = L$) of the window to ambient air at T_∞ and is characterized by the convection coefficient h. Convection and radiation emission from the inner surface may be neglected, along with radiation emission from the outer surface. Determine the temperature distribution in the quartz, expressing your result in terms of the foregoing parameters.

3.52 A truncated solid cone as shown is of circular cross section with diameter varying as $D = ae^x$, where $a = 0.8$ m and x has units of meters. The cone of length 1.8 m and thermal conductivity 8 W/m · K has uniform volumetric heat generation of 1989 W/m^3. The lateral surface of the cone is insulated, while one end surface ($x = 0$) is maintained at $T_0 = 300$°C and has a heat rate $q_x(0) = 500$ W.

Determine the temperature T_L and the heat transfer rate $q_x(L)$ at the right-end surface. *Hint:* Apply an energy balance to obtain an expression for $q_x(x)$; then apply Fourier's law to obtain $T(x)$.

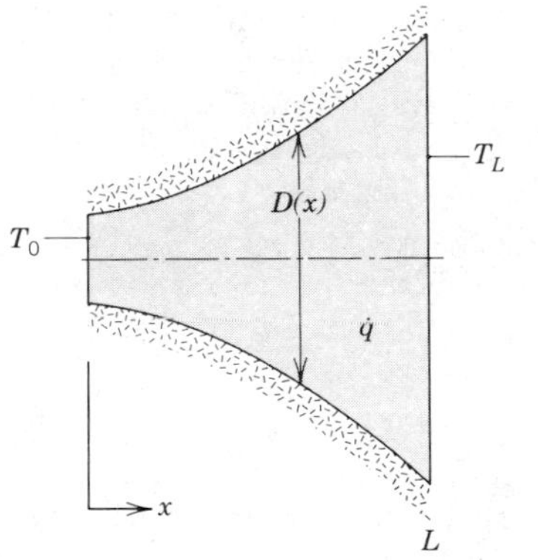

3.53 A copper cable of 30 mm diameter has an electrical resistance of 5×10^{-3} Ω/m and is used to carry an electrical current of 250 A. The cable is exposed to ambient air at 20°C, and the associated convection coefficient is 25 W/m^2 · K. What are the surface and centerline temperatures of the copper?

3.54 The cross section of a long cylindrical fuel element in a nuclear reactor is shown.

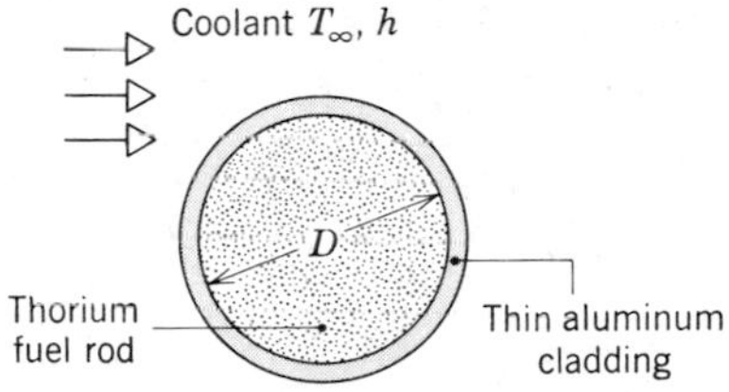

Energy generation occurs uniformly in the thorium fuel rod, which is of diameter $D = 25$ mm and is wrapped in a thin aluminum cladding. It is proposed that, under steady-state conditions, the system operate with a generation rate of $\dot{q} = 7 \times 10^8$ W/m^3 and cooling system characteristics of $T_\infty = 95$°C and $h = 7000$ W/m$^2 \cdot$ K. Is this proposal satisfactory?

3.55 A nuclear reactor fuel element consists of a solid cylindrical pin of radius r_1 and thermal conductivity k_f. The fuel pin is in good contact with a cladding material of outer radius r_2 and thermal conductivity k_c. Consider steady-state conditions for which uniform heat generation occurs within the fuel at a volumetric rate $\dot{q}$ and the outer surface of the cladding is exposed to a coolant that is characterized by a temperature T_∞ and a convection coefficient h.

(a) Obtain equations for the temperature distributions $T_f(r)$ and $T_c(r)$ in the fuel and cladding, respectively. Express your results exclusively in terms of the foregoing variables.

(b) Consider a uranium oxide fuel pin for which $k_f = 2$ W/m · K and $r_1 = 6$ mm and cladding for which $k_c = 25$ W/m·K and $r_2 = 9$ mm. If $\dot{q} = 2 \times 10^8$ W/m^3, $h = 2000$ W/m$^2 \cdot$ K, and $T_\infty = 300$ K, what is the maximum temperature in the fuel element?

3.56 A homeowner, whose water pipes have frozen during a period of cold weather, decides to melt the ice by passing an electric current I through the pipe wall. The inner and outer radii of the wall are designated as r_1 and r_2, and its electrical resistance per unit length is designated as R'_e (Ω/m). The pipe is well insulated on the outside, and during melting the ice (and water) in the pipe remain at a constant temperature T_m associated with the melting process.

(a) Assuming that steady-state conditions are reached shortly after application of the current, determine the form of the steady-state temperature distribution $T(r)$ in the pipe wall during the melting process.

(b) Develop an expression for the time t_m required to completely melt the ice. Calculate this time for $I = 100$ A, $R'_e = 0.30\ \Omega$/m, and $r_1 = 50$ mm.

3.57 A radioactive material of thermal conductivity k is cast as a solid sphere of radius r_o and placed in a liquid bath for which the temperature T_∞ and convection coefficient h are known. Heat is uniformly generated within the solid at a volumetric rate of $\dot{q}$. Obtain the steady-state radial temperature distribution in the solid, expressing your result in terms of r_o, $\dot{q}$, k, h, and T_∞.

3.58 Copper tubing is joined to the absorber of a flat plate solar collector as shown.

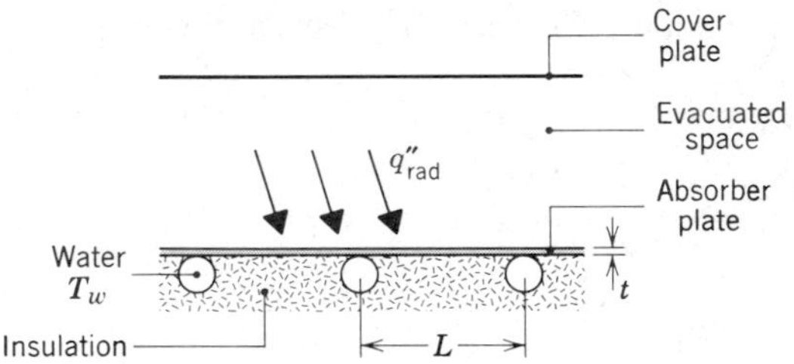

The aluminum alloy (2024-T6) absorber plate is 6 mm thick and well insulated on its bottom. The top surface of the plate is separated from a transparent cover plate by an evacuated space. The tubes are spaced a distance L of 0.20 m from each other, and water is circulated through the tubes to remove the collected energy. The water may be assumed to be at a uniform temperature of $T_w = 60°C$. Under steady-state operating conditions for which the *net* radiation heat flux to the surface is $q''_{rad} = 800$ W/m^2, what is the maximum temperature on the plate and the heat transfer rate per unit length of tube? Note that q''_{rad} represents the net effect of solar radiation absorption by the absorber plate and radiation exchange between the absorber and cover plates. You may assume the temperature of the absorber plate directly above a tube to be equal to that of the water.

3.59 Copper tubing is joined to a solar collector plate of thickness t, and the working fluid maintains the temperature of the plate above the tubes at T_o. There is a uniform net radiation heat flux q''_{rad} to the top surface of the plate, while the bottom surface is

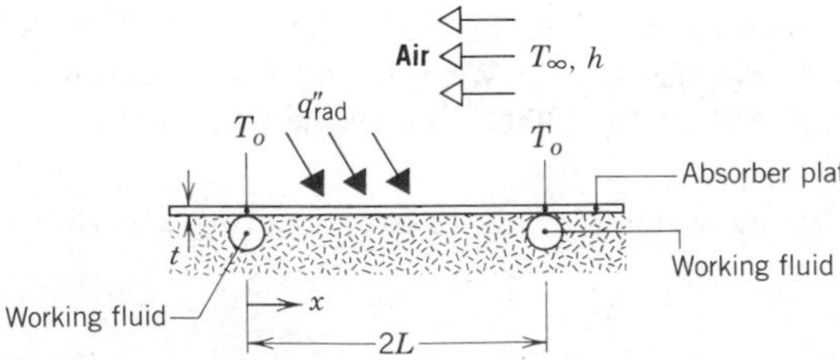

well insulated. The top surface is also exposed to a fluid at T_∞ that provides for a uniform convection coefficient h.

(a) Derive the differential equation that governs the temperature distribution $T(x)$ in the plate.

(b) Obtain a solution to the differential equation for appropriate boundary conditions.

3.60 A thin flat plate of length L, thickness t, and width $W \gg L$ is thermally joined to two large heat sinks that are maintained at a temperature T_o. The bottom of the plate is well insulated, while the net heat flux to the top surface of the plate is known to have a uniform value of q_o''.

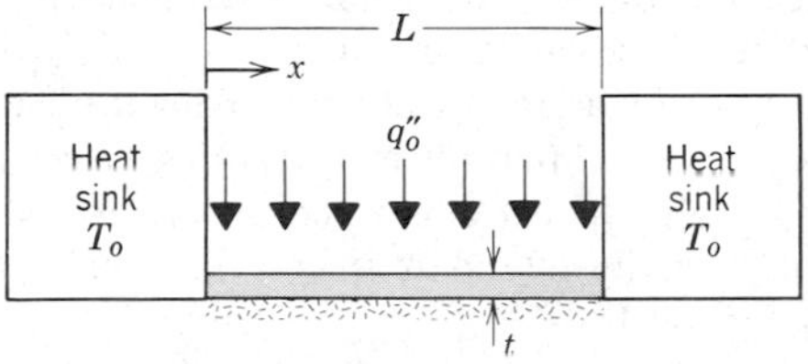

(a) Derive the differential equation that determines the steady-state temperature distribution $T(x)$ in the plate.

(b) Solve the foregoing equation for the temperature distribution, and obtain an expression for the rate of heat transfer from the plate to the heat sinks.

3.61 Consider the flat plate of Problem 3.60, but with the bottom surface no longer insulated. Convection heat transfer is now allowed to occur between this surface and a fluid at T_∞, with a convection coefficient h.

(a) Derive the differential equation that determines the steady-state temperature distribution $T(x)$ in the plate.

(b) Solve the foregoing equation for the temperature distribution, and obtain an expression for the rate of heat transfer from the plate to the heat sinks.

3.62 A thin metallic wire of thermal conductivity k, diameter D, and length $2L$ is annealed by passing an electrical current through the wire to induce a uniform volumetric heat generation $\dot{q}$. The ambient air around the wire is at a temperature T_∞, while the ends of the wire at $x = \pm L$ are also maintained at T_∞. Heat transfer from the wire to the air is characterized by the convection coefficient h. Obtain an expression for the steady-state temperature distribution $T(x)$ along the wire.

3.63 A long, circular aluminum rod is attached at one end to a heated wall and transfers heat by convection to a cold fluid.

(a) If the diameter of the rod is tripled, by how much would the rate of heat removal change?

(b) If a copper rod of the same diameter is used in place of the aluminum, by how much would the rate of heat removal change?

3.64 A brass rod 100 mm long and 5 mm in diameter extends horizontally from a casting at 200°C. The rod is in an air environment with $T_\infty = 20°\text{C}$ and $h = 30\ \text{W/m}^2 \cdot \text{K}$. What is the temperature of the rod 25, 50, and 100 mm from the casting?

3.65 Two long copper rods of diameter $D = 1$ cm are soldered together end to end, with solder having a melting point of 650°C. The rods are in air at 25°C with a convection coefficient of 10 $\text{W/m}^2 \cdot \text{K}$. What is the minimum power input needed to effect the soldering?

3.66 Circular copper rods of diameter $D = 1$ mm and length $L = 25$ mm are used to enhance heat transfer from a surface that is maintained at $T_{s,1} = 100°C$. One end of the rod is attached to this surface (at $x = 0$), while the other end ($x = 25$ mm) is joined to a second surface, which is maintained at $T_{s,2} = 0°C$. Air flowing between the surfaces (and over the rods) is also at a temperature of $T_\infty = 0°C$, and a convection coefficient of $h = 100 \text{ W/m}^2 \cdot \text{K}$ is maintained.

(a) What is the rate of heat transfer by convection from a single copper rod to the air?

(b) What is the total rate of heat transfer from a 1 m by 1 m section of the surface at 100°C, if a bundle of the rods is installed on 4-mm centers?

3.67 Consider two long, slender rods of the same diameter but different materials. One end of each rod is attached to a base surface maintained at 100°C, while the surface of the rods are exposed to ambient air at 20°C. By traversing the length of each rod with a thermocouple, it was observed that the temperatures of the rods were equal at the positions $x_A = 0.15$ m and $x_B = 0.075$ m, where x is measured from the base surface. If the thermal conductivity of rod A is known to be $k_A = 70 \text{ W/m} \cdot \text{K}$, determine the value of k_B for rod B.

3.68 The ends of two long rods of the same diameter, but different materials, are attached to a heat source at θ_b, and the rod surfaces are cooled by convection in an identical manner. The temperature distributions $\theta(x)$ for the two rods are shown.

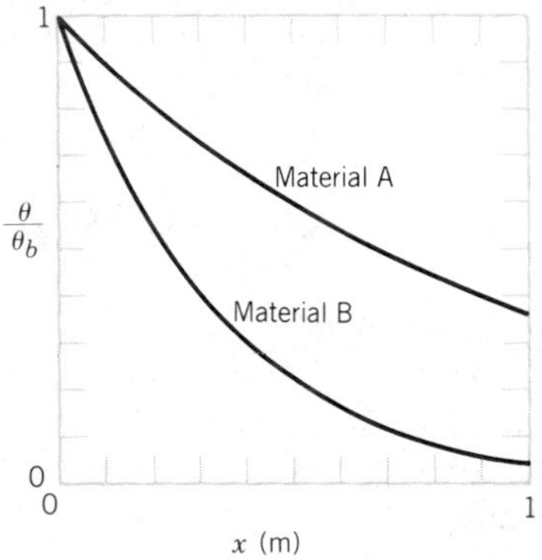

Determine k_A/k_B, the ratio of the thermal conductivities for the two rods.

3.69 An experimental arrangement for measuring the thermal conductivity of solid materials involves the use of two long rods that are equivalent in every respect, except that one is fabricated from a standard material of known thermal conductivity k_A while the other is fabricated from the material whose thermal conductivity k_B is desired. Both rods are attached at one end to a heat source of fixed temperature T_b, are exposed to a fluid of temperature T_∞, and are instrumented with thermocouples to measure the temperature at a fixed distance x_1 from the heat source. If the standard material is aluminum, with $k_A = 200 \text{ W/m} \cdot \text{K}$, and measurements reveal values of $T_A = 75°C$ and $T_B = 60°C$ at x_1 for $T_b = 100°C$ and $T_\infty = 25°C$, what is the thermal conductivity k_B of the test material?

3.70 A disk-shaped transistor, which is mounted in an insulating medium, dissipates 0.25 W during steady-state operation. To reduce the temperature of the transistor, it is proposed that a hollow copper tube be attached to the transistor as shown.

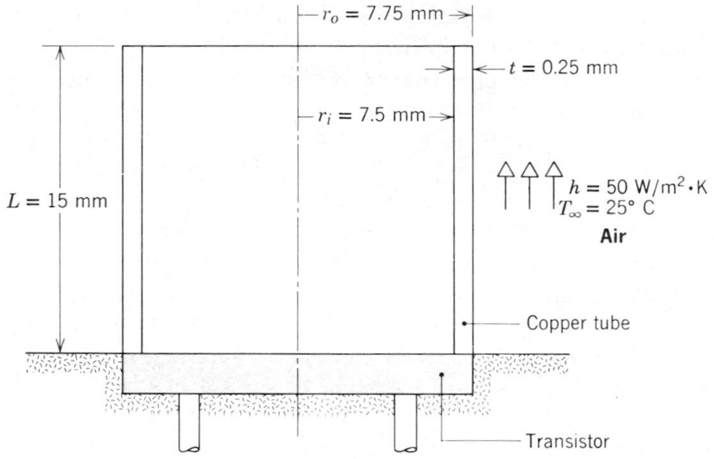

The outer surface of the tube is exposed to ambient air at $T_\infty = 25°C$ with a convection coefficient of $h = 50\ W/m^2 \cdot K$. As a first approximation, heat transfer from the interior surface of the tube and from the exposed surface of the transistor may be neglected. What is the temperature of the transistor with the fin? What is the temperature of the transistor without the fin if h and T_∞ remain the same?

3.71 Determine the percentage increase in heat transfer associated with attaching aluminum fins of rectangular profile to a plane wall. The fins are 50 mm long, 0.5 mm thick, and are equally spaced at a distance of 4 mm (250 fins/m). The convection coefficient associated with the bare wall is 40 $W/m^2 \cdot K$, while that resulting from attachment of the fins is 30 $W/m^2 \cdot K$.

3.72 Consider the use of straight, stainless steel (304) fins of rectangular and triangular profiles on a plane wall whose temperature is 100°C. The adjoining fluid is at 20°C, and the associated convection coefficient is 75 $W/m^2 \cdot K$. Each fin is 6 mm thick and 20 mm long. Compare the efficiency, the effectiveness, and the heat loss per unit width associated with the two kinds of fin.

3.73 A steel (AISI 1010) fin of rectangular cross section, 25 mm thick and 150 mm long, is attached to a plane wall whose temperature is 250°C. The temperature and convection coefficient associated with the ambient air are 20°C and 25 $W/m^2 \cdot K$, respectively. Calculate the fin heat rate per unit width of the fin by using the appropriate relation in Table 3.2 and by using the efficiency graph of Figure 3.18.

3.74 Aluminum fins of triangular profile are attached to a plane wall whose surface temperature is 250°C. The fin base thickness is 2 mm, and its length is 6 mm. The system is in ambient air at a temperature of 20°C, and the surface convection coefficient is 40 $W/m^2 \cdot K$.

(a) What are the fin efficiency and effectiveness?

(b) What is the heat dissipated per unit width by a single fin?

3.75 An annular aluminum fin of rectangular profile is attached to a circular tube having an outside diameter of 25 mm and a surface temperature of 250°C. The fin is 1 mm thick and 10 mm long, and the temperature and the convection coefficient associated with the adjoining fluid are 25°C and 25 $W/m^2 \cdot K$, respectively.

(a) What is the heat loss per fin?

(b) If 200 such fins are spaced at 5-mm increments along the tube length, what is the heat loss per meter of tube length?

3.76 Annular aluminum fins of rectangular profile are attached to a circular tube having an outside diameter of 50 mm and an outer surface temperature of 200°C. The fins are 4 mm thick and 15 mm long. The system is in ambient air at a temperature of 20°C, and the surface convection coefficient is 40 $W/m^2 \cdot K$.

(a) What are the fin efficiency and effectiveness?

(b) If there are 125 such fins per meter of tube length, what is the rate of heat transfer per unit length of tube?

3.77 Annular aluminum fins that are 2 mm thick and 15 mm long are installed on an aluminum tube of 30 mm diameter. The contact thermal resistance between a fin and the tube is known to be 2×10^{-4} $m^2 \cdot K/W$. If the tube wall is at 100°C and the adjoining fluid is at 25°C, with a convection coefficient of 75 $W/m^2 \cdot K$, what is the rate of heat transfer from a single fin? What would be the rate of heat transfer if the contact resistance could be eliminated?

3.78 Compile the forms of the heat equation and their corresponding general solutions for one-dimensional, steady-state conduction in a plane wall, a cylindrical shell, and a spherical shell with and without uniform heat generation. Compile the forms of the energy equation and their corresponding general solutions for one-dimensional, steady-state conduction in a fin of uniform cross-sectional area with and without uniform heat generation.

Four

TWO-DIMENSIONAL, STEADY-STATE CONDUCTION

To this point we have restricted our attention to conduction problems in which the temperature gradient is significant for only one coordinate direction. However, in many cases such problems are grossly oversimplified if a one-dimensional treatment is used, and it is necessary to account for multi-dimensional effects. In this chapter, we consider several techniques for treating two-dimensional systems under steady-state conditions.

4.1 ALTERNATIVE APPROACHES

Consider a long, prismatic solid in which two-dimensional conduction effects are important (Figure 4.1). With two surfaces insulated and the other surfaces maintained at different temperatures, $T_1 > T_2$, heat transfer by conduction will occur from surface 1 to 2. According to Fourier's law, Equation 2.3 or 2.4, the local heat flux in the solid is a vector that is everywhere perpendicular to lines of constant temperature (*isotherms*). The directions of the heat flux vector are represented by the *heat flow lines* of Figure 4.1, and the vector itself is the resultant of heat flux components in the x and y directions. These components are determined by Equation 2.6.

Recall that in any conduction analysis, there exist two major objectives. The first objective is to determine the temperature distribution in the medium, which for the present problem necessitates determining $T(x, y)$. This objective is generally achieved by solving the appropriate form of the heat equation. For two-dimensional, steady-state conditions with no generation and constant thermal conductivity, this form is, from Equation 2.16,

$$\frac{\partial^2 T}{\partial x^2} + \frac{\partial^2 T}{\partial y^2} = 0 \tag{4.1}$$

If Equation 4.1 can be solved for $T(x, y)$, it is then a simple matter to satisfy the second major objective, which is to determine the heat flux components q''_x and q''_y by applying the rate equations (2.6). Methods for solving Equation 4.1 include the use of *analytical*, *graphical*, and *numerical* (*finite-difference*) approaches.

The analytical method involves effecting an exact mathematical solution to

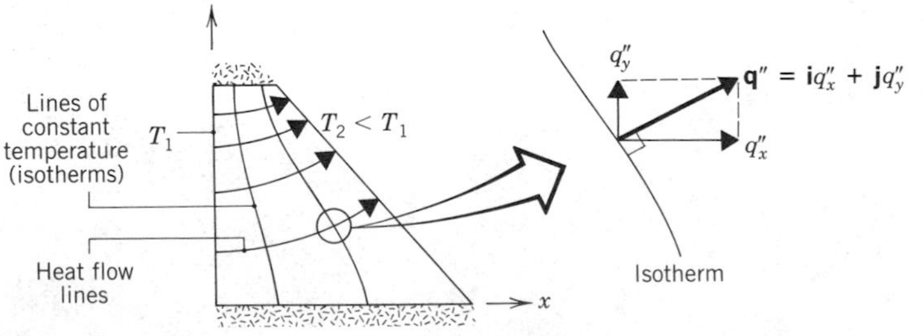

Figure 4.1 Two-dimensional conduction.

Equation 4.1. The problem is more difficult than those considered in Chapter 3, since it now involves a partial, rather than an ordinary, differential equation. Although several techniques are available for solving such equations, the solutions typically involve complicated mathematical series and functions and may be obtained for only a restricted set of simple geometries and boundary conditions [1–5]. Nevertheless, the solutions are of considerable value, since the dependent variable T is determined as a continuous function of the independent variables (x, y). Hence the solution could be used to compute the temperature at *any* point of interest in the medium. To illustrate the nature and importance of analytical techniques, an exact solution to Equation 4.1 is obtained in Section 4.2 by the method of *separation of variables.*

In contrast to the analytical methods, which provide *exact* results at *any* point, the graphical and numerical methods can provide only *approximate* results at *discrete* points. However, because the methods can accommodate complex geometries and boundary conditions, they often offer the only means by which multidimensional conduction problems can be solved. The graphical, or flux-plotting, method (Section 4.3) may be used to obtain a rough estimate of the temperature field, while the numerical method (Sections 4.4 and 4.5) may be used to obtain extremely accurate results.

4.2 THE METHOD OF SEPARATION OF VARIABLES

To appreciate how the method of separation of variables may be used to solve two-dimensional conduction problems, we consider the system of Figure 4.2. Three sides of the rectangular plate are maintained at a constant temperature T_1, while the fourth side is maintained at a constant temperature $T_2 \neq T_1$. We are interested in the temperature distribution $T(x, y)$, but to simplify the solution we introduce the transformation

$$\theta \equiv \frac{T - T_1}{T_2 - T_1} \tag{4.2}$$

Substituting Equation 4.2 into Equation 4.1, the transformed differential

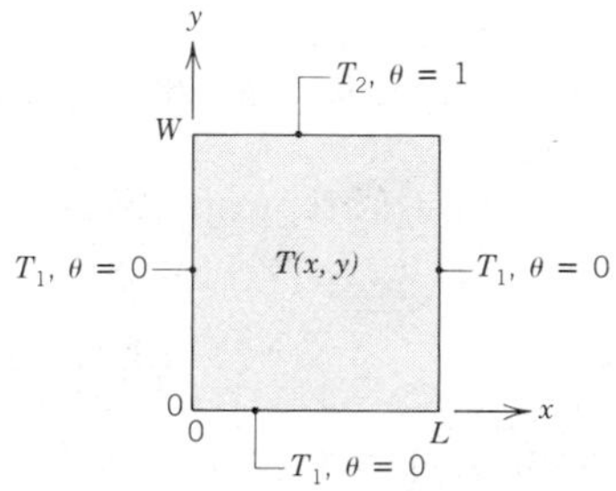

Figure 4.2 Two-dimensional conduction in a rectangular plate.

equation is then

$$\frac{\partial^2 \theta}{\partial x^2} + \frac{\partial^2 \theta}{\partial y^2} = 0 \tag{4.3}$$

Since the equation is second order in both x and y, two boundary conditions are needed for each of the coordinates. They are

$$\theta(0, y) = 0 \quad \text{and} \quad \theta(x, 0) = 0$$

$$\theta(L, y) = 0 \quad \text{and} \quad \theta(x, W) = 1$$

Note that, through the transformation of Equation 4.2, three of the four boundary conditions are now homogeneous and the value of θ is restricted to the range from 0 to 1.

We now apply the separation of variables technique by assuming that the desired solution can be expressed as the product of two functions, one of which depends only on x while the other depends only on y. That is, we assume the existence of a solution of the form

$$\theta(x, y) = X(x) \cdot Y(y) \tag{4.4}$$

Substituting into Equation 4.3 and dividing by XY, we obtain

$$-\frac{1}{X}\frac{d^2X}{dx^2} = \frac{1}{Y}\frac{d^2Y}{dy^2} \tag{4.5}$$

and it is evident that the differential equation is, in fact, separable. That is, the left-hand side of the equation depends only on x and the right-hand side depends only on y. Hence the equality can apply in general (for any x or y) only if both sides are equal to the same constant. Identifying this, as yet unknown, *separation constant* as λ^2, we then have

$$\frac{d^2X}{dx^2} + \lambda^2 X = 0 \tag{4.6}$$

$$\frac{d^2Y}{dy^2} - \lambda^2 Y = 0 \tag{4.7}$$

and the partial differential equation has been reduced to two ordinary differential equations. Note that the designation of λ^2 as a positive constant was not arbitrary. If a negative value were selected or a value of $\lambda^2 = 0$ chosen, it could be readily shown that it would be impossible to obtain a solution that satisfies the prescribed boundary conditions.

The general solutions to Equations 4.6 and 4.7 are, respectively,

$$X = C_1 \cos \lambda x + C_2 \sin \lambda x$$

$$Y = C_3 e^{-\lambda y} + C_4 e^{+\lambda y}$$

in which case the general form of the two-dimensional solution is

$$\theta = (C_1 \cos \lambda x + C_2 \sin \lambda x)(C_3 e^{-\lambda y} + C_4 e^{\lambda y}) \tag{4.8}$$

Applying the condition that $\theta(0, y) = 0$, it is evident that $C_1 = 0$. In addition from the requirement that $\theta(x, 0) = 0$, we obtain

$$C_2 \sin \lambda x (C_3 + C_4) = 0$$

which may only be satisfied if $C_3 = -C_4$. Although the requirement could also be satisfied by having $C_2 = 0$, this equality would totally eliminate the x dependence and hence would provide an unacceptable solution. If we now invoke the requirement that $\theta(L, y) = 0$, we obtain

$$C_2 C_4 \sin \lambda L (e^{\lambda y} - e^{-\lambda y}) = 0$$

The only way in which this condition may be satisfied (and still have an acceptable solution) is by requiring that λ assume discrete values for which $\sin \lambda L = 0$. These values must then be of the form

$$\lambda = \frac{n\pi}{L} \qquad n = 1, 2, 3, \ldots \tag{4.9}$$

where the integer $n = 0$ is precluded, since it provides an unacceptable solution. The desired solution may now be expressed as

$$\theta = C_2 C_4 \sin \frac{n\pi x}{L} (e^{n\pi y/L} - e^{-n\pi y/L}) \tag{4.10}$$

Combining constants and acknowledging that the new constant may depend on n, we obtain

$$\theta(x, y) = C_n \sin \frac{n\pi x}{L} \sinh \frac{n\pi y}{L}$$

where we have also used the fact that $(e^{n\pi y/L} - e^{-n\pi y/L}) = 2 \sinh (n\pi y/L)$. In the above form we have really obtained an infinite number of solutions that satisfy the original differential equation and the boundary conditions. However, since the problem is linear, a more general solution may be obtained from a superposition of the form

$$\theta(x, y) = \sum_{n=1}^{\infty} C_n \sin \frac{n\pi x}{L} \sinh \frac{n\pi y}{L} \tag{4.11}$$

To determine C_n we now apply the remaining boundary condition, which is of the form

$$\theta(x, W) = 1 = \sum_{n=1}^{\infty} C_n \sin \frac{n\pi x}{L} \sinh \frac{n\pi W}{L} \tag{4.12}$$

Although Equation 4.12 would seem to be an extremely complicated relation for evaluating C_n, a standard method is available. It involves writing an analogous infinite series expansion in terms of *orthogonal functions*. An infinite set of functions $g_1(x)$, $g_2(x)$, ..., $g_n(x)$, ... is said to be orthogonal in the domain $a \le x \le b$ if

$$\int_a^b g_m(x) g_n(x)\, dx = 0 \qquad m \ne n \tag{4.13}$$

Many functions exhibit orthogonality, including the trigonometric functions $\sin n\pi x/L$ and $\cos n\pi x/L$ for $0 \le x \le L$. Their utility in the present problem rests with the fact that any function $f(x)$ may be expressed in terms of an infinite series of orthogonal functions

$$f(x) = \sum_{n=1}^{\infty} A_n g_n(x) \tag{4.14}$$

The form of the coefficients A_n in this series may be determined by multiplying each side of the equation by $g_n(x)$ and integrating between the limits a and b.

$$\int_a^b f(x) g_n(x)\, dx = \int_a^b g_n(x) \sum_{n=1}^{\infty} A_n g_n(x)\, dx \tag{4.15}$$

However, from Equation 4.13 it is evident that all but one of the terms on the right-hand side of Equation 4.15 must be zero, leaving us with

$$\int_a^b f(x) g_n(x)\, dx = A_n \int_a^b g_n^2(x)\, dx$$

Hence

$$A_n = \frac{\int_a^b f(x) g_n(x)\, dx}{\int_a^b g_n^2(x)\, dx} \tag{4.16}$$

The properties of orthogonal functions may be used to solve Equation 4.12 for C_n by formulating an *analogous* infinite series for the appropriate form of $f(x)$. From Equation 4.12 it is evident that we should choose $f(x) = 1$ and the orthogonal function $g_n(x) = \sin n\pi x/L$. Substituting into Equation 4.16 we obtain

$$A_n = \frac{\int_0^L \sin \dfrac{n\pi x}{L}\, dx}{\int_0^L \sin^2 \dfrac{n\pi x}{L}\, dx} = \frac{2}{\pi} \frac{(-1)^{n+1} + 1}{n}$$

Hence from Equation 4.14, we have

$$1 = \sum_{n=1}^{\infty} \frac{2}{\pi} \frac{(-1)^{n+1} + 1}{n} \sin \frac{n\pi x}{L} \tag{4.17}$$

which is simply the expansion of unity in a Fourier series. Comparing Equations 4.12 and 4.17 we obtain

$$C_n = \frac{2[(-1)^{n+1} + 1]}{n\pi \sinh(n\pi W/L)} \qquad n = 1, 2, 3, \ldots \tag{4.18}$$

Substituting Equation 4.18 into Equation 4.11, we then obtain for the final solution

$$\theta(x, y) = \frac{2}{\pi} \sum_{n=1}^{\infty} \frac{(-1)^{n+1} + 1}{n} \sin \frac{n\pi x}{L} \frac{\sinh(n\pi y/L)}{\sinh(n\pi W/L)} \tag{4.19}$$

Equation 4.19 is a convergent series, from which the value of θ may be computed for any x and y. Representative results are shown in the form of isotherms for a schematic of the rectangular plate (Figure 4.3). The temperature T corresponding to a value of θ may be obtained from Equation 4.2. Exact solutions for other geometries and boundary conditions are provided in the literature.

4.3 THE GRAPHICAL METHOD

The graphical method may be employed for two-dimensional problems involving adiabatic and/or isothermal boundaries. Because the approach demands considerable patience and an artistic flair (not to mention the use of durable paper and a good eraser), it has declined in popularity over the past three decades, having been superseded largely by computer solutions based on numerical procedures. Despite its limited accuracy, however, the method still finds application as a means for obtaining a first estimate of the temperature distribution. It is also useful for developing one's intuition for the nature of the temperature field and heat flow in a system.

4.3.1 Methodology of Constructing a Flux Plot

The rationale for the graphical method comes from the fact that lines of constant temperature must be perpendicular to lines that indicate the direction of heat flow (Figure 4.1). The objective of the graphical method is one of systematically constructing such a network of isotherms and heat flow lines. This network, commonly termed a *flux plot*, is used to infer the temperature distribution and the rate of heat flow through the system.

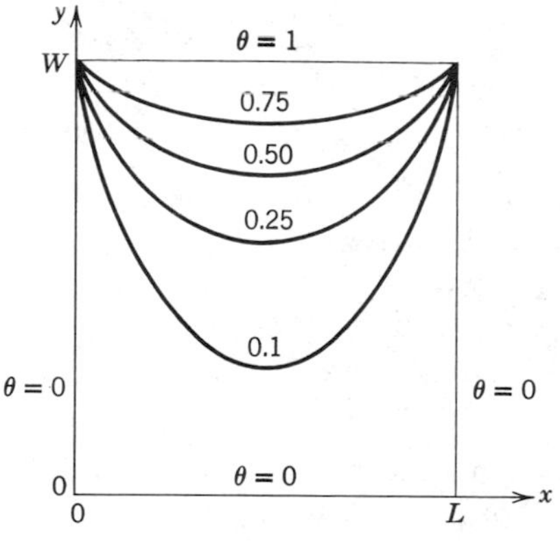

Figure 4.3 Isotherms for two-dimensional conduction in a rectangular plate.

Consider a square, two-dimensional channel whose inner and outer surfaces are maintained at T_1 and T_2, respectively. A cross section of the channel is shown in Figure 4.4*a*. A procedure for constructing the flux plot, a portion of which is shown in Figure 4.4*b*, is enumerated as follows.

1. The first step in any flux plot should be to *identify all relevant lines of symmetry*. Such lines are determined by thermal, as well as geometrical, conditions. For the square channel of Figure 4.4*a*, such lines include the designated vertical, horizontal, and diagonal lines. For this system it is therefore possible to consider only one-eighth of the configuration, as shown in Figure 4.4*b*.
2. *Lines of symmetry are adiabatic* in the sense that there can be no heat transfer in a direction perpendicular to the lines. They are therefore heat flow lines and should be treated as such. Since there is no heat flow in a direction perpendicular to a heat flow line, such a line can be termed an *adiabat*.
3. After all known lines of constant temperature associated with the system boundaries have been identified, an attempt should be made to sketch lines of constant temperature within the system. Note that *isotherms should always be perpendicular to adiabats.*
4. The heat flow lines should then be drawn with an eye toward creating a network of *curvilinear squares*. This is done by having the *heat flow lines and isotherms intersect at right angles* and by requiring that *all sides of each square be of approximately the same length.* It is often impossible to satisfy this second requirement exactly, and it is more realistic to strive for

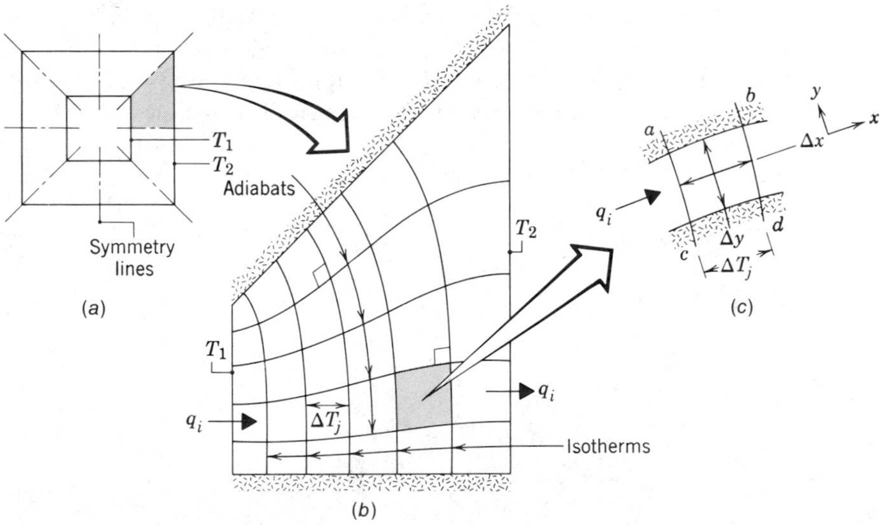

Figure 4.4 Two-dimensional conduction in a square channel of length l. (*a*) Symmetry planes. (*b*) Flux plot. (*c*) Typical curvilinear square.

equivalence between the sums of the opposite sides of each square, as shown in Figure 4.4*c*. Assigning the x coordinate to the direction of heat flow and the y coordinate to the direction normal to this flow, the requirement may be expressed as

$$\Delta x \equiv \frac{ab + cd}{2} \approx \Delta y \equiv \frac{ac + bd}{2} \tag{4.20}$$

It is difficult to create a satisfactory network of curvilinear squares in the first attempt, and numerous iterations must often be made. This trial-and-error process involves adjusting the isotherms and adiabats until satisfactory curvilinear squares are obtained for most of the network.[1] Once the flux plot has been obtained, it may be used to infer the temperature distribution in the medium. From a simple analysis, the heat transfer rate may then be obtained.

4.3.2 Determination of the Heat Transfer Rate

The rate at which energy is conducted through a *lane*, which is the region between adjoining adiabats, is designated as q_i. If the flux plot is properly constructed, the value of q_i will be the same for all lanes and the total heat transfer rate may be expressed as

$$q = \sum_{i=1}^{M} q_i = M q_i \tag{4.21}$$

where M is the *number of lanes* associated with the plot. From the curvilinear square of Figure 4.4*c* and the application of Fourier's law, q_i may be expressed as

$$q_i \approx k A_i \frac{\Delta T_j}{\Delta x} \approx k(\Delta y \cdot l) \frac{\Delta T_j}{\Delta x} \tag{4.22}$$

where ΔT_j is the temperature difference between successive isotherms, A_i is the conduction heat transfer area for the lane, and l is the length of the channel normal to the page. However, if the flux plot is properly constructed, the temperature increment is the same for all adjoining isotherms, and the overall temperature difference between boundaries, ΔT_{1-2}, may be expressed as

$$\Delta T_{1-2} = \sum_{j=1}^{N} \Delta T_j = N \Delta T_j \tag{4.23}$$

where N is the total number of temperature increments. Combining Equations 4.21 to 4.23 and recognizing that $\Delta x \approx \Delta y$ for curvilinear squares, we obtain

[1] In certain regions, such as corners, it may be impossible to approach the curvilinear square requirements. However, such difficulties generally have little effect on the overall accuracy of the results obtained from the flux plot.

Table 4.1 Conduction shape factors for selected two-dimensional systems $[q = Sk(T_1 - T_2)]$

SYSTEM	SCHEMATIC	RESTRICTIONS	SHAPE FACTOR
Isothermal sphere buried in a semiinfinite medium	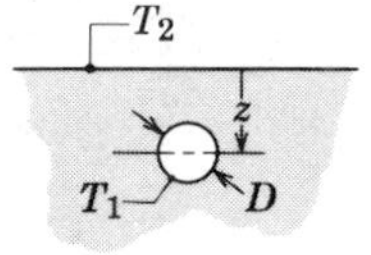	$z > D/2$	$\dfrac{2\pi D}{1 - D/4z}$
Horizontal isothermal cylinder of length L buried in a semiinfinite medium	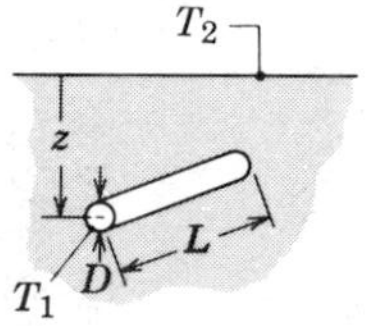	$L \gg D$	$\dfrac{2\pi L}{\cosh^{-1}(2z/D)}$
		$L \gg D$ $z > 3D/2$	$\dfrac{2\pi L}{\ln(4z/D)}$
Vertical cylinder in a semiinfinite medium	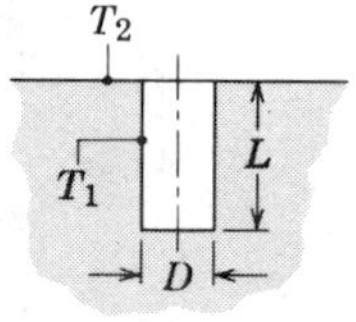	$L \gg D$	$\dfrac{2\pi L}{\ln(4L/D)}$
Conduction between two cylinders of length L in infinite medium	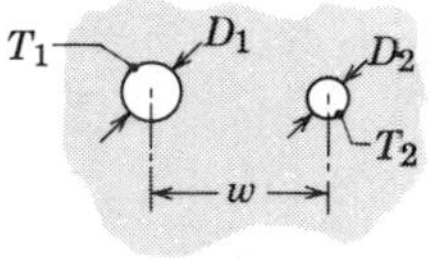	$L \gg D_1, D_2$ $L \gg w$	$\dfrac{2\pi L}{\cosh^{-1}\left(\dfrac{4w^2 - D_1^2 - D_2^2}{2D_1D_2}\right)}$
Horizontal circular cylinder of length L midway between parallel planes of equal length and infinite width	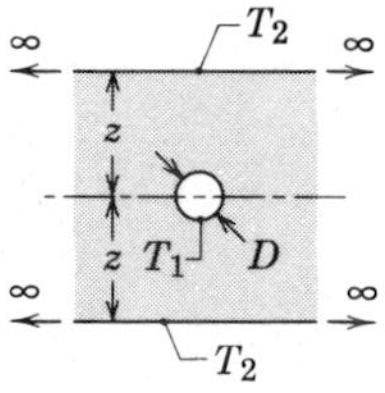	$z > D/2$	$\dfrac{2\pi L}{\ln(8z/\pi D)}$

Circular cylinder of length L in a square solid of equal length	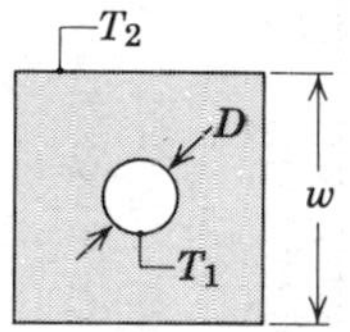	$w > D$	$\dfrac{2\pi L}{\ln(1.08\,w/D)}$
Eccentric circular cylinder of length L in a cylinder of equal length	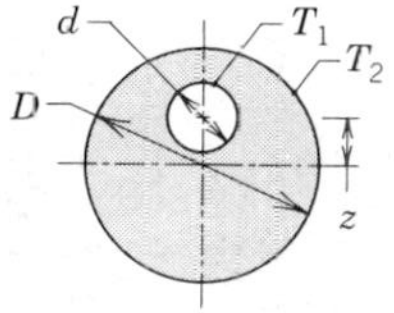	$D > d$	$\dfrac{2\pi L}{\cosh^{-1}\left(\dfrac{D^2 + d^2 - 4z^2}{2Dd}\right)}$
Conduction through the edge of adjoining walls	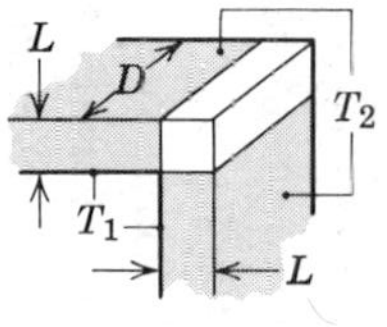	$D > L/5$	$0.54D$
Conduction through corner of three walls with a temperature difference of ΔT_{1-2} across the walls	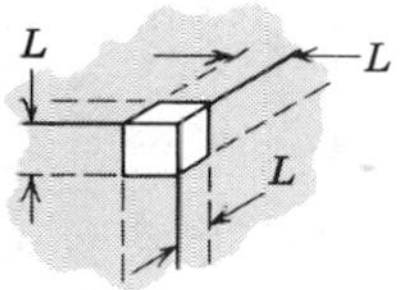	$L \ll$ length and width of wall	$0.15L$
Disk of diameter D and T_1 on a semiinfinite medium of thermal conductivity k and T_2	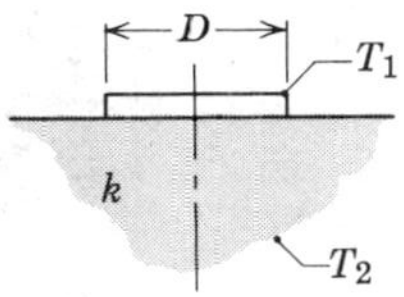	None	$2D$

$$q \approx \frac{Ml}{N} k\Delta T_{1-2} \tag{4.24}$$

The manner in which a flux plot may be used to obtain the heat transfer rate for a two-dimensional system is evident from Equation 4.24. The ratio of the number of heat flow lanes to the number of temperature increments (the value of M/N) may be obtained from the plot. Recall that specification of N is based on step 3 of the foregoing procedure, and the value, which is an integer, may be made large or small depending on the desired accuracy. The value of M is then a consequence of following step 4. Note that M is not necessarily an integer, since a fractional lane may be needed to arrive at a satisfactory network of curvilinear squares. For the network of Figure 4.4*b* it is evident that $N = 6$ and $M = 5$. Of course, as the network, or *mesh*, of curvilinear squares is made finer, N and M increase and the value of M/N should become more accurate.

4.3.3 The Conduction Shape Factor

Equation 4.24 may be used to define the *shape factor*, S, of a two-dimensional system. That is, the heat transfer rate may be expressed as

$$q = Sk\Delta T_{1-2} \tag{4.25}$$

where

$$S \equiv \frac{Ml}{N} \tag{4.26}$$

Shape factors have been obtained for numerous two-dimensional systems, and results are summarized in Table 4.1 for some common configurations. In each case, two-dimensional conduction is presumed to occur between boundaries that are maintained at uniform temperatures, with $\Delta T_{1-2} \equiv T_1 - T_2$. Results are available for many configurations not included in the table [1, 6–9].

EXAMPLE 4.1

A hole of diameter $D = 0.25$ m is drilled through the center of a solid block of square cross section with $w = 1$ m on a side. The hole is drilled along the length, $L = 2$ m, of the block, which has a thermal conductivity of $k = 150$ W/m · K. A warm fluid passing through the hole maintains an inner surface temperature of $T_1 = 75°$C, while the outer surface of the block is kept at $T_2 = 25°$C.

1. Using the flux plot method, determine the shape factor for the system.
2. What is the rate of heat transfer through the block?

SOLUTION

Known:

Dimensions and thermal conductivity of a block with a circular hole drilled along its length.

Find:

1. Shape factor.
2. Heat transfer rate for prescribed surface temperatures.

Schematic:

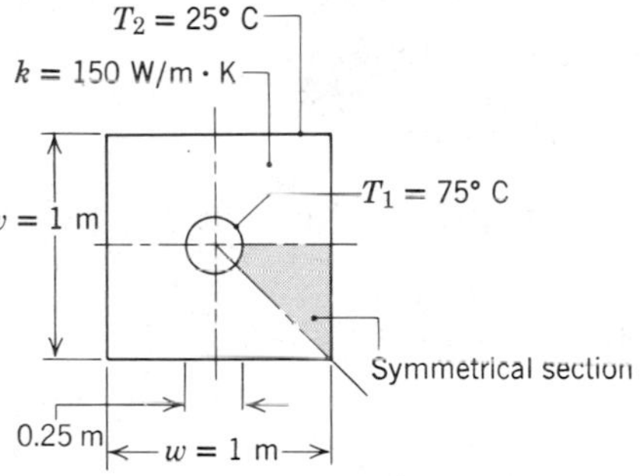

Assumptions:

1. Steady-state conditions.
2. Two-dimensional conduction.
3. Constant properties.

Analysis:

1. The flux plot may be simplified by identifying lines of symmetry and reducing the system to the one-eighth section shown in the schematic. Using a fairly coarse grid involving $N = 6$ temperature increments, the flux plot was generated. The resulting network of curvilinear squares is shown on page 142.

   With the number of heat flow lanes for the section corresponding to $M = 3$, it follows from Equation 4.26 that the shape factor for the entire block is

$$S = 8\frac{Ml}{N} = 8\frac{3 \times 2 \text{ m}}{6} = 8 \text{ m} \quad \triangleleft$$

where the factor of 8 results from the number of symmetrical sections.

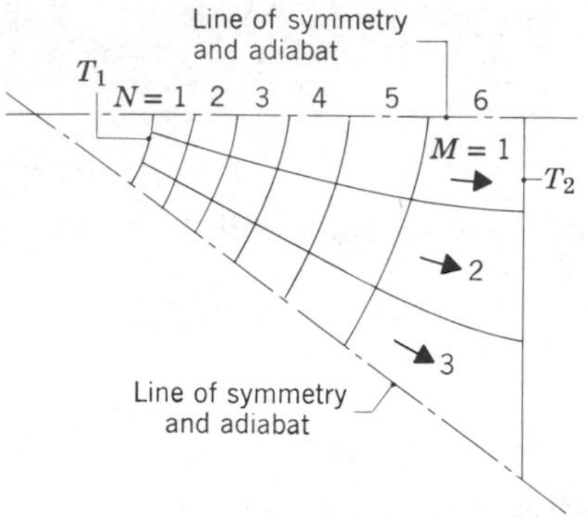

The accuracy of this result may be determined by referring to Table 4.1, where, for the prescribed system, it follows that

$$S = \frac{2\pi L}{\ln(1.08\, w/D)} = \frac{2\pi \times 2\text{ m}}{\ln(1.08 \times 1\text{ m}/0.25\text{ m})} = 8.59\text{ m}$$

Hence the result of the flux plot underpredicts the shape factor by approximately 7 percent.

2. Using $S = 8.59$ m with Equation 4.25, the heat rate is

$$q = Sk(T_1 - T_2)$$

$$q = 8.59\text{ m} \times 150\text{ W/m}\cdot\text{K}\,(75 - 25)°\text{C} = 64.4\text{ kW} \qquad \triangleleft$$

Comments:

The accuracy of the flux plot may be improved by using a finer grid (increasing the value of N).

4.4 FINITE-DIFFERENCE EQUATIONS

As discussed in Sections 4.1 and 4.2, analytical methods may be used, in certain cases, to effect exact mathematical solutions to steady, two-dimensional conduction problems. These solutions have been generated for an assortment of simple geometries and boundary conditions and are well documented in the literature [1–5]. However, more often than not, two-dimensional problems involve geometries and/or boundary conditions that preclude such solutions. In these cases, the best alternative is often one that uses *numerical*, or *finite-difference*, techniques. These techniques are particularly well suited for use with high speed digital computers.

4.4.1 The Nodal Network

In contrast to an analytical solution, which allows for temperature determination at *any* point of interest in a medium, a numerical solution enables determination of the temperature at only *discrete* points. The first step in any numerical analysis must therefore be to select these points. Referring to Figure 4.5, this is done by subdividing the medium of interest into a number of small regions and assigning to each a reference point that is at its center. The reference point is frequently termed a *nodal point* (or simply a *node*), and the aggregate of points is termed a *nodal network*, *grid* or *mesh*. The nodal points are designated by a numbering scheme that, for a two-dimensional system, may take the form shown in Figure 4.5*a*. The x and y locations are designated by the m and n indices, respectively.

It is important to note that each node represents a certain region, and its temperature is a measure of the *average* temperature of the region. For example, the temperature of the node m, n of Figure 4.5*a* may be viewed as the average temperature of the surrounding shaded area. In addition the selection of nodal points is rarely arbitrary, depending often on matters such as geometric

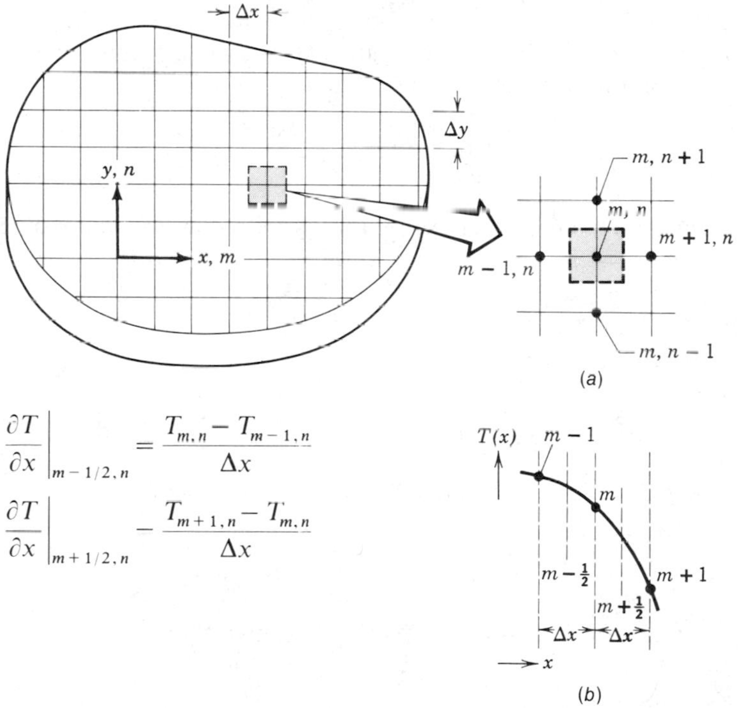

Figure 4.5 Two-dimensional conduction. (*a*) Nodal network. (*b*) Finite-difference approximation.

convenience and the desired accuracy. The numerical accuracy of calculations that are performed depends strongly on the number of designated nodal points. If this number is small (*a coarse mesh*), as the case will be if the calculations must be performed by hand, the accuracy will be limited. However, if a digital computer is used, a large number of points may be chosen (a *fine mesh*), and extremely high accuracy can be obtained.

4.4.2 Finite-Difference Form of the Heat Equation

Determination of the temperature distribution numerically dictates that an appropriate conservation equation be written for *each* of the points in the nodal network. The resulting set of equations may then be simultaneously solved for the temperature at each node. For *any interior* node of a two-dimensional system with no generation and uniform thermal conductivity, the *exact* form of the energy conservation requirement is given by the heat equation, Equation 4.1. However, if the system is characterized in terms of a nodal network, it is necessary to work with an *approximate*, or *finite-difference*, form of this equation.

A finite-difference equation which is suitable for the interior nodes of a two-dimensional system may be inferred directly from Equation 4.1. Consider the second derivative, $\partial^2 T/\partial x^2$. From Figure 4.5*b*, the value of this derivative at the m,n nodal point may be approximated as

$$\left.\frac{\partial^2 T}{\partial x^2}\right|_{m,n} \approx \frac{\partial T/\partial x|_{m+1/2,n} - \partial T/\partial x|_{m-1/2,n}}{\Delta x} \tag{4.27}$$

The temperature gradients may in turn be expressed as a function of the nodal temperatures. That is,

$$\left.\frac{\partial T}{\partial x}\right|_{m+1/2,n} \approx \frac{T_{m+1,n} - T_{m,n}}{\Delta x} \tag{4.28}$$

$$\left.\frac{\partial T}{\partial x}\right|_{m-1/2,n} \approx \frac{T_{m,n} - T_{m-1,n}}{\Delta x} \tag{4.29}$$

Substituting Equations 4.28 and 4.29 into 4.27, we obtain

$$\left.\frac{\partial^2 T}{\partial x^2}\right|_{m,n} \approx \frac{T_{m+1,n} + T_{m-1,n} - 2T_{m,n}}{(\Delta x)^2} \tag{4.30}$$

Proceeding in a similar fashion, it is readily shown that

$$\left.\frac{\partial^2 T}{\partial y^2}\right|_{m,n} \approx \frac{\partial T/\partial y|_{m,n+1/2} - \partial T/\partial y|_{m,n-1/2}}{\Delta y}$$

$$\approx \frac{T_{m,n+1} + T_{m,n-1} - 2T_{m,n}}{(\Delta y)^2} \tag{4.31}$$

Using a network for which $\Delta x = \Delta y$ and substituting Equations 4.30 and 4.31 into Equation 4.1, we obtain

$$T_{m,n+1} + T_{m,n-1} + T_{m+1,n} + T_{m-1,n} - 4T_{m,n} = 0 \tag{4.32}$$

Hence for the m,n node the heat equation, which is an *exact differential equation*, is reduced to an *approximate algebraic equation*. This approximate, *finite-difference form of the heat equation* may be applied to any interior node that is equidistant from its four neighboring nodes. It simply requires that the sum of the temperatures associated with the neighboring nodes be four times the temperature of the node of interest.

4.4.3 The Energy Balance Method

The finite-difference equation for a node may also be obtained by applying conservation of energy to a control volume about the nodal region. Since the actual direction of heat flow (into or out of the node) is often unknown, it is convenient to formulate the energy balance by *assuming* that *all* the heat flow is *into the node*. Such a condition is, of course, impossible, but if the rate equations are expressed in a manner consistent with this assumption, the correct form of the finite-difference equation is obtained. For steady-state conditions with generation, the appropriate form of Equation 1.10a is then

$$\dot{E}_{in} + \dot{E}_g = 0 \tag{4.33}$$

Consider applying Equation 4.33 to a control volume about the interior node m,n of Figure 4.6. For two-dimensional conditions, energy exchange is influenced by conduction between m,n and its four adjoining nodes, as well as by generation. Hence Equation 4.33 reduces to

$$\sum_{i=1}^{4} q_{(i)\rightarrow(m,n)} + \dot{q}(\Delta x \cdot \Delta y \cdot 1) = 0$$

where i refers to the neighboring nodes, $q_{(i)\rightarrow(m,n)}$ is the conduction rate between

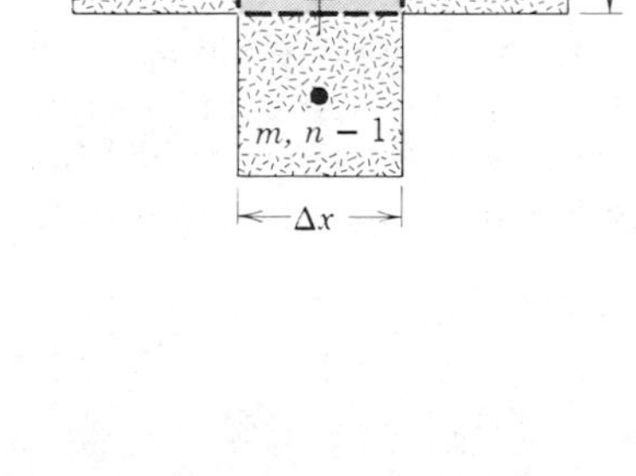

Figure 4.6 Conduction to an interior node from its adjoining nodes.

nodes, and unit depth is assumed. To evaluate the conduction rate terms, we *assume* that conduction transfer occurs exclusively through *lanes* that are oriented in either the x or y direction. Simplified forms of Fourier's law may therefore be used. For example, the rate at which energy is transferred by conduction from node $m-1,n$ to m,n may be expressed as

$$q_{(m-1,n)\to(m,n)} = k(\Delta y \cdot 1)\frac{T_{m-1,n} - T_{m,n}}{\Delta x} \tag{4.34}$$

The quantity $(\Delta y \cdot 1)$ is the heat transfer area, and the term $(T_{m-1,n} - T_{m,n})/\Delta x$ is the finite-difference approximation to the temperature gradient at the boundary between the two nodes. The remaining conduction rates may then be expressed as

$$q_{(m+1,n)\to(m,n)} = k(\Delta y \cdot 1)\frac{T_{m+1,n} - T_{m,n}}{\Delta x} \tag{4.35}$$

$$q_{(m,n+1)\to(m,n)} = k(\Delta x \cdot 1)\frac{T_{m,n+1} - T_{m,n}}{\Delta y} \tag{4.36}$$

$$q_{(m,n-1)\to(m,n)} = k(\Delta x \cdot 1)\frac{T_{m,n-1} - T_{m,n}}{\Delta y} \tag{4.37}$$

Note that in evaluating each conduction rate, we have substracted the temperature of the m,n node from the temperature of its adjoining node. This convention is necessitated by the assumption of heat flow into m,n and is consistent with the direction of the arrows shown in Figure 4.6. Substituting Equations 4.34 through 4.37 into the energy balance and remembering that $\Delta x = \Delta y$, it follows that the finite-difference equation for an interior node with generation is

$$T_{m,n+1} + T_{m,n-1} + T_{m+1,n} + T_{m-1,n} + \frac{\dot{q}(\Delta x \cdot \Delta y)}{k} - 4T_{m,n} = 0 \tag{4.38}$$

If there is no internally distributed source of energy ($\dot{q} = 0$), this expression reduces to Equation 4.32.

It is important to note that a finite-difference equation is needed for each nodal point at which the temperature is unknown. However, it is not always possible to classify all such points as interior and hence to use Equation 4.32 or 4.38. For example, the temperature may be unknown at an insulated surface or at a surface that is exposed to convective conditions. For points on such surfaces, the finite-difference equation must be obtained by applying the energy balance method.

To illustrate this method, consider the node corresponding to the internal corner of Figure 4.7. This node may be assumed to represent the three-quarter shaded section and to exchange energy by convection with an adjoining fluid at T_∞. Conduction to the nodal region (m,n) will occur along four different lanes

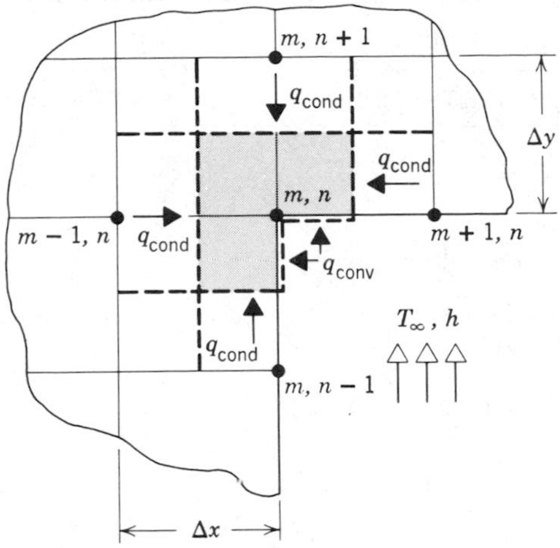

Figure 4.7 Formulation of the finite-difference equation for an internal corner of a solid with surface convection.

from neighboring nodes in the solid. The conduction heat rates q_{cond} may be expressed as

$$q_{(m-1,n)\to(m,n)} = k(\Delta y \cdot 1)\frac{T_{m-1,n} - T_{m,n}}{\Delta x} \tag{4.39}$$

$$q_{(m,n+1)\to(m,n)} = k(\Delta x \cdot 1)\frac{T_{m,n+1} - T_{m,n}}{\Delta y} \tag{4.40}$$

$$q_{(m+1,n)\to(m,n)} = k\left(\frac{\Delta y}{2} \cdot 1\right)\frac{T_{m+1,n} - T_{m,n}}{\Delta x} \tag{4.41}$$

$$q_{(m,n-1)\to(m,n)} = k\left(\frac{\Delta x}{2} \cdot 1\right)\frac{T_{m,n-1} - T_{m,n}}{\Delta y} \tag{4.42}$$

Note, that the areas for conduction from nodal regions $(m-1,n)$ and $(m,n+1)$ are proportional to Δy and Δx, respectively, whereas conduction from $(m+1,n)$ and $(m,n-1)$ occurs along lanes of half-width, $\Delta y/2$ and $\Delta x/2$, respectively.

Conditions in the nodal region m,n are also influenced by convective exchange with the fluid, and this exchange may be viewed as occurring along half-lanes in the x and y directions. The total convection rate q_{conv} may be expressed as

$$q_{(\infty)\to(m,n)} = h\left(\frac{\Delta x}{2} \cdot 1\right)(T_\infty - T_{m,n}) + h\left(\frac{\Delta y}{2} \cdot 1\right)(T_\infty - T_{m,n}) \tag{4.43}$$

Implicit in this expression is the assumption that the exposed surfaces of the corner are at a uniform temperature corresponding to the nodal temperature $T_{m,n}$. This assumption is consistent with the concept that the entire nodal region

Table 4.2 Summary of nadal finite-difference equations

CONFIGURATION	FINITE-DIFFERENCE EQUATION FOR $\Delta x = \Delta y$
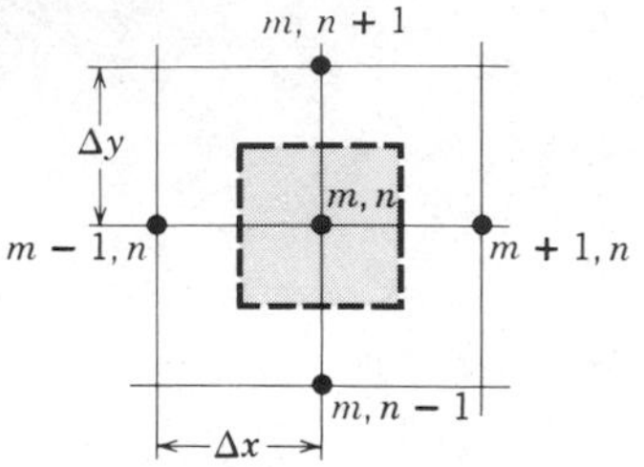	$T_{m,n+1} + T_{m,n-1} + T_{m+1,n} + T_{m-1,n} - 4T_{m,n} = 0$ (4.32) **Case 1.** Interior node.
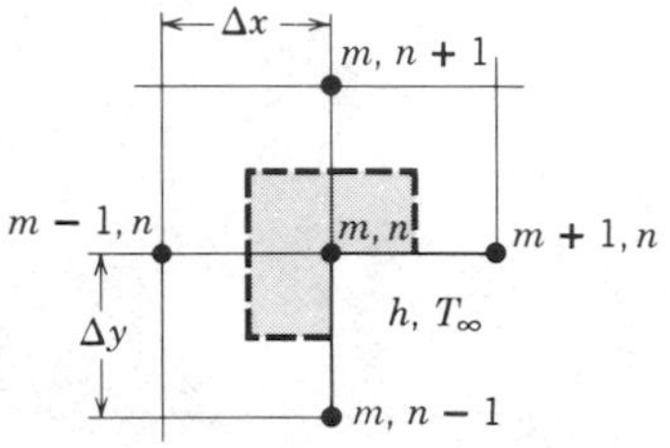	$2(T_{m-1,n} + T_{m,n+1}) + (T_{m+1,n} + T_{m,n-1}) + 2\frac{h\Delta x}{k}T_\infty - 2\left(3 + \frac{h\Delta x}{k}\right)T_{m,n} = 0$ (4.44) **Case 2.** Node at an internal corner with convection.
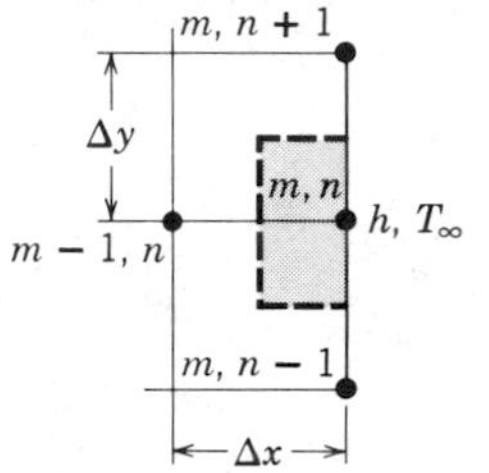	$(2T_{m-1,n} + T_{m,n+1} + T_{m,n-1}) + \frac{2h\Delta x}{k}T_\infty - 2\left(\frac{h\Delta x}{k} + 2\right)T_{m,n} = 0$ (4.45)[a] **Case 3.** Node at a plane surface with convection.
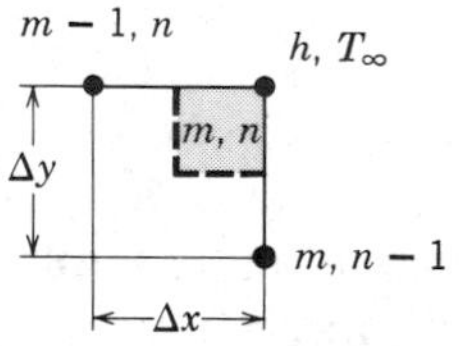	$(T_{m,n-1} + T_{m-1,n}) + 2\frac{h\Delta x}{k}T_\infty - 2\left(\frac{h\Delta x}{k} + 1\right)T_{m,n} = 0$ (4.46) **Case 4.** Node at an external corner with convection.
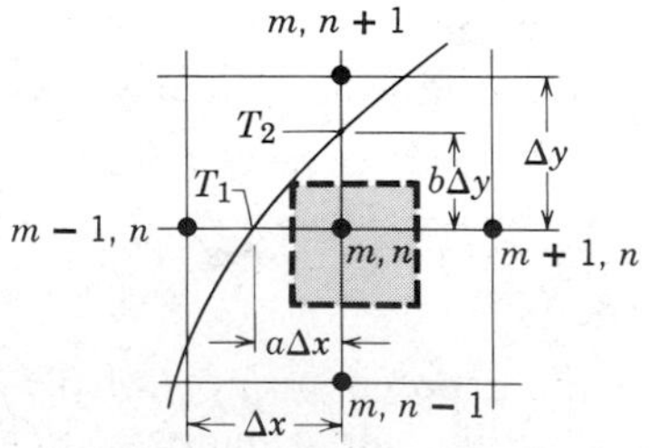	$\frac{2}{a+1}T_{m+1,n} + \frac{2}{b+1}T_{m,n-1} + \frac{2}{a(a+1)}T_1 + \frac{2}{b(b+1)}T_2 - \left(\frac{2}{a} + \frac{2}{b}\right)T_{m,n} = 0$ (4.47) **Case 5.** Node near a curved surface maintained at a nonuniform temperature.

[a]To obtain the finite-difference equation for an adiabatic surface (or surface of symmetry), simply set h equal to zero.

is characterized by a single temperature, which represents an average of the actual temperature distribution in the region. In the absence of transient, three-dimensional, and generation effects, conservation of energy, Equation 4.33, requires that the sum of Equations 4.39 to 4.43 be zero. Summing these equations and rearranging, we therefore obtain

$$T_{m-1,n} + T_{m,n+1} + \tfrac{1}{2}(T_{m+1,n} + T_{m,n-1}) + \frac{h\Delta x}{k} T_\infty - \left(3 + \frac{h\Delta x}{k}\right) T_{m,n} = 0 \tag{4.44}$$

where again the mesh is such that $\Delta x = \Delta y$.

Nodal energy balance equations pertinent to several common geometries are presented in Table 4.2. Note that the retention of equal spatial increments ($\Delta x = \Delta y$) in the presence of irregular boundaries (e.g., the curved surface of case 5 in Table 4.2) constitutes a complicating feature. Finite-difference formulations for other cases are discussed by Schneider [2] and Dusinberre [10].

EXAMPLE 4.2

Using the energy balance method, derive the finite-difference equation for the m,n nodal point located on a plane, insulated surface of a medium with uniform heat generation.

SOLUTION

Known:

Network of points adjoining a nodal point on an insulated surface.

Find: Finite-difference equation for the nodal point.

Schematic:

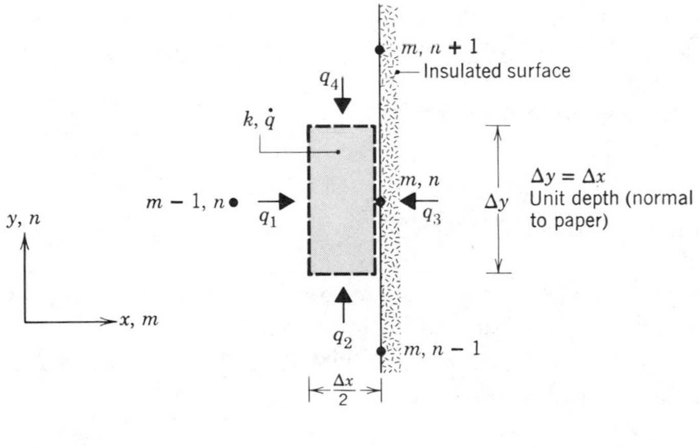

Assumptions:

1. Steady-state conditions.
2. Two-dimensional conduction.
3. Constant properties.
4. Uniform internal heat generation.

Analysis:

Applying the energy conservation requirement, Equation 4.33, to the control surface about the region $(\Delta x/2 \cdot \Delta y \cdot 1)$ associated with the m,n node, it follows that, with volumetric heat generation at a rate $\dot{q}$,

$$q_1 + q_2 + q_3 + q_4 + \dot{q}\left(\frac{\Delta x}{2} \cdot \Delta y \cdot 1\right) = 0$$

where

$$q_1 = k(\Delta y \cdot 1)\frac{T_{m-1,n} - T_{m,n}}{\Delta x}$$

$$q_2 = k\left(\frac{\Delta x}{2} \cdot 1\right)\frac{T_{m,n-1} - T_{m,n}}{\Delta y}$$

$$q_3 = 0$$

$$q_4 = k\left(\frac{\Delta x}{2} \cdot 1\right)\frac{T_{m,n+1} - T_{m,n}}{\Delta y}$$

Substituting into the energy balance and dividing by $k/2$, it follows that

$$2T_{m-1,n} + T_{m,n-1} + T_{m,n+1} - 4T_{m,n} + \frac{\dot{q}(\Delta x \cdot \Delta y)}{k} = 0 \qquad \triangleleft$$

Comments:

The same result could be obtained by using the symmetry condition, $T_{m+1,n} = T_{m-1,n}$, with the finite-difference equation (Equation 4.38) for an interior nodal point. If $\dot{q} = 0$, the desired result could also be obtained by setting $h = 0$ in Equation 4.45 (Table 4.2).

4.5 FINITE-DIFFERENCE SOLUTIONS

Once the nodal network has been established and an appropriate finite-difference equation has been written for each of the nodes, the temperature distribution may be determined. The problem reduces to one of solving a system of linear, algebraic equations. Numerous methods are available for this purpose and may be classified according to whether they are *direct* or *iterative.* Direct methods involve a fixed and predetermined number of arithmetic operations and are suitable for use when the number of equations (unknown nodal temperatures) is small. However, such methods are subject to excessive computer memory and time requirements, and it is often more efficient to use an iterative technique. Although the required number of arithmetic operations cannot be predetermined, iterative methods are characterized by reduced computer requirements and are especially appropriate when the number of equations is large.

In this section we consider matrix inversion and Gauss-Seidel iteration as examples of the direct and iterative methods, respectively. More detailed descriptions of such procedures, as well as related algorithms, may be found in the literature [2, 5, 10–14].

4.5.1 The Matrix Inversion Method

Consider a system of N finite-difference equations corresponding to N unknown temperatures. In writing these equations, it is convenient to identify the nodes by a single integer subscript, rather than by the double subscript (m,n) previously used. The procedure for setting up a matrix inversion begins by expressing the equations as

$$\begin{aligned} a_{11}T_1 + a_{12}T_2 + a_{13}T_3 + \ldots + a_{1N}T_N &= C_1 \\ a_{21}T_1 + a_{22}T_2 + a_{23}T_3 + \ldots + a_{2N}T_N &= C_2 \\ \vdots \quad\quad \vdots \quad\quad \vdots \quad\quad \vdots \quad\quad \vdots \quad &\quad \vdots \\ a_{N1}T_1 + a_{N2}T_2 + a_{N3}T_3 + \ldots + a_{NN}T_N &= C_N \end{aligned} \tag{4.48}$$

where the quantities a_{11}, a_{12}, ..., C_1, ... are known coefficients and constants involving quantities such as Δx, k, h, and T_∞. Using matrix notation, these equations may be expressed as

$$[A][T] = [C] \tag{4.49}$$

where

$$A \equiv \begin{bmatrix} a_{11} & a_{12} & \cdots & a_{1N} \\ a_{21} & a_{22} & \cdots & a_{2N} \\ \vdots & \vdots & & \vdots \\ a_{N1} & a_{N2} & \cdots & a_{NN} \end{bmatrix}, \quad T \equiv \begin{bmatrix} T_1 \\ T_2 \\ \vdots \\ T_N \end{bmatrix}, \quad C \equiv \begin{bmatrix} C_1 \\ C_2 \\ \vdots \\ C_N \end{bmatrix}$$

The *coefficient matrix* $[A]$ is square $(N \times N)$, and its *elements* are designated by a double subscript notation, for which the first and second subscripts refer to rows and columns, respectively. The matrices $[T]$ and $[C]$ have a single column and are known as *column vectors.* Typically, they are termed the *solution* and *right-hand side vectors,* respectively. If the matrix multiplication implied by the left-hand side of Equation 4.49 is performed, Equations 4.48 are obtained.

The solution vector may now be expressed as

$$[T] = [A]^{-1}[C] \tag{4.50}$$

where $[A]^{-1}$ is the inverse of $[A]$ and is defined as

$$[A]^{-1} \equiv \begin{bmatrix} b_{11} & b_{12} & \dots & b_{1N} \\ b_{21} & b_{22} & \dots & b_{2N} \\ \vdots & \vdots & & \vdots \\ b_{N1} & b_{N2} & \dots & b_{NN} \end{bmatrix}$$

Evaluating the right-hand side of Equation 4.50, it follows that

$$\begin{aligned} T_1 &= b_{11}C_1 + b_{12}C_2 + \dots + b_{1N}C_N \\ T_2 &= b_{21}C_1 + b_{22}C_2 + \dots + b_{2N}C_N \\ &\vdots \\ T_N &= b_{N1}C_1 + b_{N2}C_2 + \dots + b_{NN}C_N \end{aligned} \tag{4.51}$$

and the problem reduces to one of determining $[A]^{-1}$. That is, if $[A]$ is inverted, its elements $b_{11}, b_{12}, \dots$ may be determined and the unknown temperatures may be computed from the above expressions.

Matrix inversion may readily be performed on a digital computer by using one of numerous *canned* or *stock* programs available for this purpose. If the number of equations is small, it may also be performed by many programmable hand calculators. Hence the method provides a *convenient* means of solving two-dimensional conduction problems. Despite this convenience, however, matrix inversion is rarely numerically efficient, and it is often preferable to use an iterative numerical procedure.

EXAMPLE 4.3

A large industrial furnace is supported on a long column of fireclay brick, which is 1 m by 1 m on a side. During steady-state operation, installation is such that three surfaces of the column are maintained at 500 K while the remaining surface is exposed to an airstream for which $T_\infty = 300$ K and $h = 10$ W/m$^2\cdot$K. Using the matrix inversion method with a grid of $\Delta x = \Delta y = 0.25$ m, determine the two-dimensional temperature distribution in the column and the heat rate to the airstream per unit length of column.

SOLUTION

Known: Dimensions and surface conditions of a support column.

Find: Temperature distribution and heat rate per unit length.

Schematic:

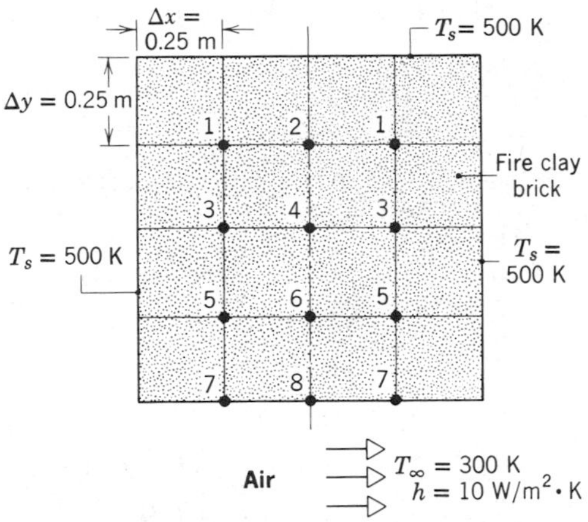

Assumptions:

1. Steady-state conditions.
2. Two-dimensional conduction.
3. Constant properties.
4. No internal heat generation.

Properties: Table A.3, fireclay brick ($T \approx 478$ K): $k = 1$ W/m·K.

Analysis:

The prescribed grid consists of 12 nodal points at which the temperature is unknown. However, the number of unknowns is reduced to 8 through symmetry, in which case the temperature of nodal points to the left of the symmetry line must equal the temperature of those to the right.

Nodes 1, 3, and 5 are interior points for which the finite-difference equations may be inferred from Equation 4.32. Hence

Node 1: $T_2 + T_3 + 1000 - 4T_1 = 0$

Node 3: $T_1 + T_4 + T_5 + 500 - 4T_3 = 0$

Node 5: $T_3 + T_6 + T_7 + 500 - 4T_5 = 0$

Equations for points 2, 4, and 6 may be obtained in a like manner or, since they lie on a symmetry adiabat, by using Equation 4.45 with $h = 0$. Hence

Node 2: $2T_1 + T_4 + 500 - 4T_2 = 0$

Node 4: $T_2 + 2T_3 + T_6 - 4T_4 = 0$

Node 6: $T_4 + 2T_5 + T_8 - 4T_6 = 0$

From Equation 4.45 and the fact that $h\Delta x/k = 2.5$, it also follows that

Node 7: $2T_5 + T_8 + 2000 - 9T_7 = 0$

Node 8: $2T_6 + 2T_7 + 1500 - 9T_8 = 0$

Having the required finite-difference equations, a matrix inversion solution may be obtained by rearranging them as follows

$$
\begin{aligned}
-4T_1 + T_2 + T_3 + 0 + 0 + 0 + 0 + 0 &= -1000\\
2T_1 - 4T_2 + 0 + T_4 + 0 + 0 + 0 + 0 &= -500\\
T_1 + 0 - 4T_3 + T_4 + T_5 + 0 + 0 + 0 &= -500\\
0 + T_2 + 2T_3 - 4T_4 + 0 + T_6 + 0 + 0 &= 0\\
0 + 0 + T_3 + 0 - 4T_5 + T_6 + T_7 + 0 &= -500\\
0 + 0 + 0 + T_4 + 2T_5 - 4T_6 + 0 + T_8 &= 0\\
0 + 0 + 0 + 0 + 2T_5 + 0 - 9T_7 + T_8 &= -2000\\
0 + 0 + 0 + 0 + 0 + 2T_6 + 2T_7 - 9T_8 &= -1500
\end{aligned}
$$

In matrix notation, following Equation 4.49, these equations are of the form $[A][T] = [C]$, where

$$
[A] = \begin{bmatrix}
-4 & 1 & 1 & 0 & 0 & 0 & 0 & 0\\
2 & -4 & 0 & 1 & 0 & 0 & 0 & 0\\
1 & 0 & -4 & 1 & 1 & 0 & 0 & 0\\
0 & 1 & 2 & -4 & 0 & 1 & 0 & 0\\
0 & 0 & 1 & 0 & -4 & 1 & 1 & 0\\
0 & 0 & 0 & 1 & 2 & -4 & 0 & 1\\
0 & 0 & 0 & 0 & 2 & 0 & -9 & 1\\
0 & 0 & 0 & 0 & 0 & 2 & 2 & -9
\end{bmatrix}
\quad
[C] = \begin{bmatrix}
-1000\\ -500\\ -500\\ 0\\ -500\\ 0\\ -2000\\ -1500
\end{bmatrix}
$$

Using a standard matrix inversion routine, it is a simple matter to find the inverse of $[A]$, $[A]^{-1}$, giving

$$[T] = [A]^{-1}[C]$$

where

$$
[T] = \begin{bmatrix} T_1\\ T_2\\ T_3\\ T_4\\ T_5\\ T_6\\ T_7\\ T_8 \end{bmatrix}
= \begin{bmatrix} 489.30\\ 485.15\\ 472.07\\ 462.01\\ 436.95\\ 418.74\\ 356.99\\ 339.05 \end{bmatrix} \; ^\circ\mathrm{C}
$$

◁

The heat rate from the column to the airstream may be computed from the expression

$$\left(\frac{q}{L}\right) = 2h\left[\left(\frac{\Delta x}{2}\right)(T_s - T_\infty) + \Delta x(T_7 - T_\infty) + \left(\frac{\Delta x}{2}\right)(T_8 - T_\infty)\right]$$

where the factor of 2 outside the brackets originates from the symmetry condition. Hence

$$\left(\frac{q}{L}\right) = 2 \times 10\ \text{W/m}^2 \cdot \text{K}\ [(0.125\ \text{m}\ (200\ \text{K})$$

$$+0.25\ \text{m}\ (56.99\ \text{K}) + 0.125\ \text{m}\ (39.05\ \text{K})] = 883\ \text{W/m}$$ ◁

Comments:

Although the foregoing temperatures comprise an exact solution to the finite-difference equations, they do not provide exact results for the physical problem. Remember that the finite-difference equations are approximations to the energy conservation requirement for the prescribed nodal points. The accuracy of the approximation may be improved by reducing the grid size (increasing the number of nodal points).

4.5.2 Gauss-Seidel Iteration

The Gauss-Seidel method is a powerful and extremely popular iterative technique. Application to the system of equations represented by Equation 4.48 is facilitated by the following procedure.

1. To whatever extent possible, the equations should be reordered to provide diagonal elements whose magnitudes are larger than those of other elements in the same row. That is, it is desirable to sequence the equations such that $|a_{11}| > |a_{12}|, |a_{13}|, \ldots, |a_{1N}|$; $|a_{22}| > |a_{21}|, |a_{23}|, \ldots, |a_{2N}|$; and so on.
2. After reordering, each of the N equations should be written in explicit form for the temperature associated with its diagonal element. Each temperature in the solution vector would then be of the form

$$T_i^{(k)} = \frac{C_i}{a_{ii}} - \sum_{j=1}^{i-1} \frac{a_{ij}}{a_{ii}} T_j^{(k)} - \sum_{j=i+1}^{N} \frac{a_{ij}}{a_{ii}} T_j^{(k-1)} \tag{4.52}$$

   where $i = 1, 2, \ldots, N$. The superscript k refers to the level of the iteration.
3. An initial ($k = 0$) value is assumed for each temperature T_i. Subsequent computations may be reduced by selecting values based on rational estimates.
4. New values of T_i are then calculated by substituting assumed ($k = 0$) or new ($k = 1$) values of T_j into the right-hand side of Equation 4.52. This step is the first iteration ($k = 1$).

5. Using Equation 4.52, the iteration procedure is continued by calculating new values of $T_i^{(k)}$ from the $T_j^{(k)}$ values of the current iteration, where $1 \le j \le i-1$, and the $T_j^{(k-1)}$ values of the previous iteration, where $i+1 \le j \le N$.
6. The iteration is terminated when a prescribed *convergence criterion* is satisfied. The criterion may be expressed as

$$|T_i^{(k)} - T_i^{(k-1)}| \le \varepsilon \tag{4.53}$$

where ε represents an uncertainty in the temperature which is considered to be acceptable.

If step 1 can be accomplished for each equation, the resulting system is said to be *diagonally dominant*, and the rate of convergence is maximized (the number of required iterations is minimized). However, convergence may also be achieved in many situations for which diagonal dominance cannot be obtained, although the rate of convergence is slowed. The manner in which new values of T_i are computed (steps 4 and 5) should also be noted. Because the T_i for a particular iteration are calculated sequentially, each value can be computed by using the *most recent estimates* of the other T_i. This feature is implicit in Equation 4.52, where the value of each unknown is updated as soon as possible, that is, for $1 \le j \le i-1$.

EXAMPLE 4.4

Consider the column of the industrial furnace described in Example 4.3. Using the same grid size, find the two-dimensional temperature distribution using the Gauss-Seidel iteration method.

SOLUTION

Find: Temperature distribution using the Gauss-Seidel iteration method.

Analysis:

Referring to the arrangement of finite-difference equations in the solution of Example 4.3, it is evident that the order is already characterized by diagonal dominance. This behavior is typical of finite-difference solutions to conduction problems. We therefore begin with step 2 and express the equations in explicit form

$$T_1^{(k)} = 0.25T_2^{(k-1)} + 0.25T_3^{(k-1)} + 250$$

$$T_2^{(k)} = 0.50T_1^{(k)} + 0.25T_4^{(k-1)} + 125$$

$$T_3^{(k)} = 0.25T_1^{(k)} + 0.25T_4^{(k-1)} + 0.25T_5^{(k-1)} + 125$$

$$T_4^{(k)} = 0.25T_2^{(k)} + 0.50T_3^{(k)} + 0.25T_6^{(k-1)}$$

$$T_5^{(k)} = 0.25T_3^{(k)} + 0.25T_6^{(k-1)} + 0.25T_7^{(k-1)} + 125$$

$$T_6^{(k)} = 0.25T_4^{(k)} + 0.50T_5^{(k)} + 0.25T_8^{(k-1)}$$

$$T_7^{(k)} = 0.2222T_5^{(k)} + 0.1111T_8^{(k-1)} + 222.22$$

$$T_8^{(k)} = 0.2222T_6^{(k)} + 0.2222T_7^{(k)} + 166.67$$

Having the finite-difference equations in the required form, the iteration procedure may be implemented by using a table which has one column for the iteration (step) number and a column for each of the nodes labeled as T_i. The calculations proceed as follows:

1. For each node, the initial temperature estimate is entered on the row for $k = 0$. Values are selected rationally to reduce the number of required iterations.
2. Using the N finite-difference equations and values of T_i from the first and second rows, the new values of T_i are calculated for the first iteration ($k = 1$). These new values are entered on the second row.
3. This procedure is repeated to calculate $T_i^{(k)}$ from the previous values of $T_i^{(k-1)}$ and the current values of $T_i^{(k)}$, until the temperature difference between iterations meets the prescribed criterion, $\varepsilon \leq 0.2$ K, at every nodal point.

k	T_1	T_2	T_3	T_4	T_5	T_6	T_7	T_8
0	480	470	440	430	400	390	370	350
1	477.5	471.3	451.9	441.3	428.0	411.8	356.2	337.3
2	480.8	475.7	462.5	453.1	436.6	413.9	355.8	337.7
3	484.6	480.6	467.6	457.4	434.3	415.9	356.2	338.2
4	487.1	482.9	469.7	459.6	435.5	417.2	356.6	338.6
5	488.1	484.0	470.8	460.7	436.1	417.9	356.7	338.8
6	488.7	484.5	471.4	461.3	436.5	418.3	356.9	338.9
7	489.0	484.8	471.7	461.6	436.7	418.5	356.9	339.0
8	489.1	485.0	471.9	461.8	436.8	418.6	356.9	339.0

Comments:

1. The results given in row 8 are in excellent agreement with those obtained by matrix inversion, although better agreement could be obtained by reducing the value of ε. However, given the approximate nature of the finite-difference equations, the results still represent approximations to the actual temperatures. The accuracy of the approximation may be improved by using a finer grid (increasing the number of nodes).
2. The Gauss-Seidel method is readily programmed on a digital computer or a hand calculator. More nodes could then be considered with little additional effort.

4.6 SUMMARY

You should now have an appreciation for the nature of a two-dimensional conduction problem and the methods that are available for its solution. When confronted with a two-dimensional problem, you should first determine whether an exact solution is known. This may be done by examining one or more of the many excellent references in which exact solutions to the heat equation are compiled [1–5]. You may also want to determine whether the shape factor is known for the system of interest [6–9]. However, conditions often are such that the use of a shape factor or an exact solution is not possible, and it is necessary to use a finite-difference solution. You should appreciate the inherent nature of the *discretization process*, and you should know how to formulate the finite-difference equations for the discrete points of a nodal network. Although you may find it convenient to solve these equations using hand calculations for a coarse mesh, you should always be prepared to use standard computer subroutines involving direct or iterative techniques.

REFERENCES

1. Carslaw, H. S., and J. C. Jaeger, *Conduction of Heat in Solids*, 2nd ed., Oxford University Press, London, 1959.
2. Schneider, P. J., *Conduction Heat Transfer*, Addison-Wesley, Reading, MA, 1955.
3. Arpaci, V. S., *Conduction Heat Transfer*, Addison-Wesley, Reading, MA, 1966.
4. Özisik, M. N., *Boundary Value Problems of Heat Conduction*, International Textbook Co., Scranton, PA, 1968.
5. Myers, G. E., *Analytical Methods in Conduction Heat Transfer*, McGraw-Hill, New York, 1971.
6. Sunderland, J. E., and K. R. Johnson, "Shape Factors for Heat Conduction Through Bodies with Isothermal or Convective Boundary Conditions," *Trans. ASHRAE*, *10*, 237–241, 1964.
7. Kutateladze, S. S., *Fundamentals of Heat Transfer*, Academic Press, NY, 1963.
8. General Electric Co. (Corporate Research and Development), *Heat Transfer Data Book*, Section 502, General Electric Company, Schenectady, New York, 1973.
9. Hahne, E., and U. Grigull, *Int. J. Heat Mass Transfer*, *18*, 751–767, 1975.
10. Dusinberre, G. M., *Heat Transfer Calculations by Finite-Difference*, International Textbook Co., Scranton, PA, 1961.
11. Adams, J. A., and D. F. Rogers, *Computer Aided Heat Transfer Analysis*, McGraw-Hill, New York, 1973.
12. Hornbeck, R. W., *Numerical Methods*, Prentice-Hall, Englewood Cliffs, NJ, 1975.
13. Roache, P. J., *Computational Fluid Dynamics*, Hermosa Publishers, Albuquerque, NM, 1976.
14. Gerald, C. F., *Applied Numerical Analysis*, Addison-Wesley, Reading, MA, 1978.

PROBLEMS

4.1 In the method of separation of variables (Section 4.2) for two-dimensional, steady-state conduction, the separation constant λ^2 in Equations 4.6 and 4.7 must be a positive constant. Show that a negative or zero value of λ^2 will result in solutions that cannot satisfy the prescribed boundary conditions.

4.2 A two-dimensional rectangular plate is subjected to the uniform temperature boundary conditions shown. Using the results of the exact solution for the heat equation presented in Section 4.2, calculate the temperature at the midpoint (1, 0.5) by considering the first five nonzero terms of the infinite series that must be evaluated. Assess the error resulting from using only the first three terms of the infinite series.

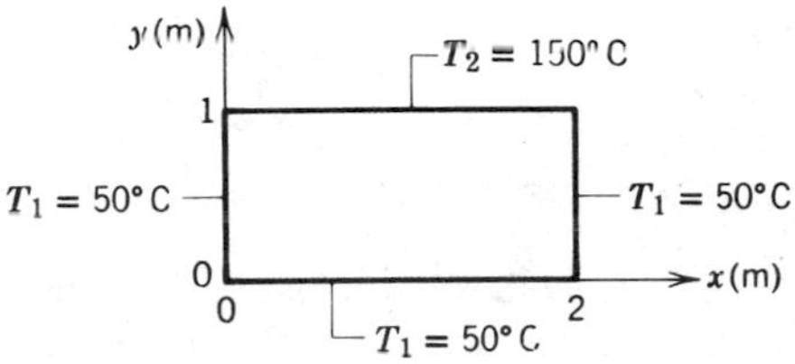

4.3 Consider the two-dimensional rectangular plate of Problem 4.2 having a thermal conductivity of 50 W/m · K. Beginning with the exact solution for the temperature distribution, derive an expression for the heat transfer rate per unit thickness to the plate along the upper surface ($0 \le x \le 2$, $y = 1$). Evaluate the heat rate considering the first five nonzero terms of the infinite series.

4.4 A two-dimensional rectangular plate is subjected to the boundary conditions shown. Derive an expression for the steady-state temperature distribution $T(x, y)$.

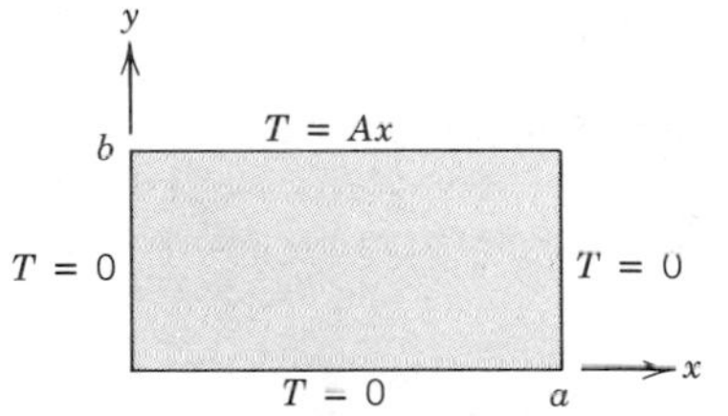

4.5 Two thin strips of materials A and B with thermal conductivities k_A and k_B, respectively, where $k_A = 2k_B$, are soldered together as shown in the sketch. The strips are insulated on their faces and exposed edges, while their ends are

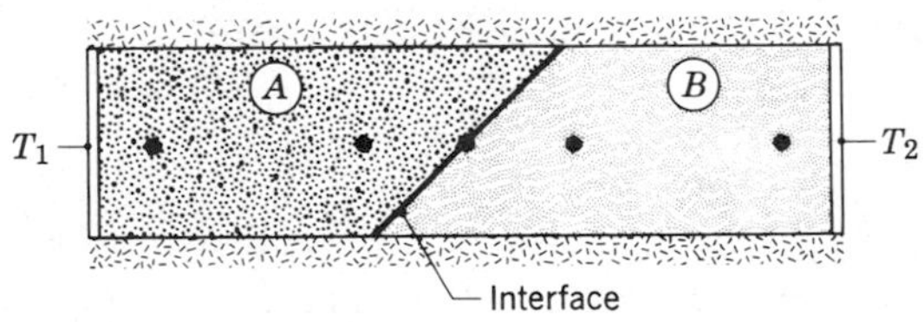

maintained at temperatures T_1 and T_2, where $T_1 > T_2$. For steady-state conditions, sketch the isotherms that pass through the points indicated in the sketch. What are the special features of the isotherms in the vicinity of the interface?

4.6 A long furnace, constructed from refractory brick with a thermal conductivity of 1.2 W/m·K, has the cross section shown, with inner and outer surface temperatures of 600 and 60°C, respectively. Determine the shape factor and the heat transfer rate per unit length using the flux plot method. Be sure to take advantage of symmetry wherever possible.

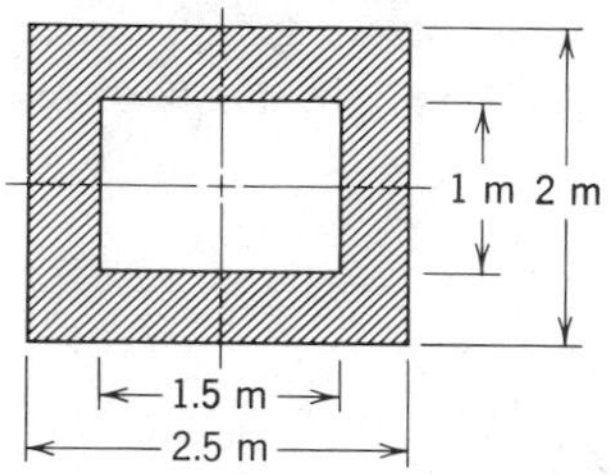

4.7 A hot pipe is embedded eccentrically as shown in a material of thermal conductivity 0.5 W/m·K. Using the flux plot method, determine the shape factor and the heat transfer per unit length when the pipe and outer surface temperatures are 150 and 35°C, respectively.

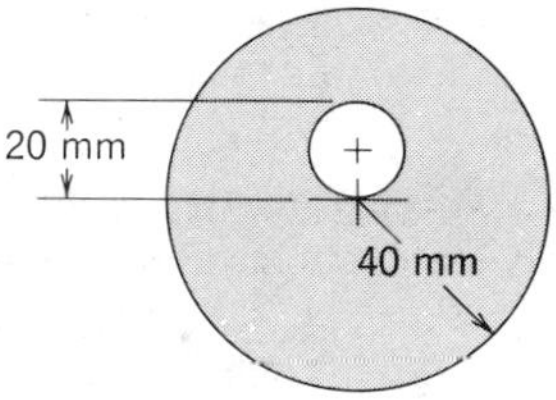

4.8 A supporting strut fabricated from a material with a thermal conductivity of 75 W/m·K has the cross section shown. The end faces are at different temperatures with $T_1 = 100°C$ and $T_2 = 0°C$, while the sides are insulated.

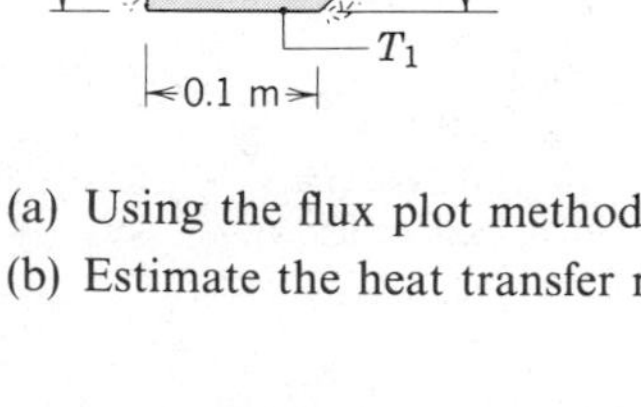

(a) Using the flux plot method, estimate the temperature at the location P.

(b) Estimate the heat transfer rate through the strut per unit length.

4.9 A hot liquid flows along a V-groove in a solid whose top and side surfaces are well insulated and whose bottom surface is in contact with a coolant.

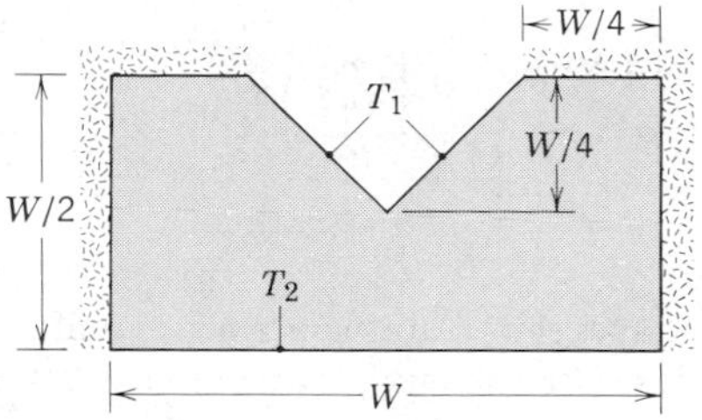

Accordingly, the V-groove surface is at a temperature T_1, which exceeds that of the bottom surface, T_2. Construct an appropriate flux plot and determine the shape factor of the system.

4.10 A plate having a thickness 6 mm and thermal conductivity of 150 W/m · K is subjected to the conditions shown. Assuming that the side and lower faces of the plate are well insulated, find the shape factor. Determine the heat rate when $T_1 = 50°C$ and $T_2 = 20°C$.

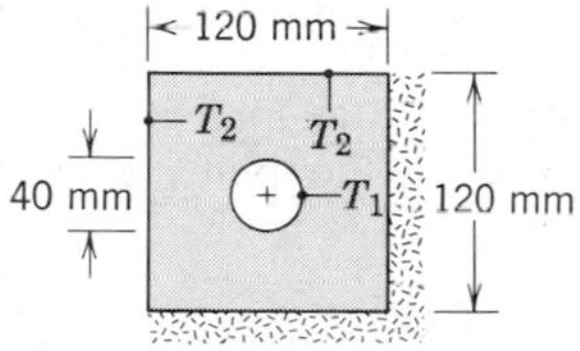

4.11 A long support column of trapezoidal cross section is well insulated on its sides, and temperatures of 100 and 0°C are maintained at its top and bottom surfaces, respectively. The column is fabricated from AISI 1010 steel, and its widths at the top and bottom surfaces are 0.3 and 0.6 m, respectively.

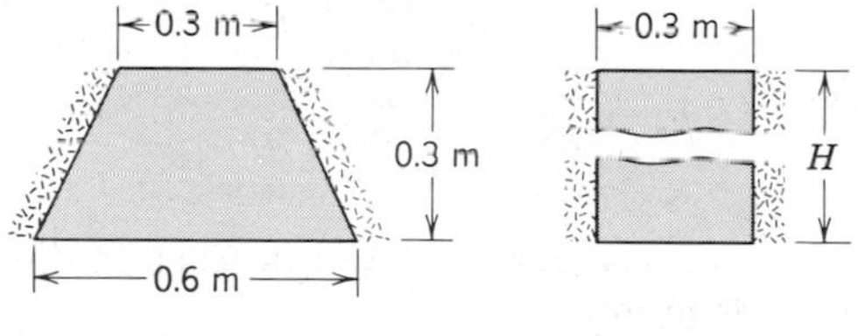

(a) Using the flux plot method, determine the heat transfer rate per unit length of the column.

(b) If the trapezoidal column is replaced by a bar of rectangular cross section 0.3 m wide and the same material, what height H must the bar be to provide an equivalent thermal resistance?

4.12 The hollow prismatic bars fabricated from plain carbon steel shown in the sketch are 1 m long with both ends insulated. For each bar, find the shape factor and then calculate the heat rate per unit length of the bar when $T_1 = 500$ K and $T_2 = 300$ K.

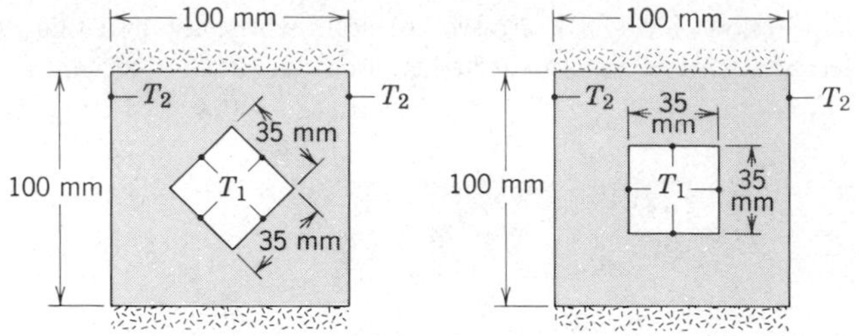

4.13 The two-dimensional, square shapes shown are maintained at uniform temperatures on portions of their boundaries. Use the flux plot method to estimate the shape factors.

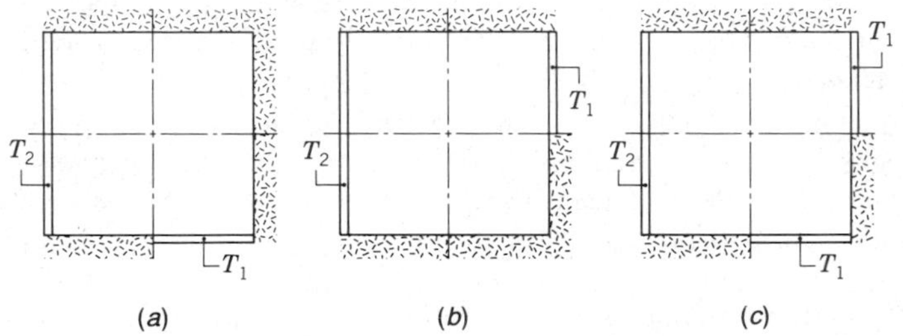

4.14 The two-dimensional circular shapes shown are maintained at uniform temperatures on portions of their boundaries. Use the flux plot method to estimate the shape factors.

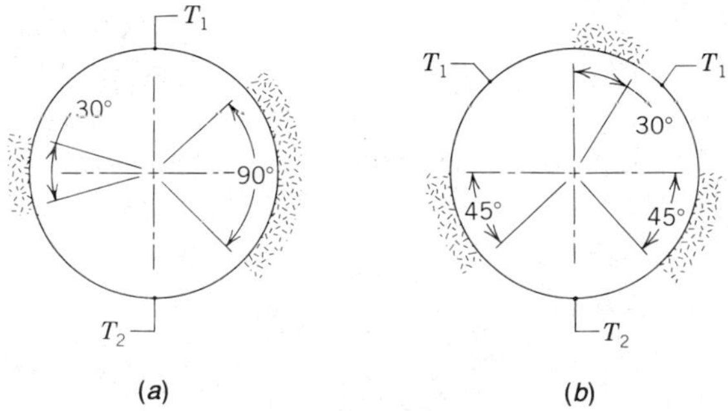

4.15 Using the thermal resistance relations developed in Chapter 3, determine shape factor expressions for the following geometries:

(a) Plane wall, cylindrical shell and spherical shell.

(b) Isothermal sphere of diameter D buried in an infinite medium.

4.16 Radioactive wastes are temporarily stored in a spherical container, the center of which is buried a distance of 10 m below the earth's surface. The outside diameter of the container is 2 m, and 500 W of heat is released as a result of the radioactive decay process. If the soil surface temperature is 20°C, what is the outside surface temperature of the container under steady-state conditions? On a sketch of the soil–container system drawn to scale, show representative isotherms and heat flow lines in the soil.

4.17 A pipeline, used for the transport of crude oil, is buried in the earth such that its centerline is a distance of 1.5 m below the surface. The pipe has an outer diameter of 0.5 m and is insulated with a layer of cellular glass 100 mm thick. What is the heat loss per unit length of pipe under conditions for which heated oil at 120°C flows through the pipe and the surface of the earth is at a temperature of 0°C?

4.18 A very long electrical conductor is buried in a large trench filled with sand ($k = 0.03$ W/m · K) to a centerline depth of 0.5 m. The conductor has an outer diameter of 25 mm, and the current flow and resistance of the cable cause a dissipation of 1 W per meter of length. The conductor is covered with an insulating sleeve of thickness 3 mm and thermal conductivity 0.01 W/m · K. Estimate the temperature at the interface between the conductor and the insulating sleeve when the temperature at the surface of the sand is 20°C.

4.19 An electrical heater 100 mm long and 5 mm in diameter is inserted into a hole drilled normal to the surface of a large block of material having a thermal conductivity of 5 W/m · K. Estimate the temperature reached by the heater when dissipating 50 W with the surface of the block at a temperature of 25°C.

4.20 Two parallel pipelines spaced 0.5 m apart are buried in soil having a thermal conductivity of 0.5 W/m · K. The pipes have outer diameters of 100 and 75 mm with surface temperatures of 175 and 5°C, respectively. Estimate the heat transfer rate per unit length between the two pipelines.

4.21 A tube of diameter 50 mm having a surface temperature of 85°C is embedded in the center plane of a concrete slab 0.1 m thick with upper and lower surfaces at 20°C.

(a) Using the appropriate tabulated relation for this configuration, find the shape factor. Determine the heat transfer rate per unit length of the tube.

(b) Using the flux plot method, estimate the shape factor and compare with the result of part (a).

4.22 Pressurized steam at 450 K flows through a long, thin-walled pipe of 0.5 m diameter. The pipe is enclosed in a concrete casing that is of square cross section and 1.5 m on a side. The axis of the pipe is centered in the casing, and the outer surfaces of the casing are maintained at 300 K. What is the heat loss per unit length of pipe?

4.23 Hot water at 85°C flows through a thin-walled copper tube of 30 mm diameter. The tube is enclosed by an eccentric cylindrical shell that is maintained at 35°C and has a diameter of 120 mm. The eccentricity, defined as the separation between the centers of the tube and shell, is 20 mm. The space between the tube and shell is filled with an insulating material having a thermal conductivity of 0.05 W/m · K. Calculate the heat loss per unit length of the tube and compare the result with the heat loss for a concentric arrangement.

4.24 A furnace of cubical shape, with external dimensions of 0.35 m, is constructed from a refractory brick (fireclay). If the wall thickness is 50 mm, the inner surface temperature is 600°C, and the outer surface temperature is 75°C, calculate the heat loss from the furnace.

4.25 A long constantan wire of 1 mm diameter is butt welded to the surface of a large copper block as shown on page 164, forming a thermocouple junction. The wire behaves as a fin, permitting heat to flow from the surface, thereby depressing the sensing junction temperature T_j below that of the block T_o. If the wire is in air at 25°C with a convection coefficient of 10 W/m^2·K, estimate the measurement error $(T_j - T_o)$ for the thermocouple, when the block is at 125°C.

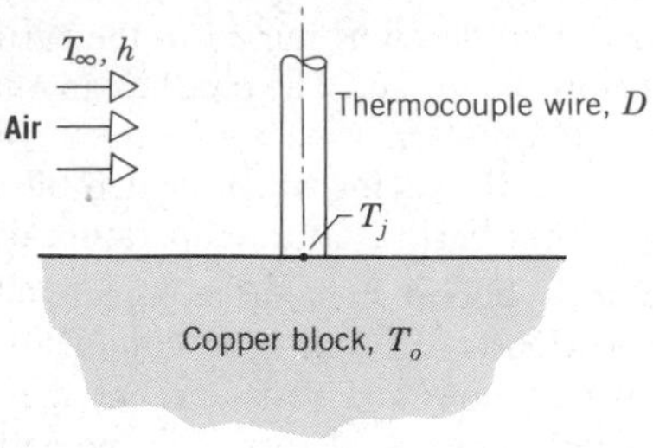

4.26 An electronic device, in the form of a disk 20 mm in diameter, dissipates 100 W when mounted flush on a large aluminum alloy (2024) block whose temperature is maintained at 27°C. The mounting arrangement is such that a contact resistance of $5 \times 10^{-5}\ \text{m}^2 \cdot \text{K/W}$ occurs at the interface between the device and the block. Calculate the temperature the device will reach, assuming that all the power generated by the device must be transferred by conduction to the block.

4.27 Consider nodal configuration 2 of Table 4.2. Derive the finite-difference equations under steady-state conditions for the following situations.

(a) The horizontal boundary of the internal corner is perfectly insulated and the vertical boundary is subjected to the convection process (h, T_∞).

(b) Both boundaries of the internal corner are perfectly insulated. How does this result compare with Equation 4.44?

4.28 Consider nodal configuration 3 of Table 4.2. Derive the finite-difference equations under steady-state conditions for the following situations.

(a) The boundary is insulated. Explain how Equation 4.45 can be modified to agree with your result.

(b) The boundary is subjected to a constant heat flux.

4.29 Consider nodal configuration 4 of Table 4.2. Derive the finite-difference equations under steady-state conditions for the following situations.

(a) The upper boundary of the external corner is perfectly insulated and the side boundary is subjected to the convection process (h, T_∞).

(b) Both boundaries of the external corner are perfectly insulated. How does this result compare with Equation 4.46?

4.30 Apply the general form of the finite-difference equation for a node near a curved surface maintained at a nonuniform temperature (case 5 of Table 4.2), to nodes 0 and 4 of the following configuration.

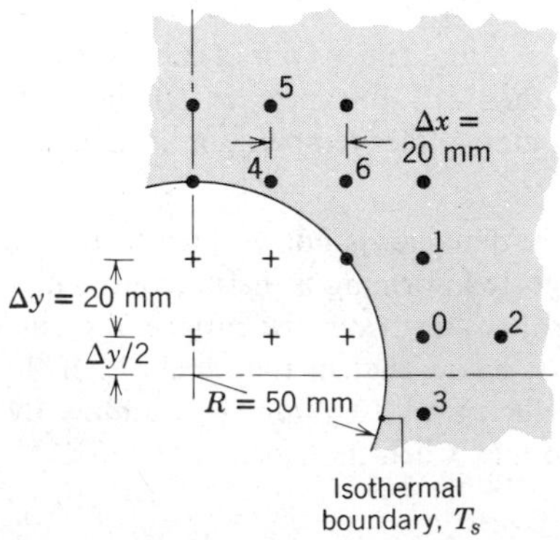

4.31 Beginning with a properly defined control volume for a node near a curved surface maintained at a nonuniform temperature (case 5 of Table 4.2), derive Equation 4.47.

4.32 Consider the nodal point arrangement in the vicinity of the isothermal surface shown.

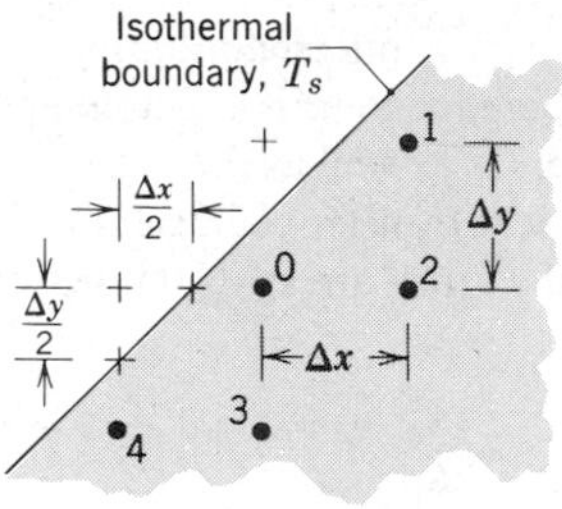

(a) Apply Equation 4.47 to the above configuration to obtain the finite-difference equation for node 0.

(b) Beginning with the properly defined control volume for node 0, derive the finite-difference equation for that node.

4.33 Derive the finite-difference equation for steady-state conditions with volumetric heat generation in the three-dimensional Cartesian coordinate system (x, y, z) for any interior node (m, n, p).

4.34 Consider heat transfer in a one-dimensional (radial) cylindrical coordinate system under steady-state conditions with volumetric heat generation.

(a) Derive the finite-difference equation for any interior node m.

(b) Derive the finite-difference equation for the node n located at the external boundary subjected to the convection process (h, T_∞).

4.35 Derive the finite-difference equations required of Problem 4.34 but for a one-dimensional (radial) spherical coordinate system.

4.36 In the two-dimensional cylindrical configuration shown, the radial (Δr) and angular $(\Delta\phi)$ spacings of the nodes are uniform. The boundary at $r = r_i$ is of uniform temperature T_i. The boundaries in the radial direction are adiabatic (insulated) and exposed to surface convection (T_∞, h), as illustrated. Derive the finite-difference equations for (a) node 2, (b) node 3, and (c) node 1.

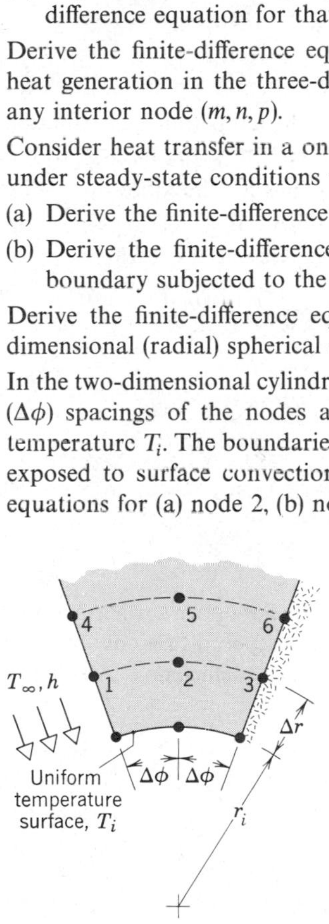

4.37 Consider a one-dimensional fin of uniform cross-sectional area, insulated at the end, $x = L$. (See Table 3.2, case B.) The temperature at the base of the fin T_b and of the adjoining fluid T_∞, as well as the heat transfer coefficient h and the thermal conductivity k, are known.

(a) Derive the finite-difference equation for any interior node m.

(b) Derive the finite-difference equation for the node n, located at the insulated end where $x = L$.

4.38 The steady-state temperature distribution in a fin of nonuniform cross-sectional area is to be determined by the finite-difference method. The temperatures at the base of the fin T_b and of the adjoining fluid T_∞, as well as the thermal conductivity k, are known. Furthermore, the convection heat transfer coefficient h, the perimeter of the fin P, and the cross-sectional area of the fin A, are known functions of x, that is, $h(x)$, $P(x)$, and $A(x)$, respectively.

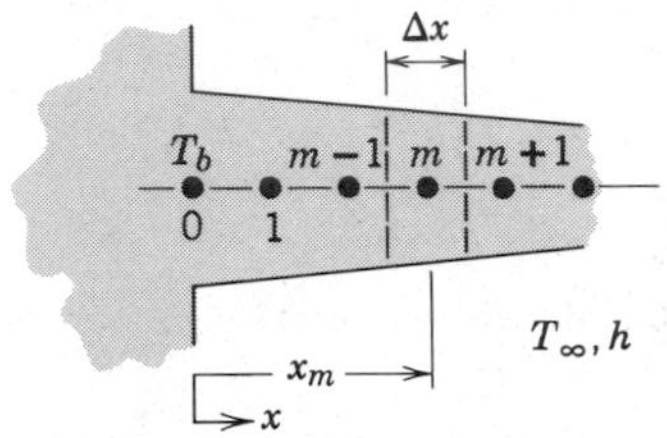

(a) Derive the nodal finite-difference equation for point m located at x_m.

(b) Describe a procedure for obtaining the temperature distribution. Indicate how you would use this distribution to determine the fin heat transfer rate, q_f.

4.39 Derive the nodal finite-difference equations for the following configurations.

(a) Node on a diagonal boundary as shown, subjected to convection with a fluid at T_∞ and a heat transfer coefficient h. Assume $\Delta x = \Delta y$.

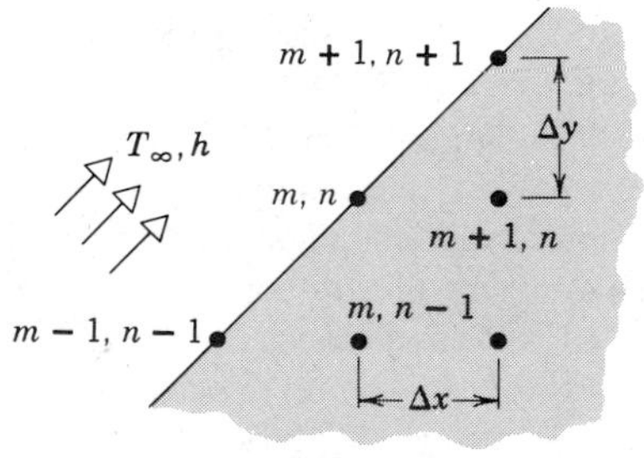

(b) Node at the tip of a cutting tool with the upper surface exposed to a constant heat flux q''_o, and the diagonal surface exposed to a convection cooling process with the fluid at T_∞ and a heat transfer coefficient h. Assume $\Delta x = \Delta y$.

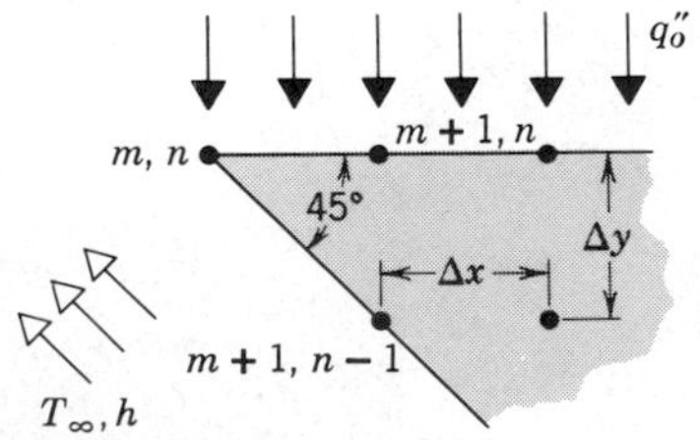

4.40 Consider the nodal point 0 located on the boundary between materials of thermal conductivity k_A and k_B as shown.

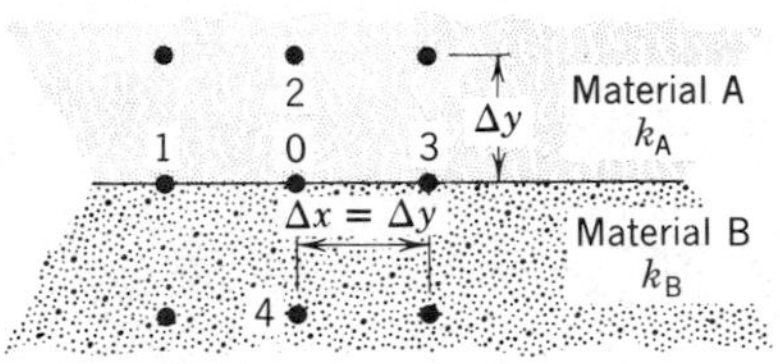

Derive the finite-difference equation under steady-state conditions, assuming no internal generation.

4.41 Steady-state temperatures (K) at three nodal points of a long rectangular rod are as shown. The rod experiences a uniform volumetric generation rate of 5×10^7 W/m^3 and has a thermal conductivity of 20 W/m · K. Two of its sides are maintained at a constant temperature of 300 K, while the others are insulated.

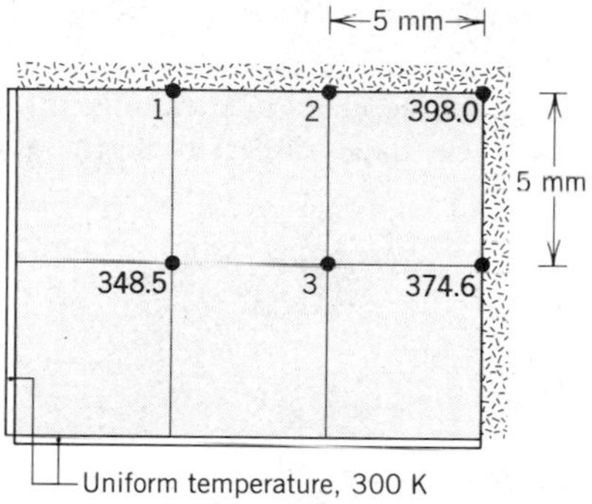

(a) Determine the temperatures at nodes 1, 2, and 3.

(b) Calculate the heat transfer rate per unit length (W/m) from the rod, using the nodal temperature values. Compare this result with the heat rate calculated from knowledge of the volumetric generation rate and the rod dimensions.

4.42 The temperatures (K) at the nodal points of a two-dimensional system are as shown. Surface B is held at a uniform temperature, while surface A is subjected to a convection boundary condition. Calculate the heat rate leaving surface A per unit thickness normal to the page. Estimate the thermal conductivity of the material.

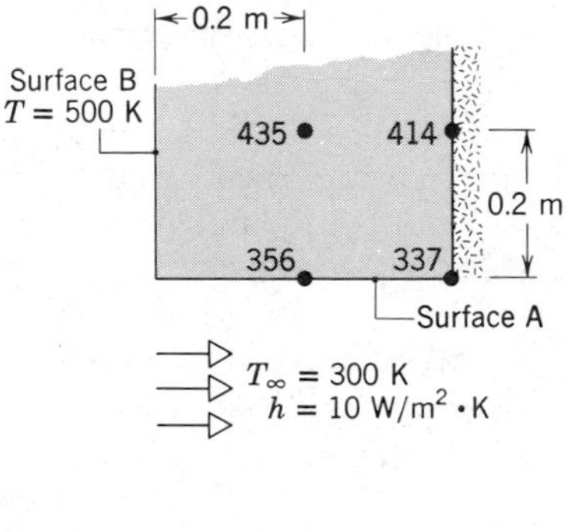

4.43 The steady-state temperatures (°C) associated with selected nodal points of a two-dimensional system having a thermal conductivity of 1.5 W/m · K are shown on the accompanying grid.

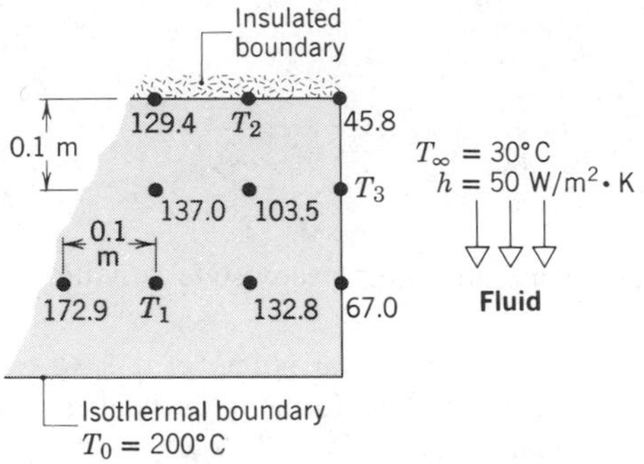

(a) Determine the temperatures at nodes 1, 2, and 3.

(b) Calculate the heat transfer rate per unit thickness normal to the page from the system to the fluid.

4.44 Assuming two-dimensional, steady-state conduction, determine the temperature at nodes 1, 2, 3, and 4 in the square shape subjected to the uniform surface temperatures shown.

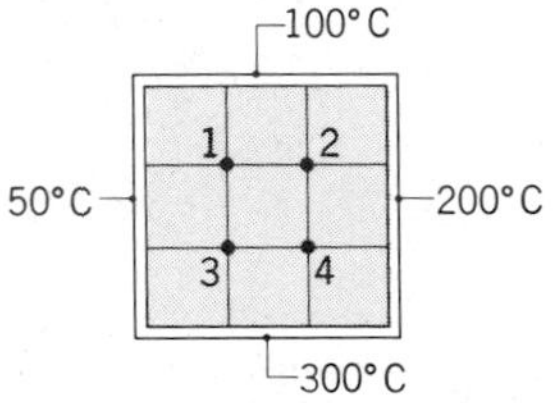

4.45 Consider a long bar of square cross section (0.8 m to the side) and of thermal conductivity 2 W/m · K. Three of these sides are maintained at a uniform temperature of 300°C. The fourth side is exposed to a fluid at 100°C for which the convection heat transfer coefficient is 10 W/m^2 · K. Using an appropriate numerical technique with a grid spacing of 0.2 m, determine the midpoint temperature and the heat transfer rate between the bar and the fluid per unit length of the bar.

4.46 A long conducting rod of rectangular cross section (20 mm × 30 mm) and thermal conductivity $k = 20$ W/m · K experiences uniform heat generation at a rate $\dot{q} = 5 \times 10^7$ W/m^3, while its surfaces are maintained at 27°C. Using a finite-difference method with a grid spacing of 5 mm, determine the temperature distribution in the rod.

4.47 A flue passing hot exhaust gases has a square cross section, 300 mm to a side. The walls are constructed of refractory brick 150 mm thick with a thermal conductivity of 0.85 W/m · K. Calculate the heat loss from the flue per unit length when the interior and exterior surfaces are maintained at 350 and 25°C, respectively. Use a grid spacing of 75 mm.

4.48 Consider the system of Problem 4.47. The interior surface is exposed to hot gases at 350°C with a convection coefficient of 100 W/m^2 · K, while the exterior surface experiences convection with air at 25°C with a convection coefficient of 5 W/m^2 · K.

Determine the heat loss from the flue per unit length by convection to the air. Compare this result with the heat gain by convection from the hot gases.

4.49 The perimeter of a plate of thickness 4 mm, thermal conductivity 2 W/m·K, and square cross section (100 mm to the side) is maintained at 600 K. One surface of the plate is insulated, while the other surface has an emissivity of 0.9 and is exposed to large surroundings at a temperature of 300 K. Neglecting temperature gradients in the direction of the plate thickness and using the Gauss-Seidel method with a grid spacing of 16.67 mm, determine the radiation heat exchange between the plate and its surroundings. *Hint:* Linearize the radiation rate equation according to Equation 1.7; after each iteration, update the value of the radiation heat transfer coefficient h_r.

4.50 A solid slab is impregnated with uniformly spaced heating pipes; the relative dimensions are shown in the diagram. Under steady-state conditions fluid passing through the pipes maintains a surface temperature $T_1 = 400$ K, while the top and bottom surfaces of the slab are maintained at $T_2 = 300$ K. Using an appropriate numerical technique with a grid spacing of $\Delta x = \Delta y = D/5$, where $D = 50$ mm, determine the temperature distribution in the slab. If the slab is constructed from granite, what is the heat loss per pipe per unit length?

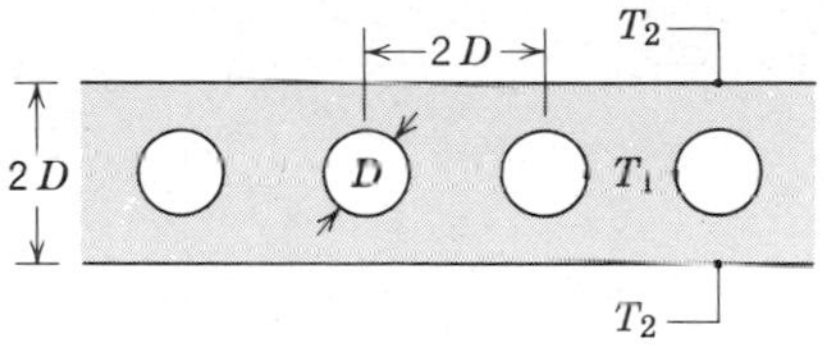

4.51 A long bar of rectangular cross section, 0.4 m × 0.6 m on a side and having a thermal conductivity of 1.5 W/m·K, is subjected to the boundary conditions shown. Two of the sides are maintained at a uniform temperature of 200°C. One of the sides is adiabatic, and the remaining side is subjected to a convection process with $T_\infty = 30$°C and $h = 50$ W/m^2·K. Using an appropriate numerical technique with a grid spacing of 0.1 m, determine the temperature distribution in the bar and the heat transfer rate between the bar and the fluid per unit length of the bar.

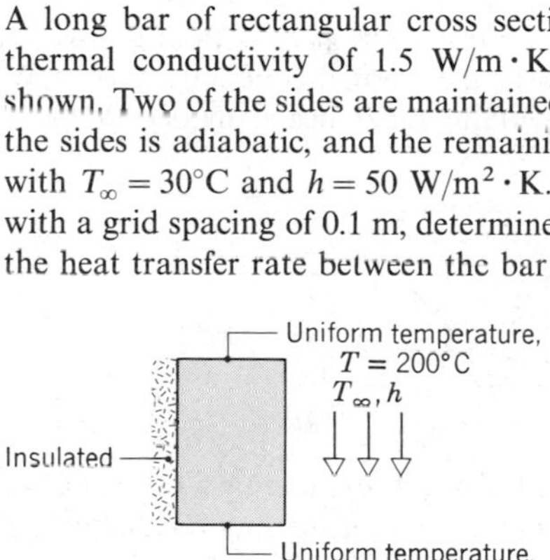

4.52 A straight fin of uniform cross section is fabricated from a material of thermal conductivity 50 W/m·K, a thickness $w = 6$ mm, and a length of $L = 48$ mm, and is

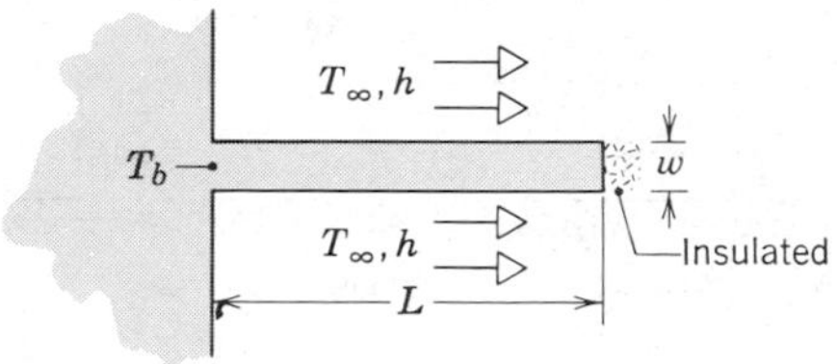

very long in the direction normal to the page. The convection heat transfer coefficient is 500 $W/m^2 \cdot K$ with an ambient air temperature of $T_\infty = 30°C$. The base of the fin is maintained at $T_b = 100°C$, while the end of the fin is insulated.

(a) Using a finite-difference method with a space increment of 4 mm, estimate the temperature distribution within the fin. Is the assumption of one-dimensional heat transfer reasonable for this fin?

(b) Estimate the fin heat transfer rate per unit length normal to the page. Compare your result with the one-dimensional, analytical solution, Equation 3.75.

4.53 The straight fin shown has a triangular profile given by the function $y = 0.2(L - x)$ and is very long in the direction normal to the page. Both top and bottom surfaces are subjected to convection with $T_\infty = 15°C$ and $h = 100$ $W/m^2 \cdot K$. The base of the fin is maintained at $T_b = 115°C$ and the fin material has a thermal conductivity of 25 $W/m \cdot K$. Assuming one-dimensional heat transfer and using a finite-difference method with a space increment of 10 mm, determine the fin heat rate and efficiency.

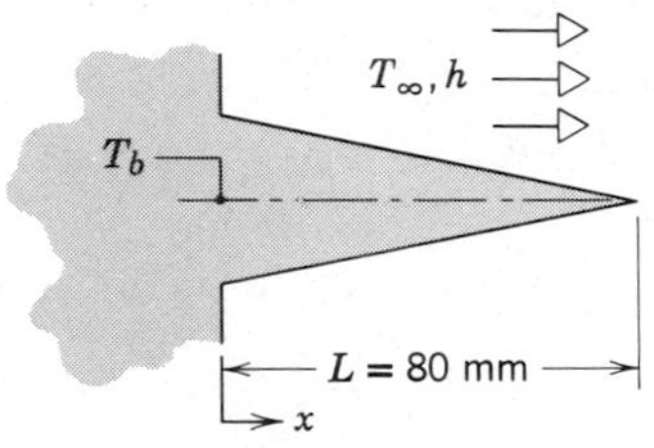

4.54 The surface of a plate and the grooves are maintained at a uniform temperature of $T_1 = 200°C$. Calculate the heat transfer rate per width of groove spacing (w) when the lower surface is at $T_2 = 20°C$, the thermal conductivity is 15 $W/m \cdot K$, and the groove spacing is 0.16 m. Use a finite-difference method with a space increment of 40 mm.

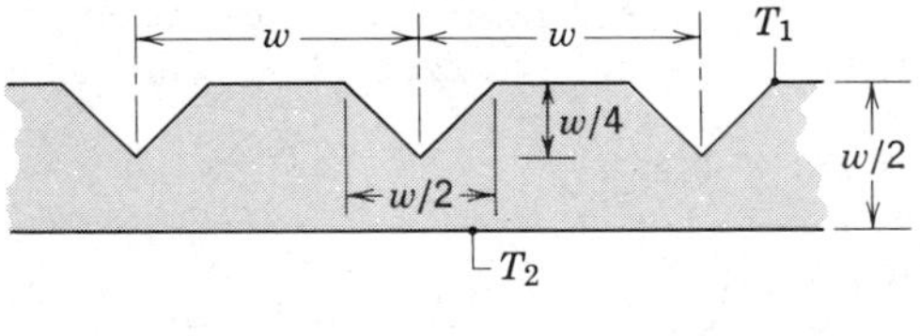

4.55 Small diameter electrical heating elements dissipating 50 W/m (length normal to the sketch) are used to heat a ceramic plate of thermal conductivity 2 $W/m \cdot K$. The upper surface of the plate is exposed to ambient air at 30°C with a convection coefficient of 100 $W/m^2 \cdot K$, while the lower surface is well insulated.

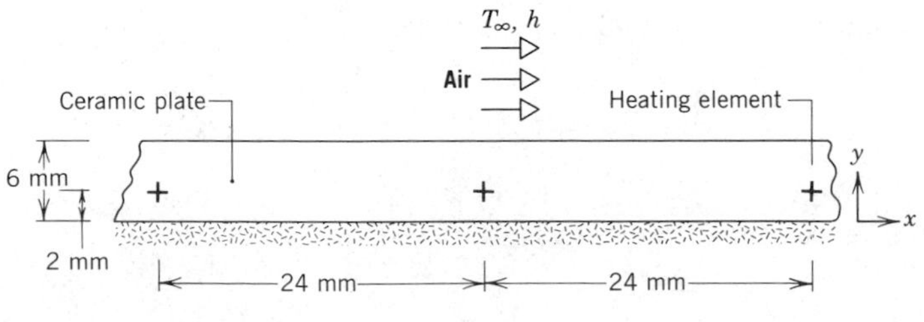

(a) Using the Gauss-Seidel method with a grid spacing of $\Delta x = 6$ mm and $\Delta y = 2$ mm, obtain the temperature distribution within the plate.

(b) Using the calculated nodal temperatures, sketch four isotherms to illustrate the temperature distribution in the plate.

(c) Calculate the heat loss by convection from the plate to the fluid. Compare this value with the element dissipation rate.

(d) What advantage, if any, is there in not making $\Delta x = \Delta y$ for this situation?

4.56 Refer to the two-dimensional rectangular plate subjected to uniform temperature boundaries as described in Problem 4.2. Using an appropriate numerical method, determine the temperature at the midpoint (1, 0.5). Use a space increment of 0.25 m for a hand calculator scheme or 0.125 m if you employ digital computer routines.

4.57 The long trapezoidal bar shown is subjected to uniform temperatures on two surfaces, while the remaining surfaces are well insulated. If the thermal conductivity of the material is 20 W/m · K, estimate the heat transfer rate per unit length of the bar using a finite-difference method. A space increment of 10 mm is suggested.

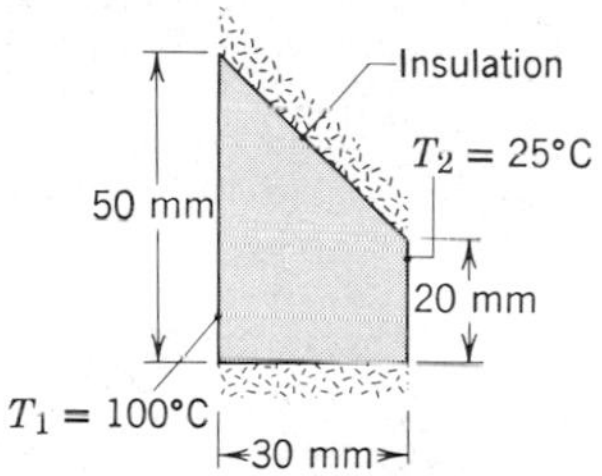

4.58 A long bar of rectangular cross section is fabricated from two materials with thermal conductivities $k_A = 15$ W/m · K and $k_B = 1$ W/m · K and with thicknesses $L_A = 50$ mm and $L_B = 100$ mm, respectively. Its width is $w = 300$ mm. Three sides of the bar are subjected to convection conditions with $T_\infty = 100°C$ and $h = 30$ W/m^2 · K. Calculate the rate at which heat must be removed per unit length of the bar by the cooling coil to maintain the bottom surface of the bar at $T_0 = 0°C$.

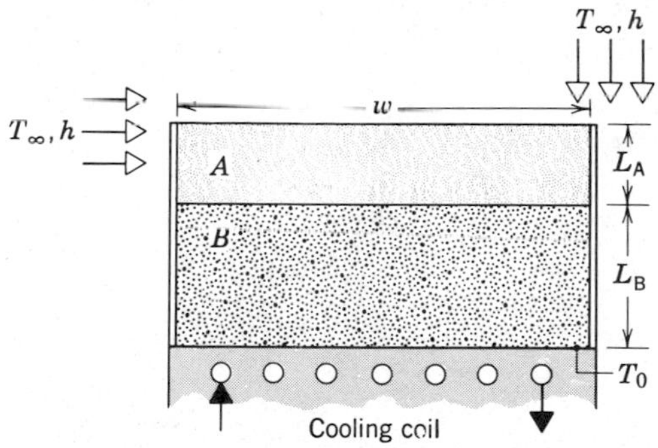

Five

TRANSIENT CONDUCTION

In our treatment of conduction we have gradually considered more complicated conditions. We began with the simple case of one-dimensional, steady-state conduction with no internal generation, and we subsequently considered complications due to multidimensional and generation effects. However, we have not yet considered situations for which conditions change with time.

We now recognize that many heat transfer problems are time dependent. Such *unsteady*, or *transient*, problems typically arise when the boundary conditions of a system are changed. For example, if the surface temperature of a system is altered, the temperature at each point in the system will also begin to change. The changes will continue to occur until a *steady-state* temperature distribution is reached. Consider a hot metal billet that is removed from a furnace and exposed to a cool airstream. Energy is transferred by convection and radiation from its surface to the surroundings. Energy transfer by conduction also occurs from the interior of the metal to the surface, and the temperature at each point in the billet decreases until a steady-state condition is reached. Such time-dependent effects occur in many industrial heating and cooling processes.

To determine the time dependence of the temperature distribution within a solid during a transient process, we could begin by solving the appropriate form of the heat equation, for example, Equation 2.13. Some cases for which solutions have been obtained are discussed in Sections 5.3 to 5.7. However, such solutions are often difficult to obtain, and where possible a simpler approach is preferred. One such approach may be used under conditions for which temperature gradients within the solid are small. It is termed the *lumped capacitance method*.

5.1 THE LUMPED CAPACITANCE METHOD

A common transient conduction problem is one in which a solid experiences a sudden change in its thermal environment. Consider a hot metal forging that is initially at a uniform temperature T_i and is quenched by immersing it in a liquid of lower temperature $T_\infty < T_i$ (Figure 5.1). If the quenching is said to begin at time $t = 0$, the temperature of the solid will decrease for time $t > 0$, until it eventually reaches T_∞. This reduction is due to convection heat transfer

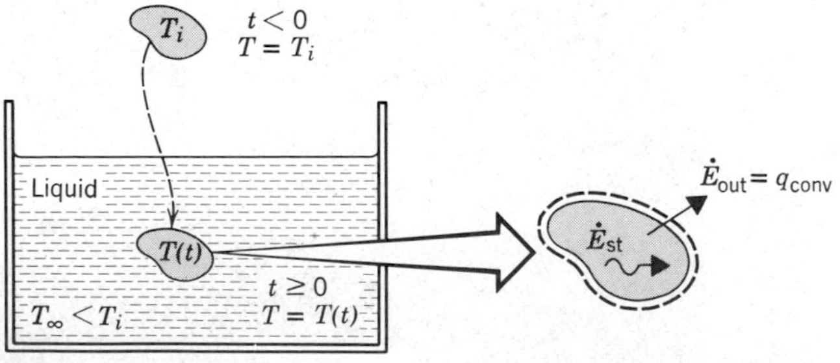

Figure 5.1 Cooling of a hot metal forging.

occurring at the solid–liquid interface. The essence of the lumped capacitance method is the assumption that the temperature of the solid is *spatially uniform* at any instant during the transient process. This assumption implies that temperature gradients within the solid are negligible.

From Fourier's law, heat conduction in the absence of a temperature gradient implies the existence of infinite thermal conductivity. Such a condition is clearly impossible. However, although the condition is never satisfied exactly, it is closely approximated if the resistance to conduction within the solid is small compared with the resistance to heat transfer between the solid and its surroundings. For now we assume that this is, in fact, the case.

In neglecting temperature gradients within the solid, we can no longer consider the problem from within the framework of the heat equation. Instead, the transient temperature response is determined by formulating an overall energy balance on the solid. This balance must relate the rate of heat loss at the surface to the rate of change of the internal energy. Applying Equation 1.10a to the control volume of Figure 5.1, this requirement takes the form

$$-\dot{E}_{\text{out}} = \dot{E}_{\text{st}} \tag{5.1}$$

or

$$-hA_s(T - T_\infty) = \rho Vc \frac{dT}{dt} \tag{5.2}$$

Introducing the temperature difference

$$\theta \equiv T - T_\infty \tag{5.3}$$

and recognizing that $(d\theta/dt) = (dT/dt)$, it follows that

$$\frac{\rho Vc}{hA_s} \frac{d\theta}{dt} = -\theta$$

Separating variables and integrating from the initial condition, for which $t = 0$ and $T(0) = T_i$, we then obtain

$$\frac{\rho Vc}{hA_s} \int_{\theta_i}^{\theta} \frac{d\theta}{\theta} = -\int_0^t dt$$

where

$$\theta_i \equiv T_i - T_\infty \tag{5.4}$$

Evaluating the integrals it follows that

$$\frac{\rho Vc}{hA_s} \ln \frac{\theta_i}{\theta} = t \tag{5.5}$$

or

$$\frac{\theta}{\theta_i} = \frac{T - T_\infty}{T_i - T_\infty} = \exp\left[-\left(\frac{hA_s}{\rho Vc}\right)t\right] \tag{5.6}$$

Equation 5.5 may be used to determine the time required for the solid to reach some temperature T, or, conversely, Equation 5.6 may be used to compute the temperature reached by the solid at some time t.

The foregoing results indicate that the difference between the solid and fluid temperatures must decay exponentially to zero as t approaches infinity. This behavior is shown in Figure 5.2. From Equation 5.6 it is also evident that the quantity $(\rho Vc/hA_s)$ may be interpreted as a *thermal time constant*. This time constant may be expressed as

$$\tau_t = \left(\frac{1}{hA_s}\right)(\rho Vc) = R_t C_t \tag{5.7}$$

where R_t is the resistance to convection heat transfer and C_t is the *lumped thermal capacitance* of the solid. Any increase in R_t or C_t will cause a solid to respond more slowly to changes in its thermal environment and will increase the time required to reach thermal equilibrium ($\theta = 0$).

It is useful to note that the foregoing behavior is analogous to the voltage decay that occurs when a capacitor is discharged through a resistor in an electrical RC circuit. Accordingly, the process may be represented by an *equivalent thermal circuit*, which is shown in Figure 5.3. With the switch closed the solid is charged to the temperature θ_i. When the switch is opened, the energy that is stored in the solid is discharged through the thermal resistance and the temperature of the solid decays with time. This analogy suggests that RC electrical circuits may be used to determine the transient behavior of thermal systems. In fact, before the advent of digital computers, RC circuits were widely used to simulate transient thermal behavior.

To determine the total energy transfer Q occurring up to some time t, we

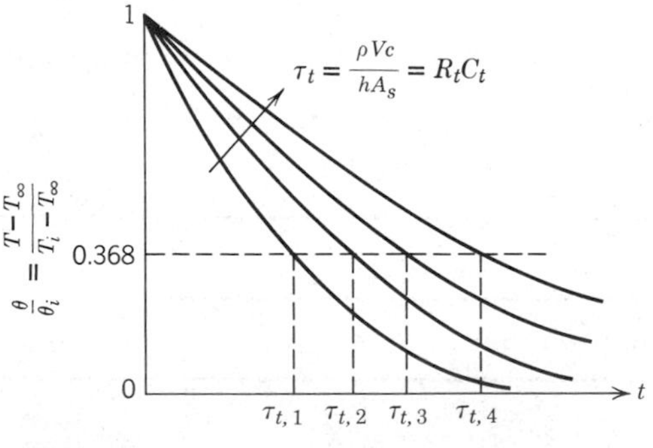

Figure 5.2 Transient temperature response of lumped capacitance solids corresponding to different thermal time constants τ_t.

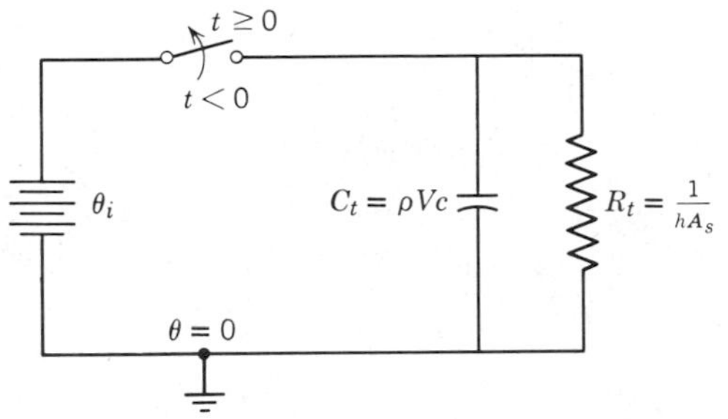

Figure 5.3 Equivalent thermal circuit for a lumped capacitance solid.

simply write

$$Q = \int_0^t q\,dt = hA_s \int_0^t \theta\,dt$$

Substituting for θ from Equation 5.6 and integrating, we obtain

$$Q = (\rho Vc)\theta_i \left[1 - \exp\left(-\frac{t}{\tau_t}\right)\right] \tag{5.8a}$$

The quantity Q is, of course, related to the change in the internal energy of the solid, and from Equation 1.10b

$$-Q = \Delta E_{st} \tag{5.8b}$$

For quenching Q is positive and the solid experiences a decrease in energy. Equations 5.5, 5.6, and 5.8a also apply to situations where the solid is heated ($\theta < 0$), in which case Q is negative and the internal energy of the solid increases.

5.2 VALIDITY OF THE LUMPED CAPACITANCE METHOD

From the foregoing results it is easy to see why there is a strong preference for using the lumped capacitance method. It is certainly the simplest and most convenient method that can be used to solve transient conduction problems. Hence it is important to determine under what conditions it may be used with reasonable accuracy.

To develop a suitable criterion consider steady-state conduction through the plane wall of area A (Figure 5.4). Although we are assuming steady-state conditions, this criterion is readily extended to transient processes. One surface is maintained at a temperature $T_{s,1}$ and the other surface is exposed to a fluid of temperature $T_\infty < T_{s,1}$. The temperature of this surface will be some intermediate value, $T_{s,2}$, for which $T_\infty < T_{s,2} < T_{s,1}$. Hence under steady-state conditions the surface energy balance, Equation 1.11, reduces to

$$\frac{kA}{L}(T_{s,1} - T_{s,2}) = hA(T_{s,2} - T_\infty)$$

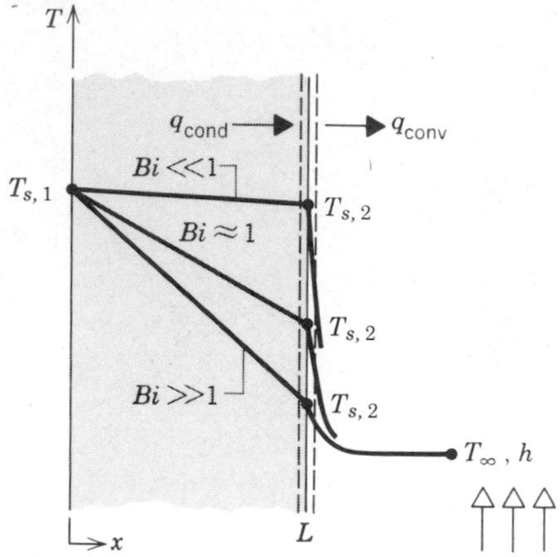

Figure 5.4 Effect of Biot number on steady-state temperature distribution in a plane wall with surface convection.

where k is the thermal conductivity of the solid. Rearranging, we then obtain

$$\frac{T_{s,1} - T_{s,2}}{T_{s,2} - T_\infty} = \frac{(L/kA)}{(1/hA)} = \frac{R_{\text{cond}}}{R_{\text{conv}}} = \frac{hL}{k} \equiv Bi \tag{5.9}$$

The quantity (hL/k) appearing in Equation 5.9 is a *dimensionless parameter*. It is termed the *Biot number*, and it plays a fundamental role in conduction problems that involve surface convection effects. According to Equation 5.9 and as illustrated in Figure 5.4, the Biot number provides a measure of the temperature drop in the solid relative to the temperature difference between the surface and the fluid. Note especially the conditions corresponding to $Bi \ll 1$. The results suggest that, for these conditions, it is reasonable to *assume* a uniform temperature distribution across a solid at any time during a transient process. This result may also be associated with interpretation of the Biot number as a ratio of thermal resistances, Equation 5.9. *If $Bi \ll 1$, the resistance to conduction within the solid is much less than the resistance to convection across the fluid boundary layer.* Hence the assumption of a uniform temperature distribution is reasonable.

We have introduced the Biot number because of its significance to transient conduction problems. Consider the plane wall of Figure 5.5, which is initially at a uniform temperature T_i and experiences convection cooling when it is immersed in a fluid of $T_\infty < T_i$. The problem may be treated as one dimensional in x, and we are interested in the temperature variation with position and time, $T(x,t)$. This variation is a strong function of the Biot number, and three conditions are shown in Figure 5.5. For $Bi \ll 1$ the temperature gradient in the solid is small and $T(x,t) \approx T(t)$. Virtually all the temperature difference is between the solid and the fluid, and the solid temperature remains nearly uniform as it decreases to T_∞. For moderate to large values of the Biot number, however, the temperature gradients within the solid are significant. Hence $T = T(x,t)$. Note that for $Bi \gg 1$, the temperature difference across the solid is now much larger than that between the surface and the fluid.

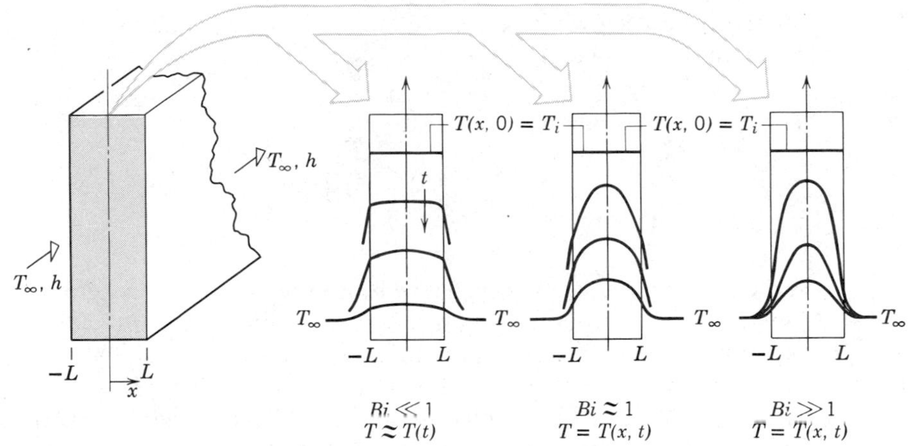

Figure 5.5 Transient temperature distribution for different Biot numbers in a plane wall symmetrically cooled by convection.

We conclude this section by emphasizing the importance of the lumped capacitance method. Its inherent simplicity renders it the preferred method for solving transient conduction problems. Hence, when confronted with such a problem, *the very first thing that one should do is calculate the Biot number.* If the following condition is satisfied

$$Bi = \frac{hL_c}{k} \leq 0.1 \tag{5.10}$$

the error associated with using the lumped capacitance method is small. The *characteristic length* $L_c \equiv V/A_s$ reduces to the half-thickness L for a plane wall of thickness $2L$ (Figure 5.5), to $r_o/2$ for a long cylinder, and to $r_o/3$ for a sphere.

Finally, we note that, with $L_c \equiv V/A_s$, the exponent of Equation 5.6 may be expressed as

$$\frac{hA_st}{\rho Vc} = \frac{ht}{\rho cL_c} = \frac{hL_c}{k}\frac{k}{\rho c}\frac{t}{L_c^2} = \frac{hL_c}{k}\frac{\alpha t}{L_c^2}$$

$\alpha = \frac{k}{\rho c}$ m^2/s

or

$$\frac{hA_st}{\rho Vc} = Bi \cdot Fo \tag{5.11}$$

where

$$Fo \equiv \frac{\alpha t}{L_c^2} \tag{5.12}$$

is termed the Fourier number. It is a *dimensionless time*, which, with the Biot number, characterizes transient conduction problems. Substituting Equation 5.11 into 5.6, we obtain

$$\frac{\theta}{\theta_i} = \frac{T - T_\infty}{T_i - T_\infty} = \exp(-Bi \cdot Fo) \tag{5.13}$$

EXAMPLE 5.1

A thermocouple junction, which may be approximated as a sphere, is to be used for temperature measurement in a gas stream. The convection coefficient between the junction surface and the gas is known to be $h = 400\ \mathrm{W/m^2 \cdot K}$, and the junction thermophysical properties are $k = 20\ \mathrm{W/m \cdot K}$, $c = 400\ \mathrm{J/kg \cdot K}$, and $\rho = 8500\ \mathrm{kg/m^3}$. Determine the junction diameter needed for the thermocouple to have a time constant of 1 s. If the junction is at 25°C and is placed in a gas stream that is at 200°C, how long will it take for the junction to reach 199°C?

SOLUTION

Known:

Thermophysical properties of thermocouple junction used to measure temperature of a gas stream.

Find:

1. Junction diameter needed for a time constant of 1 s.
2. Time required to reach 199°C in gas stream at 200°C.

Schematic:

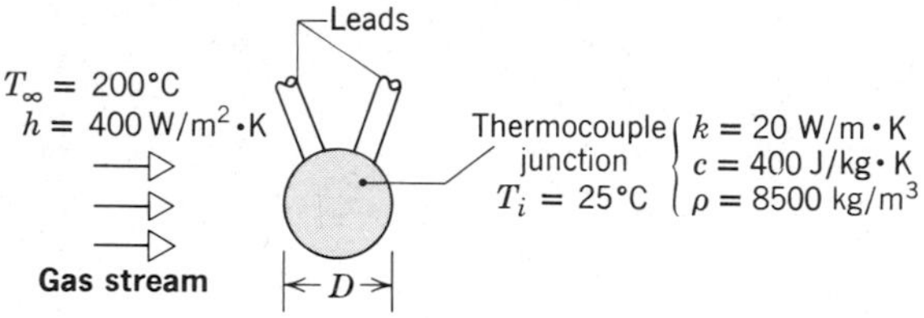

Assumptions:

1. Temperature of junction is uniform at any instant.
2. Radiation exchange with the surroundings is negligible.
3. Losses by conduction through the leads are negligible.
4. Constant properties.

Analysis:

1. Because the junction diameter is unknown, it is not possible to begin the solution by determining whether the criterion for using the lumped capacitance method, Equation 5.10, is satisfied. However, a reasonable approach is to use the method to find the diameter and to then determine whether the criterion is satisfied. From Equation 5.7 and the fact that $A_s = \pi D^2$ and $V = \pi D^3/6$ for a sphere, it follows that

$$\tau_t = \frac{1}{h\pi D^2} \times \frac{\rho \pi D^3}{6} c$$

Rearranging and substituting numerical values,

$$D = \frac{6h\tau_t}{\rho c} = \frac{6 \times 400 \text{ W/m}^2 \cdot \text{K} \times 1 \text{ s}}{8500 \text{ kg/m}^3 \times 400 \text{ J/kg} \cdot \text{K}} = 7.06 \times 10^{-4} \text{ m} \qquad \triangleleft$$

With $L_c = r_o/3$ it then follows from Equation 5.10 that

$$Bi = \frac{h(r_o/3)}{k} = \frac{400 \text{ W/m}^2 \cdot \text{K} \times 3.53 \times 10^{-4} \text{ m}}{3 \times 20 \text{ W/m} \cdot \text{K}} = 2.35 \times 10^{-4}$$

Accordingly, the lumped capacitance method may be used to an excellent approximation.

2. From Equation 5.5 the time required for the junction to reach $T = 199°\text{C}$ is

$$t = \frac{\rho(\pi D^3/6)c}{h(\pi D^2)} \ln \frac{T_i - T_\infty}{T - T_\infty} = \frac{\rho D c}{6h} \ln \frac{T_i - T_\infty}{T - T_\infty}$$

$$t = \frac{8500 \text{ kg/m}^3 \times 7.06 \times 10^{-4} \text{ m} \times 400 \text{ J/kg} \cdot \text{K}}{6 \times 400 \text{ W/m}^2 \cdot \text{K}} \ln \frac{25 - 200}{199 - 200}$$

$$t = 5.2 \text{ s} \approx 5\tau_t \qquad \triangleleft$$

Comments:

Heat losses due to radiation exchange between the junction and the surroundings and conduction through the leads would necessitate using a smaller junction diameter to achieve the desired time response.

5.3 SPATIAL EFFECTS

Situations frequently arise for which the lumped capacitance method is inappropriate, and alternative methods must be used. Regardless of the particular form of the method, we must now cope with the fact that gradients within the medium are no longer negligible.

In their most general form, transient conduction problems are described by the heat equation, Equation 2.13 for rectangular coordinates or Equations 2.18 and 2.19, respectively, for cylindrical and spherical coordinates. The solution to these partial differential equations provides the variation of temperature with both time and the spatial coordinates. However, in many problems, such as the plane wall of Figure 5.5, only one spatial coordinate is needed to describe the internal temperature distribution. With no internal generation and the assumption of constant thermal conductivity, Equation 2.13 then reduces to

$$\frac{\partial^2 T}{\partial x^2} = \frac{1}{\alpha}\frac{\partial T}{\partial t} \tag{5.14}$$

To solve Equation 5.14 for the temperature distribution $T(x,t)$, it is necessary to specify an *initial* condition and two *boundary conditions.* For the typical transient conduction problem of Figure 5.5, the initial condition is

$$T(x,0) = T_i \tag{5.15}$$

and the boundary conditions are

$$\left.\frac{\partial T}{\partial x}\right|_{x=0} = 0 \tag{5.16}$$

and

$$-k\left.\frac{\partial T}{\partial x}\right|_{x=L} = h[T(L,t) - T_\infty] \tag{5.17}$$

Equation 5.15 presumes a uniform temperature distribution at time $t=0$; Equation 5.16 reflects the *symmetry requirement* for the midplane of the wall; and Equation 5.17 describes the surface condition experienced for time $t>0$. From Equations 5.14 to 5.17, it is evident that, in addition to depending on x and t, temperatures in the wall also depend on a number of physical parameters. In particular

$$T = T(x, t, T_i, T_\infty, L, k, \alpha, h) \tag{5.18}$$

The foregoing problem may be solved analytically or numerically. These methods will be considered in subsequent sections, but first it is important to note the advantages that may be obtained by *nondimensionalizing* the governing equations. This may be done by arranging the relevant variables into suitable *groups.* Consider the dependent variable T. If the temperature difference $\theta \equiv T - T_\infty$ is divided by the *maximum possible temperature difference* $\theta_i \equiv T_i - T_\infty$, a dimensionless form of the dependent variable may be defined as

$$\theta^* \equiv \frac{\theta}{\theta_i} = \frac{T - T_\infty}{T_i - T_\infty} \tag{5.19}$$

Accordingly, θ^* must lie in the range $0 \le \theta^* \le 1$. Similarly, a dimensionless spatial coordinate may be defined as

$$x^* \equiv \frac{x}{L} \tag{5.20}$$

where L is the half-thickness of the plane wall. Similarly, a dimensionless time may be defined as

$$t^* \equiv \frac{\alpha t}{L^2} \equiv Fo \tag{5.21}$$

where t^* is equivalent to the dimensionless *Fourier number*, Equation 5.12.

Substituting the definitions of Equations 5.19 to 5.21 into Equations 5.14 to 5.17, the governing equation becomes

$$\frac{\partial^2 \theta^*}{\partial x^{*2}} = \frac{\partial \theta^*}{\partial Fo} \tag{5.22}$$

and the initial and boundary conditions become

$$\theta^*(x^*,0) = 1 \tag{5.23}$$

$$\left.\frac{\partial \theta^*}{\partial x^*}\right|_{x^*=0} = 0 \tag{5.24}$$

and

$$\left.\frac{\partial \theta^*}{\partial x^*}\right|_{x^*=1} = -Bi\theta^*(1,t^*) \tag{5.25}$$

where the Biot number is $Bi = hL/k$. In dimensionless form the functional dependence may now be expressed as

$$\theta^* = f(x^*, Fo, Bi) \tag{5.26}$$

Recall that this functional dependence, without the x^* variation, was obtained for the lumped capacitance method, as shown in Equation 5.13.

Comparing Equations 5.18 and 5.26 reveals the considerable advantage associated with casting the problem in dimensionless form. Equation 5.26 implies that *for a prescribed geometry, the transient temperature distribution is a universal function of* x^*, *Fo, and Bi.* That is, the *dimensionless solution* assumes a prescribed form that does not depend on the particular value of T_i, T_∞, L, k, α, or h. Since this generalization greatly simplifies the presentation and utilization of transient solutions, the dimensionless variables are used extensively in subsequent sections.

5.4 THE PLANE WALL WITH CONVECTION

Exact, analytical solutions to transient conduction problems have been obtained for many simplified geometries and boundary conditions and are well documented in the literature [1–4]. Several mathematical techniques, including the method of separation of variables (Section 4.2), may be used for this purpose, and typically the solution for the dimensionless temperature distribution, Equation 5.26, is in the form of an infinite series. However, except for very small values of the Fourier number, this series may be approximated by a single term and the results may be represented in a convenient graphical form.

5.4.1 Exact Solution

Consider the *plane wall* of thickness $2L$ (Figure 5.6a). If the thickness is small relative to the width and height of the wall, it is reasonable to assume that conduction occurs exclusively in the x direction. If the wall is initially at a uniform temperature, $T(x,0) = T_i$, and is suddenly immersed in a fluid of $T_\infty \neq T_i$, the resulting temperatures may be obtained by solving Equation 5.22 subject to the conditions of Equations 5.23 to 5.25. Since the convection conditions for the surfaces at $x^* = \pm 1$ are the same, the temperature distribution at any instant must be symmetrical about the midplane ($x^* = 0$). An exact solution to this problem has been obtained and is of the form [2]

$$\theta^* = \sum_{n=1}^{\infty} C_n \exp(-\zeta_n^2 Fo) \cos(\zeta_n x^*) \tag{5.27a}$$

where the coefficient C_n is

$$C_n = \frac{4 \sin \zeta_n}{2\zeta_n + \sin(2\zeta_n)} \tag{5.27b}$$

and the discrete values (*eigenvalues*) of ζ_n are positive roots of the transcendental equation

$$\zeta_n \tan \zeta_n = Bi \tag{5.27c}$$

The first four roots of this equation are given in Appendix B.3.

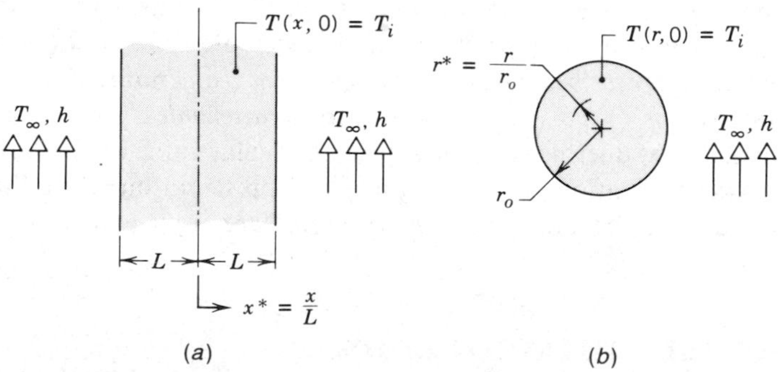

Figure 5.6 One-dimensional systems with an initial uniform temperature subjected to sudden convection conditions. (*a*) Plane wall. (*b*) Infinite cylinder or sphere.

5.4.2 Approximate Solution

It can be shown (Problem 5.16) that for values of $Fo \geq 0.2$, the infinite series solution, Equation 5.27a, can be approximated by the first term of the series. Invoking this approximation, the dimensionless form of the temperature distribution becomes

$$\theta^* = C_1 \exp(-\zeta_1^2 Fo) \cos(\zeta_1 x^*) \tag{5.28a}$$

or

$$\theta^* = \theta_o^* \cos(\zeta_1 x^*) \tag{5.28b}$$

where θ_o^* represents the midplane ($x^* = 0$) temperature

$$\theta_o^* = C_1 \exp(-\zeta_1^2 Fo) \tag{5.29}$$

An important implication of Equation 5.28b is that *the time dependence of the temperature at any location within the wall is the same as that of the midplane temperature.* The coefficients C_1 and ζ_1 are evaluated by Equations 5.27b and 5.27c, respectively, and are given in Table 5.1 for a range of Biot numbers.

5.4.3 Total Energy Transfer

In many situations it is useful to know the total energy that has left the wall up to any time t in the transient process. The conservation of energy requirement, Equation 1.10b, may be applied for the time interval bounded by the initial condition ($t = 0$) and time $t > 0$

$$E_{in} - E_{out} = \Delta E_{st} \tag{5.30}$$

Equating the energy transferred from the wall Q to E_{out} and setting $E_{in} = 0$ and $\Delta E_{st} = E(t) - E(0)$, it follows that

$$Q = -[E(t) - E(0)] \tag{5.31a}$$

or

$$Q = -\int \rho c[T(r,t) - T_i]\, dV \tag{5.31b}$$

where the integration is performed over the volume of the wall. It is convenient to nondimensionalize this result by introducing the quantity

$$Q_o = \rho c V(T_i - T_\infty) \tag{5.32}$$

which may be interpreted as the initial internal energy of the wall relative to the fluid temperature. It is also the *maximum* amount of energy transfer which could occur if the process were continued to time $t = \infty$. Hence, assuming constant properties, the ratio of the total energy transferred from the wall over the time

Table 5.1 Coefficients used in the one-term approximation to the series solutions for transient one-dimensional conduction

	PLANE WALL		INFINITE CYLINDER		SPHERE	
Bi[a]	ζ_1 (rad)	C_1	ζ_1 (rad)	C_1	ζ_1 (rad)	C_1
0.01	0.0998	1.0017	0.1412	1.0025	0.1730	1.0030
0.02	0.1410	1.0033	0.1995	1.0050	0.2445	1.0060
0.03	0.1732	1.0049	0.2439	1.0075	0.2989	1.0090
0.04	0.1987	1.0066	0.2814	1.0099	0.3450	1.0120
0.05	0.2217	1.0082	0.3142	1.0124	0.3852	1.0149
0.06	0.2425	1.0098	0.3438	1.0148	0.4217	1.0179
0.07	0.2615	1.0114	0.3708	1.0173	0.4550	1.0209
0.08	0.2791	1.0130	0.3960	1.0197	0.4860	1.0239
0.09	0.2956	1.0145	0.4195	1.0222	0.5150	1.0268
0.10	0.3111	1.0160	0.4417	1.0246	0.5423	1.0298
0.15	0.3779	1.0237	0.5376	1.0365	0.6608	1.0445
0.20	0.4328	1.0311	0.6170	1.0483	0.7593	1.0592
0.25	0.4801	1.0382	0.6856	1.0598	0.8448	1.0737
0.30	0.5218	1.0450	0.7465	1.0712	0.9208	1.0880
0.4	0.5932	1.0580	0.8516	1.0932	1.0528	1.1164
0.5	0.6533	1.0701	0.9408	1.1143	1.1656	1.1441
0.6	0.7051	1.0814	1.0185	1.1346	1.2644	1.1713
0.7	0.7506	1.0919	1.0873	1.1539	1.3525	1.1978
0.8	0.7910	1.1016	1.1490	1.1725	1.4320	1.2236
0.9	0.8274	1.1107	1.2048	1.1902	1.5044	1.2488
1.0	0.8603	1.1191	1.2558	1.2071	1.5708	1.2732
2.0	1.0769	1.1795	1.5995	1.3384	2.0288	1.4793
3.0	1.1925	1.2102	1.7887	1.4191	2.2889	1.6227
4.0	1.2646	1.2287	1.9081	1.4698	2.4556	1.7201
5.0	1.3138	1.2402	1.9898	1.5029	2.5704	1.7870
6.0	1.3496	1.2479	2.0490	1.5253	2.6537	1.8338
7.0	1.3766	1.2532	2.0937	1.5411	2.7165	1.8674
8.0	1.3978	1.2570	2.1286	1.5526	2.7654	1.8921
9.0	1.4149	1.2598	2.1566	1.5611	2.8044	1.9106
10.0	1.4289	1.2620	2.1795	1.5677	2.8363	1.9249
20.0	1.4961	1.2699	2.2881	1.5919	2.9857	1.9781
30.0	1.5202	1.2717	2.3261	1.5973	3.0372	1.9898
40.0	1.5325	1.2723	2.3455	1.5993	3.0632	1.9942
50.0	1.5400	1.2727	2.3572	1.6002	3.0788	1.9962
100.0	1.5552	1.2731	2.3809	1.6015	3.1102	1.9990

[a] $Bi = hL/k$ for the plane wall and hr_o/k for the infinite cylinder and sphere. See Figure 5.6.

interval t to the maximum possible transfer is

$$\frac{Q}{Q_o} = \int \frac{-[T(r,t) - T_i]}{T_i - T_\infty} \frac{dV}{V} = \frac{1}{V} \int (1 - \theta^*)\, dV \tag{5.33}$$

Employing the approximate form of the temperature distribution for the plane wall, Equation 5.28b, the integration prescribed by Equation 5.33 can be performed to obtain

$$\frac{Q}{Q_o} = 1 - \frac{\sin \zeta_1}{\zeta_1} \theta_o^* \tag{5.34}$$

where θ_o^* can be determined from Equation 5.29 using Table 5.1 for values of the coefficients C_1 and ζ_1.

5.4.4 Graphical Representations

Graphical representations of the *approximate* relations for the transient temperature distribution and energy transfer were first presented by Heisler [5] and Gröber et al. [6]. The graphs have been widely used for nearly four decades; in addition to offering computational convenience, they illustrate the functional dependence of the transient, dimensionless temperature distribution on the Biot and Fourier numbers.

Results for the plane wall are presented in Figures 5.7 to 5.9. Figure 5.7 may be used to obtain the *midplane* temperature of the wall, $T(0,t) = T_o(t)$, at any time during the transient process. If T_o is known for particular values of *Fo* and *Bi*, Figure 5.8 may be used to determine the corresponding temperature at any location *off the midplane*. Hence, Figure 5.8 must be used in conjunction with Figure 5.7. For example, if one wishes to determine the surface temperature ($x^* = +1$) at some time t, Figure 5.7 would first be used to determine T_o at t. Figure 5.8 would then be used to determine the surface temperature from knowledge of T_o. The procedure would be inverted if the problem were one of determining the time required for the surface to reach a prescribed temperature.

Absence of the Fourier number in Figure 5.8 implies that the time dependence of any temperature off the midplane corresponds to the time dependence of the midplane temperature. This result is, of course, a consequence of the approximation that led to Equation 5.28b and is valid for all but the earliest stages of the transient process ($Fo \geq 0.2$).

Graphical results for the energy transferred from a plane wall over the time interval t are presented in Figure 5.9. These results were generated from Equation 5.34. The dimensionless energy transfer Q/Q_o is expressed exclusively in terms of *Fo* and *Bi*.

Because the mathematical problem is precisely the same, the foregoing results may also be applied to a plane wall of thickness L, which is insulated on one side ($x^* = 0$) and experiences convective transport on the other side ($x^* = +1$). This equivalence is a consequence of the fact that, regardless of whether a symmetrical or an adiabatic requirement is prescribed at $x^* = 0$, the boundary condition is of the form $\partial\theta^*/\partial x^* = 0$.

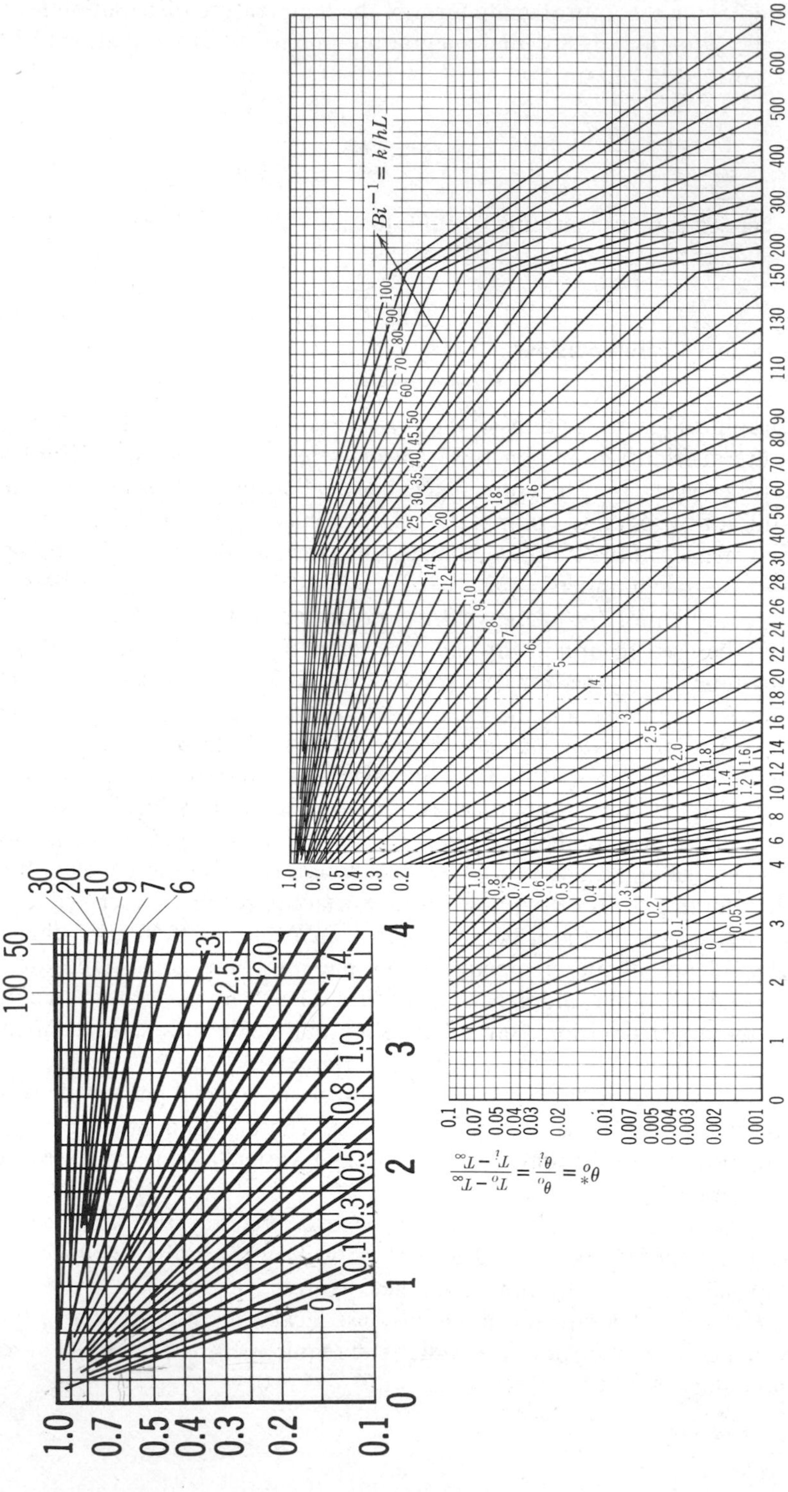

Figure 5.7 Midplane temperature as a function of time for a plane wall of thickness $2L$ [5]. Used with permission.

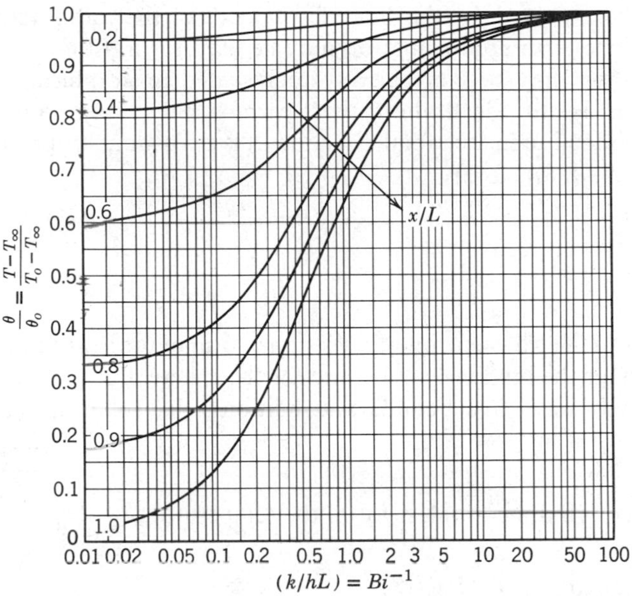

Figure 5.8 Temperature distribution in a plane wall of thickness $2L$ [5]. Used with permission.

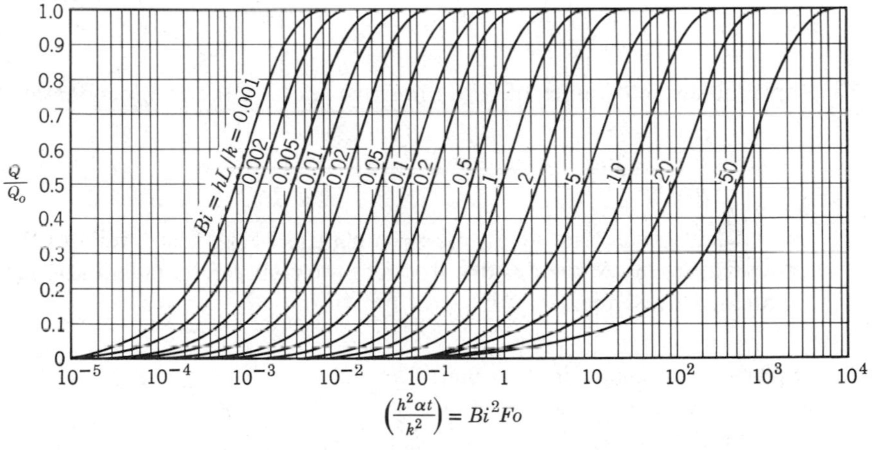

Figure 5.9 Internal energy change as a function of time for a plane wall of thickness $2L$ [6]. Adapted with permission.

5.5 RADIAL SYSTEMS WITH CONVECTION

For an infinite cylinder or sphere of radius r_o (Figure 5.6b), which is at an initial uniform temperature and experiences a change in convective conditions, results similar to those of Section 5.4 may be developed. That is, an exact series solution may be obtained for the time dependence of the radial temperature distribution; a one-term approximation may be used for most conditions; and the approximation may be conveniently represented in graphical form. The infinite cylinder is an idealization that permits the assumption of one-dimensional conduction in the radial direction. It is a reasonable approximation for cylinders having $L/r_o \geq 10$.

5.5.1 Exact Solutions

Exact solutions to the transient, one-dimensional form of the heat equation have been developed for the infinite cylinder and for the sphere. For a uniform initial temperature and convective boundary conditions, the solutions [2] are as follows.

Infinite Cylinder In dimensionless form, the temperature is

$$\theta^* = \sum_{n=1}^{\infty} C_n \exp(-\zeta_n^2 Fo) J_0(\zeta_n r^*) \tag{5.35a}$$

where

$$C_n = \frac{2}{\zeta_n} \frac{J_1(\zeta_n)}{J_0^2(\zeta_n) + J_1^2(\zeta_n)} \tag{5.35b}$$

and the discrete values of ζ_n are positive roots of the transcendental equation

$$\zeta_n \frac{J_1(\zeta_n)}{J_0(\zeta_n)} = Bi \tag{5.35c}$$

The quantities J_1 and J_0 are Bessel functions of the first kind and their values are tabulated in Appendix B.4. Roots of the transcendental equation (5.35c) are tabulated by Schneider [2].

Sphere Similarly, for the sphere

$$\theta^* = \sum_{n=1}^{\infty} C_n \exp(-\zeta_n^2 Fo) \frac{1}{\zeta_n r^*} \sin(\zeta_n r^*) \tag{5.36a}$$

where

$$C_n = \frac{4[\sin(\zeta_n) - \zeta_n \cos(\zeta_n)]}{2\zeta_n - \sin(2\zeta_n)} \tag{5.36b}$$

and the discrete values of ζ_n are positive roots of the transcendental equation

$$1 - \zeta_n \cot \zeta_n = Bi \tag{5.36c}$$

Roots of the transcendental equation are tabulated by Schneider [2].

5.5.2 Approximate Solutions

For the infinite cylinder and sphere, Heisler [5] has shown that for $Fo \geq 0.2$, the foregoing series solutions can be approximated by a single term. Hence, as for the case of the plane wall, the time dependence of the temperature at any location within the radial system is the same as that of the centerline or centerpoint.

Infinite Cylinder The one-term approximation to Equation 5.35a is

$$\theta^* = C_1 \exp(-\zeta_1^2 Fo) J_0(\zeta_1 r^*) \tag{5.37a}$$

or

$$\theta^* = \theta_o^* J_0(\zeta_1 r^*) \tag{5.37b}$$

where θ_o^* represents the centerline temperature and is of the form

$$\theta_o^* = C_1 \exp(-\zeta_1^2 Fo) \tag{5.37c}$$

Values of the coefficients C_1 and ζ_1 have been determined and are listed in Table 5.1 for a range of Biot numbers.

Sphere From Equation 5.36a, the one-term approximation is

$$\theta^* = C_1 \exp(-\zeta_1^2 Fo) \frac{1}{\zeta_1 r^*} \sin(\zeta_1 r^*) \tag{5.38a}$$

or

$$\theta^* = \theta_o^* \frac{1}{\zeta_1 r^*} \sin(\zeta_1 r^*) \tag{5.38b}$$

where θ_o^* represents the center temperature and is of the form

$$\theta_o^* = C_1 \exp(-\zeta_1^2 Fo) \tag{5.38c}$$

Values of the coefficients C_1 and ζ_1 have been determined and are listed in Table 5.1 for a range of Biot numbers.

5.5.3 Total Energy Transfer

As in Section 5.4.3, an energy balance may be performed to determine the total energy transfer from the infinite cylinder or sphere over the time interval $\Delta t = t$. Substituting from the approximate solutions, Equations 5.37b and 5.38b, and introducing Q_o from Equation 5.32, the results are as follows.

Infinite Cylinder

$$\frac{Q}{Q_o} = 1 - \frac{2\theta_o^*}{\zeta_1} J_1(\zeta_1) \tag{5.39}$$

Sphere

$$\frac{Q}{Q_o} = 1 - \frac{3\theta_o^*}{\zeta_1^3} [\sin(\zeta_1) - \zeta_1 \cos(\zeta_1)] \tag{5.40}$$

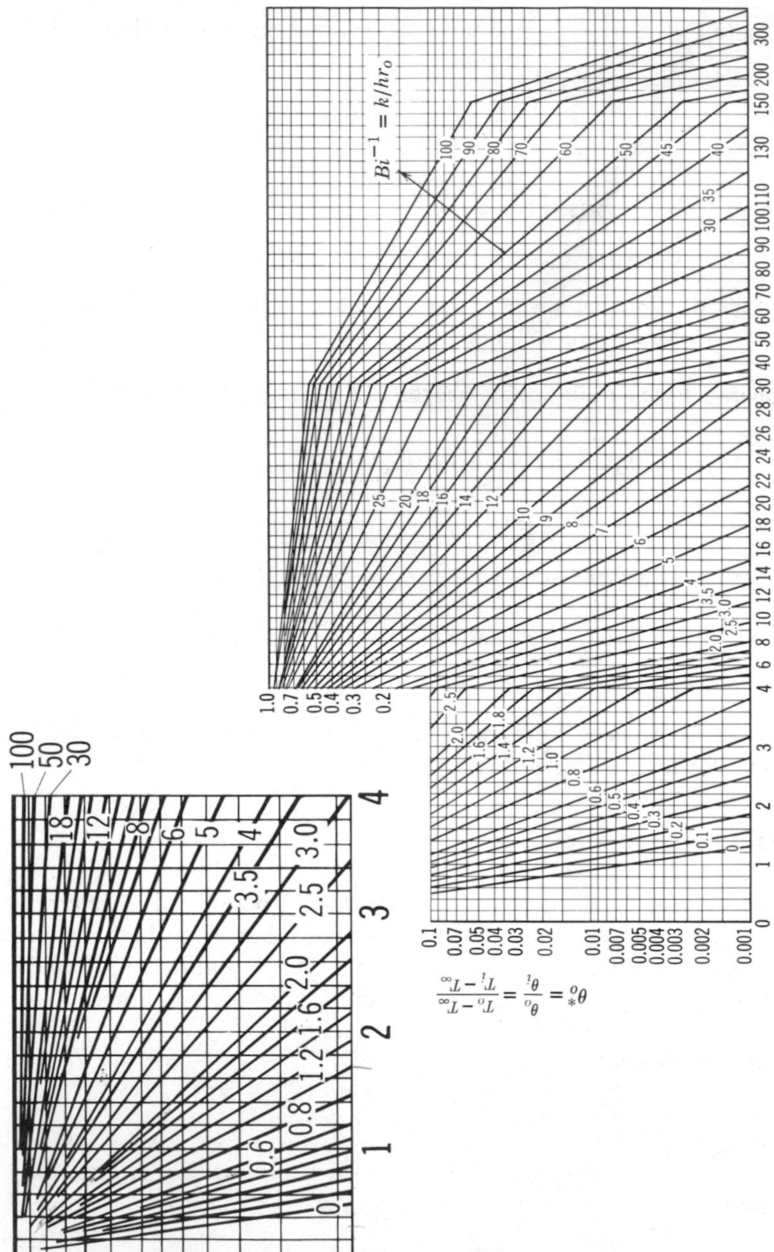

Figure 5.10 Centerline temperature as a function of time for an infinite cylinder of radius r_o [5]. Used with permission.

Values of the center temperature θ_o^* are determined from Equation 5.37c or 5.38c, using the coefficients of Table 5.1 for the appropriate system.

5.5.4 Graphical Representation

Graphical representations similar to those for the plane wall (Figures 5.7 to 5.9) have also been generated by Heisler [5] and Gröber et al. [6] for an infinite cylinder and sphere. Results for the infinite cylinder are presented in Figures 5.10 to 5.12, and those for the sphere are presented in Figures 5.13 to 5.15. *Note that,*

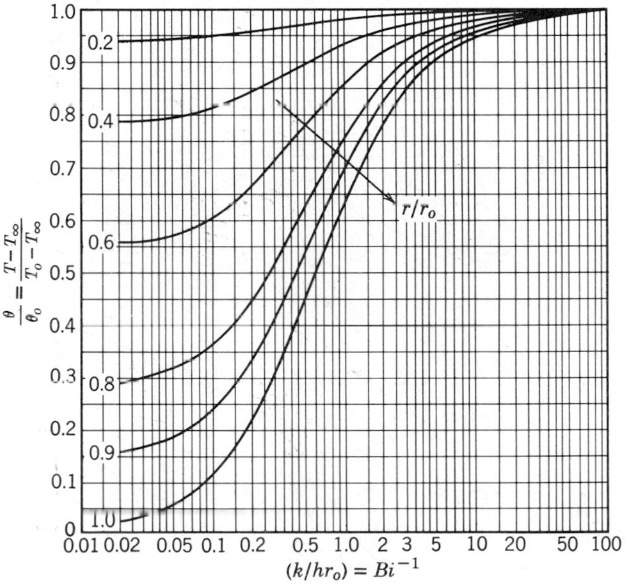

Figure 5.11 Temperature distribution in an infinite cylinder of radius r_o [5]. Used with permission.

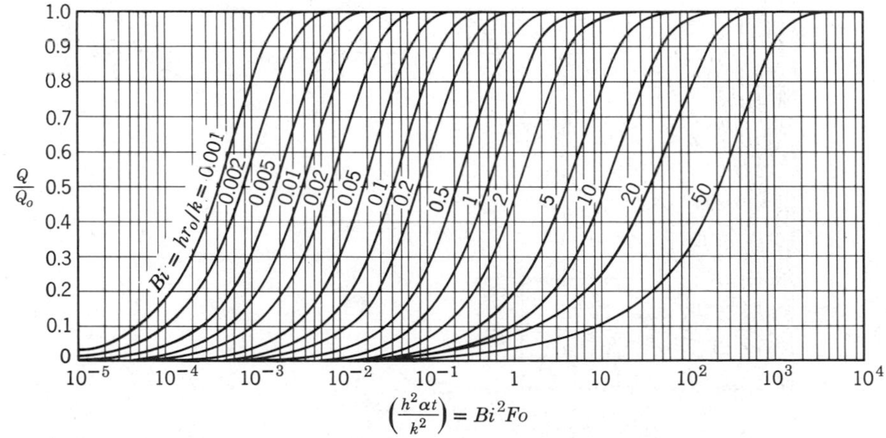

Figure 5.12 Internal energy change as a function of time for an infinite cylinder of radius r_o [6]. Adapted with permission.

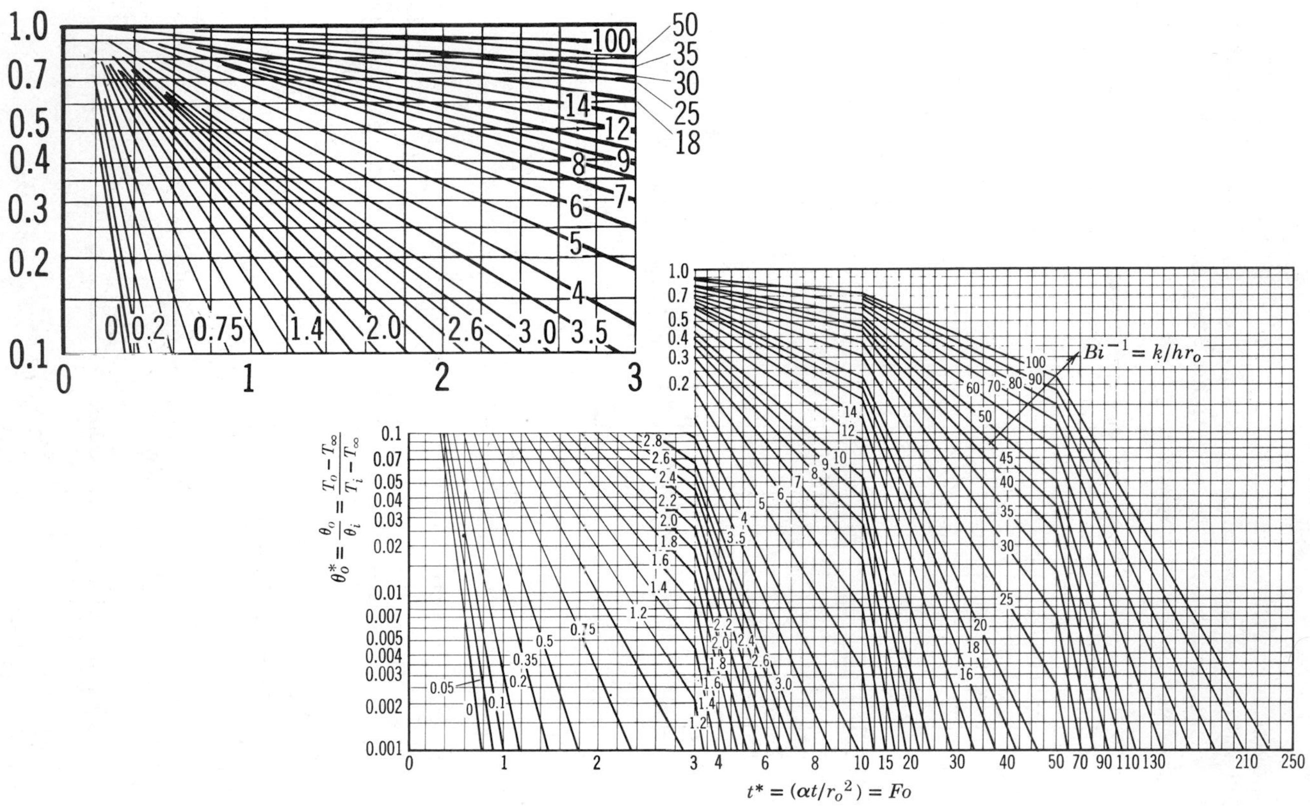

Figure 5.13 Center temperature as a function of time in a sphere of radius r_o [5]. Used with permission.

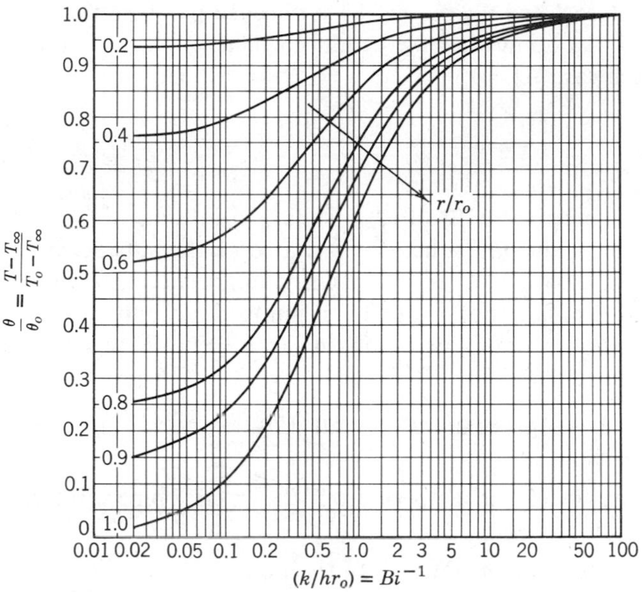

Figure 5.14 Temperature distribution in a sphere of radius r_o [5]. Used with permission.

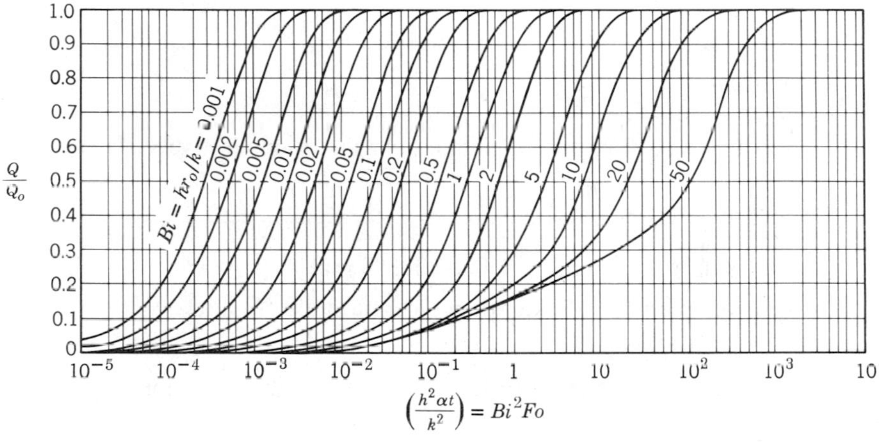

Figure 5.15 Internal energy change as a function of time for a sphere of radius r_o [6]. Adapted with permission.

with respect to the use of these figures, the Biot number is defined in terms of r_o. *In contrast recall that, for the lumped capacitance method, the characteristic length in the Biot number is* $r_o/2$ *for the cylinder and* $r_o/3$ *for the sphere.*

In closing it should be noted that the *Heisler charts* may also be used to determine the transient response of a plane wall, an infinite cylinder, or a sphere subjected to a *sudden change in surface temperature*. For such a condition it is only necessary to replace T_∞ by the prescribed surface temperature T_s and to set Bi^{-1} equal to zero. In so doing the convection coefficient is tacitly assumed to be infinite, in which case $T_\infty = T_s$.

EXAMPLE 5.2

Consider a steel pipeline (AISI 1010) that is 1 m in diameter and has a wall thickness of 40 mm. The pipe is heavily insulated on the outside, and before the initiation of flow, the walls of the pipe are at a uniform temperature of $-20°C$. With the initiation of flow, hot oil at 60°C is pumped through the pipe creating a convective surface condition corresponding to $h = 500\ W/m^2 \cdot K$ at the inner surface of the pipe.

1. What are the appropriate Biot and Fourier numbers 8 min after the initiation of flow?
2. At $t = 8$ min, what is the temperature of the exterior pipe surface covered by the insulation?
3. What is the heat flux q'' (W/m^2) to the pipe from the oil at $t = 8$ min?
4. How much energy per meter of pipe length has been transferred from the oil to the pipe at $t = 8$ min?

SOLUTION

Known:

Pipeline wall subjected to sudden change in convective surface condition.

Find:

1. Biot and Fourier numbers after 8 min.
2. Temperature of exterior pipe surface after 8 min.
3. Heat flux to the wall at 8 min.
4. Energy transferred to pipe per unit length after 8 min.

Schematic:

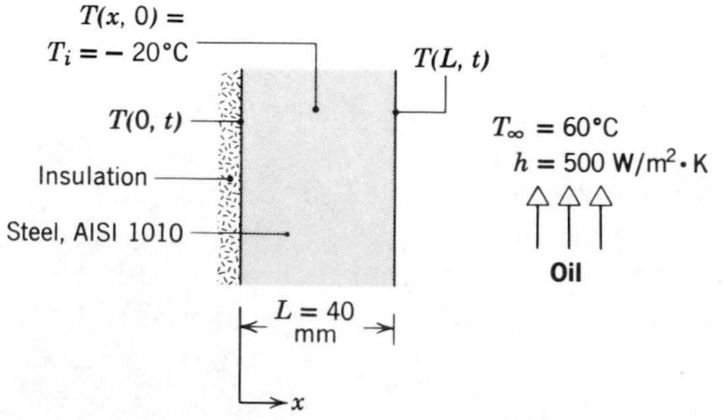

Assumptions:

1. Pipe wall can be approximated as plane wall, since thickness is much less than diameter.
2. Constant properties.
3. Outer surface of pipe is adiabatic.

Properties:

Table A.1, steel type AISI 1010 $[T = (-20 + 60)°C/2 \approx 300\ K]$: $\rho = 7823\ kg/m^3$, $c = 434\ J/kg \cdot K$, $k = 63.9\ W/m \cdot K$, $\alpha = 18.8 \times 10^{-6}\ m^2/s$.

Analysis:

1. At $t = 8$ min, the Biot and Fourier numbers are computed from Equations 5.10 and 5.12, respectively, with $L_c = L$. Hence

$$Bi = \frac{hL}{k} = \frac{500\ W/m^2 \cdot K \times 0.04\ m}{63.9\ W/m \cdot K} = 0.313 \qquad \triangleleft$$

$$Fo = \frac{\alpha t}{L^2} = \frac{18.8 \times 10^{-6}\ m^2/s \times 8\ min \times 60\ s/min}{(0.04\ m)^2} = 5.64 \qquad \triangleleft$$

2. With $Bi = 0.313$, use of the lumped capacitance method is inappropriate. However, since transient conditions in the insulated pipe wall of thickness L correspond to those in a plane wall of thickness $2L$ experiencing the same surface condition, the desired results may be obtained from the charts for the plane wall. Using Figure 5.7, with $Bi^{-1} = 3.2$,

$$\frac{\theta_o}{\theta_i} = \frac{T(0,t) - T_\infty}{T_i - T_\infty} \approx 0.22$$

Hence after 8 min, the temperature of the exterior pipe surface, which corresponds to the midplane temperature of a plane wall, is

$$T_o = T(0, 480\ s) \approx T_\infty + 0.22(T_i - T_\infty)$$

$$T_o = 60°C + 0.22(-20 - 60)°C \approx 42°C \qquad \triangleleft$$

3. Heat transfer to the inner surface at $x = L$ is by convection, and at any time t the heat flux may be obtained from Newton's law of cooling. Hence at $t = 480$ s,

$$q''_x(L, 480\ s) \equiv q''_L = h[T(L, 480\ s) - T_\infty]$$

The surface temperature $T(L, 480\ s)$ may be obtained from Figure 5.8. For the prescribed conditions

$$\frac{x}{L} = 1 \qquad \text{and} \qquad Bi^{-1} = 3.2$$

it follows that

$$\frac{\theta(L,480\,\text{s})}{\theta_o(480\,\text{s})} = \frac{T(L,480\,\text{s}) - T_\infty}{T_o(480\,\text{s}) - T_\infty} \approx 0.86$$

Hence

$$T(L,480\,\text{s}) \approx T_\infty + 0.86[T_o(480\,\text{s}) - T_\infty]$$

$$T(L,480\,\text{s}) \approx 60°\text{C} + 0.86[42 - 60]°\text{C} \approx 45°\text{C}$$

The heat flux at $t = 8$ min is then

$$q''_L = 500\ \text{W/m}^2 \cdot \text{K}\,(45 - 60)°\text{C} = -7500\ \text{W/m}^2 \qquad \triangleleft$$

4. The energy transfer to the pipe wall over the 8-min interval may be obtained from Figure 5.9 and Equation 5.32. With

$$Bi = 0.313 \qquad Bi^2 Fo = 0.55$$

it follows that

$$\frac{Q}{Q_o} \approx 0.78$$

Hence

$$Q \approx 0.78 \rho c V(T_i - T_\infty)$$

or with a volume per unit pipe length of $V' = \pi D L$,

$$Q' \approx 0.78 \rho c \pi D L(T_i - T_\infty)$$

$$Q' \approx 0.78 \times 7823\ \text{kg/m}^3 \times 434\ \text{J/kg} \cdot \text{K} \times \pi \times 1\ \text{m} \times 0.04\ \text{m}\ (-20 - 60)°\text{C}$$

$$Q' \approx -2.7 \times 10^7\ \text{J/m} \qquad \triangleleft$$

Comments:

1. The minus sign associated with q'' and Q' simply implies that the direction of heat transfer is from the oil to the pipe (into the pipe wall).
2. Since $Fo > 0.2$, the one-term approximation can be used to calculate wall temperatures and the total energy transfer. The midplane temperature can be determined from Equation 5.29

$$\theta_o^* = \frac{T_o - T_\infty}{T_i - T_\infty} = C_1 \exp(-\zeta_1^2 Fo)$$

where, with $Bi = 0.313$, $C_1 = 1.047$, and $\zeta_1 = 0.531$ rad from Table 5.1. With $Fo = 5.64$,

$$\theta_o^* = 1.047 \exp[-(0.531\ \text{rad})^2 \times 5.64] = 0.214$$

This result is in reasonable agreement with the value of 0.22 obtained from Figure 5.7. Hence,

$$T(0,8\,\text{min}) = T_\infty + \theta_o^*(T_i - T_\infty) = 60^\circ\text{C} + 0.214(-20 - 60)^\circ\text{C} = 42.9^\circ\text{C}$$

which is within 2% of the value determined from the Heisler chart.

3. Using the one-term approximation for the surface temperature, Equation 5.28b with $x^* = 1$ has the form

$$\theta^* = \theta_o^* \cos(\zeta_1)$$

$$T(L,t) = T_\infty + (T_i - T_\infty)\theta_o^* \cos(\zeta_1)$$

$$T(L,8\,\text{min}) = 60^\circ\text{C} + (-20 - 60)^\circ\text{C} \times 0.214 \times \cos(0.531\,\text{rad})$$

$$T(L,8\,\text{min}) = 45.2^\circ\text{C}$$

which is within 1% of the value determined from the Heisler chart.

4. The total energy transferred during the transient process can be determined from the result associated with the one-term approximation, Equation 5.34.

$$\frac{Q}{Q_o} = 1 - \frac{\sin(\zeta_1)}{\zeta_1}\theta_o^*$$

$$\frac{Q}{Q_o} = 1 - \frac{\sin(0.531\,\text{rad})}{0.531\,\text{rad}} \times 0.214 = 0.80$$

which is within 3% of the value determined from the Gröber chart.

EXAMPLE 5.3

A new process for treatment of a special material is to be evaluated. The material, a sphere of radius $r_o = 5$ mm, is initially in equilibrium at 400°C in a furnace. It is suddenly removed from the furnace and subjected to a two-step cooling process.

Step 1 Cooling in air at 20°C for a period of time t_a until the center temperature reaches a critical value, $T_a(0,t_a) = 335$°C. For this situation, the convective heat transfer coefficient is $h_a = 10$ W/m$^2\cdot$K.

After the sphere has reached this critical temperature, the second step is initiated.

Step 2 Cooling in a well-stirred water bath at 20°C wherein the convective heat transfer coefficient is $h_w = 6000$ W/m$^2\cdot$K.

The thermophysical properties of the material are $\rho = 3000$ kg/m^3, $k = 20$ W/m$\cdot$K, $c = 1000$ J/kg$\cdot$K, and $\alpha = 6.66 \times 10^{-6}$ m^2/s.

1. Calculate the time t_a required for step 1 of the cooling process to be completed.

2. Calculate the time t_w required during step 2 of the process for the center of the sphere to cool from 335°C (the condition at the completion of step 1) to 50°C.

SOLUTION

Known: Temperature requirements for a cooling process of a sphere.

Find:

1. Time t_a required to accomplish desired cooling in air.
2. Time t_w required to complete cooling in water bath.

Schematic:

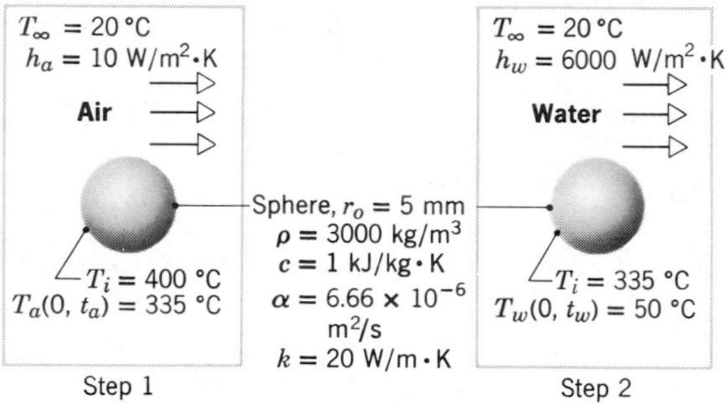

Assumptions:

1. One-dimensional conduction in r.
2. Constant properties.

Analysis:

1. To determine whether the lumped capacitance method can be used, the Biot number is calculated. From Equation 5.10, with $L_c = r_o/3$,

$$Bi = \frac{h_a r_o}{3k} = \frac{10 \text{ W/m}^2 \cdot \text{K} \times 0.005 \text{ m}}{3 \times 20 \text{ W/m} \cdot \text{K}} = 8.33 \times 10^{-4}$$

Accordingly, the lumped capacitance method may be used, and the temperature is nearly uniform throughout the sphere. From Equation 5.5 it follows that

$$t_a = \frac{\rho V c}{h_a A_s} \ln \frac{\theta_i}{\theta_a} = \frac{\rho r_o c}{3h_a} \ln \frac{T_i - T_\infty}{T_a - T_\infty}$$

where $V = (4/3)\pi r_o^3$ and $A_s = 4\pi r_o^2$. Hence

$$t_a = \frac{3000\,\text{kg/m}^3 \times 0.005\,\text{m} \times 1000\,\text{J/kg}\cdot\text{K}}{3 \times 10\,\text{W/m}^2\cdot\text{K}} \ln \frac{400-20}{335-20} = 94\,\text{s} \quad \triangleleft$$

2. To determine whether the lumped capacitance method may also be used for the second step of the cooling process, the Biot number is again calculated. In this case

$$Bi = \frac{h_w r_o}{3k} = \frac{6000\,\text{W/m}^2\cdot\text{K} \times 0.005\,\text{m}}{3 \times 20\,\text{W/m}\cdot\text{K}} = 0.50$$

and the lumped capacitance method is not appropriate. However, to an excellent approximation, the temperature of the sphere is uniform at $t = t_a$ and the Heisler charts may be used for the calculations from $t = t_a$ to $t = t_a + t_w$. Using Figure 5.13 with

$$Bi^{-1} = \frac{k}{h_w r_o} = \frac{20\,\text{W/m}\cdot\text{K}}{6000\,\text{W/m}^2\cdot\text{K} \times 0.005\,\text{m}} = 0.67$$

$$\frac{\theta_o}{\theta_i} = \frac{T_o - T_\infty}{T_i - T_\infty} = \frac{50-20}{335-20} = 0.095$$

it follows that $Fo \approx 0.80$, and

$$t_w = Fo\frac{r_o^2}{\alpha} \approx 0.80 \frac{(0.005\,\text{m})^2}{6.66 \times 10^{-6}\,\text{m}^2/\text{s}} \approx 3.0\,\text{s} \quad \triangleleft$$

Comments:

1. If the temperature distribution in the sphere at the conclusion of step 1 were not uniform, the Heisler chart could not be used for the calculations of step 2.
2. The surface temperature of the sphere at the conclusion of step 2 may be obtained from Figure 5.14. With

$$Bi^{-1} = 0.67 \qquad \text{and} \qquad \frac{r}{r_o} = 1$$

$$\frac{\theta(r_o)}{\theta_o} = \frac{T(r_o) - T_\infty}{T_o - T_\infty} \approx 0.52$$

Hence

$$T(r_o) \approx 20°\text{C} + 0.52(50-20)°\text{C} \approx 36°\text{C}$$

The variation of the center and surface temperature with time is then as follows.

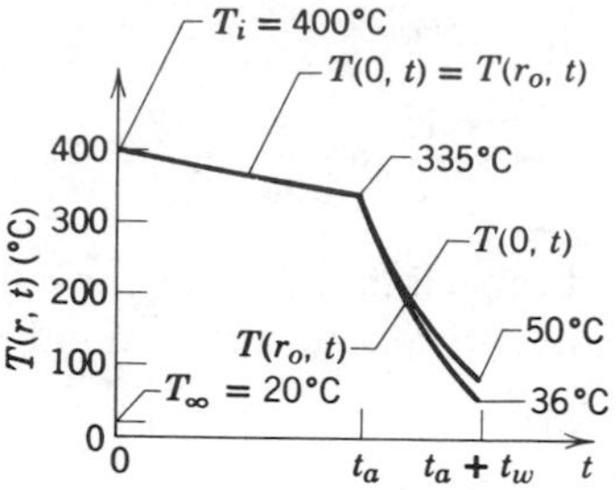

3. For the step 2 transient process in water, the one-term approximation is appropriate for determining the time t_w at which the center temperature reaches 50°C, that is, $T(0,t_w) = 50°C$. Rearranging Equation 5.38c,

$$Fo = -\frac{1}{\zeta_1^2}\ln\left[\frac{\theta_o^*}{C_1}\right] = -\frac{1}{\zeta_1^2}\ln\left[\frac{1}{C_1}\times\frac{T(0,t_w)-T_\infty}{T_i-T_\infty}\right]$$

Using Table 5.1 to obtain the coefficients for $Bi = 1/0.67 = 1.50$ ($C_1 = 1.376$ and $\zeta_1 = 1.800$ rad) and substituting appropriate temperatures, it follows that

$$Fo = -\frac{1}{(1.800\text{ rad})^2}\ln\left[\frac{1}{1.376}\times\frac{(50-20)°\text{C}}{(335-20)°\text{C}}\right] = 0.82$$

Substituting for r_o and α, it follows that $t_w = 3.1$ s, which is within 3% of the value of 3.0 s obtained from the Heisler chart.

5.6 THE SEMIINFINITE SOLID

Another simple geometry for which analytical solutions may be obtained is the *semiinfinite solid*. Since such a solid extends to infinity in all but one direction, it is characterized by a single identifiable surface (Figure 5.16). If a sudden change of conditions is imposed at this surface, transient, one-dimensional conduction will occur within the solid. The semiinfinite solid provides a *useful idealization* for many practical problems. It may be used to determine transient heat transfer effects near the surface of the earth or to approximate the transient response of a finite solid, such as a thick slab. For this second situation the approximation would be reasonable for the early portion of the transient, during which temperatures in the slab interior (well removed from the surface) are uninfluenced by the change in surface conditions.

The heat equation for transient conduction in a semiinfinite solid is given by Equation 5.14. The initial condition is prescribed by Equation 5.15, and the interior boundary condition is of the form

$$T(\infty,t) = T_i \tag{5.41}$$

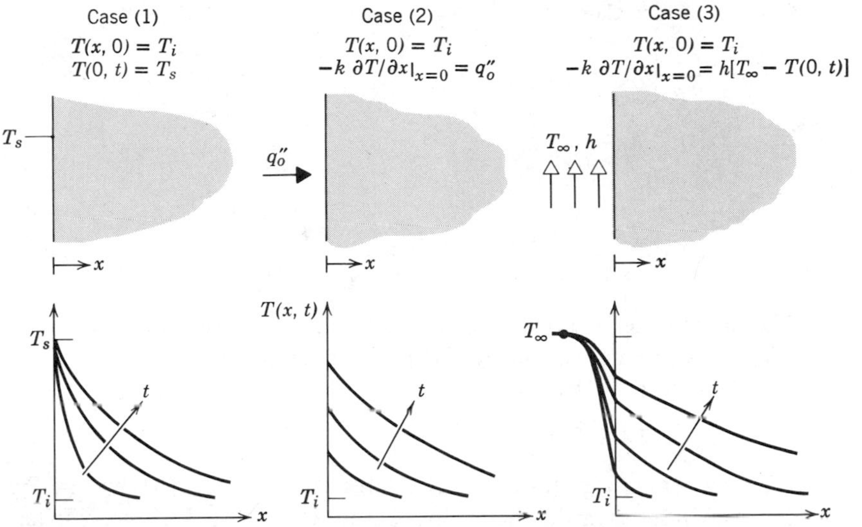

Figure 5.16 Transient temperature distributions in a semiinfinite solid for three surface conditions: constant surface temperature, constant surface heat flux, and surface convection.

Closed-form solutions have been obtained for three important surface conditions, instantaneously applied at $t = 0$ [1, 2]. These conditions are shown in Figure 5.16. They include application of a constant surface temperature $T_s \neq T_i$, application of a constant surface heat flux q''_o, and exposure of the surface to a fluid characterized by $T_\infty \neq T_i$ and the convection coefficient h. The solutions are summarized as follows.

Case 1 Constant Surface Temperature

$$T(0,t) = T_s \tag{5.42}$$

$$\frac{T(x,t) - T_s}{T_i - T_s} = \text{erf}\left(\frac{x}{2\sqrt{\alpha t}}\right) \tag{5.43}$$

$$q''_s(t) = -k\frac{\partial T}{\partial x}\bigg|_{x=0} = \frac{k(T_s - T_i)}{\sqrt{\pi \alpha t}} \tag{5.44}$$

Case 2 Constant Surface Heat Flux

$$q''_s = q''_o \tag{5.45}$$

$$T(x,t) - T_i = \frac{2q''_o(\alpha t/\pi)^{1/2}}{k}\exp\left(\frac{-x^2}{4\alpha t}\right) - \frac{q''_o x}{k}\,\text{erfc}\left(\frac{x}{2\sqrt{\alpha t}}\right) \tag{5.46}$$

Case 3 Surface Convection

$$-k\left.\frac{\partial T}{\partial x}\right|_{x=0} = h[T_\infty - T(0,t)] \tag{5.47}$$

$$\frac{T(x,t)-T_i}{T_\infty - T_i} = \operatorname{erfc}\left(\frac{x}{2\sqrt{\alpha t}}\right)$$

$$-\left[\exp\left(\frac{hx}{k}+\frac{h^2\alpha t}{k^2}\right)\right]\left[\operatorname{erfc}\left(\frac{x}{2\sqrt{\alpha t}}+\frac{h\sqrt{\alpha t}}{k}\right)\right] \tag{5.48}$$

The quantity erf w appearing in Equation 5.43 is the *Gaussian error function,* which is tabulated in Section B.1 of Appendix B. The *complementary error function,* erfc w, is defined as

$$\operatorname{erfc} w \equiv 1 - \operatorname{erf} w$$

Temperature histories for the three cases are also shown in Figure 5.16. Carefully note their distinguishing features. For case 3 the specific temperature histories computed from Equation 5.48 are plotted in Figure 5.17. Note that the curve corresponding to $h = \infty$ is equivalent to the result that would be obtained for a sudden change in the *surface temperature* to $T_s = T_\infty$. That is, for $h = \infty$ the second term on the right-hand side of Equation 5.48 goes to zero, and the result is equivalent to Equation 5.43.

EXAMPLE 5.4

In laying water mains, utilities must be concerned with the possibility of freezing during cold periods. Although the problem of determining the temperature in

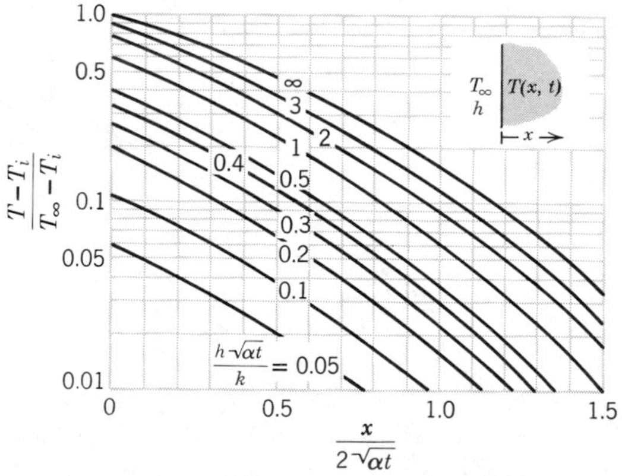

Figure 5.17 Temperature histories in a semi-infinite solid with surface convection [2]. Adapted with permission.

soil as a function of time is complicated by changing surface conditions, reasonable estimates can be based on the assumption of a constant surface temperature over a prolonged period of cold weather. What minimum burial depth x_m would you recommend to avoid freezing under conditions for which soil, initially at a uniform temperature of 20°C, is subjected to a constant surface temperature of −15°C for 60 days?

SOLUTION

Known:

Temperature below freezing imposed at the surface of soil that is initially at 20°C.

Find: The depth x_m to which the soil has frozen after 60 days.

Schematic:

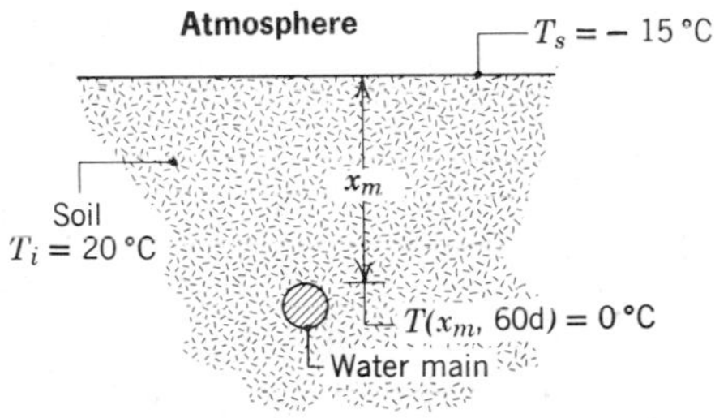

Assumptions:

1. One-dimensional conduction in x.
2. Soil is a semiinfinite medium.
3. Constant properties.

Properties:

Table A.3, soil (300 K): $\rho = 2050\ \text{kg/m}^3$, $k = 0.52\ \text{W/m}\cdot\text{K}$, $c = 1840\ \text{J/kg}\cdot\text{K}$, $\alpha = (k/\rho c) = 0.138 \times 10^{-6}\ \text{m}^2/\text{s}$.

Analysis:

The prescribed conditions correspond to those of case 1 of Figure 5.16, and the transient temperature response of the soil is governed by Equation 5.43. Hence at the time $t = 60$ days after the surface temperature change,

$$\frac{T(x_m, t) - T_s}{T_i - T_s} = \text{erf}\left(\frac{x_m}{2\sqrt{\alpha t}}\right)$$

or

$$\frac{0-(-15)}{20-(-15)} = 0.429 = \text{erf}\left(\frac{x_m}{2\sqrt{\alpha t}}\right)$$

Hence from Appendix B.1

$$\frac{x_m}{2\sqrt{\alpha t}} = 0.40$$

and

$$x_m = 0.80\sqrt{\alpha t} = 0.80(0.138 \times 10^{-6}\,\text{m}^2/\text{s} \times 60\,\text{days} \times 24\,\text{h/day}$$
$$\times\, 3600\,\text{s/h})^{1/2} = 0.68\,\text{m} \qquad \triangleleft$$

Comments:

The properties of soil are highly variable, depending on the nature of the soil and its moisture content.

5.7 MULTIDIMENSIONAL EFFECTS

Transient problems are frequently encountered for which two- and even three-dimensional effects are significant. Solution to a class of such problems can be obtained from the one-dimensional results of Sections 5.5 and 5.6.

Consider immersing the *short* cylinder of Figure 5.18, which is initially at a uniform temperature T_i, in a fluid of temperature $T_\infty \neq T_i$. Because the length and diameter are comparable, the subsequent transfer of energy by conduction will be significant for both the r and x coordinate directions. The temperature within the cylinder will therefore depend on r, x, and t.

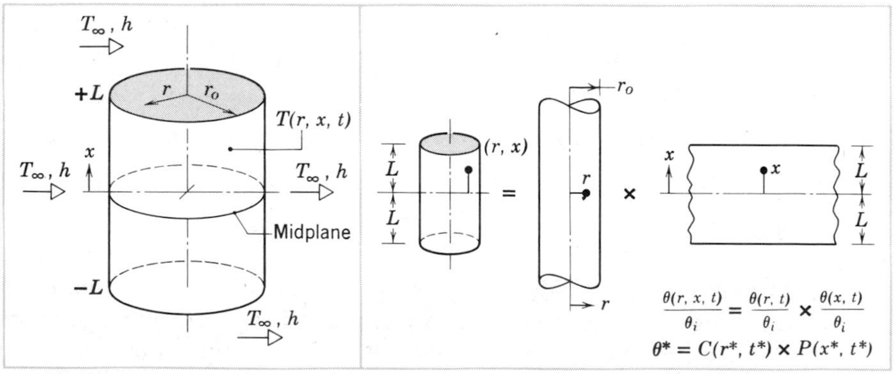

Figure 5.18 Two-dimensional, transient conduction in a short cylinder. (*a*) Geometry. (*b*) Form of the product solution.

Assuming constant properties and no generation, the appropriate form of the heat equation is, from Equation 2.18,

$$\frac{1}{r}\frac{\partial}{\partial r}\left(r\frac{\partial T}{\partial r}\right)+\frac{\partial^2 T}{\partial x^2}=\frac{1}{\alpha}\frac{\partial T}{\partial t}$$

where x has been used in place of z to designate the axial coordinate. A closed-form solution to this equation may be obtained by the separation of variables method. Although we will not consider the details of this solution, it is important to note that the end result may be expressed in the following form.

$$\frac{T(r,x,t)-T_\infty}{T_i-T_\infty}=\left.\frac{T(x,t)-T_\infty}{T_i-T_\infty}\right|_{\substack{\text{Plane}\\\text{wall}}}\cdot\left.\frac{T(r,t)-T_\infty}{T_i-T_\infty}\right|_{\substack{\text{Infinite}\\\text{cylinder}}}$$

That is, the two-dimensional solution may be expressed as a *product* of one-dimensional solutions that correspond to those for a plane wall of thickness $2L$ and an infinite cylinder of radius r_o. These solutions are available from Figures 5.7 and 5.8 for the plane wall and Figures 5.10 and 5.11 for the infinite cylinder. They are also available from the one-term approximations given by Equations 5.28 and 5.37.

Results for other multidimensional geometries are summarized in Figure 5.19. In each case the multidimensional solution is prescribed in terms of a product involving one or more of the following one-dimensional solutions.

$$S(x,t)\equiv\left.\frac{T(x,t)-T_\infty}{T_i-T_\infty}\right|_{\substack{\text{Semiinfinite}\\\text{solid}}} \tag{5.49}$$

$$P(x,t)\equiv\left.\frac{T(x,t)-T_\infty}{T_i-T_\infty}\right|_{\substack{\text{Plane}\\\text{wall}}} \tag{5.50}$$

$$C(r,t)\equiv\left.\frac{T(r,t)-T_\infty}{T_i-T_\infty}\right|_{\substack{\text{Infinite}\\\text{cylinder}}} \tag{5.51}$$

The x coordinate for the semiinfinite solid is measured from the surface, whereas for the plane wall it is measured from the midplane. In using Figure 5.19 the coordinate origins should be carefully noted. The transient, three-dimensional temperature distribution in a rectangular parallelepiped, Figure 5.19h, is then, for example, the product of three one-dimensional solutions for plane walls of thicknesses $2L_1$, $2L_2$, and $2L_3$. That is,

$$\frac{T(x_1,x_2,x_3,t)-T_\infty}{T_i-T_\infty}=P(x_1,t)\cdot P(x_2,t)\cdot P(x_3,t)$$

The distances x_1, x_2, and x_3 are all measured with respect to a rectangular coordinate system whose origin is at the center of the parallelepiped.

The amount of energy Q transferred to or from a solid during a multidimensional transient conduction process may also be determined by combining one-dimensional results, as shown by Langston [7].

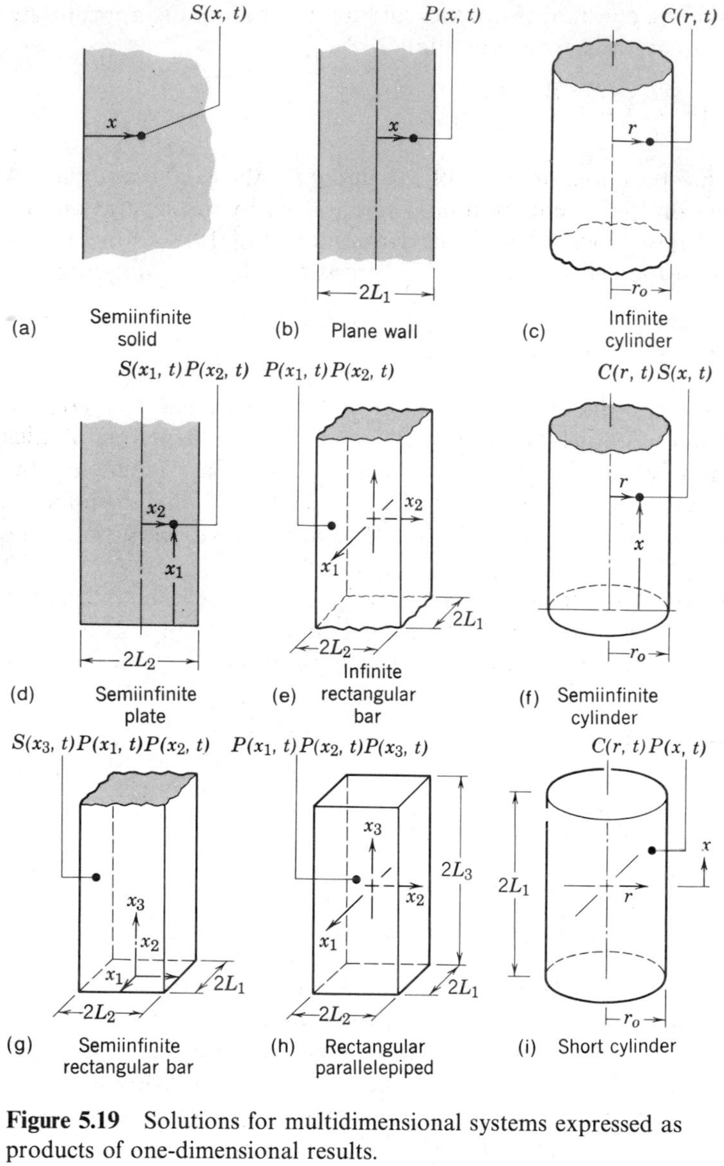

Figure 5.19 Solutions for multidimensional systems expressed as products of one-dimensional results.

EXAMPLE 5.5

In a manufacturing process stainless steel cylinders (AISI 304) initially at 600 K are quenched by submersion in an oil bath maintained at 300 K with $h = 500\ \text{W/m}^2 \cdot \text{K}$. Each cylinder is of length $2L = 60$ mm and diameter $D = 80$ mm. Consider a time 3 min into the cooling process and determine temperatures at the center of the cylinder, at the center of a circular face, and at the midplane of the side.

SOLUTION

Known:

Initial temperature and dimensions of cylinder and temperature and convection conditions of an oil bath.

Find:

Temperatures $T(r, x, t)$ after 3 min at the cylinder center, $T(0, 0, 3 \text{ min})$, at the center of a circular face, $T(0, L, 3 \text{ min})$, and at the midplane of the side, $T(r_o, 0, 3 \text{ min})$.

Schematic:

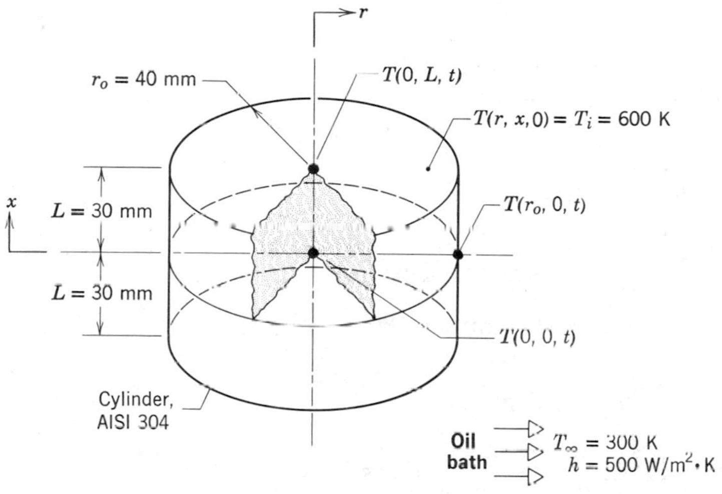

Assumptions:

1. Two-dimensional conduction in r and x.
2. Constant properties.

Properties:

Table A.1, stainless steel, AISI 304 $[T = (600 + 300)/2 = 450 \text{ K}]$: $\rho = 7900$ kg/m^3, $c = 526$ J/kg·K, $k = 17.4$ W/m·K, $\alpha = k/\rho c = 4.19 \times 10^{-6}$ m^2/s.

Analysis:

The solid steel cylinder corresponds to case i of Figure 5.19, and the temperature at any point in the cylinder may be expressed as the following product of one-dimensional solutions.

$$\frac{T(r,x,t)-T_\infty}{T_i-T_\infty}=P(x,t)C(r,t)$$

where $P(x,t)$ and $C(r,t)$ are defined by Equations 5.50 and 5.51, respectively. Accordingly, for the center of the cylinder,

$$\frac{T(0,0,3\text{ min})-T_\infty}{T_i-T_\infty}=\left.\frac{T(0,3\text{ min})-T_\infty}{T_i-T_\infty}\right|_{\substack{\text{Plane}\\\text{wall}}}\cdot\left.\frac{T(0,3\text{ min})-T_\infty}{T_i-T_\infty}\right|_{\substack{\text{Infinite}\\\text{cylinder}}}$$

Hence, for the plane wall, with

$$Bi^{-1}=\frac{k}{hL}=\frac{17.4\text{ W/m}\cdot\text{K}}{500\text{ W/m}^2\cdot\text{K}\times 0.03\text{ m}}=1.16$$

$$Fo=\frac{\alpha t}{L^2}=\frac{4.19\times 10^{-6}\text{ m}^2\text{/s}\times 180\text{ s}}{(0.03\text{ m})^2}=0.84$$

it follows from Figure 5.7 that

$$\frac{\theta_o}{\theta_i}=\left.\frac{T(0,3\text{ min})-T_\infty}{T_i-T_\infty}\right|_{\substack{\text{Plane}\\\text{wall}}}\approx 0.64$$

Similarly, for the infinite cylinder, with

$$Bi^{-1}=\frac{k}{hr_o}=\frac{17.4\text{ W/m}\cdot\text{K}}{500\text{ W/m}^2\cdot\text{K}\times 0.04\text{ m}}=0.87$$

$$Fo=\frac{\alpha t}{r_o^2}=\frac{4.19\times 10^{-6}\text{ m}^2\text{/s}\times 180\text{ s}}{(0.04\text{ m})^2}=0.47$$

it follows from Figure 5.10 that

$$\frac{\theta_o}{\theta_i}=\left.\frac{T(0,3\text{ min})-T_\infty}{T_i-T_\infty}\right|_{\substack{\text{Infinite}\\\text{cylinder}}}\approx 0.55$$

Hence, for the center of the cylinder,

$$\frac{T(0,0,3\text{ min})-T_\infty}{T_i-T_\infty}\approx 0.64\times 0.55\approx 0.35$$

$$T(0,0,3\text{ min})\approx 300\text{ K}+0.35(600-300)\text{ K}\approx 405\text{ K}$$ ◁

The temperature at the center of a circular face may be obtained from the requirement that

$$\frac{T(0,L,3\text{ min})-T_\infty}{T_i-T_\infty}=\left.\frac{T(L,3\text{ min})-T_\infty}{T_i-T_\infty}\right|_{\substack{\text{Plane}\\\text{wall}}}\cdot\left.\frac{T(0,3\text{ min})-T_\infty}{T_i-T_\infty}\right|_{\substack{\text{Infinite}\\\text{cylinder}}}$$

where, from Figure 5.8 with $(x/L) = 1$ and $Bi^{-1} = 1.16$,

$$\frac{\theta(L)}{\theta_o} = \left.\frac{T(L, 3\text{ min}) - T_\infty}{T(0, 3\text{ min}) - T_\infty}\right|_{\substack{\text{Plane}\\ \text{wall}}} \approx 0.68$$

Hence

$$\left.\frac{T(L, 3\text{ min}) - T_\infty}{T_i - T_\infty}\right|_{\substack{\text{Plane}\\ \text{wall}}} = \left.\frac{T(L, 3\text{ min}) - T_\infty}{T(0, 3\text{ min}) - T_\infty}\right|_{\substack{\text{Plane}\\ \text{wall}}} \cdot \left.\frac{T(0, 3\text{ min}) - T_\infty}{T_i - T_\infty}\right|_{\substack{\text{Plane}\\ \text{wall}}}$$

$$\left.\frac{T(L, 3\text{ min}) - T_\infty}{T_i - T_\infty}\right|_{\substack{\text{Plane}\\ \text{wall}}} \approx 0.68 \times 0.64 \approx 0.44$$

Hence

$$\frac{T(0, L, 3\text{ min}) - T_\infty}{T_i - T_\infty} \approx 0.44 \times 0.55 \approx 0.24$$

$$T(0, L, 3\text{ min}) \approx 300\text{ K} + 0.24(600 - 300)\text{ K} \approx 372\text{ K} \qquad \triangleleft$$

The temperature at the midplane of the side may be obtained from the requirement that

$$\frac{T(r_o, 0, 3\text{ min}) - T_\infty}{T_i - T_\infty} = \left.\frac{T(0, 3\text{ min}) - T_\infty}{T_i - T_\infty}\right|_{\substack{\text{Plane}\\ \text{wall}}} \cdot \left.\frac{T(r_o, 3\text{ min}) - T_\infty}{T_i - T_\infty}\right|_{\substack{\text{Infinite}\\ \text{cylinder}}}$$

where, from Figure 5.11 with $(r/r_o) = 1$ and $Bi^{-1} = 0.87$,

$$\frac{\theta(r_o)}{\theta_o} = \left.\frac{T(r_o, 3\text{ min}) - T_\infty}{T(0, 3\text{ min}) - T_\infty}\right|_{\substack{\text{Infinite}\\ \text{cylinder}}} \approx 0.61$$

Hence

$$\left.\frac{T(r_o, 3\text{ min}) - T_\infty}{T_i - T_\infty}\right|_{\substack{\text{Infinite}\\ \text{cylinder}}} = \left.\frac{T(r_o, 3\text{ min}) - T_\infty}{T(0, 3\text{ min}) - T_\infty}\right|_{\substack{\text{Infinite}\\ \text{cylinder}}} \cdot \left.\frac{T(0, 3\text{ min}) - T_\infty}{T_i - T_\infty}\right|_{\substack{\text{Infinite}\\ \text{cylinder}}}$$

$$\left.\frac{T(r_o, 3\text{ min}) - T_\infty}{T_i - T_\infty}\right|_{\substack{\text{Infinite}\\ \text{cylinder}}} \approx 0.61 \times 0.55 \approx 0.34$$

Hence

$$\frac{T(r_o, 0, 3\text{ min}) - T_\infty}{T_i - T_\infty} \approx 0.64 \times 0.34 \approx 0.22$$

$$T(r_o, 0, 3\text{ min}) \approx 300\text{ K} + 0.22(600 - 300)\text{ K} \approx 366\text{ K} \qquad \triangleleft$$

Comments:

1. Verify that the temperature at the edge of the cylinder is $T(r_o, L, 3\text{ min}) \approx 345$ K.
2. Since $Fo > 0.2$, the one-term approximations can be used to calculate the dimensionless temperatures read from the Heisler charts. For the *plane*

wall, the midplane temperature can be determined from Equation 5.29

$$\theta_o^* = \frac{\theta_o}{\theta_i} = C_1 \exp(-\zeta_1^2 Fo)$$

where, with $Bi = 0.862$, $C_1 = 1.109$ and $\zeta_1 = 0.814$ rad from Table 5.1. With $Fo = 0.84$,

$$\left.\frac{\theta_o}{\theta_i}\right|_{\substack{\text{Plane}\\ \text{wall}}} = 1.109 \exp[-(0.814 \text{ rad})^2 \times 0.84] = 0.636$$

The surface temperature can be evaluated using Equation 5.28b

$$\frac{\theta^*}{\theta_o^*} = \frac{\theta}{\theta_o} = \cos(\zeta_1 x^*)$$

with $x^* = 1$ to give

$$\frac{\theta^*(1, Fo)}{\theta_o^*} = \frac{\theta(L, t)}{\theta_o} = \cos(0.814 \text{ rad} \times 1) = 0.687$$

For the *infinite cylinder*, the centerline temperature can be determined from Equation 5.37c.

$$\theta_o^* = \frac{\theta_o}{\theta_i} = C_1 \exp(-\zeta_1^2 Fo)$$

where, with $Bi = 1.15$, $C_1 = 1.227$ and $\zeta_1 = 1.307$ from Table 5.1. With $Fo = 0.47$,

$$\left.\frac{\theta_o}{\theta_i}\right|_{\substack{\text{Infinite}\\ \text{cylinder}}} = 1.109 \exp[-(1.307 \text{ rad})^2 \times 0.47] = 0.550$$

The surface temperature can be evaluated using Equation 5.37b

$$\frac{\theta^*}{\theta_o^*} = \frac{\theta}{\theta_o} = J_0(\zeta_1 r^*)$$

with $r^* = 1$ and the value of the Bessel function determined from Table B.4,

$$\frac{\theta^*(1, Fo)}{\theta_o^*} = \frac{\theta(L, t)}{\theta_o} = J_0(1.307 \text{ rad} \times 1) = 0.616$$

The one-term approximations are in good agreement with results from the Heisler charts.

5.8 FINITE-DIFFERENCE METHODS

Analytical solutions to transient problems are restricted to simple geometries and boundary conditions, such as those considered in the preceding sections. Extensive coverage of these and other solutions is treated in the literature [1–4].

However, in many cases the geometry and/or boundary conditions preclude the use of analytical techniques, and recourse must be made to *finite-difference* methods. Such methods, introduced in Section 4.4 for steady-state conditions, are readily extended to transient problems. In this section we consider *explicit* and *implicit* forms of finite-difference solutions to transient conduction problems. More detailed treatments, as well as related algorithms, may be found in the literature [8–10].

5.8.1 Discretization of the Heat Equation: The Explicit Method

Once again consider the two-dimensional system of Figure 4.5. Under transient conditions with constant properties and no internal generation, the appropriate form of the heat equation, Equation 2.15, is

$$\frac{1}{\alpha}\frac{\partial T}{\partial t} = \frac{\partial^2 T}{\partial x^2} + \frac{\partial^2 T}{\partial y^2} \tag{5.52}$$

To obtain the finite-difference form of this equation, we may use the *central-difference* approximations to the spatial derivatives prescribed by Equations 4.30 and 4.31. Once again the m and n subscripts may be used to designate the x and y locations of *discrete nodal points*. However, in addition to being discretized in space, the problem must be discretized in time. The integer p is introduced for this purpose, where

$$t = p\Delta t \tag{5.53}$$

and the finite-difference approximation to the time derivative in Equation 5.52 is expressed as

$$\left.\frac{\partial T}{\partial t}\right|_{m,n} \approx \frac{T^{p+1}_{m,n} - T^{p}_{m,n}}{\Delta t} \tag{5.54}$$

The superscript p is used to denote the time dependence of T, and the time derivative is expressed in terms of the difference in temperatures associated with the *new* $(p+1)$ and *previous* (p) times. Hence calculations must be performed at successive times separated by the interval Δt, and just as a finite-difference solution restricts temperature determination to discrete points in space, it also restricts it to discrete points in time.

If Equation 5.54 is substituted into Equation 5.52, the nature of the finite-difference solution will then depend on the specific time at which temperatures are evaluated in the finite-difference approximations to the spatial derivatives. In the *explicit method* of solution, these temperatures are evaluated at the *previous* (p) time. Hence Equation 5.54 is considered to be a *forward-difference* approximation to the time derivative. Evaluating terms on the right-hand side of Equations 4.30 and 4.31 at p and substituting into Equation 5.52, the explicit form of the finite-difference equation for the interior node m,n is

$$\frac{1}{\alpha}\frac{T_{m,n}^{p+1}-T_{m,n}^{p}}{\Delta t}=\frac{T_{m+1,n}^{p}+T_{m-1,n}^{p}-2T_{m,n}^{p}}{(\Delta x)^2}$$

$$+\frac{T_{m,n+1}^{p}+T_{m,n-1}^{p}-2T_{m,n}^{p}}{(\Delta y)^2} \tag{5.55}$$

Solving for the nodal temperature at the new $(p+1)$ time and assuming that $\Delta x=\Delta y$, it follows that

$$T_{m,n}^{p+1}=Fo(T_{m+1,n}^{p}+T_{m-1,n}^{p}+T_{m,n+1}^{p}+T_{m,n-1}^{p})$$

$$+(1-4Fo)T_{m,n}^{p} \tag{5.56}$$

where Fo is a finite-difference form of the Fourier number

$$Fo=\frac{\alpha\Delta t}{(\Delta x)^2} \tag{5.57}$$

If the system is one-dimensional in x, the explicit form of the finite-difference equation for an interior node m reduces to

$$T_{m}^{p+1}=Fo(T_{m+1}^{p}+T_{m-1}^{p})+(1-2Fo)T_{m}^{p} \tag{5.58}$$

Equations 5.56 and 5.58 are *explicit* because *unknown* nodal temperatures for the new time are determined exclusively by *known* nodal temperatures at the previous time. Hence calculation of the unknown temperatures is straightforward. Since the temperature of each interior node is known at $t=0(p=0)$ from prescribed initial conditions, the calculations begin at $t=\Delta t(p=1)$, where Equation 5.56 or 5.58 is applied to each interior node to determine its temperature. With temperatures known for $t=\Delta t$, the appropriate finite-difference equation is then applied at each node to determine its temperature at $t=2\Delta t(p=2)$. In this way, the transient temperature distribution is obtained by *marching out in time*, using intervals of Δt.

The accuracy of the finite-difference solution may be improved by decreasing the values of Δx and Δt. Of course, the number of interior nodal points that must be considered increases with decreasing Δx, and the number of time intervals required to carry the solution to a prescribed final time increases with decreasing Δt. Hence, the computation time increases with decreasing Δx and Δt. The choice of Δx is typically based on a compromise between accuracy and computational requirements. Once this selection has been made, however, the value of Δt may not be chosen independently. It is, instead, determined by *stability* requirements.

An undesirable feature of the explicit method is that it is not unconditionally *stable*. In a transient problem, the solution for the nodal temperatures should continuously approach final (steady-state) values with increasing time. However, with the explicit method, this solution may be characterized by numerically induced oscillations, which are physically impossible. The oscillations may become unstable, causing the solution to diverge from the actual steady-state conditions. To prevent such erroneous results, the prescribed value of Δt must be maintained below a certain limit, which depends on Δx and other parameters of the system. This dependence is termed a *stability criterion*, which may be obtained mathematically [8] or demonstrated from a thermodynamic argument (see Problem 5.47). For the problems of interest in this text, *the criterion is determined by requiring that the coefficient associated with the node of interest at the previous time is greater than or equal to zero.* In general, this is done by collecting all terms involving $T^p_{m,n}$ to obtain the form of the coefficient. This result is then used to obtain a limiting relation involving *Fo*, from which the maximum allowable value of Δt may be determined. For example, with Equations 5.56 and 5.58 already expressed in the desired form, it follows that the stability criterion for a one-dimensional interior node is $(1 - 2Fo) \geq 0$, or

$$Fo \leq \tfrac{1}{2} \tag{5.59}$$

1-D interior node

and for a two-dimensional node, it is $(1-4Fo) \geq 0$, or

$$Fo \leq \tfrac{1}{4} \tag{5.60}$$

For prescribed values of Δx and α, these criteria may be used to determine upper limits to the value of Δt.

Equations 5.56 and 5.58 may also be derived by applying the energy balance method of Section 4.4.3 to a control volume about the interior node. Accounting for changes in thermal energy storage, a general form of the energy balance equation may be expressed as

$$\dot{E}_{\text{in}} + \dot{E}_g = \dot{E}_{\text{st}} \tag{5.61}$$

In the interest of adopting a consistent methodology, it is again assumed that all heat flow is *into* the node.

To illustrate application of Equation 5.61, consider the surface node of the one-dimensional system shown in Figure 5.20. To more accurately determine thermal conditions near the surface, this node has been assigned a thickness which is one-half that of the interior nodes. Assuming convection transfer from

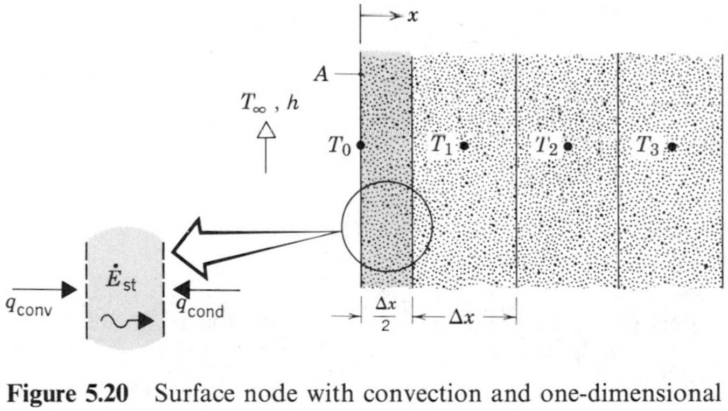

Figure 5.20 Surface node with convection and one-dimensional transient conduction.

Table 5.2 Summary of transient, two-dimensional finite-difference equations ($\Delta x = \Delta y$)

CONFIGURATION	EXPLICIT METHOD: FINITE-DIFFERENCE EQUATION		EXPLICIT METHOD: STABILITY CRITERION		IMPLICIT METHOD	
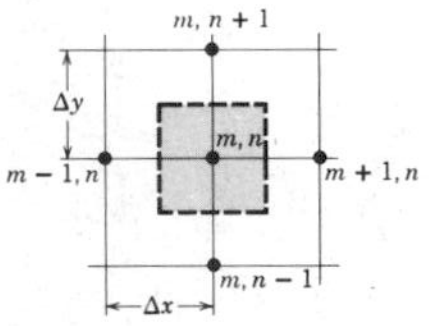	$T_{m,n}^{p+1} = Fo(T_{m+1,n}^{p} + T_{m-1,n}^{p} + T_{m,n+1}^{p} + T_{m,n-1}^{p})$ $+ (1 - 4Fo)T_{m,n}^{p}$ 1. Interior node	(5.56)	$Fo \leq \frac{1}{4}$	(5.60)	$(1 + 4Fo)T_{m,n}^{p+1} - Fo(T_{m+1,n}^{p+1} + T_{m-1,n}^{p+1}$ $+ T_{m,n+1}^{p+1} + T_{m,n-1}^{p+1}) = T_{m,n}^{p}$	(5.72)
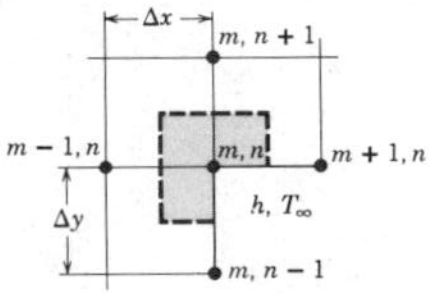	$T_{m,n}^{p+1} = \frac{2}{3} Fo(T_{m+1,n}^{p} + 2T_{m-1,n}^{p} + 2T_{m,n+1}^{p}$ $+ T_{m,n-1}^{p} + 2BiT_{\infty}) + (1 - 4Fo - \frac{4}{3} BiFo)\, T_{m,n}^{p}$ 2. Node at interior corner with convection	(5.65)	$Fo(3 + Bi) \leq \frac{3}{4}$	(5.66)	$(1 + 4Fo(1 + \frac{1}{3} Bi))\, T_{m,n}^{p+1} - \frac{2}{3} Fo$ $(T_{m+1,n}^{p+1} + 2T_{m-1,n}^{p+1} + 2T_{m,n+1}^{p+1}$ $+ T_{m,n-1}^{p+1}) = T_{m,n}^{p} + \frac{4}{3} BiFoT_{\infty}$	(5.75)
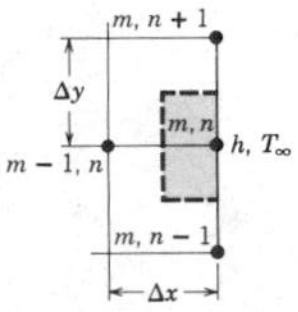	$T_{m,n}^{p+1} = Fo(2T_{m-1,n}^{p} + T_{m,n+1}^{p} + T_{m,n-1}^{p} + 2BiT_{\infty})$ $+ (1 - 4Fo - 2BiFo)T_{m,n}^{p}$ 3. Node at plane surface with convection[a]	(5.67)	$Fo(2 + Bi) \leq \frac{1}{2}$	(5.68)	$(1 + 2Fo(2 + Bi))T_{m,n}^{p+1} - Fo(2T_{m-1,n}^{p+1}$ $+ T_{m,n+1}^{p+1} + T_{m,n-1}^{p+1}) = T_{m,n}^{p} + 2BiFoT_{\infty}$	(5.76)
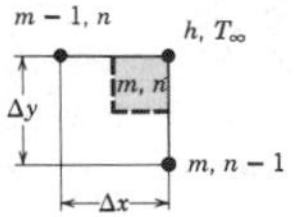	$T_{m,n}^{p+1} = 2Fo(T_{m-1,n}^{p} + T_{m,n-1}^{p} + 2BiT_{\infty})$ $+ (1 - 4Fo - 4BiFo)T_{m,n}^{p}$ 4. Node at exterior corner with convection	(5.69)	$Fo(1 + Bi) \leq \frac{1}{4}$	(5.70)	$(1 + 4Fo(1 + Bi))T_{m,n}^{p+1} - 2Fo(T_{m-1,n}^{p+1}$ $+ T_{m,n-1}^{p+1}) = T_{m,n}^{p} + 4BiFoT_{\infty}$	(5.77)

[a] To obtain the finite-difference equation and/or stability criterion for an adiabatic surface (or surface of symmetry), simply set Bi equal to zero.

an adjoining fluid and no generation, it follows from Equation 5.61 that

$$hA(T_\infty - T_0^p) + \frac{kA}{\Delta x}(T_1^p - T_0^p) = \rho c A \frac{\Delta x}{2} \frac{T_0^{p+1} - T_0^p}{\Delta t}$$

or, solving for the surface temperature at $t + \Delta t$,

$$T_0^{p+1} = \frac{2h\Delta t}{\rho c \Delta x}(T_\infty - T_0^p) + \frac{2\alpha \Delta t}{\Delta x^2}(T_1^p - T_0^p) + T_0^p$$

Recognizing that $(2h\Delta t/\rho c\Delta x) = 2(h\Delta x/k)(\alpha\Delta t/\Delta x^2) = 2BiFo$ and grouping terms involving T_0^p, it follows that

$$T_0^{p+1} = 2Fo(T_1^p + BiT_\infty) + (1 - 2Fo - 2BiFo)T_0^p \tag{5.62}$$

The finite-difference form of the Biot number is

$$Bi = \frac{h\Delta x}{k} \tag{5.63}$$

Recalling the procedure for determining the stability criterion, we require that the coefficient for T_0^p be greater than or equal to zero. Hence

$$1 - 2Fo - 2BiFo \geq 0$$

or

$$Fo(1 + Bi) \leq \tfrac{1}{2} \tag{5.64}$$

Since the complete finite-difference solution requires the use of Equation 5.58 for the interior nodes, as well as Equation 5.62 for the surface node, Equation 5.64 must be contrasted with Equation 5.59 to determine which requirement is the more stringent. Since $Bi \geq 0$, it is apparent that the limiting value of Fo for Equation 5.64 is less than that for Equation 5.59. To ensure stability for all nodes, Equation 5.64 should therefore be used to select the maximum allowable value of Fo, and hence Δt, to be used in the calculations.

Forms of the explicit finite-difference equation for several common geometries are presented in Table 5.2. Each equation may be derived by applying the energy balance method to a control volume about the corresponding node. To develop confidence in your ability to apply this method, you should attempt to verify at least one of these equations.

EXAMPLE 5.6

A fuel element of a nuclear reactor is in the shape of a plane wall of thickness $2L = 20$ mm and is convectively cooled at both surfaces, with $h = 1100\ \text{W/m}^2\cdot\text{K}$ and $T_\infty = 250°\text{C}$. At normal operating power, heat is generated uniformly within the element at a volumetric rate of $\dot{q}_1 = 10^7\ \text{W/m}^3$. A departure from the steady-state conditions associated with normal operation will occur if there is a change in the generation rate. Consider a sudden change to $\dot{q}_2 = 2 \times 10^7$ W/m^3, and use the explicit finite-difference method to determine the fuel element temperature distribution after 1.5 s. The fuel element thermal properties are $k = 30$ W/m · K and $\alpha = 5 \times 10^{-6}$ m^2/s.

SOLUTION

Known:

Conditions associated with heat generation in a rectangular fuel element with surface cooling.

Find:

Temperature distribution 1.5 s after a sudden change in operating power.

Schematic:

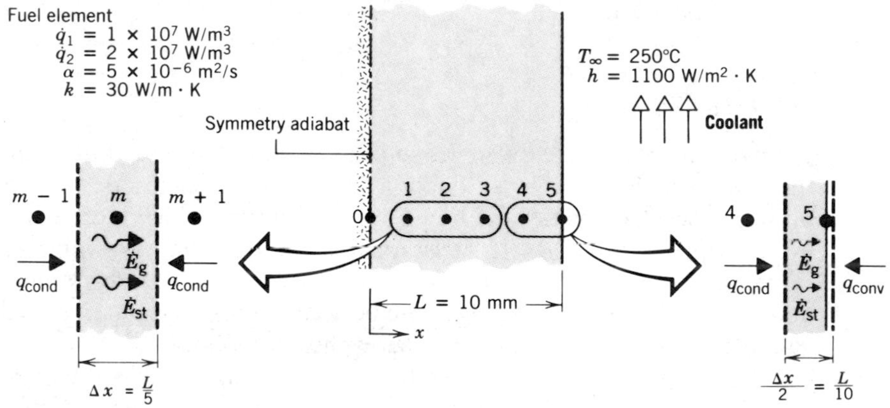

Assumptions:

1. One-dimensional conduction in x.
2. Uniform generation.
3. Constant properties.

Analysis:

A numerical solution will be obtained using a space increment of $\Delta x = 2$ mm. Since there is symmetry about the midplane, the nodal network yields six unknown nodal temperatures. Using the energy balance method, Equation

5.61, an explicit finite-difference equation may be derived for any interior node m.

$$kA\frac{T_{m-1}^p - T_m^p}{\Delta x} + kA\frac{T_{m+1}^p - T_m^p}{\Delta x} + \dot{q}A\Delta x$$

$$= \rho A \Delta x c \frac{T_m^{p+1} - T_m^p}{\Delta t}$$

Solving for T_m^{p+1} and rearranging,

$$T_m^{p+1} = Fo\left[T_{m-1}^p + T_{m+1}^p + \frac{\dot{q}(\Delta x)^2}{k}\right] + (1 - 2Fo)T_m^p \qquad (1)$$

This equation may be used for node 0, with $T_{m-1}^p = T_{m+1}^p$, as well as for nodes 1, 2, 3, and 4. Applying energy conservation to a control volume about node 5,

$$hA(T_\infty - T_5^p) + kA\frac{T_4^p - T_5^p}{\Delta x} + \dot{q}A\frac{\Delta x}{2} = \rho A\frac{\Delta x}{2}c\frac{T_5^{p+1} - T_5^p}{\Delta t}$$

or

$$T_5^{p+1} = 2Fo\left[T_4^p + BiT_\infty + \frac{\dot{q}(\Delta x)^2}{2k}\right] + (1 - 2Fo - 2BiFo)T_5^p \qquad (2)$$

Since the most restrictive stability criterion is associated with Equation 2, we select Fo from the requirement that

$$Fo(1 + Bi) \leq \tfrac{1}{2}$$

Hence, with

$$Bi = \frac{h\Delta x}{k} = \frac{1100\ \text{W/m}^2\cdot\text{K}\ (0.002\ \text{m})}{30\ \text{W/m}\cdot\text{K}} = 0.0733$$

it follows that

$$Fo \leq 0.466$$

or

$$\Delta t = \frac{Fo(\Delta x)^2}{\alpha} \leq \frac{0.466(2\times 10^{-3}\ \text{m})^2}{5\times 10^{-6}\ \text{m}^2/\text{s}} \leq 0.373\ \text{s}$$

To be well within the stability limit, we select $\Delta t = 0.3$ s, which corresponds to

$$Fo = \frac{5\times 10^{-6}\ \text{m}^2/\text{s}(0.3\ \text{s})}{(2\times 10^{-3}\ \text{m})^2} = 0.375$$

Substituting numerical values, including $\dot{q} = \dot{q}_2 = 2\times 10^7\ \text{W/m}^3$, the nodal equations become

$$T_0^{p+1} = 0.375(2T_1^p + 2.67) + 0.250T_0^p$$

$$T_1^{p+1} = 0.375(T_0^p + T_2^p + 2.67) + 0.250T_1^p$$

$$T_2^{p+1} = 0.375(T_1^p + T_3^p + 2.67) + 0.250T_2^p$$

$$T_3^{p+1} = 0.375(T_2^p + T_4^p + 2.67) + 0.250T_3^p$$

$$T_4^{p+1} = 0.375(T_3^p + T_5^p + 2.67) + 0.250T_4^p$$

$$T_5^{p+1} = 0.750(T_4^p + 19.67) + 0.195T_5^p$$

To begin the marching solution, the initial temperature distribution must be known. This distribution is given by Equation 3.41, with $\dot{q} = \dot{q}_1$. Obtaining $T_s = T_5$ from Equation 3.45,

$$T_5 = T_\infty + \frac{\dot{q}L}{h} = 250°\text{C} + \frac{10^7 \text{ W/m}^3 \times 0.01 \text{ m}}{1100 \text{ W/m}^2 \cdot \text{K}} = 340.91°\text{C}$$ 3,45

it follows that

$$T(x) = 16.67\left(1 - \frac{x^2}{L^2}\right) + 340.91°\text{C}$$ 3,41

Computed temperatures for the nodal points of interest are shown in the first row of the accompanying table.

Using the finite-difference equations, the nodal temperatures may be sequentially calculated with a time increment of 0.3 s until the desired final time is reached. The results are illustrated in rows 2 through 6 of the table and may be contrasted with the new steady-state condition (row 7), which was obtained by using Equations 3.41 and 3.45 with $\dot{q} = \dot{q}_2$.

Tabulated nodal temperatures

p	t (s)	T_0	T_1	T_2	T_3	T_4	T_5
0	0	357.58	356.91	354.91	351.58	346.91	340.91
1	0.3	358.08	357.41	355.41	352.08	347.41	341.41
2	0.6	358.58	357.91	355.91	352.58	347.91	341.88
3	0.9	359.08	358.41	356.41	353.08	348.41	342.35
4	1.2	359.58	358.91	356.91	353.58	348.89	342.82
5	1.5	360.08	359.41	357.41	354.07	349.37	343.27
∞	∞	465.15	463.82	459.82	453.15	443.82	431.82

Comments:

It is evident that at 1.5 s, the wall is in the early stages of the transient process and that many additional calculations would have to be made to reach steady-state conditions with the finite-difference solution. The computation time could be slightly reduced by using the maximum allowable time increment ($\Delta t = 0.373$ s), but with some loss of accuracy.

5.8.2 Discretization of the Heat Equation: The Implicit Method

In the *explicit* finite-difference scheme, the temperature of any node at $t + \Delta t$ may be calculated from knowledge of temperatures at the same and neighboring nodes for the *preceding time t*. Hence, determination of a nodal temperature at some time is *independent* of temperatures at other nodes for the *same time*. Although the method offers computational convenience, it suffers from limitations on the selection of Δx and Δt. For a given space increment, the time interval must be compatible with stability requirements. Frequently, this dictates the use of extremely small values of Δt, and an inordinate number of time intervals may be necessary to obtain a solution.

A reduction in the amount of computation time may often be realized by employing an *implicit*, rather than explicit, finite-difference scheme. The implicit form of a finite-difference equation may be derived by using Equation 5.54 to approximate the time derivative, while evaluating all other temperatures at the *new* $(p+1)$ time, instead of the previous (p) time. Equation 5.54 is then considered to provide a *backward-difference* approximation to the time derivative. In contrast to Equation 5.55, the implicit form of the finite-difference equation for the interior node of a two-dimensional system is then

$$\frac{1}{\alpha}\frac{T_{m,n}^{p+1} - T_{m,n}^{p}}{\Delta t} = \frac{T_{m+1,n}^{p+1} + T_{m-1,n}^{p+1} - 2T_{m,n}^{p+1}}{(\Delta x)^2} + \frac{T_{m,n+1}^{p+1} + T_{m,n-1}^{p+1} - 2T_{m,n}^{p+1}}{(\Delta y)^2} \tag{5.71}$$

Rearranging and assuming $\Delta x = \Delta y$, it follows that

$$(1 + 4Fo)T_{m,n}^{p+1} - Fo(T_{m+1,n}^{p+1} + T_{m-1,n}^{p+1} + T_{m,n+1}^{p+1} + T_{m,n-1}^{p+1}) = T_{m,n}^{p} \tag{5.72}$$

From Equation 5.72 it is evident that the *new* temperature of the m, n node depends on the *new* temperatures of its adjoining nodes, which are, in general, unknown. Hence, to determine the unknown nodal temperatures at $t + \Delta t$, the corresponding nodal equations must be *solved simultaneously*. Such a solution may be effected by using Gauss-Seidel iteration or matrix inversion, as discussed in Section 4.5. The *marching solution* would then involve simultaneously solving the nodal equations at each time $t = \Delta t, 2\Delta t, \ldots,$ until the desired final time was reached.

Although computations involving the implicit method are more complicated than those of the explicit method, the implicit formulation has the important advantage of being *unconditionally stable*. That is, the solution remains stable for all space and time intervals, in which case there are no restrictions on Δx and Δt. Since larger values of Δt may therefore be used with an implicit method, computation times may often be reduced, with little loss of accuracy.

The implicit form of a finite-difference equation may also be derived from the energy balance method. For the surface node of Figure 5.20, it is readily shown that

$$(1 + 2Fo + 2FoBi)T_0^{p+1} - 2FoT_1^{p+1} = 2FoBiT_\infty + T_0^p \tag{5.73}$$

For any interior node of Figure 5.20, it may also be shown that

$$(1 + 2Fo)T_m^{p+1} - Fo(T_{m-1}^{p+1} + T_{m+1}^{p+1}) = T_m^p \tag{5.74}$$

Forms of the implicit finite-difference equation for other common geometries are presented in Table 5.2. Each equation may be derived by applying the energy balance method.

EXAMPLE 5.7

A thick slab of copper initially at a uniform temperature of 20°C is suddenly exposed to radiation at one surface such that the net heat flux is maintained at a constant value of 3×10^5 W/m². Using the explicit and implicit finite-difference techniques with a space increment of $\Delta x = 75$ mm, determine the temperature at the irradiated surface and at an interior point that is 150 mm from the surface after 2 min have elapsed. Compare the results with those obtained from an appropriate analytical solution.

SOLUTION

Known:

Thick slab of copper, initially at a uniform temperature, is subjected to a constant net heat flux at one surface.

Find:

1. Using the explicit finite-difference method, determine temperatures at the surface and 150 mm from the surface after an elapsed time of 2 min.
2. Repeat the calculations using the implicit finite-difference method.
3. Determine the same temperatures analytically.

Schematic:

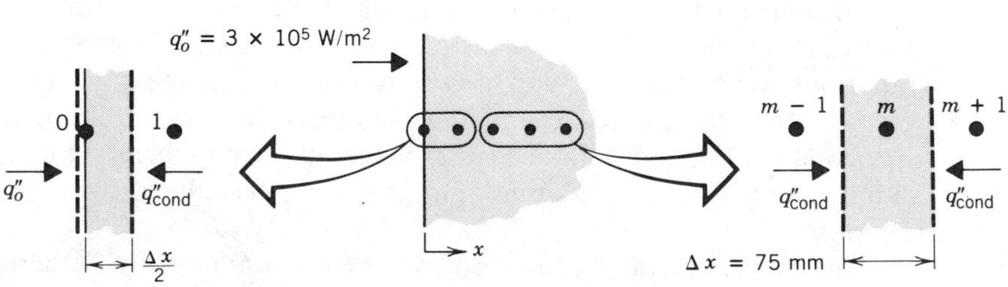

Assumptions:

1. One-dimensional conduction in x.
2. Thick slab may be approximated as a semiinfinite medium with constant surface heat flux.
3. Constant properties.

Properties:

Table A.1, pure copper (300 K): $k = 401$ W/m·K, $\alpha = 117 \times 10^{-6}$ m²/s.

Analysis:

1. An explicit form of the finite-difference equation for the surface node may be obtained by applying an energy balance to a control volume about the node.

$$q_o''A + kA\frac{T_1^p - T_0^p}{\Delta x} = \rho A\frac{\Delta x}{2}c\frac{T_0^{p+1} - T_0^p}{\Delta t}$$

or

$$T_0^{p+1} = 2Fo\left(\frac{q_o''\Delta x}{k} + T_1^p\right) + (1 - 2Fo)T_0^p$$

The finite-difference equation for any interior node is given by Equation 5.58. Both the surface and interior nodes are governed by the stability criterion

$$Fo \leq \tfrac{1}{2}$$

Noting that the finite-difference equations are simplified by choosing the maximum allowable value of Fo, we select $Fo = \frac{1}{2}$. Hence

$$\Delta t = Fo\frac{(\Delta x)^2}{\alpha} = \frac{1}{2}\frac{(0.075 \text{ m})^2}{117 \times 10^{-6} \text{ m}^2/\text{s}} = 24 \text{ s}$$

With

$$\frac{q_o''\Delta x}{k} = \frac{3 \times 10^5 \text{ W/m}^2 (0.075 \text{ m})}{401 \text{ W/m} \cdot \text{K}} = 56.1^\circ\text{C}$$

the finite-difference equations become

$$T_0^{p+1} = 56.1^\circ\text{C} + T_1^p \qquad \text{and} \qquad T_m^{p+1} = \frac{T_{m+1}^p + T_{m-1}^p}{2}$$

for the surface and interior nodes, respectively. Performing the calculations, the results are tabulated as follows.

Explicit finite-difference solution for $Fo = \frac{1}{2}$

p	t(s)	T_0	T_1	T_2	T_3	T_4
0	0	20	20	20	20	20
1	24	76.1	20	20	20	20
2	48	76.1	48.1	20	20	20
3	72	104.2	48.1	34.1	20	20
4	96	104.2	69.1	34.1	27.1	20
5	120	125.3	69.1	48.1	27.1	20

After 2 min, the surface temperature and the desired interior temperature are $T_0 = 125.3°C$ and $T_2 = 48.1°C$.

Note that calculation of identical temperatures at successive times for the same node is an idiosyncrasy of using the maximum allowable value of *Fo* with the explicit finite-difference technique. The actual physical condition is, of course, one in which the temperature changes continuously with time. The idiosyncrasy is eliminated and the accuracy of the calculations is improved by reducing the value of *Fo*.

To determine the extent to which the accuracy may be improved by reducing *Fo*, let us redo the calculations for $Fo = \frac{1}{4}(\Delta t = 12 \text{ s})$. The finite-difference equations are then of the form

$$T_0^{p+1} = \tfrac{1}{2}(56.1°\text{C} + T_1^p) + \tfrac{1}{2}T_0^p$$

$$T_m^{p+1} = \tfrac{1}{4}(T_{m+1}^p + T_{m-1}^p) + \tfrac{1}{2}T_m^p$$

and the results of the calculations are tabulated as follows.

Explicit finite-difference solution for $Fo = 1/4$

p	t (s)	T_0	T_1	T_2	T_3	T_4	T_5	T_6	T_7	T_8
0	0	20	20	20	20	20	20	20	20	20
1	12	48.1	20	20	20	20	20	20	20	20
2	24	62.1	27.0	20	20	20	20	20	20	20
3	36	72.6	34.0	21.8	20	20	20	20	20	20
4	48	81.4	40.6	24.4	20.4	20	20	20	20	20
5	60	89.0	46.7	27.5	21.3	20.1	20	20	20	20
6	72	95.9	52.5	30.7	22.6	20.4	20.0	20	20	20
7	84	102.3	57.9	34.1	24.1	20.8	20.1	20.0	20	20
8	96	108.1	63.1	37.6	25.8	21.5	20.3	20.0	20.0	20
9	108	113.7	68.0	41.0	27.6	22.2	20.5	20.1	20.0	20.0
10	120	118.9	72.6	44.4	29.6	23.2	20.8	20.2	20.0	20.0

After 2 min, the desired temperatures are $T_0 = 118.9°C$ and $T_2 = 44.4°C$. Comparing the above results with those obtained for $Fo = \frac{1}{2}$, it is clear that by reducing *Fo* we have eliminated the problem of recurring temperatures. We have also predicted greater thermal penetration (to node 6 instead of node 3). An assessment of the improvement in accuracy must await a comparison with results based on an exact solution.

2. Performing an energy balance on a control volume about the surface node, the implicit form of the finite-difference equation is

$$q''_o + k\frac{T_1^{p+1} - T_0^{p+1}}{\Delta x} = \rho\frac{\Delta x}{2}c\frac{T_0^{p+1} - T_0^p}{\Delta t}$$

or,

$$(1 + 2Fo)T_0^{p+1} - 2FoT_1^{p+1} = \frac{2\alpha q''_o\Delta t}{k\Delta x} + T_0^p$$

Arbitrarily choosing $Fo = \frac{1}{2}(\Delta t = 24 \text{ s})$, it follows that

$$2T_0^{p+1} - T_1^{p+1} = 56.1 + T_0^p$$

From Equation 5.74, the finite-difference equation for any interior node is then of the form

$$-T_{m-1}^{p+1} + 4T_m^{p+1} - T_{m+1}^{p+1} = 2T_m^p$$

Since we are dealing with a semiinfinite solid, the number of nodes is, in principle, infinite. In practice, however, the number may be limited to the nodes that are affected by the change in the boundary condition for the time period of interest. From the results of the explicit method, it is evident that we are safe in choosing nine nodes corresponding to T_0, T_1, ..., T_8. We are thereby assuming that, at $t = 120$ s, there has been no change in T_8.

We now have a set of nine equations that must be solved simultaneously for each time increment. Using the matrix inversion method, we express the equations in the form $[A][T] = [C]$, where

$$[A] = \begin{bmatrix} 2 & -1 & 0 & 0 & 0 & 0 & 0 & 0 & 0 \\ -1 & 4 & -1 & 0 & 0 & 0 & 0 & 0 & 0 \\ 0 & -1 & 4 & -1 & 0 & 0 & 0 & 0 & 0 \\ 0 & 0 & -1 & 4 & -1 & 0 & 0 & 0 & 0 \\ 0 & 0 & 0 & -1 & 4 & -1 & 0 & 0 & 0 \\ 0 & 0 & 0 & 0 & -1 & 4 & -1 & 0 & 0 \\ 0 & 0 & 0 & 0 & 0 & -1 & 4 & -1 & 0 \\ 0 & 0 & 0 & 0 & 0 & 0 & -1 & 4 & -1 \\ 0 & 0 & 0 & 0 & 0 & 0 & 0 & -1 & 4 \end{bmatrix}$$

$$[C] = \begin{bmatrix} 56.1 + T_0^p \\ 2T_1^p \\ 2T_2^p \\ 2T_3^p \\ 2T_4^p \\ 2T_5^p \\ 2T_6^p \\ 2T_7^p \\ 2T_8^p + T_9^{p+1} \end{bmatrix}$$

Note that numerical values for the components of $[C]$ are determined from previous values of the nodal temperatures. Note also how the finite-difference equation for node 8 appears in matrices $[A]$ and $[C]$.

A table of nodal temperatures may be compiled, beginning with the first row ($p = 0$) corresponding to the prescribed initial condition. To obtain nodal temperatures for subsequent times, the inverse of the coefficient matrix $[A]^{-1}$ must first be found. At each time $p + 1$, it is then multiplied by the column vector $[C]$, which is evaluated at p, to obtain the temperatures $T_0^{p+1}, T_1^{p+1}, \ldots, T_8^{p+1}$. For example, multiplying $[A]^{-1}$ by the column vector corresponding to $p = 0$,

$$[C]_{p=0} = \begin{bmatrix} 76.1 \\ 40 \\ 40 \\ 40 \\ 40 \\ 40 \\ 40 \\ 40 \\ 60 \end{bmatrix}$$

the second row of the table is obtained. Updating $[C]$, the process is repeated four more times to determine the nodal temperatures at 120 s. The desired temperatures are $T_0 = 114.7°\text{C}$ and $T_2 = 44.2°\text{C}$.

Implicit finite-difference solution for $Fo = \frac{1}{2}$

p	t (s)	T_0	T_1	T_2	T_3	T_4	T_5	T_6	T_7	T_8
0	0	20.0	20.0	20.0	20.0	20.0	20.0	20.0	20.0	20.0
1	24	52.4	28.7	22.3	20.6	20.2	20.0	20.0	20.0	20.0
2	48	74.0	39.5	26.6	22.1	20.7	20.2	20.1	20.0	20.0
3	72	90.2	50.3	32.0	24.4	21.6	20.6	20.2	20.1	20.0
4	96	103.4	60.5	38.0	27.4	22.9	21.1	20.4	20.2	20.1
5	120	114.7	70.0	44.2	30.9	24.7	21.9	20.8	20.3	20.1

3. Approximating the slab as a semiinfinite medium, the appropriate analytical expression is given by Equation 5.46, which may be applied to any point in the slab.

$$T(x,t) - T_i = \frac{2q_o''(\alpha t/\pi)^{1/2}}{k} \exp\left(-\frac{x^2}{4\alpha t}\right) - \frac{q_o'' x}{k} \operatorname{erfc}\left(\frac{x}{2\sqrt{\alpha t}}\right)$$

At the surface, this expression yields

$$T(0, 120\,\text{s}) - 20°\text{C} = \frac{2 \times 3 \times 10^5\ \text{W/m}^2}{401\ \text{W/m} \cdot \text{K}} (117 \times 10^{-6}\ \text{m}^2/\text{s} \times 120\ \text{s}/\pi)^{1/2}$$

or

$T(0, 120 \text{ s}) = 120.0°\text{C}$ ◁

At the interior point ($x = 0.15$ m)

$$T(0.15 \text{ m}, 120 \text{ s}) - 20°\text{C} = \frac{2 \times 3 \times 10^5 \text{ W/m}^2}{401 \text{ W/m}\cdot\text{K}}$$

$$\times (117 \times 10^{-6} \text{ m}^2/\text{s} \times 120 \text{ s}/\pi)^{1/2}$$

$$\times \exp\left[-\frac{(0.15 \text{ m})^2}{4 \times 117 \times 10^{-6} \text{ m}^2/\text{s} \times 120 \text{ s}}\right] - \frac{3 \times 10^5 \text{ W/m}^2 \times 0.15 \text{ m}}{401 \text{ W/m}\cdot\text{K}}$$

$$\times \left[1 - \text{erf}\left(\frac{0.15 \text{ m}}{2\sqrt{117 \times 10^{-6} \text{ m}^2/\text{s} \times 120 \text{ s}}}\right)\right]$$

or

$T(0.15 \text{ m}, 120 \text{ s}) = 45.4°\text{C}$ ◁

Comments:

1. Comparing the exact results with those obtained from the three approximate solutions, it is clear that the explicit method with $Fo = 1/4$ provides the most accurate predictions.

METHOD	$T_0 = T(0, 120 \text{ s})$	$T_2 = T(0.15 \text{ m}, 120 \text{ s})$
Explicit ($Fo = \frac{1}{2}$)	125.3	48.1
Explicit ($Fo = \frac{1}{4}$)	118.9	44.4
Implicit ($Fo = \frac{1}{2}$)	114.7	44.2
Exact	120.0	45.4

   This is not unexpected, since the corresponding value of Δt is 50% smaller than that used in the other two methods.

2. Although computations are simplified by using the maximum allowable value of Fo in the explicit method, the accuracy of the results is seldom satisfactory.
3. Note that the coefficient matrix $[A]$ is *tridiagonal*. That is, all elements are zero except those which are on, or to either side of, the main diagonal. Tridiagonal matrices are associated with one-dimensional conduction problems. In such cases the problem of solving for the unknown temperatures is greatly simplified, and stock computer programs may readily be obtained for this purpose.
4. A more general radiative heating condition would be one in which the surface is suddenly exposed to large surroundings at an elevated temperature T_{sur} (Problem 5.62). The net rate at which radiation is transferred to the surface may then be calculated from Equation 1.6.

Allowing for convection heat transfer to the surface, application of conservation of energy to the surface node yields an explicit finite-difference equation of the form

$$\varepsilon\sigma[T_{sur}^4 - (T_0^p)^4] + h(T_\infty - T_0^p) + k\frac{T_1^p - T_0^p}{\Delta x} = \rho\frac{\Delta x}{2}c\frac{T_0^{p+1} - T_0^p}{\Delta t}$$

Use of this finite-difference equation in a numerical solution is complicated by the fact that it is *nonlinear*. However, the equation may be *linearized* by introducing the radiation heat transfer coefficient h_r defined by Equation 1.8, and the finite-difference equation is

$$h_r^p(T_{sur} - T_0^p) + h(T_\infty - T_0^p) + k\frac{T_1^p - T_0^p}{\Delta x} = \rho\frac{\Delta x}{2}c\frac{T_0^{p+1} - T_0^p}{\Delta t}$$

The solution may proceed in the usual manner, although the effect of a radiative Biot number ($Bi_r \equiv h_r\Delta x/k$) must be included in the stability criterion and the value of h_r must be updated at each step in the calculations. If the implicit method is used, h_r is calculated at $p + 1$, in which case an iterative calculation must be made at each time step.

5.9 SUMMARY

Transient conduction problems are to be found in numerous engineering applications, and it is important to appreciate the different methods for dealing with such problems. There is certainly much to be said for simplicity, in which case, when confronted with a transient problem, the first thing you should do is calculate the Biot number. If this number is much less than unity, you may use the lumped capacitance method to obtain accurate results with minimal computational requirements. However, if the Biot number is not much less than unity, spatial effects must be considered, and some other method must be used. Analytical results are available in convenient graphical form for the plane wall, the infinite cylinder, and the sphere, as well as in graphical and equation form for the semiinfinite solid. You should know when and how to use these results. If geometrical complexities and/or the form of the boundary conditions preclude their use, recourse must be made to finite-difference methods. With the digital computer, such methods may be used to solve any conduction problem, regardless of complexity.

REFERENCES

1. Carslaw, H. S., and J. C. Jaeger, *Conduction of Heat in Solids*, 2nd ed., Oxford University Press, London, 1959.
2. Schneider, P. J., *Conduction Heat Transfer*, Addison-Wesley, Reading, MA, 1955.

3. Arpaci, V. S., *Conduction Heat Transfer*, Addison-Wesley, Reading, MA, 1966.
4. Özisik, M. N., *Heat Conduction*, Wiley, New York, 1980.
5. Heisler, M. P., "Temperature Charts for Induction and Constant Temperature Heating," *Trans. ASME*, *69*, 227–236, 1947.
6. Gröber, H., S. Erk, and U. Grigull, *Fundamentals of Heat Transfer*, McGraw-Hill, New York, 1961.
7. Langston, L. S., "Heat Transfer from Multidimensional Objects Using One-Dimensional Solutions for Heat Loss," *Int. J. Heat Mass Transfer*, *25*, 149–150, 1982.
8. Richtmyer, R. D., and K. W. Morton, *Difference Methods for Initial Value Problems*, Wiley-Interscience, New York, 1967.
9. Myers, G. E., *Analytical Methods in Conduction Heat Transfer*, McGraw-Hill, New York, 1971.
10. Adams, J. A., and D. F. Rogers, *Computer-Aided Heat Transfer Analysis*, McGraw-Hill, New York, 1973.

PROBLEMS

5.1 Consider a thin electrical heater attached to a plate and backed by insulation. Initially, the heater and plate are at the temperature of the ambient air, T_∞. Suddenly, the power to the heater is switched on giving rise to a constant heat flux q''_o (W/m²) at the inner surface of the plate.

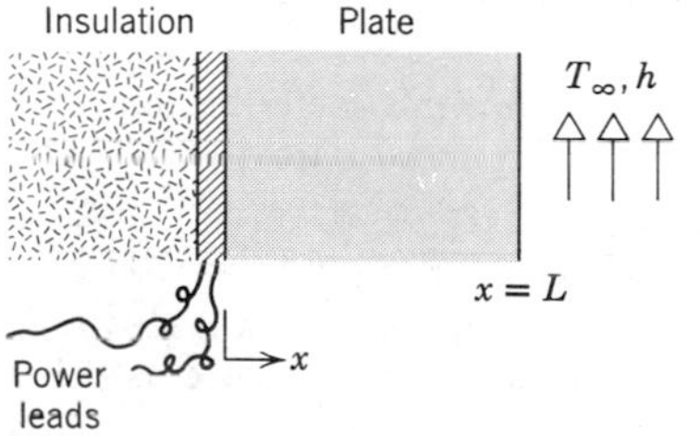

(a) Sketch and label, on T–x coordinates, the temperature distributions: initial, steady-state, and at two intermediate times.

(b) Sketch the heat flux at the outer surface $q''_x(L,t)$ as a function of time.

5.2 The inner surface of a plane wall is insulated while the outer surface is exposed to an airstream at T_∞. The wall is at a uniform temperature corresponding to that of the airstream. Suddenly, a radiation heat source is switched on applying a uniform flux q''_o to the outer surface.

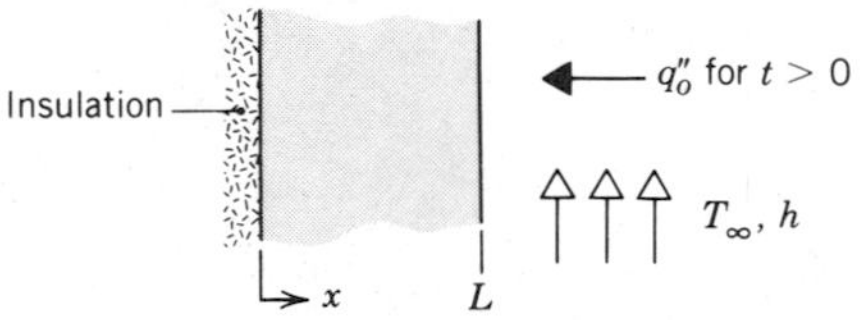

(a) Sketch and label, on $T-x$ coordinates, the temperature distributions: initial, steady-state, and at two intermediate times.

(b) Sketch the heat flux at the outer surface $q''_x(L,t)$ as a function of time.

5.3 A microwave oven operates on the principle that application of a high frequency field causes electrically polarized molecules in food to oscillate. The net effect is a *uniform generation* of thermal energy within the food, which enables it to be heated from refrigeration temperatures to 90°C in as short a time as 30 s.

Consider the process of cooking a slab of beef of thickness $2L$ in a microwave oven and compare it with cooking in a conventional oven, where *each side* of the slab is *heated by radiation* for a period of approximately 30 min. In each case the meat is to be heated from 0°C to a *minimum* temperature of 90°C. Base your comparison on a sketch of the temperature distribution at selected times for each of the cooking processes. In particular consider the time t_0 at which heating is initiated, a time t_1 during the heating process, the time t_2 corresponding to the conclusion of heating, and a time t_3 well into the subsequent cooling process.

5.4 Steel balls 12 mm in diameter are annealed by heating to 1150 K and then slowly cooling to 400 K in an air environment for which $T_\infty = 325$ K and $h = 20$ W/m$^2\cdot$K. Assuming the properties of the steel to be $k = 40$ W/m$\cdot$K, $\rho = 7800$ kg/m^3, and $c = 600$ J/kg$\cdot$K, estimate the time required for the cooling process.

5.5 The heat transfer coefficient for air flowing over a sphere is to be determined by observing the temperature–time history of a sphere fabricated from pure copper. The sphere, which is 12.7 mm in diameter, is at 66°C before it is inserted into an airstream having a temperature of 27°C. A thermocouple on the outer surface of the sphere indicates 55°C 69 s after the sphere is inserted in the airstream. Assume, and then justify, that the sphere behaves as a spacewise isothermal object and calculate the heat transfer coefficient.

5.6 A solid steel sphere (AISI 1010), 300 mm in diameter, is coated with a dielectric material layer of thickness 2 mm and thermal conductivity 0.04 W/m$\cdot$K. The coated sphere is initially at a uniform temperature of 500°C and is suddenly quenched in a large oil bath for which $T_\infty = 100$°C and $h = 3300$ W/m$^2\cdot$K. Estimate the time required for the coated sphere temperature to reach 140°C. *Hint:* Neglect the effect of energy storage in the dielectric material, since its thermal capacitance ($\rho c V$) is small compared to that of the steel sphere.

5.7 A thermal energy storage unit consists of a large rectangular channel, which is well insulated on its outer surface and encloses alternating layers of the storage material and the flow passage.

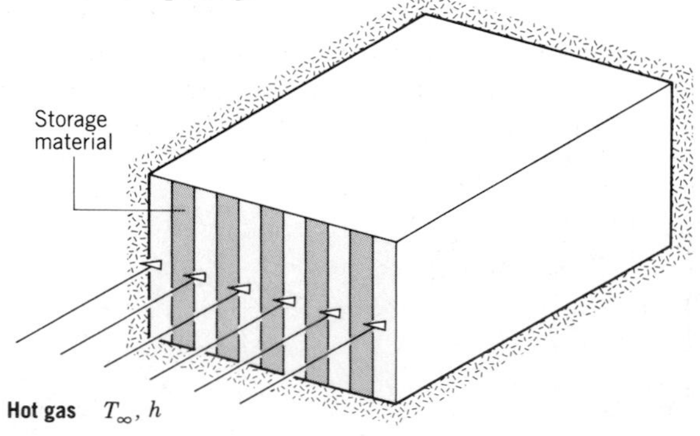

Each layer of the storage material is an aluminum slab of width $W = 0.05$ m, which is at an initial temperature of 25°C. Consider conditions for which the storage unit is charged by passing a hot gas through the passages, with the gas temperature and the convection coefficient assumed to have constant values of $T_\infty = 600°C$ and $h = 100$ $W/m^2 \cdot K$ throughout the channel. How long will it take to achieve 75% of the maximum possible energy storage? What is the temperature of the aluminum at this time?

5.8 In an industrial process requiring high d-c currents, water-jacketed copper rods, 20 mm in diameter, are used to carry the current. The water, which flows continuously between the jacket and the rod, maintains the rod temperature at 75°C during normal operation at 1000 A. The electrical resistance of the rod is known to be 0.15 Ω/m. Problems would arise if the coolant water ceased to be available (e.g. because of a valve malfunction). In such a situation heat transfer from the rod surface would diminish greatly, and the rod would eventually melt. Estimate the time required for melting to occur.

5.9 A cube of dimension L on a side is at a uniform temperature T_i and is insulated on all surfaces except for one. This surface is suddenly brought into contact with a larger, cooler object at T_o fabricated from a material with very high thermal conductivity. Using the lumped capacitance method, develop an expression for estimating the temperature of the cube as a function of time in terms of the thermophysical properties of the cube material, L, T_i, and T_o. Assume that there is negligible contact resistance between the cube and the cool surface and that the cool surface remains at T_o during the process. *Hint:* Write an energy balance for the cube approximating it as a lumped capacitance, and assume the resistance to heat transfer to be that due to conduction over a distance $L/2$ of the cube.

5.10 Consider the system of Problem 5.1 where the temperature of the plate is spacewise isothermal during the transient process.

(a) Develop an expression for the temperature of the plate as a function of time $T(t)$ in terms of q_o'', T_∞, h, L, and the plate properties ρ and c.

(b) Determine the thermal time constant and the steady-state temperature for a 12-mm-thick plate of pure copper when $T_\infty = 27°C$, $h = 50$ $W/m^2 \cdot K$, and $q_o'' = 5000$ W/m^2. Estimate the time required to reach steady-state conditions.

5.11 An electronic device, such as a power transistor mounted on a finned heat sink, can be modeled as a spatially isothermal object with internal heat generation and an external convection resistance.

(a) Consider such a system of mass M, specific heat c, and surface area A_s, which is initially in equilibrium with the environment at T_∞. Suddenly, the electronic device is energized such that a constant heat generation $\dot{E}_g$ (W) occurs. Beginning with an energy balance on the system, show that the temperature response of the device is

$$\frac{\theta}{\theta_i} = \exp\left(-\frac{t}{RC}\right)$$

where $\theta \equiv T - T(\infty)$ and $T(\infty)$ is the steady-state temperature corresponding to $t \to \infty$; $\theta_i = T_i - T(\infty)$; T_i = initial temperature of device; R = thermal resistance $1/\bar{h}A_s$; and C = thermal capacitance Mc.

(b) An electronic device, which generates 60 W of heat, is mounted on an aluminum heat sink weighing 0.31 kg and reaches a temperature of 100°C in ambient air at

20°C under steady-state conditions. If the device is initially at 20°C, what temperature will it reach 5 min after the power is switched on?

5.12 In our treatment of the lumped capacitance method (Section 5.1), the solid was exposed to convective surface conditions. Frequently applications involve surface heat transfer that is dominated by radiation, and it is interesting to determine the transient response of a lumped element to such conditions.

(a) Derive an expression for the time rate of change of the element temperature when it is suddenly exposed to a radiation exchange process characterized by the rate equation, Equation 1.6. *Hint*: the method of analysis will be similar to that used in Section 5.1 and the integral that needs evaluation is of the form

$$\int \frac{dx}{(a^4 - x^4)} = \frac{1}{4a^3} \ln \left| \frac{a + x}{a - x} \right| + \frac{1}{2a^3} \tan^{-1} \frac{x}{a}$$

(b) Develop a criterion that could be used to determine the suitability of the lumped capacitance method for the above conditions. *Hint:* Consider using Equation 1.8, which defines a radiation heat transfer coefficient.

5.13 Before being injected into a furnace, pulverized coal is preheated by passing it through a cylindrical tube whose surface is maintained at $T_{sur} = 1000°C$. The coal pellets are suspended in an airflow and are known to move with a speed of 3 m/s. If the pellets may be approximated as spheres of 1 mm diameter and it may be assumed that they are heated by radiation transfer from the tube surface, how long must the tube be to heat coal entering at 25°C to a temperature of 600°C? Is the use of the lumped capacitance method justified? *Hint:* The integral that needs evaluation is given in Problem 5.12.

5.14 A metal sphere of diameter D, which is at a uniform temperature T_i, is suddenly removed from a furnace and suspended from a fine wire in a large room with air at a uniform temperature T_∞ and the surrounding walls at a temperature T_{sur}.

(a) Assuming temperature gradients within the sphere to be negligible, develop a differential equation that governs the temperature change with time in the sphere.

(b) Neglecting heat transfer by radiation, determine the time it takes for the sphere to cool to some temperature T.

(c) Neglecting heat transfer by convection, determine the time it takes for the sphere to cool to the temperature T. *Hint:* The integral that needs evaluation is given in Problem 5.12.

(d) How would you go about determining the time required for the sphere to cool to the temperature T if both convection and radiation are of the same order of magnitude?

(e) Consider an anodized aluminum sphere ($\varepsilon = 0.75$) 50 mm in diameter, which is at an initial temperature of $T_i = 800$ K. Both the air and the surroundings are at 300 K, and the convection coefficient is 10 $W/m^2 \cdot K$. Calculate and compare the time it will take for the sphere to cool to 400 K using the results of parts b, c, and d.

5.15 Long metallic rods of circular cross section are heat treated by passing an electric current through the rods to provide uniform volumetric heat generation at a rate $\dot{q}$ (W/m^3). The rods are of diameter D and are placed in a large chamber whose walls are maintained at the same temperature T_∞ as the enclosed air. Convection from the surface of the rods to the air is characterized by the coefficient h.

(a) Assuming negligible internal resistance, derive a differential equation that could be solved to determine the rod temperature as a function of time during the heating process. Express your result in terms of the foregoing variables and appropriate physical properties of the rod material. Do not attempt to solve.

(b) Obtain an expression that could be used to determine the steady-state temperature of the rod.

(c) Neglecting radiation effects and prescribing an initial ($t = 0$) rod temperature of $T_i = T_\infty$, solve the equation in part a to obtain the transient temperature response of the rod.

5.16 Consider the series solution, Equations 5.27, for the plane wall with convection. Calculate midplane ($x^* = 0$) and surface ($x^* = 1$) temperatures θ^* for $Fo = 0.1$ and 1, using $Bi = 0.1$, 1, and 10. Consider only the first four eigenvalues. Based on these results discuss the validity of the approximate solutions, Equations 5.28 and 5.29.

5.17 Beginning with the general expression for Q/Q_o, Equation 5.33, perform the prescribed integration to obtain the result for the plane wall, Equation 5.34.

5.18 A large aluminum (2024 alloy) plate of thickness 0.15 m, initially at a uniform temperature of 300 K, is placed in a furnace having an ambient temperature of 800 K for which the convection heat transfer coefficient is estimated to be 500 $W/m^2 \cdot K$.

(a) Determine the time required for the plate midplane to reach 700 K.

(b) What is the surface temperature of the plate for this condition?

(c) Repeat the calculations if the material were stainless steel (type 304).

5.19 After a long, hard week on the books, you and your friend are ready to relax. You take a steak 50 mm thick from the freezer. How long do you have to let the good times roll before the steak has thawed? Assume that the steak is initially at $-6°C$, that it thaws when the midplane temperature reaches 4°C, and that the room temperature is 23°C with a convection heat transfer coefficient of 10 $W/m^2 \cdot K$. Treat the steak as a slab having the properties of liquid water at 0°C. Neglect the heat of fusion associated with the melting phase change.

5.20 A one-dimensional plane wall with a thickness of 0.1 m initially at a uniform temperature of 250°C is suddenly immersed in an oil bath at 30°C. Assuming the convection heat transfer coefficient for the wall in the bath is 500 $W/m^2 \cdot K$, calculate the surface temperature of the wall 9 min after immersion. The properties of the wall are $k = 50$ $W/m \cdot K$, $\rho = 7835$ kg/m^3, and $c = 465$ $J/kg \cdot K$.

5.21 Consider the thermal energy storage unit of Problem 5.7, but with a masonry material of $\rho = 1900$ kg/m^3, $c = 800$ $J/kg \cdot K$, and $k = 0.70$ $W/m \cdot K$ used in place of the aluminum. How long will it take to achieve 75% of the maximum possible energy storage? What are the maximum and minimum temperatures of the masonry at this time?

5.22 Cylindrical steel rods (AISI 1010), 50 mm in diameter, are heat treated by drawing them through an oven 5 m long in which air is maintained at 750°C. The rods enter at 50°C and achieve a centerline temperature of 600°C before leaving. For a convection coefficient of 125 $W/m^2 \cdot K$, estimate the speed at which the rods must be drawn through the oven.

5.23 Estimate the time required to cook a hot dog in boiling water. Assume that the hot dog is initially at 6°C, that the convection heat transfer coefficient is 100 $W/m^2 \cdot K$, and that the final temperature is 80°C at the centerline. Treat the hot dog as a long cylinder of 20 mm diameter having the properties: $\rho = 880\ kg/m^3$, $c = 3350\ J/kg \cdot K$, and $k = 0.52\ W/m \cdot K$.

5.24 A long rod of 60 mm diameter and thermophysical properties $\rho = 8000\ kg/m^3$, $c = 500\ J/kg \cdot K$ and $k = 50\ W/m \cdot K$ is initially at a uniform temperature and is heated in a forced convection furnace maintained at 750 K. The convection coefficient is estimated to be 1000 $W/m^2 \cdot K$. At a certain time, the surface temperature of the rod is observed to be 550 K. What is the corresponding center temperature of the rod?

5.25 A long cylinder of 30 mm diameter, initially at a uniform temperature of 1000 K, is suddenly quenched in a large, constant-temperature oil bath at 350 K. The cylinder properties are $k = 1.7\ W/m \cdot K$, $c = 1600\ J/kg \cdot K$, and $\rho = 400\ kg/m^3$, while the convection coefficient is 50 $W/m^2 \cdot K$. Calculate the time required for the surface of the cylinder to reach 500 K.

5.26 A long pyroceram rod of diameter 20 mm is clad with a very thin metallic tube for mechanical protection. The bonding between the rod and the tube has a thermal contact resistance of $R'_{t,c} = 0.12\ m \cdot K/W$.

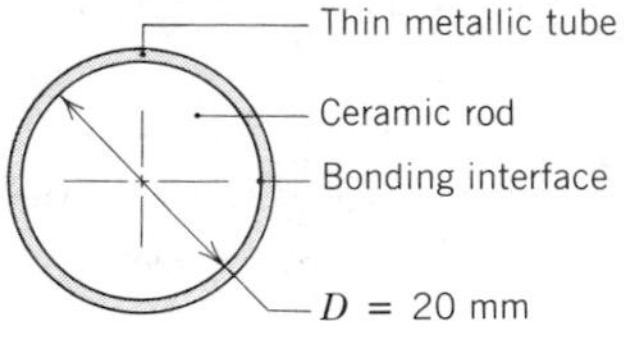

If the rod is initially at a uniform temperature of 900 K and is suddenly cooled by a fluid at $T_\infty = 300$ K and $h = 100\ W/m^2 \cdot K$, at what time will the rod centerline reach 600 K?

5.27 A long rod 40 mm in diameter, fabricated from sapphire (aluminum oxide) and initially at a uniform temperature of 800 K, is suddenly exposed to a cooling process with a fluid at 300 K having a heat transfer coefficient of 1600 $W/m^2 \cdot K$. After 35 s of exposure to the cooling process, the rod is wrapped in insulation and experiences no heat losses. What will be the temperature of the rod after a long period of time?

5.28 In heat treating to harden steel ball bearings ($c = 500\ J/kg \cdot K$, $\rho = 7800\ kg/m^3$, $k = 50\ W/m \cdot K$), it is desirable to increase the surface temperature for a short time without significantly warming the interior of the ball. This type of heating can be accomplished by sudden immersion of the ball in a molten salt bath with $T_\infty = 1300$ K and $h = 5000\ W/m^2 \cdot K$. Assume that any location within the ball whose temperature exceeds 1000 K will be hardened. Estimate the time required to harden the outer millimeter of a ball of diameter 20 mm, if its initial temperature is 300 K.

5.29 A sphere of 80 mm diameter ($k = 50\ W/m \cdot K$ and $\alpha = 1.5 \times 10^{-6}\ m^2/s$) is initially at a uniform, elevated temperature and is quenched in an oil bath maintained at 50°C. The convection coefficient for the cooling process is 1000 $W/m^2 \cdot K$. At a certain time, the surface temperature of the sphere is observed to be 150°C. What is the corresponding center temperature of the sphere?

5.30 The cold air chamber shown is proposed for quenching steel ball bearings of diameter $D = 0.2$ m and initial temperature $T_i = 400°C$. Air in the chamber is maintained at $-15°C$ by a refrigeration system, and the steel balls pass through the chamber on a conveyor belt. Optimum bearing production requires that 70% of the initial thermal energy content of the ball above $-15°C$ be removed. Radiation effects may be neglected, and the convection heat transfer coefficient within the chamber is 1000 $W/m^2 \cdot K$. Estimate the residence time of the balls within the chamber, and recommend a drive velocity of the conveyor. The following properties may be used for the steel: $k = 50$ W/m · K, $\alpha = 2 \times 10^{-5}$ m^2/s, and $c = 450$ J/kg · K.

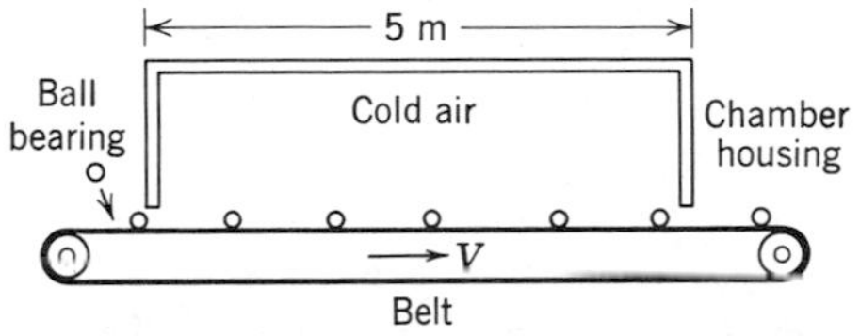

5.31 Stainless steel (AISI 304) ball bearings, which have been uniformly heated to 850°C, are hardened by quenching them in an oil bath that is maintained at 40°C. The ball diameter is 20 mm, and the convection coefficient associated with the oil bath is 1000 $W/m^2 \cdot K$.

(a) If quenching is to occur until the surface temperature of the balls reaches 100°C, how long must the balls be kept in the oil? What is the center temperature at the conclusion of the cooling period?

(b) If 10,000 balls are to be quenched per hour, what is the rate at which energy must be removed by the oil bath cooling system in order to maintain its temperature at 40°C?

5.32 A spherical hailstone that is 5 mm in diameter is formed in a high altitude cloud at $-30°C$. If the stone begins to fall through warmer air at 5°C, how long will it take before the outer surface begins to melt? What is the temperature of the stone's center at this point in time, and how much energy (J) has been transferred to the stone? A convection heat transfer coefficient of 250 $W/m^2 \cdot K$ may be assumed, and the properties of the hailstone may be taken to be those of ice.

5.33 A sphere 30 mm in diameter initially at 800 K is quenched in a large bath having a constant temperature of 320 K with a convection heat transfer coefficient of 75 $W/m^2 \cdot K$. The thermophysical properties of the sphere material are: $\rho = 400$ kg/m^3, $c = 1600$ J/kg · K, and $k = 1.7$ W/m · K.

(a) Show, in a qualitative manner on T–t coordinates, the temperatures at the center and at the surface of the sphere as a function of time.

(b) Calculate the time required for the surface of the sphere to reach 415 K.

(c) Determine the heat flux (W/m^2) at the outer surface of the sphere at the time determined in part b.

(d) Determine the energy (J) that has been lost by the sphere during the process of cooling to the surface temperature of 415 K.

(e) At the time determined by part b, the sphere is quickly removed from the bath and covered with perfect insulation, such that there is no heat loss from the surface of the sphere. What will be the temperature of the sphere after a long period of time has elapsed?

5.34 Spheres A and B are initially at 800 K, and they are simultaneously quenched in large constant temperature baths, each having a temperature of 320 K. The following parameters are associated with each of the spheres and their cooling processes.

	SPHERE A	SPHERE B
Diameter (mm)	300	30
Density (kg/m^3)	1600	400
Specific heat (kJ/kg · K)	0.400	1.60
Thermal conductivity (W/m · K)	170	1.70
Convection coefficient (W/m^2 · K)	5	50

(a) Show in a qualitative manner, on T versus t coordinates, the temperatures at the center and at the surface for each sphere as a function of time. Briefly explain the reasoning by which you determine the relative positions of the curves.

(b) Calculate the time required for the surface of each sphere to reach 415 K.

(c) Determine the energy that has been gained by each of the baths during the process of the spheres cooling to 415 K.

5.35 Two large blocks of different materials, such as copper and concrete, have been sitting in a room (23°C) for a very long time. Which of the two blocks, if either, will feel colder to the touch? Assume the blocks to be semiinfinite solids and your hand to be at a temperature of 37°C.

5.36 Asphalt pavement may achieve temperatures as high as 50°C on a hot summer day. Assume that such a temperature exists throughout the pavement, when suddenly a rainstorm reduces the surface temperature to 20°C. Calculate the total amount of energy (J/m^2) that will be transferred from the asphalt over a 30-min period in which the surface is maintained at 20°C.

5.37 An insurance company has hired you as a consultant to improve their understanding of burn injuries. They are especially interested in injuries induced when a portion of a worker's body comes into contact with machinery that is at elevated temperatures in the range of 50 to 100°C. Their medical consultant informs them that irreversible thermal injury (cell death) will occur in any living tissue that is maintained at $T \geq 48$°C for a duration $\Delta t \geq 10$ s. They want information concerning the extent of irreversible tissue damage (as measured by distance from the skin surface) as a function of the machinery temperature and the time during which contact is made between the skin and the machinery. Can you help them? Assume that living tissue has a normal temperature of 37°C, is isotropic, and has constant properties equivalent to those of liquid water.

5.38 A procedure for determining the thermal conductivity of a solid material involves embedding a thermocouple in a thick slab of the solid and measuring the response to a prescribed change in temperature at one surface. Consider an arrangement for which the thermocouple is embedded 10 mm from a surface that is suddenly brought to a temperature of 100°C by exposure to boiling water. If the initial temperature of the slab was 30°C and the thermocouple measures a temperature of 65°C, 2 min after the surface is brought to 100°C, what is its thermal conductivity?

The density and specific heat of the solid are known to be 2200 kg/m^3 and 700 J/kg · K.

5.39 An electric heater in the form of a sheet is placed in good contact with the surface of a thick slab of Bakelite having a uniform temperature of 300 K. Determine the temperature of the slab at the surface and at a depth of 25 mm, 10 min after the heater has been energized and is providing a constant heat flux to the surface of 2500 W/m^2.

5.40 A very thick slab with thermal diffusivity 5.6×10^{-6} m^2/s and thermal conductivity 20 W/m · K is initially at a uniform temperature of 325°C. Suddenly, the surface is exposed to a coolant at 15°C for which the convection heat transfer coefficient is 100 W/m^2 · K. Determine the temperatures at the surface and at a depth of 45 mm after 3 min has elapsed.

5.41 A thick oak wall initially at 25°C, is suddenly exposed to combustion products at 800°C. Determine the time of exposure required for the surface to reach the ignition temperature of 400°C, assuming the convection heat transfer coefficient between the wall and products to be 20 W/m^2 · K.

5.42 A steel (plain carbon) billet of square cross section 0.3 m × 0.3 m, initially at a uniform temperature of 30°C, is placed in a soaking oven having a temperature of 750°C. If the convection heat transfer coefficient for the heating process is 100 W/m^2 · K, how long must the billet remain in the oven before its center temperature reaches 600°C?

5.43 Fireclay brick of dimensions 0.06 m × 0.09 m × 0.20 m is removed from a kiln at 1600 K and cooled in air at 40°C with $h = 50$ W/m^2 · K. What is the temperature at the center and at the corners of the brick after 50 min of cooling?

5.44 A cylindrical copper pin 100 mm long and 50 mm in diameter is initially at a uniform temperature of 20°C. The end faces are suddenly subjected to an intense heating rate that raises them to a temperature of 500°C. At the same time, the cylindrical surface is subjected to heating by gas flow with a temperature 500°C and a heat transfer coefficient 100 W/m^2 · K.

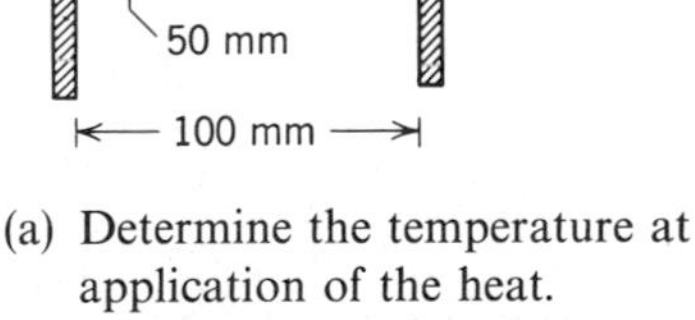

(a) Determine the temperature at the center point of the cylinder 8 s after sudden application of the heat.

(b) Considering the parameters governing the temperature distribution in transient heat diffusion problems, can any simplifying assumptions be justified in analyzing this particular problem? Explain briefly.

5.45 Recalling that your mother once said that meat should be cooked until every portion has attained a temperature of 80°C, how long will it take to cook a 2.25-kg roast? Assume that the meat is initially at 6°C and that the oven temperature is 175°C with a convection heat transfer coefficient of 15 $W/m^2 \cdot K$. Treat the roast as a cylinder with properties of liquid water, having a diameter equal to its length.

5.46 A long rod 20 mm in diameter is fabricated from alumina (polycrystalline aluminum oxide) and is initially at a uniform temperature of 850 K. The rod is suddenly exposed to fluid at 350 K with $h = 500\ W/m^2 \cdot K$. Estimate the centerline temperature of the rod after 30 s at an exposed end and at an axial distance of 6 mm from the end.

5.47 The stability criterion for the explicit method requires that the coefficient of the T_m^p term of the one-dimensional, finite-difference equation be zero or positive. Consider the situation for which the temperatures at the two neighboring nodes (T_{m-1}^p, T_{m+1}^p) are 100°C while the center node (T_m^p) is at 50°C. Show that for values of $Fo > \frac{1}{2}$, the finite-difference equation will predict a value of T_m^{p+1} that violates the second law of thermodynamics.

5.48 A thin rod of diameter D is initially in equilibrium with its surroundings, a large vacuum enclosure at temperature, T_{sur}. Suddenly an electrical current I (A) is passed through the rod having an electrical resistivity ρ_e and emissivity ε. Other pertinent thermophysical properties are identified in the sketch. Derive the transient, finite-difference equation for node m.

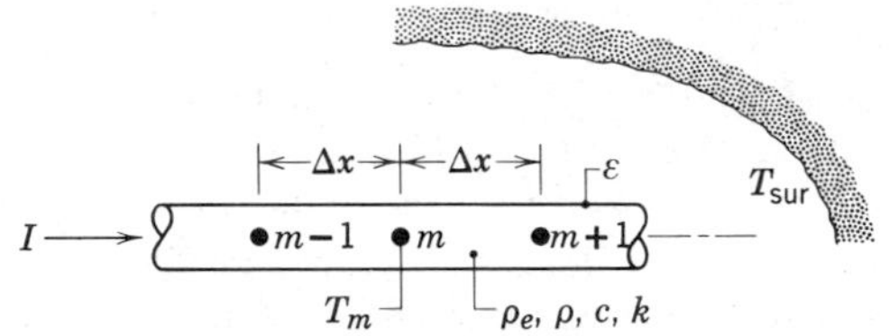

5.49 A tantalum rod of diameter 3 mm and length 120 mm is supported by two electrodes within a large vacuum enclosure. Initially the rod is in equilibrium with the electrodes and its surroundings, which are maintained at 300 K. Suddenly, an electrical current, $I = 80$ A, is passed through the rod. Assume the emissivity of the rod is 0.1 and the electrical resistivity is $95 \times 10^{-8}\ \Omega \cdot m$. Use Table A.1 to obtain the other thermophysical properties required in your solution. Use a finite-difference method with a space increment of 10 mm.

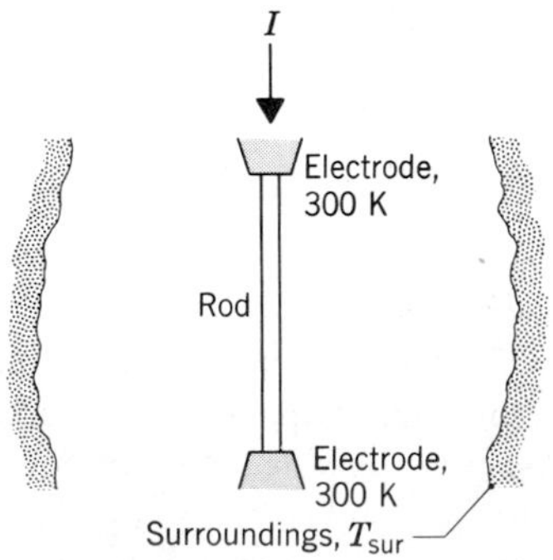

(a) Estimate the time required for the midlength of the rod to reach 1000 K.

(b) Determine the steady-state temperature distribution and estimate approximately how long it will take to reach this condition.

5.50 A one-dimensional slab of thickness $2L$ is initially at a uniform temperature T_i. Suddenly electric current is passed through the slab causing a uniform volumetric heating $\dot{q}$ (W/m^3) to be present. At the same time, both outer surfaces ($x = \pm L$) are subjected to a convection process at T_∞ with a heat transfer coefficient h.

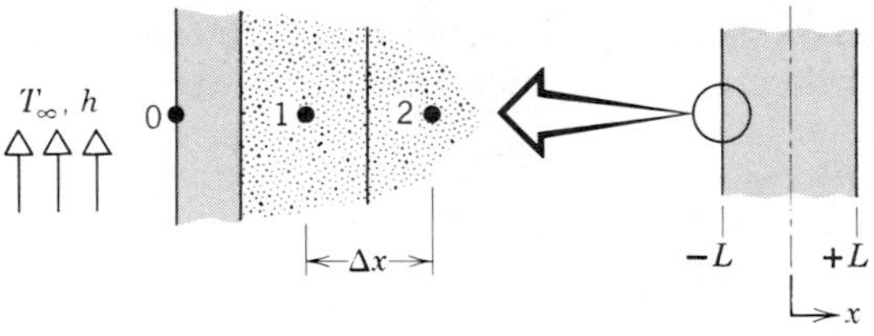

Write the finite-difference equation expressing conservation of energy for node 0 located on the outer surface at $x = -L$ as shown in the accompanying figure. Rearrange your equation and identify any important dimensionless coefficients.

5.51 A wall 0.12 m thick having a thermal diffusivity of 1.5×10^{-6} m^2/s is initially at a uniform temperature of 85°C. Suddenly one face is lowered to a temperature of 20°C while the other face is perfectly insulated. Using a numerical method with space and time increments of 30 mm and 300 s, respectively, determine the temperature distribution within the wall after 45 min has elapsed.

5.52 A large plastic casting with thermal diffusivity 6.0×10^{-7} m^2/s is removed from its mold at a uniform temperature of 150°C. The casting is then exposed to a high-velocity airstream such that the surface experiences a sudden change in temperature to 20°C. Assuming the casting approximates a semiinfinite medium and using a finite difference method with a space increment of 6 mm, estimate the temperature at a distance 18 mm from the surface after 3 min has elapsed. Verify your result by comparison with the appropriate analytical solution.

5.53 A very thick plate with thermal diffusivity 5.6×10^{-6} m^2/s and thermal conductivity 20 W/m · K is initially at a uniform temperature of 325°C. Suddenly, the surface is exposed to a coolant at 15°C for which the convection heat transfer coefficient is 100 $W/m^2 \cdot K$. Using the finite-difference method with a space increment of $\Delta x = 15$ mm and a time increment of 18 s, determine temperatures at the surface and at a depth of 45 mm after 3 min has elapsed.

5.54 Consider the fuel element of Example 5.6. Initially the element is at a uniform temperature of 250°C with no heat generation. Suddenly, the element is inserted into the reactor core causing a uniform volumetric heat generation rate of $\dot{q} = 10^8$ W/m^3. The surfaces are convectively cooled with $T_\infty = 250°C$ and $h = 1100$ $W/m^2 \cdot K$. Using the explicit method with a space increment of 2 mm, determine the temperature distribution 1.5 s after the element is inserted into the core.

5.55 A plane wall of thickness 100 mm with a uniform volumetric heat generation of $\dot{q} = 1.5 \times 10^6$ W/m^3 is exposed to convection conditions of $T_\infty = 30°C$ and $h = 1000$ $W/m^2 \cdot K$ on both surfaces. The wall is maintained under steady-state conditions when suddenly the heat generation level ($\dot{q}$) is reduced to zero. The thermal diffusivity and thermal conductivity of the wall material are 1.6×10^{-6} m^2/s and 75 W/m · K. A space increment of 10 mm is suggested.

(a) Estimate the midplane temperature 3 min after the generation has been switched off.

(b) Plot on T–x coordinates the temperature distribution obtained in part a. Show also the initial and steady-state temperature distributions for the wall.

5.56 For the conditions described in Example 5.6, use the finite-difference method to estimate the temperature at the midplane ($x = 0$) 20 s after the power level has been changed from $\dot{q}_1$ to $\dot{q}_2$.

5.57 A thin circular disk is subjected to induction heating from a coil, the effect of which is to provide a uniform heat generation within a ring section as shown.

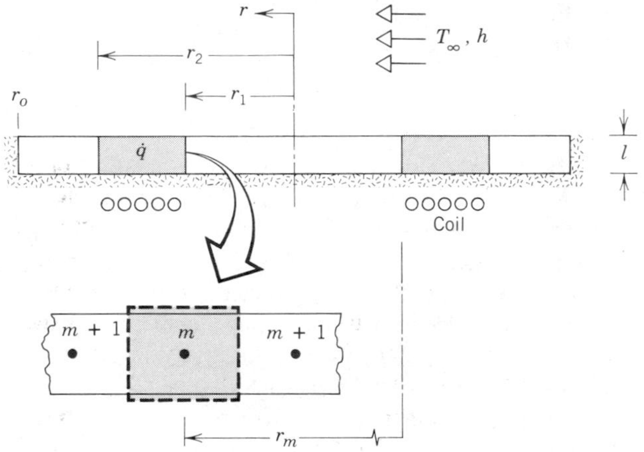

Convection occurs at the upper surface, while the lower surface is well insulated.

(a) Derive the transient, finite-difference equation for node m, which is within the region subjected to induction heating.

(b) On T–r coordinates sketch, in a qualitative manner, the steady-state temperature distribution, identifying important features.

5.58 An electrical cable, experiencing a uniform volumetric generation $\dot{q}$, is half buried in an insulating material while the upper surface is exposed to a convection process (T_∞, h).

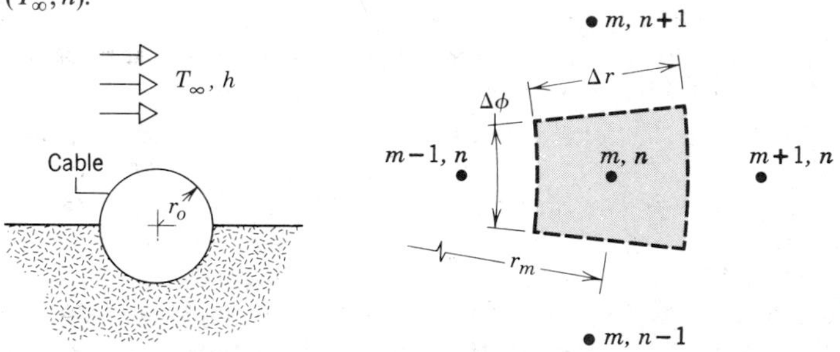

(a) Derive the explicit, finite-difference equations for an interior node (m,n), the center node $(m = 0)$, and the outer surface nodes (M,n) for the convection and insulated boundaries.

(b) Obtain the stability criterion for each of the finite-difference equations. Identify the most restrictive criterion.

5.59 One end of a stainless steel (AISI 316) rod of diameter 10 mm and length 0.16 m is inserted into a fixture maintained at 200°C. The rod, covered with an insulating sleeve, reaches a uniform temperature throughout its length. When the sleeve is removed, the rod is subjected to ambient air at 25°C such that the convection heat transfer coefficient is 30 W/m$^2\cdot$K. Using a numerical technique, estimate the time required for the midlength of the rod to reach 100°C.

5.60 The cross section of an oven wall is composed of 30-mm-thick insulation sandwiched between two thin (1.5-mm-thick) stainless steel sheets. Under steady-state conditions, the oven is operating with an inside air temperature of $T_{\infty,i} = 150°C$ and an ambient air temperature of $T_{\infty,o} = 20°C$ with $h_i = 100$ W/m$^2\cdot$K and $h_o = 10$ W/m$^2\cdot$K. When the oven heater level is changed and the fan speed changed to substantially increase air circulation within the oven, the inside surface of the oven experiences a sudden temperature change to 100°C. The insulation has a thermal conductivity of 0.03 W/m$\cdot$K and a thermal diffusivity of 7.5×10^{-7} m^2/s. In your finite-difference solution, use a space increment of 6 mm. Assume that the effect of the stainless steel sheets is negligible and that the outside convection heat transfer coefficient h_o remains unchanged. Estimate the time required for the oven wall to approximate steady-state conditions after the inner wall temperature is changed to 100°C.

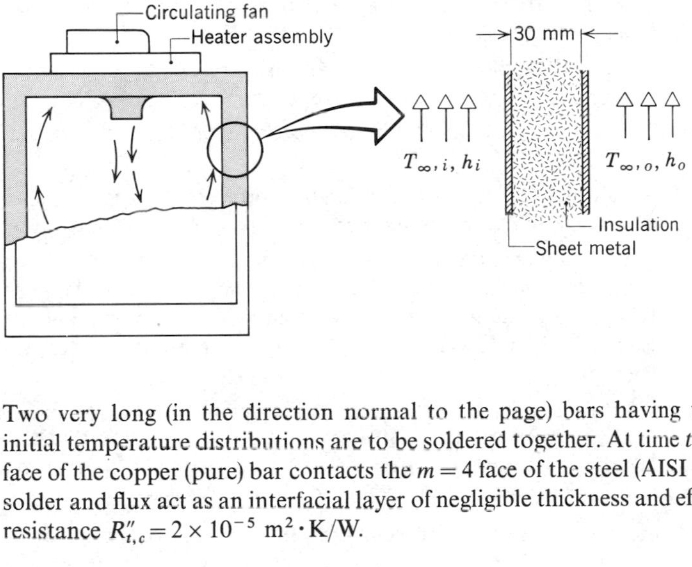

5.61 Two very long (in the direction normal to the page) bars having the prescribed initial temperature distributions are to be soldered together. At time $t = 0$, the $m = 3$ face of the copper (pure) bar contacts the $m = 4$ face of the steel (AISI 1010) bar. The solder and flux act as an interfacial layer of negligible thickness and effective contact resistance $R''_{t,c} = 2 \times 10^{-5}$ m$^2\cdot$K/W.

INITIAL TEMPERATURES (K)

n/m	1	2	3	4	5	6
1	700	700	700	1000	900	800
2	700	800	700	1000	900	800
3	700	700	700	1000	900	800

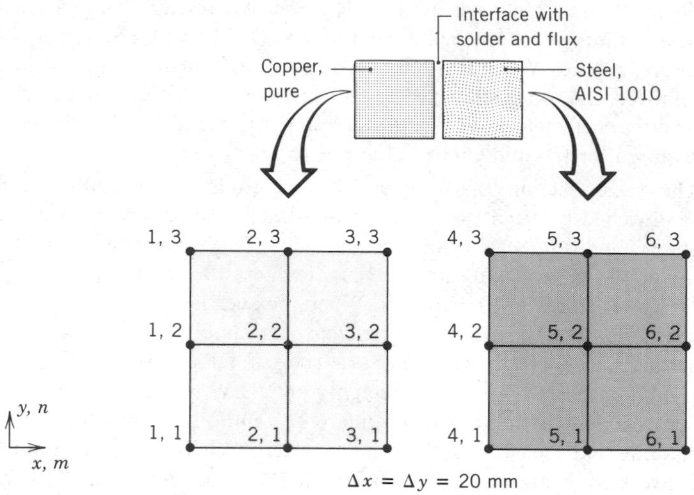

(a) Derive the explicit, finite-difference equation in terms of Fo and $Bi_c = \Delta x / kR''_{t,c}$ for $T_{4,2}$ and determine the corresponding stability criterion.

(b) Using $Fo = 0.01$, determine $T_{4,2}$ one time step after contact is made. What is Δt? Is the stability criterion satisfied?

5.62 Referring to Example 5.7, Comment 4, consider a sudden exposure of the surface to large surroundings at an elevated temperature (T_{sur}) and to convection (T_∞, h).

(a) Derive the explicit, finite-difference equation for the surface node in terms of Fo, Bi, and Bi_r.

(b) Obtain the stability criterion for the surface node. Does this criterion change with time? Is the criterion more restrictive than that for an interior node?

(c) A thick slab of material ($k = 1.5$ W/m · K, $\alpha = 7 \times 10^{-7}$ m^2/s, $\varepsilon = 0.9$), initially at a uniform temperature of 27°C, is suddenly exposed to large surroundings at 1000 K. Neglecting convection and using a space increment of 10 mm, determine temperatures at the surface and 30 mm from the surface after an elapsed time of 1 min.

5.63 Consider the system of Problem 4.47. Initially with no flue gases flowing, the walls ($\alpha = 5.5 \times 10^{-6}$ m^2/s) are at a uniform temperature of 25°C. Using the implicit, finite-difference method with a time increment of 1 h, find the temperature distribution in the wall 1, 2, 5, and 20 hours after introduction of the flue gases.

5.64 Consider the system of Problem 4.55. Initially, the ceramic plate ($\alpha = 1.5 \times 10^{-6}$ m^2/s) is at a uniform temperature of 30°C, and suddenly the electrical heating elements are energized. Using the implicit, finite-difference method, estimate the time required for the difference between the surface and initial temperatures to reach 95% of the difference for steady-state conditions. If you write a computer program, use a time increment of 2 s; otherwise use 50 s.

Six

INTRODUCTION TO CONVECTION

Thus far we have focused on heat transfer by conduction and have considered convection only to the extent that it provides a possible boundary condition for conduction problems. Recall that in Section 1.2.2 we used the term "convection" to describe energy transfer between a surface and a fluid moving over the surface. Although the mechanism of diffusion (random motion of fluid molecules) contributes to this transfer, the dominant contribution is generally made by the bulk or gross motion of fluid particles.

In our treatment of convection, we have two major objectives. In addition to obtaining a firm understanding of the physical mechanisms that underlie convection transfer, we wish to develop the means to perform convection transfer calculations. This chapter is devoted primarily to achieving the former objective. In particular, an effort has been made to concentrate many of the fundamentals in one place. Physical origins are carefully discussed, and the relevant dimensionless parameters are developed.

With conceptual foundations well established, subsequent chapters are used to develop useful tools for quantifying convection effects. Chapters 7 and 8 present methods for computing the coefficients associated with *forced convection* in *external* and *internal flow* configurations, respectively. Chapter 9 describes methods for determining these coefficients in *free convection*, and Chapter 10 considers the problem of *convection with phase change* (*boiling* and *condensation*). Chapter 11 develops methods for designing and evaluating the performance of *heat exchangers*, devices that are widely used in engineering practice to effect heat transfer between fluids.

Accordingly, we begin by developing our understanding of the nature of convection.

6.1 THE CONVECTION TRANSFER PROBLEM

Consider the flow condition of Figure 6.1*a*. A fluid of velocity V and temperature T_∞ flows over a surface of arbitrary shape and of area A_s. If the surface conditions are such that $T_s \neq T_\infty$, we know that convection heat transfer will occur. The *local heat flux* q'' may be expressed as

$$q'' = h(T_s - T_\infty) \tag{6.1}$$

where h is the *local convection coefficient*. Because flow conditions vary from point to point on the surface, both q'' and h also vary along the surface. The *total heat transfer rate* q may be obtained by integrating the local flux over the entire surface. That is,

$$q = \int_{A_s} q''\, dA_s \tag{6.2}$$

or, from Equation 6.1, assuming a uniform surface temperature,

$$q = (T_s - T_\infty) \int_{A_s} h\, dA_s \tag{6.3}$$

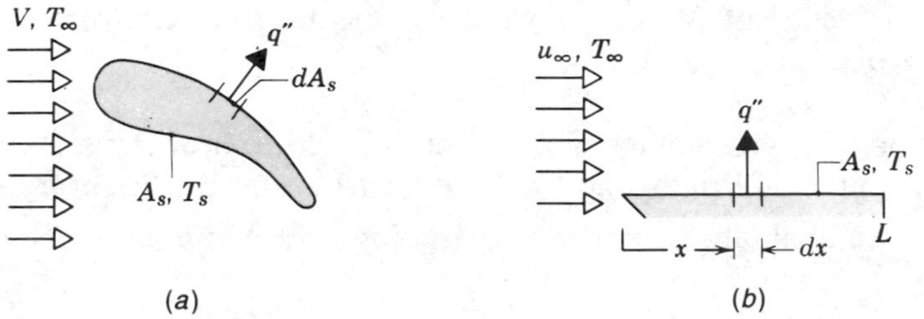

Figure 6.1 Local and total convection heat transfer effects. (*a*) Surface of arbitrary shape. (*b*) Flat plate.

Defining an *average convection coefficient* $\bar{h}$ for the entire surface, the total heat transfer rate may also be expressed as

$$q=\bar{h}A_s(T_s-T_\infty) \tag{6.4}$$

Equating Equations 6.3 and 6.4, it follows that the average and local convection coefficients are related by an expression of the form

$$\bar{h}=\frac{1}{A_s}\int_{A_s} h\, dA_s \tag{6.5}$$

Note that for the special case of flow over a flat plate (Figure 6.1*b*), h varies with the distance x from the leading edge and Equation 6.5 reduces to

$$\bar{h}=\frac{1}{L}\int_0^L h\, dx \tag{6.6}$$

The local flux and/or the total transfer rate are of paramount importance in any convection problem. These quantities may be determined from the rate equations, (6.1) and (6.4), which depend on knowledge of the local and average convection coefficients. It is for this reason that determination of these coefficients is viewed as *the problem of convection.* However, the problem is not a simple one, for in addition to depending on numerous *fluid properties* such as density, viscosity, thermal conductivity, and specific heat, the coefficients depend on the *surface geometry* and the *flow conditions.* This multiplicity of independent variables results from the fact that convection transfer is determined by the *boundary layers* that develop on the surface.

EXAMPLE 6.1

Experimental results for the local heat transfer coefficient h_x for flow over a flat plate with an extremely rough surface were found to fit the relation

$$h_x(x)=ax^{-0.1}$$

where a is a coefficient ($W/m^{1.9} \cdot K$) and x is the distance (m) from the leading edge of the plate.

1. Develop an expression for the ratio of the average heat transfer coefficient $\bar{h}_x$ for a plate of length x to the local heat transfer coefficient h_x at x.
2. Show in a qualitative manner the variation of h_x and $\bar{h}_x$ as a function of x.

SOLUTION

Known: Variation of the local heat transfer coefficient $h_x(x)$.

Find:

1. The ratio of the average heat transfer coefficient $\bar{h}(x)$ to the local value $h_x(x)$.
2. Sketch of the variation of h_x and $\bar{h}_x$ with x.

Schematic:

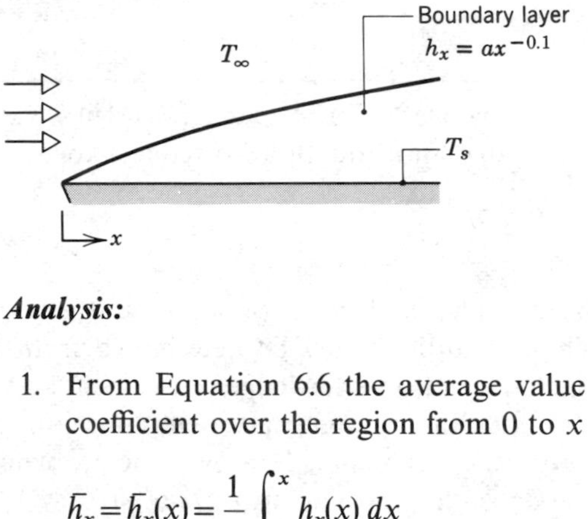

Analysis:

1. From Equation 6.6 the average value of the convection heat transfer coefficient over the region from 0 to x is

$$\bar{h}_x = \bar{h}_x(x) = \frac{1}{x}\int_0^x h_x(x)\,dx$$

Substituting the expression for the local heat transfer coefficient

$$h_x(x) = ax^{-0.1}$$

and integrating, we obtain

$$\bar{h}_x = \frac{1}{x}\int_0^x ax^{-0.1}\,dx = \frac{a}{x}\int_0^x x^{-0.1}\,dx = \frac{a}{x}\left(\frac{x^{+0.9}}{0.9}\right) = 1.11ax^{-0.1}$$

or

$$\bar{h}_x = 1.11h_x \qquad \triangleleft$$

2. The variation of h_x and $\bar{h}_x$ with x is as follows.

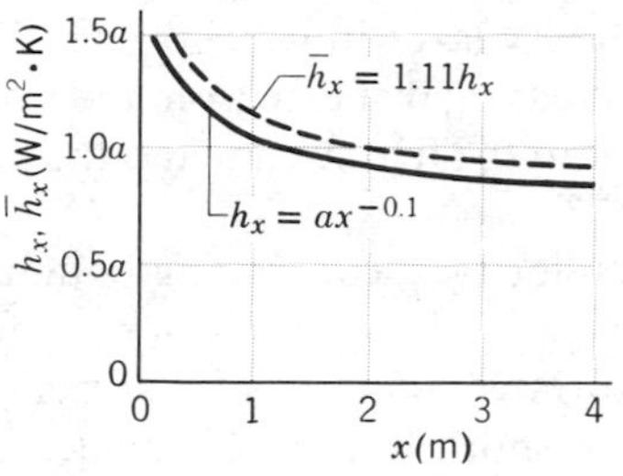

Comments:

Due to boundary layer development, both the local and average coefficients decrease with increasing distance from the leading edge. The average coefficient up to x must therefore exceed the local value at x.

6.2 THE CONVECTION BOUNDARY LAYERS

6.2.1 The Velocity Boundary Layer

To introduce the concept of a boundary layer, consider flow over the flat plate of Figure 6.2. When fluid particles make contact with the surface, they assume zero velocity. These particles then act to retard the motion of particles in the adjoining fluid layer, which act to retard the motion of particles in the next layer, and so on until, at a distance $y = \delta$ from the surface, the effect becomes negligible. This retardation of fluid motion is associated with *shear stresses* τ acting in planes that are parallel to the fluid velocity (Figure 6.2). With increasing distance y from the surface, the x velocity component of the fluid, u, must then increase until it approaches the freestream value u_∞. The subscript ∞ is used to designate conditions in the *free stream* outside the boundary layer.

The quantity δ is termed the *boundary layer thickness*, and it is formally defined as the value of y for which $u = 0.99u_\infty$. The *boundary layer velocity profile* refers to the manner in which u varies with y through the boundary layer.

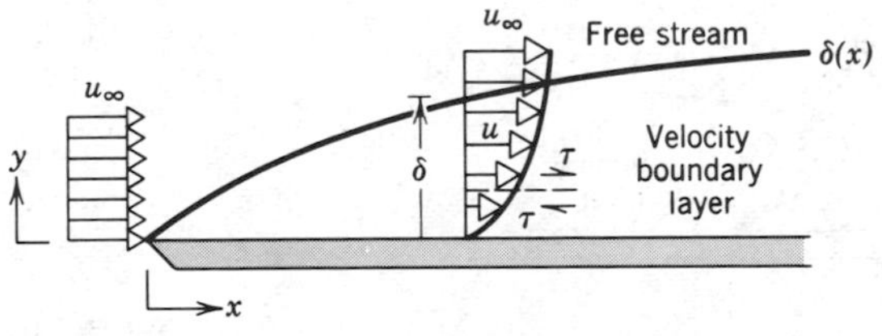

Figure 6.2 Velocity boundary layer on a flat plate.

Accordingly, the fluid flow is characterized by two distinct regions: a thin fluid layer (the boundary layer) in which velocity gradients and shear stresses are large, and a region outside the boundary layer in which velocity gradients and shear stresses are negligible. With increasing distance from the leading edge, the effects of viscosity penetrate further into the free stream and the boundary layer grows (δ increases with x).

Because it pertains to the fluid velocity, the foregoing boundary layer may be referred to more specifically as the *velocity boundary layer*. It develops whenever there is fluid flow over a surface, and it is of fundamental importance to problems involving convection transport. In fluid mechanics its significance to the engineer stems from its relation to the surface shear stress τ_s, and hence to surface frictional effects. For external flows it provides the basis for determining the local *friction coefficient*

$$C_f \equiv \frac{\tau_s}{\rho u_\infty^2/2} \tag{6.7}$$

a key dimensionless parameter from which the surface frictional drag may be determined. The surface shear stress may be evaluated from knowledge of the velocity gradient at the surface. That is,

$$\tau_s = \mu \left.\frac{\partial u}{\partial y}\right|_{y=0} \tag{6.8}$$

6.2.2 The Thermal Boundary Layer

Just as a velocity boundary layer develops when there is fluid flow over a surface, a *thermal boundary layer* must develop if the fluid freestream and surface temperatures differ. Consider flow over an isothermal flat plate (Figure 6.3). At the leading edge the *temperature profile* is uniform, with $T(y) = T_\infty$. However, fluid particles that come into contact with the plate achieve thermal equilibrium at the plate's surface temperature. In turn these particles exchange energy with those in the adjoining fluid layer, and temperature gradients develop in the fluid. The region of the fluid in which these temperature gradients exist is the thermal boundary layer, and its thickness δ_t is formally defined as the value of y for which the ratio $[(T_s - T)/(T_s - T_\infty)] = 0.99$. Note that, with increasing distance

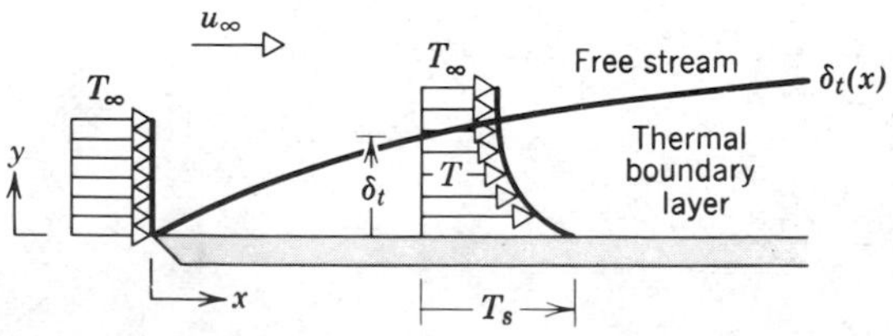

Figure 6.3 Thermal boundary layer on a flat plate.

from the leading edge, the effects of heat transfer penetrate further into the free stream and the thermal boundary layer grows.

The relationship between conditions in this boundary layer and the convection heat transfer coefficient may readily be demonstrated. At any distance from the leading edge, the *local* heat flux may be obtained by applying Fourier's law to the fluid at $y=0$. That is,

$$q_s'' = -k_f \frac{\partial T}{\partial y}\bigg|_{y=0} \tag{6.9}$$

This expression is appropriate because *at the surface, there is no fluid motion and energy transfer occurs only by conduction.* By combining Equation 6.9 with Newton's law of cooling, Equation 6.1, we then obtain

$$h = \frac{-k_f \partial T/\partial y|_{y=0}}{T_s - T_\infty} \tag{6.10}$$

Hence conditions in the thermal boundary layer, which strongly influence the wall temperature gradient $\partial T/\partial y|_{y=0}$, determine the rate of heat transfer across the boundary layer. Note that, since $(T_s - T_\infty)$ is a constant, independent of x, while δ_t increases with increasing x, temperature gradients in the boundary layer must decrease with increasing x. Accordingly, the magnitude of $\partial T/\partial y|_{y=0}$ decreases with increasing x, and it follows that q_s'' and h decrease with increasing x.

6.2.3 Significance of the Boundary Layers

The velocity boundary layer is of extent $\delta(x)$ and is characterized by the presence of velocity gradients and shear stresses. The thermal boundary layer is of extent $\delta_t(x)$ and is characterized by temperature gradients and heat transfer. For the engineer, the principal manifestations of the boundary layers are *surface friction* and *convection heat transfer*. The key boundary layer parameters are then the *friction coefficient* C_f and the *heat transfer convection coefficient* h.

For flow over any surface, there will always exist a velocity boundary layer, and hence surface friction. However, a thermal boundary layer, and hence convection heat transfer, exists only if the surface and freestream temperatures differ.

6.3 LAMINAR AND TURBULENT FLOW

An essential first step in the treatment of any convection problem is to determine whether the boundary layer is *laminar* or *turbulent*. Surface friction and the convection transfer rate depend strongly on which of these condition exists.

As shown in Figure 6.4, there are sharp differences between laminar and turbulent flow conditions. In the laminar boundary layer, fluid motion is highly

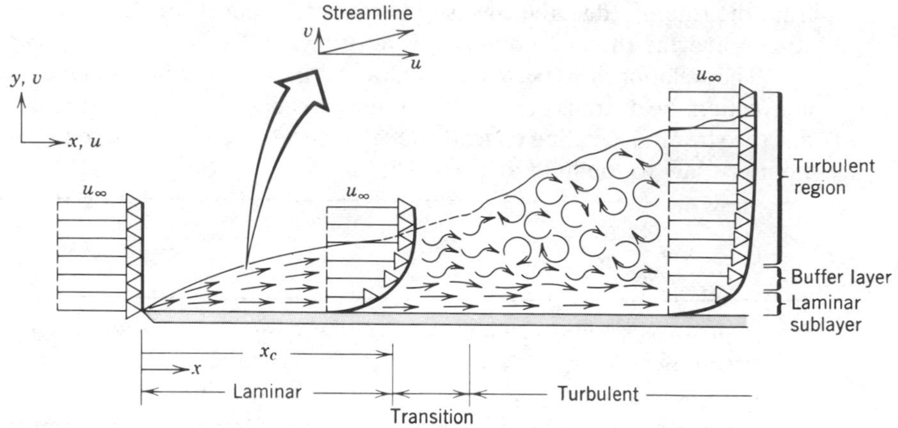

Figure 6.4 Velocity boundary layer development on a flat plate.

ordered and it is possible to identify streamlines along which particles move. Fluid motion along a streamline is characterized by velocity components in both the x and y directions. Since the velocity component v is in the direction normal to the surface, it can contribute significantly to the transfer of momentum or energy through the boundary layer. Fluid motion normal to the surface is necessitated by boundary layer growth in the x direction.

In contrast, fluid motion in the turbulent boundary layer is highly irregular and is characterized by velocity fluctuations. These fluctuations enhance the transfer of momentum and energy, hence increasing surface friction, as well as the convection heat transfer rate. Due to fluid mixing resulting from the fluctuations, turbulent boundary layer thicknesses are larger and boundary layer profiles (velocity and temperature) are flatter than in laminar flow.

The foregoing conditions are shown schematically in Figure 6.4 for velocity boundary layer development on a flat plate. The boundary layer is initially laminar, but at some distance from the leading edge, transition to turbulent flow begins to occur. Fluid fluctuations begin to develop in the *transition region*, and the boundary layer eventually becomes completely turbulent. The transition to turbulence is accompanied by significant increases in the boundary layer thicknesses, the wall shear stress, and the convection coefficient. These effects are illustrated in Figure 6.5 for the velocity boundary layer thickness δ and the local convection heat transfer coefficient h. In the turbulent boundary layer, three different regions may be delineated. We may speak of a *laminar sublayer* in which transport is dominated by diffusion and the velocity profile is nearly linear. There is an adjoining *buffer layer* in which diffusion and turbulent mixing are comparable, and there is a *turbulent zone* in which transport is dominated by turbulent mixing.

In calculating boundary layer behavior it is frequently reasonable to assume that transition begins at some location x_c. This location is determined by

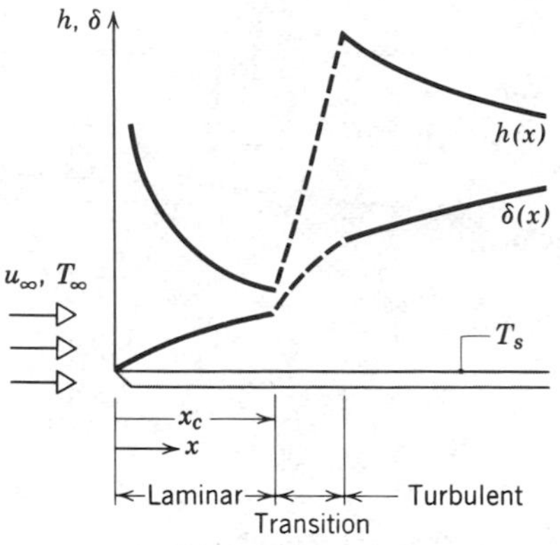

Figure 6.5 Variation of velocity boundary layer thickness δ and the local heat transfer coefficient h for flow over an isothermal flat plate.

a dimensionless grouping of variables called the *Reynolds number*,

$$Re_x \equiv \frac{\rho u_\infty x}{\mu} \tag{6.11}$$

where the characteristic length x is the distance from the leading edge. The *critical Reynolds number* is the value of Re_x for which transition begins, and for external flow it is known to vary from 10^5 to 3×10^6, depending on surface roughness, the turbulence level of the free stream and the nature of the pressure variation along the surface. A representative value of

$$Re_{x,c} = \frac{\rho u_\infty x_c}{\mu} = 5 \times 10^5 \tag{6.12}$$

is generally assumed for boundary layer calculations and, unless otherwise noted, is used for the calculations of this text.

6.4 THE CONVECTION TRANSFER EQUATIONS

We can improve our understanding of the physical effects that determine boundary layer behavior and further illustrate its relevance to convection transport by developing the equations that govern boundary layer conditions. Consider the simultaneous development of velocity and thermal boundary layers over the surface of Figure 6.6. Selection of the relative thicknesses ($\delta_t > \delta$) is arbitrary, for the moment, and the factors that influence relative boundary layer development are discussed later in this chapter. To simplify the development we assume two-dimensional, steady flow conditions for which x is in the direction along the surface and y is normal to the surface. Extension of this development to three-dimensional flows may be readily made [1–3].

For each of the boundary layers we will identify the relevant physical effects and apply the appropriate conservation laws to control volumes of infinitesimal size. Although it is not essential to follow in detail the derivation of the

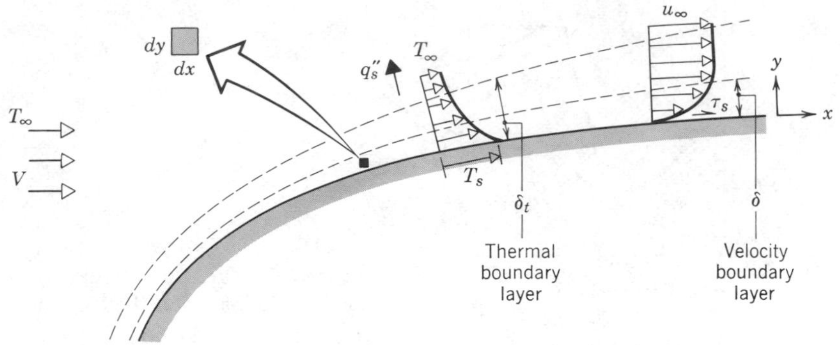

Figure 6.6 Development of the velocity and thermal boundary layers for an arbitrary surface.

conservation equations, you should attempt to develop an understanding of the underlying physical effects.

6.4.1 The Velocity Boundary Layer

One conservation law that is pertinent to the velocity boundary layer is that matter may neither be created nor destroyed. Stated in the context of the differential control volume of Figure 6.7, this law requires that, for steady flow, *the net rate at which mass enters the control volume* (inflow–outflow) *must equal zero.* Mass enters and leaves the control volume exclusively through gross fluid motion. Transport due to such motion is often referred to as *advection.* If one corner of the control volume, $dx\ dy$, is located at (x,y), the rate at which mass enters the control volume through the surface perpendicular to x may be expressed as $(\rho u)\ dy$, where ρ is the mass density and u is the x component of the velocity. Note that the control volume is of unit depth in the z direction. Since ρ and u may vary with x, the rate at which mass leaves the surface at $x + dx$ may be

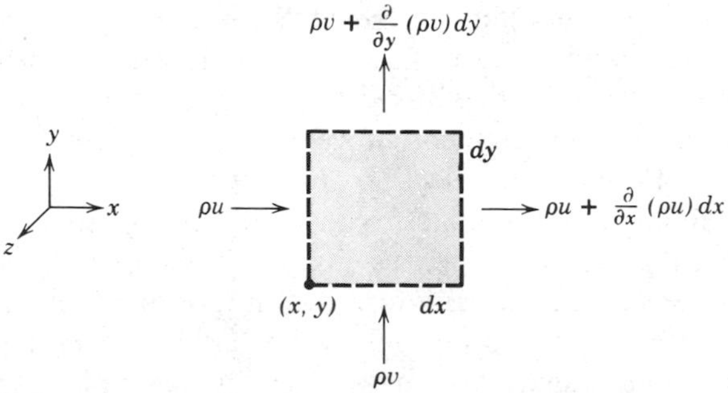

Figure 6.7 Differential control volume $(dx \cdot dy \cdot 1)$ for mass conservation in the two-dimensional velocity boundary layer.

expressed by a Taylor series expansion of the form

$$\left[(\rho u)+\frac{\partial(\rho u)}{\partial x}\,dx\right]dy$$

Using a similar result for the y direction, the conservation of mass requirement becomes

$$(\rho u)\,dy+(\rho v)\,dx-\left[\rho u+\frac{\partial(\rho u)}{\partial x}\,dx\right]dy-\left[\rho v+\frac{\partial(\rho v)}{\partial y}\,dy\right]dx=0$$

Canceling terms and dividing by $dx\,dy$, we obtain

$$\frac{\partial(\rho u)}{\partial x}+\frac{\partial(\rho v)}{\partial y}=0 \qquad (6.13)$$

Continuity Eqn.

Equation 6.13, the *continuity equation*, is a general expression of the mass conservation requirement, and it must be satisfied at every point in the velocity boundary layer.

The second fundamental law that is pertinent to the velocity boundary layer is *Newton's second law of motion.* For a differential control volume in the velocity boundary layer, this requirement states that the sum of all forces acting on the control volume must equal the net rate at which momentum leaves the control volume (outflow–inflow).

Two kinds of forces may act on the fluid in the boundary layer: *body forces*, which are proportional to the volume, and *surface forces*, which are proportional to area. Gravitational, centrifugal, magnetic, and/or electric fields may contribute to the total body force, and we designate the x and y components of this force per unit volume of fluid as X and Y, respectively. The surface forces F_s are due to the fluid static pressure, as well as to *viscous stresses.* At any point in the boundary layer, the viscous stress (a force per unit area) may be resolved into two perpendicular components, which include a *normal stress* σ_{ii} and a *shear stress* τ_{ij} (Figure 6.8).

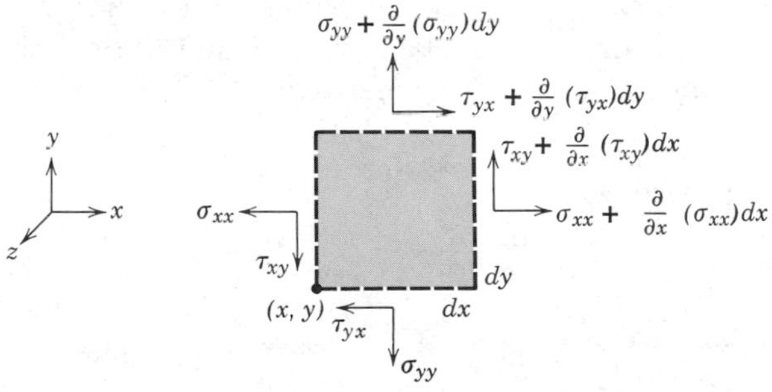

Figure 6.8 Normal and shear viscous stresses for a differential control volume $(dx\cdot dy\cdot 1)$ in the two-dimensional velocity boundary layer.

A double subscript notation is used to specify the stress components. The first subscript indicates the surface orientation by providing the direction of its outward normal, and the second subscript indicates the direction of the force component. Accordingly, for the x surface of Figure 6.8, the normal stress σ_{xx} corresponds to a force component normal to the surface, and the shear stress τ_{xy} corresponds to a force in the y direction along the surface. All the stress components shown are positive in the sense that *both* the surface normal and the force component are in the same direction. That is, they are both either in the positive coordinate direction or in the negative coordinate direction. By this convention the normal viscous stresses are *tensile* stresses. In contrast, the static pressure originates from an external force acting on the fluid in the control volume and is therefore a *compressive* stress.

Several features of the viscous stress should be noted. The associated force is between adjoining fluid elements and is a natural consequence of the fluid motion and viscosity. The surface forces of Figure 6.8 are therefore presumed to act on the fluid within the control volume and to be due to its interaction with the surrounding fluid. These stresses would vanish if the fluid velocity, or more specifically the velocity gradient, went to zero. In this respect the normal viscous stresses (σ_{xx} and σ_{yy}) must not be confused with the static pressure, which does not vanish for zero velocity.

Each of the stresses may change continuously in each of the coordinate directions. Using a Taylor series expansion for the stresses, the *net* surface force for each of the two directions may be expressed as

$$F_{s,x} = \left(\frac{\partial \sigma_{xx}}{\partial x} - \frac{\partial p}{\partial x} + \frac{\partial \tau_{yx}}{\partial y}\right) dx\, dy \tag{6.14}$$

$$F_{s,y} = \left(\frac{\partial \tau_{xy}}{\partial x} + \frac{\partial \sigma_{yy}}{\partial y} - \frac{\partial p}{\partial y}\right) dx\, dy \tag{6.15}$$

To use Newton's second law, the fluid momentum fluxes for the control volume must also be evaluated. If we focus on the x direction, the relevant fluxes are as shown in Figure 6.9. A contribution to the total x-momentum flux is made by the mass flow in each of the two directions. For example, the mass flux through the x surface (in the y–z plane) is (ρu), and the corresponding x-momentum flux is $(\rho u)u$. Similarly, the x-momentum flux due to mass flow

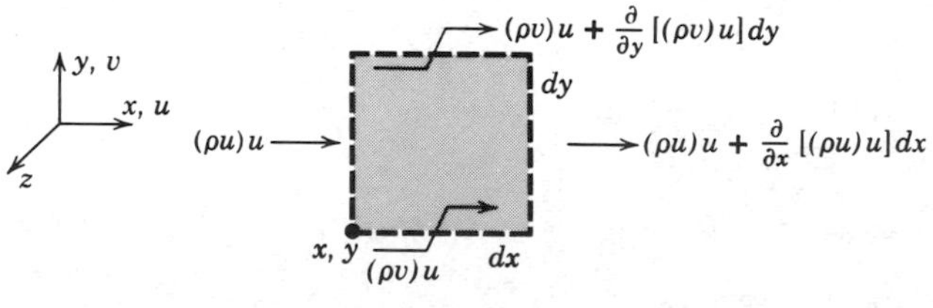

Figure 6.9 Momentum fluxes for a differential control volume ($dx \cdot dy \cdot 1$) in the two-dimensional velocity boundary layer.

through the y surface (in the x–z plane) is $(\rho v)u$. These fluxes may change in each of the coordinate directions, and the *net* rate at which x-momentum leaves the control volume is

$$\frac{\partial[(\rho u)u]}{\partial x}\,dx\,(dy)+\frac{\partial[(\rho v)u]}{\partial y}\,dy\,(dx)$$

Equating the rate of change in the x momentum of the fluid to the sum of the forces in the x direction, we then obtain

$$\frac{\partial[(\rho u)u]}{\partial x}+\frac{\partial[(\rho v)u]}{\partial y}=\frac{\partial\sigma_{xx}}{\partial x}-\frac{\partial p}{\partial x}+\frac{\partial\tau_{yx}}{\partial y}+X \tag{6.16}$$

This expression may be put in a more convenient form by expanding the derivatives on the left-hand side and substituting from the continuity equation, (6.13), in which case we obtain

$$\rho\left(u\frac{\partial u}{\partial x}+v\frac{\partial u}{\partial y}\right)=\frac{\partial}{\partial x}(\sigma_{xx}-p)+\frac{\partial\tau_{yx}}{\partial y}+X \tag{6.17}$$

A similar expression may be obtained for the y direction and is of the form

$$\rho\left(u\frac{\partial v}{\partial x}+v\frac{\partial v}{\partial y}\right)=\frac{\partial\tau_{xy}}{\partial x}+\frac{\partial}{\partial y}(\sigma_{yy}-p)+Y \tag{6.18}$$

We should not lose sight of the physics represented by Equations 6.17 and 6.18. The two terms on the left-hand side of each equation represent the *net* rate of momentum flow from the control volume. The terms on the right-hand side account for net viscous and pressure forces, as well as the body force. These equations must be satisfied at each point in the boundary layer, and with Equation 6.13 they may be solved for the velocity field.

Before a solution to the foregoing equations can be obtained, it is necessary to relate the viscous stresses to other flow variables. These stresses are associated with the deformation of the fluid and are a function of the fluid viscosity and velocity gradients. From Figure 6.10 it is evident that a *normal stress* must produce a *linear deformation* of the fluid, whereas a *shear stress* produces an *angular deformation*. Moreover, the magnitude of a stress is proportional to the

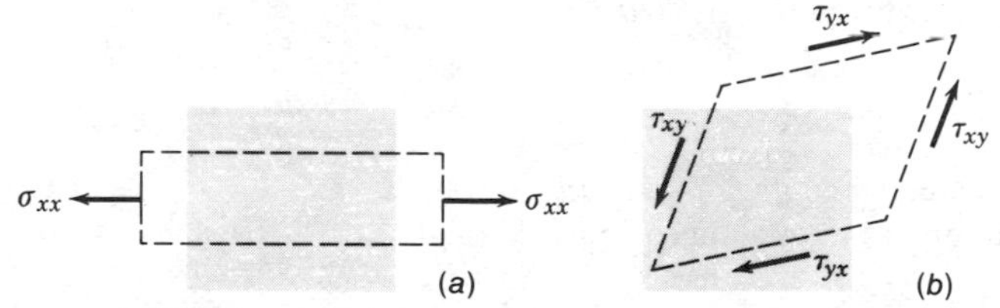

Figure 6.10 Deformations of a fluid element due to viscous stresses. (*a*) Linear deformation due to a normal stress. (*b*) Angular deformation due to shear stresses.

rate at which the deformation occurs. The deformation rate is, in turn, related to the fluid viscosity and to the velocity gradients in the flow. For a *Newtonian fluid*[1] the stresses are proportional to the velocity gradients, where the proportionality constant is the fluid viscosity. Because of its complexity, however, development of the specific relations is left to the literature [1], and we limit ourselves to a presentation of the results. In particular, it has been shown that

$$\sigma_{xx}=2\mu\frac{\partial u}{\partial x}-\frac{2}{3}\mu\left(\frac{\partial u}{\partial x}+\frac{\partial v}{\partial y}\right) \tag{6.19}$$

$$\sigma_{yy}=2\mu\frac{\partial v}{\partial y}-\frac{2}{3}\mu\left(\frac{\partial u}{\partial x}+\frac{\partial v}{\partial y}\right) \tag{6.20}$$

$$\tau_{xy}=\tau_{yx}=\mu\left(\frac{\partial u}{\partial y}+\frac{\partial v}{\partial x}\right) \tag{6.21}$$

Equation 6.13 and Equations 6.17 to 6.21 provide a complete representation of conditions in a two-dimensional velocity boundary layer, and the velocity field in the boundary layer may be determined by solving these equations. Once the velocity field is known, it is a simple matter to obtain the wall shear stress τ_s from Equation 6.8.

6.4.2 The Thermal Boundary Layer

To apply the energy conservation requirement, Equation 1.10a, to a differential control volume in the thermal boundary layer (Figure 6.11), it is necessary to first delineate the relevant physical processes. Neglecting potential energy

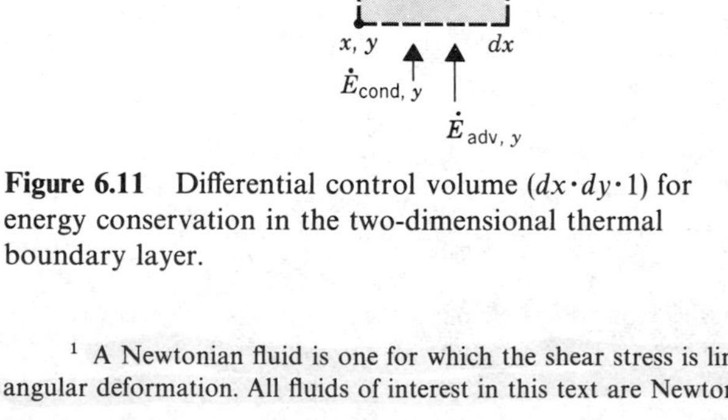

Figure 6.11 Differential control volume $(dx \cdot dy \cdot 1)$ for energy conservation in the two-dimensional thermal boundary layer.

[1] A Newtonian fluid is one for which the shear stress is linearly proportional to the rate of angular deformation. All fluids of interest in this text are Newtonian.

effects, the energy per unit mass of the fluid includes the thermal internal energy e and the kinetic energy $V^2/2$, where $V^2 \equiv u^2 + v^2$. Accordingly, thermal and kinetic energy are *advected* with the *bulk fluid* motion across the control surfaces, and for the x direction, the *net* rate at which this energy *enters* the control volume is

$$\dot{E}_{\text{adv},x} - \dot{E}_{\text{adv},x+dx} \equiv \rho u\left(e + \frac{V^2}{2}\right) dy - \left\{\rho u\left(e + \frac{V^2}{2}\right) + \frac{\partial}{\partial x}\left[\rho u\left(e + \frac{V^2}{2}\right)\right] dx\right\} dy = -\frac{\partial}{\partial x}\left[\rho u\left(e + \frac{V^2}{2}\right)\right] dx\, dy \tag{6.22}$$

Energy is also transferred across the control surface by *conduction*, and the *net* transfer of energy into the control volume is

$$\dot{E}_{\text{cond},x} - \dot{E}_{\text{cond},x+dx} = -\left(k\frac{\partial T}{\partial x}\right) dy - \left[-k\frac{\partial T}{\partial x} - \frac{\partial}{\partial x}\left(k\frac{\partial T}{\partial x}\right) dx\right] dy = \frac{\partial}{\partial x}\left(k\frac{\partial T}{\partial x}\right) dx\, dy \tag{6.23}$$

Energy may also be transferred to and from the fluid in the control volume by *work* interactions involving the *body* and *surface forces*. The *net* rate at which work is done *on* the fluid by forces in the x direction may be expressed as

$$\dot{W}_{\text{net},x} = (Xu)\, dx\, dy + \frac{\partial}{\partial x}[(\sigma_{xx} - p)u]\, dx\, dy + \frac{\partial}{\partial y}(\tau_{yx}u)\, dx\, dy \tag{6.24}$$

The first term on the right-hand side of Equation 6.24 represents the work done by the body force, and the remaining terms account for the *net* work done by the pressure and viscous forces.

Using Equations 6.22 to 6.24, as well as analogous equations for the y direction, the energy conservation requirement, Equation 1.10a, may be expressed as

$$-\frac{\partial}{\partial x}\left[\rho u\left(e + \frac{V^2}{2}\right)\right] - \frac{\partial}{\partial y}\left[\rho v\left(e + \frac{V^2}{2}\right)\right] + \frac{\partial}{\partial x}\left(k\frac{\partial T}{\partial x}\right) + \frac{\partial}{\partial y}\left(k\frac{\partial T}{\partial y}\right) + (Xu + Yv) - \frac{\partial}{\partial x}(pu) - \frac{\partial}{\partial y}(pv) + \frac{\partial}{\partial x}(\sigma_{xx}u + \tau_{xy}v) + \frac{\partial}{\partial y}(\tau_{yx}u + \sigma_{yy}v) + \dot{q} = 0 \tag{6.25}$$

where $\dot{q}$ is the rate at which energy is generated per unit volume. This expression provides a general form of the energy conservation requirement for the thermal boundary layer.

Because Equation 6.25 represents conservation of *mechanical* and *thermal* energy, it is rarely used in solving heat transfer problems. Instead, a more convenient form, which is termed the *thermal energy equation*, is obtained by multiplying Equations 6.17 and 6.18 by u and v, respectively, and subtracting the results from Equation 6.25. The desired form is [2]

$$\rho u\frac{\partial i}{\partial x}+\rho v\frac{\partial i}{\partial y}=\frac{\partial}{\partial x}\left(k\frac{\partial T}{\partial x}\right)+\frac{\partial}{\partial y}\left(k\frac{\partial T}{\partial y}\right)$$

$$+\left(u\frac{\partial p}{\partial x}+v\frac{\partial p}{\partial y}\right)+\mu\Phi+\dot{q} \tag{6.26}$$

where i is the enthalpy per unit mass of mixture

$$i=e+\frac{p}{\rho} \tag{6.27}$$

and $\mu\Phi$ is the *viscous dissipation*, defined as

$$\mu\Phi\equiv\mu\left\{\left(\frac{\partial u}{\partial y}+\frac{\partial v}{\partial x}\right)^2+2\left[\left(\frac{\partial u}{\partial x}\right)^2+\left(\frac{\partial v}{\partial y}\right)^2\right]-\frac{2}{3}\left(\frac{\partial u}{\partial x}+\frac{\partial v}{\partial y}\right)^2\right\} \tag{6.28}$$

The first term on the right-hand side of Equation 6.28 originates from the viscous shear stresses, and the remaining terms arise from the viscous normal stresses. Collectively, the terms account for the rate at which *mechanical energy is converted to thermal energy due to viscous effects in the fluid.*

EXAMPLE 6.2

One of the few situations for which *exact* solutions to the convection transfer equations may be obtained involves what is termed *parallel flow.* In this case gross fluid motion is only in one direction. Consider a special case of parallel flow involving stationary and moving plates of infinite extent separated by a distance L, with the intervening space filled by an incompressible fluid. This situation is referred to as Couette flow and occurs, for example, in a journal bearing.

1. What is the appropriate form of the continuity equation (6.13)?
2. Beginning with the momentum equation (6.17), determine the velocity distribution between the plates.
3. Beginning with the energy equation (6.26), determine the temperature distribution between the plates.
4. Consider conditions for which the fluid is engine oil with $L=3$ mm. The speed of the moving plate is $U=10$ m/s, and the temperatures of the stationary and moving plates are $T_0=10°C$ and $T_L=30°C$, respectively. Calculate the heat flux to each of the plates and determine the maximum temperature in the oil.

SOLUTION

Known: Couette flow with heat transfer.

Find:

1. Form of the continuity equation.
2. Velocity distribution.
3. Temperature distribution.
4. Surface heat fluxes and maximum temperature for prescribed conditions.

Schematic:

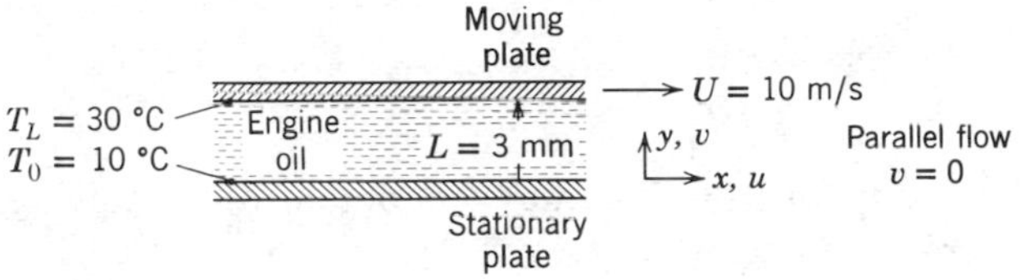

Assumptions:

1. Steady-state conditions.
2. Two-dimensional flow (no variations in z).
3. Incompressible fluid with constant properties.
4. No body forces.
5. No internal energy generation.

Properties:

Table A.5, engine oil (20°C): $\rho = 888.2\ \text{kg/m}^3$, $k = 0.145\ \text{W/m}\cdot\text{K}$, $\nu = 900 \times 10^{-6}\ \text{m}^2/\text{s}$, $\mu = \nu\rho = 0.799\ \text{N}\cdot\text{s/m}^2$.

Analysis:

1. For an incompressible fluid (constant ρ) and parallel flow ($v = 0$), Equation 6.13 reduces to

$$\frac{\partial u}{\partial x} = 0 \qquad \triangleleft$$

The important implication of this result is that, while depending on y, the x-velocity component u is independent of x. It may then be said that the velocity field is *fully developed.*

2. For two-dimensional, steady-state conditions with $v=0$, $(\partial u/\partial x)=0$ and $X=0$, Equation 6.17 reduces to

$$0=\frac{\partial}{\partial x}(\sigma_{xx}-p)+\frac{\partial \tau_{yx}}{\partial y}$$

However, in Couette flow, motion of the fluid is sustained not by the pressure gradient, $\partial p/\partial x$, but by an external force that provides for motion of the top plate relative to the bottom plate. Hence $(\partial p/\partial x)=0$. From Equation 6.19 it also follows that, with $v=(\partial u/\partial x)=0$, $\sigma_{xx}=0$, and from Equation 6.21, with $v=0$,

$$\tau_{xy}=\mu\frac{\partial u}{\partial y}$$

Accordingly, the x-momentum equation reduces to

$$\frac{\partial^2 u}{\partial y^2}=0$$

The desired velocity distribution may be obtained by solving this equation. Integrating twice, we obtain

$$u(y)=C_1 y+C_2$$

where C_1 and C_2 are the constants of integration. Applying the boundary conditions

$$u(0)=0 \qquad \text{and} \qquad u(L)=U$$

it follows that $C_2=0$ and $C_1=U/L$. The velocity distribution is then

$$u(y)=\frac{y}{L}U$$ ◁

3. The energy equation, (6.26), may be simplified for the prescribed conditions. In particular for two-dimensional, steady-state conditions with $v=0$, $(\partial u/\partial x)=0$, $(\partial p/\partial x)=0$, and $\dot{q}=0$, it follows that

$$\rho u\frac{\partial i}{\partial x}=\frac{\partial}{\partial x}\left(k\frac{\partial T}{\partial x}\right)+\frac{\partial}{\partial y}\left(k\frac{\partial T}{\partial y}\right)+\mu\left(\frac{\partial u}{\partial y}\right)^2$$

However, because the top and bottom plates are at uniform temperatures, the temperature field must also be fully developed, in which case $(\partial T/\partial x)=0$. Hence, since the enthalpy is a function of temperature and pressure, $i(T,p)$, its longitudinal gradient may be expressed as

$$\frac{\partial i}{\partial x}=\frac{\partial i}{\partial T}\frac{\partial T}{\partial x}+\frac{\partial i}{\partial p}\frac{\partial p}{\partial x}=0$$

For constant properties the appropriate form of the energy equation is then

$$0 = k\frac{\partial^2 T}{\partial y^2} + \mu\left(\frac{\partial u}{\partial y}\right)^2$$

The desired temperature distribution may be obtained by solving this equation. Rearranging and substituting for the velocity distribution,

$$k\frac{d^2 T}{dy^2} = -\mu\left(\frac{du}{dy}\right)^2 = -\mu\left(\frac{U}{L}\right)^2$$

Integrating twice, we obtain

$$T(y) = -\frac{\mu}{2k}\left(\frac{U}{L}\right)^2 y^2 + C_3 y + C_4$$

The constants of integration may be obtained from the boundary conditions

$$T(0) = T_0 \qquad \text{and} \qquad T(L) = T_L$$

in which case

$$C_4 = T_0 \qquad \text{and} \qquad C_3 = \frac{T_L - T_0}{L} + \frac{\mu}{2k}\frac{U^2}{L}$$

and

$$T(y) = T_0 + \frac{\mu}{2k}U^2\left[\frac{y}{L} - \left(\frac{y}{L}\right)^2\right] + (T_L - T_0)\frac{y}{L} \qquad \triangleleft$$

4. Knowing the temperature distribution, the surface heat fluxes may be obtained by applying Fourier's law. Hence

$$q''_y = -k\frac{dT}{dy} = -k\left[\frac{\mu}{2k}U^2\left(\frac{1}{L} - \frac{2y}{L^2}\right) + \frac{T_L - T_0}{L}\right]$$

At the bottom and top surfaces, respectively, it follows that

$$q''_0 = -\frac{\mu U^2}{2L} - \frac{k}{L}(T_L - T_0) \qquad \text{and} \qquad q''_L = +\frac{\mu U^2}{2L} - \frac{k}{L}(T_L - T_0)$$

Hence, for the prescribed numerical values,

$$q''_0 = -\frac{0.799\ \text{N}\cdot\text{s/m}^2 \times 100\ \text{m}^2/\text{s}^2}{2 \times 3 \times 10^{-3}\ \text{m}} - \frac{0.145\ \text{W/m}\cdot\text{K}}{3 \times 10^{-3}\ \text{m}}(30 - 10)^\circ\text{C}$$

$$q''_0 = -13{,}315\ \text{W/m}^2 - 967\ \text{W/m}^2 = -14.3\ \text{kW/m}^2 \qquad \triangleleft$$

$$q''_L = +13{,}315\ \text{W/m}^2 - 967\ \text{W/m}^2 = 12.3\ \text{kW/m}^2 \qquad \triangleleft$$

The location of the maximum temperature in the oil may be found from the requirement that

$$\frac{dT}{dy} = \frac{\mu}{2k}U^2\left(\frac{1}{L} - \frac{2y}{L^2}\right) + \frac{T_L - T_0}{L} = 0$$

Solving for y it follows that

$$y_{\max} = \left[\frac{k}{\mu U^2}(T_L - T_0) + \frac{1}{2} \right] L$$

or for the prescribed conditions

$$y_{\max} = \left[\frac{0.145 \text{ W/m}\cdot\text{K}}{0.799 \text{ N}\cdot\text{s/m}^2 \times 100 \text{ m}^2/\text{s}^2}(30 - 10)^\circ\text{C} + \frac{1}{2} \right] L = 0.536L$$

Substituting the value of $y_{\max}$ into the expression for $T(y)$, it follows that

$$T_{\max} = 89.3^\circ\text{C}$$ ◁

Comments:

1. Due to the strong effect of viscous dissipation for the prescribed conditions, the maximum temperature occurs in the oil and there is heat transfer to the hot, as well as to the cold, plate. The temperature distribution is a function of the velocity of the moving plate, and the effect is shown schematically below.

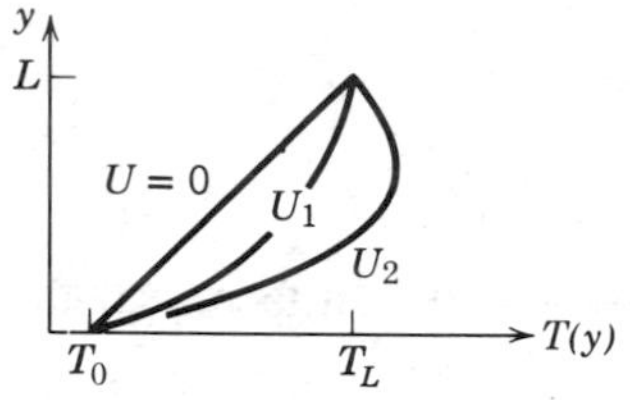

   For velocities less than U_1 the maximum temperature corresponds to that of the hot plate. For $U = 0$ there is no viscous dissipation, and the temperature distribution is linear.

2. The properties were evaluated at $\bar{T} = (T_L + T_0)/2 = 20^\circ$C, which is *not* a good measure of the average oil temperature. For more precise calculations, the properties should be evaluated at a more appropriate value of the average temperature (e.g., $\bar{T} \approx 55^\circ$C), and the calculations should be repeated.

6.5 APPROXIMATIONS AND SPECIAL CONDITIONS

The equations of the preceding section provide a complete account of the physical processes that influence conditions in the velocity and thermal boundary layers. Rarely, however, do all the terms need to be considered and it is customary to work with simplified forms of the equations. The usual situation is one in which the two-dimensional boundary layer can be characterized as:

steady (time-independent), *incompressible* (ρ is constant), having *constant properties* (k, μ, etc.) and *negligible body forces* ($X = Y = 0$), and *without energy generation* ($\dot{q} = 0$).

Additional simplifications may be made by invoking what are known as the *boundary layer approximations.* Because the boundary layer thicknesses are typically very small, the following inequalities are known to apply.

Velocity Boundary Layer:

$$u \gg v$$

$$\frac{\partial u}{\partial y} \gg \frac{\partial u}{\partial x}, \frac{\partial v}{\partial y}, \frac{\partial v}{\partial x}$$

Thermal Boundary Layer:

$$\frac{\partial T}{\partial y} \gg \frac{\partial T}{\partial x}$$

That is, the velocity component in the direction along the surface is much larger than that normal to the surface, and gradients normal to the surface are much larger than those along the surface. The normal stresses given by Equations 6.19 and 6.20 are then negligible, and the single relevant shear stress component of Equation 6.21 reduces to

$$\tau_{xy} = \tau_{yx} = \mu\left(\frac{\partial u}{\partial y}\right) \tag{6.29}$$

Moreover, the conduction rate for the y direction is much larger than that for the x direction.

With the foregoing simplifications and approximations, the overall continuity equation, (6.13), and the x-momentum equation, (6.17), reduce to

$$\frac{\partial u}{\partial x} + \frac{\partial v}{\partial y} = 0 \qquad \text{Continuity} \tag{6.30}$$

$$u\frac{\partial u}{\partial x} + v\frac{\partial u}{\partial y} = -\frac{1}{\rho}\frac{\partial p}{\partial x} + \nu\frac{\partial^2 u}{\partial y^2} \qquad \text{x-momentum} \tag{6.31}$$

In addition, from an order-of-magnitude analysis that uses the velocity boundary layer approximations [1, 3], it may be shown that the y-momentum equation, (6.18), reduces to

$$\frac{\partial p}{\partial y} = 0 \qquad \text{y-momentum} \tag{6.32}$$

That is, the *pressure does not vary in the direction normal to the surface.* Hence, the pressure in the boundary layer depends only on x and is equal to the pressure in the free stream outside the boundary layer. The form of $p(x)$, which depends on the surface geometry, may then be obtained from a separate consideration of flow conditions in the free stream. Hence, as far as Equation 6.31 is concerned, $(\partial p/\partial x) = (dp/dx)$, and the pressure gradient may be treated as a known quantity.

With the foregoing simplifications and the assumption that the terms $u(\partial p/\partial x)$ and $v(\partial p/\partial y)$ are negligible, the energy equation, (6.26), reduces to

$$u\frac{\partial T}{\partial x} + v\frac{\partial T}{\partial y} = \alpha\frac{\partial^2 T}{\partial y^2} + \frac{\nu}{c_p}\left(\frac{\partial u}{\partial y}\right)^2 \qquad (6.33)$$

The last term on the right-hand side of Equation 6.33 is what remains of the viscous dissipation, Equation 6.28. In most situations this term may be neglected relative to those that account for advection (the left-hand side of the equation) and conduction (the first term on the right-hand side). In fact it is only for sonic flows or the high speed motion of lubricating oils that viscous dissipation may not be neglected.

Equations 6.30, 6.31, and 6.33 may be solved to determine the spatial variations of u, v and T in the boundary layers. For incompressible, constant property flow, Equations 6.30 and 6.31 are *uncoupled* from 6.33. That is, Equations 6.30 and 6.31 may be solved for the *velocity field*, $u(x,y)$ and $v(x,y)$, to the exclusion of Equation 6.33. From knowledge of $u(x,y)$, the velocity gradient $(\partial u/\partial y)_{y=0}$ could then be evaluated, and the wall shear stress could be obtained from Equation 6.8. In contrast, through the appearance of u and v in Equation 6.33, the temperature and velocity fields are *coupled*. Hence, $u(x,y)$ and $v(x,y)$ must be known before Equation 6.33 may be solved for $T(x,y)$. Once $T(x,y)$ has been obtained, the convection heat transfer coefficient may be determined from Equation 6.10. It follows that this coefficient depends strongly on the velocity field.

Because boundary layer solutions generally involve mathematics beyond the scope of this text, our treatment of such solutions is restricted to the analysis of parallel flow over an isothermal flat plate (Section 7.2 and Appendix C). However, many analytical solutions are discussed in advanced texts on convection [4–6], and detailed boundary layer solutions may be obtained by using numerical (finite-difference) techniques [7, 8].

We have not developed the boundary layer equations solely for the purpose of obtaining solutions. In fact, we have been motivated primarily by two other considerations. One major motivation has been to cultivate an appreciation for the different physical processes that occur in the boundary layers. These processes will, of course, affect wall friction, as well as energy transfer into the boundary layers. A second motivation comes from the fact that the equations may be used to identify key *boundary layer similarity parameters*, as well as an important *analogy* between *momentum* and *heat* transfer.

6.6 BOUNDARY LAYER SIMILARITY: THE NORMALIZED CONVECTION TRANSFER EQUATIONS

If we examine Equations 6.31 and 6.33 more carefully, we note a strong similarity. In fact, if the pressure gradient appearing in Equation 6.31 and the viscous dissipation term of Equation 6.33 are negligible, the two equations are of the same form. *Each equation is characterized by advection terms on the left-hand side and a diffusion term on the right-hand side.* This situation describes *lowspeed, forced-convection flows,* which are found in many engineering applications and will occupy much of our attention in this text. Implications of this similarity may be developed in a rational manner by first *nondimensionalizing* the governing equations.

6.6.1 Boundary Layer Similarity Parameters

The boundary layer equations are normalized by first defining dimensionless independent variables of the form

$$x^* \equiv \frac{x}{L} \quad \text{and} \quad y^* = \frac{y}{L} \tag{6.34}$$

where L is some *characteristic length* for the surface of interest (e.g., the length of a flat plate). Moreover, dependent dimensionless variables may also be defined as

$$u^* \equiv \frac{u}{V} \quad \text{and} \quad v^* \equiv \frac{v}{V} \tag{6.35}$$

where V is the velocity upstream of the surface (Figure 6.6), and as

$$T^* \equiv \frac{T - T_s}{T_\infty - T_s} \tag{6.36}$$

Equations 6.34 to 6.36 may be substituted into Equations 6.31 and 6.33 to obtain the dimensionless forms of the conservation equations shown in Table 6.1. Note that viscous dissipation has been neglected and that $p^* = (p/\rho V^2)$ is a dimensionless pressure. The boundary conditions required to solve the equations are also shown in the table.

From the form of Equations 6.37 and 6.38, two *similarity parameters* may be inferred. Similarity parmeters are important because they allow us to apply results obtained for a surface experiencing one set of conditions to *geometrically*

Table 6.1 The convection transfer equations and their boundary conditions in nondimensional form

BOUNDARY LAYER	CONSERVATION EQUATION	BOUNDARY CONDITIONS: WALL	BOUNDARY CONDITIONS: FREESTREAM	SIMILARITY PARAMETER(S)
Velocity	$u^*\frac{\partial u^*}{\partial x^*} + v^*\frac{\partial u^*}{\partial y^*} = -\frac{dp^*}{dx^*} + \frac{\nu}{VL}\frac{\partial^2 u^*}{\partial y^{*2}}$ (6.37)	$u^*(x^*,0)=0$, $v^*(x^*,0)=0$	$u^*(x^*,\infty)=\frac{u_\infty(x^*)}{V}$ (6.39)	Re_L
Thermal	$u^*\frac{\partial T^*}{\partial x^*} + v^*\frac{\partial T^*}{\partial y^*} = \frac{\alpha}{VL}\frac{\partial^2 T^*}{\partial y^{*2}}$ (6.38)	$T^*(x^*,0)=0$	$T^*(x^*,\infty)=1$ (6.40)	Re_L, Pr

similar surfaces experiencing entirely different conditions. These conditions may vary, for example, with the nature of the fluid, the fluid velocity, and/or with the size of the surface (as determined by L).

Beginning with Equation 6.37, we note that the quantity ν/VL is a *dimensionless group* whose reciprocal is termed the Reynolds number.

Reynolds Number:

$$Re_L \equiv \frac{VL}{\nu} \tag{6.41}$$

From Equation 6.38 we also note that the term α/VL is a dimensionless group that may be expressed as $(\nu/VL)(\alpha/\nu) = (Re_L)^{-1}(\alpha/\nu)$. The ratio of properties α/ν is also dimensionless, and its reciprocal is referred to as the Prandtl number.

Prandtl Number:

$$Pr \equiv \frac{\nu}{\alpha} \tag{6.42}$$

Using Equations 6.41 and 6.42 with the boundary layer equations (6.37) and (6.38), and including the dimensionless form of the continuity equation (6.30), the complete set of boundary layer equations becomes

$$\frac{\partial u^*}{\partial x^*} + \frac{\partial v^*}{\partial y^*} = 0 \tag{6.43}$$

Continuity

$$u^* \frac{\partial u^*}{\partial x^*} + v^* \frac{\partial u^*}{\partial y^*} = -\frac{dp^*}{dx^*} + \frac{1}{Re_L} \frac{\partial^2 u^*}{\partial y^{*2}} \tag{6.44}$$

Momentum

$$u^* \frac{\partial T^*}{\partial x^*} + v^* \frac{\partial T^*}{\partial y^*} = \frac{1}{Re_L Pr} \frac{\partial^2 T^*}{\partial y^{*2}} \tag{6.45}$$

Energy

6.6.2 Functional Form of the Solutions

The preceding equations are extremely useful from the standpoint of suggesting how important boundary layer results may be simplified and generalized. The momentum equation, (6.44), suggests that, although conditions in the velocity boundary layer depend on the fluid properties ρ and μ, the velocity V, and the length scale L, this dependence may be simplified by grouping these variables in the form of the Reynolds number. We therefore anticipate that the solution to Equation 6.44 will be of the functional form

$$u^* = f_1\left(x^*, y^*, Re_L, \frac{dp^*}{dx^*}\right) \tag{6.46}$$

Note that the pressure distribution $p^*(x^*)$ depends on the surface geometry and may be obtained independently by considering flow conditions in the free stream. Hence the appearance of dp^*/dx^* in Equation 6.46 represents the influence of geometry on the velocity distribution.

From Equation 6.8, the shear stress at the surface, $y^*=0$, may be expressed as

$$\tau_s=\mu\frac{\partial u}{\partial y}\bigg|_{y=0}=\left(\frac{\mu V}{L}\right)\frac{\partial u^*}{\partial y^*}\bigg|_{y^*=0}$$

and from Equations 6.7 and 6.41 it follows that the friction coefficient is

$$C_f=\frac{\tau_s}{\rho V^2/2}=\frac{2}{Re_L}\frac{\partial u^*}{\partial y^*}\bigg|_{y^*=0} \tag{6.47}$$

From Equation 6.46 we also know that

$$\frac{\partial u^*}{\partial y^*}\bigg|_{y^*=0}=f_2\left(x^*, Re_L, \frac{dp^*}{dx^*}\right)$$

Hence *for a prescribed geometry* Equation 6.47 may be expressed as

$$C_f=\frac{2}{Re_L}f_2(x^*, Re_L) \tag{6.48}$$

The significance of this result should not be overlooked. Equation 6.48 states that the friction coefficient, a dimensionless parameter of considerable importance to the engineer, may be expressed exclusively in terms of a dimensionless space coordinate and the Reynolds number. Hence for a prescribed geometry we expect the function that relates C_f to x^* and Re_L to be *universally* applicable. That is, we expect it to apply to different fluids and over a wide range of values for V and L.

A similar result may be obtained for the convection heat transfer coefficient. Intuitively, we might anticipate that h depends on the fluid properties (k, c_p, μ, and ρ), the fluid velocity V, the length scale L, and the surface geometry. However, Equation 6.45 suggests the manner in which this dependence may be simplified. In particular the solution to this equation may be expressed in the form

$$T^*=f_3\left(x^*, y^*, Re_L, Pr, \frac{dp^*}{dx^*}\right) \tag{6.49}$$

where the dependence on dp^*/dx^* originates from the influence of the fluid motion (u^* and v^*) on the thermal conditions. Once again the term dp^*/dx^* represents the effect of surface geometry. From the definition of the convection coefficient, Equation 6.10, and the dimensionless variables, Equations 6.34 and 6.36, we also obtain

$$h=-\frac{k_f}{L}\frac{(T_\infty-T_s)}{(T_s-T_\infty)}\frac{\partial T^*}{\partial y^*}\bigg|_{y^*=0}=+\frac{k_f}{L}\frac{\partial T^*}{\partial y^*}\bigg|_{y^*=0}$$

This expression suggests defining a dependent dimensionless parameter termed the Nusselt number.

Nusselt Number:

$$Nu \equiv \frac{hL}{k_f} = + \left.\frac{\partial T^*}{\partial y^*}\right|_{y^*=0} \tag{6.50}$$

This parameter is equal to the dimensionless temperature gradient at the surface, and it provides a measure of the convection heat transfer occurring at the surface. From Equation 6.49 it follows that, *for a prescribed geometry,*

$$Nu = f_4(x^*, Re_L, Pr) \tag{6.51}$$

The Nusselt number is to the thermal boundary layer what the friction coefficient is to the velocity boundary layer. Equation 6.51 implies that for a given geometry, the Nusselt number must be some *universal function* of x^*, Re_L, and Pr. If this function were known, it could be used to compute the value of Nu for different fluids and for different values of V and L. From knowledge of Nu, the local convection coefficient h may be found and the *local* heat flux may then be computed from Equation 6.1. Moreover, since the *average* heat transfer coefficient is obtained by integrating over the surface of the body, it must be independent of the spatial variable x^*. Hence the functional dependence of the *average* Nusselt number is

$$\overline{Nu} = \frac{\bar{h}L}{k_f} = f_5(Re_L, Pr) \tag{6.52}$$

From the foregoing development we have obtained the relevant dimensionless parameters for a low speed, forced-convection boundary layer. We have done so by nondimensionalizing the differential equations that describe the physical processes occurring within the boundary layers. An alternative approach could have involved the use of dimensional analysis in the form of the Buckingham pi theorem [9]. However, the success of this method depends on one's ability to select, largely from intuition, the various parameters that influence a problem. For example, knowing beforehand that $\bar{h} = f(k, c_p, \rho, \mu, V, L)$, one could use the Buckingham pi theorem to obtain Equation 6.52. However, having begun with the differential form of the conservation equations, we have eliminated the guesswork and have established the similarity parameters in a rigorous fashion.

The value of an expression such as Equation 6.52 should be fully appreciated. It states that convection heat transfer results, whether obtained theoretically or experimentally, can be represented in terms of three dimensionless groups, instead of the original seven parameters. The convenience afforded by such simplifications is evident. Moreover, once the form of the functional dependence of Equation 6.52 has been obtained for a particular surface geometry, let us say from laboratory measurements, it is known to be *universally* applicable. That is, it may be applied for different fluids, velocities, and length scales, as long as the assumptions implicit in the originating boundary layer equations remain valid (e.g., negligible viscous dissipation and body forces).

EXAMPLE 6.3

Experimental tests on a portion of the turbine blade shown indicate a heat flux to the blade of $q'' = 95{,}000\ \text{W/m}^2$. To maintain a steady-state surface temperature of 800°C, heat transferred to the blade is removed by circulating a coolant inside the blade.

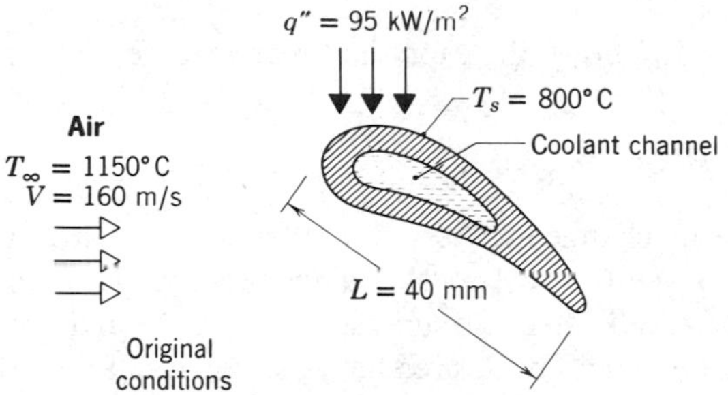

1. Determine the heat flux to the blade if its temperature is reduced to 700°C by increasing the coolant flow.
2. Determine the heat flux at the same dimensionless location for a similar turbine blade having a chord length of $L = 80$ mm, when the blade operates in an airflow at $T_\infty = 1150$°C and $V = 80$ m/s, with $T_s = 800$°C.

SOLUTION

Known: Operating conditions of an internally cooled turbine blade.

Find:

1. Heat flux to the blade when the surface temperature is reduced.
2. Heat flux to a larger turbine blade of the same shape with reduced air velocity.

Schematic:

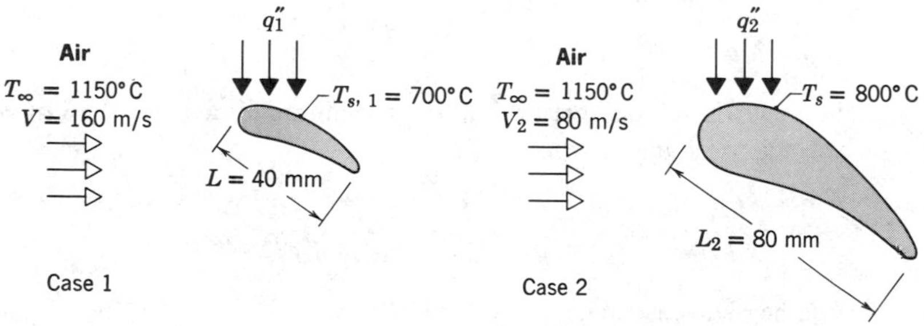

Assumptions:

1. Steady-state conditions.
2. Constant air properties.

Analysis:

1. From Equation 6.51 it follows that, for the prescribed geometry,

$$Nu = \frac{hL}{k} = f_4(x^*, Re_L, Pr)$$

Hence, since there is no change in x^*, Re_L, or Pr associated with a change in T_s for constant properties, the local Nusselt number is unchanged. Moreover, since L and k are also unchanged, the local convection coefficient remains the same. The desired heat flux for case 1 may then be obtained from Newton's law of cooling

$$q_1'' = h_1(T_\infty - T_s)_1$$

where

$$h_1 = h = \frac{q''}{(T_\infty - T_s)}$$

Hence

$$q_1'' = q'' \frac{(T_\infty - T_s)_1}{(T_\infty - T_s)} = 95{,}000\ \text{W/m}^2 \frac{(1150 - 700)°\text{C}}{(1150 - 800)°\text{C}} = 122{,}000\ \text{W/m}^2 \quad \triangleleft$$

2. To determine the heat flux associated with the larger blade and the reduced airflow (case 2), we first note that, although L has increased by a factor of 2, the velocity has decreased by the same factor and the Reynolds number has not changed. That is,

$$Re_{L,2} = \frac{V_2 L_2}{\nu} = \frac{VL}{\nu} = Re_L$$

Accordingly, since x^* and Pr are also unchanged, the local Nusselt number remains the same.

$$Nu_2 = Nu$$

Because the characteristic length is different, however, the convection coefficient changes, where

$$\frac{h_2 L_2}{k} = \frac{hL}{k} \quad \text{or} \quad h_2 = h\frac{L}{L_2} = \frac{q''}{(T_\infty - T_s)} \frac{L}{L_2}$$

The heat flux is then

$$q_2'' = h_2(T_\infty - T_s) = q'' \frac{(T_\infty - T_s)}{(T_\infty - T_s)} \frac{L}{L_2}$$

$$q_2'' = 95{,}000 \text{ W/m}^2 \times \frac{0.04 \text{ m}}{0.08 \text{ m}} = 47{,}500 \text{ W/m}^2 \qquad \triangleleft$$

Comments:

If the Reynolds number for the two situations of part 2 differed (i.e., $Re_{L,2} \neq Re_L$), the heat flux q_2'' could be obtained only if the particular form of the function f_4 were known. Such forms are provided for many different shapes in subsequent chapters.

6.7 PHYSICAL SIGNIFICANCE OF THE DIMENSIONLESS PARAMETERS

All of the foregoing dimensionless parameters have physical interpretations that relate to conditions in the boundary layers. Consider the *Reynolds number Re* (Equation 6.41), which may be interpreted as the *ratio of inertia to viscous forces* in the velocity boundary layer. For a differential control volume in this boundary layer, inertia forces are associated with an increase in the momentum flux of fluid moving through the control volume. From Equation 6.16, it is evident that these forces are of the form $\partial[(\rho u)u]/\partial x$, in which case an order of magnitude approximation gives $F_I \approx \rho V^2/L$. Similarly, the net shear force is of the form $\partial\tau_{yx}/\partial y = \partial[\mu(\partial u/\partial y)]/\partial y$ and may be approximated as $F_s \approx \mu V/L^2$. Therefore the ratio of forces is

$$\frac{F_I}{F_s} \approx \frac{\rho V^2/L}{\mu V/L^2} = \frac{\rho V L}{\mu} = Re_L$$

We therefore expect inertia forces to dominate for large values of *Re* and viscous forces to dominate for small *Re*.

The above result has several important implications. Recall that the Reynolds number determines the existence of laminar or turbulent flow. In any flow there exist small disturbances that can be amplified to produce turbulent conditions. However, for small *Re*, viscous forces are sufficiently large relative to inertia forces to prevent this amplification. Hence laminar flow is maintained. With increasing *Re* viscous effects become progressively less important relative to inertia effects, and small disturbances may be amplified to a point where transition occurs. We should also expect the magnitude of the Reynolds number to influence the velocity boundary layer thickness δ. With increasing *Re* at a fixed location on a surface, we expect viscous forces to become less influential relative to inertia forces. Hence the effects of viscosity do not penetrate as far into the freestream, and the value of δ diminishes.

The physical interpretation of the *Prandtl number* follows from its definition as a ratio of the momentum diffusivity ν to the thermal diffusivity α. The Prandtl number provides a *measure of the relative effectiveness of momentum and energy transport by diffusion in the velocity and thermal boundary layers*, respectively. From Table A.4 we see that the Prandtl number of gases is near unity, in which case energy and momentum transfer by diffusion are comparable. In a liquid metal (Table A.7), $Pr \ll 1$ and the energy diffusion rate greatly exceeds the momentum diffusion rate. The opposite is true for oils (Table A.5), for which $Pr \gg 1$. From this interpretation it follows that the value of Pr strongly influences the relative growth of the velocity and thermal boundary layers. In fact for laminar boundary layers (in which transport by diffusion is *not* overshadowed by turbulent mixing), it is reasonable to expect that

$$\frac{\delta}{\delta_t} \approx Pr^n \tag{6.53}$$

where n is a positive exponent. Hence for a gas $\delta_t \approx \delta$; for a liquid metal $\delta_t \gg \delta$; for an oil $\delta_t \ll \delta$.

Table 6.2 lists the dimensionless groups that appear frequently in the heat transfer literature. The list includes groups already considered, as well as those yet to be introduced for special convection problems. As a new group is confronted, its definition and interpretation should be committed to memory. Note that the *Grashof number* provides a measure of the ratio of buoyancy forces to viscous forces in the velocity boundary layer. Its role in free convection (Chapter 9) is much the same as that of the Reynolds number in forced convection. The *Eckert number* provides a measure of the kinetic energy of the flow relative to the enthalpy difference across the thermal boundary layer. It plays an important role in high speed flows for which viscous dissipation is significant [3]. Note also that, although similar in form, the Nusselt and Biot numbers differ in both definition and interpretation. Whereas the Nusselt number is defined in terms of the thermal conductivity of the fluid, the Biot number is based on the solid thermal conductivity, Equation 5.9.

6.8 MOMENTUM AND HEAT TRANSFER (REYNOLDS) ANALOGY

As engineers, our interest in boundary layer behavior is directed principally toward the dimensionless parameters C_f and Nu. It is from knowledge of these parameters that we may compute the wall shear stress and the convection heat transfer rate. It is therefore understandable that an expression that relates C_f and Nu can be a useful tool in convection analysis. Such an expression is available in the form of a boundary layer analogy.

If two or more processes are governed by dimensionless equations of the same form, the processes are said to be analogous. From Table 6.1, we note that, if $dp^*/dx^* = 0$ and $Pr = 1$, Equations 6.37 and 6.38 are of precisely the same form. Each equation is composed of analogous advection and diffusion terms, and the coefficient of the diffusion term is the reciprocal Reynolds number. Moreover,

Table 6.2 Selected dimensionless groups of heat transfer

GROUP	DEFINITION	INTERPRETATION
Biot number (Bi)	$\frac{hL}{k_s}$	Ratio of the internal thermal resistance of a solid to the boundary layer thermal resistance.
Bond number (Bo)	$\frac{g(\rho_l-\rho_v)L^2}{\sigma}$	Ratio of gravitational and surface tension forces.
Coefficient of friction (C_f)	$\frac{\tau_s}{\rho V^2/2}$	Dimensionless surface shear stress.
Eckert number (Ec)	$\frac{V^2}{c_p(T_s-T_\infty)}$	Kinetic energy of the flow relative to the boundary layer enthalpy difference.
Fourier number (Fo)	$\frac{\alpha t}{L^2}$	Ratio of the heat conduction rate to the rate of thermal energy storage in a solid. Dimensionless time.
Friction factor (f)	$\frac{\Delta p}{(L/D)(\rho u_m^2/2)}$	Dimensionless pressure drop for internal flow.
Grashof number (Gr_L)	$\frac{g\beta(T_s-T_\infty)L^3}{\nu^2}$	Ratio of buoyancy to viscous forces.
Colburn j factor (j_H)	$St Pr^{2/3}$	Dimensionless heat transfer coefficient.
Jakob number (Ja)	$\frac{c_p(T_s-T_{sat})}{h_{fg}}$	Ratio of sensible to latent energy absorbed during liquid–vapor phase change.
Nusselt number (Nu_L)	$\frac{hL}{k_f}$	Dimensionless temperature gradient at the surface.
Peclet number (Pe_L)	$\frac{VL}{\alpha}=Re_L Pr$	Dimensionless independent heat transfer parameter.
Prandtl number (Pr)	$\frac{c_p\mu}{k}=\frac{\nu}{\alpha}$	Ratio of the momentum and thermal diffusivities.
Reynolds number (Re_L)	$\frac{VL}{\nu}$	Ratio of the inertia and viscous forces.
Stanton number (St)	$\frac{h}{\rho V c_p}=\frac{Nu_L}{Re_L Pr}$	Modified Nusselt number.
Weber number (We)	$\frac{\rho V^2 L}{\sigma}$	Ratio of inertia to surface tension forces.

since $u_\infty = V$ if $dp^*/dx^* = 0$, the boundary conditions, Equations 6.39 and 6.40, are equivalent. With governing differential equations and boundary conditions of the same form, it follows that dimensionless relations that govern velocity boundary layer behavior must be the same as those that govern the thermal boundary layer. Hence the boundary layer velocity and temperature profiles must be of the same functional form.

Recalling the discussion of Section 6.6.2, key features of which are summarized in Table 6.3, the most important results of the momentum and heat transfer analogy may be obtained. From the foregoing paragraph, it follows that f_1 of Equation 6.46 must be of the same form as f_3 of Equation 6.49. From Equations 6.47 and 6.50 it also follows that the dimensionless velocity and temperature gradients at the surface, and therefore the values of C_f and Nu, are analogous. That is, f_2 of Equation 6.48 is of the same form as f_4 of Equation 6.51. Accordingly, friction and heat transfer relations for a particular geometry are interchangeable. We conclude that

$$C_f \frac{Re_L}{2} = Nu \tag{6.54}$$

Replacing Nu by the *Stanton number St*,

$$St \equiv \frac{h}{\rho V c_p} = \frac{Nu}{Re Pr} \tag{6.55}$$

the analogy takes the form

$$\frac{C_f}{2} = St \tag{6.56}$$

Equation 6.56 is known as the *Reynolds analogy*, and it relates the key engineering parameters of the velocity and thermal boundary layers. If the velocity parameter is known, the analogy may be used to obtain the heat transfer parameter, and vice versa. However, there are numerous restrictions associated with using this result. In addition to relying on the validity of the boundary layer approximations, the accuracy of Equation 6.56 depends on having $Pr \approx 1$ and $dp^*/dx^* \approx 0$. However, it has been shown that the analogy

Table 6.3 Functional relations pertinent to the Reynolds analogy

FLUID FLOW	HEAT TRANSFER
$u^* = f_1\left(x^*, y^*, Re_L, \frac{dp^*}{dx^*}\right)$ (6.46)	$T^* = f_3\left(x^*, y^*, Re_L, Pr, \frac{dp^*}{dx^*}\right)$ (6.49)
$C_f = \frac{2}{Re_L} \frac{\partial u^*}{\partial y^*}\bigg\rvert_{y^*=0}$ (6.47)	$Nu = + \frac{\partial T^*}{\partial y^*}\bigg\rvert_{y^*=0}$ (6.50)
$C_f = \frac{2}{Re_L} f_2(x^*, Re_L)$ (6.48)	$Nu = f_4(x^*, Re_L, Pr)$ (6.51)
—	$\overline{Nu} = f_5(Re_L, Pr)$ (6.52)

may be applied over a wide range of Pr, if a Prandtl number correction is added. In particular, the *modified Reynolds*, or *Chilton-Colburn*, analogy [10, 11] has the form

$$\frac{C_f}{2} = StPr^{2/3} \equiv j_H \qquad 0.6 < Pr < 60 \tag{6.57}$$

where j_H is the *Colburn j factor* for heat transfer. For laminar flow, Equation 6.57 is only appropriate when $dp^*/dx^* \approx 0$; but in turbulent flow, conditions are less sensitive to the effect of pressure gradients and these equations remain approximately valid. If the analogy is applicable at every point on a surface, it may be applied to the surface average coefficients.

6.9 THE EFFECTS OF TURBULENCE

At this point we recognize that turbulent conditions characterize numerous flows of practical interest. In fact the practicing engineer deals more often with turbulent flows than with laminar flows. It is well known that small disturbances associated with distortions in the fluid streamlines of a laminar flow can eventually lead to turbulent conditions. These disturbances may originate from the free stream, or they may be induced by surface roughness. The onset of turbulence depends on whether these disturbances are amplified or attenuated in the direction of fluid flow, which in turn depends on the ratio of the inertia to viscous forces (the Reynolds number). Recall that if the Reynolds number is small, inertia forces are small relative to viscous forces. The naturally occurring disturbances are then dissipated, and the flow remains laminar. For a large Reynolds number, however, the inertia forces are sufficiently large to amplify the disturbances, and a transition to turbulence occurs. In Section 6.3 we observed that the critical Reynolds number Re_c required for transition is approximately 5×10^5 for flow over a flat plate.

The onset of turbulence is associated with the existence of *random fluctuations* in the fluid, and at least on a small scale, the flow is inherently *unsteady*. This behavior is shown in Figure 6.12, which plots the variation of an

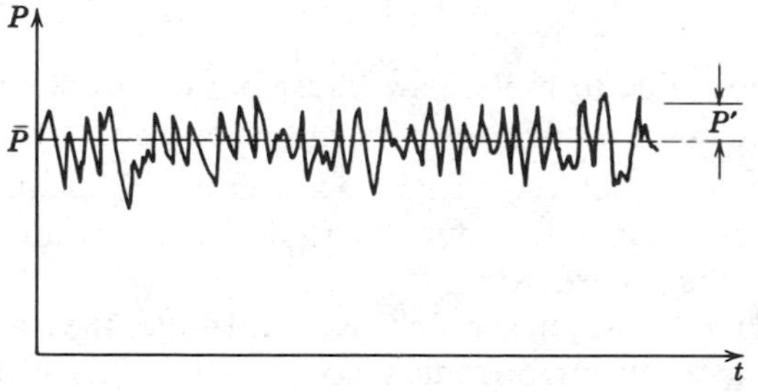

Figure 6.12 Property variation with time at some point in a turbulent boundary layer.

arbitrary flow property P as a function of time at some location in a turbulent boundary layer. The property P could be a velocity component or the fluid temperature, and at any instant it may be represented as the sum of a *time-mean* value $\bar{P}$ and a fluctuating component P'. The average is taken over a time that is large compared with the period of a typical fluctuation, and if $\bar{P}$ is independent of time, the time-mean flow is said to be *steady*.

The existence of turbulent flow can be advantageous in the sense of providing increased heat transfer rates. However the motion is extremely complicated and difficult to describe theoretically. Although the boundary layer conservation equations developed in preceding sections remain applicable, the dependent variables (u,v,T) must be interpreted as *instantaneous values*, and it is impossible to predict their exact variation with time. In the practical sense, however, this inability to determine the variation of the instantaneous properties P with time is not a serious restriction, since the engineer is generally concerned only with the time-mean properties $\bar{P}$. Equations of the form $P=\bar{P}+P'$ may be substituted for each of the flow variables into the boundary layer equations. For steady, incompressible, constant property flow, using well-established time-averaging procedures [1, 12], the following forms of the x-momentum and energy conservation equations may be obtained:

$$\rho\left(\bar{u}\frac{\partial \bar{u}}{\partial x}+\bar{v}\frac{\partial \bar{u}}{\partial y}\right)=-\frac{dp}{dx}+\frac{\partial}{\partial y}\left(\mu\frac{\partial \bar{u}}{\partial y}-\rho\overline{u'v'}\right) \tag{6.58}$$

$$\rho c_p\left(\bar{u}\frac{\partial \bar{T}}{\partial x}+\bar{v}\frac{\partial \bar{T}}{\partial y}\right)=\frac{\partial}{\partial y}\left(k\frac{\partial \bar{T}}{\partial y}-\rho c_p\overline{v'T'}\right) \tag{6.59}$$

The equations are like those for the laminar boundary layer, except for the presence of additional terms of the form $\overline{a'b'}$. These terms account for the effect of the turbulent fluctuations on momentum and energy transport.

On the basis of the foregoing results it is customary to speak of a *total* shear stress and a *total* heat flux, which are defined as

$$\tau_{\text{tot}}=\left(\mu\frac{\partial \bar{u}}{\partial y}-\rho\overline{u'v'}\right) \tag{6.60}$$

$$q''_{\text{tot}}=-\left(k\frac{\partial \bar{T}}{\partial y}-\rho c_p\overline{v'T'}\right) \tag{6.61}$$

and consist of contributions due to molecular diffusion and turbulent mixing. From the form of these equations we see how momentum and heat transfer rates are enhanced by the existence of turbulence. The term $\rho\overline{u'v'}$ appearing in Equation 6.60 represents the momentum flux due to the turbulent fluctuations, and it is often termed the *Reynolds stress*.

A simple conceptual model, which is widely used, attributes the transport of momentum and heat in a turbulent boundary layer to the motion of *eddies*, small portions of fluid in the boundary layer that move about for a short time before losing their identity. Due to this motion, the transport of momentum and energy is greatly enhanced. The notion of transport by eddies has prompted the introduction of a transport coefficient, which is defined as the *eddy diffusivity for momentum transfer* ε_M, and has the form

$$\rho\varepsilon_M \frac{\partial \bar{u}}{\partial y} \equiv -\rho\overline{u'v'} \tag{6.62}$$

Hence the total shear stress may be expressed as

$$\tau_{\text{tot}} = \rho\left(\nu + \varepsilon_M\right)\frac{\partial \bar{u}}{\partial y} \tag{6.63}$$

Similarly, the *eddy diffusivity for heat transfer* ε_H, may be defined by the relation

$$\varepsilon_H \frac{\partial \bar{T}}{\partial y} \equiv -\overline{v'T'} \tag{6.64}$$

in which case

$$q''_{\text{tot}} = -\rho c_p(\alpha + \varepsilon_H)\frac{\partial \bar{T}}{\partial y} \tag{6.65}$$

In the region of a turbulent boundary layer removed from the surface (the core region), the eddy diffusivities are much larger than the molecular diffusivities. The enhanced mixing associated with this condition has the effect of making velocity and temperature profiles more uniform in the core. This behavior is shown in Figure 6.13 for laminar and turbulent velocity boundary layers corresponding to the same freestream velocity. Accordingly, the velocity gradient at the surface, and therefore the surface shear stress, is much larger for the turbulent boundary layer than for the laminar boundary layer. In like manner, it may be argued that the surface temperature gradient, and therefore the heat transfer rate, is much larger for turbulent than for laminar flow. Because of this enhancement of the convection heat transfer rate, it is desirable to have turbulent flow conditions in many engineering applications. However, the increase in wall shear stress will always have the adverse effect of increasing pump or fan power requirements. A fundamental problem in performing a turbulent boundary layer analysis involves determining the eddy diffusivities as

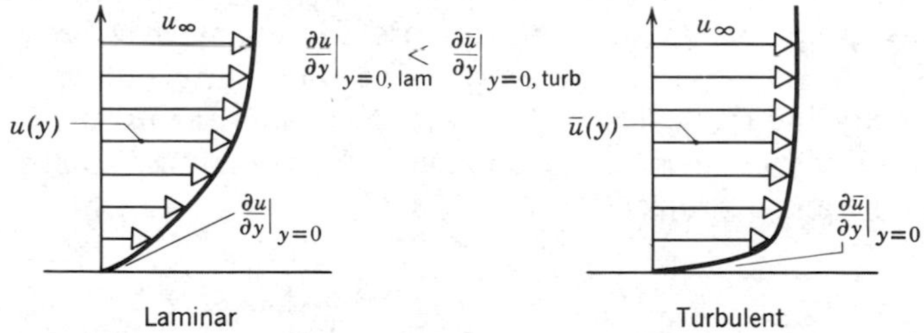

Figure 6.13 Comparison of laminar and turbulent velocity boundary layer profiles for the same freestream velocity.

a function of the mean properties of the flow. Unlike the molecular diffusivities, which are strictly fluid properties, the eddy diffusivities depend strongly on the nature of the flow and vary from point to point in a boundary layer. The problem of determining the eddy diffusivities continues to attract many researchers in fluid mechanics. Although numerous turbulence models have been proposed [13, 14], there is, to date, no formulation that holds for all flow conditions.

6.10 THE CONVECTION COEFFICIENTS

In this chapter we have attempted to develop the fundamentals of convection transport phenomena. In the process, however, it is hoped that you have not lost sight of what remains *the problem of convection.* Our primary objective is still one of developing means to determine the convection coefficient h. Although this coefficient may be obtained by solving the boundary layer equations, it is only for simple flow situations that such solutions are readily effected. The more practical approach frequently involves calculating h from empirical equations of the form given by Equation 6.51. The particular form of these equations is obtained by *correlating* measured convection heat transfer results in terms of appropriate dimensionless groups. It is this approach that is emphasized in the chapters that follow.

6.11 SUMMARY

In this chapter we have developed the mathematical and physical basis of convection transport, and the ideas are essential to anyone wishing to deal rationally with convection processes. To test your comprehension of the material, you should challenge yourself with appropriate questions. What are the velocity and thermal boundary layers? Under what conditions do they develop? And why are they of interest to the engineer? How do laminar and turbulent boundary layers differ, and how does one determine whether a particular boundary layer is laminar or turbulent? Several processesses affect momentum and energy transfer in a boundary layer. What are they? How are they represented mathematically? What are the boundary layer approximations, and in what way do they alter the conservation equations? What are the relevant dimensionless groups for the boundary layers? How may they be physically interpreted? How will the use of these groups facilitate convection calculations? How are velocity and thermal boundary layer behaviors analogous? How may the effects of turbulence be treated in a boundary layer analysis? Finally, what is *the* convection transfer problem?

REFERENCES

1. Schlichting, H., *Boundary Layer Theory*, 6th ed., McGraw-Hill, New York, 1968.
2. Bird, R. B., W. E. Stewart, and E. N. Lightfoot, *Transport Phenomena*, Wiley, New York, 1960, Chapters 10 and 18.

3. Eckert, E. R. G., and R. M. Drake, Jr., *Analysis of Heat and Mass Transfer*, McGraw-Hill, New York, 1973, pp. 270–275.
4. Kays, W. M., and M. E. Crawford, *Convective Heat and Mass Transfer*, McGraw-Hill, New York, 1980.
5. Shah, R. K., and A. L. London, *Laminar Flow Forced Convection in Ducts*, Academic Press, New York, 1978.
6. Burmeister, L. C., *Convective Heat Transfer*, Wiley-Interscience, New York, 1983.
7. Roache, P. J., *Computational Fluid Dynamics*, Hermosa Publishers, Albuquerque, N.M., 1976.
8. Patankar, S. V., *Numerical Heat Transfer and Fluid Flow*, Hemisphere Publishing Corp., New York, 1980.
9. Fox, R. W., and A. T. McDonald, *Introduction to Fluid Mechanics*, 3rd ed., Wiley, New York, 1985.
10. Colburn, A. P., *Trans. AIChE*, *29*, 174, 1933.
11. Chilton, T. H., and A. P. Colburn, *Ind. Eng. Chem.*, *26*, 1183, 1934.
12. Hinze, J. O., *Turbulence*, 2nd ed., McGraw-Hill, New York, 1975.
13. Launder, B. E., and D. B. Spalding, *Mathematical Models of Turbulence*, Academic Press, New York, 1972.
14. Bradshaw, P., T. Cebeci, and J. Whitelaw, *Engineering Calculational Methods for Turbulent Flow*, Academic Press, New York, 1981.

PROBLEMS

6.1 For flow over a flat plate, the local heat transfer coefficient h_x is known to vary as $x^{-1/2}$, where x is the distance from the leading edge ($x=0$) of the plate. What is the ratio of the average coefficient between the leading edge and some location x on the plate to the local coefficient at x?

6.2 For laminar free convection from a heated vertical surface, the local convection coefficient may be expressed as $h_x = Cx^{-1/4}$, where h_x is the coefficient at a distance x from the leading edge of the surface and the quantity C, which depends on the fluid properties, is independent of x. Obtain an expression for the ratio $\bar{h}_x/h_x$, where $\bar{h}_x$ is the average coefficient between the leading edge ($x=0$) and the x location. Sketch the variation of h_x and $\bar{h}_x$ with x.

6.3 A circular, hot gas jet at T_∞ is directed normal to a circular plate that has radius r_o and is maintained at a uniform temperature T_s. Gas flow over the plate is axisymmetric, causing the local convection coefficient to have a radial dependence of the form $h(r) = a + br^n$, where a, b, and n are constants. Determine the rate of heat transfer to the plate, expressing your result in terms of T_∞, T_s, r_o, a, b, and n.

6.4 In flow over a surface, velocity and temperature profiles are of the forms

$$u(y) = Ay + By^2 - Cy^3 \qquad \text{and} \qquad T(y) = D + Ey + Fy^2 - Gy^3$$

where the coefficients A through G are constants. Obtain expressions for the friction coefficient C_f and the convection coefficient h in terms of u_∞, T_∞, and appropriate profile coefficients and fluid properties.

6.5 Water at $T_\infty = 25°C$ flows over one of the surfaces of a steel wall (AISI 1010) whose temperature is $T_{s,1} = 40°C$. The wall is 0.35 m thick, and its other surface temperature is $T_{s,2} = 100°C$. For steady-state conditions, what is the convection coefficient associated with the water flow? What are the temperature gradients in the wall and in the water that is in contact with the wall? Sketch the temperature distribution in the wall and in the adjoining water.

6.6 In a particular application involving airflow over a heated surface, the boundary layer temperature distribution may be approximated as

$$\frac{T-T_s}{T_\infty-T_s}=1-\exp\left(-Pr\frac{u_\infty y}{\nu}\right)$$

where y is the distance normal to the surface and the Prandtl number $Pr=c_p\mu/k=0.7$ is a dimensionless fluid property. If $T_\infty=400$ K, $T_s=300$ K, and $u_\infty/\nu=5000$ m^{-1}, what is the surface heat flux?

6.7 A fan that can provide air speeds up to 50 m/s is to be used in a low speed wind tunnel with atmospheric air at 25°C. If one wishes to use the wind tunnel to study flat plate boundary layer behavior up to Reynolds numbers of $Re_x=10^8$, what is the minimum plate length? At what distance from the leading edge would transition occur, if the critical Reynolds number is $Re_{x,c}=5\times10^5$?

6.8 Assuming a transition Reynolds number of 5×10^5, determine the distance from the leading edge of a flat plate at which transition will occur for each of the following fluids when $u_\infty=1$ m/s: atmospheric air, water, engine oil, and mercury. In each case the fluid temperature is 27°C.

6.9 Consider the control volume shown for the special case of steady-state conditions with $v=0$, $T=T(y)$, and $\rho=$constant.

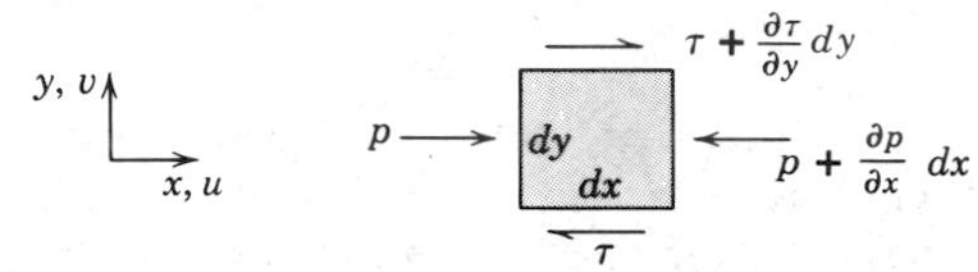

(a) Prove that $u=u(y)$ if $v=0$ everywhere.

(b) Derive the x-momentum equation and simplify it as much as possible.

(c) Derive the energy equation and simplify it as much as possible.

6.10 Consider a lightly loaded journal bearing using oil having the constant properties $\mu=10^{-2}$ kg/s·m and $k=0.15$ W/m·K. If the journal and the bearing are each maintained at a temperature of 40°C, what is the maximum temperature in the oil when the journal is rotating at 10 m/s?

6.11 Consider a lightly loaded journal bearing using oil having the constant properties $\rho=800$ kg/m^3, $\nu=10^{-5}$ m^2/s, and $k=0.13$ W/m·K. The journal diameter is 75 mm, the clearance is 0.25 mm, and the bearing operates at 3600 rpm.

(a) Determine the temperature distribution in the oil film, assuming that there is no heat transfer into the journal and that the bearing surface is maintained at 75°C.

(b) What is the rate of heat transfer from the bearing, and how much power is needed to rotate the journal?

6.12 Consider two large (infinite) parallel plates separated by 5 mm. One plate is stationary, while the other plate is moving at 200 m/s. Both plates are maintained at 27°C. Consider cases for which the plates are separated by water and by air, respectively.

(a) For each of the two fluids, what is the force per unit surface area required to maintain the above condition? What is the corresponding power requirement?

(b) What is the viscous dissipation associated with each of the two fluids?

(c) What is the maximum temperature in each of the two fluids?

6.13 A judgment concerning the influence of viscous dissipation in forced-convection heat transfer may be made by calculating the quantity $PrEc$, where the Prandtl number $Pr = c_p\mu/k$ and the Eckert number $Ec = U^2/c_p\Delta T$ are *dimensionless groups*. The characteristic velocity and temperature difference of the problem are designated as U and ΔT, respectively. If $PrEc \ll 1$, dissipation effects may be neglected. Consider Couette flow for which one plate moves at 10 m/s and a temperature difference of 25°C is maintained between the plates. Evaluating properties at 27°C, determine the value of $PrEc$ for air, water, and engine oil. What is the value of $PrEc$ for air, if the plate is moving at the sonic velocity?

6.14 Consider Couette flow for which the moving plate is maintained at a uniform temperature and the stationary plate is insulated. Determine the temperature of the insulated plate, expressing your result in terms of fluid properties and the temperature and speed of the moving plate. Obtain an expression for the heat flux at the moving plate.

6.15 Consider Couette flow with heat transfer as described in Example 6.2.

(a) Rearrange the temperature distribution to obtain the dimensionless form

$$\theta(\eta) = \eta[1 + \tfrac{1}{2}PrEc(1-\eta)]$$

where $\theta \equiv [T(y) - T_0]/[T_L - T_0]$ and $\eta = y/L$. The *dimensionless groups* are the Prandtl number $Pr = \mu c_p/k$ and the Eckert number $Ec = U^2/c_p(T_L - T_0)$.

(b) Derive an expression that prescribes the conditions under which there will be no heat transfer to the upper plate.

(c) Derive an expression for the heat transfer rate to the lower plate for the conditions identified in part b.

6.16 Consider the problem of steady, incompressible laminar flow between two stationary, infinite parallel plates maintained at different temperatures.

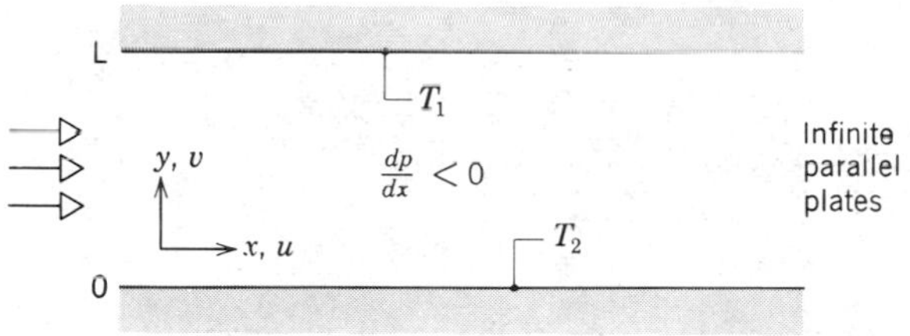

Referred to as *Poiseuille flow* with heat transfer, this special case of parallel flow is one for which the x velocity component is finite, but the y and z components (v and w) are zero.

(a) What is the form of the continuity equation for this case? In what way is the flow *fully developed*?

(b) What forms do the x- and y-momentum equations take? What is the form of the velocity profile? Note that, unlike Couette flow, fluid motion between the plates is now sustained by a finite pressure gradient. How is this pressure gradient related to the maximum fluid velocity?

(c) Assuming viscous dissipation to be significant and recognizing that conditions must be thermally fully developed, what is the appropriate form of the energy equation? Solve this equation for the temperature distribution. What is the heat flux at the upper ($y=L$) surface? *Hint*: In evaluating $\partial i/\partial x$ in Equation 6.26, substitute from Equation 6.27 and use the fact that the internal energy is a function of temperature and density, $e(T,\rho)$.

6.17 Consider the conservation equations, (6.13), (6.17), and (6.26).

(a) Identify each equation and briefly describe the physical significance of its terms.

(b) Identify the approximations and special conditions that are made to reduce these expressions to the boundary layer equations, (6.30), (6.31), and (6.33).

(c) Comparing Equations 6.31 and 6.33, identify conditions for which the equations have the same form. Comment on the existence of a heat and momentum transfer analogy.

6.18 To a good approximation, the dynamic viscosity μ, thermal conductivity k, and specific heat c_p are independent of pressure. In what manner do the kinematic viscosity ν and thermal diffusivity α vary with pressure for an incompressible liquid and for a perfect gas? Determine ν and α of air at 350 K for pressures of 1 and 10 atm.

6.19 The Prandtl number (Pr) is a dimensionless parameter used in convection analysis and is defined as $\mu c_p/k$. Using the thermophysical properties μ, c_p, and k from the appropriate tables, calculate the Prandtl number for these fluids: air at 1 atm and 300 K, engine oil at 60°C, water at 373 K, and liquid sodium at 93°C.

6.20 Forced air at $T_\infty = 25°C$ and $V = 10$ m/s is used to cool electronic elements on a circuit board. One such element is a chip, 4 mm × 4 mm, located 120 mm from the leading edge of the board. Experiments have revealed that flow over the board is disturbed by the elements and that convection heat transfer is correlated by an expression of the form

$$Nu_x = 0.04\, Re_x^{0.85}\, Pr^{1/3}$$

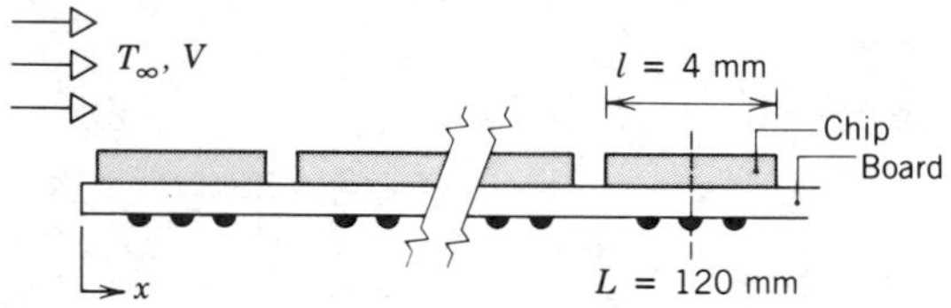

Estimate the surface temperature of the chip if it is dissipating 30 mW.

6.21 An object of irregular shape has a characteristic length of $L=1$ m and is maintained at a uniform surface temperature of $T_s=400$ K. When placed in atmospheric air at a temperature of $T_\infty=300$ K and moving with a velocity of $V=100$ m/s, the average heat flux from the surface to the air is 20,000 W/m². If a second object of the same shape, but with a characteristic length of $L=5$ m, is maintained at a surface temperature of $T_s=400$ K and placed in atmospheric air at $T_\infty=300$ K, what will the value of the average convection coefficient be if the air velocity is $V=20$ m/s?

6.22 Experiments have shown that, for airflow at $T_\infty = 35°C$ and $V_1 = 100$ m/s, the rate of heat transfer from a turbine blade of characteristic length $L_1 = 0.15$ m and surface temperature $T_{s,1} = 300°C$ is $q_1 = 1500$ W. What would be the heat transfer rate from a second turbine blade of characteristic length $L_2 = 0.3$ m operating at $T_{s,2} = 400°C$ in airflow of $T_\infty = 35°C$ and $V_2 = 50$ m/s? The surface area of the blade may be assumed to be directly proportional to its characteristic length.

6.23 Experimental measurements of the convection heat transfer coefficient for a square bar in cross flow, as shown below, resulted in these values:

$\bar{h}_1 = 50\ \text{W/m}^2\cdot\text{K}$ when $V_1 = 20$ m/s
$\bar{h}_2 = 40\ \text{W/m}^2\cdot\text{K}$ when $V_2 = 15$ m/s

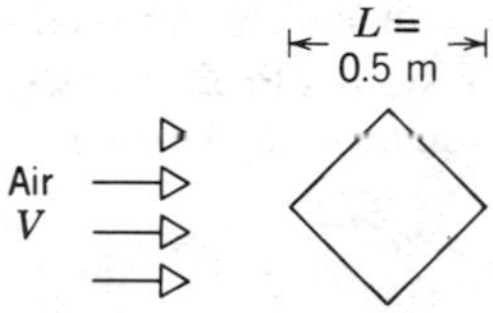

Assume that the functional form of the Nusselt number is $\overline{Nu} = C Re^m Pr^n$, where C, m, and n are constants.

(a) What will be the convection heat transfer coefficient for a similar bar with $L = 1$ m when $V = 15$ m/s?

(b) What will be the convection heat transfer coefficient for a similar bar with $L = 1$ m when $V = 30$ m/s?

(c) Would your results be the same if the side of the bar, rather than its diagonal, were used as the characteristic length?

6.24 Experimental results for heat transfer over a flat plate with an extremely rough surface were found to be correlated by an expression of the form

$$Nu_x = 0.04 Re_x^{0.9}\ Pr^{1/3}$$

where Nu_x is the local value of the Nusselt number at a position x measured from the leading edge of the plate. Obtain an expression for the ratio of the average heat transfer coefficient $\bar{h}_x$ to the local coefficient h_x.

6.25 Consider conditions for which a fluid with a freestream velocity of $V = 1$ m/s flows over a surface with a characteristic length of $L = 1$ m, providing an average convection heat transfer coefficient of $\bar{h} = 100\ \text{W/m}^2\cdot\text{K}$. Calculate the dimensionless parameters $\overline{Nu}_L$, Re_L, Pr, and $\bar{j}_H$ for the following fluids: air, engine oil, mercury, and water. Assume the fluids to be at 300 K.

6.26 For flow over a flat plate of length L, the local heat transfer coefficient h_x is known to vary as $x^{-1/2}$, where x is the distance from the leading edge of the plate. What is the ratio of the average Nusselt number for the entire plate, $\overline{Nu}_L$, to the local Nusselt number at $x = L$, Nu_L?

6.27 For laminar boundary layer flow over a flat plate with air at 20°C and 1 atm, the thermal boundary layer thickness δ_t is approximately 13% larger than the velocity boundary layer thickness δ. Determine the ratio δ/δ_t if the fluid is ethylene glycol under the same flow conditions.

6.28 Sketch the variation of the velocity and thermal boundary layer thicknesses with distance from the leading edge of a flat plate for the laminar flow of air, water, engine oil, and mercury. For each case assume a mean fluid temperature of 300 K.

6.29 A thin flat plate that is 0.2 m × 0.2 m on a side is oriented parallel to an atmospheric airstream having a velocity of 40 m/s. The air is at a temperature of $T_\infty = 20°C$, while the plate is maintained at $T_s = 120°C$. The air flows over the top and bottom surfaces of the plate, and measurement of the drag force reveals a value of 0.075 N. What is the rate of heat transfer from both sides of the plate to the air?

6.30 Atmospheric air is in parallel flow ($u_\infty = 15$ m/s, $T_\infty = 15°C$) over a flat heater surface that is to be maintained at 140°C. The heater surface area is 0.25 m^2, and the airflow is known to induce a drag force of 0.25 N on the heater. What is the electrical power needed to maintain the prescribed surface temperature?

6.31 For flow over a flat plate with an extremely rough surface, convection heat transfer effects are known to be correlated by the expression of Problem 6.24. For airflow at 50 m/s, what is the surface shear stress at $x = 1$ m from the leading edge of the plate? Assume the air to be at 300 K.

Seven

EXTERNAL FLOW

In this chapter we focus on the problem of computing heat transfer rates to or from a surface in *external flow*. In such a flow boundary layers develop freely, without constraints imposed by adjacent surfaces. Accordingly, there will always exist a region of the flow outside the boundary layer in which velocity and temperature gradients are negligible. Examples include fluid motion over a flat plate (inclined or parallel to the freestream velocity) and flow over curved surfaces such as a sphere, cylinder, airfoil, or turbine blade.

For the moment we confine our attention to problems of *low speed, forced convection*, with *no phase change* occurring within the fluid. In *forced* convection the relative motion between the fluid and the surface is maintained by external means, such as a fan or a pump, and not by buoyancy forces due to temperature gradients in the fluid (*natural* convection). *Internal flows, natural convection*, and *convection with phase change* are treated in Chapters 8, 9, and 10, respectively.

Our primary objective is to determine convection coefficients for different flow geometries. In particular, we wish to obtain specific forms of the functions that represent these coefficients. By nondimensionalizing the boundary layer conservation equations in Chapter 6, we found that the local and average convection coefficients may be correlated by equations of the form

$$Nu_x = f_4(x^*, Re_x, Pr) \tag{6.51}$$

$$\overline{Nu}_x = f_5(Re_x, Pr) \tag{6.52}$$

The subscript x has been added to emphasize our interest in conditions at a particular location on the surface. The overbar indicates an average from $x^* = 0$, where the boundary layer begins to develop, to the location of interest. Recall that *the problem of convection* is one of obtaining these functions. There are two approaches that we could take, one theoretical and the other experimental.

The *experimental*, or *empirical, approach* involves performing heat transfer measurements under controlled laboratory conditions and correlating the data in terms of appropriate dimensionless parameters. A general discussion of the approach is provided in Section 7.1. It has been applied to many different geometries and flow conditions. Important results are presented in Sections 7.2 to 7.7.

The *theoretical approach* involves solving the boundary layer equations for a particular geometry. For example, obtaining the temperature profile T^* from such a solution, Equation 6.50 could then be used to evaluate the local Nusselt number Nu_x, and therefore the local convection coefficient h_x. With knowledge of how h_x varies over the surface, Equation 6.5 may then be used to determine the average convection coefficient $\bar{h}_x$, and therefore the Nusselt number $\overline{Nu}_x$. In Section 7.2.1 this approach is illustrated by using the *similarity method* to obtain an *exact solution* of the boundary layer equations for a flat plate in parallel, laminar flow. An *approximate solution* to the same problem is obtained in Appendix C by using the *integral method*.

7.1 THE EMPIRICAL METHOD

The manner in which a convection heat transfer correlation may be obtained experimentally is illustrated in Figure 7.1. If a prescribed geometry, such as the flat plate in parallel flow, is heated electrically to maintain $T_s > T_\infty$, convection heat transfer occurs from the surface to the fluid. It would be a simple matter to measure T_s and T_∞, as well as the electrical power $E \cdot I$, which is equal to the total heat transfer rate q. The convection coefficient $\bar{h}_L$, which is an average associated with the entire plate, could then be computed from Newton's law of cooling, Equation 6.4. Moreover, from knowledge of the characteristic length L and the fluid properties, the Nusselt, Reynolds, and Prandtl numbers could be computed from their definitions, Equations 6.50, 6.41, and 6.42, respectively.

The foregoing procedure could be repeated for a variety of test conditions. We could vary the velocity u_∞ and the plate length L, as well as the nature of the fluid, using, for example, air, water, and engine oil, which have substantially different Prandtl numbers. We would then be left with many different values of the Nusselt number corresponding to a wide range of Reynolds and Prandtl numbers, and the results could be plotted on a *log-log* scale, as shown in Figure 7.2*a*. Each symbol represents a unique set of test conditions. As is often the case, the results associated with a given fluid, and hence a fixed Prandtl number, fall close to a straight line, which may be represented by an algebraic expression of the form

$$\overline{Nu}_L = C Re_L^m Pr^n \tag{7.1}$$

Since the values of C, m, and n are independent of the nature of the fluid, the family of straight lines corresponding to different Prandtl numbers can be collapsed to a single line by plotting the results in terms of the ratio $\overline{Nu}_L/Pr^n$, as shown in Figure 7.2*b*.

Because Equation 7.1 is inferred from experimental measurements, it is termed an *empirical correlation*. However, the specific values of the coefficient C and the exponents m and n vary with the nature of the surface geometry and the type of flow.

We will use expressions of the form given by Equation 7.1 for many special cases, and it is important to note that the assumption of *constant fluid properties*

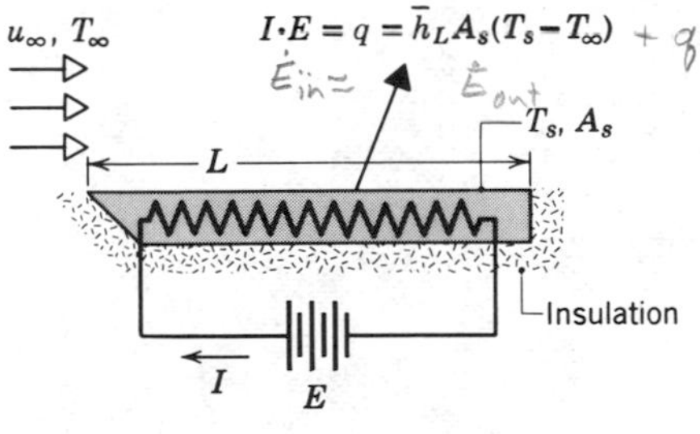

Figure 7.1 Experiment for measuring the average convection heat transfer coefficient $\bar{h}_L$.

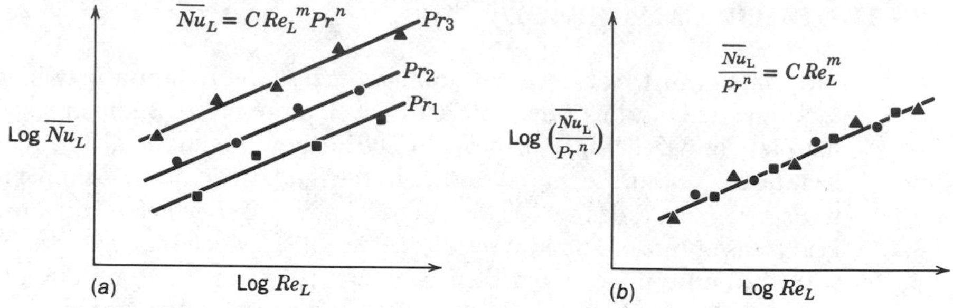

Figure 7.2 Dimensionless representation of convection heat transfer measurements.

is often implicit in the results. However, we know that the fluid properties vary with temperature across the boundary layer and that this variation can certainly influence the heat transfer rate. This influence may be handled in one of two ways. In one method, Equation 7.1 is used with all properties evaluated at a mean boundary layer temperature T_f, termed the *film temperature.*

$$T_f \equiv \frac{T_s + T_\infty}{2} \tag{7.2}$$

The alternate method is to evaluate all properties at T_∞ and to multiply the right-hand side of Equation 7.1 by an additional parameter to account for the property variations. The parameter is commonly of the form $(Pr_\infty/Pr_s)^r$ or $(\mu_\infty/\mu_s)^r$, where the subscripts ∞ and s designate evaluation of the properties at the freestream and surface temperatures, respectively. Both methods are used in the results that follow.

7.2 THE FLAT PLATE IN PARALLEL FLOW

Despite its simplicity, parallel flow over a flat plate (Figure 7.3) occurs in numerous engineering applications. Moreover, this geometry is often a reasonable approximation for flow over slightly contoured surfaces, such as airfoils or turbine blades.

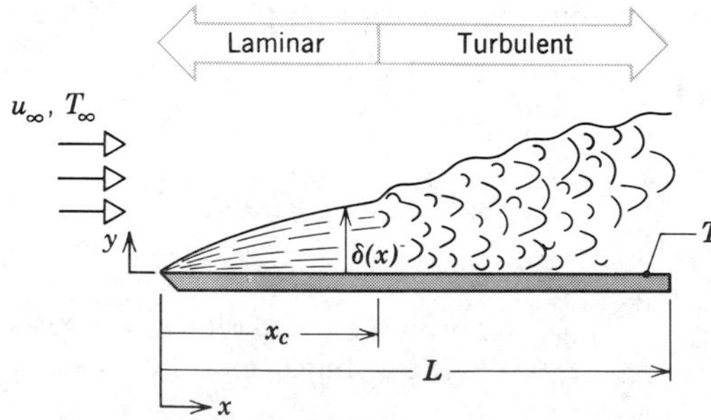

Figure 7.3 The flat plate in parallel flow.

7.2.1 Laminar Flow: A Similarity Solution

The major convection parameters may be obtained by solving the appropriate form of the boundary layer equations. Assuming *steady, incompressible, laminar* flow with *constant fluid properties* and *negligible viscous dissipation,* and recognizing that $dp/dx=0$, the boundary layer equations, (6.30), (6.31), and (6.33), reduce to

Continuity:

$$\frac{\partial u}{\partial x}+\frac{\partial v}{\partial y}=0 \tag{7.3}$$

Momentum:

$$u\frac{\partial u}{\partial x}+v\frac{\partial u}{\partial y}=\nu\frac{\partial^2 u}{\partial y^2} \tag{7.4}$$

Energy:

$$u\frac{\partial T}{\partial x}+v\frac{\partial T}{\partial y}=\alpha\frac{\partial^2 T}{\partial y^2} \tag{7.5}$$

Solution of these equations is simplified by the fact that, for constant properties, conditions in the velocity (hydrodynamic) boundary layer are independent of temperature. Hence, we may begin by solving the hydrodynamic problem, Equations 7.3 and 7.4, to the exclusion of Equation 7.5. Once the hydrodynamic problem has been solved, a solution to Equation 7.5, which depends on u and v, may be obtained.

The hydrodynamic solution follows the method of Blasius [1, 2]. The velocity components are defined in terms of a stream function $\psi(x,y)$,

$$u\equiv\frac{\partial \psi}{\partial y} \quad \text{and} \quad v\equiv-\frac{\partial \psi}{\partial x} \tag{7.6}$$

such that Equation 7.3 is automatically satisfied, and hence is no longer needed. New dependent and independent variables, f and η, respectively, are then defined such that

$$f(\eta)\equiv\frac{\psi}{u_\infty\sqrt{\nu x/u_\infty}} \tag{7.7}$$

$$\eta\equiv y\sqrt{u_\infty/\nu x} \tag{7.8}$$

As we will find, use of these variables simplifies matters by reducing the partial differential equation (7.4) to an ordinary differential equation.

The Blasius solution is termed a *similarity solution,* and η is a *similarity variable.* This terminology is used because, despite the growth of the boundary layer with distance x from the leading edge, the velocity profile, u/u_∞, remains *geometrically similar*. This similarity is of the functional form

$$\frac{u}{u_\infty}=\phi\left(\frac{y}{\delta}\right)$$

where δ is the boundary layer thickness. Assuming this thickness to vary as $(\nu x/u_\infty)^{1/2}$, it follows that

$$\frac{u}{u_\infty}=\phi(\eta) \tag{7.9}$$

Hence, the velocity profile is assumed to be uniquely determined by the similarity variable η, which depends on both x and y.

From Equations 7.6 through 7.8 we obtain

$$u=\frac{\partial\psi}{\partial y}=\frac{\partial\psi}{\partial\eta}\frac{\partial\eta}{\partial y}=u_\infty\sqrt{\nu x/u_\infty}\frac{df}{d\eta}\sqrt{u_\infty/\nu x}=u_\infty\frac{df}{d\eta} \tag{7.10}$$

and

$$v=-\frac{\partial\psi}{\partial x}=-\left(u_\infty\sqrt{\nu x/u_\infty}\frac{\partial f}{\partial x}+\frac{u_\infty}{2}\sqrt{\nu/u_\infty x}f\right)$$

$$=\tfrac{1}{2}\sqrt{\nu u_\infty/x}\left(\eta\frac{df}{d\eta}-f\right) \tag{7.11}$$

By differentiating the velocity components, it may also be shown that

$$\frac{\partial u}{\partial x}=-\frac{u_\infty}{2x}\eta\frac{d^2f}{d\eta^2} \tag{7.12}$$

$$\frac{\partial u}{\partial y}=u_\infty\sqrt{u_\infty/\nu x}\frac{d^2f}{d\eta^2} \tag{7.13}$$

$$\frac{\partial^2 u}{\partial y^2}=\frac{u_\infty^2}{\nu x}\frac{d^3f}{d\eta^3} \tag{7.14}$$

Substituting these expressions into Equation 7.4, we then obtain

$$2\frac{d^3f}{d\eta^3}+f\frac{d^2f}{d\eta^2}=0 \tag{7.15}$$

Hence the hydrodynamic boundary layer problem is reduced to one of solving a nonlinear, third-order ordinary differential equation. The appropriate boundary conditions are

$$u(x,0)=v(x,0)=0 \qquad \text{and} \qquad u(x,\infty)=u_\infty$$

or, in terms of the similarity variables,

$$\left.\frac{df}{d\eta}\right|_{\eta=0}=f(0)=0 \qquad \text{and} \qquad \left.\frac{df}{d\eta}\right|_{\eta=\infty}=1 \tag{7.16}$$

The solution to Equation 7.15, subject to the conditions of Equation 7.16, is not easily obtained. In fact, a closed-form solution does not exist. However, an accurate solution may still be obtained by a series expansion [2] or by numerical methods [3], and selected results are presented in Table 7.1. Several important results may be extracted from this table. First, to a good approximation, $(u/u_\infty)=0.99$ for $\eta=5.0$. Defining the boundary layer thickness δ as that value of y for which $(u/u_\infty)=0.99$, it follows from Equation 7.8 that

$$\delta=\frac{5.0}{\sqrt{u_\infty/\nu x}}=\frac{5x}{\sqrt{Re_x}} \tag{7.17}$$

From Equation 7.17 it is clear that δ increases with increasing x and ν but decreases with increasing u_∞ (the larger the freestream velocity, the *thinner* the boundary layer). In addition, from Equation 7.13 the wall shear stress may be expressed as

$$\tau_s=\mu\frac{\partial u}{\partial y}\bigg|_{y=0}=\mu u_\infty\sqrt{u_\infty/\nu x}\,\frac{d^2 f}{d\eta^2}\bigg|_{\eta=0}$$

Hence, from Table 7.1

$$\tau_s=0.332u_\infty\sqrt{\rho\mu u_\infty/x}$$

Table 7.1 Flat plate laminar boundary layer functions [3]

$\eta=y\sqrt{\frac{u_\infty}{\nu x}}$	f	$\frac{df}{d\eta}=\frac{u}{u_\infty}$	$\frac{d^2f}{d\eta^2}$
0	0	0	0.332
0.4	0.027	0.133	0.331
0.8	0.106	0.265	0.327
1.2	0.238	0.394	0.317
1.6	0.420	0.517	0.297
2.0	0.650	0.630	0.267
2.4	0.922	0.729	0.228
2.8	1.231	0.812	0.184
3.2	1.569	0.876	0.139
3.6	1.930	0.923	0.098
4.0	2.306	0.956	0.064
4.4	2.692	0.976	0.039
4.8	3.085	0.988	0.022
5.2	3.482	0.994	0.011
5.6	3.880	0.997	0.005
6.0	4.280	0.999	0.002
6.4	4.679	1.000	0.001
6.8	5.079	1.000	0.000

The *local* friction coefficient is then

$$C_{f,x} \equiv \frac{\tau_{s,x}}{\rho u_\infty^2/2} = 0.664 Re_x^{-1/2} \tag{7.18}$$

From knowledge of conditions in the velocity boundary layer, the energy equation may now be solved. To solve Equation 7.5 we nondimensionalize the temperature according to Equation 6.36, and we assume a similarity solution of the form $T^* = T^*(\eta)$. Making the necessary substitutions, Equation 7.5 reduces to

$$\frac{d^2 T^*}{d\eta^2} + \frac{Pr}{2} f \frac{dT^*}{d\eta} = 0 \tag{7.19}$$

Note the dependence of the thermal solution on hydrodynamic conditions through appearance of the variable f in Equation 7.19. The appropriate boundary conditions are

$$T^*(0) = 0 \qquad \text{and} \qquad T^*(\infty) = 1 \tag{7.20}$$

Subject to the conditions of Equation 7.20, Equation 7.19 has been solved numerically for different values of the Prandtl number [4]. One important consequence of this solution is that, for $Pr \gtrsim 0.6$, results for the surface temperature gradient $dT^*/d\eta|_{\eta=0}$ may be correlated by the following relation:

$$\left.\frac{dT^*}{d\eta}\right|_{\eta=0} = 0.332 Pr^{1/3}$$

Expressing the local convection coefficient as

$$h_x = \frac{q_s''}{T_s - T_\infty} = -\frac{T_\infty - T_s}{T_s - T_\infty} k \left.\frac{\partial T^*}{\partial y}\right|_{y=0}$$

$$= k \left(\frac{u_\infty}{\nu x}\right)^{1/2} \left.\frac{dT^*}{d\eta}\right|_{\eta=0}$$

it follows that the *local* Nusselt number is of the form

$$Nu_x \equiv \frac{h_x x}{k} = 0.332 Re_x^{1/2} Pr^{1/3} \qquad Pr \gtrsim 0.6 \tag{7.21}$$

Doesn't apply to liquid metals

From the solution to Equation 7.19, it also follows that the ratio of the velocity to thermal boundary layer thickness is

$$\frac{\delta}{\delta_t} \approx Pr^{1/3} \tag{7.22}$$

where δ is given by Equation 7.17.

The foregoing results may be used to compute important *laminar* boundary layer parameters for any $0 < x < x_c$, where x_c is the distance from the leading edge at which transition begins. Generally, this distance may be computed from the transition criterion given by Equation 6.12. Note that Equations 7.18 and 7.21 imply that $\tau_{s,x}$ and h_x are, in principle, infinite at the leading edge and decrease as $x^{-1/2}$ in the flow direction. Equation 7.22 also implies that for values of *Pr* close to unity, which is the case for most gases, the velocity and thermal boundary layers experience nearly identical growth.

From the foregoing local results, the average boundary layer parameters may be determined. With the average friction coefficient defined as

$$\bar{C}_{f,x} \equiv \frac{\bar{\tau}_{s,x}}{\rho u_\infty^2/2} \tag{7.23}$$

where

$$\bar{\tau}_{s,x} \equiv \frac{1}{x}\int_0^x \tau_{s,x}\, dx$$

the form of $\tau_{s,x}$ may be substituted from Equation 7.18 and the integration performed to obtain

$$\bar{C}_{f,x} = 1.328 Re_x^{-1/2} \tag{7.24}$$

Moreover, from Equations 6.6 and 7.21, the *average* heat transfer coefficient for laminar flow is

$$\bar{h}_x = \frac{1}{x}\int_0^x h_x\, dx = 0.332\left(\frac{k}{x}\right) Pr^{1/3}\left(\frac{u_\infty}{\nu}\right)^{1/2}\int_0^x \frac{dx}{x^{1/2}}$$

Integrating and substituting from Equation 7.21, it follows that

$$\bar{h}_x = 2h_x$$

Hence

$$\overline{Nu}_x \equiv \frac{\bar{h}_x x}{k} = 0.664 Re_x^{1/2} Pr^{1/3} \qquad Pr \gtrsim 0.6 \qquad \text{Laminar Flow} \tag{7.25}$$

Note that, if the flow is laminar over the entire surface, the subscript x may be replaced by L, and Equations 7.24 and 7.25 may be used to predict average conditions for the entire surface.

From the foregoing expressions we see that, for laminar flow over a flat plate, the *average* friction and convection coefficients from the leading edge up to a point x on the surface are *twice* the *local* coefficients at that point. We also note that, in using these expressions, the effect of variable properties can be treated by evaluating all properties at the *film temperature*, Equation 7.2.

For fluids of small Prandtl number, namely *liquid metals*, Equation 7.21 does not apply. However, for this case the thermal boundary layer development is much more rapid than that of the velocity layer ($\delta_t \gg \delta$), and it is reasonable to assume uniform velocity ($u = u_\infty$) throughout the thermal boundary layer. From a solution to the thermal boundary layer equation based on this assumption [5], it may then be shown that

$$Nu_x = 0.565 Pe_x^{1/2} \qquad Pr \lesssim 0.05,\ Pe_x \gtrsim 100 \tag{7.26}$$

liquid metals

where $Pe_x \equiv Re_x Pr$ is the *Peclet number* (Table 6.2). Heat transfer in liquid metals is of considerable current interest, particularly as it relates to nuclear reactor cooling. Despite the corrosive and reactive nature of such liquids, their unique properties (low melting point and vapor pressure, as well as high thermal capacity and conductivity) render them attractive as coolants in applications requiring high heat transfer rates.

7.2.2 Turbulent flow

From experiment [6] it is known that the *local* friction coefficient for Reynolds numbers between 5×10^5 and 10^7 is well correlated by an expression of the form

$$C_{f,x} = 0.0592 Re_x^{-1/5} \qquad 5 \times 10^5 < Re_x < 10^7 \tag{7.27}$$

The expression may also be used to within 15% accuracy for values of Re_x up to 10^8. Moreover, it is known that, to a reasonable approximation, the velocity boundary layer thickness may be expressed as

$$\delta = 0.37 x Re_x^{-1/5} \tag{7.28}$$

Comparing these results with those for the laminar boundary layer, Equations 7.17 and 7.18, we see that turbulent boundary layer growth is much more rapid (δ varies as $x^{4/5}$, in contrast to $x^{1/2}$ for laminar flow) and that the decay in the friction coefficient is more gradual ($x^{-1/5}$ vs. $x^{-1/2}$). For turbulent flow, boundary layer development is influenced strongly by random fluctuations in the fluid, and not by molecular diffusion. Hence relative boundary layer growth does not depend on the value of Pr, and Equation 7.28 may be used to obtain the thermal, as well as the velocity, boundary layer thickness. That is, for turbulent flow, $\delta \approx \delta_t$.

Using Equation 7.27 with the modified Reynolds, or Chilton-Colburn, analogy, Equation 6.57, the *local* Nusselt number for turbulent flow is

$$Nu_x = St Re_x Pr = 0.0296 Re_x^{4/5} Pr^{1/3} \qquad 0.6 < Pr < 60 \tag{7.29}$$

Due to enhanced mixing, the turbulent boundary layer grows more rapidly than the laminar boundary layer and is characterized by larger friction and convection coefficients.

Expressions for the average coefficients could now be determined using the procedures of the preceding section. Since the turbulent boundary layer is generally preceded by a laminar boundary layer, however, it is convenient to first consider *mixed* boundary layer conditions.

7.2.3 Mixed Boundary Layer Conditions

For laminar flow over the entire plate, Equations 7.24 and 7.25 may be used to compute the average coefficients. Moreover, if transition occurs toward the rear of the plate, for example, in the range $0.95 \lesssim (x_c/L) \leq 1$, these equations may be used to compute the average coefficients to a reasonable approximation. However, when transition occurs sufficiently upstream of the rear edge, $(x_c/L) \lesssim 0.95$, the surface average coefficients will be influenced by conditions in both the laminar and turbulent boundary layers.

In the mixed boundary layer situation (Figure 7.3), Equation 6.6 may be used to obtain the average convection heat transfer coefficient for the entire plate. Integrating over the laminar region $(0 \leq x \leq x_c)$ and then over the turbulent region $(x_c < x \leq L)$, this equation may be expressed as

$$\bar{h}_L = \frac{1}{L}\left(\int_0^{x_c} h_{\mathrm{lam}}\, dx + \int_{x_c}^{L} h_{\mathrm{turb}}\, dx\right)$$

Turbulent: $h_L = \dfrac{1}{\frac{1}{L}\int_0^L \frac{1}{h_x}\,dx}$

where it is assumed that transition occurs abruptly at $x = x_c$. Substituting from Equations 7.21 and 7.29, for h_{lam} and h_{turb}, respectively, we obtain

$$\bar{h}_L = \left(\frac{k}{L}\right)\left[0.332\left(\frac{u_\infty}{\nu}\right)^{1/2}\int_0^{x_c}\frac{dx}{x^{1/2}} + 0.0296\left(\frac{u_\infty}{\nu}\right)^{4/5}\int_{x_c}^{L}\frac{dx}{x^{1/5}}\right]Pr^{1/3}$$

Integrating, we then obtain

$$\overline{Nu}_L = [0.664 Re_{x,c}^{1/2} + 0.037(Re_L^{4/5} - Re_{x,c}^{4/5})]\, Pr^{1/3}$$

or,

$$\overline{Nu}_L = (0.037 Re_L^{4/5} - A) Pr^{1/3} \tag{7.30}$$

where the constant A is determined by the value of the critical Reynolds number $Re_{x,c}$. That is,

$$A = 0.037 Re_{x,c}^{4/5} - 0.664 Re_{x,c}^{1/2} \tag{7.31}$$

If the typical transition Reynolds number of $Re_{x,c} = 5 \times 10^5$ is assumed, Equation 7.30 reduces to

$$\overline{Nu}_L = (0.037 Re_L^{4/5} - 871) Pr^{1/3} \tag{7.32}$$

Assume $Re_{x,c} = 5\times10^5$

$$\begin{bmatrix} 0.6 < Pr < 60 \\ 5 \times 10^5 < Re_L \lesssim 10^8 \\ Re_{x,c} = 5 \times 10^5 \end{bmatrix}$$

where the bracketed relations indicate the range of applicability. In a similar manner it may be shown that for the friction coefficient

$$\bar{C}_{f,L} = \frac{0.074}{Re_L^{1/5}} - \frac{1742}{Re_L} \tag{7.33}$$

$$\begin{bmatrix} 5 \times 10^5 < Re_L \lesssim 10^8 \\ Re_{x,c} = 5 \times 10^5 \end{bmatrix}$$

It is useful to note that, in situations for which $L \gg x_c$ ($Re_L \gg Re_{x,c}$), $A \ll 0.037 Re_L^{4/5}$ and to a reasonable approximation Equation 7.32 reduces to

$$\overline{Nu}_L = 0.037 Re_L^{4/5} Pr^{1/3} \tag{7.34}$$

Similarly, it follows that

$$\bar{C}_{f,L} = 0.074 Re_L^{-1/5} \tag{7.35}$$

Use of these results is also appropriate when a turbulent boundary layer exists over the entire plate. Such a condition may be realized by *tripping* the boundary layer at the leading edge, using a fine wire or some other turbulence promoter.

All of the foregoing correlations require evaluation of the fluid properties at the film temperature, Equation 7.2. An alternative approach for treating the temperature dependence of the fluid properties has been suggested by Whitaker [7] on the basis of measurements by Zhukauskas [8]. More recently, Churchill [9, 10] has critically reviewed theoretical and experimental results available in the literature and has suggested a single correlating equation that applies over the complete range of *Re* and *Pr*. Different constants are suggested for use with the equation, where specific values depend on the existence of uniform surface temperature or heat flux and whether one is interested in local or average conditions.

7.2.4 Special Cases

All the foregoing Nusselt number expressions are restricted to situations for which the surface temperature T_s is uniform. A common exception involves existence of an *unheated starting length* ($T_s = T_\infty$) upstream of a heated section ($T_s \neq T_\infty$). As shown in Figure 7.4, velocity boundary layer development begins at $x = 0$, while thermal boundary layer development begins at $x = \xi$. Hence there is no heat transfer occurring for $0 \le x \le \xi$. Through use of an integral boundary layer solution [5], it has been found that, for laminar flow,

$$Nu_x = \frac{Nu_x|_{\xi=0}}{[1 - (\xi/x)^{3/4}]^{1/3}} \tag{7.36}$$

where $Nu_x|_{\xi=0}$ is given by Equation 7.21. Note that, in both Nu_x and $Nu_x|_{\xi=0}$, the characteristic length x is measured from the leading edge of the unheated

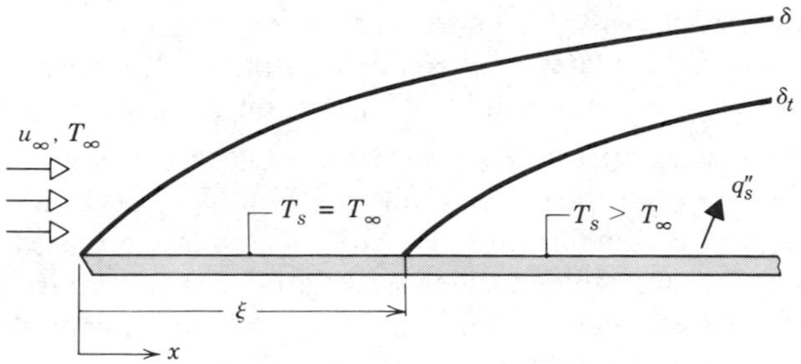

Figure 7.4 Flat plate in parallel flow with unheated starting length.

starting length. It has also been found that, for turbulent flow,

$$Nu_x = \frac{Nu_x|_{\xi=0}}{[1-(\xi/x)^{9/10}]^{1/9}} \tag{7.37}$$

where $Nu_x|_{\xi=0}$ is given by Equation 7.29. Equations 7.36 and 7.37 apply for $x>\xi$, and to obtain average Nusselt numbers for $\xi<x<L$, they would have to be integrated numerically.

In heat transfer problems, it is also possible to have a uniform surface heat flux, rather than a uniform temperature, imposed at the plate. Integral methods [5] may again be used to show that, for laminar flow,

$$Nu_x = 0.453 Re_x^{1/2} Pr^{1/3} \tag{7.38}$$

while for turbulent flow

$$Nu_x = 0.308 Re_x^{4/5} Pr^{1/3} \tag{7.39}$$

wrong $Nu_x = 0.0308\, Re_x^{4/5}\, Pr^{1/3}$

Hence the Nusselt number is 36 and 4% larger than the constant surface temperature result for laminar and turbulent flow, respectively. Correction for the effect of an unheated starting length may be made by using Equations 7.38 and 7.39 with Equations 7.36 and 7.37, respectively. If the heat flux is known, the convection coefficient may be used to determine the local surface temperature

$$T_s(x) = T_\infty + \frac{q_s''}{h_x} \tag{7.40}$$

Although the equations of this section are suitable for most engineering calculations, in practice they rarely provide exact values for the convection coefficients. Conditions vary according to freestream turbulence and surface roughness, and errors as large as 25% may be incurred by using the expressions. A detailed description of freestream turbulence effects is provided by Blair [11].

7.3 METHODOLOGY FOR A CONVECTION CALCULATION

Although we have discussed only correlations for parallel flow over a flat plate, selection and application of a convection correlation for *any flow situation* are facilitated by following a few simple rules.

1. *Become immediately cognizant of the flow geometry.* Does the problem involve flow over a flat plate, a sphere, or a cylinder? The specific form of the convection correlation depends, of course, on the geometry.
2. *Specify the appropriate reference temperature and evaluate the pertinent fluid properties at that temperature.* For moderate boundary layer temperature differences, the film temperature, Equation 7.2, may be used for this purpose. However, we will consider correlations that require property evaluation at the freestream temperature and include a property ratio to account for the nonconstant property effect.
3. *Calculate the Reynolds number.* Boundary layer conditions are strongly influenced by this parameter. If the geometry is a flat plate in parallel flow, determine whether the flow is laminar or turbulent.
4. *Decide whether a local or surface average coefficient is required.* Recall that for constant surface temperature, the local coefficient is used to determine the flux at a particular point on the surface; whereas the average coefficient determines the transfer rate for the entire surface.

Having done the above, you will have sufficient information to select the appropriate correlation.

EXAMPLE 7.1

Air at a pressure of 6 kN/m^2 and a temperature of 300°C flows with a velocity of 10 m/s over a flat plate of length 0.5 m. Estimate the cooling rate per unit width of the plate needed to maintain it at a surface temperature of 27°C.

SOLUTION

Known: Airflow over an isothermal flat plate.

Find: Cooling rate per unit width of the plate, q' (W/m).

Schematic:

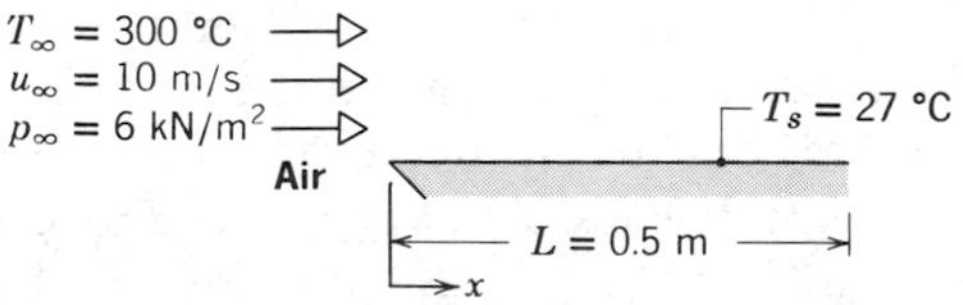

Assumptions:

1. Steady-state conditions.
2. Negligible radiation effects.

Properties:

Table A.4, air ($T_f = 437$ K, $p = 1$ atm): $\nu = 30.84 \times 10^{-6}$ m²/s, $k = 36.4 \times 10^{-3}$ W/m·K, $Pr = 0.687$. Properties such as k, Pr, and μ may be assumed to be independent of pressure to an excellent approximation. However, for a gas the kinematic viscosity, $\nu = \mu/\rho$, will vary with pressure through its dependence on density. From the perfect gas law, $\rho = p/RT$, it follows that the ratio of kinematic viscosities for a gas at the same temperature but at different pressures, p_1 and p_2, is $(\nu_1/\nu_2) = (p_2/p_1)$. Hence the kinematic viscosity of air at 437 K and $p_\infty = 6 \times 10^3$ N/m² is

$$\nu = 30.84 \times 10^{-6}\ \text{m}^2/\text{s} \times \frac{1.0133 \times 10^5\ \text{N/m}^2}{6 \times 10^3\ \text{N/m}^2} = 5.21 \times 10^{-4}\ \text{m}^2/\text{s}$$

Analysis:

For a plate of unit width, it follows from Newton's law of cooling that the rate of convection heat transfer *to* the plate is

$$q' = \bar{h} L (T_\infty - T_s)$$

To determine the appropriate convection correlation for computing $\bar{h}$, the Reynolds number must first be determined

$$Re_L = \frac{u_\infty L}{\nu} = \frac{10\ \text{m/s} \times 0.5\ \text{m}}{5.21 \times 10^{-4}\ \text{m}^2/\text{s}} = 9597$$

Hence the flow is laminar over the entire plate, and the appropriate correlation is given by Equation 7.25

$$\overline{Nu}_L = 0.664 Re_L^{1/2} Pr^{1/3} = 0.664\,(9597)^{1/2}\,(0.687)^{1/3} = 57.4$$

The average convection coefficient is then

$$\bar{h} = \frac{\overline{Nu}_L k}{L} = \frac{57.4 \times 0.0364\ \text{W/m}\cdot\text{K}}{0.5\ \text{m}} = 4.18\ \text{W/m}^2\cdot\text{K}$$

and the required cooling rate per unit width of plate is

$$q' = 4.18\ \text{W/m}^2\cdot\text{K} \times 0.5\ \text{m}\ (300 - 27)^\circ\text{C} = 570\ \text{W/m} \qquad \triangleleft$$

Comments:

The results of Table A.4 apply to gases at atmospheric pressure. Except for the kinematic viscosity, mass density, and thermal diffusivity, they may generally be used at other pressures without correction. The kinematic viscosity and the thermal diffusivity for a pressure other than 1 atm may be obtained by dividing the tabulated value by the pressure in atmospheres.

EXAMPLE 7.2

A flat plate of width $w = 1$ m is maintained at a uniform surface temperature, $T_s = 230°C$, by using independently controlled, electrical strip heaters, each of which is 50 mm long. If atmospheric air at 25°C flows over the plate at a velocity of 60 m/s, at what heater is the electrical input a maximum? What is the value of this input?

SOLUTION

Known: Airflow over a flat plate with segmented heaters.

Find: Maximum heater power requirement.

Schematic:

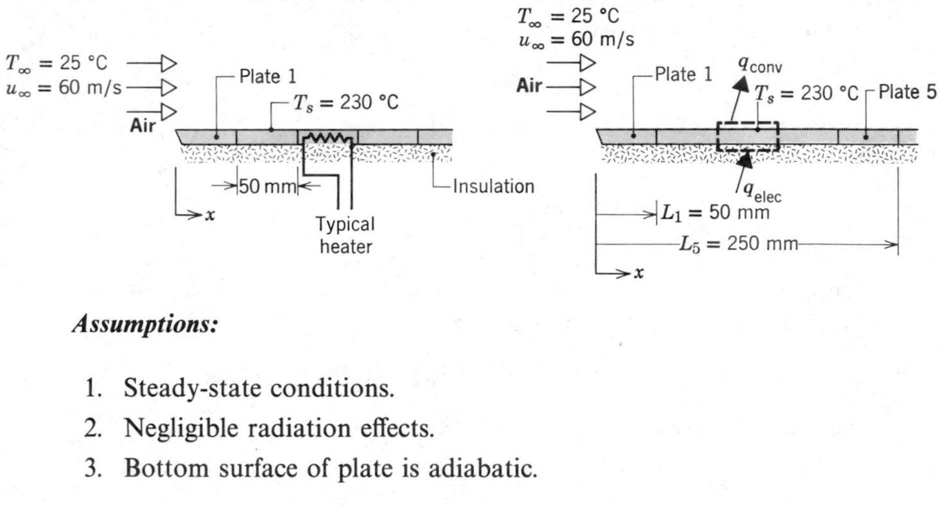

Assumptions:

1. Steady-state conditions.
2. Negligible radiation effects.
3. Bottom surface of plate is adiabatic.

Properties:

Table A.4, air ($T_f = 400$ K, $p = 1$ atm): $\nu = 26.41 \times 10^{-6}$ m²/s, $k = 0.0338$ W/m·K, $Pr = 0.690$.

Analysis:

The location of the heater requiring the maximum electrical power may be determined by first finding the point of boundary layer transition. The Reynolds number based on the length L_1 of the first heater is

$$Re_1 = \frac{u_\infty L_1}{\nu} = \frac{60 \text{ m/s} \times 0.05 \text{ m}}{26.41 \times 10^{-6} \text{ m}^2\text{/s}} = 1.14 \times 10^5$$

If the transition Reynolds number is assumed to be $Re_{x,c} = 5 \times 10^5$, it follows that transition will occur on the fifth heater, or more specifically at

$$x_c = \frac{\nu}{u_\infty} Re_{x,c} = \frac{26.41 \times 10^{-6}\ \text{m}^2/\text{s}}{60\ \text{m/s}} 5 \times 10^5 = 0.22\ \text{m}$$

The heater requiring the maximum electrical power is that for which the average convection coefficient is largest. From knowledge of how the local convection coefficient varies with distance from the leading edge, we conclude that there are three possibilities:

1. Heater 1, since it corresponds to the largest local, laminar convection coefficient.
2. Heater 5, since it corresponds to the largest local, turbulent convection coefficient.
3. Heater 6, since turbulent conditions exist over the entire heater.

For each of these heaters, conservation of energy requires that

$$q_{\text{elec}} = q_{\text{conv}}$$

For the first heater,

$$q_{\text{conv},1} = \bar{h}_1 L_1 w (T_s - T_\infty)$$

where $\bar{h}_1$ is determined from Equation 7.25

$$\overline{Nu}_1 = 0.664 Re_1^{1/2} Pr^{1/3} = 0.664(1.14 \times 10^5)^{1/2}(0.69)^{1/3} = 198$$

Hence

$$\bar{h}_1 = \frac{\overline{Nu}_1 k}{L_1} = \frac{198 \times 0.0338\ \text{W/m}\cdot\text{K}}{0.05\ \text{m}} = 134\ \text{W/m}^2\cdot\text{K}$$

and

$$q_{\text{conv},1} = 134\ \text{W/m}^2\cdot\text{K}\ (0.05 \times 1)\ \text{m}^2\ (230 - 25)^\circ\text{C} = 1370\ \text{W}$$

The power requirement for the fifth heater may be obtained by subtracting the total heat loss associated with the first four heaters from that associated with the first five heaters. Accordingly,

$$q_{\text{conv},5} = \bar{h}_{1-5} L_5 w (T_s - T_\infty) - \bar{h}_{1-4} L_4 w (T_s - T_\infty)$$

$$q_{\text{conv},5} = (\bar{h}_{1-5} L_5 - \bar{h}_{1-4} L_4)\, w (T_s - T_\infty)$$

The value of $\bar{h}_{1-4}$ may be obtained from Equation 7.25, where

$$\overline{Nu}_4 = 0.664 Re_4^{1/2} Pr^{1/3}$$

With $Re_4 = 4Re_1 = 4.56 \times 10^5$,

$$\overline{Nu}_4 = 0.664(4.56 \times 10^5)^{1/2}(0.69)^{1/3} = 396$$

Hence

$$\bar{h}_{1-4} = \frac{\overline{Nu}_4 k}{L_4} = \frac{396 \times 0.0338 \text{ W/m}\cdot\text{K}}{0.2 \text{ m}} = 67 \text{ W/m}^2\cdot\text{K}$$

In contrast, the fifth heater is characterized by mixed boundary layer conditions, and $\bar{h}_{1-5}$ must be obtained from Equation 7.32. With $Re_5 = 5Re_1 = 5.70 \times 10^5$,

$$\overline{Nu}_5 = (0.037 Re_5^{4/5} - 871) Pr^{1/3}$$

$$\overline{Nu}_5 = [0.037(5.70 \times 10^5)^{4/5} - 871](0.69)^{1/3} = 546$$

Hence

$$\bar{h}_{1-5} = \frac{\overline{Nu}_5 k}{L_5} = \frac{546 \times 0.0338 \text{ W/m}\cdot\text{K}}{0.25 \text{ m}} = 74 \text{ W/m}^2\cdot\text{K}$$

The rate of heat transfer from the fifth heater is then

$$q_{\text{conv},5} = (74 \text{ W/m}^2\cdot\text{K} \times 0.25 \text{ m} - 67 \text{ W/m}^2\cdot\text{K} \times 0.20 \text{ m}) \text{ 1 m } (230 - 25)°\text{C}$$

$$q_{\text{conv},5} = 1050 \text{ W}$$

Similarly, the power requirement for the sixth heater may be obtained by subtracting the total heat loss associated with the first five heaters from that associated with the first six heaters. Hence

$$q_{\text{conv},6} = (\bar{h}_{1-6} L_6 - \bar{h}_{1-5} L_5)\, w (T_s - T_\infty)$$

where $\bar{h}_{1-6}$ may be obtained from Equation 7.32. With $Re_6 = 6Re_1 = 6.84 \times 10^5$,

$$\overline{Nu}_6 = [0.037(6.84 \times 10^5)^{4/5} - 871](0.69)^{1/3} = 753$$

Hence

$$\bar{h}_{1-6} = \frac{\overline{Nu}_6 k}{L_6} = \frac{753 \times 0.0338 \text{ W/m}\cdot\text{K}}{0.30 \text{ m}} = 85 \text{ W/m}^2\cdot\text{K}$$

and

$$q_{\text{conv},6} = (85 \text{ W/m}^2\cdot\text{K} \times 0.30 \text{ m} - 74 \text{ W/m}^2\cdot\text{K} \times 0.25 \text{ m}) \text{ 1 m } (230 - 25)°\text{C}$$

$$q_{\text{conv},6} = 1440 \text{ W} \qquad \triangleleft$$

Hence $q_{\text{conv},6} > q_{\text{conv},1} > q_{\text{conv},5}$, and the sixth plate has the largest power requirement.

Comments:

An alternative method of finding the convection heat transfer rate from a particular plate involves estimating an average local convection coefficient for the surface. For example, Equation 7.29 could be used to evaluate the local convection coefficient at the midpoint of the sixth plate. With $x = 0.275$ m, $Re_x = 6.27 \times 10^5$, $Nu_x = 1136$, and $h_x = 140\ \text{W/m}^2 \cdot \text{K}$, the convection heat transfer rate from the sixth plate is

$$q_{conv,6} = h_x (L_6 - L_5)\, w (T_s - T_\infty)$$

$$q_{conv,6} = 140\ \text{W/m}^2 \cdot \text{K}\ (0.30\text{–}0.25)\ \text{m} \times 1\ \text{m}\ (230 - 25)^\circ\text{C} = 1440\ \text{W}$$

This procedure may be used only when the variation of the local convection coefficient with distance is gradual, such as in turbulent flow. It could lead to significant error when used for a surface that experiences transition.

7.4 THE CYLINDER IN CROSS FLOW

7.4.1 Flow Considerations

Another common external flow involves fluid motion normal to the axis of a circular cylinder. As shown in Figure 7.5, the freestream fluid is brought to rest at the *forward stagnation point*, with an accompanying rise in pressure. From this point, the pressure decreases with increasing x, the streamline coordinate, and the boundary layer develops under the influence of a *favorable pressure gradient* $(dp/dx < 0)$. However, the pressure must eventually reach a minimum, and toward the rear of the cylinder further boundary layer development occurs in the presence of an *adverse pressure gradient* $(dp/dx > 0)$.

In Figure 7.5 the distinction between the upstream velocity V and the freestream velocity u_∞ should be noted. Unlike conditions for the flat plate in parallel flow, these velocities differ, with u_∞ now depending on the distance x from the stagnation point. From Euler's equation for an inviscid flow [12], $u_\infty(x)$ must exhibit behavior opposite to that of $p(x)$. That is, from $u_\infty = 0$ at the stagnation point, the fluid accelerates because of the favorable pressure gradient

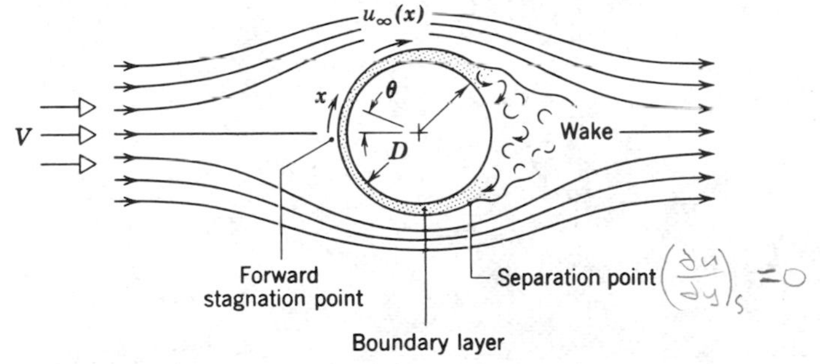

Figure 7.5 Boundary layer formation and separation on a circular cylinder in cross flow.

($du_\infty/dx > 0$ when $dp/dx < 0$), reaches a maximum velocity when $dp/dx = 0$, and decelerates as a result of the adverse pressure gradient ($du_\infty/dx < 0$ when $dp/dx > 0$). As the fluid decelerates, the velocity gradient at the surface $\partial u/\partial y|_{y=0}$ eventually becomes zero (Figure 7.6). At this location, termed the *separation point*, fluid near the surface lacks sufficient momentum to overcome the pressure gradient, and continued downstream movement is impossible. Since the oncoming fluid also precludes flow back upstream, *boundary layer separation* must occur. This is a condition for which the boundary layer detaches from the surface, and a *wake* is formed in the downstream region. Flow in this region is characterized by vortex formation and is highly irregular. The *separation point* is the location for which $(\partial u/\partial y)_s = 0$.

It is important to note that the occurrence of *boundary layer transition*, which depends on the Reynolds number, strongly influences the position of the separation point. For the circular cylinder the characteristic length is the diameter, and the Reynolds number is defined as

$$Re_D \equiv \frac{\rho V D}{\mu} = \frac{VD}{\nu}$$

Since the momentum of fluid in a turbulent boundary layer is larger than in the laminar boundary layer, it is reasonable to expect transition to delay the occurrence of separation. If $Re_D \lesssim 2 \times 10^5$, the boundary layer remains laminar, and separation occurs at $\theta \approx 80°$ (Figure 7.7). However, if $Re_D \gtrsim 2 \times 10^5$, boundary layer transition occurs, and separation is delayed to $\theta \approx 140°$.

The foregoing processes strongly influence the drag force F_D acting on the cylinder. This force has two components, one of which is due to the boundary layer surface shear stress (*friction drag*). The other component is due to a pressure differential in the flow direction resulting from formation of the wake (*form*, or *pressure*, *drag*). A dimensionless *drag coefficient* C_D may be defined as

$$C_D \equiv \frac{F_D}{A_f(\rho V^2/2)} \tag{7.41}$$

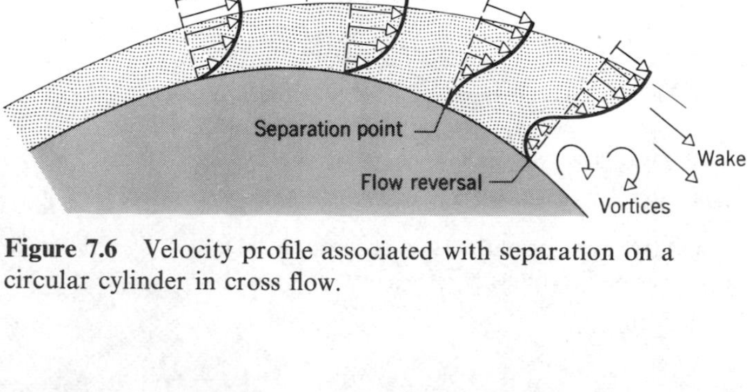

Figure 7.6 Velocity profile associated with separation on a circular cylinder in cross flow.

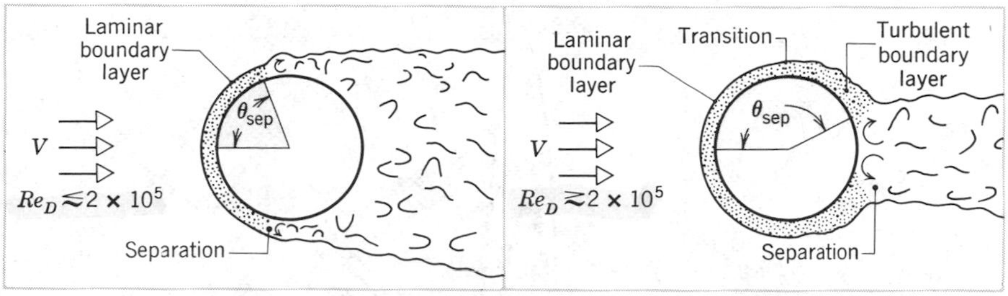

Figure 7.7 The effect of turbulence on separation.

where A_f is the cylinder frontal area (the area projected perpendicular to the freestream velocity). The drag coefficient is a function of Reynolds number and results are presented in Figure 7.8. For $Re_D < 2$ separation effects are negligible, and conditions are dominated by friction drag. However, with increasing Reynolds number, the effect of separation, and therefore form drag, becomes more important. The large reduction in C_D that occurs for $Re_D > 2 \times 10^5$ is due to boundary layer transition, which delays separation, thereby reducing the extent of the wake region and the magnitude of the form drag.

7.4.2 Convection Heat Transfer

Because of complexities associated with flow over a cylinder, emphasis is placed on the use of experimental methods to determine heat transfer effects. Experimental results for the variation of the local Nusselt number with θ are shown

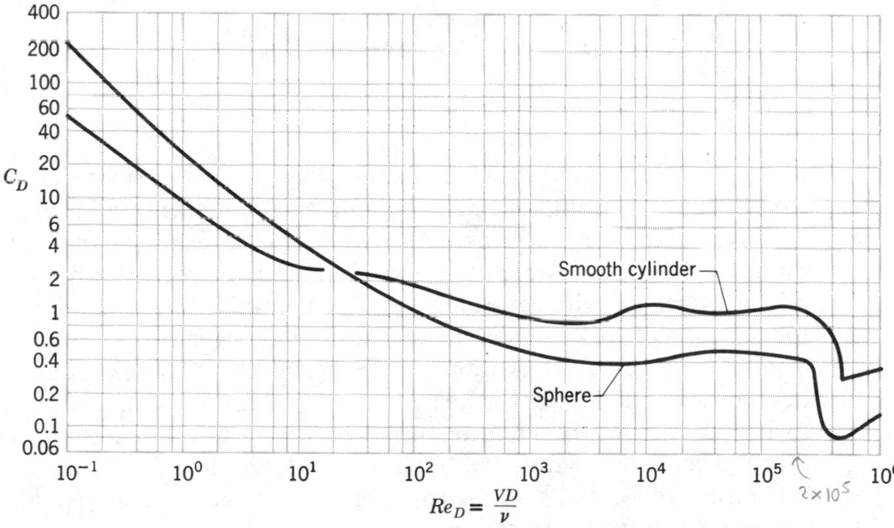

Figure 7.8 Drag coefficients for a smooth, circular cylinder in cross flow and for a sphere [6]. Adapted with permission.

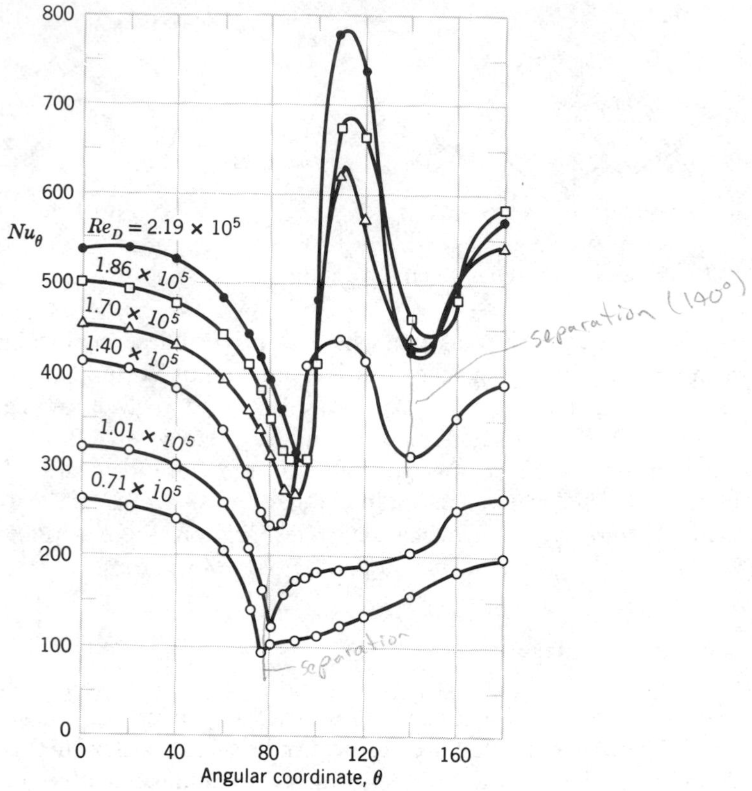

Figure 7.9 Local Nusselt number for airflow normal to a circular cylinder. Adapted with permission from W. H. Giedt, *Trans. ASME*, *71*, 375, 1949.

in Figure 7.9 for the cylinder in a cross flow of air. Consider the results for $Re_D \lesssim 10^5$. Starting at the stagnation point, Nu_θ decreases with increasing θ due to laminar boundary layer development. However, a minimum is reached at $\theta \approx 80°$. At this point separation occurs, and Nu_θ increases with θ due to the mixing associated with vortex formation in the wake. In contrast for $Re_D \gtrsim 10^5$ the variation of Nu_θ with θ is characterized by two minima. The decline in Nu_θ from the value at the stagnation point is again due to laminar boundary layer development, but the sharp increase that occurs between 80 and 100° is now due to boundary layer transition to turbulence. With further development of the turbulent boundary layer, Nu_θ again begins to decline. Eventually separation occurs ($\theta \approx 140°$), and Nu_θ increases as a result of the considerable mixing associated with the wake region.

From the standpoint of engineering calculations, we are more interested in overall average conditions. The empirical correlation due to Hilpert [13]

$$\overline{Nu}_D \equiv \frac{\bar{h}D}{k} = C Re_D^m Pr^{1/3} \qquad \text{@ } T_f \qquad \text{Table 7.2, Table 7.3} \tag{7.42}$$

is widely used, where the constants C and m are listed in Table 7.2. Equation 7.42 may also be used for *gas flow* over cylinders of noncircular cross section, with the characteristic length D and the constants obtained from Table 7.3. In working with Equation 7.42 all properties are evaluated at the film temperature.

More recently, alternative correlations have been suggested for the circular cylinder in cross flow [7, 16, 17]. The correlation due to Zhukauskas [16] is of the form

$$\overline{Nu}_D = C Re_D^m Pr^n (Pr_\infty / Pr_s)^{1/4} \qquad \text{@ } T_\infty \; (Pr_s \text{ @ } T_s) \qquad \text{Table 7.4} \tag{7.43}$$

$$\begin{bmatrix} 0.7 < Pr < 500 \\ 1 < Re_D < 10^6 \end{bmatrix}$$

Table 7.2 Constants of Equation 7.42 for the circular cylinder in cross flow [13, 14]

Re_D	C	m
0.4–4	0.989	0.330
4–40	0.911	0.385
40–4000	0.683	0.466
4000–40,000	0.193	0.618
40,000–400,000	0.027	0.805

Table 7.3 Constants of Equation 7.42 for noncircular cylinders in cross flow of a gas [15]

GEOMETRY	Re_D	C	m
Square			
V → ◇ D	$5 \times 10^3 - 10^5$	0.246	0.588
V → □ D	$5 \times 10^3 - 10^5$	0.102	0.675
Hexagon			
V → ⬡ D	$5 \times 10^3 - 1.95 \times 10^4$	0.160	0.638
	$1.95 \times 10^4 - 10^5$	0.0385	0.782
V → ⬢ D	$5 \times 10^3 - 10^5$	0.153	0.638
Vertical plate			
V → ▯ D	$4 \times 10^3 - 1.5 \times 10^4$	0.228	0.731

where all properties are evaluated at T_∞, except Pr_s, which is evaluated at T_s. Values of C and m are listed in Table 7.4. If $Pr \leq 10$, $n = 0.37$; if $Pr > 10$, $n = 0.36$. Most recently, Churchill and Bernstein [17] have proposed a single comprehensive equation that covers the entire range of Re_D for which data are available, as well as a wide range of Pr. The equation is recommended for all $Re_D Pr > 0.2$ and has the form

$$\overline{Nu}_D = 0.3 + \frac{0.62 Re_D^{1/2} Pr^{1/3}}{[1 + (0.4/Pr)^{2/3}]^{1/4}} \left[1 + \left(\frac{Re_D}{28{,}200}\right)^{5/8}\right]^{4/5} \quad Re_D Pr > 0.2 \qquad (7.44)$$

where all properties are evaluated at the film temperature.

Again we caution the reader not to view any of the foregoing correlations as sacrosanct. Each correlation is reasonable over a certain range of conditions, but for most engineering calculations one should not expect accuracy to much better than 25%. Because they are based on more recent results encompassing a wide range of conditions, Equations 7.43 and 7.44 are used for the calculations of this text. A detailed review of the many correlations that have been developed for the circular cylinder is provided by Morgan [18].

Table 7.4 Constants of Equation 7.43 for the circular cylinder in cross flow [16]

Re_D	C	m	$Pr \leq 10$	$Pr > 10$
1–40	0.75	0.4	$n = 0.37$	$n = 0.36$
40–1000	0.51	0.5		
$10^3 - 2 \times 10^5$	0.26	0.6		
$2 \times 10^5 - 10^6$	0.076	0.7		

EXAMPLE 7.3

Experiments have been conducted on a metallic cylinder 12.7 mm in diameter and 94 mm long. The cylinder is heated internally by an electrical heater and is subjected to a cross flow of air in a low speed wind tunnel. Under a specific set of operating conditions for which the freestream air velocity and temperature were maintained at $V = 10$ m/s and 26.2°C, respectively, the heater power dissipation was measured to be $P = 46$ W, while the average cylinder surface temperature was determined to be $T_s = 128.4$°C. It is estimated that 15% of the power dissipation is lost through the cumulative effect of surface radiation and conduction through the endpieces.

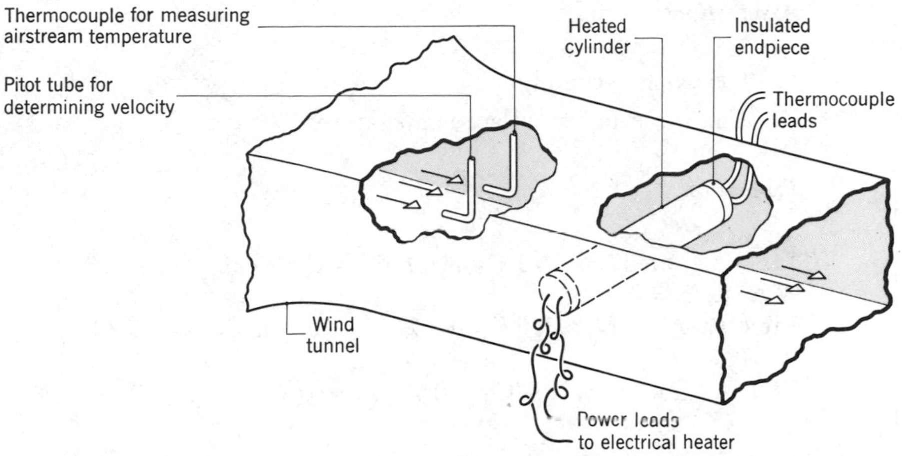

1. Determine the convection heat transfer coefficient from the experimental observations.
2. Compare the experimental result with the convection coefficient computed from an appropriate correlation.

SOLUTION

Known: Operating conditions for a heated cylinder.

Find:

1. Convection coefficient associated with the operating conditions.
2. Convection coefficient from an appropriate correlation.

Schematic:

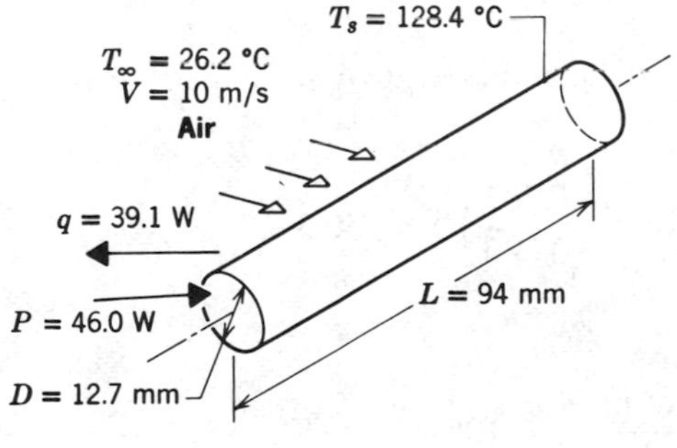

Assumptions:

1. Steady-state conditions.
2. Uniform cylinder surface temperature.

Properties:

Table A.4, air ($T_\infty = 26.2°\text{C} \approx 300$ K): $\nu = 15.89 \times 10^{-6}$ m²/s, $k = 26.3 \times 10^{-3}$ W/m·K, $Pr = 0.707$.
Table A.4, air ($T_f \approx 350$ K): $\nu = 20.92 \times 10^{-6}$ m²/s, $k = 30 \times 10^{-3}$ W/m·K, $Pr = 0.700$.
Table A.4, air ($T_s = 128.4°\text{C} = 401$ K): $Pr = 0.690$.

Analysis:

1. The convection heat transfer coefficient may be determined from the data by using Newton's law of cooling. That is,

$$\bar{h} = \frac{q}{A(T_s - T_\infty)}$$

With $q = 0.85P$ and $A = \pi DL$, it follows that

$$\bar{h} = \frac{0.85 \times 46 \text{ W}}{\pi \times 0.0127 \text{ m} \times 0.094 \text{ m } (128.4 - 26.2)°\text{C}} = 102 \text{ W/m}^2 \cdot \text{K} \qquad \triangleleft$$

2. Working with the Zhukauskas relation, Equation 7.43,

$$\overline{Nu}_D = C Re_D^m Pr^n \left(\frac{Pr_\infty}{Pr_s}\right)^{1/4}$$

all properties, except Pr_s, are evaluated at T_∞. Accordingly,

$$Re_D = \frac{VD}{\nu} = \frac{10 \text{ m/s} \times 0.0127 \text{ m}}{15.89 \times 10^{-6} \text{ m}^2/\text{s}} = 7992$$

Hence, from Table 7.4, $C = 0.26$ and $m = 0.6$. Also, since $Pr < 10$, $n = 0.37$. It follows that

$$\overline{Nu}_D = 0.26(7992)^{0.6}(0.707)^{0.37}\left(\frac{0.707}{0.690}\right)^{0.25} = 50.5$$

Hence

$$\bar{h} = \overline{Nu}_D \frac{k}{D} = 50.5 \frac{0.0263 \text{ W/m} \cdot \text{K}}{0.0127 \text{ m}} = 105 \text{ W/m}^2 \cdot \text{K} \qquad \triangleleft$$

Comments:

1. Using the Churchill relation, Equation 7.44,

$$\overline{Nu}_D = 0.3 + \frac{0.62 Re_D^{1/2} Pr^{1/3}}{[1+(0.4/Pr)^{2/3}]^{1/4}}\left[1+\left(\frac{Re_D}{28{,}200}\right)^{5/8}\right]^{4/5}$$

With all properties evaluated at T_f, $Pr = 0.70$ and

$$Re_D = \frac{VD}{\nu} = \frac{10\ \text{m/s} \times 0.0127\ \text{m}}{20.92 \times 10^{-6}\ \text{m}^2/\text{s}} = 6071$$

Hence the Nusselt number and convection coefficient are

$$\overline{Nu}_D = 0.3 + \frac{0.62(6071)^{1/2}(0.70)^{1/3}}{[1+(0.4/0.70)^{2/3}]^{1/4}}\left[1+\left(\frac{6071}{28{,}200}\right)^{5/8}\right]^{4/5} = 49.1$$

$$\bar{h} = \overline{Nu}_D \frac{k}{D} = 49.1 \frac{0.030\ \text{W/m}\cdot\text{K}}{0.0127\ \text{m}} = 115.9\ \text{W/m}^2\cdot\text{K}$$

Alternatively, from the Hilpert correlation, Equation 7.42,

$$\overline{Nu}_D = C Re_D^m Pr^{1/3}$$

With all properties evaluated at the film temperature, $Re_D = 6071$ and $Pr = 0.70$. Hence, from Table 7.2, $C = 0.193$ and $m = 0.618$. The Nusselt number and convection coefficient are then

$$\overline{Nu}_D = 0.193(6071)^{0.618}(0.700)^{0.333} = 37.3$$

$$\bar{h} = \overline{Nu}_D \frac{k}{D} = 37.3 \frac{0.030\ \text{W/m}\cdot\text{K}}{0.0127\ \text{m}} = 88\ \text{W/m}^2\cdot\text{K}$$

2. Uncertainties associated with measuring the air velocity, estimating the heat loss from cylinder ends, and averaging the cylinder surface temperature, which varies axially and circumferentially, render the experimental result accurate to no better than 15%. Accordingly, calculations based on each of the three correlations are within the experimental uncertainty of the measured result.
3. Recognize the importance of using the proper temperature when evaluating fluid properties.

7.5 THE SPHERE

Boundary layer effects associated with flow over a sphere are much like those for the circular cylinder, with transition and separation both playing prominent roles. Results for the drag coefficient, which is defined by Equation 7.41, are presented in Figure 7.8.

Numerous heat transfer correlations have been proposed, and Whitaker [7] recommends an expression of the form

$$\overline{Nu}_D = 2 + (0.4Re_D^{1/2} + 0.06Re_D^{2/3})Pr^{0.4}\left(\frac{\mu_\infty}{\mu_s}\right)^{1/4} \tag{7.45}$$

$$\left[\begin{array}{l} 0.71 < Pr < 380 \\ 3.5 < Re_D < 7.6 \times 10^4 \\ 1.0 < (\mu_\infty/\mu_s) < 3.2 \end{array}\right]$$

The correlation is accurate to within $\pm 30\%$ for the range of parameter values listed. Note that all properties except μ_s should be evaluated at T_∞.

A special case of convection heat transfer from spheres relates to transport from freely falling liquid drops. Although the correlation of Ranz and Marshall [19] is often used

$$\overline{Nu}_D = 2 + 0.6Re_D^{1/2}Pr^{1/3} \tag{7.46}$$

it has been suggested [20] that the expression be modified to account for the effects of droplet oscillations and distortion. The modified correlation is

$$\overline{Nu}_D = 2 + 0.6Re_D^{1/2}Pr^{1/3}\left[25\left(\frac{x}{D}\right)^{-0.7}\right] \tag{7.47}$$

where x is the falling distance measured from rest and the properties are evaluated at T_∞.

EXAMPLE 7.4

The decorative plastic film on a copper sphere 10 mm in diameter is cured in an oven at 75°C. Upon removal from the oven, the sphere is subjected to an airstream at 1 atm and 23°C having a velocity of 10 m/s. Estimate how long it will take to cool the sphere to 35°C.

SOLUTION

Known: Sphere cooling in an airstream.

Find: Time t required to cool from $T_i = 75°C$ to $T(t) = 35°C$.

Schematic:

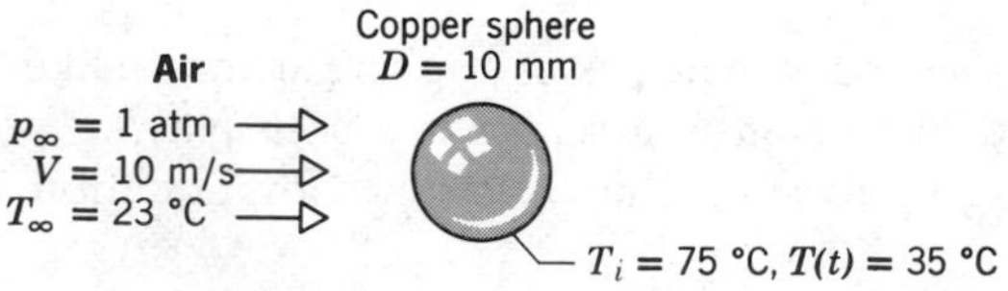

Assumptions:

1. Negligible thermal resistance and capacitance for the plastic film.
2. Spatially isothermal sphere.
3. Negligible radiation effects.

Properties:

Table A.1, copper ($T \approx 328$ K): $\rho = 8933$ kg/m^3, $k = 399$ W/m·K, $c_p = 387$ J/kg·K.
Table A.4, air ($T_\infty = 296$ K): $\mu = 181.6 \times 10^{-7}$ N·s/m^2, $\nu = 15.36 \times 10^{-6}$ m^2/s, $k = 0.0258$ W/m·K, $Pr = 0.709$.
Table A.4, air ($T_s \approx 328$ K): $\mu = 197.8 \times 10^{-7}$ N·s/m^2.

Analysis:

The time required to complete the cooling process may be obtained from results for a lumped capacitance. In particular, from Equations 5.4 and 5.5

$$t = \frac{\rho V c_p}{\bar{h} A_s} \ln \frac{T_i - T_\infty}{T - T_\infty}$$

or, with $V = \pi D^3/6$ and $A_s = \pi D^2$

$$t = \frac{\rho c_p D}{6\bar{h}} \ln \frac{T_i - T_\infty}{T - T_\infty}$$

From Equation 7.45

$$\overline{Nu}_D = 2 + (0.4 Re_D^{1/2} + 0.06 Re_D^{2/3}) Pr^{0.4} \left(\frac{\mu_\infty}{\mu_s}\right)^{1/4}$$

where

$$Re_D = \frac{VD}{\nu} = \frac{10 \text{ m/s} \times 0.01 \text{ m}}{15.36 \times 10^{-6} \text{ m}^2\text{/s}} = 6510$$

Hence the Nusselt number and convection coefficient are

$$\overline{Nu}_D = 2 + [0.4(6510)^{1/2} + 0.06(6510)^{2/3}](0.709)^{0.4} \times \left(\frac{181.6 \times 10^{-7} \text{ N·s/m}^2}{197.8 \times 10^{-7} \text{ N·s/m}^2}\right)^{1/4} = 47.4$$

$$\bar{h} = \overline{Nu}_D \frac{k}{D} = 47.4 \frac{0.0258 \text{ W/m·K}}{0.01 \text{ m}} = 122 \text{ W/m}^2\text{·K}$$

The time required for cooling is then

$$t = \frac{8933 \text{ kg/m}^3 \times 387 \text{ J/kg·K} \times 0.01 \text{ m}}{6 \times 122 \text{ W/m}^2\text{·K}} \ln \left(\frac{75 - 23}{35 - 23}\right) = 69.2 \text{ s} \qquad \triangleleft$$

Comments:

1. The validity of the lumped capacitance method may be determined by calculating the Biot number. From Equation 5.10

$$Bi = \frac{\bar{h}L_c}{k_s} = \frac{\bar{h}(r_o/3)}{k_s} = \frac{122\ \text{W/m}^2\cdot\text{K} \times 0.005\ \text{m}/3}{399\ \text{W/m}\cdot\text{K}} = 5.1 \times 10^{-4}$$

and the criterion is satisfied.

2. Remember that, although having similar definitions, the Nusselt number is defined in terms of the thermal conductivity of the fluid, whereas the Biot number is defined in terms of the thermal conductivity of the solid.

7.6 FLOW ACROSS BANKS OF TUBES

Heat transfer to or from a bank (or bundle) of tubes in cross flow is relevant to numerous industrial applications, such as steam generation in a boiler or air cooling in the coil of an air conditioner. The geometric arrangement is shown schematically in Figure 7.10. Typically, one fluid moves over the tubes, while a second fluid at a different temperature passes through the tubes. In this section

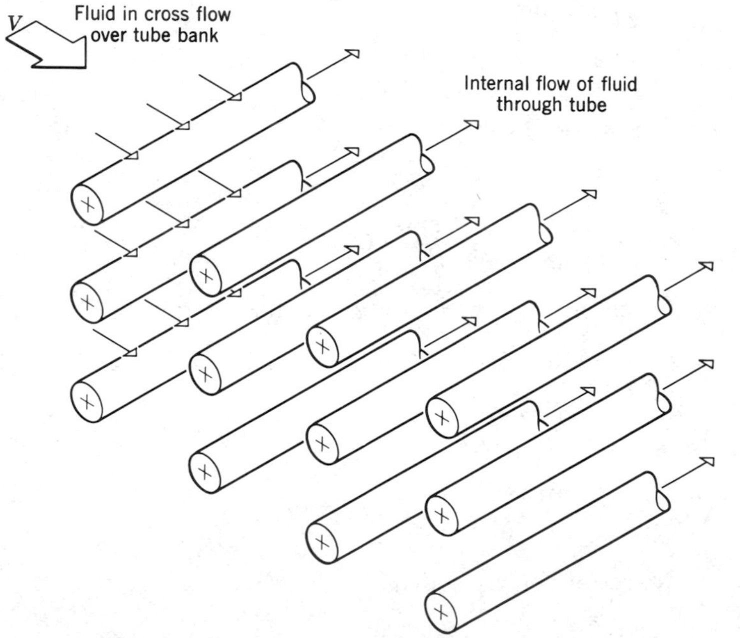

Figure 7.10 Schematic of a tube bank in cross flow.

we are specifically interested in the convection heat transfer associated with cross flow over the tubes.

The tube rows of a bank are either *staggered* or *aligned* in the direction of the fluid velocity V (Figure 7.11). The configuration is characterized by the tube diameter D, and by the *transverse pitch* S_T and *longitudinal pitch* S_L, measured between tube centers. Flow conditions within the bank are dominated by boundary layer separation effects and by wake interactions, which in turn influence convection heat transfer.

The heat transfer coefficient associated with a tube is determined by its position in the bank. The coefficient for a tube in the first row is approximately equal to that for a single tube in cross flow, whereas larger heat transfer coefficients are associated with tubes of the inner rows. The tubes of the first few rows act as turbulence grids, which increase the heat transfer coefficient for tubes in the following rows. In most configurations, however, heat transfer conditions stabilize, such that little change occurs in the convection coefficient for a tube beyond the fourth or fifth row.

Generally, we wish to know the *average* heat transfer coefficient for the *entire* tube bundle. For airflow across tube bundles comprised of *10 or more rows* ($N_L \geq 10$), Grimison [21] has obtained a correlation of the form

$$\overline{Nu}_D = C_1 Re_{D,\max}^m \qquad \begin{bmatrix} N_L \geq 10 \\ 2000 < Re_{D,\max} < 40{,}000 \\ Pr = 0.7 \end{bmatrix} \tag{7.48}$$

where C_1 and m are listed in Table 7.5 and

$$Re_{D,\max} \equiv \frac{\rho V_{\max} D}{\mu} \tag{7.49}$$

It has become common practice to extend this result to other fluids through

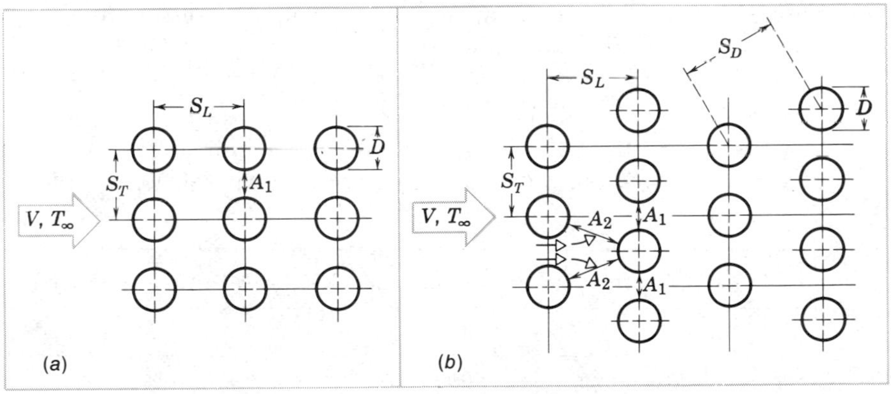

Figure 7.11 Tube arrangements in a bank. (*a*) Aligned. (*b*) Staggered.

Table 7.5 Constants of Equations 7.48 and 7.50 for airflow over a tube bank of 10 or more rows [21]

	S_T/D							
	1.25		1.5		2.0		3.0	
S_L/D	C_1	m	C_1	m	C_1	m	C_1	m
Aligned								
1.25	0.348	0.592	0.275	0.608	0.100	0.704	0.0633	0.752
1.50	0.367	0.586	0.250	0.620	0.101	0.702	0.0678	0.744
2.00	0.418	0.570	0.299	0.602	0.229	0.632	0.198	0.648
3.00	0.290	0.601	0.357	0.584	0.374	0.581	0.286	0.608
Staggered								
0.600	—	—	—	—	—	—	0.213	0.636
0.900	—	—	—	—	0.446	0.571	0.401	0.581
1.000	—	—	0.497	0.558	—	—	—	—
1.125	—	—	—	—	0.478	0.565	0.518	0.560
1.250	0.518	0.556	0.505	0.554	0.519	0.556	0.522	0.562
1.500	0.451	0.568	0.460	0.562	0.452	0.568	0.488	0.568
2.000	0.404	0.572	0.416	0.568	0.482	0.556	0.449	0.570
3.000	0.310	0.592	0.356	0.580	0.440	0.562	0.428	0.574

insertion of the factor $1.13Pr^{1/3}$, in which case

$$\overline{Nu}_D = 1.13 C_1 Re_{D,\max}^m Pr^{1/3} \tag{7.50}$$

$$\begin{bmatrix} N_L \geq 10 \\ 2000 < Re_{D,\max} < 40{,}000 \\ Pr \geq 0.7 \end{bmatrix}$$

All properties appearing in these equations are evaluated at the film temperature. If $N_L < 10$ a correction factor may be applied such that

$$\overline{Nu}_D|_{(N_L<10)} = C_2 \overline{Nu}_D|_{(N_L \geq 10)} \tag{7.51}$$

where C_2 is given in Table 7.6.

Note that the Reynolds number $Re_{D,\max}$ for the foregoing correlations is based on the *maximum fluid velocity* occurring within the tube bank. For the

Table 7.6 Correction factor C_2 of Equation 7.51 for $N_L < 10$ [22]

N_L	1	2	3	4	5	6	7	8	9
Aligned	0.64	0.80	0.87	0.90	0.92	0.94	0.96	0.98	0.99
Staggered	0.68	0.75	0.83	0.89	0.92	0.95	0.97	0.98	0.99

aligned arrangement, V_{max} occurs at the transverse plane A_1 of Figure 7.11*a*, and from the mass conservation requirement for an incompressible fluid

$$V_{max} = \frac{S_T}{S_T - D} V \tag{7.52}$$

For the staggered configuration, the maximum velocity may occur at either the transverse plane A_1 or the diagonal plane A_2 of Figure 7.11*b*. It will occur at A_2 if the rows are spaced such that

$$2(S_D - D) < (S_T - D)$$

The factor of 2 results from the bifurcation experienced by the fluid moving from the A_1 to the A_2 planes. Hence V_{max} occurs at A_2 if

$$S_D = \left[S_L^2 + \left(\frac{S_T}{2}\right)^2 \right]^{1/2} < \frac{S_T + D}{2}$$

in which case it is given by

$$V_{max} = \frac{S_T}{2(S_D - D)} V \tag{7.53}$$

If V_{max} occurs at A_1 for the staggered configuration, it may again be computed from Equation 7.52.

More recent results have been obtained [7, 16], and Zhukauskas [16] has proposed a correlation of the form

$$\overline{Nu}_D = C Re_{D,max}^m Pr^{0.36} \left(\frac{Pr_\infty}{Pr_s}\right)^{1/4} \tag{7.54}$$

$$\left[\begin{array}{l} N_L \geq 20 \\ 0.7 < Pr < 500 \\ 1000 < Re_{D,max} < 2 \times 10^6 \end{array}\right]$$

where all properties, except Pr_s, are evaluated at T_∞ and the constants, C and m, are listed in Table 7.7. This result correlates data very well and may be used for

Table 7.7 Constants of Equation 7.54 for the tube bank in cross flow [16]

CONFIGURATION	$Re_{D,max}$	C	m
Aligned	$10^3 - 2 \times 10^5$	0.27	0.63
Staggered ($S_T/S_L < 2$)	$10^3 - 2 \times 10^5$	$0.35(S_T/S_L)^{1/5}$	0.60
Staggered ($S_T/S_L > 2$)	$10^3 - 2 \times 10^5$	0.40	0.60
Aligned	$2 \times 10^5 - 2 \times 10^6$	0.021	0.84
Staggered	$2 \times 10^5 - 2 \times 10^6$	0.022	0.84

tube banks having $N_L < 20$. In fact accuracy to within 25% is retained for N_L as low as 4.

In using Equation 7.50 or 7.54 to determine the rate of heat transfer for a tube bank, it is important to acknowledge that the fluid in cross flow may experience a large change in temperature as it moves through the bank. Hence the heat transfer rate could be significantly overpredicted by using $\Delta T = T_s - T_\infty$ as the temperature difference in Newton's law of cooling. As the fluid moves through the bank, its temperature approaches T_s and ΔT decreases. In Chapter 11 the appropriate form of ΔT is shown to be the *log mean temperature difference*, which may be expressed as

$$\Delta T_{\mathrm{lm}} = \frac{(T_s - T_i) - (T_s - T_o)}{\ln\left[(T_s - T_i)/(T_s - T_o)\right]} \tag{7.55}$$

where T_i and T_o are temperatures of the fluid as it enters and leaves the bank, respectively. The outlet temperature, which is needed to determine ΔT_{lm}, may be estimated from

$$\frac{T_s - T_o}{T_s - T_i} = \exp\left(-\frac{\pi D N \bar{h}}{\rho V N_T S_T c_p}\right) \tag{7.56}$$

where N is the total number of tubes in the bank and N_T is the number of tubes in the transverse plane. Once ΔT_{lm} is known, the heat transfer rate per unit length of the tubes may be computed from

$$q' = N(\bar{h}\pi D \Delta T_{\mathrm{lm}}) \tag{7.57}$$

Note that neither Equation 7.50 nor 7.54 may be used in the low Reynolds number range, $Re_{D,\max} \lesssim 1000$. Heat transfer results for this range are provided by Zhukauskas [16] and Bergelin et al. [23].

We close by recognizing that there is generally as much interest in the pressure drop associated with flow across a tube bank as in the overall heat transfer rate. The power required to move the fluid across the bank is often a major operating expense and is directly proportional to the pressure drop. The most recent and reliable results are those of Zhukauskas [16], where the pressure drop is given by

$$\Delta p = N_L \chi \left(\frac{\rho V_{\max}^2}{2}\right) f \tag{7.58}$$

The friction factor f and the correction factor χ are plotted in Figures 7.12 and 7.13. Figure 7.12 pertains to a square, in-line tube arrangement for which the dimensionless longitudinal and transverse pitches $P_L \equiv S_L/D$ and $P_T \equiv S_T/D$, respectively, are equal. The correction factor χ, plotted in the insert, is used to apply the results to other in-line arrangements. Similarly, Figure 7.13 applies to a staggered arrangement of tubes in the form of an equilateral triangle ($S_T = S_D$), and the correction factor enables extension of the results to other staggered arrangements. Note that the Reynolds number appearing in Figures 7.12 and 7.13 is based on the maximum fluid velocity V_{max}.

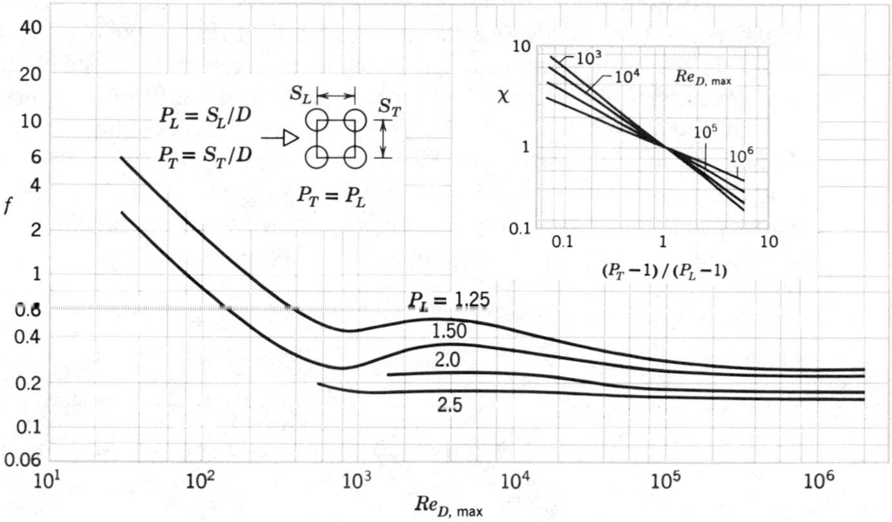

Figure 7.12 Friction factor f and correction factor χ for Equation 7.58. In-line tube bundlc arrangement [16]. Used with permission.

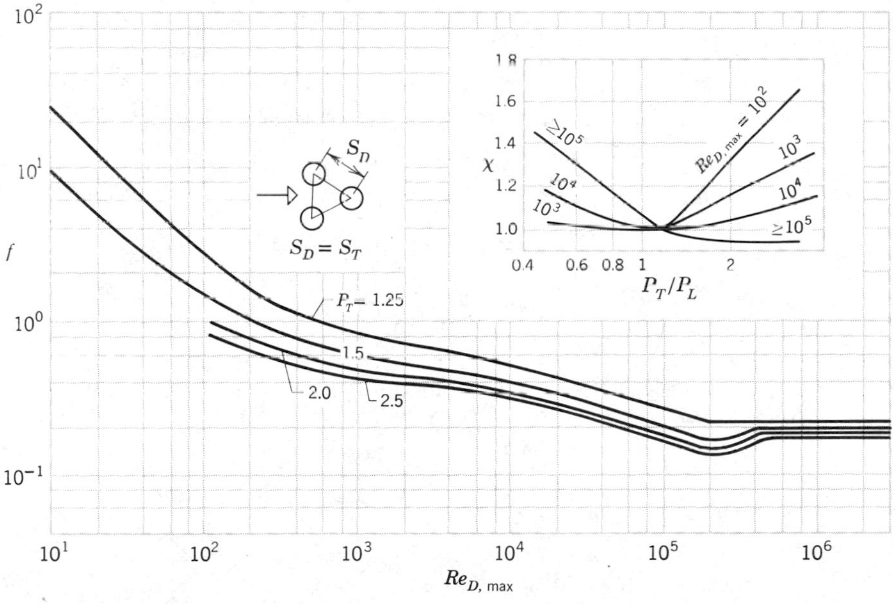

Figure 7.13 Friction factor f and correction factor χ for Equation 7.58. Staggered tube bundle arrangement [16]. Used with permission.

EXAMPLE 7.5

Pressurized water is often available at elevated temperatures and may be used for space heating or industrial process applications. In such cases it is customary to use a tube bundle in which the water is passed through the tubes, while air is passed in cross flow over the tubes. Consider a staggered arrangement for which the tube outside diameter is 16.4 mm and the longitudinal and transverse pitches S_L and S_T are 34.3 and 31.3 mm. There are seven rows of tubes in the airflow direction and eight tubes per row. Under typical operating conditions the cylinder surface temperature is 70°C, while the air upstream temperature and velocity are 15°C and 6 m/s, respectively. Determine the air-side convection coefficient and the rate of heat transfer for the tube bundle. What is the air-side pressure drop?

SOLUTION

Known: Geometry and operating conditions of a tube bank.

Find:

1. Air-side convection coefficient and heat rate.
2. Pressure drop.

Schematic:

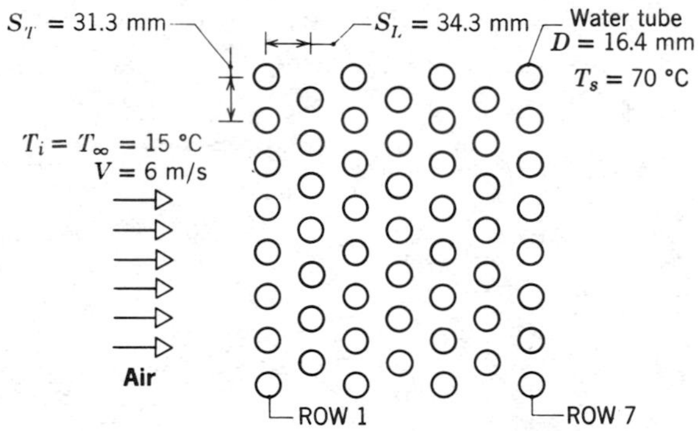

Assumptions:

1. Steady-state conditions.
2. Negligible radiation effects.

Properties:

Table A.4, air ($T_\infty = 15°C$): $\rho = 1.217\ \text{kg/m}^3$, $c_p = 1007\ \text{J/kg·K}$, $\nu = 14.82 \times 10^{-6}\ \text{m}^2/\text{s}$, $k = 0.0253\ \text{W/m·K}$, $Pr = 0.710$.
Table A.4, air ($T_s = 70°C$): $Pr = 0.701$.
Table A.4, air ($T_f = 43°C$): $\nu = 17.4 \times 10^{-6}\ \text{m}^2/\text{s}$, $k = 0.0274\ \text{W/m·K}$, $Pr = 0.705$.

Analysis:

1. From Equation 7.54 the air-side Nusselt number is

$$\overline{Nu}_D = C Re_{D,\max}^m Pr^{0.36} \left(\frac{Pr_\infty}{Pr_s} \right)^{1/4}$$

Since $S_D = [S_L^2 + (S_T/2)^2]^{1/2} = 37.7$ mm is greater than $(S_T + D)/2$, the maximum velocity occurs on the transverse plane, A_1 of Figure 7.11. Hence from Equation 7.52

$$V_{\max} = \frac{S_T}{S_T - D} V$$

$$V_{\max} = \frac{31.3\ \text{mm}}{(31.3 - 16.4)\ \text{mm}} 6\ \text{m/s} = 12.6\ \text{m/s}$$

With

$$Re_{D,\max} = \frac{V_{\max} D}{\nu} = \frac{12.6\ \text{m/s} \times 0.0164\ \text{m}}{14.82 \times 10^{-6}\ \text{m}^2/\text{s}} = 13{,}943$$

and

$$\frac{S_T}{S_L} = \frac{31.3\ \text{mm}}{34.3\ \text{mm}} = 0.91 < 2$$

it follows from Table 7.7 that

$$C = 0.35 \left(\frac{S_T}{S_L} \right)^{1/5} = 0.34 \qquad \text{and} \qquad m = 0.60$$

Hence

$$\overline{Nu}_D = 0.34(13{,}943)^{0.60}(0.71)^{0.36} \left(\frac{0.710}{0.701} \right)^{0.25} = 92.5$$

and

$$\bar{h} = \overline{Nu}_D \frac{k}{D} = 92.5 \times \frac{0.0253\ \text{W/m·K}}{0.0164\ \text{m}} = 142.7\ \text{W/m}^2\text{·K} \qquad \triangleleft$$

From Equation 7.56

$$T_s - T_o = (T_s - T_i)\exp\left(-\frac{\pi D N \bar{h}}{\rho V N_T S_T c_p}\right)$$

$$T_s - T_o = (55°\text{C})\exp\left(-\frac{\pi(0.0164\text{ m})\,56\,(142.7\text{ W/m}^2\cdot\text{K})}{1.217\text{ kg/m}^3(6\text{ m/s})\,8\,(0.0313\text{ m})1007\text{ J/kg}\cdot\text{K}}\right)$$

$$T_s - T_o = 44.0°\text{C}$$

Hence from Equations 7.55 and 7.57

$$\Delta T_{\text{lm}} = \frac{(T_s - T_i) - (T_s - T_o)}{\ln\,[(T_s - T_i)/(T_s - T_o)]} = \frac{(55-44)°\text{C}}{\ln\,(55/44)} = 49.3°\text{C}$$

and

$$q' = N(\bar{h}\pi D \Delta T_{\text{lm}}) = 56\pi \times 142.7\text{ W/m}^2\cdot\text{K} \times 0.0164\text{ m} \times 49.3°\text{C}$$

$$q' = 20.3\text{ kW/m} \qquad \triangleleft$$

2. The pressure drop may be obtained from Equation 7.58

$$\Delta p = N_L \chi \left(\frac{\rho V_{\max}^2}{2}\right) f$$

With $Re_{D,\max} = 13{,}943$, $P_T = (S_T/D) = 1.91$, and $(P_T/P_L) = 0.91$, it follows from Figure 7.13 that

$$\chi \approx 1.04 \qquad \text{and} \qquad f \approx 0.35$$

Hence with $N_L = 7$

$$\Delta p = 7 \times 1.04 \left[\frac{1.217\text{ kg/m}^3\,(12.6\text{ m/s})^2}{2}\right] 0.35$$

$$\Delta p = 246\text{ N/m}^2 = 2.46 \times 10^{-3}\text{ bars} \qquad \triangleleft$$

Comments:

1. With properties evaluated at T_f, it is readily found that $Re_{D,\max} = 11{,}876$. With $S_T/D \approx 2$ and $S_L/D \approx 2$, it follows from Tables 7.5 and 7.6 that $C_1 = 0.482$, $m = 0.556$, and $C_2 = 0.97$. From Equations 7.50 and 7.51, the Nusselt number is then $\overline{Nu}_D = 86.7$, and $\bar{h} = 144.8\text{ W/m}^2\cdot\text{K}$. Values of $\bar{h}$ obtained from Equations 7.50 and 7.54 therefore agree to within 2%, which is well within their uncertainties.
2. If $\Delta T_i \equiv T_s - T_i$ had been used in lieu of ΔT_{lm} in Equation 7.57, the heat rate would have been overpredicted by 12%.

7.7 PACKED BEDS

Gas flow through a *packed bed* of solid particles (Figure 7.14) is relevant to many industrial processes, including the transfer and storage of thermal energy, heterogeneous catalytic reactions, and drying. The position of the particles in a *packed bed* is *fixed.* In contrast, a *fluidized bed* is one for which the particles are in motion as a result of advection with the fluid.

For a packed bed a large amount of heat transfer surface area can be contained in a small volume, and the irregular flow that exists in the voids of the bed enhances transport through turbulent mixing. Many correlations that have been developed for different particle shapes, sizes, and packing densities are described in the literature [24, 25]. One such correlation, which has been recommended for gas flow in a bed of spheres, is of the form

$$\varepsilon \bar{j}_H = 2.06 Re_D^{-0.575} \qquad \begin{bmatrix} Pr \approx 0.7 \\ 90 \leq Re_D \leq 4000 \end{bmatrix} \tag{7.59}$$

where $\bar{j}_H$ is the Colburn *j* factor defined by Equation 6.57. The Reynolds number $Re_D = VD/\nu$ is defined in terms of the sphere diameter and the upstream velocity V that would exist in the empty channel without the packing. The quantity ε is the *porosity*, or *void fraction*, of the bed (volume of void space per unit volume of bed), and its value typically ranges from 0.30 to 0.50. The correlation may be applied to packing materials other than spheres by multiplying the right-hand side by an appropriate correction factor. For a bed of uniformly sized cylinders, with length-to-diameter ratio of 1, the factor is 0.79; for a bed of cubes it is 0.71.

7.8 SUMMARY

In this chapter we have compiled convection correlations for a variety of external flow conditions. For simple surface geometries these results may be derived from a boundary layer analysis, but in most cases they are obtained from generalizations based on experiment. The correlations of this chapter may be used to estimate convection transfer rates in many practical problems. You should know when and how to use the various expressions, and you should be familiar with the general methodology of a convection calculation. To facilitate their use, the correlations are summarized in Table 7.8.

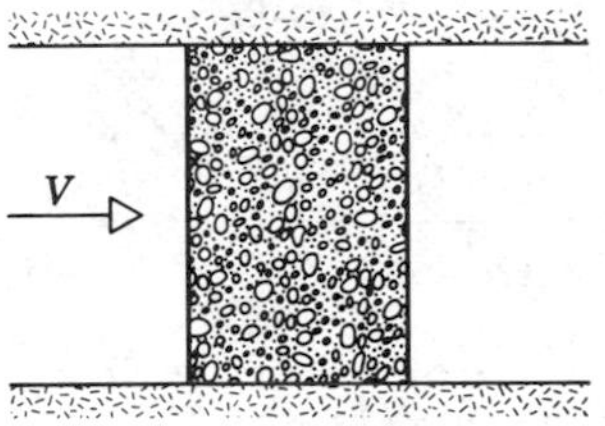

Figure 7.14 Gas flow through a packed bed of solid particles.

Table 7.8 Summary of convection heat transfer correlations for external flow[a]

CORRELATION		GEOMETRY	CONDITIONS
$\delta = 5x/Re_x^{-1/2}$	(7.17)	Flat plate	Laminar, T_f
$C_{f,x} = 0.664 Re_x^{-1/2}$	(7.18)	Flat plate	Laminar, local, T_f
$Nu_x = 0.332 Re_x^{1/2} Pr^{1/3}$	(7.21)	Flat plate	Laminar, local, T_f, $Pr \gtrsim 0.6$
$\delta_t = \delta Pr^{-1/3}$	(7.22)	Flat plate	Laminar, T_f
$\bar{C}_{f,x} = 1.328 Re_x^{-1/2}$	(7.24)	Flat plate	Laminar, average, T_f
$\overline{Nu}_x = 0.664 Re_x^{1/2} Pr^{1/3}$	(7.25)	Flat plate	Laminar, average, T_f, $Pr \gtrsim 0.6$
$Nu_x = 0.565 Pe_x^{1/2}$	(7.26)	Flat plate	Laminar, local, T_f, $Pr \lesssim 0.05$
$C_{f,x} = 0.0592 Re_x^{-1/5}$	(7.27)	Flat plate	Turbulent, local, T_f, $Re_x \lesssim 10^8$
$\delta = 0.37 x Re_x^{-1/5}$	(7.28)	Flat plate	Turbulent, local, T_f, $Re_x \lesssim 10^8$
$Nu_x = 0.0296 Re_x^{4/5} Pr^{1/3}$	(7.29)	Flat plate	Turbulent, local, T_f, $Re_x \lesssim 10^8$, $0.6 \lesssim Pr \lesssim 60$
$\bar{C}_{f,L} = 0.074 Re_L^{-1/5} - 1742 Re_L^{-1}$	(7.33)	Flat plate	Mixed, average, T_f, $Re_{x,c} = 5 \times 10^5$, $Re_L \lesssim 10^8$
$\overline{Nu}_L = (0.037 Re_L^{4/5} - 871) Pr^{1/3}$	(7.32)	Flat plate	Mixed, average, T_f, $Re_{x,c} = 5 \times 10^5$, $Re_L \lesssim 10^8$, $0.6 < Pr < 60$
$\overline{Nu}_D = C Re_D^m Pr^{1/3}$ (Tables 7.2 and 7.3)	(7.42)	Cylinder	Average, T_f, $0.4 < Re_D < 4 \times 10^5$, $Pr \gtrsim 0.7$
$\overline{Nu}_D = C Re_D^m Pr^n (Pr_\infty/Pr_s)^{1/4}$ (Table 7.4)	(7.43)	Cylinder	Average, T_∞, $1 < Re_D < 10^6$, $0.7 < Pr < 500$
$\overline{Nu}_D = 0.3 + \{0.62 Re_D^{1/2} Pr^{1/3}/[1 + (0.4/Pr)^{2/3}]^{1/4}\}[1 + (Re_D/28{,}200)^{5/8}]^{4/5}$	(7.44)	Cylinder	Average, T_f, $Re_D Pr > 0.2$
$\overline{Nu}_D = 2 + (0.4 Re_D^{1/2} + 0.06 Re_D^{2/3}) Pr^{0.4} \cdot (\mu_\infty/\mu_s)^{1/4}$	(7.45)	Sphere	Average, T_∞, $3.5 < Re_D < 7.6 \times 10^4$, $0.71 < Pr < 380$, $1.0 < (\mu_\infty/\mu_s) < 3.2$
$\overline{Nu}_D = 2 + 0.6 Re_D^{1/2} Pr^{1/3} \cdot [25(x/D)^{-0.7}]$	(7.47)	Falling drop	Average, T_∞
$\overline{Nu}_D = 1.13 C_1 Re_{D,\max}^m Pr^{1/3}$ (Tables 7.5, 7.6)	(7.50)	Tube bank	Average, T_f, $2 \times 10^3 < Re_{D,\max} < 4 \times 10^4$, $Pr \geq 0.7$
$\overline{Nu}_D = C Re_{D,\max}^m Pr^{0.36} (Pr_\infty/Pr_s)^{1/4}$ (Table 7.7)	(7.54)	Tube bank	Average, T_∞, $1000 < Re_D < 2 \times 10^6$, $0.7 < Pr < 500$
$\varepsilon \bar{j}_H = 2.06 Re_D^{-0.575}$	(7.59)	Packed bed of spheres	Average, T_∞, $90 \leq Re_D \leq 4000$, $Pr \approx 0.7$

[a]Correlations in this table pertain to isothermal surfaces; for special cases see Section 7.2.4.

REFERENCES

1. Blasius, H., *Z. Math. Phys.*, *56*, 1, 1908. English translation in National Advisory Committee for Aeronautics Technical Memo No. 1256.
2. Schlichting, H., *Boundary Layer Theory*, 4th ed., McGraw-Hill, New York, 1960.
3. Howarth, L., *Proc. R. Soc. Lond.*, *A*, *164*, 1938, p. 547.
4. Pohlhausen, E., *Z. Angew. Math. Mech.*, *1*, 1921, p. 115.
5. Kays, W. M., and M. E. Crawford, *Convective Heat and Mass Transfer*, McGraw-Hill, New York, 1980.
6. Schlichting, H., *Boundary Layer Theory*, 6th ed., McGraw-Hill, New York, 1968.
7. Whitaker, S., *AIChE J.*, *18*, 361, 1972.
8. Zhukauskas, A., and A. B. Ambrazyavichyus, *Int. J. Heat Mass Transfer*, *3*, 305, 1961.
9. Churchill, S. W., and H. Ozoe, *J. Heat Transfer*, *95*, 78, 1973.
10. Churchill, S. W., *AIChE J.*, *22*, 264, 1976.
11. Blair, M. F., *J. Heat Transfer*, *105*, 33 and 41, 1983.
12. Fox, R. W., and A. T. McDonald, *Introduction to Fluid Mechanics*, 3rd ed., Wiley, New York, 1985.
13. Hilpert, R., *Forsch. Geb. Ingenieurwes.*, *4*, 215, 1933.
14. Knudsen, J. D., and D. L. Katz, *Fluid Dynamics and Heat Transfer*, McGraw-Hill, New York, 1958.
15. Jakob, M., *Heat Transfer*, Vol. 1, Wiley, New York, 1949.
16. Zhukauskas, A., "Heat Transfer from Tubes in Cross Flow", in J. P. Hartnett and T. F. Irvine, Jr., Eds., *Advances in Heat Transfer*, Vol. 8, Academic Press, New York, 1972.
17. Churchill, S. W., and M. Bernstein, *J. Heat Transfer*, *99*, 300, 1977.
18. Morgan, V. T., "The Overall Convective Heat Transfer from Smooth Circular Cylinders," in T. F. Irvine, Jr. and J. P. Hartnett, Eds., *Advances in Heat Transfer*, Vol. 11, Academic Press, New York, 1975.
19. Ranz, W., and W. Marshall, *Chem. Eng. Prog.*, *48*, 141, 1952.
20. Yao, S.-C., and V. E. Schrock, ASME Publication 75-WA/HT-37, Winter Annual Meeting, Houston, November 30–December 4, 1975.
21. Grimison, E. D., *Trans. ASME*, *59*, 583, 1937.
22. Kays, W. M., and R. K. Lo, Stanford University Technical Report No. 15, 1952.
23. Bergelin, O. P., G. A. Brown, and S. C. Doberstein, *Trans. ASME*, *74*, 953, 1952.
24. Bird, R. B., W. E. Stewart, and E. N. Lightfoot, *Transport Phenomena*, Wiley, New York, 1960.
25. Jakob, M., *Heat Transfer*, Vol. 2, Wiley, New York, 1957.

PROBLEMS

7.1 Consider the following fluids at a film temperature of 300 K in parallel flow over a flat plate with velocity of 1 m/s: atmospheric air, water, engine oil, and mercury. For each fluid, determine the velocity and thermal boundary layer thicknesses at a distance of 40 mm from the leading edge.

7.2 Consider atmospheric air at 25°C in parallel flow at 5 m/s over both surfaces of a 1-m-long flat plate maintained at 75°C. Determine the boundary layer thickness, the surface shear stress, and the heat flux at the trailing edge. Determine the drag force on the plate and the total heat transfer from the plate, each per unit width of the plate.

7.3 Engine oil at 100°C and a velocity of 0.1 m/s flows over both surfaces of a 1-m-long flat plate maintained at 20°C. Determine:

(a) The velocity and thermal boundary layer thicknesses at the trailing edge.

(b) The local heat flux and surface shear stress at the trailing edge.

(c) The total drag force and heat transfer per unit width of the plate.

7.4 A common flow geometry, which includes the flat plate in parallel flow as a special case, is the *wedge*.

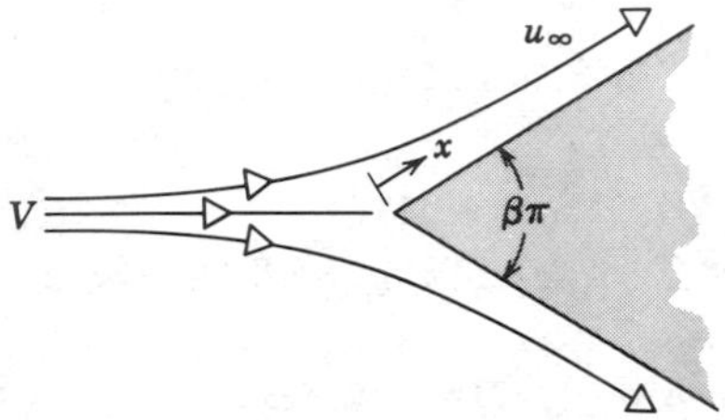

From a potential flow solution, it is known that the velocity at the edge of the boundary layer increases with x according to the relation $u_\infty = Vx^m$, where $m = \beta/(2-\beta)$ and $\beta\pi$ is the wedge angle. From a laminar boundary layer solution, it is also known that the local Nusselt number may be expressed as $Nu_x = C_1 Re_x^{1/2}$ where $Re_x \equiv (u_\infty x/\nu)$ and C_1 is a known function of Pr and β, as tabulated below.

		Pr				
		0.7	0.8	1.0	5.0	10.0
β	m	C_1				
0	0	0.292	0.307	0.332	0.585	0.730
0.2	0.111	0.331	0.348	0.378	0.669	0.851
0.5	0.333	0.384	0.403	0.440	0.792	1.013
1.0	1.000	0.496	0.523	0.570	1.043	1.344

(a) Comment on the nature of flow conditions at $x=0$ for $\beta>0$. How does u_∞ vary with x for $\beta=1$?

(b) Obtain the ratio of the average convection coefficient $\bar{h}_x$ to the local coefficient h_x for $\beta=0.5$ and for $\beta=1.0$.

(c) It is common practice to approximate wedge flow heat transfer by using convection correlations associated with the flat plate in parallel flow. Comment on the accuracy of such an approximation by computing the ratio $\bar{h}_{x,\beta>0}/\bar{h}_{x,\beta=0}$ at $x=1$ m for airflow over wedges of $\beta=0.5$ and $\beta=1.0$.

7.5 Consider a liquid metal ($Pr \ll 1$), with freestream conditions u_∞ and T_∞, in parallel flow over an isothermal flat plate at T_s. Assuming that $u=u_\infty$ throughout the thermal boundary layer, write the corresponding form of the boundary layer energy equation. Applying appropriate initial ($x=0$) and boundary conditions, solve this equation for the boundary layer temperature field $T(x,y)$. Use the result to obtain an expression for the local Nusselt number Nu_x. (*Hint:* This problem is analogous to one-dimensional heat transfer in a semiinfinite medium with a sudden change in surface temperature.)

7.6 Consider the velocity boundary layer profile for flow over a flat plate to be of the form $u=C_1+C_2y$. Applying appropriate boundary conditions, obtain an expression for the velocity profile in terms of the boundary layer thickness δ and the freestream velocity u_∞. Using the integral form of the boundary layer momentum equation (Appendix C), obtain expressions for the boundary layer thickness and the local friction coefficient, expressing your result in terms of the local Reynolds number. Compare your results with those obtained from the exact solution (Section 7.2.1) and the integral solution with a cubic profile (Appendix C).

7.7 Do Problem 7.6 for a velocity boundary layer profile of the form $u=C_1+C_2 \sin(C_3y)$.

7.8 Consider a steady, turbulent boundary layer on an isothermal flat plate of temperature T_s. The boundary layer is "tripped" at the leading edge, $x=0$, by a fine wire. Assume constant physical properties and velocity and temperature profiles of the form

$$\frac{u}{u_\infty}=\left(\frac{y}{\delta}\right)^{1/7} \quad \text{and} \quad \frac{T-T_\infty}{T_s-T_\infty}=1-\left(\frac{y}{\delta_t}\right)^{1/7}$$

(a) From experiment it is known that the surface shear stress is related to the boundary layer thickness by an expression of the form

$$\tau_s=0.0228\rho u_\infty^2\left(\frac{u_\infty\delta}{\nu}\right)^{-1/4}$$

Beginning with the momentum integral equation (Appendix C), show that

$$\frac{\delta}{x}=0.376Re_x^{-1/5}$$

Determine the average friction coefficient $\bar{C}_{f,x}$.

(b) Beginning with the energy integral equation, obtain an expression for the local Nusselt number Nu_x, and use this result to evaluate the average Nusselt number, $\overline{Nu}_x$.

7.9 An electric air heater consists of a horizontal array of thin metal strips that are each 10 mm long in the direction of an airstream that is in parallel flow over the top of the strips. Each strip is 0.2 m wide, and 25 strips are arranged side by side, forming a continuous and smooth surface over which the air flows at 2 m/s. During operation each strip is maintained at 500°C and the air is at 25°C. What is the rate of convection heat transfer from the first strip? The fifth strip? The tenth strip? All the strips?

7.10 Do Problem 7.3 for atmospheric air at $T_\infty=25$°C and $u_\infty=25$ m/s, with the plate maintained at $T_s=125$°C.

7.11 Consider atmospheric air at 25°C and a velocity of 25 m/s flowing over both surfaces of a 1-m-long flat plate that is maintained at 125°C. Determine the rate of heat transfer per unit width from the plate for values of the critical Reynolds number corresponding to 10^5, 5×10^5, and 10^6.

7.12 Consider air at 27°C and 1 atm in parallel flow over an isothermal, 1-m-long flat plate with a velocity of 10 m/s. Plot the variation of the local heat transfer coefficient with distance along the plate. What is the value of the average coefficient?

7.13 Consider water at 27°C in parallel flow over an isothermal, 1-m-long flat plate with a velocity of 2 m/s. Plot the variation of the local heat transfer coefficient with distance along the plate. What is the value of the average coefficient?

7.14 Air at a pressure and temperature of 1 atm and 15°C, respectively, is in parallel flow at a velocity of 10 m/s over a 3-m-long flat plate that is heated to a uniform temperature of 140°C.

(a) What is the average heat transfer coefficient for the plate?

(b) What is the local heat transfer coefficient at the midpoint of the plate?

(c) Sketch qualitatively the manner in which the heat flux varies along the length of the plate.

7.15 Explain under what conditions the total rate of heat transfer from an isothermal flat plate of dimensions $L \times 2L$ would be the same, independent of whether parallel flow over the plate is directed along the side of length L or $2L$. With a critical Reynolds number of 5×10^5, for what values of Re_L would the total heat transfer be independent of orientation?

7.16 The surface of a 1.5-m-long flat plate is maintained at 40°C, and water at a temperature of 4°C and a velocity of 0.6 m/s flows over the surface.

(a) Using the film temperature T_f for evaluation of the properties, calculate the heat transfer rate per unit width of the plate q' (W/m).

(b) Calculate the error in q' that would be incurred in part a if the thermophysical properties of the water were evaluated at the freestream temperature and the same empirical correlation were used.

(c) In part a, if a wire were placed near the leading edge of the plate to induce turbulence over its entire length, what would be the heat transfer rate?

7.17 Air at a pressure and temperature of 1 atm and 50°C, respectively, is in parallel flow over the top surface of a flat plate that is heated to a uniform temperature of 100°C. The plate has a length of 0.20 m (in the flow direction) and a width of 0.10 m. The Reynolds number based on the plate length is 40,000. What is the rate of heat transfer from the plate to the air? If the freestream velocity of the air is doubled and the pressure is increased to 10 atm, what is the rate of heat transfer?

7.18 A thin, flat plate of length $L = 1$ m separates two airstreams that are in parallel flow over opposite surfaces of the plate. One airstream has a temperature of $T_{\infty,1} = 200$°C and a velocity of $u_{\infty,1} = 60$ m/s, while the other airstream has a temperature of $T_{\infty,2} = 25$°C and a velocity of $u_{\infty,2} = 10$ m/s. What is the heat flux between the two streams at the midpoint of the plate?

7.19 Consider a rectangular fin that is used to cool a motorcycle engine. The fin is 0.15 m long and at a temperature of 250°C, while the motorcycle is moving at 80 km/h in air at 27°C. The air is in parallel flow over both surfaces of the fin, and turbulent flow conditions may be assumed to exist throughout. What is the rate of heat removal per unit width of fin?

7.20 Consider the wing of an aircraft as a flat plate of 2.5 m length in the flow direction.

The plane is moving at a speed of 100 m/s in air that is at a pressure of 0.7 bar and a temperature of $-10°C$. If the top surface of the wing absorbs solar radiation at a rate of 800 W/m^2, estimate its steady-state temperature. Assume the wing to be of solid construction and to have a single, uniform temperature.

7.21 The top surface of a heated compartment consists of very smooth (A) and highly roughened (B) portions, and the surface is placed in an atmospheric airstream.

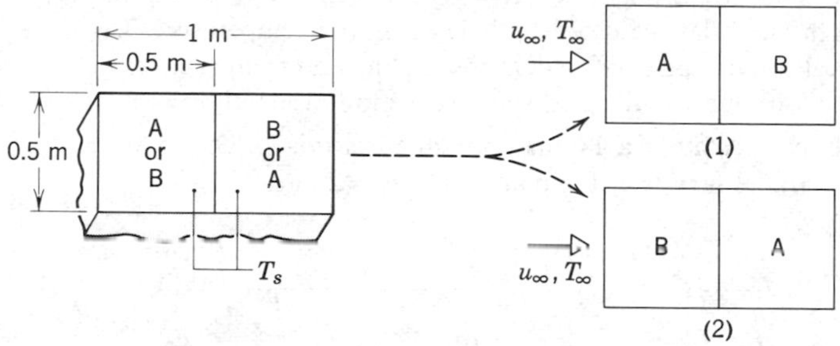

In the interest of minimizing total convection heat transfer from the surface, which orientation, (1) or (2), is preferred? If $T_s = 100°C$, $T_\infty = 20°C$, and $u_\infty = 20$ m/s, what is the convection heat transfer from the entire surface for this orientation?

7.22 Two rooms are arranged on the side of a building. The wall of room A is 13 m long and 3.5 m high, while that of room B is 7 m long and 3.5 m high. Both walls are 0.25 m thick and have an effective thermal conductivity of 1 W/m·K.

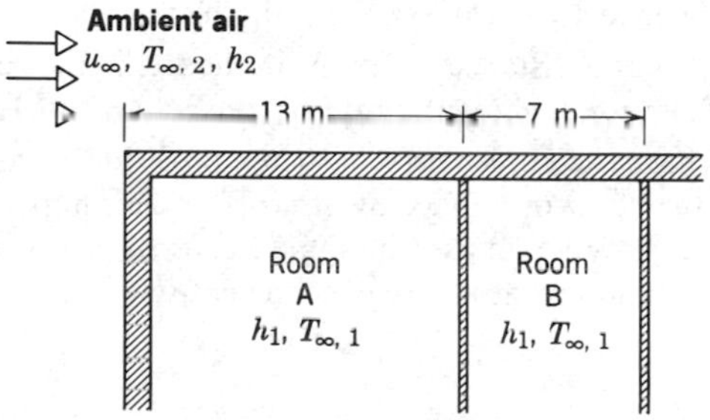

Consider conditions for which the ambient air is in parallel flow over the outer surface of the wall with $u_\infty = 7$ m/s and $T_{\infty,2} = -20°C$. The inside air temperature is maintained at $T_{\infty,1} = 20°C$, and the inside convection coefficient is $h_1 = 5$ W/m^2·K. Estimate the rate at which heat is lost through the 13- and 7-m-long walls of rooms A and B, respectively. Turbulent boundary layer conditions may be assumed for the ambient airflow over the entire wall.

7.23 As a means of supplying fresh water to arid regions of the world, it has been advocated that icebergs be towed from Antarctica or Arctica. Icebergs that are considered to be best suited for towing are relatively broad and flat. Consider an iceberg that is 1 km long by 0.5 km wide and of depth $D = 0.25$ km. It is proposed that this iceberg be towed at 1 km/h in the direction of its length for 6000 km through water whose average temperature (over the trip) is 10°C. As a first approximation, the interaction of the iceberg with its surroundings may be assumed to be dominated by conditions at the bottom (1 km × 0.5 km) surface. The latent heat of fusion of ice is 3.34×10^5 J/kg.

(a) What is the average recession (melting) rate, dD/dt, at the bottom surface?

(b) What is the power required to move the iceberg at the designated speed?

(c) If towing costs amount to \$1/kW·h of power requirement, what is the minimum cost of fresh water at the destination?

7.24 Consider atmospheric air at $u_\infty = 2$ m/s and $T_\infty = 300$ K in parallel flow over an isothermal flat plate of length $L = 1$ m and temperature $T_s = 350$ K. Compute the local convection coefficient at the leading and trailing edges of the heated plate with and without an unheated starting section of length $\xi = 1$ m.

7.25 The cover plate of a flat plate solar collector is at 15°C, while ambient air at 10°C is in parallel flow over the plate, with $u_\infty = 2$ m/s.

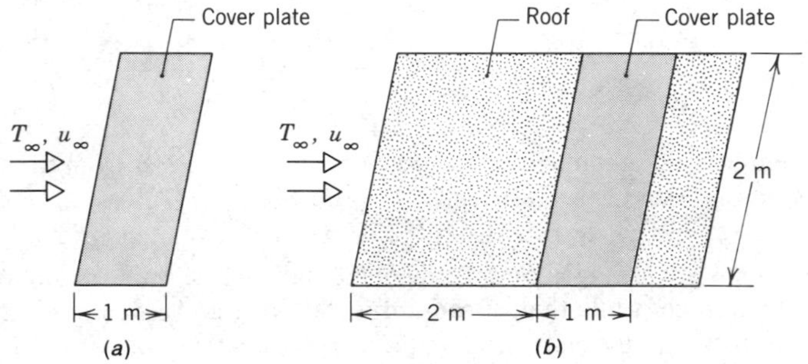

(a) What is the rate of convective heat loss from the plate?

(b) If the plate is installed at a distance of 2 m from the leading edge of a roof and flush with the roof surface, what is the rate of convective heat loss?

7.26 An array of 10 silicon chips, each of length $L = 10$ mm on a side, is insulated on one surface and cooled on the opposite surface by atmospheric air in parallel flow with $T_\infty = 24$°C and $u_\infty = 40$ m/s. When in use, the same electrical power is dissipated in each chip, maintaining a uniform heat flux over the entire cooled surface.

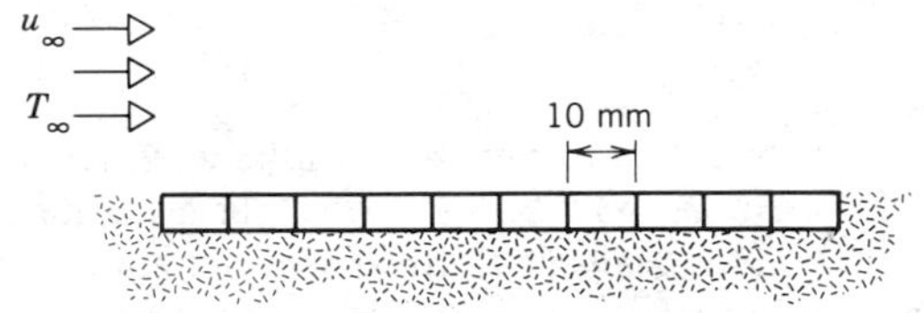

If the temperature of each chip may not exceed 80°C, what is the maximum allowable power per chip? What is the maximum allowable power if a turbulence promoter is used to trip the boundary layer at the leading edge? Would it be preferable to orient the array normal, instead of parallel, to the airflow?

7.27 A square (10 mm × 10 mm) silicon chip is insulated on one side and cooled on the opposite side by atmospheric air in parallel flow at $u_\infty = 20$ m/s and $T_\infty = 24$°C. When in use, electrical power dissipation within the chip maintains a uniform heat flux at the cooled surface. If the chip temperature may not exceed 80°C at any point on its surface, what is the maximum allowable power? What is the maximum allowable power if the chip is flush mounted in a substrate that provides for an unheated starting length of 20 mm?

7.28 Calculate the rate of heat transfer per unit length for a 10-mm-diameter cylinder maintained at 50°C with the following fluids in cross flow over the cylinder at $T_\infty = 20°C$ and $V = 5$ m/s: atmospheric air, saturated water, and engine oil.

7.29 A circular pipe of 25 mm outside diameter is placed in an airstream at 25°C and 1 atm pressure. The air moves in cross flow over the pipe at 15 m/s, while the outer surface of the pipe is maintained at 100°C. What is the drag force exerted on the pipe per unit length? What is the rate of heat transfer from the pipe per unit length?

7.30 Do Problem 7.29 for a pipe diameter and temperature of 30 mm and 90°C and water flow at 25°C and 3 m/s.

7.31 A circular cylinder of 25 mm diameter is initially at 150°C and is quenched by immersion in a 80°C oil bath, which moves at a velocity of 2 m/s in cross flow over the cylinder. What is the initial rate of heat loss per unit length of the cylinder?

7.32 Consider atmospheric air at 25°C and flowing at a velocity of 15 m/s. What is the rate of heat transfer per unit length from the following surfaces, each at 75°C, when the air is in cross flow over the surface?

(a) A circular cylinder 10 mm in diameter.

(b) A square cylinder 10 mm long on a side.

(c) A vertical plate 10 mm high.

7.33 A fine wire of diameter D is positioned across a passage to determine flow velocity from heat transfer characteristics. Current is passed through the wire to heat it, and the heat is disspated to the flowing fluid by convection. The resistance of the wire is determined from electrical measurements, and the temperature is known from the resistance.

(a) For a fluid of arbitrary Prandtl number, develop an expression for its velocity in terms of the difference between the temperature of the wire and the freestream temperature of the fluid.

(b) What is the velocity of an airstream at 1 atm and 25°C, if a wire of 0.5 mm diameter achieves a temperature of 40°C while dissipating 35 W/m?

7.34 A horizontal copper rod 10 mm in diameter and 100 mm long is inserted in the airspace between surfaces of an electronic device to enhance heat dissipation. The ends of the rod are at 90°C, while air at 25°C is in cross flow over the cylinder with a velocity of 25 m/s. What is the temperature at the midplane of the rod? What is the rate of heat transfer from the rod?

7.35 An uninsulated steam pipe is used to transport high temperature steam from one building to another. The pipe is of 0.5 m diameter, has a surface temperature of 150°C, and is exposed to ambient air at −10°C. The air moves in cross flow over the pipe with a velocity of 5 m/s. What is the heat loss per unit length of pipe?

7.36 Assume that a person can be approximated as a cylinder of 0.3 m diameter and 1.8 m height with a surface temperature of 24°C. Calculate the body heat loss while this person is subjected to a 15 m/s wind whose temperature is −5°C.

7.37 Water at 20°C flows over a 20-mm-diameter sphere with a velocity of 5 m/s. The surface of the sphere is at 60°C. What is the drag force on the sphere? What is the rate of heat transfer from the sphere?

7.38 Air at 25°C flows over a 10-mm-diameter sphere with a velocity of 25 m/s, while the surface of the sphere is maintained at 75°C. What is the drag force on the sphere? What is the rate of heat transfer from the sphere?

7.39 Atmospheric air at 25°C and a velocity of 0.5 m/s flows over a 50-W incandescent bulb whose surface temperature is at 140°C. The bulb may be approximated as a sphere 50 mm in diameter. What is the rate of heat loss by convection to the air?

7.40 In manufacturing lead shot, molten lead droplets fall from a tower through cool air into a tank of water. The intent is to have the lead solidify before hitting the water. Consider each pellet to be a sphere 3 mm in diameter, moving at the terminal velocity (for which the drag force equals the gravitational force) throughout the descent. If pellets at the melting point are to be converted from the molten to the solid state in air at 15°C, what is the necessary height of the tower above the tank? The latent heat of fusion of lead is 2.45×10^4 J/kg.

7.41 Copper spheres 20 mm in diameter are quenched by being dropped into a tank of water that is maintained at 280 K. The spheres may be assumed to reach the terminal velocity on impact and to drop freely through the water. Estimate the terminal velocity by equating the drag and gravitational forces acting on the sphere. What is the approximate height of the water tank needed to cool the spheres from an initial temperature of 360 K to a center temperature of 320 K?

7.42 For the conditions of Problem 7.41, determine the terminal velocity and tank height if engine oil at 300 K, rather than water, is used as the coolant.

7.43 Repeat Example 7.5 for a more compact tube bank in which the longitudinal and transverse pitches are $S_L = S_T = 20.5$ mm. All other conditions remain the same.

7.44 A preheater involves the use of condensing steam at 100°C on the inside of a bank of tubes to heat air that enters at 1 atm and 25°C. The air moves at 5 m/s in cross flow over the tubes. Each tube is 1 m long and has an outside diameter of 10 mm. The bank consists of 196 tubes in a square, aligned array for which $S_T = S_L = 15$ mm. What is the total rate of heat transfer to the air? What is the pressure drop associated with the airflow?

7.45 A tube bank uses an aligned arrangement of 10-mm-diameter tubes with $S_T = S_L = 20$ mm. There are 10 rows of tubes with 50 tubes in each row. Consider an application for which cold water flows through the tubes, maintaining the outer surface temperature at 27°C, while flue gases at 427°C and a velocity of 5 m/s are in cross flow over the tubes. The properties of the flue gas may be approximated as those of atmospheric air at 427°C. What is the total rate of heat transfer per unit length of the tubes in the bank?

7.46 A tube bank uses an aligned arrangement of 30-mm-diameter tubes with $S_T = S_L = 60$ mm and a tube length of 1 m. There are 10 tube rows in the flow direction ($N_L = 10$) and 7 tubes per row ($N_T = 7$). Air with upstream conditions of $T_\infty = 27$°C and $V = 15$ m/s is in cross flow over the tubes, while a tube wall temperature of 100°C is maintained by steam condensation within the tubes. Determine the temperature of air leaving the tube bank, the pressure drop across the bank, and the fan power requirement.

7.47 The use of rock pile thermal energy storage systems has been considered for solar energy and industrial process heat applications. A particular system involves a cylindrical container, 2 m long by 1 m in diameter, in which nearly spherical rocks of 0.03 m diameter are packed. The bed has a void space of 0.42, and the density and specific heat of the rock are $\rho = 2300$ kg/m^3 and $c_p = 879$ J/kg·K, respectively. Consider conditions for which atmospheric air is supplied to the rock pile at a steady flow rate of 1 kg/s and a temperature of 90°C. The air flows in the axial direction through the container. If the rock is at a temperature of 25°C, what is the total rate of heat transfer from the air to the rock pile?

Eight

INTERNAL FLOW

Having acquired the means to compute convection transfer rates for external flow, we now consider the convection transfer problem for *internal flow*. Recall that an external flow is one for which boundary layer development on a surface is allowed to continue without external constraints, as for the flat plate of Figure 6.4. In contrast an internal flow, such as flow in a pipe, is one for which the fluid is *confined* by a surface. Hence the boundary layer is unable to to develop without eventually being constrained. The internal flow configuration represents a convenient geometry for the heating and cooling of fluids used in the chemical processing, environmental control, and energy conversion areas.

We begin by considering velocity (hydrodynamic) effects pertinent to internal flows, focusing on certain unique features of boundary layer development. Thermal boundary layer effects will then be considered, and an overall energy balance will be applied to determine fluid temperature variations in the flow direction. Empirical correlations for estimating the convection heat transfer coefficient will then be presented for a variety of internal flow conditions.

8.1 HYDRODYNAMIC CONSIDERATIONS

When considering external flow, it is only necessary to ask whether this flow is laminar or turbulent. However, for an internal flow we must also be concerned with the existence of *entrance* and *fully developed* regions.

8.1.1 Flow Conditions

Consider laminar flow in a circular tube of radius r_o (Figure 8.1), where fluid enters the tube with a uniform velocity. We know that, when the fluid makes contact with the surface, viscous effects become important and a boundary layer develops with increasing x. This development occurs at the expense of a shrinking inviscid flow region and concludes with boundary layer merger at the centerline. The distance from the entrance at which this merger occurs is termed the *hydrodynamic entry length*, $x_{\text{fd},h}$. For $x > x_{\text{fd},h}$ the boundary layer, and

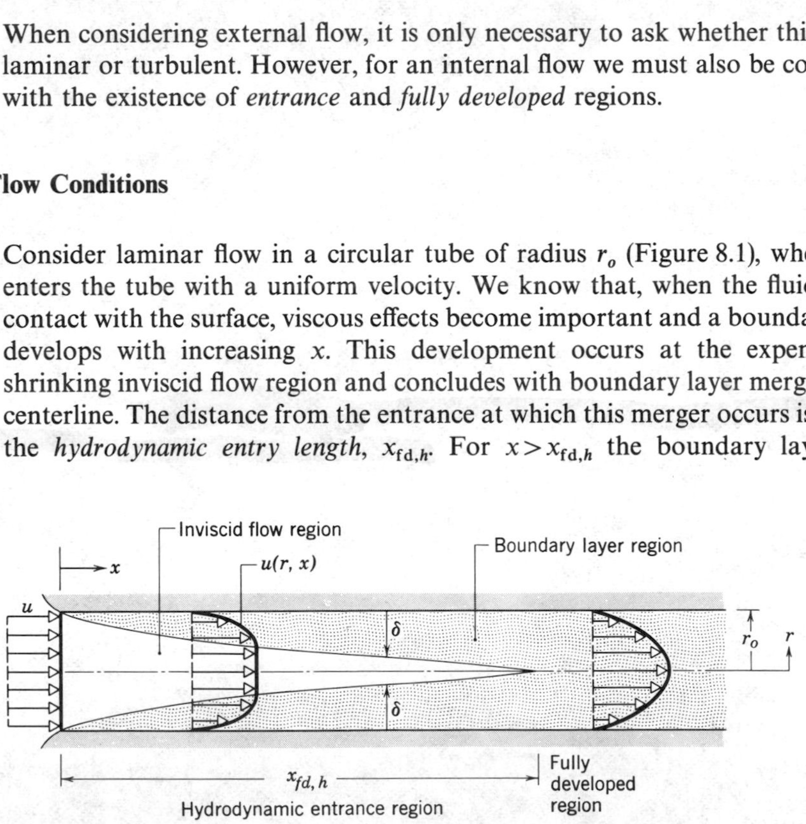

Figure 8.1 Laminar, hydrodynamic boundary layer development in a circular tube.

therefore the effect of viscosity, extends over the entire cross section, and the flow is said to be *fully developed.*

When dealing with internal flows, it is important to be cognizant of the extent of the entry region, which depends on whether the flow is laminar or turbulent. The Reynolds number for flow in a circular tube is defined as

$$Re_D \equiv \frac{\rho u_m D}{\mu} \tag{8.1}$$

where u_m is the mean fluid velocity over the tube cross section and D is the tube diameter. The critical Reynolds number for which transition from laminar to turbulent flow begins is

$$Re_{D,c} \approx 2300 \tag{8.2}$$

and the flow is fully turbulent for $Re_D \gtrsim 4000$. For laminar flow ($Re_D \lesssim 2300$), the hydrodynamic entry length may be obtained from an expression of the form [1]

$$\left(\frac{x_{\text{fd},h}}{D}\right)_{\text{lam}} \approx 0.05 Re_D \tag{8.3}$$

Although there is no satisfactory general expression for the entry length in turbulent flow, we know that it is independent of Reynolds number and that, as a first approximation [2],

$$10 \lesssim \left(\frac{x_{\text{fd},h}}{D}\right)_{\text{turb}} \lesssim 60 \tag{8.4}$$

For the purposes of this text, we shall assume fully developed turbulent flow for $(x/D) > 10$.

8.1.2 The Mean Velocity

Because the velocity varies over the cross section and there is no well-defined free stream, it is necessary to work with a mean velocity u_m when dealing with internal flows. This velocity is defined such that, when multiplied by the fluid density ρ and the cross-sectional area of the tube A_c, it provides the rate of mass flow through the tube. Hence

$$\dot{m} = \rho u_m A_c \tag{8.5}$$

For steady, incompressible flow in a tube of uniform cross-sectional area, $\dot{m}$ and u_m are constants independent of x. From Equations 8.1 and 8.5 it is evident that for flow in a *circular tube* ($A_c = \pi D^2/4$), the Reynolds number reduces to

$$Re_D = \frac{4\dot{m}}{\pi D \mu} \tag{8.6}$$

Since the mass flow rate may also be expressed as the integral of the mass flux (ρu) over the cross section

$$\dot{m} = \int_{A_c} \rho u(r,x) \, dA_c \tag{8.7}$$

it follows that, for *incompressible* flow in a *circular* tube,

$$u_m = \frac{\int_{A_c} \rho u(r,x) \, dA_c}{\rho A_c} = \frac{2\pi\rho}{\rho \pi r_o^2} \int_0^{r_o} u(r,x) r \, dr = \frac{2}{r_o^2} \int_0^{r_o} u(r,x) r \, dr \tag{8.8}$$

The foregoing expression may be used to determine u_m at any axial location x from knowledge of the velocity profile $u(r)$ at that location.

8.1.3 Velocity Profile in the Fully Developed Region

The form of the velocity profile may readily be determined for the *laminar flow* of an *incompressible, constant property fluid* in the *fully developed region* of a *circular tube.* An important feature of hydrodynamic conditions in the fully developed region is that both the radial velocity component v and the gradient of the axial velocity component ($\partial u/\partial x$) are everywhere zero.

$$v = 0 \quad \text{and} \quad \left(\frac{\partial u}{\partial x}\right) = 0 \tag{8.9}$$

F-d region (laminar flow)

Hence the axial velocity component depends only on r, $u(x,r) = u(r)$.

The radial dependence of the axial velocity may be obtained by solving the appropriate form of the x-momentum equation. This form is determined by first recognizing that, for the conditions of Equation 8.9, the net momentum flux is everywhere zero in the fully developed region. Hence the momentum conservation requirement reduces to a simple balance between shear and pressure forces in the flow. For the annular differential element of Figure 8.2, this force balance may be expressed as

$$-\tau_r(2\pi r \, dx) + \left\{\tau_r(2\pi r \, dx) + \frac{d}{dr}[\tau_r(2\pi r \, dx)] \, dr\right\}$$

$$+ p(2\pi r \, dr) - \left\{p(2\pi r \, dr) + \frac{d}{dx}[p(2\pi r \, dr)] \, dx\right\} = 0$$

which reduces to

$$+\frac{d}{dr}(r\tau_r) = r\frac{dp}{dx} \tag{8.10}$$

Substituting from Newton's law of viscosity, Equation 6.8, which may be expressed as

$$\tau_r = \mu \frac{du}{dr} \tag{8.11}$$

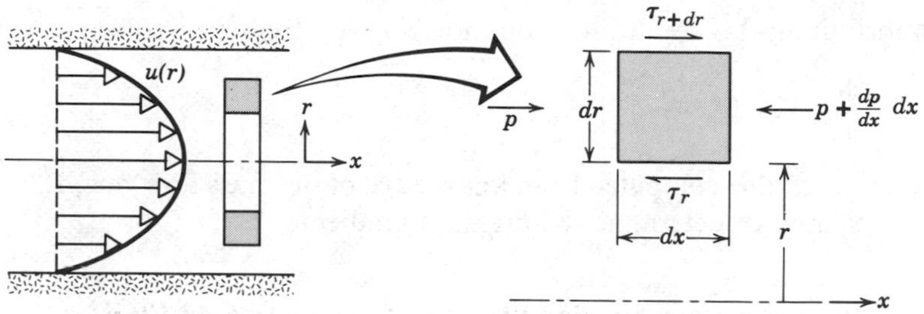

Figure 8.2 Force balance on a differential element for laminar, fully developed flow in a circular tube.

Equation 8.10 becomes

$$\frac{\mu}{r}\frac{d}{dr}\left(r\frac{du}{dr}\right)=\frac{dp}{dx} \tag{8.12}$$

Since the axial pressure gradient is independent of r, Equation 8.12 may be solved by integrating twice to obtain

$$r\frac{du}{dr}=\frac{1}{\mu}\left(\frac{dp}{dx}\right)\frac{r^2}{2}+C_1$$

and

$$u(r)=\frac{1}{\mu}\left(\frac{dp}{dx}\right)\frac{r^2}{4}+C_1\ln r+C_2$$

The integration constants may be determined by invoking the boundary conditions

$$u(r_o)=0 \qquad \text{and} \qquad \left.\frac{\partial u}{\partial r}\right|_{r=0}=0$$

which, respectively, impose the requirements of zero slip at the tube surface and radial symmetry about the centerline. It is a simple matter to evaluate the constants, and it follows that

$$u(r)=-\frac{1}{4\mu}\left(\frac{dp}{dx}\right)r_o^2\left[1-\left(\frac{r}{r_o}\right)^2\right] \tag{8.13}$$

Hence the fully developed velocity profile is *parabolic.* Note that the pressure gradient must always be negative.

The foregoing result may be used to determine the mean velocity of the flow. Substituting Equation 8.13 into Equation 8.8 and integrating, we obtain

$$u_m=-\frac{r_o^2}{8\mu}\frac{dp}{dx} \tag{8.14}$$

Substituting this result into Equation 8.13, the velocity profile is then

$$\frac{u(r)}{u_m} = 2\left[1 - \left(\frac{r}{r_o}\right)^2\right] \qquad (8.15)$$

Since u_m can be computed from knowledge of the mass flow rate, Equation 8.14 can be used to determine the pressure gradient.

8.1.4 Pressure Gradient and Friction Factor in Fully Developed Flow

Since the pressure drop needed to sustain an internal flow determines pump or fan power requirements, the engineer is frequently interested in this parameter. To determine the pressure drop, it is convenient to work with the *Moody* (or Darcy) *friction factor*, which is a dimensionless parameter defined as

$$f \equiv \frac{-(dp/dx)D}{\rho u_m^2/2} \qquad (8.16)^{1}$$

Substituting Equations 8.1 and 8.14 into 8.16, it follows that, for fully developed laminar flow,

$$f = \frac{64}{Re_D} \qquad (8.19)$$

F-d laminar flow

For fully developed turbulent flow, the analysis is much more complicated, and we must ultimately rely on the use of experimental results. Friction factors for a wide Reynolds number range are presented in the *Moody diagram* of Figure 8.3. In addition to depending on Reynolds number, the friction factor is also a function of the tube surface condition. It is a minimum for *smooth* surfaces and increases with increasing surface roughness, *e*. Correlations that reasonably approximate the smooth surface condition are of the form

$$f = 0.316 Re_D^{-1/4} \qquad Re_D \lesssim 2 \times 10^4 \qquad (8.20)$$

$$f = 0.184 Re_D^{-1/5} \qquad Re_D \gtrsim 2 \times 10^4 \qquad (8.21)$$

Turbulent flow
smooth pipes

[1] This quantity is not to be confused with the *friction coefficient*, sometimes called the Fanning friction factor, which is defined as

$$C_f \equiv \frac{\tau_s}{\rho u_m^2/2} \qquad (8.17)$$

Since the shear force of the fluid on the surface is equal and opposite to the force of the surface on the fluid, the surface shear stress may be expressed as $\tau_s = -\mu(du/dr)_{r=r_o}$. It follows from Equation 8.13 that

(8.18)

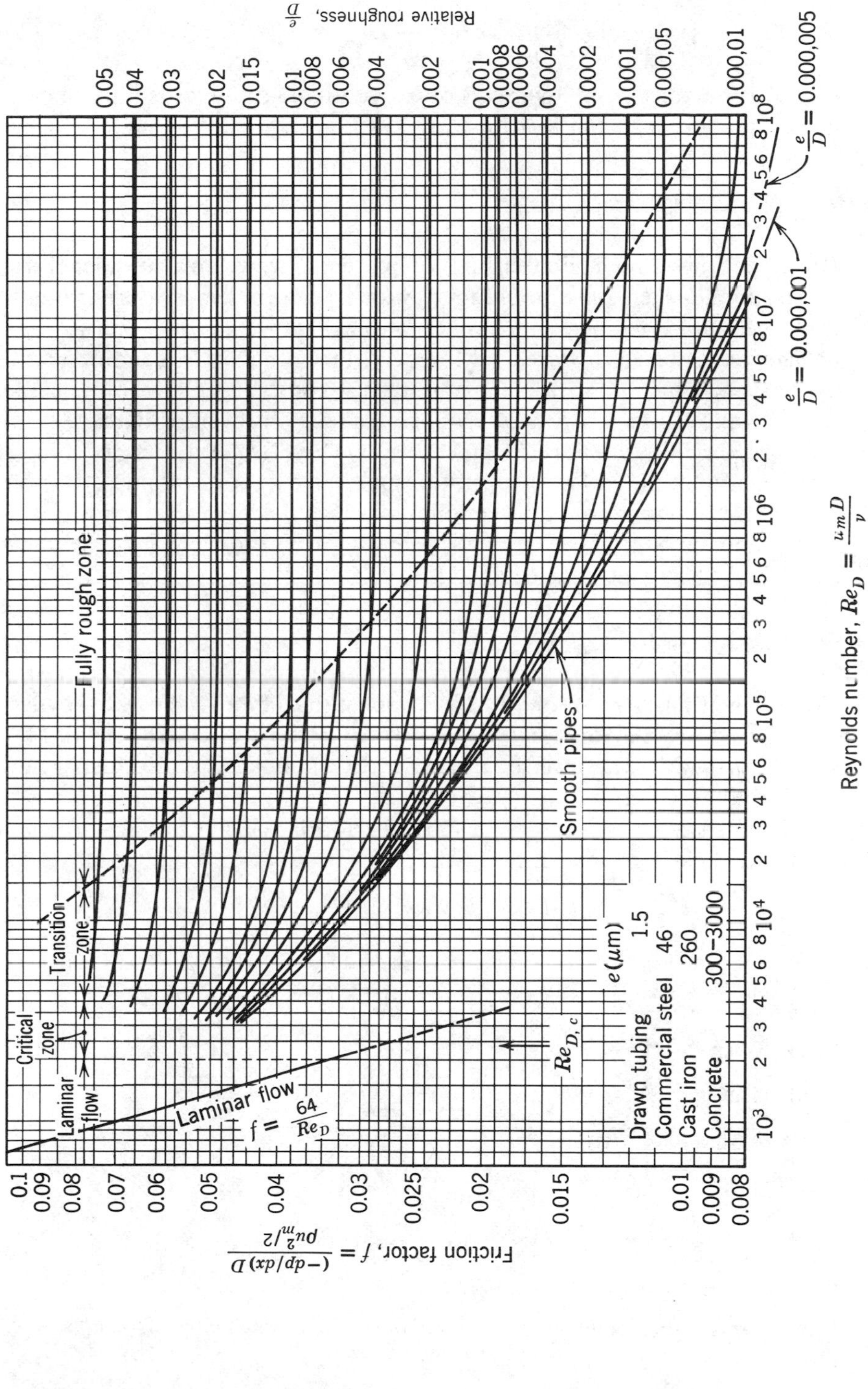

Figure 8.3 Friction factor for fully developed flow in a circular tube [3]. Used with permission.

Note that f, hence dp/dx, is a constant in the fully developed region. From Equation 8.16 the pressure drop, $\Delta p = p_1 - p_2$, associated with fully developed flow from the axial position x_1 to x_2 may then be expressed as

$$\Delta p = -\int_{p_1}^{p_2} dp = f\frac{\rho u_m^2}{2D}\int_{x_1}^{x_2} dx = f\frac{\rho u_m^2}{2D}(x_2 - x_1) \tag{8.22}$$

where f is obtained from Figure 8.3 or from Equation 8.19 for laminar flow and from Equations 8.20 and 8.21 for turbulent flow in smooth tubes.

8.2 THERMAL CONSIDERATIONS

Having reviewed the fluid mechanics of internal flow, we now consider thermal effects. If fluid enters the tube of Figure 8.4 at a uniform temperature that differs from the surface temperature, convection heat transfer occurs and a *thermal boundary layer* begins to develop. Moreover, if the tube *surface* condition is fixed by imposing either a uniform temperature (T_s = constant) or a uniform heat flux (q_s'' = constant), a *thermally fully developed condition* is eventually reached. The shape of the fully developed temperature profile $T(r, x)$ differs according to whether a uniform surface temperature or heat flux is maintained. For both surface conditions, however, the amount by which fluid temperatures exceed the entrance temperature increases with increasing x. For laminar flow the *thermal entry length* may be expressed as [2].

$$\left(\frac{x_{fd,t}}{D}\right)_{lam} \approx 0.05 Re_D Pr \tag{8.23}$$

Comparing Equations 8.3 and 8.23 it is evident that, if $Pr > 1$, the hydrodynamic boundary layer develops more rapidly than the thermal boundary layer ($x_{fd,h} < x_{fd,t}$), while the inverse is true for $Pr < 1$. For extremely high Prandtl number fluids, such as oils ($Pr \gtrsim 100$), $x_{fd,h}$ is very much smaller than $x_{fd,t}$ and it is reasonable to assume a fully developed velocity profile throughout the thermal entry region. In contrast, for turbulent flow, conditions are nearly

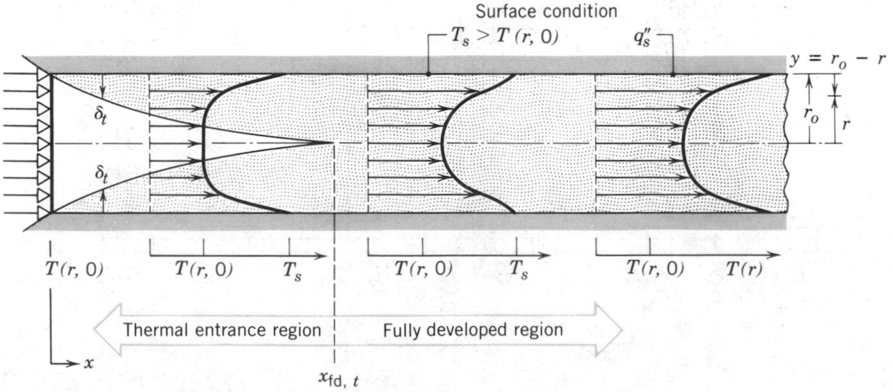

Figure 8.4 Thermal boundary layer development in a heated circular tube.

independent of Prandtl number, and to a first approximation, we shall assume $(x_{fd,t}/D) = 10$.

Thermal conditions in the fully developed region are characterized by several interesting and useful features. Before we can consider these features (Section 8.2.3), however, it is necessary to introduce the concept of a mean temperature and the appropriate form of Newton's law of cooling.

8.2.1 The Mean Temperature

Just as the absence of a freestream velocity requires use of a mean velocity to describe an internal flow, the absence of a fixed freestream temperature necessitates using a *mean temperature.* The *mean* (or *bulk*) *temperature* of the fluid at a given cross section is defined in terms of the *thermal* energy transported by the fluid as it moves past the cross section. The rate at which this transport occurs, $\dot{E}_t$, may be obtained by integrating the product of the mass flux (ρu) and the internal energy per unit mass $(c_v T)$ over the cross section. That is,

$$\dot{E}_t = \int_{A_c} \rho u c_v T \, dA_c \tag{8.24}$$

Hence if a mean temperature is defined such that

$$\dot{E}_t \equiv \dot{m} c_v T_m \tag{8.25}$$

we obtain

$$T_m = \frac{\int_{A_c} \rho u c_v T \, dA_c}{\dot{m} c_v} \tag{8.26}$$

For *incompressible flow* in a *circular tube* with constant c_v, it follows from Equations 8.5 and 8.26 that

$$T_m = \frac{2}{u_m r_o^2} \int_0^{r_o} u T r \, dr \tag{8.27}$$

It is important to note that, when multiplied by the mass flow rate and the specific heat, T_m provides the rate at which thermal energy is transported with the fluid as it moves along the tube.

8.2.2 Newton's Law of Cooling

The mean temperature T_m is a convenient reference temperature for internal flows, playing much the same role as the freestream temperature T_∞ for external flows. Accordingly, Newton's law of cooling may be expressed as

$$q_s'' = h(T_s - T_m) \tag{8.28}$$

where h is the *local* convection heat transfer coefficient. However, there is an essential difference between T_m and T_∞. Whereas T_∞ is a constant in the flow direction, T_m must vary in this direction. That is, dT_m/dx is never equal to zero if heat transfer is occurring. The value of T_m increases with x if heat transfer is from the surface to the fluid ($T_s > T_m$); it decreases with x if the opposite is true ($T_s < T_m$).

8.2.3 Fully Developed Conditions

Since the existence of convection heat transfer between the surface and the fluid dictates that the fluid temperature must continue to change with x, one might legitimately question whether fully developed thermal conditions could ever be reached. The situation is certainly different from the hydrodynamic case, for which $(\partial u/\partial x)=0$ in the fully developed region. In contrast, if there is heat transfer, (dT_m/dx), as well as $(\partial T/\partial x)$ at any radius r, is not zero. Accordingly, the temperature profile $T(r)$ is continuously changing with x, and it would seem that a fully developed condition could never be reached. This apparent contradiction may be reconciled by working with a dimensionless form of the temperature.

Recall that analyses have often been simplified by working with dimensionless temperature differences, as for transient conduction (Chapter 5) and the energy conservation equation (Chapter 6). Introducing a dimensionless temperature difference of the form $(T_s - T)/(T_s - T_m)$, conditions for which this ratio becomes independent of x are known to exist [2]. That is, although the temperature profile $T(r)$ continues to change with x, the *relative* shape of the profile no longer changes, and the flow is said to be *thermally fully developed.* The requirement for such a condition is formally stated as

$$\frac{\partial}{\partial x}\left[\frac{T_s(x)-T(r,x)}{T_s(x)-T_m(x)}\right]_{\text{fd},t}=0 \qquad (8.29)$$

Thermally F-d

where T_s is the tube surface temperature, T is the local fluid temperature, and T_m is the mean temperature of the fluid over the cross section of the tube.

The condition given by Equation 8.29 is eventually reached in a tube for which there is either a *uniform surface heat flux* (q_s'' is constant) *or* a *uniform surface temperature* (T_s is constant). These surface conditions arise in many engineering applications. For example, a constant surface heat flux would exist if the tube wall were heated electrically or if the outer surface were uniformly irradiated. In contrast a constant surface temperature would exist if a phase change (due to boiling or condensation) were occurring at the outer surface. It is impossible to *simultaneously* impose the conditions of constant surface heat flux and constant surface temperature. If q_s'' is constant, T_s must vary with x; conversely, if T_s is constant, q_s'' must vary with x.

Several important features of thermally developed flow may be inferred from Equation 8.29. Since the temperature ratio is independent of x, the derivative of this ratio with respect to r must also be independent of x. Evaluating this derivative at the tube surface (note that T_s and T_m are constants insofar as differentiation with respect to r is concerned), we then obtain

$$\frac{\partial}{\partial r}\left(\frac{T_s - T}{T_s - T_m}\right)\bigg|_{r=r_o} = \frac{-\partial T/\partial r|_{r=r_o}}{T_s - T_m} \neq f(x)$$

Substituting for $\partial T/\partial r$ from Fourier's law, which, from Figure 8.4, is of the form

$$q_s'' = -k\frac{\partial T}{\partial y}\bigg|_{y=0} = k\frac{\partial T}{\partial r}\bigg|_{r=r_o}$$

and for q_s'' from Newton's law of cooling, Equation 8.28, we obtain

$$\frac{h}{k} \neq f(x) \tag{8.30}$$

Hence *in the thermally fully developed flow* of a fluid *with constant properties*, the *local convection coefficient is a constant, independent of x.*

Equation 8.29 is not satisfied in the entrance region, where h varies with x, as shown in Figure 8.5. Because the thermal boundary layer thickness is zero at the tube entrance, the convection coefficient is extremely large at $x = 0$. However, h decays rapidly as the thermal boundary layer develops, until the constant value associated with fully developed conditions is reached.

Additional simplifications are associated with the special case of *uniform surface heat flux.* Since both h and q_s'' are constant in the fully developed region, it follows from Equation 8.28 that

$$\frac{dT_s}{dx}\bigg|_{\text{fd},t} = \frac{dT_m}{dx}\bigg|_{\text{fd},t} \qquad q_s'' = \text{constant} \tag{8.31}$$

If we expand Equation 8.29 and solve for $\partial T/\partial x$, it also follows that

$$\frac{\partial T}{\partial x}\bigg|_{\text{fd},t} = \frac{dT_s}{dx}\bigg|_{\text{fd},t} - \frac{(T_s - T)}{(T_s - T_m)}\frac{dT_s}{dx}\bigg|_{\text{fd},t} + \frac{(T_s - T)}{(T_s - T_m)}\frac{dT_m}{dx}\bigg|_{\text{fd},t} \tag{8.32}$$

Substituting from Equation 8.31, we then obtain

$$\frac{\partial T}{\partial x}\bigg|_{\text{fd},t} = \frac{dT_m}{dx}\bigg|_{\text{fd},t} \qquad q_s'' = \text{constant} \tag{8.33}$$

Hence the axial temperature gradient is independent of the radial location. For

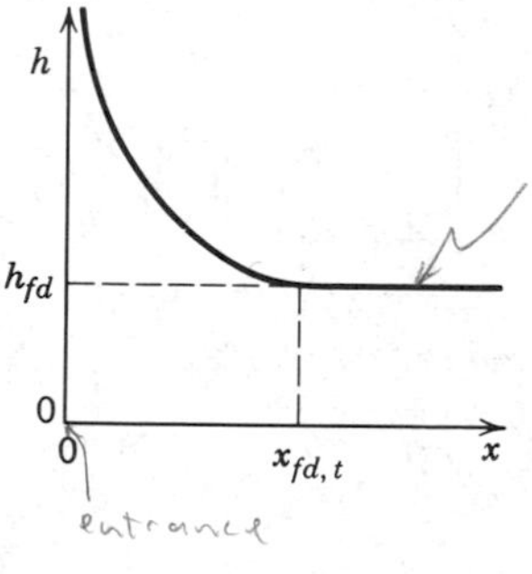

Figure 8.5 Axial variation of the convection heat transfer coefficient for flow in a tube.

the case of *constant surface temperature* ($dT_s/dx = 0$), it also follows from Equation 8.32 that

$$\left.\frac{\partial T}{\partial x}\right|_{\mathrm{fd},t} = \frac{(T_s - T)}{(T_s - T_m)} \left.\frac{dT_m}{dx}\right|_{\mathrm{fd},t} \qquad T_s = \text{constant} \tag{8.34}$$

in which case the value of $\partial T/\partial x$ depends on the radial coordinate.

From the foregoing results, it is evident that the mean temperature is a very important variable for internal flows. To describe such flows, it is essential that its variation with x be known. This variation may be obtained by applying an *overall energy balance* to the flow.

EXAMPLE 8.1

For flow of a liquid metal through a circular tube, the velocity and temperature profiles at a particular axial location may be approximated as being uniform and parabolic, respectively. That is, $u(r) = C_1$ and $T(r) - T_s = C_2[1 - (r/r_o)^2]$, where C_1 and C_2 are constants. What is the value of the Nusselt number Nu_D at this location?

SOLUTION

Known:

Form of the velocity and temperature profiles at a particular axial location for flow in a circular tube.

Find: Nusselt number at the prescribed location.

Schematic:

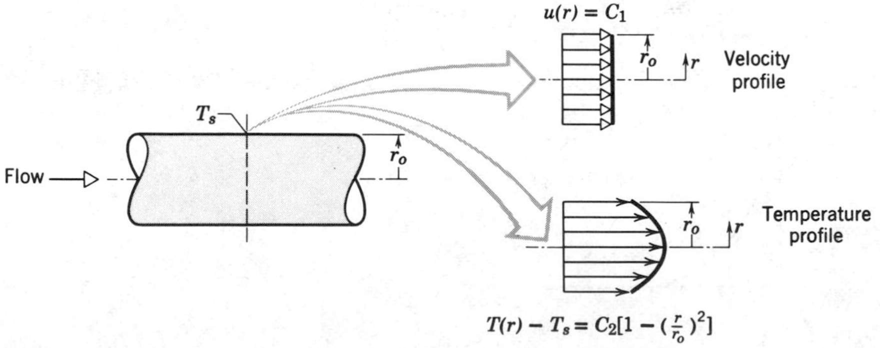

Assumptions: Incompressible, constant property flow.

Analysis:

The Nusselt number may be obtained by first determining the convection coefficient, which, from Equation 8.28, is given as

$$h=\frac{q_s''}{T_s-T_m}$$

From Equation 8.27, the mean temperature is then

$$T_m=\frac{2}{u_m r_o^2}\int_0^{r_o} uTr\,dr=\frac{2C_1}{u_m r_o^2}\int_0^{r_o}\left\{T_s+C_2\left[1-\left(\frac{r}{r_o}\right)^2\right]\right\}r\,dr$$

or, since $u_m=C_1$ from Equation 8.8,

$$T_m=\frac{2}{r_o^2}\int_0^{r_o}\left\{T_s+C_2\left[1-\left(\frac{r}{r_o}\right)^2\right]\right\}r\,dr$$

$$T_m=\frac{2}{r_o^2}\left[T_s\frac{r^2}{2}+C_2\frac{r^2}{2}-\frac{C_2}{4}\frac{r^4}{r_o^2}\right]\Bigg|_0^{r_o}$$

$$T_m=\frac{2}{r_o^2}\left(T_s\frac{r_o^2}{2}+\frac{C_2}{2}r_o^2-\frac{C_2}{4}r_o^2\right)=T_s+\frac{C_2}{2}$$

The heat flux may be obtained from Fourier's law, in which case,

$$q_s''=k\frac{\partial T}{\partial r}\bigg|_{r=r_o}=-kC_2 2\frac{r}{r_o^2}\bigg|_{r=r_o}=-2C_2\frac{k}{r_o}$$

Hence

$$h=\frac{q_s''}{T_s-T_m}=\frac{-2C_2(k/r_o)}{-C_2/2}=\frac{4k}{r_o}$$

and

$$Nu_D=\frac{hD}{k}=\frac{(4k/r_o)\times 2r_o}{k}=8 \qquad \triangleleft$$

8.3 THE ENERGY BALANCE

8.3.1 General Considerations

Because the flow in a tube is completely enclosed, an energy balance may be applied to determine how the mean temperature $T_m(x)$ varies with position along the tube and how the total convection heat transfer q_{conv} is related to the difference in temperatures at the tube inlet and outlet. Consider the tube flow of Figure 8.6. Fluid moves at a constant flow rate $\dot{m}$, and convection heat transfer

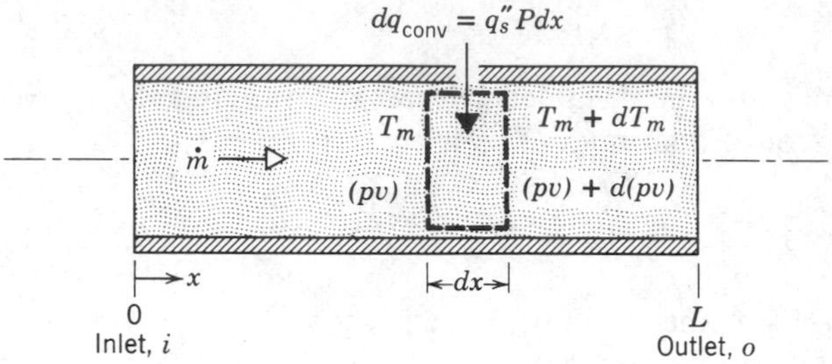

Figure 8.6 Control volume for internal flow in a tube.

occurs at the inner surface. Typically, fluid kinetic and potential energy changes, as well as energy transfer by conduction in the axial direction, are negligible. Hence if no shaft work is done by the fluid as it moves through the tube, the only significant effects will be those associated with *thermal energy changes* and with *flow work*. Flow work is performed to move fluid through a control surface [4] and, per unit mass of fluid, may be expressed as the product of the fluid pressure p and specific volume $v(v = 1/\rho)$.

Applying conservation of energy, Equation 1.10a, to the differential control volume of Figure 8.6 and recalling the definition of the mean temperature, Equation 8.25, we obtain

$$dq_{\text{conv}} + \dot{m}(c_v T_m + pv) - \left[\dot{m}(c_v T_m + pv) + \dot{m}\frac{d(c_v T_m + pv)}{dx} dx\right] = 0$$

or

$$dq_{\text{conv}} = \dot{m} d(c_v T_m + pv) \tag{8.35}$$

That is, the rate of convection heat transfer to the fluid must equal the rate at which the fluid thermal energy increases plus the net rate at which work is done in moving the fluid through the control volume. If the fluid is assumed to be a *perfect gas* ($pv = RT_m$, $c_p = c_v + R$) and c_p is assumed to be constant, Equation 8.35 reduces to

$$dq_{\text{conv}} = \dot{m} c_p \, dT_m \tag{8.36}$$

This expression may also be used to a good approximation for *incompressible liquids*. In this case $c_v = c_p$, and since v is very small, $d(pv)$ is generally much less than $d(c_v T_m)$.[2] Accordingly, Equation 8.36 again follows from Equation 8.35.

A special form of Equation 8.36 relates to conditions for the *entire* tube. In particular, integrating from the tube inlet i to the outlet o, it follows that

$$q_{\text{conv}} = \dot{m} c_p (T_{m,o} - T_{m,i}) \tag{8.37}$$

[2] The only exception arises when the pressure gradient is extremely large. This situation occurs when $\dot{m}$ is very large and/or A_c is very small (see Problem 8.6).

where q_{conv} is the total tube heat transfer rate. This simple overall energy balance relates three important thermal variables (q_{conv}, $T_{m,o}$, $T_{m,i}$). *It is a general expression that applies regardless of the nature of the surface thermal condition or tube flow conditions.*

Equation 8.36 may be cast in a convenient form by expressing the rate of convection heat transfer to the differential element as $dq_{\text{conv}} = q_s'' P\, dx$, where P is the surface perimeter ($P = \pi D$ for a circular tube). Substituting from Equation 8.28, it follows that

$$\frac{dT_m}{dx} = \frac{q_s'' P}{\dot{m} c_p} = \frac{P}{\dot{m} c_p} h(T_s - T_m) \tag{8.38}$$

This expression is an extremely useful result, from which the axial variation of T_m may be determined. If $T_s > T_m$, heat is transferred to the fluid and T_m increases with x; if $T_s < T_m$, the opposite is true.

The manner in which quantities on the right-hand side of Equation 8.38 vary with x should be noted. Although P may vary with x, the most common situation is one in which it is a constant (a tube of constant cross-sectional area). Hence, the quantity ($P/\dot{m}c_p$) is a constant. In the fully developed region, the convection coefficient h is also constant, although it varies with x in the entrance region (Figure 8.5). Finally, although T_s may be constant, T_m must always vary with x (except for the trivial case of no heat transfer, $T_s = T_m$).

The solution to Equation 8.38 for $T_m(x)$ depends on the surface thermal condition. Recall that the two special cases of interest include *constant surface heat flux* and *constant surface temperature.* It is common to find one of these conditions existing to a reasonable approximation.

8.3.2 Constant Surface Heat Flux

For this condition we first note that it is a simple matter to determine the total heat transfer rate q_{conv}. Since q_s'' is independent of x, it follows that

$$q_{\text{conv}} = q_s''(P \cdot L) \tag{8.39}$$

This expression could be used with Equation 8.37 to determine the fluid temperature change, $T_{m,o} - T_{m,i}$.

For constant q_s'' it also follows that the right-hand side of Equation 8.38 is a constant independent of x. Hence

$$\frac{dT_m}{dx} = \frac{q_s'' P}{\dot{m} c_p} \neq f(x) \tag{8.40}$$

Integrating from $x = 0$, it follows that

$$T_m(x) = T_{m,i} + \frac{q_s'' P}{\dot{m} c_p} x \qquad q_s'' = \text{constant} \tag{8.41}$$

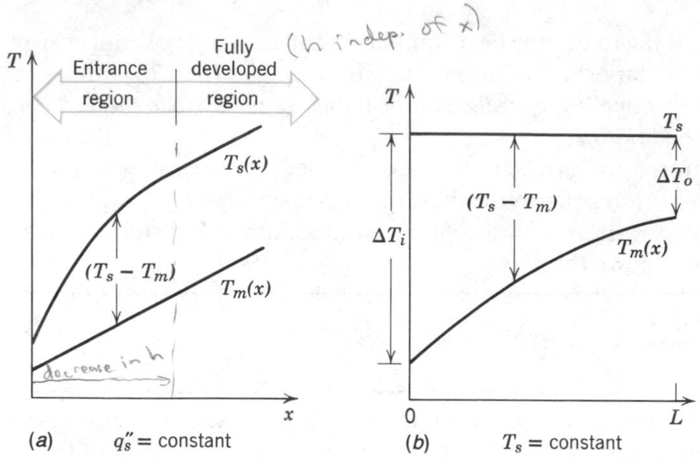

Figure 8.7 Axial temperature variations for heat transfer in a tube. (*a*) Constant surface heat flux. (*b*) Constant surface temperature.

Accordingly, the mean temperature varies *linearly* with x along the tube (Figure 8.7*a*). Moreover, from Equation 8.28 and Figure 8.5 we also expect the temperature difference $(T_s - T_m)$ to vary with x, as shown in Figure 8.7*a*. This difference is initially small (due to the large value of h at the entrance) but increases with increasing x as a result of the decrease in h that occurs as the boundary layer develops. However, in the fully developed region we know that h is independent of x. Hence from Equation 8.28 it follows that $(T_s - T_m)$ must also be independent of x in this region. Note from Equations 8.31 and 8.33 that the axial temperature gradient assumes a fixed value, regardless of radial location in the fully developed region.

EXAMPLE 8.2

A system for heating water from an inlet temperature of $T_{m,i} = 20°C$ to an outlet temperature of $T_{m,o} = 60°C$ involves passing the water through a thick-walled tube of inner and outer diameters of 20 and 40 mm. The outer surface of the tube is well insulated, and electrical heating within the wall provides for a uniform generation rate of $\dot{q} = 10^6$ W/m^3.

1. For a water mass flow rate of $\dot{m} = 0.1$ kg/s, how long must the tube be to achieve the desired outlet temperature?
2. If the inner surface temperature of the tube is $T_s = 70°C$ at the outlet, what is the local convection heat transfer coefficient at the outlet?

SOLUTION

Known:

Internal flow through a thick-walled tube having uniform heat generation.

Find:

1. Length of tube needed to achieve the desired outlet temperature.
2. Local convection coefficient at the outlet.

Schematic:

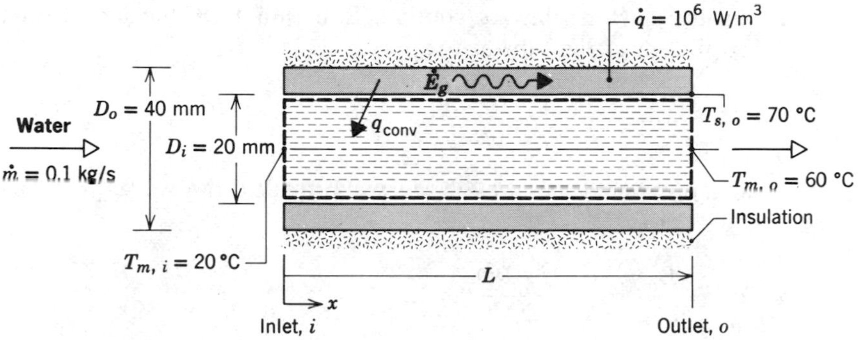

Assumptions:

1. Steady-state conditions.
2. Uniform heat generation.
3. Negligible potential energy, kinetic energy, and flow work changes.
4. Constant properties.
5. Adiabatic outer tube surface.

Properties:

Table A.6, water ($\bar{T}_m = 313$ K): $c_p = 4.179$ kJ/kg·K.

Analysis:

1. Since the outer surface of the tube is adiabatic, the rate at which energy is generated within the tube wall must equal the rate at which it is convected to the water.

   $$\dot{E}_g = q_{\text{conv}}$$

   With

   $$\dot{E}_g = \dot{q}\frac{\pi}{4}(D_o^2 - D_i^2)L$$

   it follows from Equation 8.37 that

   $$\dot{q}\frac{\pi}{4}(D_o^2 - D_i^2)L = \dot{m}c_p(T_{m,o} - T_{m,i})$$

or

$$L = \frac{4\dot{m}c_p}{\pi(D_o^2 - D_i^2)\dot{q}}(T_{m,o} - T_{m,i})$$

$$L = \frac{4 \times 0.1 \text{ kg/s} \times 4179 \text{ J/kg}\cdot\text{K}}{\pi(0.04^2 - 0.02^2) \text{ m}^2 \times 10^6 \text{ W/m}^3}(60-20)°\text{C} = 17.7 \text{ m} \quad \triangleleft$$

2. From Newton's law of cooling, Equation 8.28, the local convection coefficient at the tube exit is

$$h_o = \frac{q_s''}{T_{s,o} - T_{m,o}}$$

Uniform heat generation in the wall provides for a constant surface heat flux, with

$$q_s'' = \frac{\dot{E}_g}{\pi D_i L} = \frac{\dot{q}}{4}\frac{(D_o^2 - D_i^2)}{D_i}$$

$$q_s'' = \frac{10^6 \text{ W/m}^3}{4}\frac{(0.04^2 - 0.02^2) \text{ m}^2}{0.02 \text{ m}} = 1.5 \times 10^4 \text{ W/m}^2$$

Hence

$$h_o = \frac{1.5 \times 10^4 \text{ W/m}^2}{(70-60)°\text{C}} = 1500 \text{ W/m}^2\cdot\text{K} \quad \triangleleft$$

Comments:

1. If conditions are fully developed over the entire tube, the local convection coefficient and the temperature difference, $T_s - T_m$, are independent of x. Hence $h = 1500$ W/m$^2\cdot$K and $(T_s - T_m) = 10°$C over the entire tube, and the inner surface temperature at the tube inlet is $T_{s,i} = 30°$C.
2. The required tube length L could have been computed by applying the expression for $T_m(x)$, Equation 8.41, at $x = L$.

8.3.3 Constant Surface Temperature

Results for the total heat transfer rate and the axial distribution of the mean temperature are entirely different for the *constant surface temperature* condition. Defining ΔT as $T_s - T_m$, Equation 8.38 may be expressed as

$$\frac{dT_m}{dx} = -\frac{d(\Delta T)}{dx} = \frac{P}{\dot{m}c_p}h\Delta T$$

Separating variables and integrating from the tube inlet to the outlet,

$$\int_{\Delta T_i}^{\Delta T_o} \frac{d(\Delta T)}{\Delta T} = -\frac{P}{\dot{m}c_p}\int_0^L h\,dx$$

or,

$$\ln\frac{\Delta T_o}{\Delta T_i} = -\frac{PL}{\dot{m}c_p}\left(\frac{1}{L}\int_0^L h\,dx\right)$$

From the definition of the average convection heat transfer coefficient given by Equation 6.5, it follows that

$$\ln\frac{\Delta T_o}{\Delta T_i} = -\frac{PL}{\dot{m}c_p}\bar{h}_L \qquad T_s = \text{constant} \tag{8.42a}$$

where $\bar{h}_L$, or simply $\bar{h}$, is the average value of h for the entire tube. Rearranging,

$$\frac{\Delta T_o}{\Delta T_i} = \frac{T_s - T_{m,o}}{T_s - T_{m,i}} = \exp\left(-\frac{PL}{\dot{m}c_p}\bar{h}\right) \qquad T_s = \text{constant} \tag{8.42b}$$

Had we integrated from the tube inlet to some axial position x within the tube, we would have obtained the similar, but more general, result that

$$\frac{T_s - T_m(x)}{T_s - T_{m,i}} = \exp\left(-\frac{Px}{\dot{m}c_p}\bar{h}\right) \qquad T_s = \text{constant} \tag{8.43}$$

where $\bar{h}$ is now the average value of h from the tube inlet to x. This result suggests that the temperature difference $(T_s - T_m)$ *decays exponentially* with distance along the tube axis. The axial surface and mean temperature distributions are therefore as shown in Figure 8.7*b*.

Determination of an expression for the total heat transfer rate q_{conv} is complicated by the exponential nature of the temperature decay. Expressing Equation 8.37 in the form

$$q_{\text{conv}} = \dot{m}c_p[(T_s - T_{m,i}) - (T_s - T_{m,o})] = \dot{m}c_p(\Delta T_i - \Delta T_o)$$

and substituting for $\dot{m}c_p$ from Equation 8.42a, we obtain

$$q_{\text{conv}} = \bar{h}A_s\Delta T_{\text{lm}} \qquad T_s = \text{constant} \tag{8.44}$$

where A_s is the tube surface area ($A_s = P \cdot L$) and ΔT_{lm} is the *log mean temperature difference.*

$$\Delta T_{\text{lm}} \equiv \frac{\Delta T_o - \Delta T_i}{\ln(\Delta T_o/\Delta T_i)} = \frac{(T_s - T_{mo}) - (T_s - T_{mi})}{\ln[(T_s - T_{mo})/(T_s - T_{mi})]} \tag{8.45}$$

Equation 8.44 is a form of Newton's law of cooling for the entire tube, and ΔT_{lm} is the appropriate *average* of the temperature difference over the tube length.

The logarithmic nature of this average temperature difference [in contrast, e.g., to an *arithmetic mean temperature* difference of the form $\Delta T_{am} = (\Delta T_i + \Delta T_o)/2$] is due to the exponential nature of the temperature decay.

Before concluding this section, it is important to note that in many applications, it is the temperature of an *external* fluid that is fixed, rather than the tube surface temperature. In such cases, it is readily shown that the results of this section may still be used, if T_s is replaced by T_∞ (the freestream temperature of the external fluid) and $\bar{h}$ is replaced by $\bar{U}$ (the overall heat transfer coefficient).

EXAMPLE 8.3

Steam condensing on the outer surface of a thin-walled circular tube of 50 mm diameter and 6 m length maintains a uniform surface temperature of 100°C. Water flows through the tube at a rate of $\dot{m} = 0.25$ kg/s, and its inlet and outlet temperatures are $T_{m,i} = 15°C$ and $T_{m,o} = 57°C$. What is the average convection coefficient associated with the water flow?

SOLUTION

Known:

Flow rate and inlet and outlet temperatures of water flowing through a tube of prescribed dimensions and surface temperature.

Find: Average convection heat transfer coefficient.

Schematic:

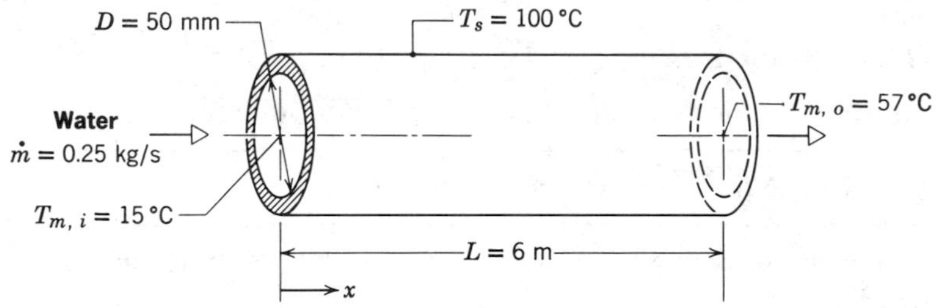

Assumptions:

1. Constant surface temperature T_s.
2. Negligible kinetic energy, potential energy, and flow work changes.
3. Constant properties.

Properties:

Table A.6, water (36°C): $c_p = 4178$ J/kg·K.

Analysis:

Combining the energy balance, Equation 8.37, with the rate equation, (8.44), the average convection coefficient is given by

$$\bar{h} = \frac{\dot{m}c_p}{\pi DL}\frac{(T_{m,o} - T_{m,i})}{\Delta T_{\text{lm}}}$$

From Equation 8.45

$$\Delta T_{\text{lm}} = \frac{(T_s - T_{m,o}) - (T_s - T_{m,i})}{\ln\,[(T_s - T_{m,o})/(T_s - T_{m,i})]}$$

$$\Delta T_{\text{lm}} = \frac{(100 - 57) - (100 - 15)}{\ln\,[(100 - 57)/(100 - 15)]} = 61.6°\text{C}$$

Hence

$$\bar{h} = \frac{0.25\ \text{kg/s} \times 4178\ \text{J/kg}\cdot\text{K}}{\pi \times 0.05\ \text{m} \times 6\ \text{m}}\frac{(57 - 15)°\text{C}}{61.6°\text{C}}$$

or

$\bar{h} = 756\ \text{W/m}^2\cdot\text{K}$ ◁

Comments:

1. In this case use of an arithmetic mean temperature difference, $\Delta T_{\text{am}} = T_s - (T_{m,i} + T_{m,o})/2 = 64°\text{C}$, in place of the log mean temperature difference, $\Delta T_{\text{lm}} = 61.6°\text{C}$, would have been a reasonable approximation.
2. If conditions were fully developed over the entire tube, the local convection coefficient would be everywhere 756 $\text{W/m}^2\cdot\text{K}$.

8.4 CONVECTION CORRELATIONS: LAMINAR FLOW IN CIRCULAR TUBES

To use many of the foregoing results, the convection coefficients must be known. In this section we outline the manner in which such coefficients may be obtained theoretically for laminar flow in a circular tube. In subsequent sections we consider empirical correlations pertinent to turbulent flow in a circular tube, as well as to flows in tubes of noncircular cross section.

8.4.1 The Fully Developed Region

The problem of laminar flow in a circular tube has been treated theoretically, and the results may be used to determine the convection coefficients. At any point in the tube the boundary layer approximations (Section 6.5) may be applied, and for constant properties, the energy equation is

$$u\frac{\partial T}{\partial x}+v\frac{\partial T}{\partial r}=\frac{\alpha}{r}\frac{\partial}{\partial r}\left(r\frac{\partial T}{\partial r}\right) \tag{8.46}$$

This equation, which applies for cylindrical coordinates, is of the same form as the boundary layer equation, (6.33), which was developed in rectangular coordinates, except that viscous dissipation has been neglected. The terms on the left-hand side of Equation 8.46 account for net energy transfer by gross fluid motion, and the term on the right accounts for net energy transfer by conduction in the radial direction.

The solution to Equation 8.46 is readily obtained for the *fully developed region.* In this region the velocity boundary layer approximations are satisfied exactly. That is, $v=0$ and $(\partial u/\partial x)=0$, in which case the axial velocity component is given by the parabolic profile of Equation 8.15. Moreover, for the case of *constant surface heat flux*, the thermal boundary layer approximation is also satisfied exactly. That is, $(\partial^2 T/\partial x^2)=0$. Substituting for the axial temperature gradient from Equation 8.33 and for the axial velocity component from Equation 8.15, the energy equation, Equation 8.46, reduces to

$$\frac{1}{r}\frac{d}{dr}\left(r\frac{dT}{dr}\right)=\frac{2u_m}{\alpha}\left(\frac{dT_m}{dx}\right)\left[1-\left(\frac{r}{r_o}\right)^2\right] \qquad q_s''=\text{constant} \tag{8.47}$$

where $(2u_m/\alpha)(dT_m/dx)$ is a constant. Separating variables and integrating twice, we obtain the following expression for the radial temperature distribution:

$$T(r)=\frac{2u_m}{\alpha}\left(\frac{dT_m}{dx}\right)\left[\frac{r^2}{4}-\frac{r^4}{16r_o^2}\right]+C_1\ln r+C_2$$

The constants of integration may be evaluated by applying appropriate boundary conditions. From the requirement that the temperature remain finite at $r=0$, it follows that $C_1=0$. From the requirement that $T(r_o)=T_s$, where T_s varies with x, it also follows that

$$C_2=T_s-\frac{2u_m}{\alpha}\left(\frac{dT_m}{dx}\right)\left(\frac{3r_o^2}{16}\right)$$

Accordingly, for the fully developed region with constant surface heat flux, the temperature profile is of the form

$$T(r)=T_s-\frac{2u_m r_o^2}{\alpha}\left(\frac{dT_m}{dx}\right)\left[\frac{3}{16}+\frac{1}{16}\left(\frac{r}{r_o}\right)^4-\frac{1}{4}\left(\frac{r}{r_o}\right)^2\right] \tag{8.48}$$

From knowledge of the temperature profile, all other thermal parameters may be determined. For example, if the velocity and temperature profiles,

Equations 8.15 and 8.48, respectively, are substituted into Equation 8.27 and the integration over r is performed, the mean temperature is found to be

$$T_m = T_s - \frac{11}{48}\left(\frac{u_m r_o^2}{\alpha}\right)\left(\frac{dT_m}{dx}\right) \tag{8.49}$$

From Equation 8.40, where $P = \pi D$ and $\dot{m} = \rho u_m(\pi D^2/4)$, we then obtain

$$T_m - T_s = -\frac{11}{48}\frac{q_s'' D}{k} \tag{8.50}$$

Combining Newton's law of cooling, Equation 8.28, and 8.50, it follows that

$$h = \frac{48}{11}\left(\frac{k}{D}\right)$$

or

$$Nu_D \equiv \frac{hD}{k} = 4.36 \qquad q_s'' = \text{constant} \tag{8.51}$$

(laminar, F-d conditions)

Hence in a *circular tube* characterized by *uniform surface heat flux* and *laminar, fully developed conditions*, the *Nusselt number is a constant*, independent of Re_D, Pr, and axial location.

For *laminar, fully developed conditions* with a *constant surface temperature*, the velocity boundary layer approximations are again satisfied exactly, and the thermal boundary layer approximation $(\partial^2 T/\partial x^2) \ll (\partial^2 T/\partial r^2)$ is often reasonable. Substituting for the velocity profile from Equation 8.15 and for the axial temperature gradient from Equation 8.34, the energy equation becomes

$$\frac{1}{r}\frac{d}{dr}\left(r\frac{dT}{dr}\right) = \frac{2u_m}{\alpha}\left(\frac{dT_m}{dx}\right)\left[1 - \left(\frac{r}{r_o}\right)^2\right]\frac{T_s - T}{T_s - T_m} \qquad T_s = \text{constant} \tag{8.52}$$

A solution to this equation may be obtained by an iterative procedure that involves making successive approximations to the temperature profile. The resulting profile is not described by a simple algebraic expression, but the resulting Nusselt number may be shown to be of the form [2]

$$Nu_D = 3.66 \qquad T_s = \text{constant} \tag{8.53}$$

(laminar, F-d conditions)

Note that in using Equation 8.51 or 8.53 to determine h, the thermal conductivity should be evaluated at T_m.

EXAMPLE 8.4

One concept used for solar energy collection involves placing a tube at the focal point of a parabolic reflector and passing a fluid through the tube. The net effect of this arrangement *may be approximated* as one of creating a

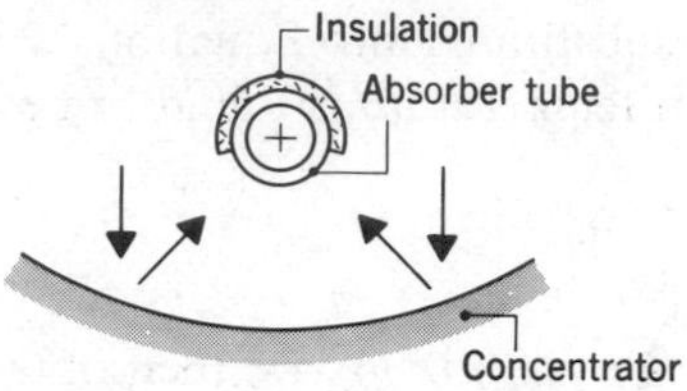

condition of uniform heating at the surface of the tube. That is, the resulting heat flux to the fluid, q_s'', may be assumed to be a constant along the axis of the tube. Consider operation with a tube of diameter $D = 60$ mm on a sunny day for which $q_s'' = 2000\ \text{W/m}^2$.

1. If water enters the tube at $\dot{m} = 0.01$ kg/s and $T_{m,i} = 20°\text{C}$, what tube length L is required to obtain an exit temperature of 80°C?
2. What is the surface temperature at the outlet of the tube, where fully developed conditions may be assumed to exist?

SOLUTION

Known: Internal flow with uniform surface heat flux.

Find:

1. Length of tube L to achieve required heating.
2. Tube wall temperature $T_s(L)$ at the outlet section, $x = L$.

Schematic:

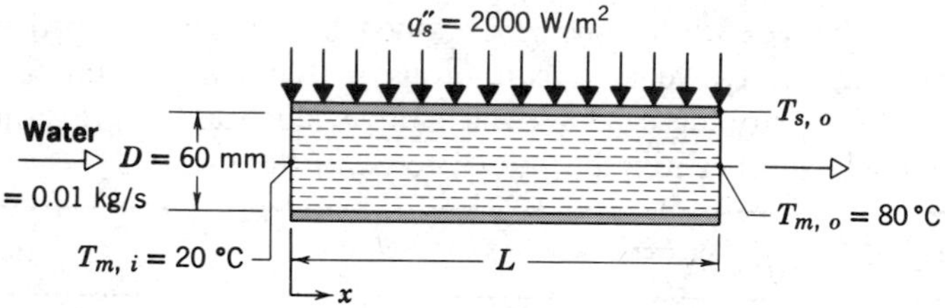

Assumptions:

1. Steady-state conditions.
2. Incompressible flow with constant properties.
3. Negligible kinetic and potential energy and flow work changes.
4. Fully developed conditions at tube outlet.

Properties:

Table A.6, water ($\bar{T}_m = 323$ K): $c_p = 4181$ J/kg·K.
Table A.6, water ($T_{m,o} = 353$ K): $k = 0.670$ W/m·K, $\mu = 352 \times 10^{-6}$ N·s/m², $Pr = 2.2$.

Analysis:

1. For constant surface heat flux, Equation 8.39 may be used with the energy balance, Equation 8.37, to obtain

$$A_s = \pi D L = \frac{\dot{m} c_p (T_{m,o} - T_{m,i})}{q_s''}$$

$$L = \frac{\dot{m} c_p}{\pi D q_s''}(T_{m,o} - T_{m,i})$$

Hence

$$L = \frac{0.01 \text{ kg/s} \times 4181 \text{ J/kg}\cdot\text{K}}{\pi \times 0.060 \text{ m} \times 2000 \text{ W/m}^2}(80 - 20)°\text{C} = 6.65 \text{ m} \qquad \triangleleft$$

2. The surface temperature at the outlet may be obtained from Newton's law of cooling, Equation 8.28, where

$$T_{s,o} = \frac{q_s''}{h} + T_{m,o}$$

To find the local convection coefficient at the tube outlet, the nature of the flow condition must first be established. From Equation 8.6

$$Re_D = \frac{4\dot{m}}{\pi D \mu} = \frac{4 \times 0.01 \text{ kg/s}}{\pi \times 0.060 \text{ m} \times 352 \times 10^{-6} \text{ N}\cdot\text{s/m}^2} = 603$$

Hence the flow is laminar. With the assumption of fully developed conditions, the appropriate heat transfer correlation is then

$$Nu_D = \frac{hD}{k} = 4.36$$

and

$$h = 4.36\frac{k}{D} = 4.36\frac{0.670 \text{ W/m}\cdot\text{K}}{0.06 \text{ m}} = 48.7 \text{ W/m}^2\cdot\text{K}$$

The surface temperature at the tube outlet is then

$$T_{s,o} = \frac{2000 \text{ W/m}^2}{48.7 \text{ W/m}^2\cdot\text{K}} + 80°\text{C} = 121°\text{C} \qquad \triangleleft$$

Comments:

For the conditions of this problem, $(x_{fd}/D) = 0.05 Re_D Pr = 66.3$, and $L/D = 110$. Hence the assumption of fully developed conditions is justified.

8.4.2 The Entry Region

The solution to the energy equation, (8.46), for the entry region is considerably more difficult to obtain, since velocity and temperature now depend on x as well as r. Even if the radial advection term is neglected, the axial temperature gradient $\partial T/\partial x$ may no longer be simplified through Equation 8.33 or 8.34. However, two different entry length solutions have been obtained. The simplest solution is for the *thermal entry length problem*, and it is based on assuming that thermal conditions develop in the presence of a *fully developed velocity profile*. Such a situation would exist if the location at which heat transfer begins were preceded by an *unheated starting length*. It could also be assumed to a reasonable approximation for large Prandtl number fluids, such as oils. Even in the absence of an unheated starting length, velocity boundary layer development would occur far more rapidly than thermal boundary layer development, and a thermal entry length approximation could be made. In contrast, the *combined* (thermal and velocity) *entry length problem* corresponds to the case for which the temperature and velocity profiles develop simultaneously.

Solutions have been obtained for both entry length conditions [2], and selected results are presented in Figure 8.8. The Nusselt numbers are, in principle, infinite

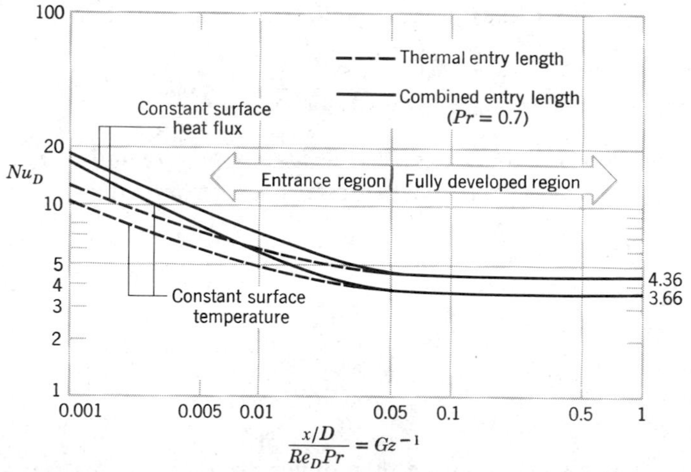

Figure 8.8 Local Nusselt number obtained from entry length solutions for laminar flow in a circular tube [2]. Adapted with permission.

at $x=0$ and decay to their asymptotic (fully developed) values with increasing x. When plotted against the dimensionless parameter $x/(DRe_DPr)$, which is the reciprocal of the *Graetz number*, $Gz_D \equiv (D/x)Re_DPr$, the manner in which Nu_D varies with Gz_D^{-1} is independent of Pr for the thermal entry length problem. In contrast, for the combined entry length problem, results depend on the Prandtl number and have been presented only for $Pr=0.7$. However, these results are representative of conditions for most gases. Note also that fully developed conditions are reached for $[(x/D)/Re_DPr] \approx 0.05$.

For the *constant surface temperature* condition, it is desirable to know the *average* convection coefficient for use with Equation 8.44. Kays [6] presents a correlation attributed to Hausen [7], which is of the form

$$\overline{Nu}_D = 3.66 + \frac{0.0668(D/L)Re_DPr}{1+0.04[(D/L)Re_DPr]^{2/3}} \tag{8.54}$$

where $\overline{Nu}_D \equiv \bar{h}D/k$. Because this result presumes a thermal entry length, it is not generally applicable. For the combined entry length, a suitable correlation, due to Sieder and Tate [8], is of the form

$$\overline{Nu}_D = 1.86\left(\frac{Re_DPr}{L/D}\right)^{1/3}\left(\frac{\mu}{\mu_s}\right)^{0.14} \tag{8.55}$$

$$\left[\begin{array}{l} T_s = \text{constant} \\ 0.48 < Pr < 16{,}700 \\ 0.0044 < (\mu/\mu_s) < 9.75 \end{array}\right]$$

The correlation has been recommended by Whitaker [9] for values of $\{[(Re_D/Pr)/(L/D)]^{1/3}(\mu/\mu_s)^{0.14}\} \gtrsim 2$. Below this limit fully developed conditions encompass much of the tube, and Equation 8.53 may be used to a good approximation. Note that all properties appearing in Equations 8.54 and 8.55, except μ_s, should be evaluated at the average value of the mean temperature, $\bar{T}_m \equiv (T_{m,i}+T_{m,o})/2$.

The subject of laminar flow in ducts has been studied extensively, and numerous results are available for a variety of duct cross sections and surface conditions. These results have been compiled in a monograph by Shah and London [10].

8.5 CONVECTION CORRELATIONS: TURBULENT FLOW IN CIRCULAR TUBES

Since the analysis of turbulent flow conditions is a good deal more involved, greater emphasis is placed on determining empirical correlations. A classical expression for computing the *local* Nusselt number for *fully developed turbulent flow* in a *smooth circular tube* is due to Colburn [11] and may be obtained from the Chilton-Colburn analogy. Substituting Equation 6.57 into Equation 8.18, the analogy is of the form

$$\frac{C_f}{2} = \frac{f}{8} = St Pr^{2/3} = \frac{Nu_D}{Re_D Pr} Pr^{2/3} \tag{8.56}$$

Substituting for the friction factor from Equation 8.21, the *Colburn equation* is then

$$Nu_D = 0.023 Re_D^{4/5} Pr^{1/3} \tag{8.57}$$

The *Dittus-Boelter equation* [12] is a slightly different and preferred version of the above result and is of the form

$$Nu_D = 0.023 Re_D^{4/5} Pr^n \tag{8.58}$$

Prop's, eval. at T_m

where $n = 0.4$ for heating ($T_s > T_m$) and 0.3 for cooling ($T_s < T_m$). These equations have been confirmed experimentally for the range of conditions

$$\begin{bmatrix} 0.7 \le Pr \le 160 \\ Re_D \gtrsim 10{,}000 \\ L/D \gtrsim 10 \end{bmatrix}$$

The equations should be used only for small to moderate temperature differences, $T_s - T_m$, with all the properties evaluated at T_m. For flows characterized by large property variations, the following equation due to Sieder and Tate [8] is recommended.

$$Nu_D = 0.027 Re_D^{4/5} Pr^{1/3} \left(\frac{\mu}{\mu_s}\right)^{0.14} \tag{8.59}$$

flows w/large prop. variations

$$\begin{bmatrix} 0.7 \le Pr \le 16{,}700 \\ Re_D \gtrsim 10{,}000 \\ L/D \gtrsim 10 \end{bmatrix}$$

where all properties except μ_s are evaluated at T_m. Note that, to a good approximation, the foregoing correlations may be applied for both the constant surface temperature and heat flux conditions. Although improved accuracy may often be obtained by using more recent results [13], the correlations suffice for most engineering calculations.

Since the average Nusselt number, $\overline{Nu}_D$, for the entire tube depends on the nature of the inlet conditions, as well as on the Reynolds and Prandtl numbers, generally applicable correlations are not available. However, to a first approximation, the foregoing equations may be used to obtain a reasonable estimate of

$\overline{Nu}_D$ if $L/D \gtrsim 10$. When determining $\overline{Nu}_D$, all fluid properties should be evaluated at the arithmetic average of the mean temperature, $\bar{T}_m \equiv (T_{m,i} + T_{m,o})/2$.

The foregoing correlations are restricted to *smooth surface* conditions. Heat transfer is enhanced by surface roughness, and the Nusselt number may be estimated by applying the Chilton-Colburn analogy, Equation 8.56, to the friction factor data of Figure 8.3 or from available correlations [14]. In fact, heat transfer enhancement is often promoted by using internal ribs, fins, or corrugations, as well as by surface roughening [15–17].

Finally, the foregoing correlations do not apply to liquid metals ($3 \times 10^{-3} \lesssim Pr \lesssim 5 \times 10^{-2}$). For fully developed turbulent flow ($L/D \gtrsim 10$) in smooth circular tubes with constant surface heat flux, Skupinski et al. [18] recommend a correlation of the form

$$Nu_D = 4.82 + 0.0185 Pe_D^{0.827} \qquad q_s'' = \text{constant} \qquad (8.60)$$

$$\begin{bmatrix} 3.6 \times 10^3 < Re_D < 9.05 \times 10^5 \\ 10^2 < Pe_D < 10^4 \end{bmatrix}$$

Similarly, for constant surface temperature Seban and Shimazaki [19] recommend the following correlation for $Pe_D > 100$

$$Nu_D = 5.0 + 0.025 Pe_D^{0.8} \qquad T_s = \text{constant} \qquad (8.61)$$

Extensive data and additional correlations are available in the literature [20].

EXAMPLE 8.5

Hot air flows with a mass rate of $\dot{m} = 0.050$ kg/s through an uninsulated sheet metal duct of diameter $D = 0.15$ m, which is in the crawlspace of a house. The hot air enters at 103°C and, after a distance of $L = 5$ m, cools to 77°C. The heat transfer coefficient between the duct outer surface and the ambient air at $T_\infty = 0$°C is known to be $h_o = 6$ W/m$^2 \cdot$K.

1. Calculate the heat loss (W) from the duct over the length L.
2. Determine the heat flux and the duct surface temperature at $x = L$.

SOLUTION

Known: Hot air flowing in a duct.

Find:

1. Heat loss from the duct over the length L, q (W).
2. Heat flux and surface temperature at $x = L$.

Schematic:

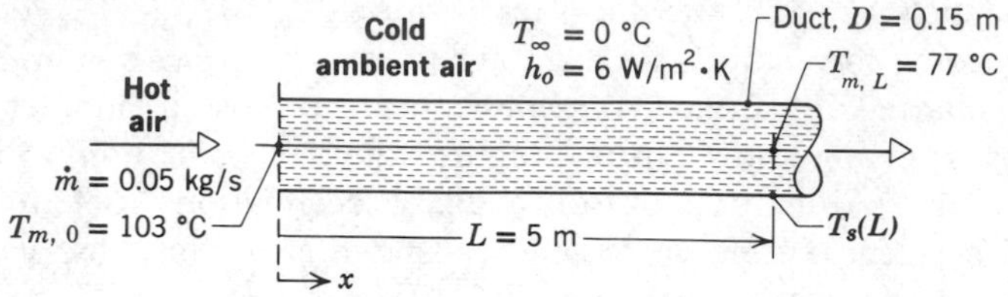

Assumptions:

1. Steady-state conditions.
2. Constant properties.
3. Perfect gas behavior.
4. Negligible kinetic and potential energy changes.
5. Negligible duct wall thermal resistance.
6. Uniform convection coefficient at outer surface of duct.

Properties:

Table A.4, air ($\bar{T}_m = 363$ K): $c_p = 1010$ J/kg·K.
Table A.4, air ($T_{m,L} = 350$ K): $k = 0.030$ W/m·K, $\mu = 208 \times 10^{-7}$ N·s/m^2, $Pr = 0.70$.

Analysis:

1. From the energy balance for the entire tube, Equation 8.37,

   $$q = \dot{m} c_p (T_{m,L} - T_{m,0})$$

   $$q = 0.05 \text{ kg/s} \times 1010 \text{ J/kg·K } (77 - 103)\text{°C} = -1313 \text{ W} \quad \triangleleft$$

2. An expression for the heat flux at $x = L$ may be inferred from the resistance network

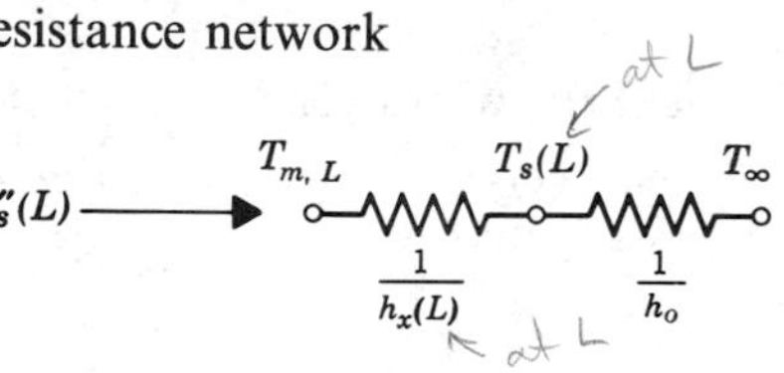

$q = \frac{\Delta T}{R_t} = \frac{\Delta T}{1/hA}$

$q'' = \frac{\Delta T}{R_t} = \frac{\Delta T}{1/h}$

where $h_x(L)$ is the inside convection heat transfer coefficient at $x = L$. Hence

$$q_s''(L) = \frac{T_{m,L} - T_\infty}{1/h_x(L) + 1/h_o}$$

The inside convection coefficient may be obtained from knowledge of the Reynolds number. From Equation 8.6,

$$Re_D = \frac{4\dot{m}}{\pi D \mu} = \frac{4 \times 0.05 \text{ kg/s}}{\pi \times 0.15 \text{ m} \times 208 \times 10^{-7} \text{ N} \cdot \text{s/m}^2} = 20{,}404$$

Hence the flow is turbulent. Moreover, with $L/D = 5/0.15 = 33.3$, it is reasonable to assume fully developed conditions at $x = L$. Hence from Equation 8.58, with $n = 0.3$,

$$Nu_D = \frac{h_x(L)D}{k} = 0.023 Re_D^{4/5} Pr^{0.3}$$

$$Nu_D = 0.023(20{,}404)^{4/5}(0.70)^{0.3} = 57.9$$

$$h_x(L) = Nu_D \frac{k}{D} = 57.9 \frac{0.030 \text{ W/m} \cdot \text{K}}{0.15 \text{ m}} = 11.6 \text{ W/m}^2 \cdot \text{K}$$

Hence

$$q_s''(L) = \frac{(77 - 0)^\circ\text{C}}{[(1/11.6) + (1/6.0)] \text{ m}^2 \cdot \text{K/W}} = 304.5 \text{ W/m}^2$$

Referring back to the network, it also follows that

$$q_s''(L) = \frac{T_{m,L} - T_{s,L}}{1/h_x(L)}$$

in which case

$$T_{s,L} = T_{m,L} - \frac{q_s''(L)}{h_x(L)}$$

$$T_{s,L} = 77^\circ\text{C} - \frac{304.5 \text{ W/m}^2}{11.6 \text{ W/m}^2 \cdot \text{K}} = 50.7^\circ\text{C} \qquad \triangleleft$$

Comments:

1. In using the energy balance of part 1 for the entire tube, properties (in this case, only c_p) are evaluated at $\bar{T}_m = (T_{m,0} + T_{m,L})/2$. However, in using the correlation for a local heat transfer coefficient, Equation 8.58, properties are evaluated at the local mean temperature, $T_{m,L} = 77^\circ$C.
2. This problem is characterized neither by constant surface temperature nor by constant surface heat flux. It would therefore be erroneous to presume that the total heat loss from the tube is given by $q_s''(L)\pi DL = 717$ W. This result is substantially less than the actual heat loss of 1313 W because $q_s''(x)$ decreases with increasing x. This decrease in $q_s''(x)$ is due to reductions in both $h_x(x)$ and $[T_m(x) - T_\infty]$ with increasing x.

8.6 CONVECTION CORRELATIONS: NONCIRCULAR TUBES

OMIT

Although we have thus far restricted our consideration to internal flows of circular cross section, many engineering applications involve convection transport in *noncircular tubes.* At least to a first approximation, however, many of the circular tube results may be applied by using *an effective diameter* as the characteristic length. It is termed the *hydraulic diameter* and is defined as

$$D_h \equiv \frac{4A_c}{P} \tag{8.62}$$

(for calc, Re_D + Nu_D)

where A_c and P are the *flow* cross-sectional area and the *wetted perimeter,* respectively [21]. It is this diameter that should be used in calculating parameters such as Re_D and Nu_D.

For turbulent flow, which still occurs if $Re_D \gtrsim 2300$, it is reasonable to use the correlations of Section 8.5 for $Pr \gtrsim 0.7$. However, in a noncircular tube the

Table 8.1 Nusselt numbers for fully developed laminar flow in tubes of different cross section

CROSS SECTION	$\frac{b}{a}$	$Nu_D \equiv \frac{hD_h}{k}$ (constant q_s'')	(constant T_s)
	—	4.36	3.66
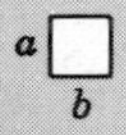	1.0	3.61	2.98
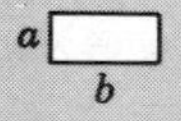	1.43	3.73	3.08
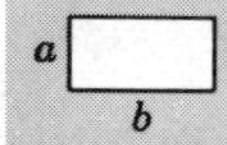	2.0	4.12	3.39
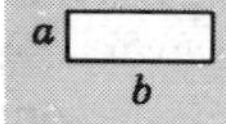	3.0	4.79	3.96
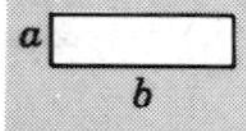	4.0	5.33	4.44
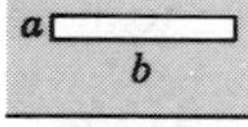	8.0	6.49	5.60
	∞	8.23	7.54
	—	3.00	2.35

Used with permission from W. M. Kays and M. E. Crawford, *Convection Heat and Mass Transfer*, McGraw-Hill, New York, 1980.

convection coefficients vary around the periphery, approaching zero in the corners. Hence in using a circular tube correlation, the coefficient is presumed to be an average over the perimeter.

For laminar flow, the use of circular tube correlations is less accurate, particularly with cross sections characterized by sharp corners. For such cases the Nusselt number corresponding to fully developed conditions may be obtained from Table 8.1.

Although the foregoing procedures are generally satisfactory, exceptions do exist. Detailed discussions of heat transfer in noncircular tubes are provided by Kays and Perkins [14] and Shah and London [10, 22].

OMIT

8.7 THE CONCENTRIC TUBE ANNULUS

Many internal flow problems involve heat transfer in a *concentric tube annulus* (Figure 8.9). Fluid passes through the space (annulus) formed by the concentric tubes, and convection heat transfer may occur to or from both the inner and outer tube surfaces. It is possible to independently specify the heat flux or temperature, that is, the thermal condition, at each of these surfaces. In any case the heat flux at each surface may be computed from expressions of the form

$$q_i'' = h_i(T_{s,i} - T_m) \tag{8.63}$$

$$q_o'' = h_o(T_{s,o} - T_m) \tag{8.64}$$

Note that separate convection coefficients are associated with the inner and outer surfaces. The corresponding Nusselt numbers are of the form

$$Nu_i \equiv \frac{h_i D_h}{k} \tag{8.65}$$

$$Nu_o \equiv \frac{h_o D_h}{k} \tag{8.66}$$

where, from Equation 8.62, the hydraulic diameter D_h is

$$D_h = \frac{4(\pi/4)(D_o^2 - D_i^2)}{\pi D_o + \pi D_i} = D_o - D_i \tag{8.67}$$

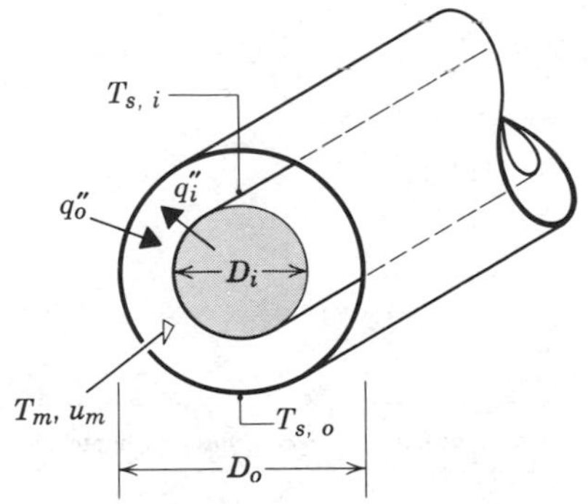

Figure 8.9 The concentric tube annulus.

For the case of fully developed laminar flow with one surface insulated and the other surface at a constant temperature, Nu_i or Nu_o may be obtained from Table 8.2. Note that in such cases we would only be interested in the convection coefficient associated with the isothermal (nonadiabatic) surface.

If uniform heat flux conditions exist at both surfaces, the Nusselt numbers may be computed from expressions of the form

$$Nu_i = \frac{Nu_{ii}}{1-(q_o''/q_i'')\theta_i^*} \tag{8.68}$$

$$Nu_o = \frac{Nu_{oo}}{1-(q_i''/q_o'')\theta_o^*} \tag{8.69}$$

The influence coefficients (Nu_{ii}, Nu_{oo}, θ_i^*, and θ_o^*) appearing in these equations may be obtained from Table 8.3. Note that q_i'' and q_o'' may be positive or negative, depending on whether heat transfer is to or from the fluid, respectively. Moreover, situations may arise for which the values of h_i and h_o are negative. Such results, when used with the sign convention implicit in Equations 8.63 and 8.64, reveal the relative magnitudes of T_s and T_m.

Table 8.2 Nusselt number for fully developed laminar flow in a circular tube annulus with one surface insulated and the other at constant temperature

D_i/D_o	Nu_i	Nu_o
0	—	3.66
0.05	17.46	4.06
0.10	11.56	4.11
0.25	7.37	4.23
0.50	5.74	4.43
1.00	4.86	4.86

Used with permission from W. M. Kays and H. C. Perkins, *Handbook of Heat Transfer*, Chapter 7, W. M. Rohsenow and J. P. Hartnett, Eds., McGraw-Hill, New York, 1972.

Table 8.3 Influence coefficients for fully developed laminar flow in a circular tube annulus with uniform heat flux maintained at both surfaces

D_i/D_o	Nu_{ii}	Nu_{oo}	θ_i^*	θ_o^*
0	—	4.364	∞	0
0.05	17.81	4.792	2.18	0.0294
0.10	11.91	4.834	1.383	0.0562
0.20	8.499	4.833	0.905	0.1041
0.40	6.583	4.979	0.603	0.1823
0.60	5.912	5.099	0.473	0.2455
0.80	5.58	5.24	0.401	0.299
1.00	5.385	5.385	0.346	0.346

Used with permission from W. M. Kays and H. C. Perkins, *Handbook of Heat Transfer*, Chapter 7, W. M. Rohsenow and J. P. Hartnett, Eds., McGraw-Hill, New York, 1972.

For fully developed turbulent flow the influence coefficients are a function of the Reynolds and Prandtl numbers [14]. However, to a first approximation the inner and outer convection coefficients may be assumed to be equal, and they may be evaluated by using the hydraulic diameter, Equation 8.67, with the Dittus-Boelter equation (8.58).

Table 8.4 Summary of convection correlations for internal flow.[a,b]

CORRELATION		CONDITIONS
$f = \frac{64}{Re_D}$	(8.19)	Laminar, fully developed
$Nu_D = 4.36$	(8.51)	Laminar, fully developed, constant q_s'', $Pr \gtrsim 0.6$
$Nu_D = 3.66$	(8.53)	Laminar, fully developed, constant T_s, $Pr \gtrsim 0.6$
$\overline{Nu}_D = 3.66 + \frac{0.0668(D/L)Re_D Pr}{1 + 0.04[(D/L)Re_D Pr]^{2/3}}$	(8.54)	Laminar, thermal entry length ($Pr \gg 1$ or an unheated starting length), constant T_s
or,		
$\overline{Nu}_D = 1.86\left(\frac{Re_D Pr}{L/D}\right)^{1/3}\left(\frac{\mu}{\mu_s}\right)^{0.14}$	(8.55)	Laminar, combined entry length, $[(Re_D Pr/L/D)^{1/3}(\mu/\mu_s)^{0.14}] \gtrsim 2$, constant T_s, $0.48 < Pr < 16{,}700$, $0.0044 < (\mu/\mu_s) < 9.75$
$f = 0.316 Re_D^{-1/4}$	(8.20)[c]	Turbulent, fully developed, $Re_D \lesssim 2 \times 10^4$
$f = 0.184 Re_D^{-1/5}$	(8.21)[c]	Turbulent, fully developed, $Re_D \gtrsim 2 \times 10^4$
$Nu_D = 0.023 Re_D^{4/5} Pr^{1/3}$	(8.57)[d]	Turbulent, fully developed, $0.7 \leq Pr \leq 160$, $Re_D \gtrsim 10{,}000$, $L/D \gtrsim 10$
or, $Nu_D = 0.023 Re_D^{4/5} Pr^n$	(8.58)[d]	Turbulent, fully developed, $0.7 \leq Pr \leq 160$, $Re_D \gtrsim 10{,}000$, $L/D \gtrsim 10$, $n = 0.4$ for $T_s > T_m$ and $n = 0.3$ for $T_s < T_m$
or, $Nu_D = 0.027 Re_D^{4/5} Pr^{1/3}\left(\frac{\mu}{\mu_s}\right)^{0.14}$	(8.59)[d]	Turbulent, fully developed, $0.7 \leq Pr \leq 16{,}700$, $Re_D \gtrsim 10{,}000$, $L/D \gtrsim 10$
$Nu_D = 4.82 + 0.0185(Re_D Pr)^{0.827}$	(8.60)	Liquid metals, turbulent, fully developed, constant q_s'', $3.6 \times 10^3 < Re_D < 9.05 \times 10^5$, $10^2 < Pe_D < 10^4$
$Nu_D = 5.0 + 0.025(Re_D Pr)^{0.8}$	(8.61)	Liquid metals, turbulent, fully developed, constant T_s, $Pe_D > 100$

[a] Properties in Equations 8.51, 8.53, 8.57, 8.58, 8.59, 8.60, and 8.61 are based on T_m; properties in Equations 8.19, 8.20, and 8.21 are based on $T_f \equiv (T_s + T_m)/2$; properties in Equations 8.54 and 8.55 are based on $\bar{T}_m \equiv (T_{m,i} + T_{m,o})/2$.

[b] $Re_D \equiv D_h u_m/\nu$; $D_h \equiv 4A_c/P$; $u_m \equiv \dot{m}/\rho A_c$.

[c] Equations 8.20 and 8.21 pertain to smooth tubes. For rough tubes the Chilton-Colburn analogy, Equation 8.56, should be used with the results of Figure 8.3.

[d] As a first approximation, Equation 8.57, 8.58, or 8.59 may be used to evaluate the average Nusselt number $\overline{Nu}_D$ over the entire tube length, if $(L/D) \gtrsim 10$. The properties should then be evaluated at the average of the mean temperature, $\bar{T}_m \equiv (T_{m,i} + T_{m,o})/2$.

8.8 SUMMARY

Internal flow is encountered in numerous applications, and it is important to appreciate its unique features. What is the nature of fully developed flow, and how does it differ from flow in the entry region? How does the Prandtl number influence boundary layer development in the entry region? How do thermal conditions in the fluid depend on the surface condition? For example, how do the surface and mean temperatures vary with x for the case of uniform surface heat flux? Or, how do the mean temperature and the surface heat flux vary for the case of uniform surface temperature?

You must be able to perform engineering calculations that involve an energy balance and appropriate convection correlations. The methodology involves first determining whether the flow is laminar or turbulent and then establishing the length of the entry region. After it has been decided whether local conditions (at a particular axial location) or average conditions (for the entire tube) are of interest, the convection correlation may be selected and used with the appropriate form of the energy balance to solve the problem. A summary of the correlations is provided in Table 8.4.

Several features that complicate internal flows have not been considered in this chapter. For example, a situation may exist for which there is a prescribed axial variation in T_s or q_s'', rather than uniform surface conditions. Among other things, such a variation would preclude the existence of a fully developed region. There may also exist surface roughness effects, circumferential heat flux or temperature variations, widely varying fluid properties, or transition flow conditions. A complete discussion of these effects is provided by Kays and Perkins [14].

REFERENCES

1. Langhaar, H. L., *J. Appl. Mech.*, *64*, A-55, 1942.
2. Kays, W. M., and M. E. Crawford, *Convective Heat and Mass Transfer*, McGraw-Hill, New York, 1980.
3. Moody, L. F., *Trans. ASME*, *66*, 671, 1944.
4. Wark, K., *Thermodynamics*, 4th ed., McGraw-Hill, New York, 1983.
5. Bird, R. B., W. E. Stewart, and E. N. Lightfoot, *Transport Phenomena*, Wiley, New York, 1960.
6. Kays, W. M., *Trans. ASME*, *77*, 1265, 1955.
7. Hausen, H., *Z. VDI. Beih. Verfahrenstech*, *4*, 91, 1943.
8. Sieder, E. N., and G. E. Tate, *Ind. Eng. Chem.*, *28*, 1429, 1936.
9. Whitaker, S., *AIChE J.*, *18*, 361, 1972.
10. Shah, R. K., and A. L. London, *Laminar Flow Forced Convection in Ducts*, Academic Press, New York, 1978.
11. Colburn, A. P., *Trans. AIChE*, *29*, 174, 1933.
12. Dittus, F. W., and L. M. K. Boelter, University of California, Berkeley, Publications on Engineering, Vol. 2, 1930, p. 443.

13. Petukhov, B. S., in T. F. Irvine and J. P. Hartnett, Eds., *Advances in Heat Transfer*, Vol. 6, Academic Press, New York, 1970.
14. Kays, W. M., and H. C. Perkins, in *Handbook of Heat Transfer*, W. M. Rohsenow and J. P. Hartnett, Eds., McGraw-Hill, New York, 1972, Chapter 7.
15. Bergles, A. E., "Survey and Evaluation of Techniques to Augment Convective Heat and Mass Transfer," *Progress in Heat and Mass Transfer*, Vol. 1, Pergamon Press, New York, 1969, pp. 331–424.
16. Bergles, A. E., "Principles of Heat Transfer Augmentation," *Heat Exchangers, Thermal-Hydraulic Fundamentals and Design*, Hemisphere Publishing Corp., New York, 1981, pp. 819–842.
17. Bergles, A. E., V. Nirmalan, G. H. Junkhan, and R. L. Webb, *Bibliography on Augmentation of Convective Heat and Mass Transfer*, II, HTL-31, ISU-ERI-Ames-84257, 1983.
18. Skupinski, E. S., J. Tortel, and L. Vautrey, *Int. J. Heat Mass Transfer*, *8*, 937, 1965
19. Seban, R. A., and T. T. Shimazaki, *Trans. ASME*, *73*, 803, 1951.
20. Stein, R., in J. P. Hartnett and T. F. Irvine, Eds., *Advances in Heat Transfer*, Vol. 3, Academic Press, New York, 1966.
21. Fox, R. W., and A. T. McDonald, *Introduction to Fluid Mechanics*, 3rd ed., Wiley, New York, 1978.
22. Shah, R. K., and A. L. London, *J. Heat Trans.*, *96*, 159, 1974.

PROBLEMS

8.1 Fully developed conditions are known to exist for water flowing through a 25-mm-diameter tube at 0.01 kg/s and 27°C. What is the maximum velocity of the water in the tube? What is the pressure gradient associated with the flow?

8.2 What is the pressure drop associated with water at 27°C flowing with a mean velocity of 0.2 m/s through a 600-m-long cast iron pipe of 0.15 m inside diameter?

8.3 Water at 27°C flows with a mean velocity of 1 m/s through a 1-km-long cast iron pipe of 0.25 m inside diameter.

(a) Determine the pressure drop over the pipe length and the corresponding pump power requirement, if the pipe surface is clean.

(b) The pipe surface roughness is increased by 25% due to contamination. What is the new pressure drop and pump power requirement.

8.4 Consider a 25-mm-diameter circular tube through which liquid mercury, water, or engine oil at 27°C may flow at a rate of 0.03 kg/s. Determine the velocity, the hydrodynamic entry length, and the thermal entry length for each of the fluids.

8.5 Compare the thermal and velocity entry lengths for oil, water, and mercury flowing through a 25-mm-diameter tube with a mean velocity and temperature of $u_m = 5$ mm/s and $T_m = 27$°C, respectively.

8.6 Water at a flow rate of 2 kg/s enters a long section of pipe with a temperature of 25°C and a pressure of 100 bars. The pipe wall is heated such that 10^5 W is transferred to the water as it flows through the pipe.

(a) If the water leaves the pipe with a pressure of 2 bars, what is its outlet temperature?

(b) What value of the outlet temperature would be obtained if Equation 8.37 were used for the calculation?

8.7 Water enters a tube at 27°C with a flow rate of 450 kg/h. The heat transfer from the tube wall to the fluid is given as q' (W/m) $= ax$, where the coefficient a is 20 W/m^2 and x (m) is the axial distance from the tube entrance. Beginning with a properly defined differential control volume in the tube, derive an expression for the temperature distribution $T_m(x)$ of the water. What is the outlet temperature of the water for a heated section 30 m long?

8.8 Consider the flow in a circular tube as shown. Within the test section length (between 1 and 2) a constant heat flux q_s'' is maintained.

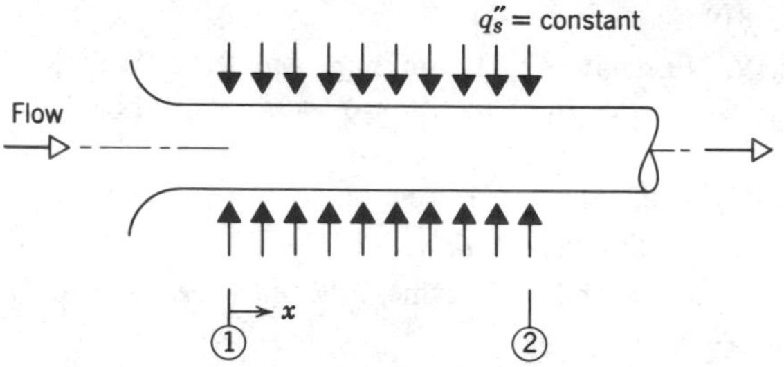

(a) For the two cases identified, sketch, qualitatively, the surface temperature $T_s(x)$ and the fluid mean temperature $T_m(x)$, as a function of distance along the test section x. Case A: flow is hydrodynamically and thermally fully developed. Case B: flow is not developed.

(b) Assuming that the surface flux q_s'' and the inlet mean temperature $T_{m,1}$ are identical for both cases, will the exit mean temperature $T_{m,2}$ for case A be greater than, equal to, or less than $T_{m,2}$ for case B? Briefly explain why.

8.9 Consider a cylindrical nuclear fuel rod of length L and diameter D that is encased within a concentric tube. Pressurized water flows through the annular region between the rod and the tube at a rate $\dot{m}$, and the outer surface of the tube is well insulated. Heat generation occurs within the fuel rod, and the volumetric generation rate is known to vary sinusoidally with distance along the rod. That is, $\dot{q}(x) = \dot{q}_o \sin(\pi x/L)$, where $\dot{q}_o$ (W/m^3) is a constant. A uniform convection coefficient h may be assumed to exist between the surface of the rod and the water.

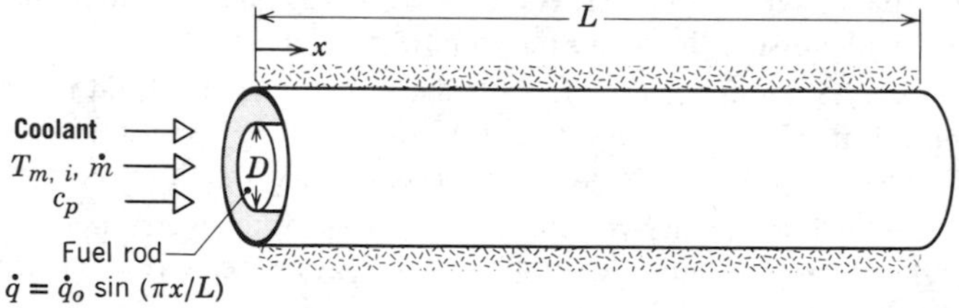

(a) Obtain expressions for the local heat flux $q''(x)$ and the total heat transfer q from the fuel rod to the water.

(b) Obtain an expression for the variation of the mean temperature $T_m(x)$ of the water with distance x along the tube.

(c) Obtain an expression for the variation of the rod surface temperature $T_s(x)$ with distance x along the tube. Develop an expression for the x location at which this temperature is maximized.

8.10 A flat plate solar collector is used to heat atmospheric air flowing through a rectangular channel, as shown. The bottom surface of the channel is well insulated, while the top surface is subjected to a uniform heat flux q''_o, which is due to the net effect of solar radiation absorption and heat exchange between the absorber and cover plates.

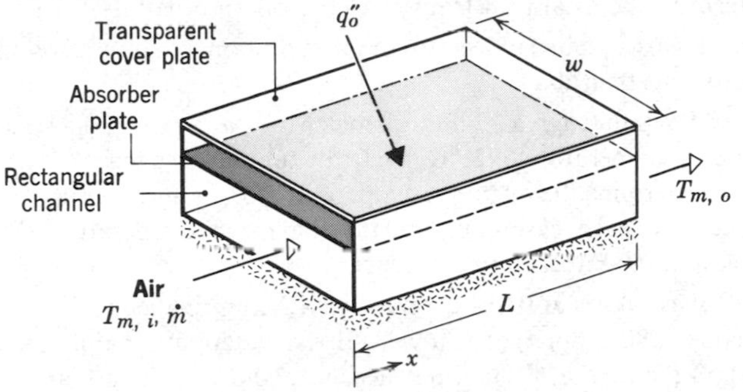

(a) Beginning with an appropriate differential control volume, obtain an equation that could be used to determine the mean air temperature $T_m(x)$ as a function of distance along the channel. Solve this equation to obtain an expression for the mean temperature of the air leaving the collector.

(b) With air inlet conditions of $\dot{m}=0.1$ kg/s and $T_{m,i}=40°C$, what is the air outlet temperature if $L=3$ m, $w=1$ m, and $q''_o=700$ W/m^2? The specific heat of air is $c_p=1008$ J/kg·K.

8.11 Atmospheric air enters the heated section of a circular tube at a flow rate of 0.005 kg/s and a temperature of 20°C. The tube is of diameter $D=50$ mm, and fully developed conditions with $h=25$ W/m^2·K exist over the entire length of $L=3$ m.

(a) For the case of uniform surface heat flux at $q''_s=1000$ W/m^2, determine the total heat transfer rate q and the mean temperature of the air leaving the tube $T_{m,o}$. What is the value of the surface temperature at the tube inlet $T_{s,i}$ and outlet $T_{s,o}$? Sketch the axial variation of T_s and T_m. On the same figure, sketch (qualitatively) the axial variation of T_s and T_m for the more realistic case in which the local convection coefficient varies with x.

(b) If the surface heat flux varies linearly with x, such that q''_s (W/m^2)$=500x$ (m), what are the values of q, $T_{m,o}$, $T_{s,i}$, and $T_{s,o}$? Sketch the axial variation of T_s and T_m. On the same figure, sketch (qualitatively) the axial variation of T_s and T_m for the more realistic case in which the local convection coefficient varies with x.

8.12 Slug flow is an idealized tube flow condition for which the velocity is assumed to be uniform over the entire tube cross section. For the case of laminar slug flow with a uniform surface heat flux, determine the form of the fully developed temperature distribution $T(r)$ and the Nusselt number Nu_D.

8.13 Engine oil at a rate of 0.02 kg/s flows through a 3-mm-diameter tube 30 m long. The oil has an inlet temperature of 60°C, while the tube wall temperature is maintained at 100°C by steam condensing on its outer surface.

(a) Estimate the average heat transfer coefficient for internal flow of the oil.

(b) Determine the outlet temperature of the oil.

8.14 Oil is heated by flowing through a circular tube of diameter $D=50$ mm and length $L=25$ m and whose outer surface is maintained at 150°C. If the flow rate and inlet temperature of the oil are 0.5 kg/s and 20°C, respectively, what is the oil outlet temperature? What is the total heat transfer rate for the tube?

8.15 Engine oil flows through a 25-mm-diameter, 10-m-long tube at a rate of 0.5 kg/s. The oil enters the tube at 25°C, while the tube surface is maintained at 100°C.

(a) Determine the total heat transfer to the oil and the oil outlet temperature.

(b) Repeat part a, but subject to the assumption of fully developed conditions throughout the tube.

8.16 Engine oil flows through a 25-mm-diameter tube at a rate of 0.5 kg/s. The oil enters the tube at a temperature of 25°C, while the tube surface temperature is maintained at 100°C. Determine the oil outlet temperature for a 5-m-long tube and for a 100-m-long tube. For each case compare the log mean temperature difference to the arithmetic mean temperature difference.

8.17 Ethylene glycol flows at 0.01 kg/s through a 3-mm-diameter, thin-walled tube. The tube is coiled and submerged in a well-stirred water bath maintained at 25°C. If the fluid enters the tube at 85°C, what heat rate and tube length are required for the fluid to leave at 35°C?

8.18 A 25-mm-diameter circular tube, whose outer surface is maintained at 100°C, is used to heat water flowing at 1 kg/s from 30 to 70°C. How long must the tube be?

8.19 Water flowing at 2 kg/s through a 40-mm-diameter tube is to be heated from 25 to 75°C by maintaining the tube surface temperature at 100°C. What is the required tube length?

8.20 Water flows at 2 kg/s through a 40-mm-diameter tube 4 m long. If the water enters the tube at 25°C and the surface temperature is 90°C, what is the temperature of the water as it leaves the tube? What is the rate of heat transfer to the water?

8.21 A thick-walled, stainless steel (AISI 316) pipe of inside and outside diameters $D_i=20$ mm and $D_o=40$ mm is heated electrically to provide a uniform heat generation rate of $\dot{q}=10^6$ W/m^3. The outer surface of the pipe is insulated, while water flows through the pipe at a rate of $\dot{m}=0.1$ kg/s.

(a) If the water inlet temperature is $T_{m,i}=20$°C and the desired outlet temperature is $T_{m,o}=40$°C, what is the required pipe length?

(b) What are the location and value of the maximum pipe temperature?

8.22 Atmospheric air enters a 10-m-long, 150-mm-diameter uninsulated heating duct at 60°C and 0.04 kg/s. If the duct surface temperature is approximately constant at $T_s=15$°C, what is the outlet air temperature and what is the duct heat loss?

8.23 Liquid mercury at 0.5 kg/s is to be heated from 300 to 400 K by passing it through a 50-mm-diameter tube whose surface is maintained at 450 K. Calculate the required tube length by using an appropriate liquid metal convection heat transfer correlation. Compare your result with that which would have been obtained by using a correlation appropriate for $Pr \gtrsim 0.7$.

8.24 The surface of a 50-mm-diameter, thin-walled tube is maintained at 100°C. In one case air is in cross flow over the tube with temperature and velocity of 25°C and 30 m/s, respectively. In another case air is in fully developed flow through the tube with a temperature of 25°C and a mean velocity of 30 m/s. Compare the heat flux from the tube to the air for the two cases.

8.25 A device that recovers heat from high temperature combustion products involves passing the combustion gas between parallel plates, each of which is maintained at 350 K by water flow on the opposite surface. The plate separation is 40 mm, and the gas flow is fully developed. The gas may be assumed to have the properties of atmospheric air, and its mean temperature and velocity are 1000 K and 60 m/s, respectively.

(a) What is the heat flux at the plate surface?

(b) If a third plate, 20 mm thick, is suspended midway between the original plates, what is the surface heat flux for the original plates? Assume the temperature and *flow rate* of the gas to be unchanged and radiation effects to be negligible.

8.26 Consider a horizontal, thin-walled circular tube of diameter $D = 0.025$ m submerged in a container of *n*-octadecane (paraffin), which is used to store thermal energy. As hot water flows through the tube, heat is transferred to the paraffin, converting it from the solid to liquid state at the phase change temperature of $T_\infty = 27.4°\text{C}$.

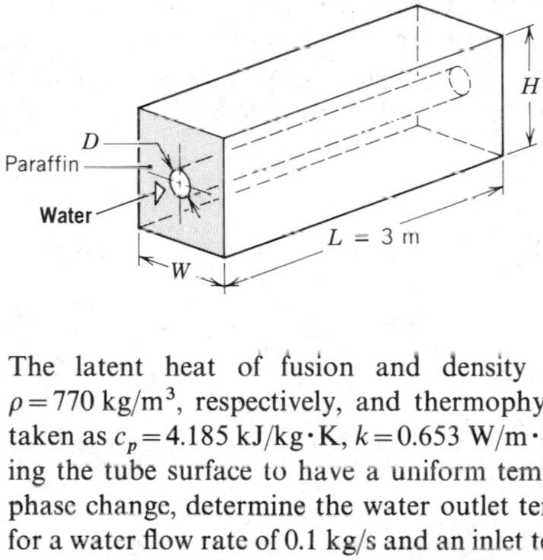

The latent heat of fusion and density of paraffin are $h_{sf} = 244$ kJ/kg and $\rho = 770$ kg/m^3, respectively, and thermophysical properties of the water may be taken as $c_p = 4.185$ kJ/kg·K, $k = 0.653$ W/m·K, and $\mu = 467 \times 10^{-6}$ kg/s·m. Assuming the tube surface to have a uniform temperature corresponding to that of the phase change, determine the water outlet temperature and total heat transfer rate for a water flow rate of 0.1 kg/s and an inlet temperature of 60°C. If $H = W = 0.25$ m, how long would it take to completely liquefy the paraffin, starting from an initial state for which all the paraffin is solid and at 27.4°C?

8.27 In a gas-fired water heater, heat is transferred from high temperature products of combustion flowing through long tubes to water moving over the tubes.

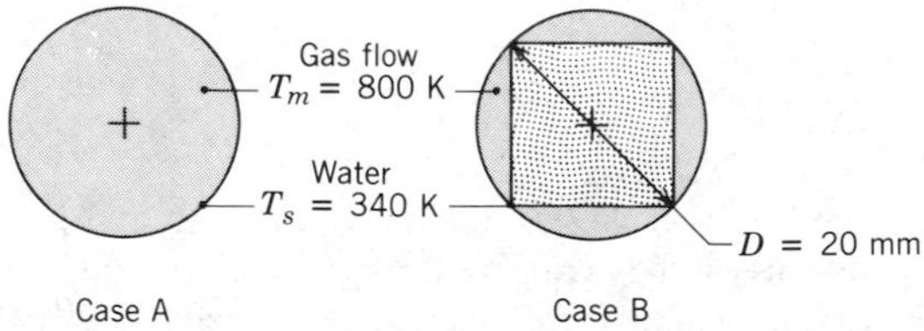

(a) Operating the heater in the conventional fashion (Case A), gas flow through the tubes is fully developed and unobstructed. If the gas flow rate and temperature are 0.01 kg/s and 800 K, respectively, and the surface temperature is 340 K, what is the rate of heat transfer to the water per unit length of tube? The gas properties may be approximated to be those of air at 1 atm.

(b) It is proposed to increase heater performance by inserting a square rod in the tubes (Case B). For the conditions of Case A ($\dot{m}=0.01$ kg/s, $T_m=800$ K, and $T_s=340$ K), what is the rate of heat transfer to the water per unit length of tube?

8.28 Freon is being transported at 0.1 kg/s through a Teflon tube of inside diameter $D_i=25$ mm and outside diameter $D_o=28$ mm, while atmospheric air at $V=25$ m/s and 300 K is in cross flow over the tube. What is the heat transfer per unit length of tube to Freon at 240 K?

8.29 Oil at 150°C flows slowly through a long, thin-walled pipe of 30 mm inner diameter. The pipe is suspended in a room in which the air temperature is 20°C and the convection coefficient at the outer tube surface is 11 W/m$^2\cdot$K. Estimate the heat loss per unit length of tube.

8.30 Exhaust gases from a wire processing oven are discharged into a tall stack, and the gas and stack surface temperatures at the outlet of the stack must be estimated. Knowledge of the outlet gas temperature $T_{m,o}$ is useful for predicting the dispersion of effluents in the thermal plume, while knowledge of the outlet stack surface temperature $T_{s,o}$ provides an indication of whether condensation of the gas products will occur.

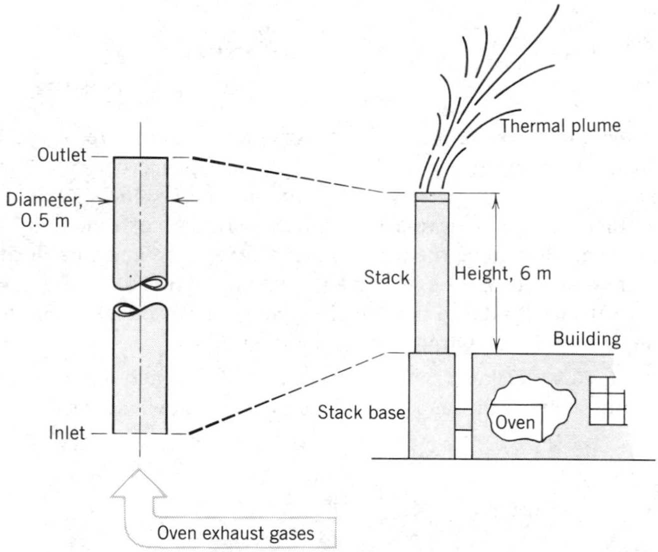

The thin-walled, cylindrical stack has diameter and height of 0.5 and 6.0 m, respectively. The exhaust gas flow rate is 0.5 kg/s, and the inlet temperature is 600°C. The ambient air temperature and wind velocity are 4°C and 5 m/s, respectively. Approximating the thermophysical properties of the gas as those of atmospheric air, estimate the outlet gas and stack surface temperatures for the given conditions.

8.31 Consider a thin-walled, metallic tube of length $L=1$ m and inside diameter $D_i=3$ mm. Water enters the tube at $\dot{m}=0.015$ kg/s and $T_{m,i}=97°C$.

(a) What is the outlet temperature of the water if the tube surface temperature is maintained at 27°C?

(b) If a 0.5-mm-thick layer of insulation of $k=0.05$ W/m·K is applied to the tube and its outer surface is maintained at 27°C, what is the outlet temperature of the water?

(c) If the outer surface of the insulation is no longer maintained at 27°C but is allowed to exchange heat by free convection with ambient air at 27°C, what is the outlet temperature of the water? The free convection heat transfer coefficient is 5 W/m²·K.

8.32 The problem of heat losses from a fluid moving through a buried pipeline has received considerable attention. Practical applications include the Alaskan pipeline, as well as power plant steam and water distribution lines. Consider a steel pipe of diameter D that is used to transport oil flowing at a rate $\dot{m}_o$ through a cold region. The pipe is covered with a layer of insulation of thickness t and thermal conductivity k_i and is buried in soil to a depth z (distance from the soil surface to the pipe centerline). Each section of pipe is of length L and extends between pumping stations in which the oil is heated to ensure low viscosity, and hence low pump power requirements. The temperature of the oil entering the pipe from a pumping station and the temperature of the ground above the pipe are designated as $T_{m,i}$ and T_s, respectively, and are presumed to be known.

Consider conditions for which the oil (o) properties may be approximated as $\rho_o=900$ kg/m³, $c_{p,o}=2000$ J/kg·K, $\nu_o=8.5\times10^{-4}$ m²/s, $k_o=0.140$ W/m·K, $Pr_o=10^4$; the oil flow rate is $\dot{m}_o=500$ kg/s; and the pipe diameter is 1.2 m.

(a) Expressing your results in terms of D, L, z, t, $\dot{m}_o$, $T_{m,i}$, and T_s, as well as the appropriate oil (o), insulation (i), and soil (s) properties, obtain all the expressions needed to estimate the temperature $T_{m,o}$ of the oil leaving the pipe.

(b) If $T_s=-40°C$, $T_{m,i}=120°C$, $t=0.15$ m, $k_i=0.05$ W/m·K, $k_s=0.5$ W/m·K, $z=3$ m, and $L=100$ km, what is the value of $T_{m,o}$? What is the total rate of heat transfer from a section of the pipeline?

8.33 Water flows at 0.25 kg/s through a thin-walled, 40-mm-diameter tube that is 4 m long. The water enters at 30°C and is heated by hot gases moving in cross flow over the tube with $V=100$ m/s and $T_\infty=225°C$. Estimate the outlet temperature of the water. The gas properties may be approximated to be those of atmospheric air.

8.34 You are designing an operating room heat exchange device that cools blood (bypassed from a human) from 40 to 30°C by passing the fluid through a coiled tube sitting in a vat of water–ice mixture. The volumetric flow rate (V) is 10^{-4} m³/min; the tube diameter (D) is 2.5 mm; and the $T_{m,i}$ and $T_{m,o}$ represent the inlet and outlet temperatures of the blood, respectively.

(a) At what temperature would you evaluate the fluid properties in determining $\bar{h}$ for the entire tube length?

(b) If the properties of blood evaluated at the temperature for part a are: $\rho=1000$ kg/m³, $\nu=7\times10^{-7}$ m²/s, $k=0.5$ W/m·K, and $c_p=4.0$ kJ/kg·K, what is the Prandtl number for the blood?

(c) Is the blood flow laminar or turbulent?

(d) Neglecting all entrance effects and assuming fully developed conditions, calculate the value of $\bar{h}$ for heat transfer from the blood.

(e) What is the total heat rate lost from the blood as it passes through the tube?

(f) When free convection effects on the outside of the tube are included, the average overall heat transfer coefficient $\bar{U}$ between the blood and the ice–water mixture can be approximated as 300 $W/m^2 \cdot K$. Determine the tube length L required to obtain the outlet temperature $T_{m,o}$.

8.35 Pressurized water at $T_{m,i} = 200°C$ is pumped at $\dot{m} = 2$ kg/s from a power plant to a nearby industrial user through a thin-walled, round pipe of inside diameter $D = 1$ m. The pipe is covered with a layer of insulation of thickness $t = 0.15$ m and thermal conductivity $k = 0.05$ W/m·K. The pipe, which is of length $L = 500$ m, is exposed to a cross flow of air at $T_\infty = -10°C$ and $V = 4$/ms. Obtain a differential equation that could be used to solve for the variation of the mixed mean temperature of the water $T_m(x)$ with the axial coordinate. As a first approximation, the internal flow may be assumed to be fully developed throughout the pipe. Express your results in terms of $\dot{m}$, V, T_∞, D, t, k, and appropriate water (w) and air (a) properties. Evaluate the heat loss per unit length of the pipe at the inlet. What is the mean temperature of the water at the outlet?

8.36 Air at 3×10^{-4} kg/s and 27°C enters a rectangular duct that is 1 m long and 4 mm × 16 mm on a side. A uniform heat flux of 600 W/m^2 is imposed on the duct surface. What is the temperature of the air and of the duct surface at the outlet?

8.37 Air at 4×10^{-4} kg/s and 27°C enters a triangular duct that is 20 mm on a side and 2 m long. The duct surface is maintained at 100°C. Assuming fully developed flow throughout the duct, determine the air outlet temperature.

8.38 Air at 1 atm and 285 K enters a 2-m-long rectangular duct with cross section 75 mm × 150 mm. The duct is maintained at a constant surface temperature of 400 K, and the air mass flow rate is 0.10 kg/s. Determine the heat transfer rate from the duct to the air and the air outlet temperature.

8.39 Referring to Figure 8.9, consider conditions in an annulus whose outer surface is insulated ($q''_o = 0$) and which has a uniform heat flux q''_i at the inner surface. Fully developed, laminar flow may be assumed to exist.

(a) Determine the velocity profile $u(r)$ in the annular region.

(b) Determine the temperature profile $T(r)$ and obtain an expression for the Nusselt number Nu_i associated with the inner surface.

8.40 Consider a concentric tube annulus for which the inner and outer diameters are 25 and 50 mm. Water enters the annular region at 0.04 kg/s and 25°C. If the inner tube wall is heated electrically at a rate (per unit length) of $q' = 4000$ W/m, while the outer tube wall is insulated, how long must the tubes be for the water to achieve an outlet temperature of 85°C? What is the inner tube surface temperature at the outlet, where fully developed conditions may be assumed?

8.41 It is common practice to recover waste heat from oil- or gas-fired furnaces by using the exhaust gases to preheat the combustion air. A device commonly used for this purpose consists of a concentric pipe arrangement for which the exhaust gases are passed through the inner pipe, while the cooler combustion air flows through an annular passage around the pipe.

Consider conditions for which there is a uniform heat transfer rate per unit length, $q'_i = 1.25 \times 10^5$ W/m, from the exhaust gases to the pipe inner surface, while air flows through the annular passage at a rate of $\dot{m}_a = 2.1$ kg/s. The thin-walled inner pipe is of diameter $D_i = 2$ m, while the outer pipe, which is well insulated from

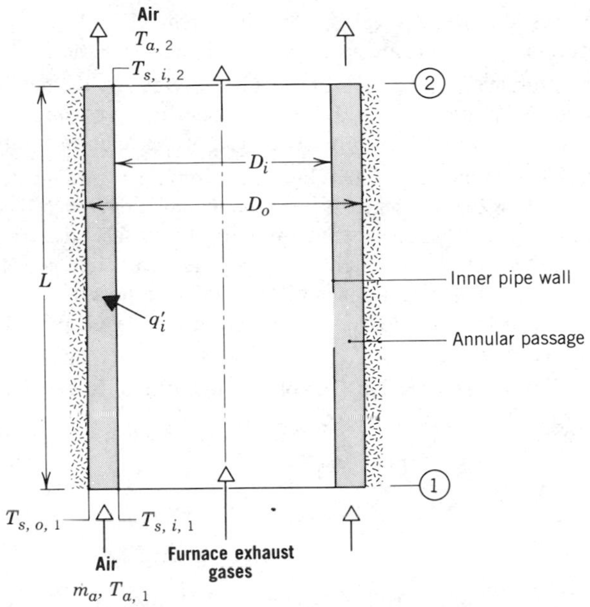

the surroundings, is of diameter $D_o = 2.05$ m. The air properties may be taken to be $c_p = 1030$ J/kg·K, $\mu = 270 \times 10^{-7}$ N·s/m^2, $k = 0.041$ W/m·K, and $Pr = 0.68$.

(a) If air enters at $T_{a,1} = 300$ K and $L = 7$ m, what is the air outlet temperature, $T_{a,2}$?

(b) If the air flow is fully developed throughout the annular region, what is the temperature of the inner pipe at the inlet ($T_{s,i,1}$) and outlet ($T_{s,i,2}$) sections of the device? What is the outer surface temperature $T_{s,o,1}$ at the inlet?

8.42 An industrial process heat application involves passing ethylene glycol through a concentric tube annulus, which is itself enclosed in a transparent glass tube. Under sunny skies, reflectors are situated such that the blackened outer surface of the annulus is uniformly irradiated to heat the ethylene glycol as it passes through the system. The space between the glass cover and the absorbing surface is evacuated, and the fluid flow is fully developed throughout the annulus. An electrically heated rod is used to heat the fluid during the evening hours, as well as during cloudy days.

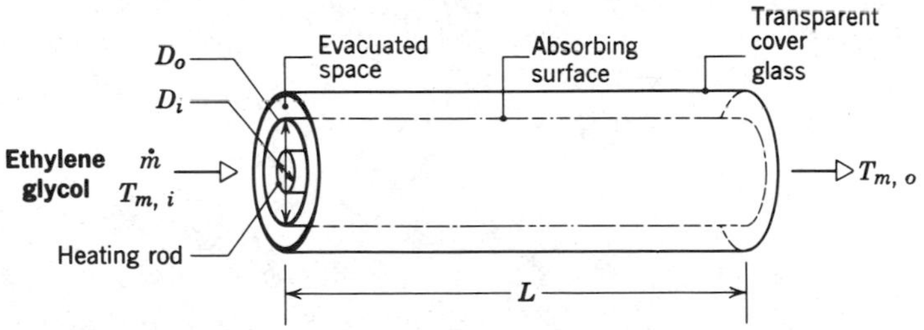

Consider a system for which $D_i = 10$ mm, $D_o = 100$ mm, and $L = 20$ m. The ethylene glycol enters at a rate of $\dot{m} = 0.25$ kg/s and a temperature of $T_{m,i} = 300$ K.

(a) Under sunny skies, there is a net radiation heat flux of $q_o'' = 3000$ W/m^2 uniformly distributed on the outer surface of the annulus, and the heating rod is not operational. What is the outlet fluid temperature $T_{m,o}$? What is the temperature of the absorbing surface at the inlet and outlet?

(b) What is the surface heat flux q_i'' that must be provided by the heating rod during the evening to maintain the fluid outlet temperature predicted in part a? What would be the surface temperature of the rod at the inlet and outlet?

8.43 Water at $\dot{m}=0.02$ kg/s and $T_{m,i}=20°C$ enters an annular region formed by an inner tube of diameter $D_i=25$ mm and an outer tube of diameter $D_o=100$ mm. Saturated steam flows through the inner tube, maintaining its surface at a uniform temperature of $T_{s,i}=100°C$, while the outer surface of the outer tube is well insulated. If fully developed conditions may be assumed throughout the annulus, how long must the system be to provide an outlet water temperature of 75°C? What is the heat flux from the inner tube at the outlet?

8.44 For the conditions of the preceding problem, how long must the annulus be if the water flow rate is 0.30 kg/s instead of 0.02 kg/s?

Nine

FREE CONVECTION

In preceding chapters we considered convection transfer in fluid flows that originate from an *external forcing* condition. For example, fluid motion can be induced by a fan or a pump, or it can result from propulsion of a solid through the fluid. In the presence of a temperature gradient, *forced-convection* heat transfer will then occur.

It is now time to consider situations for which there is no *forced* velocity, and yet convection currents exist within the fluid. Such situations are referred to as *free* or *natural convection*, and they originate when a *body force* acts on a fluid in which there are *density gradients*. The net effect is a *buoyancy force*, which induces free convection currents. In the most common case, the density gradient is due to a temperature gradient, and the body force is due to the gravitational field.

Since free convection flow velocities are generally much smaller than those associated with forced convection, the corresponding convection transfer rates are also smaller. It is perhaps tempting to therefore attach less significance to free convection processes. This temptation should be resisted. In many systems involving multimode heat transfer effects, free convection provides the largest resistance to heat transfer and therefore plays an important role in the design or performance of the system. Moreover, when it is desirable to minimize heat transfer rates or to minimize operating costs, free convection is often preferred to forced convection.

There are, of course, many applications. Free convection strongly influences heat transfer from pipes and transmission lines, as well as from various electronic devices. It is important in transferring heat from electric baseboard heaters or steam radiators to room air and in dissipating heat from the coil of a refrigeration unit to the surrounding air. It is also relevant to the environmental sciences, where it is responsible for oceanic and atmospheric motions, as well as related heat transfer processes.

9.1 PHYSICAL CONSIDERATIONS

In free convection fluid motion is due to buoyancy forces within the fluid, while in forced convection it is externally imposed. *Buoyancy is due to the combined presence of a fluid density gradient* and *a body force that is proportional to density.* In practice the body force is usually *gravitational*, although it may be a centrifugal force in rotating fluid machinery or a Coriolis force in atmospheric and oceanic rotational motions. There are also several ways in which a mass density gradient may arise in a fluid, but in the most common situation it is due to the presence of a temperature gradient. We know that the density of gases and liquids depends on temperature, generally decreasing (due to fluid expansion) with increasing temperature ($\partial\rho/\partial T < 0$).

In this text we focus on free convection problems in which the density gradient is due to a temperature gradient and the body force is gravitational. However, the presence of a fluid density gradient in a gravitational field does not ensure the existence of free convection currents. Consider the conditions of Figure 9.1. A fluid is enclosed by two large, horizontal plates of different temperature ($T_1 \neq T_2$). In case (*a*) the temperature of the lower plate exceeds that of the upper plate, and the density decreases in the direction of the gravitational force. This condition is *unstable*, and free convection currents must exist. The gravitational force on the denser fluid in the upper layers exceeds that acting on the lighter fluid in the lower layers, and the designated circulation pattern will exist. The heavier fluid will descend, being warmed in the process, while the lighter fluid will rise, cooling as it moves. However, this condition does not characterize case (*b*), for which $T_1 > T_2$ and the density no longer decreases in the direction of the gravitational force. Conditions are now *stable*, and there is no bulk fluid motion. In case (*a*) heat transfer occurs from the bottom to the top surface by free convection; for case (*b*) heat transfer (from top to bottom) occurs by conduction.

Free convection flows may be classified according to whether the flow is bounded by a surface. In the absence of an adjoining surface, *free boundary flows* may occur in the form of a *plume* or a *buoyant jet* (Figure 9.2). A plume is associated with fluid rising from a submerged heated object. Consider the heated wire of Figure 9.2*a*, which is immersed in an *extensive, quiescent* fluid.[1] Fluid that is heated by the wire rises due to buoyancy forces, entraining fluid from the quiescent region. Although the width of the plume increases with distance from the wire, the plume itself will eventually dissipate due to viscous effects and a reduction in the buoyancy force caused by cooling of the fluid in the plume. The distinction between a plume and a buoyant jet is generally made on the basis of the *initial* fluid velocity. This velocity is zero for the plume, but finite for the buoyant jet. Figure 9.2*b* shows a heated fluid being discharged as a horizontal jet into a quiescent medium of lower temperature. The vertical motion that the jet begins to assume is due to the buoyancy force. Such a condition occurs when warm water from the condenser of a central power station is discharged into a

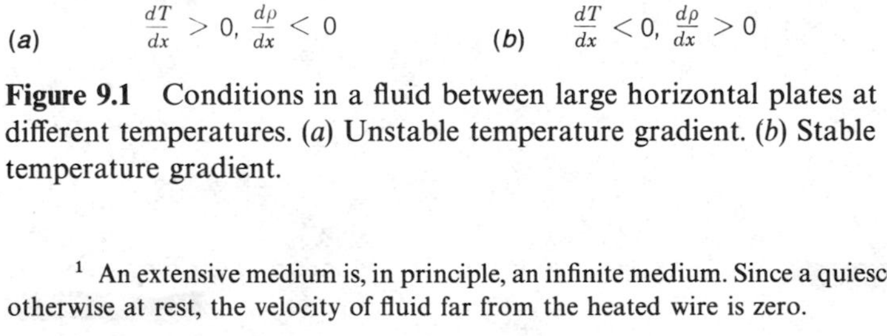

Figure 9.1 Conditions in a fluid between large horizontal plates at different temperatures. (*a*) Unstable temperature gradient. (*b*) Stable temperature gradient.

[1] An extensive medium is, in principle, an infinite medium. Since a quiescent fluid is one that is otherwise at rest, the velocity of fluid far from the heated wire is zero.

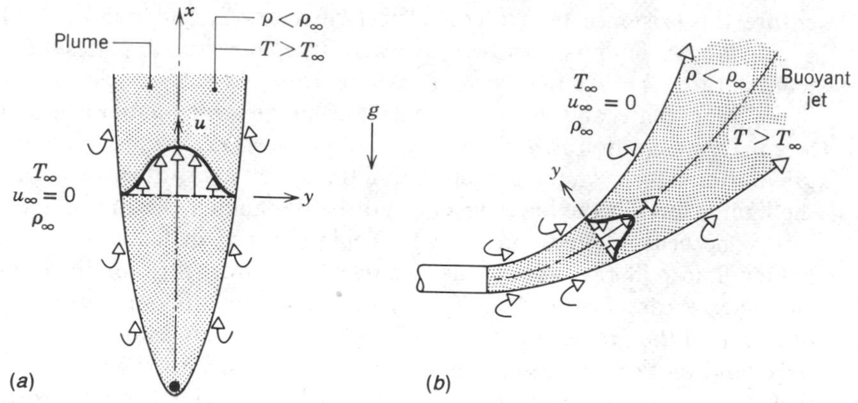

Figure 9.2 Buoyancy-driven free boundary layer flows in an extensive, quiescent medium. (*a*) Plume formation above a heated wire. (*b*) Buoyant jet associated with a heated discharge.

reservoir of cooler water. Free boundary flows are discussed by Gebhart [1] and in considerable detail by Jaluria [2].

In this text we focus on free convection flows bounded by a surface, and a classical example relates to boundary layer development on a heated vertical plate (Figure 9.3). The plate is immersed in an extensive, quiescent fluid, and with $T_s > T_\infty$, the density of fluid close to the plate is less than that of fluid that is further removed. Buoyancy forces therefore induce a free convection boundary layer in which the heated fluid rises vertically, entraining fluid from the quiescent region. The resulting velocity distribution is unlike that associated with forced-convection boundary layers. In particular the velocity is zero as $y \rightarrow \infty$, as well as at $y=0$. A free convection boundary layer also develops if $T_s < T_\infty$. In this case, however, fluid motion is downward.

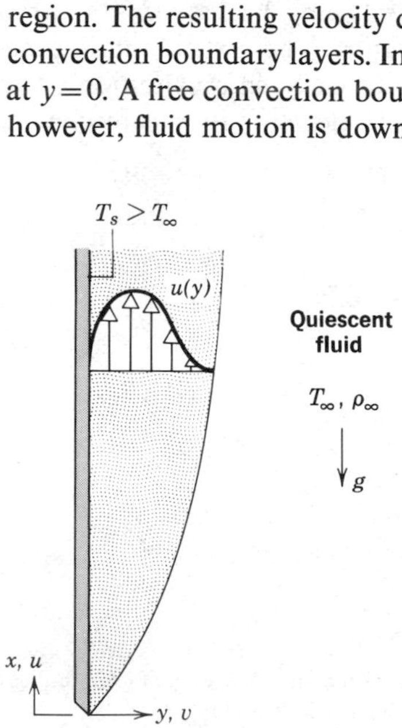

Figure 9.3 Boundary layer development on a heated vertical plate.

9.2 THE GOVERNING EQUATIONS

As for forced convection, the equations that describe momentum and energy transfer in free convection originate from the related conservation principles. Moreover, the specific processes related to free convection are much like those that dominate in forced convection. Inertia and viscous forces remain important, as does energy transfer by advection and diffusion. The major difference between the two flows is that, in free convection, a major role is played by buoyancy forces. It is such forces that, in fact, sustain the flow.

Consider a laminar boundary layer flow (Figure 9.3) that is driven by buoyancy forces. Assume steady, two-dimensional, constant property conditions in which the gravity force acts in the negative x direction. Also, with one exception, assume the fluid to be incompressible. The exception involves accounting for the effect of variable density in the buoyancy force (the so-called *Boussinesq approximation*), since it is this variation that induces fluid motion. Finally, assume that the boundary layer approximations are valid.

With the foregoing simplifications the x-momentum equation, (6.17), reduces to the boundary layer equation (6.31), except that the body force term X is retained. If the only contribution to this force is made by gravity, the body force per unit volume is $X=-\rho g$, where g is the local acceleration due to gravity. The appropriate form of the x-momentum equation is then

$$u\frac{\partial u}{\partial x}+v\frac{\partial u}{\partial y}=-\frac{1}{\rho}\frac{\partial p}{\partial x}-g+\nu\frac{\partial^2 u}{\partial y^2} \tag{9.1}$$

The above equation may be couched in a more convenient form by first noting that, if there is no body force in the y direction, $(\partial p/\partial y)=0$ from the y-momentum equation (6.32). Hence the x-pressure gradient at any point *in* the boundary layer must equal the pressure gradient in the quiescent region *outside* the boundary layer. However, in this region $u=v=0$ and Equation 9.1 reduces to

$$\frac{\partial p}{\partial x}=-\rho_\infty g \tag{9.2}$$

Substituting Equation 9.2 into 9.1, we then obtain the following equation

$$u\frac{\partial u}{\partial x}+v\frac{\partial u}{\partial y}=\frac{g}{\rho}(\rho_\infty-\rho)+\nu\frac{\partial^2 u}{\partial y^2} \tag{9.3}$$

which must apply at every point in the free convection boundary layer.

The first term on the right-hand side of Equation 9.3 is the buoyancy force, and flow originates because the density ρ is a variable. The origin of this variation may be made more explicit by introducing the *volumetric thermal expansion coefficient*

$$\beta=-\frac{1}{\rho}\left(\frac{\partial \rho}{\partial T}\right)_p \tag{9.4}$$

This *thermodynamic* property of the fluid provides a measure of the amount by which the density changes in response to a change in temperature at constant pressure. If it is expressed in the following approximate form,

$$\beta \approx -\frac{1}{\rho}\frac{\rho_\infty-\rho}{T_\infty-T}$$

it follows that

$$(\rho_\infty-\rho)\approx\rho\beta(T-T_\infty)$$

Substituting into Equation 9.3, the x-momentum equation then becomes

$$u\frac{\partial u}{\partial x}+v\frac{\partial u}{\partial y}=g\beta(T-T_\infty)+\nu\frac{\partial^2 u}{\partial y^2} \tag{9.5}$$

where it is now apparent how the buoyancy force, which drives the flow, is related to the temperature difference.

Since buoyancy effects are confined to the momentum equation, the mass and energy conservation equations are unchanged from forced convection. Equations 6.30 and 6.33 may then be used to complete the problem formulation. The set of governing equations is then

$$\frac{\partial u}{\partial x}+\frac{\partial v}{\partial y}=0 \tag{9.6}$$

$$u\frac{\partial u}{\partial x}+v\frac{\partial u}{\partial y}=g\beta(T-T_\infty)+\nu\frac{\partial^2 u}{\partial y^2} \tag{9.7}$$

$$u\frac{\partial T}{\partial x}+v\frac{\partial T}{\partial y}=\alpha\frac{\partial^2 T}{\partial y^2} \tag{9.8}$$

Note that viscous dissipation has been neglected in the energy equation, (9.8), an assumption that is certainly reasonable for the small velocities associated with free convection. In the mathematical sense the appearance of the buoyancy term in Equation 9.7 complicates matters. No longer may the hydrodynamic problem, given by Equations 9.6 and 9.7, be uncoupled from and solved to the exclusion of the thermal problem, given by Equation 9.8. The solution to the momentum equation depends on knowledge of T and hence on the solution to the energy equation. Equations 9.6 to 9.8 are therefore strongly coupled and must be solved simultaneously.

Free convection effects obviously depend on the expansion coefficient β. The manner in which β is obtained depends on the nature of the fluid. For a perfect gas, $\rho=p/RT$ and

$$\beta=-\frac{1}{\rho}\left(\frac{\partial\rho}{\partial T}\right)_p=\frac{1}{\rho}\frac{p}{RT^2}=\frac{1}{T} \tag{9.9}$$

where T is the *absolute* temperature. For liquids and nonideal gases, β must be obtained from appropriate property tables (Appendix A).

9.3 SIMILARITY CONSIDERATIONS

Let us now consider the dimensionless parameters that govern free convective flow and heat transfer. As for forced convection (Chapter 6), the parameters may be obtained by nondimensionalizing the governing equations. Introducing

$$x^* \equiv \frac{x}{L} \qquad y^* \equiv \frac{y}{L}$$

$$u^* \equiv \frac{u}{u_0} \qquad v^* \equiv \frac{v}{u_0} \qquad T^* \equiv \frac{T - T_\infty}{T_s - T_\infty}$$

where L is a characteristic length and u_0 is an *arbitrary* reference velocity,[2] the x-momentum and energy equations, (9.7) and (9.8), reduce to

$$u^* \frac{\partial u^*}{\partial x^*} + v^* \frac{\partial u^*}{\partial y^*} = \frac{g\beta(T_s - T_\infty)L}{u_0^2} T^* + \frac{1}{Re_L} \frac{\partial^2 u^*}{\partial y^{*2}} \tag{9.10}$$

$$u^* \frac{\partial T^*}{\partial x^*} + v^* \frac{\partial T^*}{\partial y^*} = \frac{1}{Re_L Pr} \frac{\partial^2 T^*}{\partial y^{*2}} \tag{9.11}$$

The dimensionless parameter in the first term on the right-hand side of Equation 9.11 is a direct consequence of the buoyancy force. However, because it is expressed in terms of the unknown reference velocity u_0, it is inconvenient in its present form. It is therefore customary to work with an alternative form that is obtained from multiplying by $Re_L^2 = (u_0 L/\nu)^2$. The result is termed the *Grashof number*, Gr_L.

$$Gr_L \equiv \frac{g\beta(T_s - T_\infty)L}{u_0^2} \left(\frac{u_0 L}{\nu}\right)^2 = \frac{g\beta(T_s - T_\infty)L^3}{\nu^2} \tag{9.12}$$

The Grashof number plays the same role in free convection that the Reynolds number plays in forced convection. Recall that the *Reynolds number* provides a measure of the *ratio of* the *inertial to viscous forces* acting on a fluid element. In contrast, the *Grashof number* indicates the *ratio of* the *buoyancy force to* the *viscous force* acting on the fluid.

Although Equations 9.10 to 9.12 prompt us to expect heat transfer correlations of the form $Nu_L = f(Re_L, Gr_L, Pr)$, it is important to note that such correlations are pertinent only when forced- and free-convection effects are comparable. For such cases an external flow is superposed on the buoyancy-driven flow, and there exists a well-defined forced-convection velocity. Generally, the combined effects of free and forced convection must be considered when $(Gr_L/Re_L^2) \sim 1$. If the inequality $(Gr_L/Re_L^2) \ll 1$ is satisfied, free convection effects may be neglected and $Nu_L = f(Re_L, Pr)$. Conversely, if $(Gr_L/Re_L^2) \gg 1$, forced-

[2] Since freestream conditions are quiescent in free convection, there is no logical external reference velocity (V or u_∞), as in forced convection.

convection effects may be neglected and $Nu_L = f(Gr_L, Pr)$. In the strict sense, a free convection flow is induced solely by buoyancy forces, in which case there is no well-defined forced-convection velocity and $(Gr_L/Re_L^2) = \infty$.

9.4 LAMINAR FREE CONVECTION ON A VERTICAL SURFACE

Numerous solutions to the laminar free convection boundary layer equations have been obtained, and a special case that has received much attention involves free convection from an isothermal vertical surface in an extensive quiescent medium (Figure 9.3). For this geometry Equations 9.6 to 9.8 must be solved subject to boundary conditions of the form[3]

$$y = 0: \qquad u = v = 0 \qquad T = T_s$$

$$y \to \infty: \qquad u \to 0 \qquad T \to T_\infty$$

A similarity solution to the foregoing problem has been obtained by Ostrach [3]. The solution involves transforming variables by introducing a *similarity parameter* of the form

$$\eta \equiv \frac{y}{x}\left(\frac{Gr_x}{4}\right)^{1/4} \tag{9.13}$$

and representing the velocity components in terms of a stream function defined as

$$\psi(x,y) \equiv f(\eta)\left[4\nu\left(\frac{Gr_x}{4}\right)^{1/4}\right] \tag{9.14}$$

With the foregoing definition of the stream function, the x-velocity component may be expressed as

$$u = \frac{\partial \psi}{\partial y} = \frac{\partial \psi}{\partial \eta}\frac{\partial \eta}{\partial y} = 4\nu\left(\frac{Gr_x}{4}\right)^{1/4} f'(\eta)\frac{1}{x}\left(\frac{Gr_x}{4}\right)^{1/4}$$

$$= \frac{2\nu}{x} Gr_x^{1/2} f'(\eta) \tag{9.15}$$

where primed quantities indicate differentiation with respect to η. Hence $f'(\eta) \equiv df/d\eta$. Evaluating the y-velocity component $v = -\partial\psi/\partial x$ in a similar fashion and introducing the dimensionless temperature

$$T^* \equiv \frac{T - T_\infty}{T_s - T_\infty} \tag{9.16}$$

[3] The boundary layer approximations are assumed in using Equations 9.6 to 9.8. However, the approximations are valid only for $(Gr_x Pr) \gtrsim 10^4$. Below this value (close to the leading edge), the boundary layer thickness is too large relative to the characteristic length x to ensure the validity of the approximations.

the three original partial differential equations, (9.6) to (9.8), may then be reduced to two ordinary differential equations of the form

$$f''' + 3ff'' - 2(f')^2 + T^* = 0 \tag{9.17}$$

$$T^{*\prime\prime} + 3Pr\, f\, T^{*\prime} = 0 \tag{9.18}$$

where f and T^* are functions of only η and the double and triple primes, respectively, refer to second and third derivatives with respect to η. Note that f is the key dependent variable for the velocity boundary layer and that the continuity equation, (9.6), is automatically satisfied through introduction of the stream function.

The transformed boundary conditions required to solve the momentum and energy equations, (9.17) and (9.18), are of the form

$$\eta = 0: \qquad f = f' = 0 \qquad T^* = 1$$

$$\eta \to \infty: \qquad f' \to 0 \qquad T^* \to 0$$

A numerical solution has been obtained by Ostrach [3], and selected results are shown in Figure 9.4. Note that the x-velocity component u may readily be obtained from Figure 9.4a through the use of Equation 9.15. Note also that, through the definition of the similarity parameter η, Figure 9.4 may be used to obtain values of u and T for any value of x and y.

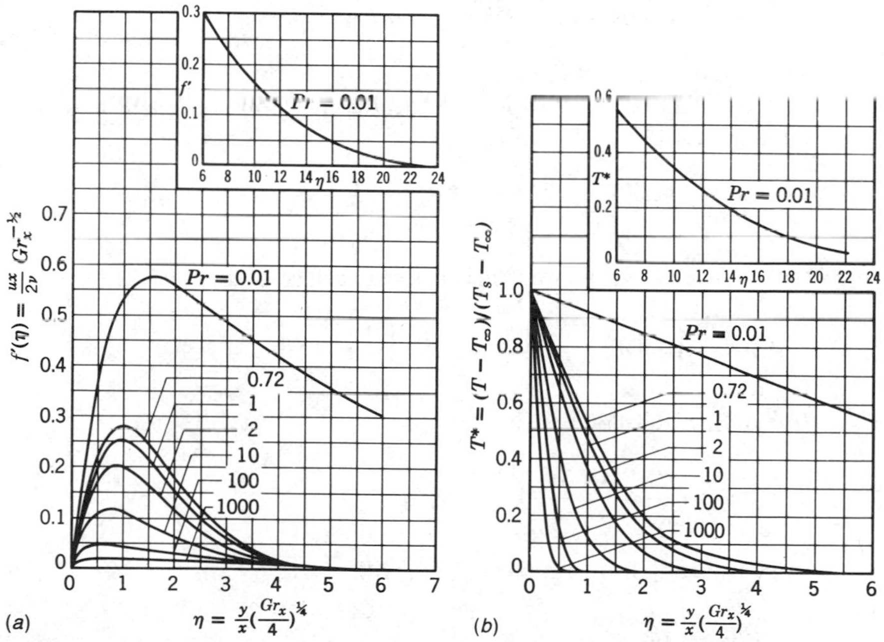

Figure 9.4 Laminar, free convection boundary layer conditions on an isothermal, vertical surface. (a) Velocity profiles. (b) Temperature profiles [3].

Figure 9.4*b* may also be used to infer the appropriate form of the heat transfer correlation. Using Newton's law of cooling for the local convection coefficient *h*, the local Nusselt number may be expressed as

$$Nu_x = \frac{hx}{k} = \frac{[q_s''/(T_s - T_\infty)]x}{k}$$

Using Fourier's law to obtain q_s'' and expressing the surface temperature gradient in terms of η, Equation 9.13, and T^*, Equation 9.16, it follows that

$$q_s'' = -k\frac{\partial T}{\partial y}\bigg|_{y=0} = -\frac{k}{x}(T_s - T_\infty)\left(\frac{Gr_x}{4}\right)^{1/4}\frac{dT^*}{d\eta}\bigg|_{\eta=0}$$

Hence

$$Nu_x = \frac{hx}{k} = -\left(\frac{Gr_x}{4}\right)^{1/4}\frac{dT^*}{d\eta}\bigg|_{\eta=0} = \left(\frac{Gr_x}{4}\right)^{1/4} g(Pr) \tag{9.19}$$

which acknowledges that the dimensionless temperature gradient at the surface is a function of the Prandtl number, $g(Pr)$. This dependence is evident from Figure 9.4*b* and has been determined numerically for selected values of Pr [3]. The results have been correlated to within 0.5% by an interpolation formula of the form [4]

$$g(Pr) = \frac{0.75Pr^{1/2}}{(0.609 + 1.221Pr^{1/2} + 1.238Pr)^{1/4}} \tag{9.20}$$

which applies for $0 \leq Pr \leq \infty$.

Using Equation 9.19 for the local convection coefficient and substituting for the local Grashof number

$$Gr_x = \frac{g\beta(T_s - T_\infty)x^3}{\nu^2}$$

the average convection coefficient for a surface of length L is then

$$\bar{h} = \frac{1}{L}\int_0^L h\,dx = \frac{k}{L}\left[\frac{g\beta(T_s - T_\infty)}{4\nu^2}\right]^{1/4} g(Pr)\int_0^L \frac{dx}{x^{1/4}}$$

Integrating, it follows that

$$\overline{Nu}_L = \frac{\bar{h}L}{k} = \frac{4}{3}\left(\frac{Gr_L}{4}\right)^{1/4} g(Pr) \tag{9.21}$$

or substituting from Equation 9.19, with $x = L$,

$$\overline{Nu}_L = \frac{4}{3}Nu_L \tag{9.22}$$

The foregoing results apply regardless of whether $T_s > T_\infty$ or $T_s < T_\infty$. If $T_s < T_\infty$, conditions are inverted from those of Figure 9.3. The leading edge is at the top of the plate, and positive x is defined in the direction of the gravity force.

9.5 THE EFFECTS OF TURBULENCE

It is important to note that free convection boundary layers are not restricted to laminar flow. All free convection flows of interest in this chapter originate from a *thermal instability*. That is, warmer, lighter fluid moves vertically upward relative to cooler, heavier fluid. However, as with forced convection, *hydrodynamic instabilities* may also arise. That is, disturbances in the flow may be amplified, leading to transition from laminar to turbulent flow. This process is shown schematically in Figure 9.5 for a heated vertical plate.

Transition in a free convection boundary layer depends on the relative magnitude of the buoyancy and viscous forces in the fluid. It is customary to correlate its occurrence in terms of the *Rayleigh number*, which is simply the product of the Grashof and Prandtl numbers. For vertical plates the critical Rayleigh number is

$$Ra_{x,c} = Gr_{x,c}Pr = \frac{g\beta(T_s - T_\infty)x^3}{\nu\alpha} \approx 10^9 \qquad (9.23)$$

An extensive discussion of stability and transition effects is provided by Gebhart [5, 6].

As in forced convection, transition to turbulence has a strong effect on heat transfer. Hence the results of the foregoing section apply only if $Ra_L \lesssim 10^9$. To obtain appropriate correlations for turbulent flow, heavy reliance is placed on experimental results.

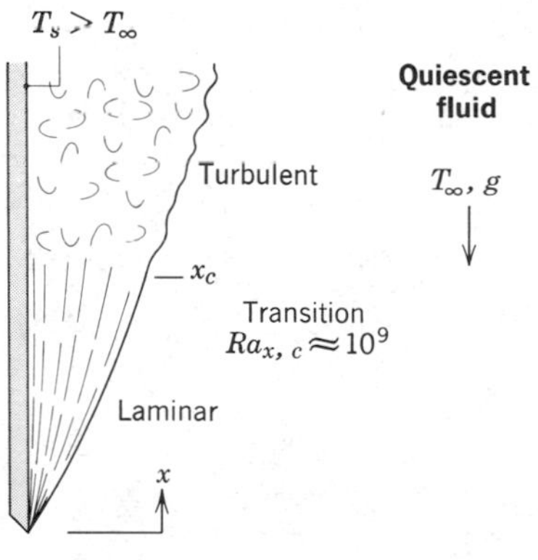

Figure 9.5 Free convection boundary layer transition on a vertical plate.

EXAMPLE 9.1

Consider a vertical plate that is 0.25 m long and at a temperature of 70°C. The plate is suspended in air that is at 25°C. Estimate the boundary layer thickness at the trailing edge of the plate if the air is quiescent. How does this thickness compare with that which would exist if the air were flowing over the plate at a freestream velocity of 5 m/s?

SOLUTION

Known: Vertical plate is in quiescent air at a lower temperature.

Find:

Boundary layer thickness at trailing edge. Compare with thickness corresponding to an air speed of 5 m/s.

Schematic:

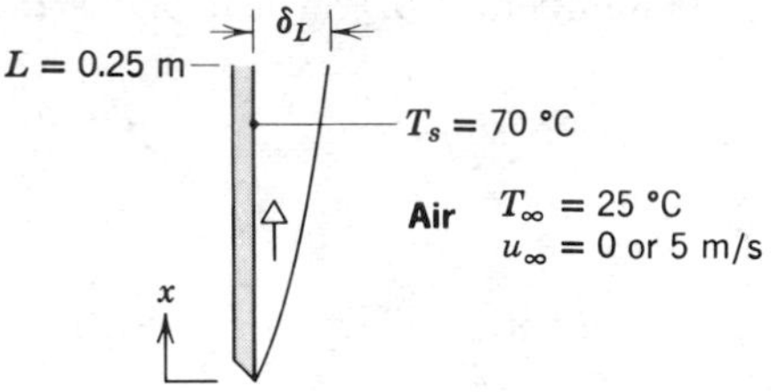

Assumptions:

1. Constant properties.
2. Buoyancy effects are negligible when $u_\infty = 5$ m/s.

Properties:

Table A.4, air ($T_f = 320.5$ K): $\nu = 17.95 \times 10^{-6}$ m²/s, $Pr = 0.7$, $\beta = T_f^{-1} = 3.12 \times 10^{-3}$ K^{-1}.

Analysis:

For the quiescent air, Equation 9.12 gives

$$Gr_L = \frac{g\beta(T_s - T_\infty)L^3}{\nu^2}$$

$$= \frac{9.8 \text{ m/s}^2 \times (3.12 \times 10^{-3} \text{ K}^{-1})(70 - 25)°\text{C}(0.25 \text{ m})^3}{(17.95 \times 10^{-6} \text{ m}^2/\text{s})^2} = 6.69 \times 10^7$$

Hence $Ra_L = Gr_L Pr = 4.68 \times 10^7$ and, from Equation 9.23, the free convection boundary layer is laminar. The analysis of Section 9.4 is therefore applicable. From the results of Figure 9.4, it follows that, for $Pr = 0.7$, $\eta \approx 6.0$ at the edge of the boundary layer; that is, at $y = \delta$. Hence

$$\delta_L \approx \frac{6L}{(Gr_L/4)^{1/4}} = \frac{6(0.25\text{ m})}{(1.67 \times 10^7)^{1/4}} = 0.024\text{ m} \qquad \triangleleft$$

For airflow at $u_\infty = 5$ m/s

$$Re_L = \frac{u_\infty L}{\nu} = \frac{(5\text{ m/s}) \times 0.25\text{ m}}{17.95 \times 10^{-6}\text{ m}^2/\text{s}} = 6.97 \times 10^4$$

and the boundary layer is laminar. Hence from Equation 7.17

$$\delta_L \approx \frac{5L}{Re_L^{1/2}} = \frac{5(0.25\text{ m})}{(6.97 \times 10^4)^{1/2}} = 0.0047\text{ m} \qquad \triangleleft$$

Comments:

1. Boundary layer thicknesses are typically larger for free convection than for forced convection.
2. $(Gr_L/Re_L^2) = 0.014 \ll 1$, and the assumption of negligible buoyancy effects for $u_\infty = 5$ m/s is justified.

9.6 EMPIRICAL CORRELATIONS: EXTERNAL FREE CONVECTION FLOWS

In this section we summarize appropriate empirical correlations that have been developed for common *immersed* (external flow) geometries. The correlations are suitable for most engineering calculations and are generally of the form

$$\overline{Nu}_L = \frac{\bar{h}L}{k} = C Ra_L^n \tag{9.24}$$

where the Rayleigh number,

$$Ra_L = Gr_L Pr = \frac{g\beta(T_s - T_\infty)L^3}{\nu\alpha} \tag{9.25}$$

is based on the characteristic length L of the geometry. Typically, $n = \frac{1}{4}$ and $\frac{1}{3}$ for laminar and turbulent flows, respectively. For turbulent flow it then follows that $\bar{h}_L$ is independent of L. Note that all properties are evaluated at the film temperature, $T_f \equiv (T_s + T_\infty)/2$. The specific correlations that follow are summarized in Table 9.2 (Section 9.6.5).

9.6.1 The Vertical Plate

Equations of the form given by Equation 9.24 have been developed for the vertical plate [7–9] and are plotted in Figure 9.6. The coefficient C and the exponent n depend on the Rayleigh number range; for Rayleigh numbers less than 10^4, the Nusselt number should be obtained directly from the figure.

More recently, Churchill and Chu [10] have recommended a correlation that may be applied over the *entire* range of Ra_L. It is of the form

$$\overline{Nu}_L = \left\{0.825 + \frac{0.387 Ra_L^{1/6}}{[1+(0.492/Pr)^{9/16}]^{8/27}}\right\}^2 \tag{9.26}$$

Although Equation 9.26 is suitable for most engineering calculations, slightly better accuracy may be obtained for laminar flow by using [10]

$$\overline{Nu}_L = 0.68 + \frac{0.670 Ra_L^{1/4}}{[1+(0.492/Pr)^{9/16}]^{4/9}} \qquad 0 < Ra_L < 10^9 \tag{9.27}$$

It is important to recognize that the foregoing results have been obtained for an isothermal plate (constant T_s). If the surface condition is, instead, one of uniform heat flux (constant q_s''), the temperature difference $(T_s - T_\infty)$ will vary with x, increasing from a value of zero at the leading edge. An approximate

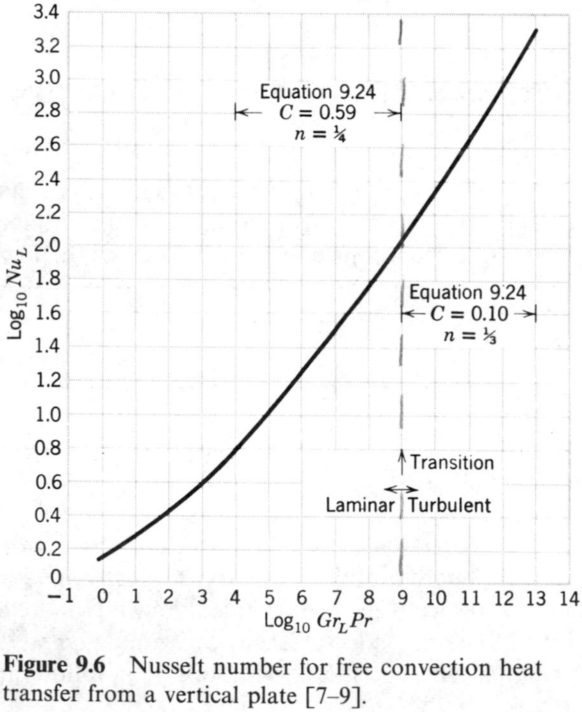

Figure 9.6 Nusselt number for free convection heat transfer from a vertical plate [7–9].

procedure for determining this variation may be based on results [10, 11] which show that $\overline{Nu}_L$ correlations obtained for the isothermal plate may still be used to an excellent approximation, if $\overline{Nu}_L$ and Ra_L are defined in terms of the temperature difference at the midpoint of the plate, $\Delta T_{L/2} = T_s(L/2) - T_\infty$. Hence, with $\bar{h} \equiv q_s''/\Delta T_{L/2}$, a correlation such as Equation 9.27 could be used (in a trial-and-error solution) to determine $\Delta T_{L/2}$ and thus the midpoint surface temperature $T_s(L/2)$. If it is assumed that $Nu_x \propto Ra_x^{1/4}$ over the entire plate, it follows that

$$\frac{q_s'' x}{k \Delta T} \propto \Delta T^{1/4} x^{3/4}$$

or

$$\Delta T \propto x^{1/5}$$

Hence, the temperature difference at any x is

$$\Delta T_x \approx \frac{x^{1/5}}{(L/2)^{1/5}} \Delta T_{L/2} = 1.15 \left(\frac{x}{L}\right)^{1/5} \Delta T_{L/2} \qquad (9.28)$$

A more detailed discussion of constant heat flux results is provided by Churchill [12].

The foregoing results may also be applied to *vertical* cylinders of height L, if the boundary layer thickness δ is much less than the cylinder diameter D. This condition is known to be satisfied [6, 13] when

$$\frac{D}{L} \gtrsim \frac{35}{Gr_L^{1/4}}$$

Cebeci [14] and Minkowycz and Sparrow [15] present results for slender, vertical cylinders not meeting the above condition, where transverse curvature influences boundary layer development and enhances the rate of heat transfer.

EXAMPLE 9.2

A glass-door firescreen, used to reduce exfiltration of room air through a chimney, has a height of 0.71 m and a width of 1.02 m and reaches a temperature of 232°C. If the room temperature is 23°C, estimate the convection heat rate from the fireplace to the room.

SOLUTION

Known: Glass screen situated in fireplace opening.

Find: Heat transfer by convection between screen and room air.

Schematic:

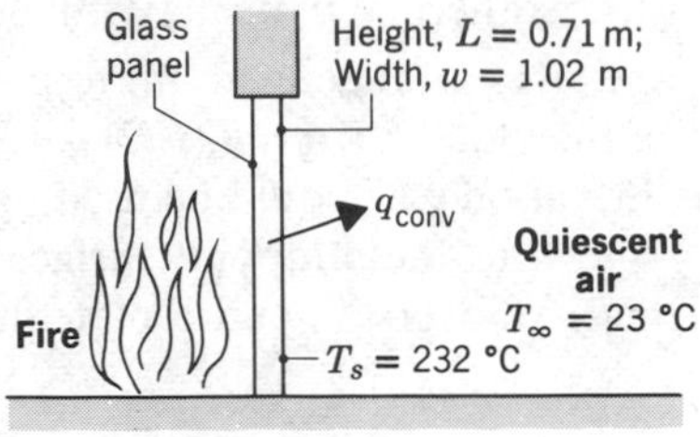

Assumptions:

1. Screen is at a uniform temperature T_s.
2. Room air is quiescent.

Properties:

Table A.4, air ($T_f = 400$ K): $k = 33.8 \times 10^{-3}$ W/m·K, $\nu = 26.4 \times 10^{-6}$ m²/s, $\alpha = 38.3 \times 10^{-6}$ m²/s, $Pr = 0.690$, $\beta = (1/T_f) = 0.0025$ K^{-1}.

Analysis:

The rate of heat transfer by free convection from the panel to the room is given by Newton's law of cooling

$$q = \bar{h}A_s(T_s - T_\infty)$$

where $\bar{h}$ may be obtained from knowledge of the Rayleigh number. Using Equation 9.25,

$$Ra_L = \frac{g\beta(T_s - T_\infty)L^3}{\alpha\nu}$$

$$= \frac{9.8\ \text{m/s}^2 \times 1/400\ \text{K}\ (232 - 23)^\circ\text{C} \times (0.71\ \text{m})^3}{38.3 \times 10^{-6}\ \text{m/s}^2 \times 26.4 \times 10^{-6}\ \text{m/s}^2} = 1.813 \times 10^9$$

and from Equation 9.23 it follows that transition to turbulence occurs on the panel. The appropriate correlation is then given by Equation 9.26.

$$\overline{Nu}_L = \left\{0.825 + \frac{0.387 Ra_L^{1/6}}{[1 + (0.492/Pr)^{9/16}]^{8/27}}\right\}^2$$

$$\overline{Nu}_L = \left\{0.825 + \frac{0.387(1.813 \times 10^9)^{1/6}}{[1 + (0.492/0.690)^{9/16}]^{8/27}}\right\}^2 = 147$$

Hence

$$\bar{h} = \frac{\overline{Nu}_L \cdot k}{L} = \frac{147 \times 33.8 \times 10^{-3}\ \text{W/m}\cdot\text{K}}{0.71\ \text{m}} = 7.00\ \text{W/m}^2\cdot\text{K}$$

and

$$q = 7.0\ \text{W/m}^2\cdot\text{K}\ (1.02\times 0.71)\ \text{m}^2\ (232-23)°\text{C} = 1060\ \text{W} \qquad \triangleleft$$

Comments:

1. If $\bar{h}$ were computed from Equation 9.24, with $C = 0.10$ and $n = \frac{1}{3}$, we would obtain $\bar{h} = 5.8\ \text{W/m}^2\cdot\text{K}$ and the heat transfer prediction would be approximately 20% lower than the foregoing result. This difference is within the uncertainty normally associated with using such correlations.
2. Radiation heat transfer effects are often significant relative to free convection. Using Equation 1.6 and assuming $\varepsilon = 1.0$ for the glass surface and $T_{sur} = 23°\text{C}$, the net rate of radiation heat transfer between the glass and the surroundings is

   $$q_{rad} = \varepsilon A_s \sigma (T_s^4 - T_{sur}^4)$$

   $$q_{rad} = 1(1.02\times 0.71)\ \text{m}^2 \times 5.67\times 10^{-8}\ \text{W/m}^2\cdot\text{K}^4\ (505^4 - 296^4)\ \text{K}^4$$

   $$q_{rad} = 2355\ \text{W}$$

   Hence in this case radiation heat transfer exceeds free convection heat transfer by more than a factor of 2.

9.6.2 The Horizontal Plate

For this geometry the specific form of the correlation depends on whether the plate is *warmer* or *cooler* than the surrounding fluid and on whether it is *facing upward* or *downward.* Although correlations suggested by McAdams [7] are widely used, improved accuracy may be obtained by altering the form of the characteristic length on which the correlations are based [16, 17]. In particular with the characteristic length defined as

$$L \equiv \frac{A_s}{P} \tag{9.29}$$

where A_s and P are the plate surface area and perimeter, respectively, the recommended correlations are

Upper Surface of Heated Plate or Lower Surface of Cooled Plate:

$$\overline{Nu}_L = 0.54 Ra_L^{1/4} \qquad 10^4 \lesssim \text{Ra}_L \lesssim 10^7 \tag{9.30}$$

$$\overline{Nu}_L = 0.15 Ra_L^{1/3} \qquad 10^7 \lesssim \text{Ra}_L \lesssim 10^{11} \tag{9.31}$$

Lower Surface of Heated Plate or Upper Surface of Cooled Plate:

$$\overline{Nu}_L = 0.27 Ra_L^{1/4} \qquad 10^5 \lesssim Ra_L \lesssim 10^{10} \tag{9.32}$$

EXAMPLE 9.3

Airflow through a long rectangular heating duct of width and height 0.75 and 0.30 m, respectively, maintains the outer duct surface at 45°C. If the duct is uninsulated and exposed to air at 15°C in the crawlspace beneath a home, what is the heat loss from the duct per meter of length?

SOLUTION

Known: Surface temperature of a long rectangular duct.

Find: Heat loss from duct per meter of length.

Schematic:

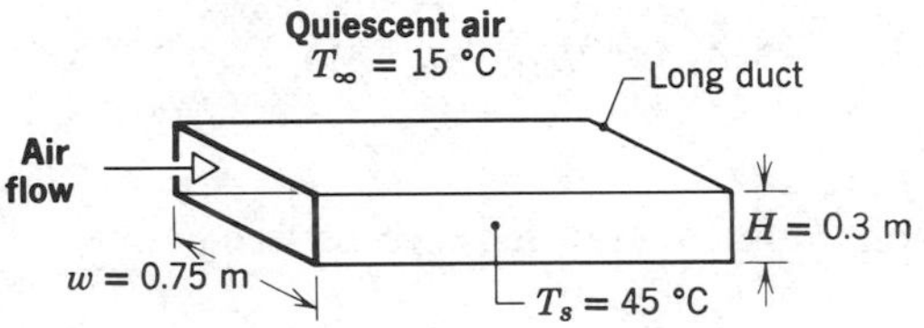

Assumptions:

1. Ambient air is quiescent.
2. Surface radiation effects are negligible.

Properties:

Table A.4, air ($T_f = 303$ K): $\nu = 16.2 \times 10^{-6}$ m^2/s, $\alpha = 22.9 \times 10^{-6}$ m^2/s, $k = 0.0265$ W/m·K, $\beta = 0.0033$ K^{-1}, $Pr = 0.71$.

Analysis:

Surface heat loss is by free convection from the vertical sides and the horizontal top and bottom. From Equation 9.25

$$Ra_L = \frac{g\beta(T_s - T_\infty)L^3}{\nu\alpha} = \frac{(9.8 \text{ m/s}^2)(0.0033 \text{ K}^{-1})(30 \text{ K})L^3(\text{m}^3)}{(16.2 \times 10^{-6} \text{ m/s}^2)(22.9 \times 10^{-6} \text{ m/s}^2)}$$

$$Ra_L = 2.62 \times 10^9 L^3$$

For the two sides, $L = H = 0.3$ m. Hence $Ra_L = 7.07 \times 10^7$. The free convection boundary layer is therefore laminar, and from Equation 9.27

$$\overline{Nu}_L = 0.68 + \frac{0.670 Ra_L^{1/4}}{[1 + (0.492/Pr)^{9/16}]^{4/9}}$$

The convection coefficient associated with the sides is then

$$\bar{h}_s = \frac{k}{H}\overline{Nu}_L$$

$$\bar{h}_s = \frac{0.0265\ \text{W/m}\cdot\text{K}}{0.3\ \text{m}}\left\{0.68 + \frac{0.670(7.07\times 10^7)^{1/4}}{[1+(0.492/0.71)^{9/16}]^{4/9}}\right\} = 4.23\ \text{W/m}^2\cdot\text{K}$$

For the top and bottom, $L=(A_s/P)\approx(w/2)=0.375$ m. Hence $Ra_L=1.38\times 10^8$, and from Equations 9.31 and 9.32, respectively,

$$\bar{h}_t = \frac{k}{w/2}\times 0.15 Ra_L^{1/3} = \frac{0.0265\ \text{W/m}\cdot\text{K}}{0.375\ \text{m}}$$

$$\times 0.15(1.38\times 10^8)^{1/3} = 5.47\ \text{W/m}^2\cdot\text{K}$$

$$\bar{h}_b = \frac{k}{w/2}\times 0.27 Ra_L^{1/4} = \frac{0.0265\ \text{W/m}\cdot\text{K}}{0.375\ \text{m}}$$

$$\times 0.27(1.38\times 10^8)^{1/4} = 2.07\ \text{W/m}^2\cdot\text{K}$$

The rate of heat loss per unit length of duct is then

$$q' = 2q'_s + q'_t + q'_b = (2\bar{h}_s\cdot H + \bar{h}_t\cdot w + \bar{h}_b\cdot w)(T_s - T_\infty)$$

$$q' = (2\times 4.23\times 0.3 + 5.47\times 0.75 + 2.07\times 0.75)(45-15)\ \text{W/m}$$

$$q' = 246\ \text{W/m}$$ ◁

Comments:

Even in this situation, radiation effects are significant. From Equation 1.6 with ε assumed to be unity and $T_{sur}=288$ K, $q'_{rad}=398$ W/m.

9.6.3 The Inclined Plate

As a first approximation, Churchill and Chu [10] recommended using Equations 9.26 and 9.27 for surfaces that are inclined up to $\theta=60°$ from the vertical. For laminar flow Equation 9.27 may be used if g is replaced by $g\cos\theta$ in evaluating Ra_L. Equation 9.26 may be used for turbulent flow without this modification. More accurate representations may be obtained from the literature [12, 18, 19].

9.6.4 The Long Horizontal Cylinder

This important geometry has been studied extensively, and many existing correlations have been reviewed by Morgan [20]. Based on this review Morgan suggests using an expression of the form

Table 9.1 Constants of Equation 9.33 for free convection on a horizontal circular cylinder [20]

Ra_D	C	n
$10^{-10}-10^{-2}$	0.675	0.058
$10^{-2}-10^{2}$	1.02	0.148
$10^{2}-10^{4}$	0.850	0.188
$10^{4}-10^{7}$	0.480	0.250
$10^{7}-10^{12}$	0.125	0.333

$$\overline{Nu}_D = \frac{\bar{h}D}{k} = CRa_D^n \qquad (9.33)$$

Long Horizontal Cylinder

where C and n are given in Table 9.1. Churchill and Chu [21] have recommended a single correlation for a wide Rayleigh number range.

$$\overline{Nu}_D = \left\{0.60 + \frac{0.387Ra_D^{1/6}}{[1+(0.559/Pr)^{9/16}]^{8/27}}\right\}^2 \qquad 10^{-5} < Ra_D < 10^{12} \qquad (9.34)$$

EXAMPLE 9.4

A horizontal, high-pressure steam pipe of 0.1 m outside diameter passes through a large room whose walls and air are at 23°C. The pipe has an outside surface temperature of 165°C and an emissivity of $\varepsilon = 0.85$. Estimate the heat loss from the pipe per unit length.

SOLUTION

Known:

Surface temperature of a horizontal steam pipe passing through a room.

Find: Heat loss from the pipe per unit length, q' (W/m).

Schematic:

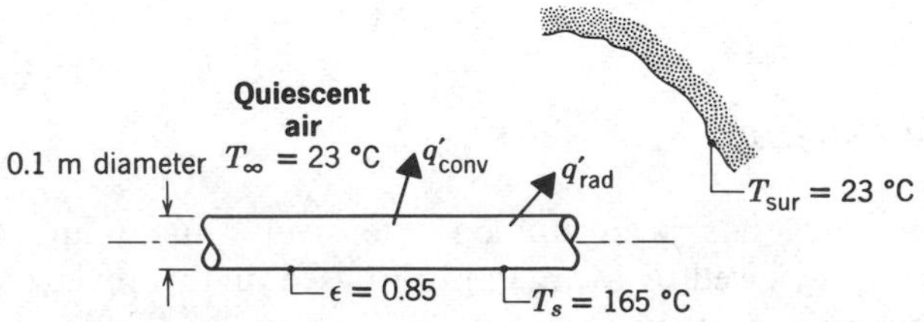

Assumptions:

1. Pipe surface area is small compared to surroundings.
2. Room air is quiescent.

Properties:

Table A.4, air ($T_f = 367$ K): $k = 0.0313$ W/m·K, $\nu = 22.8 \times 10^{-6}$ m^2/s, $\alpha = 32.8 \times 10^{-6}$ m^2/s, $Pr = 0.697$, $\beta = 2.725 \times 10^{-3}$ K^{-1}.

Analysis:

The total heat loss per unit length of pipe is

$$q' = q'_{conv} + q'_{rad} = \bar{h}\pi D(T_s - T_\infty) + \varepsilon\pi D\sigma(T_s^4 - T_{sur}^4)$$

The convection coefficient may be obtained from Equation 9.34

$$\overline{Nu}_D = \left\{0.60 + \frac{0.387 Ra_D^{1/6}}{[1 + (0.559/Pr)^{9/16}]^{8/27}}\right\}^2$$

where

$$Ra_D = \frac{g\beta(T_s - T_\infty)D^3}{\nu\alpha}$$

$$Ra_D = \frac{9.8 \text{ m/s}^2 \times 2.725 \times 10^{-3} \text{ 1/K } (165 - 23)°\text{C}(0.1 \text{ m})^3}{22.8 \times 10^{-6} \text{ m}^2/\text{s} \times 32.8 \times 10^{-6} \text{ m}^2/\text{s}} = 5.073 \times 10^6$$

Hence

$$\overline{Nu}_D = \left\{0.60 + \frac{0.387(5.073 \times 10^6)^{1/6}}{[1 + (0.559/0.697)^{9/16}]^{8/27}}\right\}^2 = 23.3$$

and

$$\bar{h} = \frac{k}{D}\overline{Nu}_D = \frac{0.0313 \text{ W/m·K}}{0.1 \text{ m}} \times 23.3 = 7.29 \text{ W/m}^2\text{·K}$$

The total heat loss is then

$$q' = 7.29 \text{ W/m}^2\text{·K } (\pi \times 0.1 \text{ m})(165 - 23)°\text{C}$$

$$+ 0.85(\pi \times 0.1 \text{ m})5.67 \times 10^{-8} \text{ W/m}^2\text{·K}^4 \ (438^4 - 296^4) \text{ K}^4$$

$$q' = (325 + 441) \text{ W/m} = 766 \text{ W/m}$$ ◁

Comments:

Equation 9.33 could also be used to estimate the Nusselt number, with the result that $\overline{Nu}_D = 22.8$.

Table 9.2 Summary of free convection empirical correlations for immersed geometries

GEOMETRY	RECOMMENDED CORRELATION	RESTRICTIONS
1. Vertical plates[a] 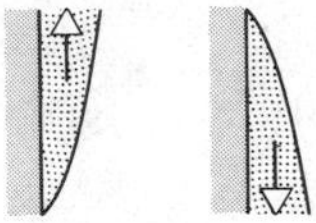	Equation 9.26	None
2. Horizontal plates (a) Hot surface up or cold surface down 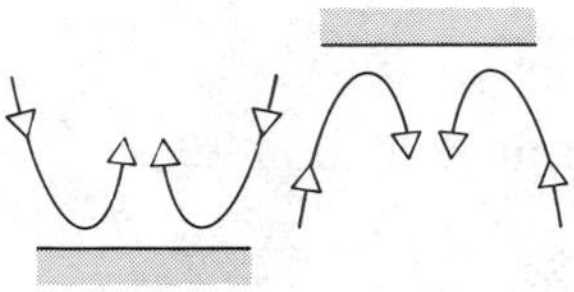	Equation 9.30 Equation 9.31	$10^4 \lesssim Ra_L \lesssim 10^7$ $10^7 \lesssim Ra_L \lesssim 10^{11}$
(b) Cold surface up or hot surface down 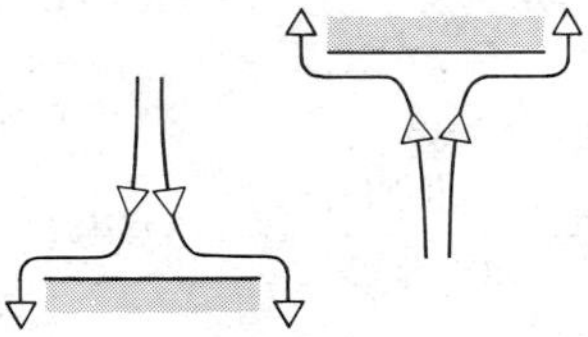	Equation 9.32	$10^5 \lesssim Ra_L \lesssim 10^{10}$
3. Inclined plates 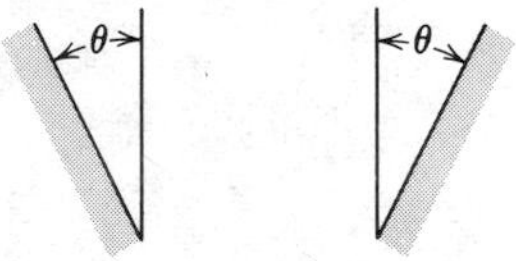	Equation 9.26 Equation 9.27	$Ra_L \gtrsim 10^9$ $\theta < 60°$ $Ra_L \lesssim 10^9$ $\theta < 60°$ $g \rightarrow g \cos\theta$
4. Horizontal cylinder 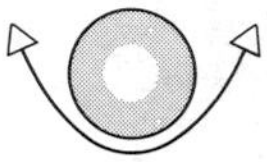	Equation 9.34	$10^{-5} < Ra_D < 10^{12}$
5. Sphere 	Equation 9.35	$Ra_D \lesssim 10^{11}$ $Pr \geq 0.7$

[a]The correlation may be applied to a vertical cylinder if $(D/L) \gtrsim (35/Gr_L^{1/4})$.

9.6.5 Spheres

The following correlation due to Churchill [12] is recommended for spheres in fluids of $Pr \geq 0.7$ and for $Ra_D \lesssim 10^{11}$.

$$\overline{Nu}_D = 2 + \frac{0.589 Ra_D^{1/4}}{[1 + (0.469/Pr)^{9/16}]^{4/9}} \tag{9.35}$$

Recommended correlations from this section are summarized in Table 9.2. Results for other immersed geometries and special conditions are presented in the comprehensive review by Churchill [12].

9.7 EMPIRICAL CORRELATIONS: ENCLOSURES

The correlations of the preceding section pertain to free convection between a surface and an extensive fluid medium. However, engineering applications frequently involve heat transfer between surfaces that are at different temperatures and are separated by an *enclosed* fluid. In this section we present correlations that are pertinent to the most common geometries.

9.7.1 Rectangular Cavities

The rectangular cavity, shown schematically in Figure 9.7, has been studied extensively and comprehensive reviews of both experimental and theoretical results are available in the literature [22, 23]. The tilt angle τ between the heated and cooled surfaces of the cavity and the horizontal plane strongly influences the heat flux across the cavity. This flux may be expressed as

$$q'' = h(T_1 - T_2) \tag{9.36}$$

The horizontal cavity heated from below ($\tau = 0$) has been considered by many investigators. It is known that when

$$Ra_L = \frac{g\beta(T_1 - T_2)L^3}{\alpha\nu} > 1708$$

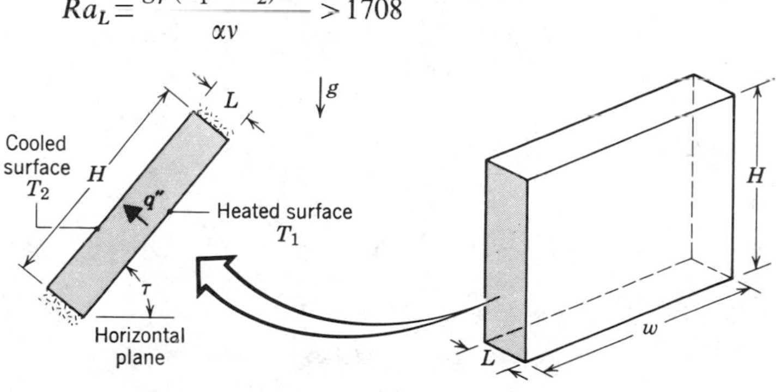

Figure 9.7 Free convection in a rectangular cavity.

conditions are thermally unstable and there is heat transfer by free convection. As a first approximation, the convection coefficient may be obtained from a correlation proposed by Globe and Dropkin [24]

$$\overline{Nu}_L = \frac{\bar{h}L}{k} = 0.069 Ra_L^{1/3} Pr^{0.074} \qquad 3\times 10^5 < Ra_L < 7\times 10^9 \tag{9.37}$$

where all properties are evaluated at the average temperature, $\bar{T} \equiv (T_1 + T_2)/2$. More detailed correlations, which apply over a wider range of Ra_L, have been proposed [25, 26]. In each case, however, the ratios L/H and L/w were small, and the effects of lateral boundaries on heat transfer were negligible. For $Ra_L < 1708$ or for $\tau = 180°$, heat transfer between the surfaces occurs by conduction and $\overline{Nu}_L = 1$.

In the vertical rectangular cavity ($\tau = 90°$), the vertical surfaces are heated and cooled, whereas the horizontal surfaces are adiabatic. For aspect ratios in the range $1 < (H/L) < 10$, the following correlations have been suggested [23]:

$$\overline{Nu}_L = 0.22\left(\frac{Pr}{0.2+Pr}Ra_L\right)^{0.28}\left(\frac{H}{L}\right)^{-1/4} \tag{9.38}$$

$$\begin{bmatrix} 2 < (H/L) < 10 \\ Pr < 10^5 \\ Ra_L < 10^{10} \end{bmatrix}$$

$$\overline{Nu}_L = 0.18\left(\frac{Pr}{0.2+Pr}Ra_L\right)^{0.29} \tag{9.39}$$

$$\begin{bmatrix} 1 < (H/L) < 2 \\ 10^{-3} < Pr < 10^5 \\ 10^3 < (Ra_L Pr)/(0.2+Pr) \end{bmatrix}$$

For larger aspect ratios, the following correlations have been proposed [27]:

$$\overline{Nu}_L = 0.42 Ra_L^{1/4} Pr^{0.012} (H/L)^{-0.3} \qquad \begin{bmatrix} 10 < (H/L) < 40 \\ 1 < Pr < 2\times 10^4 \\ 10^4 < Ra_L < 10^7 \end{bmatrix} \tag{9.40}$$

$$\overline{Nu}_L = 0.046 Ra_L^{1/3} \qquad \begin{bmatrix} 1 < (H/L) < 40 \\ 1 < Pr < 20 \\ 10^6 < Ra_L < 10^9 \end{bmatrix} \tag{9.41}$$

Convection coefficients computed from the foregoing expressions are to be used with Equation 9.36. Again, all properties are evaluated at the mean temperature, $(T_1+T_2)/2$.

Many publications have appeared in recent years concerning free convection heat transfer in inclined, rectangular spaces, with particular attention given to solar collector applications [28–33]. For large aspect ratios, $(H/L)\gtrsim 12$, and tilt angles less than the critical value τ^* given in Table 9.3, the following correlation due to Hollands et al. [33] is in excellent agreement with available data.

$$\overline{Nu}_L=1+1.44\left[1-\frac{1708}{Ra_L\cos\tau}\right]^{\bullet}\left[1-\frac{1708(\sin 1.8\tau)^{1.6}}{Ra_L\cos\tau}\right]+\left[\left(\frac{Ra_L\cos\tau}{5830}\right)^{1/3}-1\right]^{\bullet}\quad\begin{bmatrix}(H/L)\gtrsim 12\\ 0<\tau\lesssim\tau^*\end{bmatrix}\qquad(9.42)$$

The notation [ ]• implies that, if the quantity in brackets is negative, it must be set equal to zero. For small aspect ratios Catton [23] suggests that reasonable results may be obtained from a correlation of the form

$$\overline{Nu}_L=\overline{Nu}_L(\tau=0)\left[\frac{\overline{Nu}_L(\tau=90)}{\overline{Nu}_L(\tau=0)}\right]^{\tau/\tau^*}(\sin\tau^*)^{(\tau/4\tau^*)}\quad\begin{bmatrix}(H/L)\lesssim 12\\ 0<\tau\leq\tau^*\end{bmatrix}\qquad(9.43)$$

Beyond the critical tilt angle, the following correlations due to Ayyaswamy and Catton [28] and Arnold et al. [31], respectively, have been recommended [23] for all aspect ratios (H/L).

$$\overline{Nu}_L=\overline{Nu}_L(\tau=90^\circ)(\sin\tau)^{1/4}\qquad \tau^*<\tau<90^\circ\qquad(9.44)$$

$$Nu_L=1+[Nu_L(\tau=90^\circ)-1]\sin\tau\qquad 90^\circ<\tau<180^\circ\qquad(9.45)$$

9.7.2 Concentric Cylinders

Free convection heat transfer in the annular space between *long*, horizontal concentric cylinders (Figure 9.8) has been considered by Raithby and Hollands [34]. The heat transfer rate per unit length of cylinder (W/m) may be expressed as

$$q'=\frac{2\pi k_{eff}}{\ln(D_o/D_i)}(T_i-T_o)\qquad(9.46)$$

where the *effective thermal conductivity* k_{eff} is the thermal conductivity that a stationary fluid should have to transfer the same amount of heat as the moving

Table 9.3 Critical angle for inclined rectangular cavities

(H/L)	1	3	6	12	>12
τ^*	25	53	60	67	70

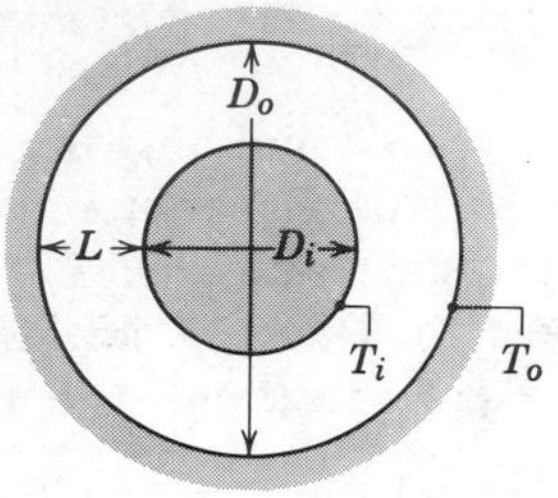

Figure 9.8 Free convection heat transfer in the annular space between long concentric cylinders or concentric spheres.

fluid. The suggested correlation for k_{eff} is of the form

$$\frac{k_{eff}}{k} = 0.386\left(\frac{Pr}{0.861 + Pr}\right)^{1/4}(Ra_c^*)^{1/4} \tag{9.47}$$

where

$$Ra_c^* = \frac{[\ln (D_o/D_i)]^4}{L^3(D_i^{-3/5} + D_o^{-3/5})^5} Ra_L \tag{9.48}$$

Equation 9.47 may be used for the range $10^2 \lesssim Ra_c^* \lesssim 10^7$. A more detailed correlation, which accounts for cylinder eccentricity effects, has been developed by Kuehn and Goldstein [35].

9.7.3 Concentric Spheres

Raithby and Hollands [34] have also considered free convection heat transfer between concentric spheres (Figure 9.8) and express the total heat transfer rate as

$$q = k_{eff}\pi\left(\frac{D_i D_o}{L}\right)(T_i - T_o) \tag{9.49}$$

The effective thermal conductivity is

$$\frac{k_{eff}}{k} = 0.74\left(\frac{Pr}{0.861 + Pr}\right)^{1/4}(Ra_s^*)^{1/4} \tag{9.50}$$

where

$$Ra_s^* = \left[\frac{L}{(D_o D_i)^4}\frac{Ra_L}{(D_i^{-7/5} + D_o^{-7/5})^5}\right] \tag{9.51}$$

The result may be used to a reasonable approximation for $10^2 \lesssim Ra_s^* \lesssim 10^4$.

EXAMPLE 9.5

A long tube of 0.1 m diameter is maintained at 120°C by passing steam through its interior. A radiation shield is installed concentric to the tube with an air gap of 10 mm. If the shield is at 35°C, estimate the heat transfer by free convection from the tube per unit length. What is the heat loss if the space between the tube and the shield is filled with glass-fiber blanket insulation?

SOLUTION

Known:

Temperatures and diameters of a steam tube and a concentric radiation shield.

Find:

1. Heat loss per unit length of tube.
2. Heat loss if air space is filled with glass-fiber blanket insulation.

Schematic:

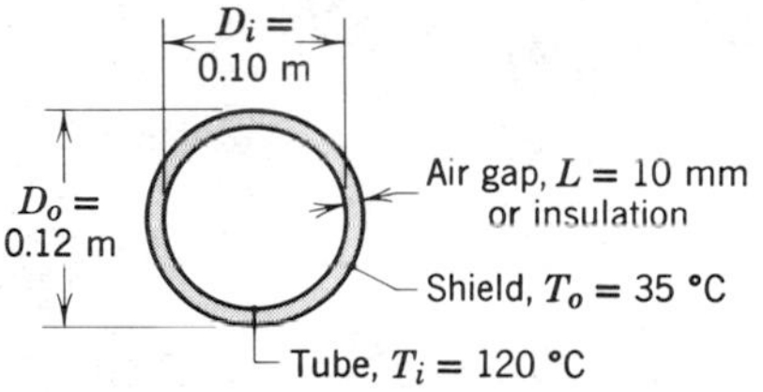

Assumptions:

1. Radiation heat transfer effects may be neglected.
2. Negligible contact resistance with insulation.

Properties:

Table A.4, air $[T=(T_i+T_o)/2=350\text{ K}]$: $k=0.030$ W/m·K, $\nu=20.92\times10^{-6}$ m²/s, $\alpha=29.9\times10^{-6}$ m²/s, $Pr=0.70$, $\beta=0.00285\text{ K}^{-1}$.
Table A.3, insulation, glass-fiber ($T\approx300$ K): $k=0.038$ W/m·K.

Analysis:

1. From Equation 9.46 the heat loss by free convection is

$$q'=\frac{2\pi k_{\text{eff}}}{\ln(D_o/D_i)}(T_i-T_o)$$

where k_{eff} may be obtained from Equations 9.47 and 9.48. With

$$Ra_L=\frac{g\beta(T_i-T_o)L^3}{\nu\alpha}$$

$$Ra_L = \frac{9.8 \text{ m/s}^2 \times 0.00285 \text{ K}^{-1} \times (120-35)°\text{C}(0.01 \text{ m})^3}{20.92 \times 10^{-6} \text{ m}^2/\text{s} \times 29.9 \times 10^{-6} \text{ m}^2/\text{s}} = 3795$$

it follows that

$$Ra_c^* = \frac{[\ln (D_o/D_i)]^4 Ra_L}{L^3(D_i^{-3/5} + D_o^{-3/5})^5}$$

$$Ra_c^* = \frac{[\ln (0.12 \text{ m}/0.10 \text{ m})]^4 3795}{(0.01 \text{ m})^3[(0.10 \text{ m})^{-3/5} + (0.12 \text{ m})^{-3/5}]^5} = 171$$

Hence

$$\frac{k_{\text{eff}}}{k} = 0.386\left(\frac{Pr}{0.861 + Pr}\right)^{1/4}(Ra_c^*)^{1/4}$$

$$\frac{k_{\text{eff}}}{k} = 0.386\left(\frac{0.70}{0.861 + 0.70}\right)^{1/4}(171)^{1/4} = 1.14$$

The effective thermal conductivity is then

$$k_{\text{eff}} = 1.14(0.030 \text{ W/m}\cdot\text{K}) = 0.0343 \text{ W/m}\cdot\text{K}$$

and the heat loss is

$$q' = \frac{2\pi(0.0343 \text{ W/m}\cdot\text{K})}{\ln (0.12 \text{ m}/0.10 \text{ m})}(120-35)°\text{C} = 100 \text{ W/m} \quad \triangleleft$$

2. With insulation in the space between the tube and the shield, heat loss is by conduction and from Equation 3.26

$$q' = \frac{2\pi k(T_i - T_o)}{\ln (D_o/D_i)}$$

$$q' = \frac{2\pi(0.038 \text{ W/m}\cdot\text{K})}{\ln (0.12 \text{ m}/0.10 \text{ m})}(120-35)°\text{C} = 111 \text{ W/m} \quad \triangleleft$$

Comments:

Although there is slightly more heat loss by conduction through the insulation than by free convection across the air space, the total heat loss across the air space may exceed that through the insulation due to the effects of radiation. The heat loss due to radiation may be minimized by using a radiation shield of low emissivity, and the means for calculating the loss will be developed in Chapter 13.

9.8 COMBINED FREE AND FORCED CONVECTION

In dealing with forced convection (Chapters 6–8), we ignored the effects of free convection. This was, of course, an assumption; for, as we now know, free convection will always exist when there is an unstable temperature gradient. Similarly, in the preceding sections of this chapter, we assumed that forced-convection effects were negligible. It is now time to acknowledge that situations may arise for which free and forced convection effects are comparable, in which case it is inappropriate to neglect either process. In Section 9.3 we indicated that free convection is negligible if $(Gr_L/Re_L^2) \ll 1$ and that forced convection is negligible if $(Gr_L/Re_L^2) \gg 1$. The *combined free* and *forced* (or *mixed*) *convection* regime is then one for which $(Gr_L/Re_L^2) \sim 1$.

The effect of buoyancy on heat transfer in a forced flow is strongly influenced by the direction of the buoyancy force relative to that of the flow. Three special cases that have been studied extensively correspond to buoyancy-induced and forced motions having the same direction (*assisting* flow), opposite directions (*opposing* flow), and perpendicular directions (*transverse* flow). Upward and downward forced motions over a heated vertical plate are examples of assisting and opposing flows, respectively. Examples of transverse flow include horizontal motion over a heated cylinder, sphere, or horizontal plate. In assisting and transverse flows, buoyancy acts to enhance the rate of heat transfer associated with pure forced convection; in opposing flows it acts to decrease this rate.

Experimental and theoretical mixed convection heat transfer results for external and internal flows have been reviewed [36–39], and a correlating equation of the form

$$Nu^n = Nu_F^n \pm Nu_N^n \tag{9.52}$$

has been recommended. The Nusselt numbers Nu_F and Nu_N are determined from correlations for pure forced and natural (free) convection, respectively, and specific correlations have been recommended for various flow geometries. The plus sign on the right-hand side of Equation 9.52 applies for assisting and transverse flows; the minus sign applies for opposing flow. The best correlation of data is generally obtained for $n = 3$, although values of 7/2 and 4 are better suited for transverse flows involving horizontal plates and cylinders (or spheres), respectively.

9.9 SUMMARY

In this chapter we have considered convective flows that originate in part, or exclusively, from buoyancy forces, and we have introduced the dimensionless parameters needed to characterize such flows. You should be able to discern when free convection effects are important and to quantify the associated heat transfer rates. An assortment of empirical correlations has been provided for this purpose.

REFERENCES

1. Gebhart, B., "Buoyancy-Induced Fluid Motions Characteristic of Applications in Technology," *J. Fluids Eng.*, *101*, 5, 1979.
2. Jaluria, Y., *Natural Convection Heat and Mass Transfer*, Pergamon Press, New York, 1980.
3. Ostrach, S., "An Analysis of Laminar Free Convection Flow and Heat Transfer About a Flat Plate Parallel to the Direction of the Generating Body Force," National Advisory Committee for Aeronautics Report 1111, 1953.
4. LeFevre, E. J., "Laminar Free Convection from a Vertical Plane Surface," *Proceedings of the Ninth International Congress on Applied Mechanics, Brussels*, Vol. 4, 1956, p. 168.
5. Gebhart, B., "External Natural Convection Flow," *Appl. Mech. Rev.*, *22*, 691, 1969.
6. Gebhart, B., *Heat Transfer*, 2nd ed., McGraw-Hill, New York, 1971, Chapter 8.
7. McAdams, W. H., *Heat Transmission*, 3rd ed., McGraw-Hill, New York, 1954, Chapter 7.
8. Warner, C. Y., and V. S. Arpaci, "An Experimental Investigation of Turbulent Natural Convection in Air at Low Pressure for a Vertical Heated Flat Plate," *Int. J. Heat Mass Transfer*, *11*, 397, 1968.
9. Bayley, F. J., "An Analysis of Turbulent Free Convection Heat Transfer," *Proc. Inst. Mech. Eng.*, *169*, 361, 1955.
10. Churchill, S. W., and H. H. S. Chu, "Correlating Equations for Laminar and Turbulent Free Convection from a Vertical Plate," *Int. J. Heat Mass Transfer*, *18*, 1323, 1975.
11. Sparrow, E. M., and J. L. Gregg, "Laminar Free Convection from a Vertical Flat Plate," *Trans. ASME*, *78*, 435, 1956.
12. Churchill, S. W., "Free Convection Around Immersed Bodies," in E. U. Schlünder, Ed.-in-Chief, *Heat Exchanger Design Handbook*, Hemisphere Publishing Corp., New York, 1983, Section 2.5.7.
13. Sparrow, E. M., and J. L. Gregg, "Laminar Free Convection Heat Transfer from the Outer Surface of a Vertical Circular Cylinder," *Trans. ASME*, *78*, 1823, 1956.
14. Cebeci, T., "Laminar-Free-Convective-Heat Transfer from the Outer Surface of a Vertical Slender Circular Cylinder," *Proceedings of the Fifth International Heat Transfer Conference*, Paper NC1.4, 1974, pp. 15–19.
15. Minkowycz, W. J., and E. M. Sparrow, "Local Nonsimilar Solutions for Natural Convection on a Vertical Cylinder," *J. Heat Transfer*, *96*, 178, 1974.
16. Goldstein, R. J., E. M. Sparrow, and D. C. Jones, "Natural Convection Mass Transfer Adjacent to Horizontal Plates," *Int. J. Heat Mass Transfer*, *16*, 1025, 1973.
17. Lloyd, J. R., and W. R. Moran, "Natural Convection Adjacent to Horizontal Surfaces of Various Planforms," ASME Paper 74-WA/HT-66, 1974.

18. Vliet, G. C., "Natural Convection Local Heat Transfer on Constant-Heat-Flux Inclined Surfaces," *Trans. ASME, 91C*, 511, 1969.
19. Fujii, T., and H. Imura, "Natural Convection Heat Transfer from a Plate with Arbitrary Inclination," *Int. J. Heat Mass Transfer, 15*, 755, 1972.
20. Morgan, V. T., "The Overall Convective Heat Transfer from Smooth Circular Cylinders," in T. F. Irvine and J. P. Hartnett, Eds., *Advances in Heat Transfer*, Vol. 11, Academic Press, New York, 1975, pp. 199–264.
21. Churchill, S. W., and H. H. S. Chu, "Correlating Equations for Laminar and Turbulent Free Convection from a Horizontal Cylinder," *Int. J. Heat Mass Transfer, 18*, 1049, 1975.
22. Ostrach, S., "Natural Convection in Enclosures," in J. P. Hartnett and T. F. Irvine, Eds., *Advances in Heat Transfer*, Vol. 8, Academic Press, New York, 1972, pp. 161–227.
23. Catton, I., "Natural Convection in Enclosures," *Proceedings of the Sixth International Heat Transfer Conference, Toronto*, Vol. 6, 1978, p. 13.
24. Globe, S., and D. Dropkin, "Natural Convection Heat Transfer in Liquids Confined Between Two Horizontal Plates," *J. Heat Transfer, 81C*, 24, 1959.
25. Hollands, K. G. T., G. D. Raithby, and L. Konicek, "Correlation Equations for Free Convection Heat Transfer in Horizontal Layers of Air and Water," *Int. J. Heat Mass Transfer, 18*, 879, 1975.
26. Churchill, S. W., "Free Convection in Layers and Enclosures," in E. U. Schlünder, Ed.-in-Chief, *Heat Exchanger Design Handbook*, Hemisphere Publishing Corp., New York, 1983, Section 2.5.8.
27. MacGregor, R. K., and A. P. Emery, "Free Convection Through Vertical Plane Layers: Moderate and High Prandtl Number Fluids," *J. Heat Transfer, 91*, 391, 1969.
28. Ayyaswamy, P. S., and I. Catton, "The Boundary-Layer Regime for Natural Convection in a Differentially Heated, Tilted Rectangular Cavity," *J. Heat Transfer, 95*, 543, 1973.
29. Catton, I., P. S. Ayyaswamy, and R. M. Clever, "Natural Convection Flow in a Finite, Rectangular Slot Arbitrarily Oriented with Respect to the Gravity Vector," *Int. J. Heat Mass Transfer, 17*, 173, 1974.
30. Clever, R. M., "Finite Amplitude Longitudinal Convection Rolls in an Inclined Layer," *J. Heat Transfer, 95*, 407, 1973.
31. Arnold, J. N., I. Catton, and D. K. Edwards, "Experimental Investigation of Natural Convection in Inclined Rectangular Regions of Differing Aspect Ratios," ASME Paper 75-HT-62, 1975.
32. Buchberg, H., I. Catton, and D. K. Edwards, "Natural Convection in Enclosed Spaces — A Review of Application to Solar Energy," *J. Heat Transfer, 98*, 182, 1976.
33. Hollands, K. G. T., S. E. Unny, G. D. Raithby, and L. Konicek, "Free Convective Heat Transfer Across Inclined Air Layers," *J. Heat Transfer, 98*, 189, 1976.
34. Raithby, G. D., and K. G. T. Hollands, "A General Method of Obtaining Approximate Solutions to Laminar and Turbulent Free Convection Problems," in T. F. Irvine and J. P. Hartnett, Eds., *Advances in Heat Transfer*, Vol. 11, Academic Press, New York, 1975, pp. 265–315.
35. Kuehn, T. H., and R. J. Goldstein, "Correlating Equations for Natural Convection

Heat Transfer Between Horizontal Circular Cylinders," *Int. J. Heat Mass Transfer*, *19*, 1127, 1976.
36. Churchill, S. W., "A Comprehensive Correlating Equation for Laminar, Assisted, Forced and Free Convection," *AIChE J.*, *23*, 10, 1977.
37. Churchill, S. W., "Combined Free and Forced Convection Around Immersed Bodies," in E. U. Schlünder, Ed.-in-Chief, *Heat Exchanger Design Handbook*, Hemisphere Publishing Corp., New York, 1983, Section 2.5.9.
38. Churchill, S. W., "Combined Free and Forced Convection in Channels, in E. U. Schlünder, Ed.-in-Chief, *Heat Exchanger Design Handbook*, Hemisphere Publishing Corp., New York, 1983, Section 2.5.10.
39. Hieber, C. A., "Laminar Mixed Convection in an Isothermal Horizontal Tube: Correlation of Heat Transfer Data," *Int. J. Heat Mass Transfer*, *25*, 1737, 1982.

PROBLEMS

9.1 Using the tabulated values of density for water in Table A.6, calculate the volumetric thermal expansion coefficient at 300 K from its definition, Equation 9.4, and compare your result with the tabulated value.

9.2 Consider an object having a characteristic length of 0.25 m and a situation for which the temperature difference is 25°C. Using thermophysical properties evaluated at 350 K, calculate the Grashof number for air, hydrogen, water, and ethylene glycol. Assume that the pressure is 1 atm.

9.3 Consider an object of characteristic length 0.01 m and a situation for which the temperature difference is 30°C. Evaluating thermophysical properties at the prescribed conditions, determine the Rayleigh number for the following fluids: air (1 atm, 400 K), helium (1 atm, 400 K), glycerin (285 K), and water (310 K).

9.4 A solid object is to be cooled by submerging it in a quiescent fluid, and the associated free convection coefficient is given by $\bar{h} = C\Delta T^{1/4}$, where C is a constant and $\Delta T = T - T_\infty$. Invoking the lumped capacitance approximation, obtain an expression for the time required for the object to cool from an initial temperature T_i to a final temperature T_f.

9.5 The heat transfer rate due to free convection from a vertical surface, 1 m high and 0.6 m wide, to quiescent air that is 20 K colder than the surface is known. What is the ratio of the heat transfer rate for that situation to the rate corresponding to a vertical surface, 0.6 m high and 1 m wide, when the quiescent air is 20 K warmer than the surface. Neglect heat transfer by radiation and any influence of temperature on the relevant thermophysical properties of air.

9.6 The heat transfer rate per unit length due to free convection from a horizontal tube is 200 W/m when its surface is maintained at 65°C and the ambient air is 25°C. Estimate the heat transfer rate per unit length when the tube surface is maintained at 145°C. Neglect heat transfer by radiation and any influence of temperature on the relevant thermophysical properties of air.

9.7 Consider a large vertical plate with a uniform surface temperature of 130°C suspended in quiescent air at 25°C and atmospheric pressure.

(a) Estimate the boundary layer thickness at a location 0.25 m measured from the lower edge.

(b) What is the maximum velocity in the boundary layer at this location and at what position in the boundary layer does the maximum occur?

(c) Using the similarity solution result, Equation 9.19, determine the heat transfer coefficient for the location 0.25 m measured from the lower edge.

(d) At what location on the plate measured from the lower edge will the boundary layer become turbulent?

9.8 A number of thin plates are to be cooled by vertically suspending them in a water bath at a temperature of 20°C. If the plates are initially at 54°C and have a length of 0.15 m, what is the minimum spacing permitted such that no interference will occur between their free convection boundary layers?

9.9 A square aluminum plate 5 mm thick and 200 mm on a side is cooled while vertically suspended in quiescent air at 40°C. Determine the average heat transfer coefficient for the plate when its temperature is 15°C by using: results from the similarity solution to the boundary layer equations, and an empirical correlation.

9.10 Consider an array of vertical rectangular fins, which is to be used to cool an electronic device mounted in quiescent, atmospheric air at $T_\infty = 27°C$. Each fin is of length and height $L = 20$ mm and $H = 150$ mm, and operates at an approximately uniform temperature of $T_s = 77°C$.

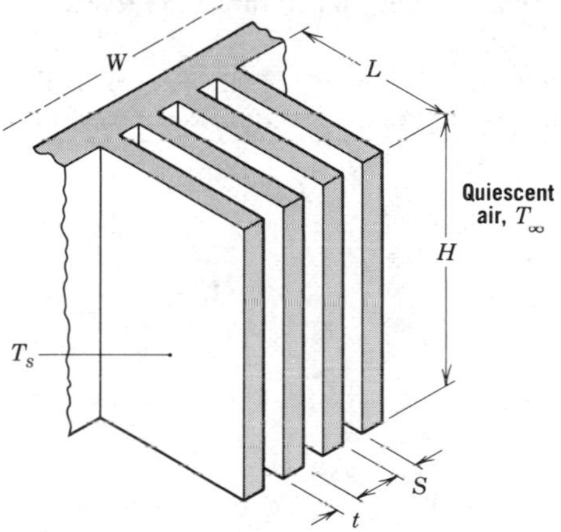

(a) Briefly describe why there exists an optimum fin spacing S. Estimate the optimum value of S for the prescribed conditions.

(b) For the optimal value of S and a fin thickness of $t = 1.5$ mm, estimate the rate of heat transfer from the fins for an array of width $W = 355$ mm.

9.11 Determine the average convection heat transfer coefficient for the 2.5-m-high vertical walls of a home having interior air and wall surface temperatures of (a) 20 and 10°C, respectively, and (b) 27 and 37°C, respectively.

9.12 Beginning with the free convection correlation of the form given by Equation 9.24, show that for air at atmospheric pressure and a film temperature of 400 K, the average heat transfer coefficient for a vertical plate can be expressed as

$$\bar{h}_L = 1.40\left(\frac{\Delta T}{L}\right)^{1/4} \qquad 10^4 < Ra_L < 10^9$$

$$\bar{h}_L = 0.98\Delta T^{1/3} \qquad 10^9 < Ra_L < 10^{13}$$

9.13 A household oven door 0.5 m high and 0.7 m wide reaches an average surface temperature of 32°C during operation. Estimate the heat loss to the room with ambient air and surroundings at 22°C.

9.14 An aluminum alloy (2024) plate, heated to a uniform temperature of 227°C, is allowed to cool while vertically suspended in a room where the ambient air and surroundings are at 27°C. The plate is 0.3 m square and 15 mm thick and has an emissivity of 0.25.

(a) Develop an expression for the time rate of change of the plate temperature assuming the temperature to be uniform at any time.

(b) Determine the initial rate of cooling (K/s) when the plate temperature is 227°C.

(c) Justify the uniform plate temperature assumption.

9.15 The plate described in Problem 9.14 has been used in an experiment to determine the free convection heat transfer coefficient. At an instant of time when the plate temperature was 127°C, the time rate of change of this temperature was observed to be −0.0465 K/s. What is the corresponding free convection heat transfer coefficient? Compare this result with an estimate based on a standard empirical correlation.

9.16 Determine the maximum allowable uniform heat flux that may be imposed at a wall heating panel 1 m high if the maximum temperature is not to exceed 37°C when the ambient air temperature is 25°C.

9.17 The components of a vertical circuit board, 150 mm on a side, dissipate 5 W. The back surface is well insulated, and the front surface is exposed to quiescent air at 27°C.

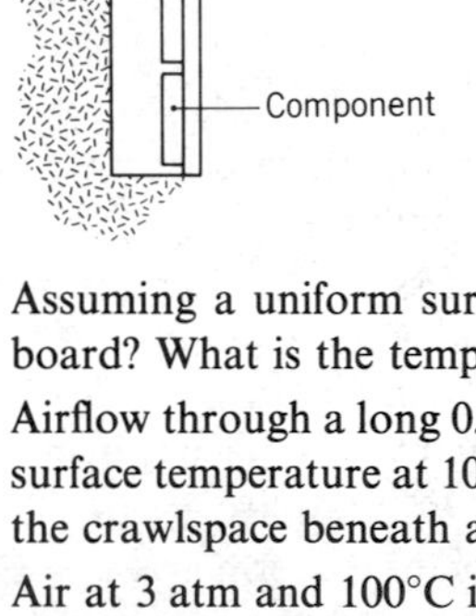

Assuming a uniform surface heat flux, what is the maximum temperature of the board? What is the temperature of the board for an isothermal surface condition?

9.18 Airflow through a long 0.2 m square air conditioning duct maintains the outer duct surface temperature at 10°C. If the duct is uninsulated and exposed to air at 35°C in the crawlspace beneath a house, what is the heat gain per unit length of the duct?

9.19 Air at 3 atm and 100°C is discharged from a compressor into a vertical receiver of height 2.5 m and diameter 0.75 m. Estimate the receiver wall temperature and the heat transfer to the ambient air at 25°C. Neglect radiation and heat losses from the ends. Assume that the receiver wall is at a uniform temperature and that heat transfer at the inner and outer surfaces is by free convection.

9.20 An electrical heater in the form of a horizontal disk of 400 mm diameter is used to heat the bottom of a tank filled with engine oil at 5°C. Calculate the power required to maintain the heater surface temperature at 70°C.

9.21 Consider a horizontal straight fin fabricated from plain carbon steel, 6 mm thick and 100 mm long. Estimate the fin heat rate per unit width when the base is maintained at 150°C and the surrounding atmospheric air is quiescent and at 25°C.

9.22 A circular grill of diameter 0.25 m and emissivity 0.9 is maintained at a constant surface temperature of 130°C. What electrical power is required when the room air and surroundings are at 24°C?

9.23 A plate 1 m × 1 m, inclined at an angle of 45° with the horizontal, is exposed to the environment, which provides a net radiation heat flux of 300 W/m^2. If the back side of the plate is well insulated, estimate the temperature the plate reaches when the ambient air is still and at 0°C.

9.24 A horizontal rod 5 mm in diameter is immersed in water maintained at 18°C. If the rod surface temperature is 56°C, estimate the free convection heat transfer rate per unit length of the rod.

9.25 A horizontal uninsulated steam pipe passes through a large room whose walls and ambient air are at 300 K. The pipe of 150 mm diameter has an emissivity of 0.85 and an outer surface temperature of 400 K. Calculate the heat loss per unit length from the pipe.

9.26. Beer in cans 150 mm long and 60 mm in diameter is initially at 27°C and is to be cooled by placement in a refrigerator compartment at 4°C. In the interest of maximizing the cooling rate, should the cans be laid horizontally or vertically in the compartment? As a first approximation, neglect heat transfer from the ends.

9.27 Consider Problem 8.26. A more realistic solution would account for the resistance to heat transfer due to free convection in the paraffin during melting. Assuming the tube surface to have a uniform temperature of 55°C and the paraffin to be an infinite, quiescent liquid, determine the convection coefficient associated with the outer surface. Using this result and recognizing that the tube surface temperature is not known, determine the water outlet temperature, the total heat transfer rate, and the time required to completely liquefy the paraffin, for the prescribed conditions. Thermophysical properties associated with the liquid state of the paraffin are $k = 0.15$ W/m·K, $\beta = 8 \times 10^{-4}$ K^{-1}, $\rho = 770$ kg/m^3, $\nu = 5 \times 10^{-6}$ m^2/s, and $\alpha = 8.85 \times 10^{-8}$ m^2/s.

9.28 A horizontal pipe of 70 mm outside diameter passes through a room in which the air and the walls are at 25°C. The outer surface has an emissivity of 0.8 and is maintained at 150°C by steam passing through the pipe. What is the heat loss per unit length of the pipe?

9.29 A horizontal tube of 12.5 mm diameter with an outer surface temperature of 240°C is located in a room with an air temperature of 20°C. Estimate the heat transfer rate per unit length of the tube due to free convection.

9.30 Saturated steam at 4 bars absolute with a mean velocity of 3 m/s flows through a tube whose inner and outer diameters are 55 and 65 mm, respectively. The heat transfer coefficient for the steam flow is known to be 11,000 $W/m^2 \cdot K$. The tube is covered with a 25-mm-thick layer of 85% magnesia insulation and is exposed to atmospheric air at 25°C. Estimate the heat transfer rate by free convection to the room per unit length of the tube. If the steam is saturated at the inlet of the tube, estimate its quality at the outlet of a tube 30 m long.

9.31 An electrical cable of 25 mm diameter has a heat dissipation rate of 30 W/m. If the ambient air temperature is 27°C, estimate the surface temperature of the cable.

9.32 An electrical immersion heater, 10 mm in diameter and 300 mm long, is rated at 550 W. If the heater is horizontally positioned in a large tank of water at 20°C, estimate its surface temperature. Estimate the surface temperature if the heater is accidentally operated in air at 20°C.

9.33 A sphere of 25 mm diameter contains an embedded electrical heater. Calculate the power required to maintain the surface temperature at 94°C when the sphere is exposed to a quiescent medium at 20°C for (a) air at atmospheric pressure, (b) water, and (c) ethylene glycol.

9.34 Under steady-state operation the surface temperature of a small 20-W incandescent light bulb is 125°C when the room air and the walls are at 25°C. Approximating the bulb as a sphere 40 mm in diameter with a surface emissivity of 0.8, what is the rate of heat transfer from the surface of the bulb to the surroundings?

9.35 A building window pane that has a height of 1.2 m and a width of 0.8 m is separated from the ambient air by a storm window of the same height and width. The thickness of the air space between the two windows is 0.06 m. If the building and storm windows are at temperatures of 20 and −10°C, respectively, what is the rate of heat loss by free convection across the air space?

9.36 The absorber plate and the adjoining cover plate of a flat plate solar collector are at 70 and 35°C, respectively, and are separated by an air space of 0.05 m. What is the rate of free convection heat transfer per unit surface area between the two plates if they are inclined at an angle of 60° from the horizontal?

9.37 In a solar collector 1 m wide × 3 m long, the glass cover plate at an average temperature of 29°C is spaced 60 mm from the absorber plate at an average temperature of 50°C. Estimate the convection heat loss from the absorber plate to the glass when the collector is positioned horizontally. What is the convection heat loss if the spacing is reduced to 10 mm?

9.38 A rectangular cavity consists of two parallel plates, 0.5 m × 0.5 m, separated by 50 mm, with the lateral boundaries insulated. The heated plate is maintained at 325 K and the cooled plate at 275 K. Estimate the heat flux between the surfaces for three orientations of the cavity using the notation of Figure 9.7: vertical, $\tau = 90°$, horizontal with $\tau = 0°$, and horizontal with $\tau = 180°$.

9.39 Consider a horizontal flat roof section having the same dimensions as a vertical wall section. For both sections, the surfaces exposed to the air gap are at 18°C (inside) and −10°C (outside).

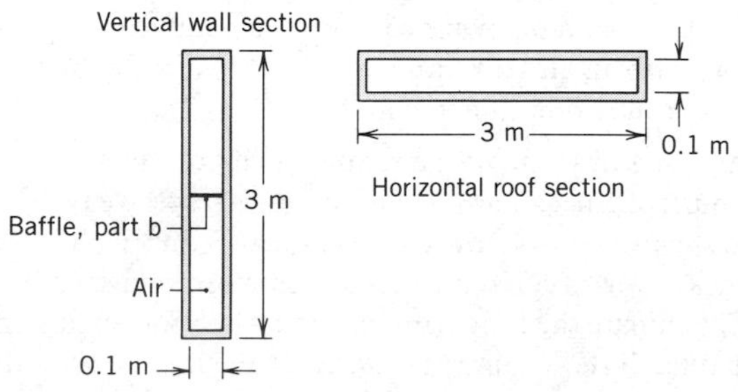

(a) Estimate the ratio of the convection heat rate for the horizontal section to that of the vertical section.

(b) What effect will inserting a baffle at the midheight of the vertical section have on the convection heat rate for that section?

9.40 The top surface (0.5 m × 0.5 m) of an oven is at 60°C for a particular operating condition when the room air is at 23°C. To reduce heat loss from the oven and to minimize burn hazard, it is proposed to create a 50-mm air space by adding a cover plate.

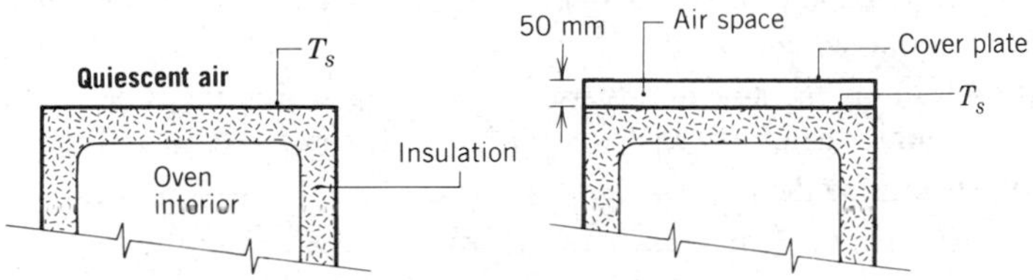

Assuming the same oven surface temperature T_s for both situations, estimate the reduction in the convection heat loss resulting from installation of the cover plate. What is the temperature of the cover plate?

9.41 Consider the rectangular cavity of Problem 9.38. Determine the heat flux between the two surfaces for tilt angles of 45° and 75°.

9.42 A solar collector design consists of an inner tube enclosed concentrically within an outer tube that is transparent to solar radiation. The tubes are thin walled, with inner and outer diameters of 0.10 and 0.15 m, respectively. The annular space between the tubes is completely enclosed and filled with air at atmospheric pressure. Under operating conditions for which the inner and outer tube surface temperatures are 70 and 30°C, respectively, what is the convective heat loss per meter of tube length across the air space?

9.43 The annulus formed by two concentric, horizontal tubes with inner and outer diameters of 50 and 75 mm, respectively, is filled with water. If the inner and outer surfaces are maintained at 300 and 350 K, respectively, estimate the convection heat transfer rate per unit length of the tubes.

9.44 The surfaces of two long, horizontal, concentric thin-walled tubes having radii of 100 and 125 mm are maintained at 300 and 400 K, respectively. If the annular space is pressurized with nitrogen at 5 atm, estimate the convection heat transfer rate per unit length of the tubes.

9.45 The surfaces of two concentric spheres having radii of 75 and 100 mm are maintained at 325 and 275 K, respectively. If the space between the spheres is filled with air at 3 atm, estimate the convection heat transfer rate.

9.46 Water at a temperature 17°C and a mass rate of 0.012 kg/s is to be used for maintaining a small plate (on which a special sensor is to be mounted) at a fixed temperature. The plate is situated within a hot air environment at a temperature of 235°C. The tube is horizontal and 1 m long. Fabricated from a plastic with a thermal conductivity of 0.05 W/m·K, the tube has an inner diameter $D_i = 1.4$ mm and an outer diameter $D_o = 3.2$ mm.

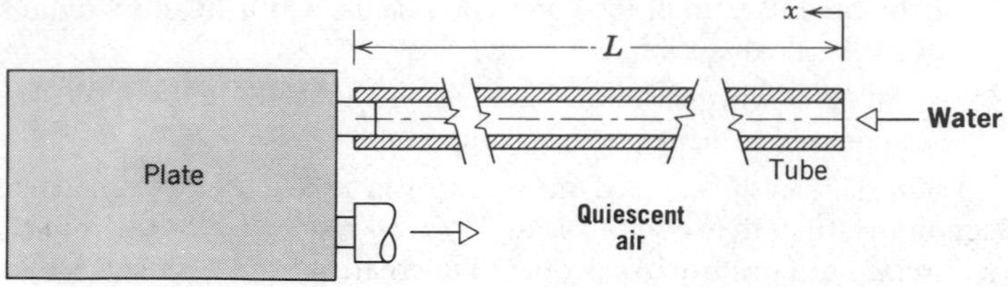

(a) Assuming that the average outer surface temperature of the tube is 120°C, estimate the heat transfer coefficient for free convection between the tube and the ambient air.

(b) Assuming the flow and thermal conditions within the tube are fully developed, estimate the heat transfer coefficient between the tube and the water.

(c) Determine the overall heat transfer coefficient based on the outer tube area.

(d) Estimate the temperature of the water at the end of the tube ($x = 1$ m).

9.47 Water at 35°C with a velocity of 0.05 m/s flows over a horizontal, 50-mm-diameter cylinder maintained at a uniform surface temperature of 20°C. Do you anticipate that heat transfer by free convection will be significant? What would be the situation if the fluid were air at atmospheric pressure?

9.48 According to experimental results for parallel airflow over a uniform temperature, heated vertical plate, the effect of free convection on the heat transfer convection coefficient will be 5% when $Gr_L/Re_L^2 = 0.08$. Consider a heated vertical plate 0.3 m long, maintained at a surface temperature of 60°C in atmospheric air at 25°C. What is the minimum vertical velocity required of the airflow such that free convection effects will be less than 5% of the heat transfer rate?

9.49 A vertical array of circuit boards 150 mm in height is to be air cooled such that the board temperature does not exceed 60°C when the ambient temperature is 25°C.

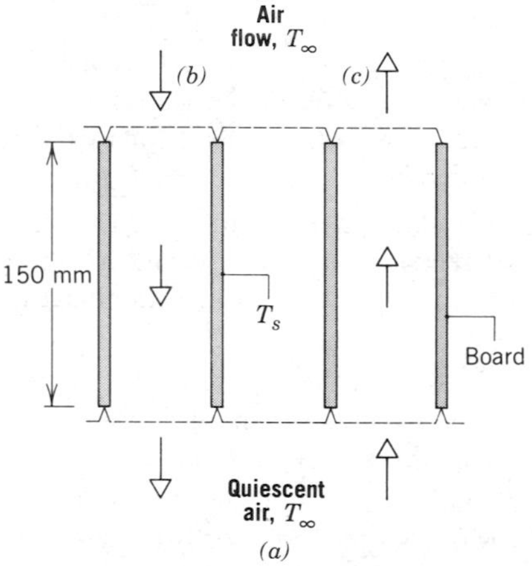

Assuming isothermal surface conditions, determine the allowable electrical power dissipation per board for the following cooling arrangements.

(a) Free convection only (no forced air flow).

(b) Airflow with a downward velocity of 0.6 m/s.

(c) Airflow with an upward velocity of 0.3 m/s.

(d) Airflow with a velocity (upward or downward) of 5 m/s.

9.50 A horizontal 100-mm-diameter pipe passing hot oil is to be used in the design of an industrial water heater. Based on a typical water draw rate, the velocity over the pipe is 0.5 m/s. The hot oil maintains the outer surface temperature at 85°C and the water temperature is 37°C.

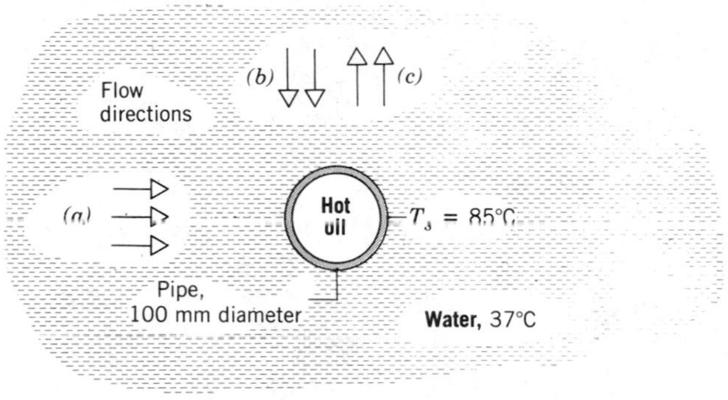

Investigate the effect of flow direction on the heat rate (W/m) for (a) horizontal, (b) downward, and (c) upward flow.

9.51 Panels (250 mm × 250 mm) with a decorative, highly reflective plastic finish are cured in an oven at 125°C and cooled in quiescent air at 29°C. Quality considerations dictate that the panels remain horizontal and that the cooling rate be controlled. To increase productivity in the plant, it is proposed to replace the batch cooling method with a conveyor system having a velocity of 0.5 m/s.

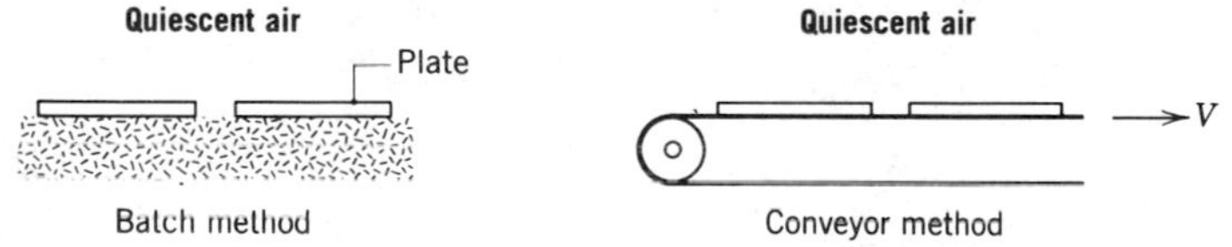

Compare the initial (immediately after leaving the oven) convection heat transfer rates for the two methods.

Ten

BOILING AND CONDENSATION

In this chapter we focus on convection processes associated with the change in phase of a fluid. In particular, we consider important processes that can occur at a solid–liquid interface, that is, *boiling* and *condensation*. For these cases *latent* heat effects associated with the phase change are significant. The change from the liquid to the vapor state due to boiling is sustained by heat transfer from the solid surface; conversely, condensation of a vapor to the liquid state results in heat transfer to the solid surface.

Since they involve fluid motion, boiling and condensation are classified as forms of the convection mode of heat transfer. However, they are characterized by unique features. Because there is a phase change, heat transfer to or from the fluid can occur without influencing the fluid temperature. In fact, through boiling or condensation, large heat transfer rates may be achieved with small temperature differences. In addition to the *latent heat* h_{fg}, two other parameters are important in characterizing the processes, namely, the *surface tension* σ between the liquid–vapor interface and the *density difference* between the two phases. This difference induces a *buoyancy force*, which is proportional to $g(\rho_l - \rho_v)$. Due to combined latent heat and buoyancy-driven flow effects, boiling and condensation heat transfer coefficients and rates are generally much larger than those characteristic of convection heat transfer without phase change.

Many engineering problems involve boiling and condensation. For example, both processes are essential to all closed-loop power and refrigeration cycles. In a power cycle, pressurized liquid is converted to vapor in a *boiler*. After expansion in a turbine, the vapor is restored to its liquid state in a *condenser*, whereupon it is pumped to the boiler to repeat the cycle. Evaporators, in which the boiling process occurs, and condensers are also essential components in vapor–compression refrigeration cycles. The rational design of such components dictates that the associated phase change processes be well understood.

10.1 DIMENSIONLESS PARAMETERS IN BOILING AND CONDENSATION

In our treatment of boundary layer phenomena (Section 6.6), we nondimensionalized the governing equations to identify relevant dimensionless groups. This approach enhanced our understanding of related physical mechanisms and suggested simplified procedures for generalizing and representing heat transfer results.

Since it is difficult to develop governing equations for boiling and condensation processes, it is easier to obtain the appropriate dimensionless parameters by using the Buckingham pi theorem [1]. For either process, the convection coefficient could depend on the difference between the surface and saturation temperatures, $\Delta T = |T_s - T_{sat}|$, the body force arising from the liquid–vapor density difference $g(\rho_l - \rho_v)$, the latent heat h_{fg}, the surface tension σ, a characteristic length L, and the thermophysical properties of the liquid or vapor: ρ, c_p, k, μ. That is,

$$h = h[\Delta T, g(\rho_l - \rho_v), h_{fg}, \sigma, L, \rho, c_p, k, \mu] \tag{10.1}$$

Since there are 10 variables in 5 dimensions (m, kg, s, J, K), there will be $(10-5)=5$ pi groups, which can be expressed in the following forms:

$$\frac{hL}{k}=f\left[\frac{\rho g(\rho_l-\rho_v)L^3}{\mu^2}, \frac{c_p\Delta T}{h_{fg}}, \frac{\mu c_p}{k}, \frac{g(\rho_l-\rho_v)L^2}{\sigma^2}\right] \tag{10.2a}$$

or, defining the dimensionless groups,

$$Nu_L=f\left[\frac{\rho g(\rho_l-\rho_v)L^3}{\mu^2}, Ja, Pr, Bo\right] \tag{10.2b}$$

The Nusselt and Prandtl numbers are familiar from our earlier single-phase convection analyses; the new dimensionless parameters are the Jakob number Ja, the Bond number Bo, and a parameter that is without a name but bears a strong resemblance to the Grashof number (see Equation 9.12). This unnamed parameter represents the effect of buoyancy-induced fluid motion on heat transfer. The Jakob number, named after Max Jakob in honor of his early work on phase change phenomena, is the ratio of the maximum sensible energy absorbed by the liquid (vapor) to the latent energy absorbed by the liquid (vapor) during condensation (boiling). In many applications, the sensible energy is much less than the latent energy and the Jakob number has a small numerical value. The Bond number is the ratio of the gravitational body force to the surface tension force. In subsequent sections, we delineate the role of each of the foregoing parameters in boiling and condensation.

10.2 BOILING MODES

When evaporation occurs at a solid–liquid interface, it is termed *boiling*. The process occurs when the temperature of the surface T_s exceeds the saturation temperature corresponding to the liquid pressure T_{sat}. Heat is transferred from the solid surface to the liquid, and the appropriate form of Newton's law of cooling is

$$q_s''=h(T_s-T_{sat})=h\Delta T_e \tag{10.3}$$

where $\Delta T_e \equiv T_s-T_{sat}$ is termed the *excess temperature*. The process is characterized by the formation of vapor bubbles, which grow and subsequently detach from the surface. Vapor bubble growth and dynamics depend, in a complicated manner, on the excess temperature, on the nature of the surface, and on thermophysical properties of the fluid, such as its surface tension. In turn, the dynamics of vapor bubble formation affect fluid motion near the surface and therefore strongly influence the heat transfer coefficient.

Boiling may occur under various conditions. For example, the term *pool boiling* refers to a situation in which the liquid is quiescent and its motion near the surface is due to free convection and to mixing induced by bubble growth and detachment. In contrast, for *forced-convection boiling*, fluid motion is induced by external means, as well as by free convection and bubble-induced mixing. Boiling may also be classified according to whether it is *subcooled* or *saturated.* In subcooled (or local) boiling, the temperature of the liquid is below the saturation temperature and bubbles formed at the surface eventually condense in the liquid. In contrast the temperature of the liquid slightly exceeds the saturation temperature in *saturated boiling.* Bubbles formed at the surface are then propelled through the liquid by buoyancy forces, eventually escaping from a free surface.

10.3 POOL BOILING

Saturated pool boiling, as shown in Figure 10.1, has been studied extensively. Although there is a sharp decline in the liquid temperature close to the solid surface, the temperature through most of the liquid remains slightly above saturation. Bubbles generated at the liquid–solid interface therefore rise to and are transported across the liquid–vapor interface. An appreciation for the underlying physical mechanisms may be obtained by examining the *boiling curve.*

10.3.1 The Boiling Curve

Nukiyama [2] was the first to identify different regimes of pool boiling using the apparatus of Figure 10.2. The heat flux from a horizontal *nichrome* wire to the saturated water was determined by measuring the current flow I and potential drop E. The wire temperature was determined from knowledge of the manner in which its electrical resistance varied with temperature. This arrangement is referred to as *power-controlled* heating, wherein the wire temperature T_s (hence the excess temperature ΔT_e) is the dependent variable and the power setting (hence the heat flux q_s'') is the independent variable. Following the arrows of the *heating curve* of Figure 10.3, it is evident that as power is applied, the heat flux increases, at first slowly and then very rapidly, with excess temperature. Nukiyama observed that boiling, as evidenced by the presence of bubbles, did

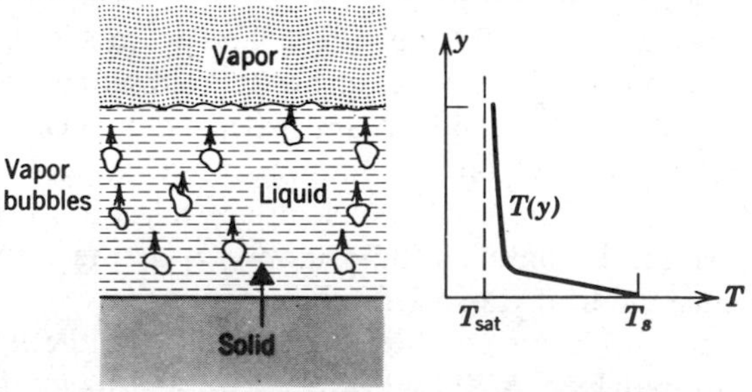

Figure 10.1 Temperature distribution in saturated pool boiling with a liquid–vapor interface.

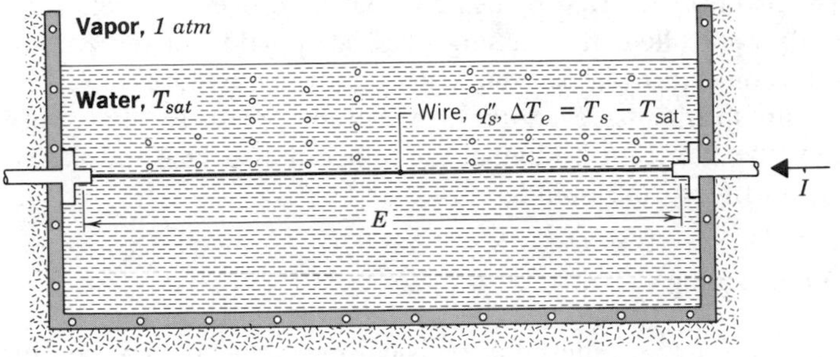

Figure 10.2 Nukiyama's power-controlled heating apparatus for demonstrating the boiling curve.

not begin until $\Delta T_e \approx 5°C$. With further increase in power, the heat flux increased to very high levels until *suddenly*, for a value slightly larger than q''_{max}, the wire temperature jumped to the melting point and burnout occurred. However, repeating the experiment with a *platinum* wire having a higher melting point (2045 vs. 1500 K). Nukiyama was able to maintain heat fluxes above the maximum value q''_{max} without burnout. When he subsequently reduced the power, the variation of ΔT_e with q''_s followed the *cooling curve* of Figure 10.3. When the heat flux reached the minimum point q''_{min}, a further decrease in power caused the excess temperature to abruptly drop and the process followed the original heating curve back to the saturation point.

Nukiyama believed that the hysteresis effect of Figure 10.3 was a consequence of the power-controlled method of heating, wherein ΔT_e is a dependent

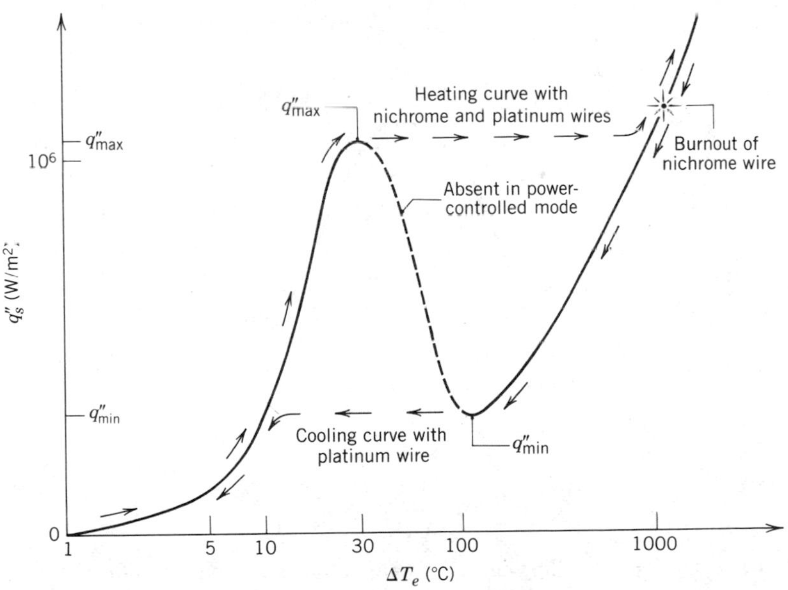

Figure 10.3 Nukiyama's boiling curve for saturated water.

variable. He also believed that by using a heating process wherein ΔT_e could be independently controlled, the missing (dashed) portion of the curve could be obtained. His conjecture was subsequently confirmed by Drew and Mueller [3]. By condensing steam inside a tube at different pressures, they were able to control the value of ΔT_e for boiling at the tube outer surface and thereby obtain the missing portion of the boiling curve.

10.3.2 Modes of Pool Boiling

An appreciation for the underlying physical mechanisms may be obtained by examining the different modes, or regimes, of pool boiling. These regimes are identified in the boiling curve of Figure 10.4. The specific curve pertains to water, although similar trends characterize the behavior of other fluids. From Equation 10.3 we note that q_s'' depends on the convection coefficient h, as well as on the excess temperature ΔT_e. Different boiling regimes may be delineated according to the value of ΔT_e.

Free Convection Boiling Free convection boiling is said to exist if $\Delta T_e \leq \Delta T_{e,A}$, where $\Delta T_{e,A} \approx 5°C$. In this regime there is insufficient vapor in contact with the liquid phase to cause boiling at the saturation temperature. As the excess

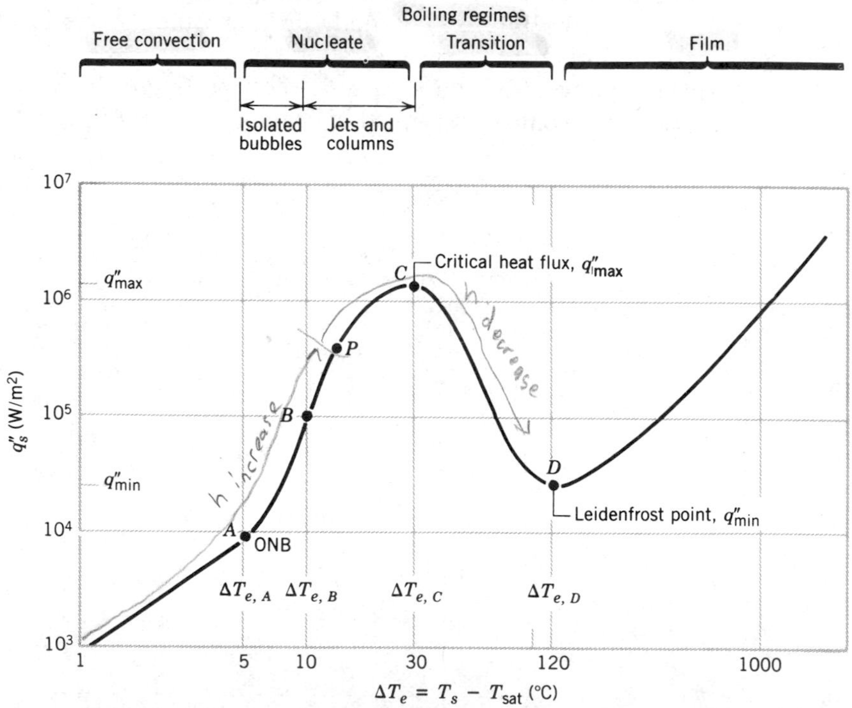

Figure 10.4 Typical boiling curve for water at one atmosphere: surface heat flux q_s'' as a function of excess temperature $\Delta T_e \equiv T_s - T_{sat}$.

temperature is increased, bubble inception will eventually occur, but below point A (referred to as the *onset of nucleate boiling*, ONB), fluid motion is determined principally by free convection effects. According to whether the flow is laminar or turbulent, h varies as ΔT_e to the $\frac{1}{4}$ or $\frac{1}{3}$ power, respectively, in which case q_s'' varies as ΔT_e to the $\frac{5}{4}$ or $\frac{4}{3}$ power.

Nucleate Boiling Nucleate boiling exists in the range, $\Delta T_{e,A} \leq \Delta T_e \leq \Delta T_{e,C}$, where $\Delta T_{e,C} \approx 30°\text{C}$. In this range, two different flow regimes may be distinguished. In the region A–B, *isolated bubbles* form at nucleation sites and separate from the surface, as illustrated in the sketch of Figure 10.2. This separation induces considerable fluid mixing near the surface, substantially increasing h and q_s''. In this regime most of the heat exchange is through direct transfer from the surface to liquid in motion at the surface, and not through the vapor bubbles rising from the surface. As ΔT_e is increased beyond $\Delta T_{e,B}$, more nucleation sites become active and increased bubble formation causes bubble interference and coalescence. In the region B–C, the vapor escapes as *jets* or *columns*, which subsequently merge into slugs of vapor. This condition is illustrated in Figure 10.5*a*. Interference between the densely populated bubbles inhibits the motion of liquid near the surface. Point P of Figure 10.4 corresponds to an inflection point in the boiling curve at which the heat transfer coefficient is a maximum. At this point h begins to decrease with increasing ΔT_e, although q_s'', which is the product of h and ΔT_e, continues to increase. This trend results because for $\Delta T_e > \Delta T_{e,P}$, the relative increase in ΔT_e exceeds the relative reduction in h. At point C, however, further increase in ΔT_e is exactly balanced by the reduction in h. The maximum heat flux, $q_{s,c}'' = q_{max}''$, is usually termed the *critical heat flux* and is generally in excess of 1 MW/m^2. At the point of this maximum, considerable vapor is being formed, making it difficult for liquid to continuously wet the surface.

Because high heat transfer rates and convection coefficients are associated with small values of the excess temperature, it is desirable to operate many engineering devices in the nucleate boiling regime. The magnitude of the convection coefficient may be inferred by using Equation 10.3 with the boiling curve of Figure 10.4. Dividing q_s'' by ΔT_e, it is evident that convection coefficients in excess of 10^4 W/m$^2 \cdot$K are characteristic of this regime. These values are considerably larger than those normally corresponding to convection with no phase change.

Transition Boiling The region corresponding to $\Delta T_{e,C} \leq \Delta T_e \leq \Delta T_{e,D}$, where $\Delta T_{e,D} \approx 120°\text{C}$, is termed *transition boiling, unstable film boiling*, or *partial film boiling*. Bubble formation is now so rapid that a vapor film or blanket begins to form on the surface. At any point on the surface, conditions may oscillate between film and nucleate boiling, but the fraction of the total surface covered by the film increases with increasing ΔT_e. Because the thermal conductivity of the vapor is much less than that of the liquid, the value of h (and q_s'') must then decrease with increasing ΔT_e.

(a)

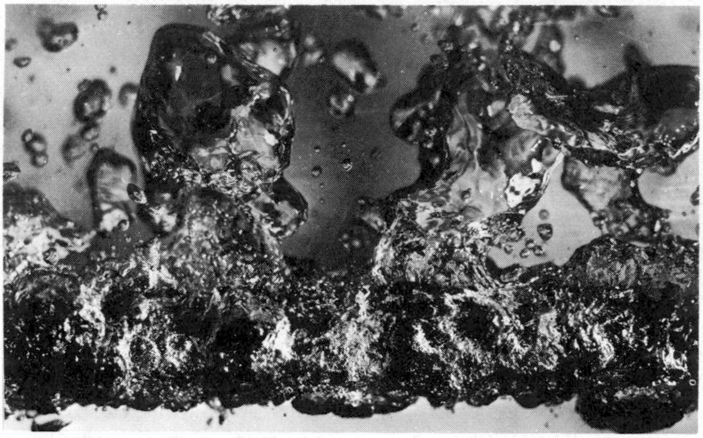

(b)

(c)

Figure 10.5 Boiling of methanol on a horizontal tube. (a) Nucleate boiling in the jets and columns regime. (b) Transition boiling. (c) Film boiling. Photographs courtesy of Professor J. W. Westwater, University of Illinois at Champaign-Urbana.

Film Boiling Film boiling exists for $\Delta T_e \geq \Delta T_{e,D}$. At point D of the boiling curve, referred to as the *Leidenfrost point*, the heat flux is a minimum, $q''_{s,D} = q''_{\min}$, and the surface is completely covered by a *vapor blanket*. Heat transfer from the surface to the liquid occurs by conduction through the vapor. It was Leidenfrost who in 1756 observed that water droplets supported by the vapor film slowly boil away as they move about a hot surface. As the surface temperature is increased, radiation through the vapor film becomes significant and the heat flux increases with increasing ΔT_e.

The nature of the vapor formation and bubble dynamics associated with nucleate boiling, transition boiling, and film boiling is shown in Figure 10.5. The photographs were obtained for the boiling of methanol on a horizontal tube.

Although the foregoing discussion of the boiling curve assumes that control may be maintained over T_s, it is important to remember the Nukiyama experiment and that many applications involve controlling q''_s (e.g., in a nuclear reactor or in an electric resistance heating device). Consider starting at some point P on Figure 10.6 and gradually increasing q''_s. The value of ΔT_e, hence the value of T_s, will also increase, following the boiling curve to point C. However, any increase in q''_s beyond this point will induce a sharp departure from the boiling curve in which surface conditions change abruptly from $\Delta T_{e,C}$ to $\Delta T_{e,E} \equiv T_{s,E} - T_{\text{sat}}$. Because $T_{s,E}$ generally exceeds the melting point of the solid, destruction or failure of the system may occur. For this reason point C is often termed the *burnout point* or the *boiling crisis*, and accurate knowledge of the critical flux, $q''_{s,C} \equiv q''_{\max}$, is important. We may want to operate a heat transfer surface close to this value, but rarely would we want to exceed it.

10.4 POOL BOILING CORRELATIONS

From the shape of the boiling curve and the fact that various physical mechanisms characterize the different regimes, it is no surprise that a multiplicity of heat transfer correlations exist for the boiling process. For the region

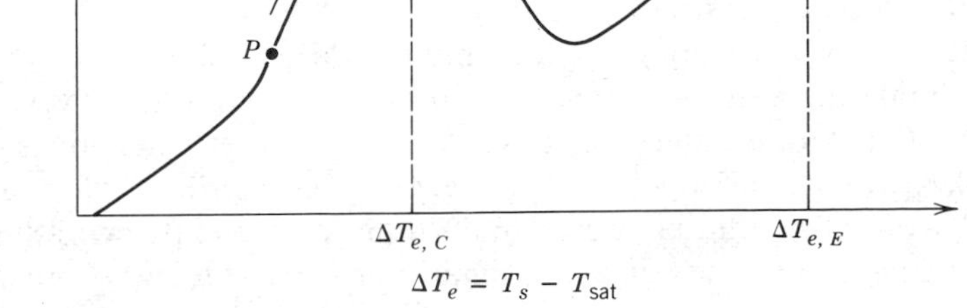

Figure 10.6 Onset of the boiling crisis.

below $\Delta T_{e,A}$ of the boiling curve (Figure 10.4), appropriate free convection correlations from Chapter 9 can be used to estimate heat transfer coefficients and heat rates. In this section we review some of the more widely used correlations for nucleate and film boiling.

10.4.1 Nucleate Pool Boiling

The analysis of nucleate boiling requires prediction of the number of surface nucleation sites and the rate at which bubbles originate from each site. While mechanisms associated with this boiling regime have been extensively studied, complete and reliable mathematical models have yet to be developed. Yamagata et al. [4] were the first to show the influence of nucleation sites on the heat rate through the relation

$$q_s'' = C \Delta T_e^a n^b \tag{10.4}$$

where n is the site density (active nucleation sites per unit area) and the exponents are approximately $a = 1.2$ and $b = \frac{1}{3}$. Although C and n vary considerably from one fluid–surface combination to another, it has been found that, for most commercial surfaces, $n \propto \Delta T_e^5$ or ΔT_e^6. Hence from Equation 10.4 it follows that q_s'' is approximately proportional to ΔT_e^3.

The foregoing dependence of q_s'' on ΔT_e characterizes the first and most useful correlation for nucleate boiling, which was developed by Rohsenow [5]

$$q_s'' = \mu_l h_{fg} \left[\frac{g(\rho_l - \rho_v)}{\sigma} \right]^{1/2} \left(\frac{c_{p,l} \Delta T_e}{C_{s,f} h_{fg} Pr_l^n} \right)^3 \tag{10.5}$$

where the subscripts l and v, respectively, denote the saturated liquid and vapor states. Appearance of the surface tension σ (N/m) follows from the significant effect this fluid property has on bubble formation and development (see Problem 10.3). The coefficient $C_{s,f}$ and exponent n depend on the surface–liquid combination, and representative values are presented in Table 10.1. Values for other liquid–surface combinations may be obtained from the literature [6, 7]. Values of the surface tension, as well as the latent heat of vaporization, for water are presented in Table A.6 and for selected fluids in Table A.5. Values for other liquids may be obtained from any recent edition of the *Handbook of Chemistry and Physics*. It is evident that the empirical correlation, Equation 10.5, does not explicitly contain the dimensionless parameters identified by Equations 10.2. Note, however, that the Jakob number, $Ja = c_{p,l} \Delta T_e / h_{fg}$, and the Prandtl number are present and that $q_s'' \propto Ja^3 Pr^{-3n}$.

The Rohsenow correlation applies only for clean surfaces, and when it is used to estimate the heat flux, errors can amount to $\pm 100\%$. However, since $\Delta T_e \propto (q_s'')^{1/3}$, this error is reduced by a factor of 3 when the expression is used to estimate ΔT_e from knowledge of q_s''. Note also that, since $q_s'' \propto h_{fg}^{-2}$ and h_{fg} decreases markedly with increasing saturation pressure (temperature), the nucleate boiling heat flux will increase significantly as the liquid is pressurized.

Table 10.1 Values of $C_{s,f}$ for various fluid–surface combinations [5–7]

FLUID–SURFACE COMBINATION	$C_{s,f}$	n
Water–copper		
Scored	0.0068	1.0
Polished	0.0130	1.0
Water–stainless steel		
Chemically etched	0.0130	1.0
Mechanically polished	0.0130	1.0
Ground and polished	0.0060	1.0
Water–brass	0.0060	1.0
Water–nickel	0.006	1.0
Water–platinum	0.0130	1.0
n-Pentane–copper		
Polished	0.0154	1.7
Lapped	0.0049	1.7
Benzene–chromium	0.101	1.7
Ethyl alcohol–chromium	0.0027	1.7

10.4.2 Critical Heat Flux for Nucleate Pool Boiling

We recognize that the critical heat flux $q''_{s,C}=q''_{max}$ represents an important point on the boiling curve. We may wish to operate a boiling process close to this point, but we appreciate the danger of dissipating heat in excess of this amount. Kutateladze [8], through dimensional analysis, and Zuber [9], through a stability analysis, obtained an expression of the form

$$q''_{max}=\frac{\pi}{24}h_{fg}\rho_v\left[\frac{\sigma g(\rho_l-\rho_v)}{\rho_v^2}\right]^{1/4}\left(\frac{\rho_l+\rho_v}{\rho_l}\right)^{1/2} \tag{10.6}$$

which, as a first approximation, is independent of surface material and is only weakly dependent on geometry. Replacing the Zuber constant $(\pi/24)-0.131$ by an experimental value of 0.149 [10] and approximating the last term in parentheses by unity, Equation 10.6 becomes

$$q''_{max}=0.149h_{fg}\rho_v\left[\frac{\sigma g(\rho_l-\rho_v)}{\rho_v^2}\right]^{1/4} \tag{10.7}$$

In principle, since this expression applies to a heater surface of infinite extent, there is no characteristic length; in practice, however, the expression is applicable if the characteristic length is large compared to the mean bubble diameter parameter D_b (see Problem 10.16). If the ratio of the heater characteristic length to D_b, which is the Bond number, is less than 3, a correction factor must be applied to Equation 10.7. Lienhard and Dhir [11] have developed such correction factors for different geometries, including plates, cylinders, spheres, and vertically and horizontally oriented ribbons.

It is important to note that the critical heat flux depends strongly on pressure, mainly through the pressure dependence of surface tension and the heat of vaporization. Cichelli and Bonilla [12] have experimentally demonstrated that the peak flux increases with pressure up to one-third of the critical pressure, after which it falls to zero at the critical pressure.

10.4.3 Minimum Heat Flux

The transition boiling regime is of little practical interest, since it may be obtained only by controlling the surface heater temperature. While no adequate theory has been developed for this regime, conditions can be characterized by periodic, *unstable* contact between the liquid and the heated surface. However, the upper limit of this regime is of interest because it corresponds to formation of a *stable* vapor blanket or film and to a minimum heat flux condition. If the heat flux drops below this minimum, the film will collapse, causing the surface to cool and nucleate boiling to be reestablished.

Zuber [9] used stability theory to derive the following expression for the minimum heat flux, $q''_{s,D} = q''_{\min}$, from a large horizontal plate.

$$q''_{\min} = C\rho_v h_{fg}\left[\frac{g\sigma(\rho_l - \rho_v)}{(\rho_l + \rho_v)^2}\right]^{1/4} \tag{10.8}$$

The constant $C = 0.09$ has been experimentally determined by Berenson [13]. This result is accurate to approximately 50% for most fluids at moderate pressures but provides poorer estimates at higher pressures [14]. A similar result has been obtained for horizontal cylinders [15].

10.4.4 Film Pool Boiling

At excess temperatures beyond the Leidenfrost point, a continuous vapor film blankets the surface and there is no contact between the liquid phase and the surface. Because conditions in the stable vapor film bear a strong resemblance to those of laminar film condensation (Section 10.7), it is customary to base film boiling correlations on results obtained from condensation theory. One such result, which applies to film boiling on a cylinder or sphere of diameter D, is of the form

$$\overline{Nu}_D = \frac{\bar{h}_{\text{conv}} D}{k_v} = C\left[\frac{g(\rho_l - \rho_v)h'_{fg}D^3}{\nu_v k_v (T_s - T_{\text{sat}})}\right]^{1/4} \tag{10.9}$$

The correlation constant C is 0.62 for horizontal cylinders [16] and 0.67 for spheres [11], and the corrected latent heat is $h'_{fg} = h_{fg} + 0.4c_{p,v}(T_s - T_{\text{sat}})$. Vapor properties are evaluated at the film temperature, $T_f = (T_s + T_{\text{sat}})/2$.

At elevated surface temperatures ($T_s \gtrsim 300°C$), radiation heat transfer across the vapor film becomes significant. Since radiation acts to increase the film thickness, it is not reasonable to assume that the radiative and convective processes are simply additive. Bromley [16] investigated film boiling from the outer surface of horizontal tubes and suggested calculating the total heat transfer coefficient from a transcendental equation of the form

$$\bar{h}^{4/3} = \bar{h}_{\text{conv}}^{4/3} + \bar{h}_{\text{rad}}\bar{h}^{1/3} \tag{10.10a}$$

If $\bar{h}_{\text{rad}} < \bar{h}_{\text{conv}}$, a simpler form may be used.

$$\bar{h} = \bar{h}_{\text{conv}} + \tfrac{3}{4}\bar{h}_{\text{rad}} \tag{10.10b}$$

The effective radiation coefficient $\bar{h}_{\text{rad}}$ may be approximated as

$$\bar{h}_{\text{rad}} = \frac{\varepsilon\sigma(T_s^4 - T_{\text{sat}}^4)}{T_s - T_{\text{sat}}} \tag{10.11}$$

where ε is the emissivity of the solid (Table A.8) and σ is the Stefan-Boltzmann constant.

Note that the analogy between film boiling and film condensation does not hold for small surfaces with high curvature because of the large disparity between vapor and liquid film thicknesses for the two processes. The analogy is also questionable for a vertical surface, although satisfactory predictions have been obtained for limited conditions.

10.4.5 Parametric Effects on Pool Boiling

In this section we briefly consider other parameters that can affect pool boiling, confining our attention to the gravitational field, liquid subcooling, and solid surface conditions.

The influence of the *gravitational field* on boiling must be considered in applications involving space travel and rotating machinery. This influence is evident from appearance of the gravitational acceleration g in the foregoing expressions. Siegel [17], in his review of low gravity effects, confirms that the $g^{1/4}$ dependence in Equations 10.7, 10.8, and 10.9 (for the maximum and minimum heat fluxes and for film boiling) is correct for values of g as low as 0.10 m^2/s. For nucleate boiling, however, evidence indicates that the heat flux is nearly independent of gravity, which is in contrast to the $g^{1/2}$ dependence of Equation 10.5. Above-normal gravitational forces show similar effects, although near the ONB, gravity can influence bubble-induced convection.

If liquid below the saturation temperature is continuously fed to a pool boiling system, without introducing forced-convection effects, the liquid is said to be *subcooled*, where $\Delta T_{\text{sub}} \equiv T_{\text{sat}} - T_l$. In the natural convection regime, the heat flux increases typically as $(T_s - T_l)^{5/4}$ or $(\Delta T_e + \Delta T_{\text{sub}})^{5/4}$. In contrast, for nucleate boiling the influence of subcooling is considered to be negligible, although the maximum and minimum heat fluxes (q''_{max} and q''_{min}, respectively) are known to increase linearly with ΔT_{sub}. For film boiling, the influence of ΔT_{sub} on heat transfer is considerable, but little information on the effect is available in the literature.

The influence of *surface roughness* (by machining, grooving, scoring, or sandblasting) on the maximum and minimum heat fluxes and on film boiling is negligible. However, as demonstrated by Berensen [18], increased surface roughness can cause nearly an order of magnitude increase in heat flux for the nucleate boiling regime. This influence can be appreciated by considering the Yamagata correlation of Equation 10.4. As Figure 10.7 illustrates, a roughened surface will have numerous cavities, which serve to trap vapor and thereby provide more and larger sites for bubble growth. It follows that the nucleation site density for a rough surface can be substantially larger than that for a smooth surface. However, under prolonged boiling, the effects of surface roughness generally diminish, indicating that the new, large sites created by roughening are not stable sources of vapor entrapment.

Special surface arrangements that provide stable *augmentation* (*enhancement*) of nucleate boiling are available commercially and have been reviewed by Webb [19]. *Enhancement surfaces* are of two types: (1) coatings of very porous material formed by sintering, brazing, flame spraying, electrolytic deposition, or foaming; and (2) mechanically machined or formed double-reentrant cavities to ensure continuous vapor trapping (see Figure 10.8). Such surfaces provide for continuous renewal of vapor at the nucleation sites and heat transfer augmentation by more than an order of magnitude. Active augmentation techniques,

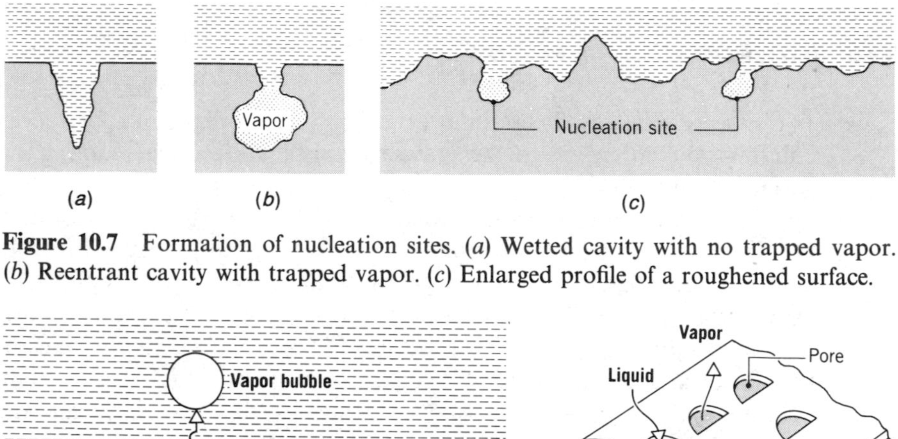

Figure 10.7 Formation of nucleation sites. (*a*) Wetted cavity with no trapped vapor. (*b*) Reentrant cavity with trapped vapor. (*c*) Enlarged profile of a roughened surface.

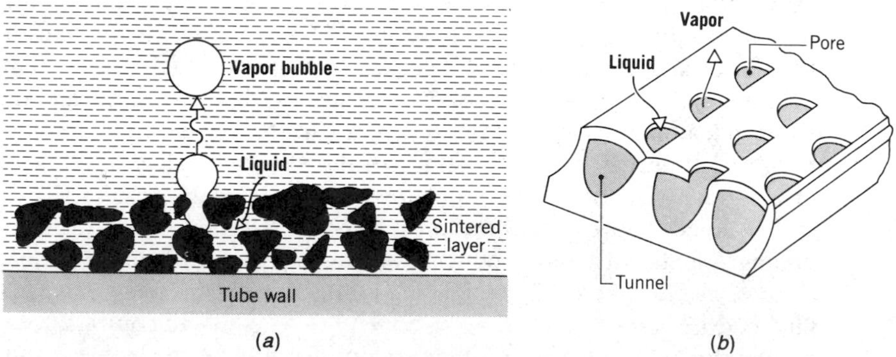

Figure 10.8 Typical structured, enhancement surfaces for augmentation of nucleate boiling. (*a*) Sintered metallic coating. (*b*) Mechanically formed double-reentrant cavity.

such as surface wiping and rotation, surface vibration, fluid vibration, and electrostatic fields, have also been reviewed by Bergles [20, 21]. However, because such techniques complicate the boiling system and, in many instances, impair reliability, they have found little practical application.

EXAMPLE 10.1

The bottom of a copper pan, 0.3 m in diameter, is maintained at 118°C by an electric heater. Estimate the power required to boil water in this pan. What is the evaporation rate? Estimate the critical heat flux.

SOLUTION

Known: Water boiling in copper pan of prescribed surface temperature.

Find:

1. Power required by electric heater to cause boiling.
2. Rate of water evaporation due to boiling.
3. Critical heat flux corresponding to the burnout point.

Schematic:

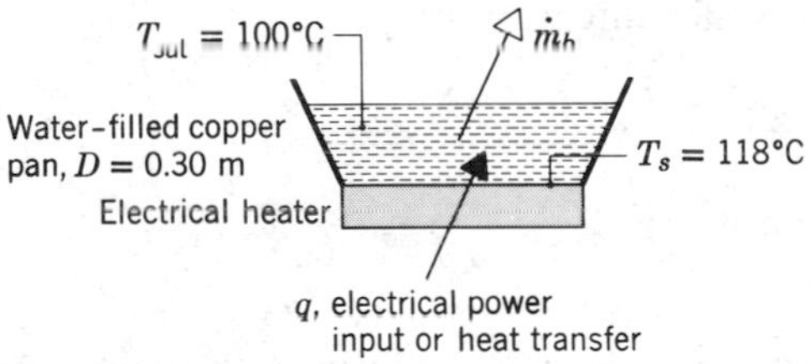

Assumptions:

1. Steady-state conditions.
2. Water exposed to standard atmospheric pressure, 1.01 bar.
3. Water at uniform temperature $T_{sat} = 100°C$.
4. Pan bottom surface polished copper.

Properties:

Table A.6, saturated water, liquid (100°C): $\rho_l = 1/v_f = 957.9\ \text{kg/m}^3$, $c_{p,l} = c_{p,f} = 4.217\ \text{kJ/kg}\cdot\text{K}$, $\mu_l = \mu_f = 279 \times 10^{-6}\ \text{N}\cdot\text{s/m}^2$, $Pr_l = Pr_f = 1.76$, $h_{fg} = 2257\ \text{kJ/kg}$, $\sigma = 58.9 \times 10^{-3}\ \text{N/m}$.
Table A.6, saturated water, vapor (100°C): $\rho_v = 1/v_g = 0.5955\ \text{kg/m}^3$.

Analysis:

1. From knowledge of the saturation temperature T_{sat} of water boiling at 1 atm and the temperature of the heated copper surface T_s, the excess temperature ΔT_e is

$$\Delta T_e \equiv T_s - T_{sat} = (118 - 100)°\text{C} = 18°\text{C}$$

According to the boiling curve of Figure 10.4, nucleate pool boiling will occur and the recommended correlation for estimating the heat transfer rate per unit area of plate surface is given by Equation 10.5.

$$q_s'' = \mu_l h_{fg} \left[\frac{g(\rho_l - \rho_v)}{\sigma}\right]^{1/2} \left(\frac{c_{p,l}\Delta T_e}{C_{s,f} h_{fg} Pr_l^n}\right)^3$$

The values of $C_{s,f}$ and n corresponding to the polished copper surface–water combination are determined from the experimental results of Table 10.1, where $C_{s,f} = 0.0130$ and $n = 1.0$. Substituting numerical values, the boiling heat flux is

$$q_s'' = 279 \times 10^{-6}\ \text{N·s/m}^2 \times 2257 \times 10^3\ \text{J/kg}$$

$$\times \left[\frac{9.8\ \text{m/s}^2\ (957.9 - 0.5955)\text{kg/m}^3}{58.9 \times 10^{-3}\ \text{N/m}}\right]^{1/2}$$

$$\times \left(\frac{4.217 \times 10^3\ \text{J/kg·K} \times 18°\text{C}}{0.0130 \times 2257 \times 10^3\ \text{J/kg} \times 1.76}\right)^3 = 789\ \text{kW/m}^2$$

Hence the boiling heat transfer rate is

$$q_s = q_s'' \times A = q_s'' \times \frac{\pi D^2}{4}$$

$$q_s = 7.89 \times 10^5\ \text{W/m}^2 \times \frac{\pi(0.30\ \text{m})^2}{4} = 55.8\ \text{kW}$$ ◁

2. Under steady-state conditions all heat addition to the pan will result in water evaporation from the pan. Hence

$$q_s = \dot{m}_b h_{fg}$$

where $\dot{m}_b$ is the rate at which water evaporates from the free surface to the room. It follows that

$$\dot{m}_b = \frac{q_s}{h_{fg}} = \frac{5.58 \times 10^4\ \text{W}}{2257 \times 10^3\ \text{J/kg}} = 0.0247\ \text{kg/s} = 89\ \text{kg/h}$$ ◁

3. The critical heat flux for nucleate pool boiling can be estimated from Equation 10.7.

$$q_{max}'' = 0.149 h_{fg} \rho_v \left[\frac{\sigma g(\rho_l - \rho_v)}{\rho_v^2}\right]^{1/4}$$

Substituting the appropriate numerical values

$$q''_{max} = 0.149 \times 2257 \times 10^3\ \text{J/kg} \times 0.5955\ \text{kg/m}^3$$

$$\times \left[\frac{58.9 \times 10^{-3}\ \text{N/m} \times 9.8\ \text{m/s}^2 (957.9 - 0.5955)\ \text{kg/m}^3}{(0.5955)^2\ (\text{kg/m}^3)^2}\right]^{1/4}$$

$$q''_{max} = 1.26\ \text{MW/m}^2 \quad \triangleleft$$

Comments:

1. The critical heat flux, $q''_{max} = 1.26\ \text{MW/m}^2$, represents the maximum heat flux for boiling water at normal atmospheric pressure. Operation of the heater at $q''_s = 0.789\ \text{MW/m}^2$ is therefore below the critical condition.
2. Using Equation 10.8, the minimum heat flux at the Leidenfrost point is $q''_{min} = 18.9\ \text{kW/m}^2$. Note from Figure 10.4 that, for this condition, $\Delta T_e \approx 120°\text{C}$.

EXAMPLE 10.2

A metal-clad heating element of 6 mm diameter and emissivity $\varepsilon = 1$ is horizontally immersed in a water bath. The surface temperature of the metal is 255°C under steady-state boiling conditions. Estimate the power dissipation per unit length for the heater.

SOLUTION

Known: Boiling from outer surface of horizontal cylinder in water.

Find: Power dissipation per unit length for the cylinder q'_s.

Schematic:

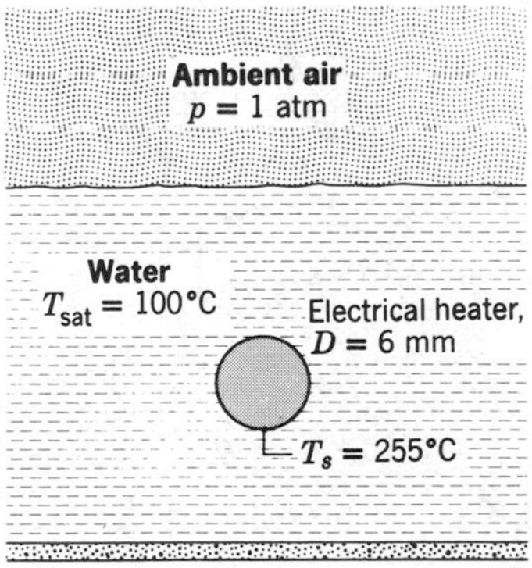

Assumptions:

1. Steady-state conditions.
2. Water exposed to standard atmospheric pressure and at uniform temperature, T_{sat}.

Properties:

Table A.6, saturated water, liquid (100°C): $\rho_l = 1/v_f = 957.9$ kg/m³, $h_{fg} = 2257$ kJ/kg.
Table A.6, saturated water, vapor ($T_f \approx 450$ K): $\rho_v = 1/v_g = 4.808$ kg/m³, $c_{p,v} = c_{p,g} = 2.56$ kJ/kg·K, $k_v = k_g = 0.0331$ W/m·K, $\mu_v = \mu_g = 14.85 \times 10^{-6}$ N·s/m².

Analysis:

The excess temperature

$$\Delta T_e = T_s - T_{\text{sat}} = (255 - 100)°\text{C} = 155°\text{C}$$

According to the boiling curve of Figure 10.4, film pool boiling conditions are approached, in which case heat transfer is due to both convection and radiation. The heat transfer rate follows from Equation 10.3, written on a per unit length basis for a cylindrical surface of diameter D.

$$q'_s = q''_s \pi D = \bar{h} \pi D \Delta T_e$$

The heat transfer coefficient $\bar{h}$ is calculated from Equation 10.10a

$$\bar{h}^{4/3} = \bar{h}_{\text{conv}}^{4/3} + \bar{h}_{\text{rad}} \bar{h}^{1/3}$$

where the convection and radiation heat transfer coefficients follow from Equations 10.9 and 10.11, respectively. For the convection coefficient:

$$h_{\text{conv}} = 0.62 \left[\frac{k_v^3 \rho_v (\rho_l - \rho_v) g (h_{fg} + 0.4 c_{p,v} \Delta T_e)}{\mu_v D \Delta T_e} \right]^{1/4}$$

$$h_{\text{conv}} = 0.62$$

$$\times \left[\frac{(0.0331)^3 (\text{W/m}\cdot\text{K})^3 \times 4.808 \text{ kg/m}^3 \ (957.9 - 4.808) \text{ kg/m}^3 \times 9.8 \text{ m/s}^2}{1} \right.$$

$$\left. \times \frac{(2257 \times 10^3 \text{ J/kg} + 0.4 \times 2.56 \times 10^3 \text{ J/kg}\cdot\text{K} \times 155°\text{C})}{14.85 \times 10^{-6} \text{ N}\cdot\text{s/m}^2 \times 6 \times 10^{-3} \text{ m} \times 155°\text{C}} \right]^{1/4}$$

$$h_{\text{conv}} = 453 \text{ W/m}^2\cdot\text{K}$$

For the radiation heat transfer coefficient

$$h_{\text{rad}} = \frac{\varepsilon \sigma (T_s^4 - T_{\text{sat}}^4)}{T_s - T_{\text{sat}}}$$

$$h_{rad} = \frac{5.67\times10^{-8}\ \text{W/m}^2\cdot\text{K}^4\ (528^4-373^4)\ \text{K}^4}{(528-373)\ \text{K}} = 21.3\ \text{W/m}^2\cdot\text{K}$$

Solving Equation 10.10a by trial and error,

$$\bar{h}^{4/3} = 453^{4/3} + 21.3\bar{h}^{1/3}$$

it follows that

$$\bar{h} = 469\ \text{W/m}^2\cdot\text{K}$$

Hence the heat transfer rate per unit length of heater element is

$$q'_s = 469\ \text{W/m}^2\cdot\text{K}\times\pi\times6\times10^{-3}\ \text{m}\times155^\circ\text{C} = 1.37\ \text{kW/m} \quad \triangleleft$$

Comments:

Equation 10.10b is appropriate for estimating $\bar{h}$ from values of $\bar{h}_{conv}$ and $\bar{h}_{rad}$ and gives the identical numerical result.

Omit

10.5 FORCED-CONVECTION BOILING

In *pool boiling* fluid flow is due exclusively to the buoyancy-driven motion of bubbles originating from the heated surface. In contrast, for *forced-convection boiling*, flow is due to a directed (bulk) motion of the fluid, as well as to buoyancy effects. Conditions depend strongly on geometry, which may involve *external* flow over heated plates and cylinders or *internal* (duct) flow. Internal, forced-convection boiling is commonly referred to as *two-phase flow* and is characterized by rapid changes from liquid to vapor in the flow direction.

10.5.1 External Forced-Convection Boiling

For external flow over a heated plate, the heat flux can be estimated by standard forced-convection correlations up to the inception of boiling. As the temperature of the heated plate is increased, nucleate boiling will occur, causing the heat flux to increase. If vapor generation is not extensive and the liquid is subcooled, Bergles and Rohsenow [7, 22] suggest a method for estimating the total heat flux in terms of components associated with pure forced convection and pool boiling.

Both forced convection and subcooling are known to increase the critical heat flux q''_{max} for nucleate boiling. Experimental values as high as 35 MW/m^2 (compared with 1.3 MW/m^2 for pool boiling of water at 1 atm) have been reported [23]. For a liquid of velocity V moving in cross flow over a cylinder of diameter D, Lienhard and Eichhorn [24] have developed the following expressions for low and high velocity flows.

Low Velocity:

$$\frac{q''_{max}}{\rho_v h_{fg} V} = \frac{1}{\pi}\left[1 + \left(\frac{4}{We_D}\right)^{1/3}\right] \tag{10.12}$$

High Velocity:

$$\frac{q''_{max}}{\rho_v h_{fg} V} = \frac{(\rho_l/\rho_v)^{3/4}}{169\pi} + \frac{(\rho_l/\rho_v)^{1/2}}{19.2\pi We_D^{1/3}} \tag{10.13}$$

The Weber number We_D is the ratio of inertia to surface tension forces and has the form

$$We_D \equiv \frac{\rho_v V^2 D}{\sigma} \tag{10.14}$$

The high and low velocity regions, respectively, are determined by whether the heat flux parameter $q''_{max}/\rho_v h_{fg} V$ is less than or greater than $[(0.275/\pi)(\rho_v/\rho_l)^{1/2} + 1]$. In most cases, Equations 10.12 and 10.13 correlate q''_{max} data within 20%.

10.5.2 Two-Phase Flow

Internal forced-convection boiling is associated with bubble formation at the inner surface of a heated tube through which a liquid is flowing. Bubble growth and separation are strongly influenced by the flow velocity, and hydrodynamic effects are significantly different from those corresponding to pool boiling. The process is complicated by the existence of different two-phase flow patterns that preclude the development of generalized theories.

Consider flow development in the heated tube of Figure 10.9. Heat transfer to the subcooled liquid that enters the tube is initially by forced convection and may be predicted from the correlations of Chapter 8. However, boiling is soon initiated, with bubbles appearing at the surface growing and being carried into the mainstream of the liquid. There is a sharp increase in the convection heat transfer coefficient associated with this *bubbly flow regime*. As the volume fraction of the vapor increases, individual bubbles coalesce to form plugs or slugs of vapor. This *slug-flow regime* is followed by an *annular-flow* regime in which the liquid forms a film. This film moves along the inner surface, while vapor moves at a faster velocity through the core of the tube. The heat transfer coefficient continues to increase through the *bubbly flow* and much of the *annular-flow* regimes. However, dry spots eventually appear on the inner surface, at which point the convection coefficient begins to decrease. The *transition regime* is characterized by the growth of the dry spots, until the surface is completely dry and all remaining liquid is in the form of droplets appearing in the vapor core. The convection coefficient continues to decrease through this regime. There is little change in this coefficient through the *mist-flow* regime, which persists until all of the droplets are converted to vapor. The vapor is then *superheated* by forced convection from the surface.

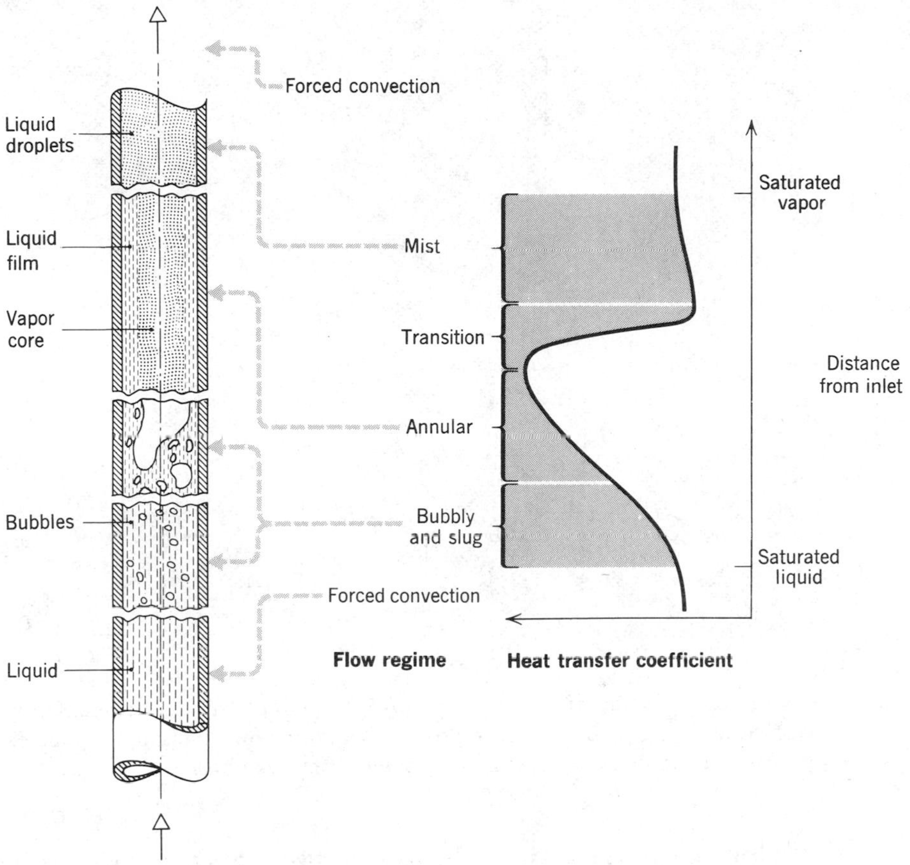

Figure 10.9 Flow regimes for forced-convection boiling inside a tube.

Discussion of the many correlations that have been developed to quantify the foregoing two-phase flow phenomena is left to the literature [7, 25–28].

10.6 CONDENSATION: PHYSICAL MECHANISMS

Condensation occurs when the temperature of a vapor is reduced below its saturation temperature. In industrial equipment, the process commonly results from contact between the vapor and a cool *surface* (Figure 10.10*a*, *b*). The latent energy of the vapor is released, heat is transferred to the surface, and the condensate is formed. Other common modes are *homogeneous* condensation (Figure 10.10*c*), where vapor condenses out as droplets suspended in a gas phase to form a fog, and *direct contact* condensation (Figure 10.10*d*), which occurs when vapor is brought into contact with a cold liquid. In this chapter we consider only surface condensation.

As shown in Figure 10.10*a*, *b*, condensation may occur in one of two possible ways, depending on the condition of the surface. The dominant form of

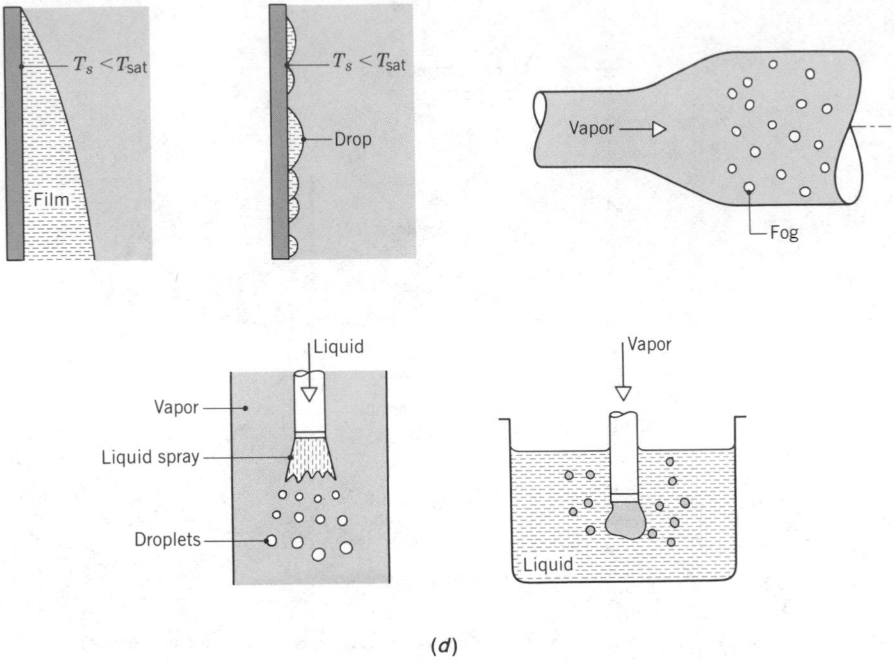

Figure 10.10 Modes of condensation. (*a*) Film and (*b*) dropwise condensation on a surface. (*c*) Homogeneous condensation or fog formation resulting from increased pressure due to expansion. (*d*) Direct contact condensation.

condensation is one in which a liquid film covers the entire condensing surface, and under the action of gravity the film flows continuously from the surface. *Film condensation* is generally characteristic of clean uncontaminated surfaces. However, if the surface is coated with a substance that inhibits wetting, it is possible to maintain *dropwise condensation.* The drops form in cracks, pits, and cavities on the surface and may grow and coalesce through condensation. Typically, more than 90% of the surface is covered by drops, ranging from a few micrometers in diameter to agglomerations visible to the naked eye. The droplets flow from the surface as a result of the action of gravity. Film and dropwise condensation of steam on a vertical copper surface are shown in Figure 10.11. A thin coating of cupric oleate was applied to the left-hand portion of the surface to promote the dropwise condensation. A thermocouple probe of 1 mm diameter extends across the photograph.

Regardless of whether it is in the form of a film or droplets, the condensate provides a resistance to heat transfer between the vapor and the surface. Because this resistance increases with condensate thickness, which increases in the flow direction, it is desirable to use short vertical surfaces or horizontal cylinders in situations involving film condensation. Most condensers therefore consist of horizontal tube bundles through which a liquid coolant flows and around which

(*a*) (*b*)

Figure 10.11 Condensation on a vertical surface. (*a*) Dropwise. (*b*) Film. Photograph courtesy of Professor J. W. Westwater, University of Illinois at Champaign Urbana.

the vapor to be condensed is circulated. In terms of maintaining high condensation and heat transfer rates, droplet formation is superior to that of film formation. In dropwise condensation most of the heat transfer is through drops less than 100 μm in diameter, and transfer rates that are more than an order of magnitude larger than those associated with film condensation may be achieved. It is therefore common practice to use surface coatings that inhibit wetting and hence stimulate dropwise condensation. Silicones, Teflon, and an assortment of waxes and fatty acids are often used for this purpose. However, such coatings gradually lose their effectiveness because of oxidation, fouling, or outright removal, and film condensation eventually occurs.

Although it is desirable to achieve dropwise condensation in industrial applications, it is often difficult to maintain this condition. For this reason and because the convection coefficients for film condensation are smaller than those for the dropwise case, condenser design calculations are often based on the assumption of film condensation. In the remaining sections of this chapter, we therefore focus on film condensation effects and only briefly mention the limited results available for dropwise condensation.

Omit

10.7 LAMINAR FILM CONDENSATION ON A VERTICAL PLATE

As shown in Figure 10.12, there may be several complicating features associated with film condensation. The film originates at the top of the plate and flows downward under the influence of gravity. The thickness δ and the condensate mass flow rate $\dot{m}$ increase with increasing x as a result of continuous condensation at the liquid–vapor interface, which is at T_{sat}. There is then heat transfer from this interface through the film to the surface, which is maintained at $T_s < T_{sat}$. In the most general case the vapor may be superheated ($T_{v,\infty} > T_{sat}$) and may be part of a mixture containing one or more noncondensable gases. Moreover, there exists a finite shear stress at the liquid–vapor interface, contributing to a velocity gradient in the vapor, as well as in the film [29, 30].

Despite the complexities associated with film condensation, useful results may be obtained by making assumptions that originated with an analysis by Nusselt [31].

1. Laminar flow and constant properties are assumed for the liquid film.
2. The gas is assumed to be a pure vapor and at a uniform temperature equal to T_{sat}. With no temperature gradient in the vapor, heat transfer to the liquid–vapor interface can occur only by condensation at the interface, not by conduction from the vapor.
3. The shear stress at the liquid–vapor interface is assumed to be negligible, in which case $\partial u/\partial y|_{y=\delta}=0$. With this assumption and the foregoing assump-

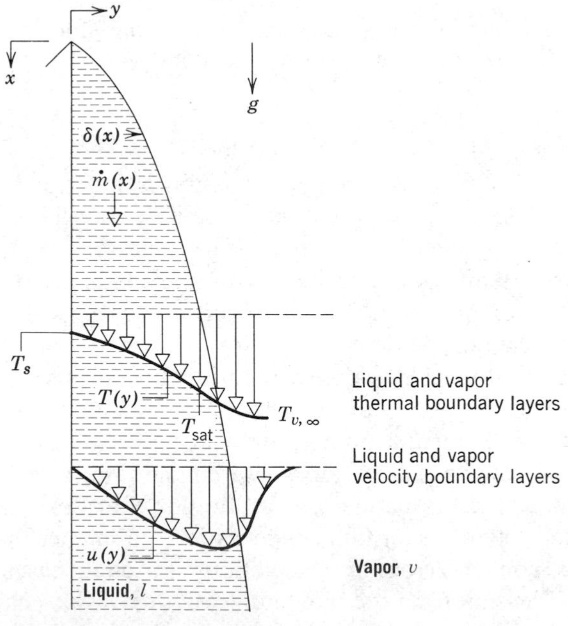

Figure 10.12 Boundary layer effects related to film condensation on a vertical surface.

tion of a uniform vapor temperature, there is then no need to consider the vapor velocity or thermal boundary layers shown in Figure 10.12.

4. Momentum and energy transfer by advection in the condensate film are assumed to be negligible. This assumption is reasonable by virtue of the low velocities associated with the film. It follows that heat transfer across the film occurs only by conduction, in which case the liquid temperature distribution is linear.

Film conditions resulting from these assumptions are shown in Figure 10.13.

From the fourth approximation, momentum advection terms may be neglected, and the x-momentum equation, (6.31), may be expressed as

$$\frac{\partial^2 u}{\partial y^2} = \frac{1}{\mu_l}\frac{dp}{dx} - \frac{X}{\mu_l} \tag{10.15}$$

where the body force X, introduced in Equation 6.17, has been retained. The body force within the film is equal to $\rho_l g$, and the pressure gradient may be approximated in terms of conditions outside the film. That is, invoking the boundary layer approximation that $(\partial p/\partial y) \approx 0$, it follows that $(dp/dx) \approx \rho_v g$. The momentum equation may therefore be expressed as

$$\frac{\partial^2 u}{\partial y^2} = -\frac{g}{\mu_l}(\rho_l - \rho_v) \tag{10.16}$$

Integrating twice and applying boundary conditions of the form $u(0) = 0$ and $\partial u/\partial y|_{y=\delta} = 0$, the velocity profile in the film becomes

$$u(y) = \frac{g(\rho_l - \rho_v)\delta^2}{\mu_l}\left[\frac{y}{\delta} - \frac{1}{2}\left(\frac{y}{\delta}\right)^2\right] \tag{10.17}$$

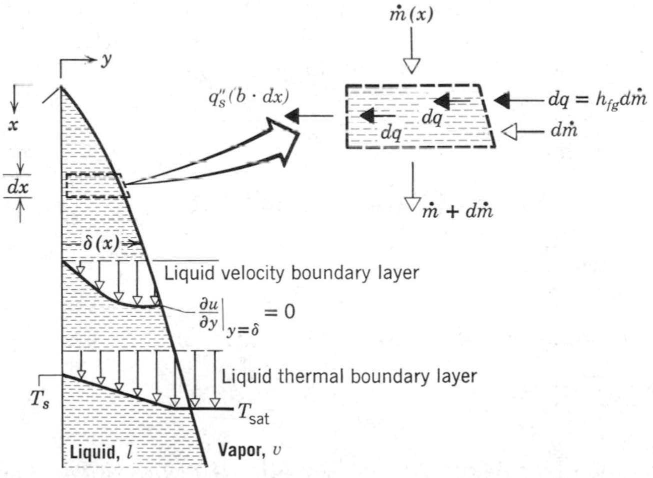

Figure 10.13 Boundary layer conditions associated with Nusselt's analysis for a vertical plate of width b.

From this result the condensate mass flow rate per unit width $\Gamma(x)$ may be obtained in terms of an integral involving the velocity profile

$$\frac{\dot{m}(x)}{b} = \int_0^{\delta(x)} \rho_l u(y)\, dy \equiv \Gamma(x) \tag{10.18}$$

Substituting from Equation 10.17, it follows that

$$\Gamma(x) = \frac{g\rho_l(\rho_l - \rho_v)\delta^3}{3\mu_l} \tag{10.19}$$

The specific variation of δ, and hence of Γ, with x may be obtained by first applying the conservation of energy requirement to the differential element shown in Figure 10.13. At a portion of the liquid–vapor interface of unit width and length dx, the rate of heat transfer into the film dq must equal the rate of energy release due to condensation at the interface. Hence

$$dq = h_{fg} d\dot{m} \tag{10.20}$$

Since advective effects are neglected, it also follows that the rate of heat transfer across the interface must equal the rate of heat transfer to the surface. Hence

$$dq = q_s''(b \cdot dx) \tag{10.21}$$

Since the liquid temperature distribution is linear, Fourier's law may be used to express the surface heat flux as

$$q_s'' = \frac{k_l(T_{\text{sat}} - T_s)}{\delta} \tag{10.22}$$

Combining Equations 10.18 and 10.20 to 10.22, we then obtain

$$\frac{d\Gamma}{dx} = \frac{k_l(T_{\text{sat}} - T_s)}{\delta h_{fg}} \tag{10.23}$$

Differentiating Equation 10.19 we also obtain

$$\frac{d\Gamma}{dx} = \frac{g\rho_l(\rho_l - \rho_v)\delta^2}{\mu_l}\frac{d\delta}{dx} \tag{10.24}$$

Combining Equations 10.23 and 10.24, it follows that

$$\delta^3 d\delta = \frac{k_l\mu_l(T_{\text{sat}} - T_s)}{g\rho_l(\rho_l - \rho_v)h_{fg}}\, dx$$

Integrating from $x = 0$, where $\delta = 0$, to any x location of interest on the surface,

$$\delta(x) = \left[\frac{4k_l\mu_l(T_{\text{sat}} - T_s)x}{g\rho_l(\rho_l - \rho_v)h_{fg}}\right]^{1/4} \tag{10.25}$$

This result may then be substituted into Equation 10.19 to obtain $\Gamma(x)$.

An improvement to the foregoing result for $\delta(x)$ was made by Nusselt [31] and Rohsenow [32], who showed that, with the inclusion of thermal advection effects, a term is added to the latent heat of vaporization. In lieu of h_{fg}, Rohsenow recommended using a modified latent heat of the form $h'_{fg}=h_{fg}+0.68\ c_{p,l}(T_{sat}-T_s)$ or, in terms of the Jakob number,

$$h'_{fg}=h_{fg}(1+0.68\,Ja) \tag{10.26}$$

The surface heat flux may also be expressed as

$$q''_s=h_x(T_{sat}-T_s) \tag{10.27}$$

Substituting from Equation 10.22, the local convection coefficient is then of the form

$$h_x=\frac{k_l}{\delta} \tag{10.28}$$

or, from Equation 10.25, with h_{fg} replaced by h'_{fg},

$$h_x=\left[\frac{g\rho_l(\rho_l-\rho_v)k_l^3h'_{fg}}{4\mu_l(T_{sat}-T_s)x}\right]^{1/4} \tag{10.29}$$

Since h_x depends on $x^{-1/4}$, it follows that the average convection coefficient for the entire plate is

$$\bar{h}_L=\frac{1}{L}\int_0^L h_x\,dx=\frac{4}{3}h_L$$

or

$$\bar{h}_L=0.943\left[\frac{g\rho_l(\rho_l-\rho_v)k_l^3h'_{fg}}{\mu_l(T_{sat}-T_s)L}\right]^{1/4} \tag{10.30}$$

The average Nusselt number then has the form

$$\overline{Nu}_L=\frac{\bar{h}_LL}{k_l}=0.943\left[\frac{\rho_lg(\rho_l-\rho_v)h'_{fg}L^3}{\mu_lk_l(T_{sat}-T_s)}\right]^{1/4} \tag{10.31}$$

In using this equation all liquid properties should be evaluated at the film temperature, $T_f=(T_{sat}+T_s)/2$, and h_{fg} should be evaluated at T_{sat}.

A more detailed boundary layer analysis of film condensation on a vertical plate has been performed by Sparrow and Gregg [29]. Their results, confirmed by Chen [33], indicate that errors associated with using Equation 10.31 are less than 3% for $Ja\leq 0.1$ and $1\leq Pr\leq 100$. Dhir and Lienhard [34] have also shown that Equation 10.31 may be used for inclined plates, if g is replaced by $g\cdot\cos\theta$, where θ is the angle between the vertical and the surface. However, it must be used with caution for large values of θ and does not apply if θ is $\pi/2$. The expression may be used for condensation on the inner or outer surface of a vertical tube of radius R, if R is much greater than δ.

The total heat transfer to the surface may be obtained by using Equation 10.30 with the following form of Newton's law of cooling:

$$q=\bar{h}_LA(T_{sat}-T_s) \tag{10.32}$$

The total condensation rate may then be determined from the relation

$$\dot{m} = \frac{q}{h'_{fg}} = \frac{\bar{h}_L A(T_{\text{sat}} - T_s)}{h'_{fg}} \tag{10.33}$$

Equations 10.32 and 10.33 are generally applicable to any surface geometry, although the form of $\bar{h}_L$ will vary according to geometry and flow conditions.

10.8 TURBULENT FILM CONDENSATION

As for all previous convection phenomena, turbulent flow conditions may exist in film condensation. Consider the vertical surface of Figure 10.14*a*. The transition criterion may be expressed in terms of a Reynolds number defined as

$$Re_\delta \equiv \frac{4\Gamma}{\mu_l} \tag{10.34}$$

With the condensate mass flow rate given by $\dot{m} = \rho_l u_m b\delta$, the Reynolds number may be expressed as

$$Re_\delta = \frac{4\dot{m}}{\mu_l b} = \frac{4\rho_l u_m \delta}{\mu_l} \tag{10.35}$$

where u_m is the average velocity in the film and δ, the film thickness, is the characteristic length. As in the case of single-phase boundary layers, the Reynolds number is an indicator of flow conditions. As shown in Figure 10.14*b*, for $Re_\delta \lesssim 30$ the film is laminar and wave free. For increased Re_δ, ripples or

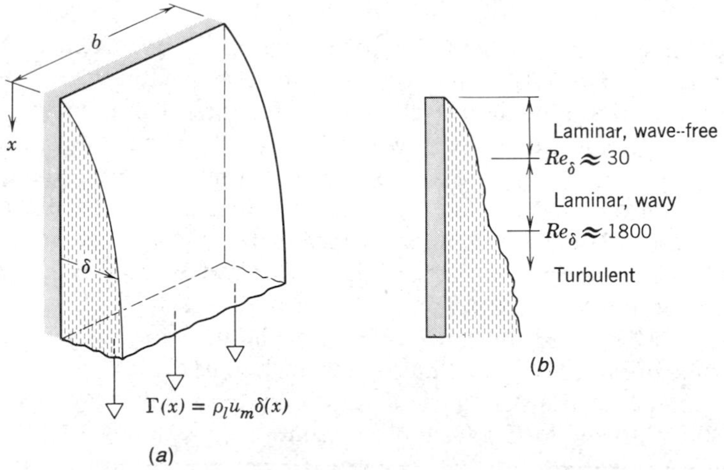

Figure 10.14 Film condensation on a vertical plate. (*a*) Condensate rate for plate of width *b*. (*b*) Flow regimes.

waves form on the condensate film, and at $Re_\delta \approx 1800$, the transition from laminar to turbulent flow is complete.

For the wave-free laminar region ($Re_\delta \lesssim 30$), Equations 10.34 and 10.19 may be combined to yield

$$Re_\delta = \frac{4g\rho_l(\rho_l - \rho_v)\delta^3}{3\mu_l^2} \tag{10.36}$$

Assuming $\rho_l \gg \rho_v$ and substituting first from Equation 10.25 and then from (10.30), Equation 10.36 can be expressed in terms of a modified Nusselt number.

$$\frac{\bar{h}_L(\nu_l^2/g)^{1/3}}{k_l} = 1.47 Re_\delta^{-1/3} \qquad Re_\delta \lesssim 30 \tag{10.37}$$

In the laminar wavy region, Kutateladze [35] recommends a correlation of the form

$$\frac{\bar{h}_L(\nu_l^2/g)^{1/3}}{k_l} = \frac{Re_\delta}{1.08 Re_\delta^{1.22} - 5.2} \qquad 30 \lesssim Re_\delta \lesssim 1800 \tag{10.38}$$

and for the turbulent region, Labuntsov [36] recommends

$$\frac{\bar{h}_L(\nu_l^2/g)^{1/3}}{k_l} = \frac{Re_\delta}{8750 + 58 Pr^{-0.5}(Re_\delta^{0.75} - 253)} \qquad Re_\delta \gtrsim 1800 \tag{10.39}$$

Graphical representation of the foregoing correlations is provided in Figure 10.15, and the trends have been experimentally verified by Gregorig et al. [37] for water over the range $1 < Re_\delta < 7200$. However, conditions associated with the transition to turbulence ($Re_\delta \sim 1800$) are somewhat uncertain and are the subject of current research.

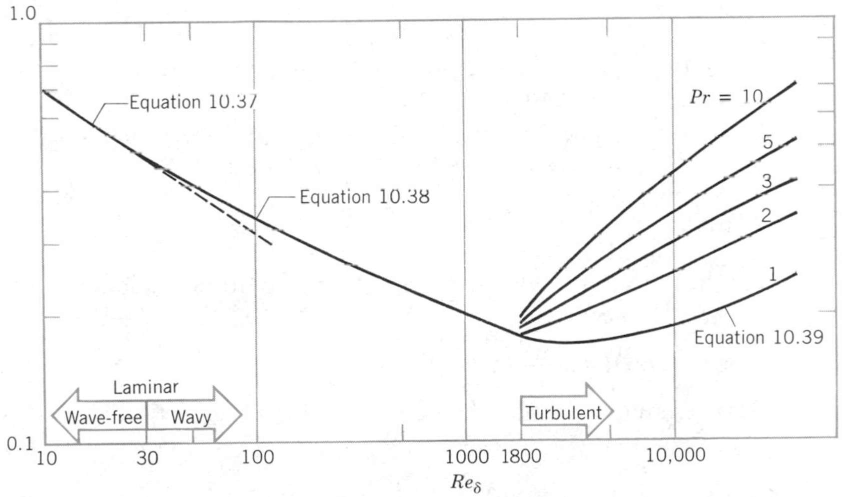

Figure 10.15 Modified Nusselt number for condensation on a vertical plate.

EXAMPLE 10.3

The outer surface of a vertical tube, which is 1 m long and has an outer diameter of 80 mm, is exposed to saturated steam at atmospheric pressure and is maintained at 50°C by the flow of cool water through the tube. What is the rate of heat transfer to the coolant, and what is the rate at which steam is condensed at the surface?

SOLUTION

Known:

Dimensions and temperature of a vertical tube experiencing condensation of steam at its outer surface.

Find: Heat transfer and condensation rates.

Schematic:

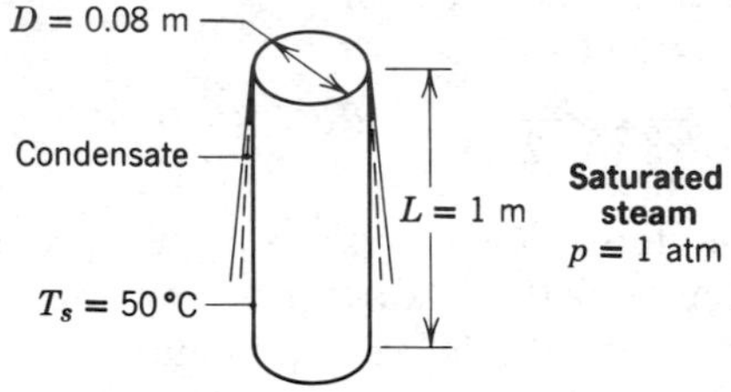

Assumptions: Laminar film condensation on a vertical surface.

Properties:

Table A.6, saturated vapor ($p = 1.0133$ bars): $T_{\text{sat}} = 100°\text{C}$, $\rho_v = (1/v_g) = 0.596$ kg/m^3, $h_{fg} = 2257$ kJ/kg.
Table A.6, saturated water ($T_f = 75°\text{C}$): $\rho_l = (1/v_f) = 975$ kg/m^3, $\mu_l = 375 \times 10^{-6}$ N·s/m^2, $k_l = 0.668$ W/m·K, $c_{p,l} = 4193$ J/kg·K.

Analysis:

The heat transfer rate may be determined from Equation 10.32, where $A = \pi DL$. Hence

$$q = \bar{h}_L(\pi DL)(T_{\text{sat}} - T_s)$$

Having assumed laminar film condensation on a vertical surface, it follows from Equation 10.30 that

$$\bar{h}_L = 0.943\left[\frac{g\rho_l(\rho_l - \rho_v)k_l^3 h'_{fg}}{\mu_l(T_{\text{sat}} - T_s)L}\right]^{1/4}$$

With

$$Ja = \frac{c_{p,l}(T_{\text{sat}} - T_s)}{h_{fg}} = \frac{4193 \text{ J/kg·K } (100 - 50) \text{ K}}{2257 \text{ kJ/kg}} = 0.0929$$

it follows that

$$h'_{fg} = h_{fg}(1 + 0.68Ja) = 2257 \text{ kJ/kg } (1.0632) = 2400 \text{ kJ/kg}$$

Hence

$$\bar{h}_L = 0.943 \times \{[9.8 \text{ m/s}^2 \times 975 \text{ kg/m}^3 \ (975 - 0.596) \text{ kg/m}^3 \ (0.668 \text{ W/m·K})^3 2.400 \times 10^6 \text{ J/kg}]/[375 \times 10^{-6} \text{ kg/s·m } (100 - 50) \text{ K} \times 1 \text{ m}]\}^{1/4}$$

$$\bar{h}_L = 4094 \text{ W/m}^2\text{·K}$$

and

$$q = 4094 \text{ W/m}^2\text{·K} \times \pi \times 0.08 \text{ m} \times 1 \text{ m } (100 - 50) \text{ K} = 51{,}446 \text{ W} \qquad \triangleleft$$

From Equation 10.33 the condensation rate is then

$$\dot{m} = \frac{q}{h'_{fg}} = \frac{51{,}446 \text{ W}}{2.400 \times 10^6 \text{ J/kg}} = 0.0214 \text{ kg/s} \qquad \triangleleft$$

Comments:

The assumption of laminar film conditions may be checked by calculating Re_δ from Equation 10.35.

$$Re_\delta = \frac{4\dot{m}}{\mu_l b} = \frac{4 \times 0.0214 \text{ kg/s}}{375 \times 10^{-6} \text{ kg/s·m } \pi(0.08 \text{ m})} = 908$$

Since $30 < Re_\delta < 1800$, a significant portion of the condensate is in the wavy-laminar region. Hence the wave-free laminar assumption may be poor.

For the wavy-laminar region, Equation 10.38 may be used with the specified surface dimensions and temperature. Expressing $\bar{h}_L$ in terms of the Reynolds number by combining Equations 10.33 and 10.35,

$$\bar{h}_L = \frac{\dot{m}h'_{fg}}{A(T_{\text{sat}} - T_s)} = \frac{Re_\delta(\mu_l b)h'_{fg}}{4A(T_{\text{sat}} - T_s)}$$

Equation 10.38 may be expressed as

$$\frac{Re_\delta \mu_l b h'_{fg}}{4A(T_{\text{sat}} - T_s)} = \frac{Re_\delta}{1.08Re_\delta^{1.22} - 5.2} \times \frac{k_l}{(\nu_l^2/g)^{1/3}}$$

which can be solved to obtain a Reynolds number of $Re_\delta = 1173$. It follows from Equations 10.38, 10.32, and 10.33, respectively, that $\bar{h}_L = 5285 \text{ W/m}^2\text{·K}$, $q = 66{,}418$ W, and $\dot{m} = 0.0276$ kg/s. Hence estimates of the heat and condensate rates based on the wavy-laminar correlation exceed by 29% those based on Nusselt's analysis.

Note that using Equation 10.25, with the corrected latent heat, the film thickness at the bottom of the tube $\delta(L)$ for the wave-free laminar assumption is

$$\delta(L) = \left[\frac{4k_l\mu_l(T_{\text{sat}} - T_s)L}{g\rho_l(\rho_l - \rho_v)h'_{fg}}\right]^{1/4}$$

$$\delta(L) = \left[\frac{4 \times 0.668 \text{ W/m}\cdot\text{K} \times 375 \times 10^{-6} \text{ kg/s}\cdot\text{m } (100-50) \text{ K} \times 1 \text{ m}}{9.8 \text{ m/s}^2 \times 975 \text{ kg/m}^3 \text{ } (975 - 0.596) \text{ kg/m}^3 \times 2.400 \times 10^6 \text{ J/kg}}\right]^{1/4}$$

$$\delta(L) = 2.18 \times 10^{-4} \text{ m} = 0.218 \text{ mm}$$

Hence $\delta(L) \ll (D/2)$ and use of the vertical plate correlation for a vertical cylinder is justified.

10.9 FILM CONDENSATION ON RADIAL SYSTEMS

The Nusselt analysis may be extended to laminar film condensation on the outer surface of a sphere and a horizontal tube (Figure 10.16*a*, *b*), and the average convection coefficient may be expressed as [34]

$$\bar{h}_D = C\left[\frac{g\rho_l(\rho_l - \rho_v)k_l^3 h'_{fg}}{\mu_l(T_{\text{sat}} - T_s)D}\right]^{1/4} \tag{10.40}$$

where $C = 0.815$ for the sphere and 0.729 for the tube.

For a vertical tier of N horizontal tubes, Figure 10.16*c*, the average (over the N tubes) may be expressed as

$$\bar{h}_{D,N} = 0.729\left[\frac{g\rho_l(\rho_l - \rho_v)k_l^3 h'_{fg}}{N\mu_l(T_{\text{sat}} - T_s)D}\right]^{1/4} \tag{10.41}$$

That is, $\bar{h}_{D,N} = \bar{h}_D N^{-1/4}$, where $\bar{h}_D$ is the heat transfer coefficient for the first (upper) tube. Such an arrangement is often used in condenser design. The reduction in $\bar{h}$ with increasing N may be attributed to an increase in the average film thickness for each successive tube. Equations 10.40 and 10.41 are generally in agreement with or slightly lower than experimental results for pure vapors. Departures may be attributed to ripples in the liquid surface for the single horizontal tube. For the tube bank, it is assumed that the condensate falls in a continuous sheet (Figure 10.16*c*), and two effects are neglected: heat transfer to the condensate sheet between the tubes and momentum gain as the sheet falls freely under gravity. These effects enhance heat transfer, and Chen [38] has

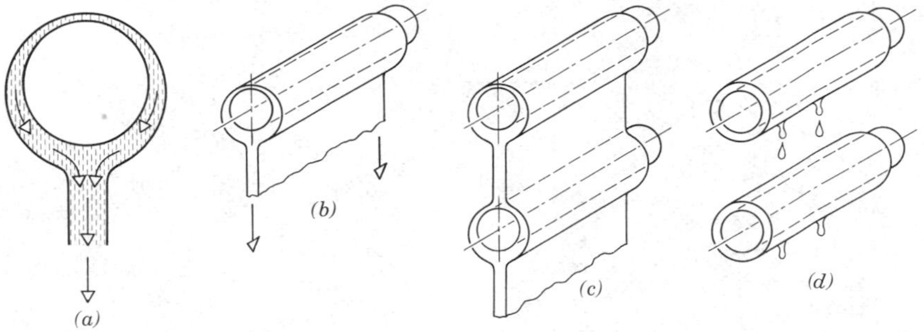

Figure 10.16 Film condensation on (*a*) a sphere, (*b*) a single horizontal tube, (*c*) a vertical tier of horizontal tubes with a continuous condensate sheet, and (*d*) with dripping condensate.

accounted for their influence in terms of the Jakob number and the number of tubes. For $Ja < 0.1$, however, heat transfer is enhanced by less than 15%. Despite this correction, experimental results tend to be higher than the predictions. A plausible explanation for the discrepancy is that, rather than flow as a continuous sheet, the condensate drips from tube to tube, as illustrated in Figure 10.16*d*. Dripping reduces the sheet thickness and promotes turbulence, thereby enhancing heat transfer.

If the length-to-diameter ratio exceeds $1.8 \tan\theta$ [39], the foregoing equations may be applied to inclined tubes by replacing g with $g \cos\theta$, where the angle θ is measured from the horizontal position. In the presence of noncondensable gases, however, the convection coefficient will be less than predictions based on the foregoing correlations.

EXAMPLE 10.4

A steam condenser consists of a square array of 400 tubes, each 6 mm in diameter. If the tubes are exposed to saturated steam at a pressure of 0.15 bar and the tube surface temperature is maintained at 25°C, what is the rate at which steam is condensed per unit length of the tubes?

SOLUTION

Known:

Configuration and surface temperature of condenser tubes exposed to saturated steam at 0.15 bar.

Find: Condensation rate per unit length of tubes.

Schematic:

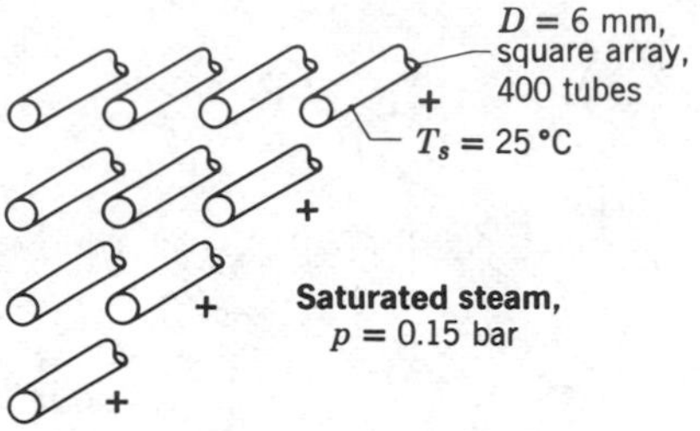

Assumptions:

1. Negligible concentration of noncondensable gases in the steam.
2. Laminar film condensation on the tubes.

Properties:

Table A.6, saturated vapor ($p=0.15$ bar): $T_{sat}=327\text{ K}=54°\text{C}$, $\rho_v=(1/v_g)=0.098\text{ kg/m}^3$, $h_{fg}=2373$ kJ/kg.
Table A.6, saturated water ($T_f=312.5$ K): $\rho_l=(1/v_f)=992\text{ kg/m}^3$, $\mu_l=663\times10^{-6}\text{ N}\cdot\text{s/m}^2$, $k_l=0.631\text{ W/m}\cdot\text{K}$, $c_{p,l}=4178\text{ J/kg}\cdot\text{K}$.

Analysis:

The condensation rate for a single tube of the array may be obtained from Equation 10.33, where for a unit length of the tube,

$$\dot{m}'_1=\frac{q'_1}{h'_{fg}}=\frac{\bar{h}_D(\pi D)(T_{sat}-T_s)}{h'_{fg}}$$

From Equation 10.41

$$\bar{h}_{D,N}=0.729\left[\frac{g\rho_l(\rho_l-\rho_v)k_l^3h'_{fg}}{N\mu_l(T_{sat}-T_s)D}\right]^{1/4}$$

or with $N=20$, $Ja=0.051$, and $h'_{fg}=2455$ J/kg,

$$\bar{h}_{D,N}=0.729$$
$$\times\{[9.8\text{ m/s}^2\times992\text{ kg/m}^3\,(992-0.098)\text{ kg/m}^3\,(0.631\text{ W/m}\cdot\text{K})^3 2.455$$
$$\times10^6\text{ J/kg}]/[20\times663\times10^{-6}\text{ kg/s}\cdot\text{m}\,(54-25)\text{ K}\times0.006\text{ m}]\}^{1/4}$$
$$\bar{h}_D=5188\text{ W/m}^2\cdot\text{K}$$

Hence for a single tube

$$\dot{m}'_1=\frac{5188\text{ W/m}^2\cdot\text{K}\,(\pi\times0.006\text{ m})(54-25)\text{ K}}{2.455\times10^6\text{ J/kg}}=1.16\times10^{-3}\text{ kg/s}\cdot\text{m}$$

For the complete array, the condensation rate per unit length is then

$$\dot{m}' = N^2 \dot{m}_1' = 400 \times 1.16 \times 10^{-3} \text{ kg/s}\cdot\text{m} = 0.464 \text{ kg/s}\cdot\text{m} \quad \triangleleft$$

Comments:

Since $Ja < 0.1$, Equation 10.41 provides a reliable estimate of the average heat transfer coefficient.

10.10 FILM CONDENSATION IN HORIZONTAL TUBES

Condensers used for refrigeration and air conditioning systems generally involve vapor condensation inside horizontal or vertical tubes. Conditions within the tube are complicated and depend strongly on the velocity of the vapor flowing through the tube. If this velocity is small, condensation occurs in the manner depicted by Figure 10.17*a* for a horizontal tube. That is, the condensate flow is from the upper portion of the tube to the bottom, from whence it flows in a longitudinal direction with the vapor. For low vapor velocities such that

$$Re_{v,i} = \left(\frac{\rho_v u_{m,v} D}{\mu_v}\right)_i < 35{,}000$$

where i refers to the tube inlet, Chato [40] recommends an expression of the form

$$\bar{h}_D = 0.555 \left[\frac{g \rho_l (\rho_l - \rho_v) k_l^3 h'_{fg}}{\mu_l (T_{\text{sat}} - T_s) D}\right]^{1/4} \tag{10.42}$$

where, for this case, the modified latent heat is

$$h'_{fg} \equiv h_{fg} + \tfrac{3}{8} c_{p,l} (T_{\text{sat}} - T_s) \tag{10.43}$$

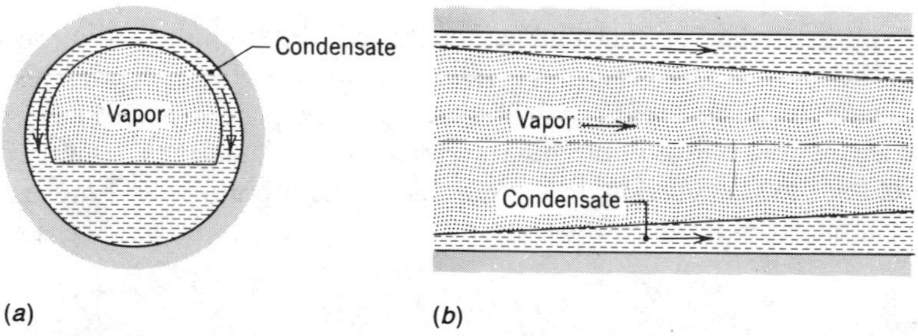

Figure 10.17 Film condensation in a horizontal tube. (*a*) Cross section of condensate flow for low vapor velocities. (*b*) Longitudinal section of condensate flow for large vapor velocities.

At higher vapor velocities the two-phase flow regime becomes annular in nature, as shown in Figure 10.17*b*. The vapor occupies the core of the annulus, diminishing in diameter as the thickness of the outer condensate layer increases in the flow direction. Results for this flow condition are provided by Rohsenow [41].

10.11 DROPWISE CONDENSATION

Typically, heat transfer coefficients for dropwise condensation are an order of magnitude larger than those for film condensation. In fact, in heat exchanger applications for which dropwise condensation is promoted, other thermal resistances may be significantly larger than that due to condensation; therefore, reliable correlations for the condensation process are not needed.

Of the many surface–fluid systems studied [42], most of the data are for steam condensation on well-promoted copper surfaces. Near atmospheric pressure [43], the data are correlated by an expression of the form

$$\bar{h}_{dc} = 51{,}104 + 2044T_s \qquad 22°\text{C} < T_s < 100°\text{C} \tag{10.44}$$

$$\bar{h}_{dc} = 255{,}510 \qquad 100°\text{C} < T_s \tag{10.45}$$

where the heat transfer coefficient has units of $\text{W/m}^2 \cdot \text{K}$.

The effect of noncondensable vapors in the steam can be very important and has been studied by Shade and Mikic [44]. In addition, if the condensing surface material does not conduct as well as copper or silver, its thermal resistance becomes a factor. Since all of the heat is transferred to the drops, which are very small and widely distributed over the surface, heat flow lines within the surface material near the active areas of condensation will *crowd*, inducing a *constriction* resistance. This effect has been studied by Hannemann and Mikic [45].

10.12 SUMMARY

It is apparent that boiling and condensation are complicated processes for which the existence of generalized relations is somewhat limited. This chapter has identified the essential physical features of the processes and has presented correlations suitable for approximate engineering calculations. However, a great deal of additional information is available in the literature, and much of it has been summarized in several extensive reviews [7, 14, 23, 25–28, 41, 43, 46–49].

REFERENCES

1. Fox, R. W., and A. T. McDonald, *Introduction to Fluid Mechanics*, 3rd ed., Wiley, New York, 1985.
2. Nukiyama, S., "The Maximum and Minimum Values of Heat Transmitted from

Metal to Boiling Water Under Atmospheric Pressure," *J. Japan Soc. Mech. Eng.*, *37*, 367–374, 1934. (Translation: *Int. J. Heat Mass Transfer*, *9*, 1419–1433, 1966.)

3. Drew, T. B., and C. Mueller, "Boiling," *Trans. AIChE*, *33*, 449, 1937.
4. Yamagata, K., F. Kirano, K. Nishiwaka, and H. Matsuoka, "Nucleate Boiling of Water on the Horizontal Heating Surface," *Mem. Fac. Eng. Kyushu*, *15*, 98, 1955.
5. Rohsenow, W. M., "A Method of Correlating Heat Transfer Data for Surface Boiling Liquids," *Trans. ASME*, *74*, 969, 1952.
6. Vachon, R. I., G. H. Nix, and G. E. Tanger, "Evaluation of Constants for the Rohsenow Pool-Boiling Correlation," *J. Heat Transfer*, *90*, 239, 1968.
7. Rohsenow, W. M., "Boiling," in W. M. Rohsenow and J. P. Hartnett, Eds., *Handbook of Heat Transfer*, McGraw-Hill, New York, 1973, Chapter 13.
8. Kutateladze, S. S., "On the Transition to Film Boiling Under Natural Convection," *Kotloturbostroenie*, *3*, 10, 1948
9. Zuber, N., "On the Stability of Boiling Heat Transfer," *Trans. ASME*, *80*, 711, 1958.
10. Lienhard, J. H., V. K. Dhir, and D. M. Riherd, "Peak Pool Boiling Heat Flux Measurements on Finite Horizontal Flat Plates," *J. Heat Transfer*, *95*, 477–482, 1973.
11. Lienhard, J. H., and V. K. Dhir, "Extended Hydrodynamic Theory of the Peak and Minimum Pool Boiling Heat Fluxes," National Aeronautics and Space Administration Rep. NASA-CR-2270, July 1973.
12. Cichelli, M. T., and C. F. Bonilla, "Heat Transfer to Liquids Boiling Under Pressure," *Trans. AIChE*, *41*, 755, 1945.
13. Berenson, P. J., "Film Boiling Heat Transfer for a Horizontal Surface," *J. Heat Transfer*, *83*, 351–358, 1961.
14. Hahne, E., and U. Grigull, *Heat Transfer in Boiling*, Hemisphere/Academic Press, New York, 1977.
15. Lienhard, J. H., and P. T. Y. Wong, "The Dominant Unstable Wavelength and Minimum Heat Flux During Film Boiling on a Horizontal Cylinder," *J. Heat Transfer*, *86*, 220–226, 1964.
16. Bromley, L. A., "Heat Transfer in Stable Film Boiling," *Chem. Eng. Prog.*, *46*, 221, 1950.
17. Siegel, R., "Effect of Reduced Gravity on Heat Transfer," *Adv. Heat Transfer*, *4*, 143–228, 1967.
18. Berensen, P. J., "Experiments on Pool Boiling Heat Transfer," *Int. J. Heat Mass Transfer*, *5*, 985–999, 1962.
19. Webb, R. L., "The Evolution of Enhanced Surface Geometries for Nucleate Boiling," *Heat Transfer Eng.*, *12*, 46–49, 1981, and "Nucleate Boiling on Porous Coated Surfaces," *Heat Transfer Eng.*, *4*, 71–82, 1983.
20. Bergles, A. E., "Enhancement of Heat Transfer," *Heat Transfer 1978*, *6*, pp. 89–108, Hemisphere Publishing Corp., New York, 1978.
21. Bergles, A. E., "Augmentation of Boiling and Evaporation," in E. U. Schlünder, Ed.-in-Chief, *Heat Exchanger Design Handbook*, Vol. 2, Hemisphere Publishing Corp., 1983, Section 2.7.9.
22. Bergles, A. E., and W. H. Rohsenow, "The Determination of Forced-Convection Surface Boiling Heat Transfer, *J. Heat Transfer*, *86*, 365–372, 1964.
23. van Stralen, S., and R. Cole, *Boiling Phenomena*, McGraw-Hill/Hemisphere, New York, 1979.

24. Lienhard, J. H., and R. Eichhorn, "Peak Boiling Heat Flux on Cylinders in a Cross Flow," *Int. J. Heat Mass Transfer*, *19*, 113, 1966.
25. Tong, L. S., *Boiling Heat Transfer and Two-Phase Flow*, Wiley, New York, 1965.
26. Collier, J. G., *Convective Boiling and Condensation*, McGraw-Hill, New York, 1972.
27. Griffith, P., "Two-Phase Flow," in W. M. Rohsenow and J. P. Hartnett, Eds., *Handbook of Heat Transfer*, McGraw-Hill, New York, 1973, Chapter 14.
28. Ginoux, J. N., *Two-Phase Flow and Heat Transfer*, McGraw-Hill/Hemisphere, New York, 1978.
29. Sparrow, E. M., and J. L. Gregg, "A Boundary Layer Treatment of Laminar Film Condensation," *J. Heat Transfer*, *81*, 13, 1959.
30. Koh, J. C. Y., E. M. Sparrow, and J. P. Hartnett, "The Two-Phase Boundary Layer in Laminar Film Condensation," *Int. J. Heat Mass Transfer*, *2*, 69, 1961.
31. Nusselt, W., "Die Oberflachenkondensation des Wasserdampfes," *Z. Ver. Deut. Ing.*, *60*, 541, 1916.
32. Rohsenow, W. M., "Heat Transfer and Temperature Distribution in Laminar Film Condensation," *Trans, ASME*, *78*, 1645–1648, 1956.
33. Chen, M. M., "An Analytical Study of Laminar Film Condensation. Part 1. Flat Plate," *J. Heat Transfer*, *83*, 48–54, 1961.
34. Dhir, V. K., and J. H. Lienhard, "Laminar Film Condensation on Plane and Axisymmetric Bodies in Non-uniform Gravity," *J. Heat Transfer*, *93*, 97–100, 1971.
35. Kutateladze, S. S., *Fundamentals of Heat Transfer*, Academic Press, New York, 1963.
36. Labuntsov, D. A., "Heat Transfer in Film Condensation of Pure Steam on Vertical Surfaces and Horizontal Tubes," *Teploeneroetika*, *4*, 72–80, 1957.
37. Gregorig, R., J. Kern, and K. Turek, "Improved Correlation of Film Condensation Data Based on a More Rigorous Application of Similarity Parameters," *Wärme-Stoffübertragung*, *7*, 1–13, 1974.
38. Chen, M. M., "An Analytical Study of Laminar Film Condensation. Part 2. Single and Multiple Horizontal Tubes," *J. Heat Transfer*, *83*, 55–60, 1961.
39. Selin, G., "Heat Transfer by Condensing Pure Vapours Outside Inclined Tubes," *International Developments in Heat Transfer*, Part 2, International Heat Transfer Conference, University of Colorado, ASME, New York, 1961, pp. 278–289.
40. Chato, J. C., "Laminar Condensation Inside Horizontal and Inclined Tubes," *J. ASHRAE*, *4*, 52, 1962.
41. Rohsenow, W. M., "Film Condensation," in W. M. Rohsenow and J. P. Hartnett, Eds., *Handbook of Heat Transfer*, McGraw-Hill, New York, 1973, Chapter 12A.
42. Tanner, D. W., D. Pope, C. J. Potter, and D. West, "Heat Transfer in Dropwise Condensation of Low Pressures in the Absence and Presence of Noncondensable Gas," *Int. J. Heat Mass Transfer*, *11*, 181, 1968.
43. Griffith, P., "Dropwise Condensation," in E. U. Schlünder, Ed.-in-Chief, *Heat Exchanger Design Handbook*, Vol. 2, Hemisphere Publishing Corp., New York, 1983, Section 2.6.5.
44. Shade, R., and Mikic, B., "The Effects of Noncondensable Gases on Heat Transfer During Dropwise Condensation," AIChE 67th Annual Meeting, Paper 67b, Washington, D.C., 1974.
45. Hannemann, R., and B. Mikic, "An Experimental Investigation into the Effect of

Surface Thermal Conductivity on the Rate of Heat Transfer in Dropwise Condensation," *Int. J. Heat Mass Transfer*, *19*, 1309, 1976.

46. Collier, J. G., "Pool Boiling," in E. U. Schlünder, Ed.-in-Chief, *Heat Exchanger Design Handbook*, Vol. 2, Hemisphere Publishing Corp., New York, 1983, Section 2.7.2.
47. Butterworth, D., "Filmwise Condensation," in D. Butterworth and G. F. Hewitt, Eds., *Two-Phase Flow and Heat Transfer*, Oxford University Press, London, 1977, pp. 426–462.
48. Butterworth, D., "Film Condensation of Pure Vapor," in E. U. Schlünder, Ed.-in-Chief, *Heat Exchanger Design Handbook*, Vol. 2, Hemisphere Publishing Corp., New York, 1983, Section 2.6.2.
49. Rose, J. W., "Dropwise Condensation Theory," *Int. J. Heat Mass Transfer*, *24*, 191–194, 1981.
50. Jakob, M. and G. A. Hawkins, *Elements of Heat Transfer*, 3rd ed., Wiley, New York, 1957.
51. Lienhard, J. H., and V. K. Dhir, "Hydrodynamic Prediction of Peak Boiling Heat Fluxes from Finite Bodies," *J. Heat Transfer*, *95*, 152–158, 1973.

PROBLEMS

10.1 Show that, for water at 1 atm pressure with $T_s - T_{sat} = 10°C$, the Jakob number is much less than unity. What is the physical significance of this result? Verify that this conclusion applies to other fluids.

10.2 The surface of a horizontal, 20-mm-diameter cylinder is maintained at an excess temperature of 5°C in saturated water at 1 atm. Estimate the heat flux using the appropriate free convection correlation and compare your result with the boiling curve of Figure 10.4. For nucleate boiling, estimate the maximum value of the heat transfer coefficient from the boiling curve.

10.3 The role of surface tension in bubble formation can be demonstrated by considering a spherical bubble of pure saturated vapor in *mechanical* and *thermal* equilibrium with its superheated liquid.

(a) Beginning with an appropriate free body diagram of the bubble, perform a force balance to obtain an expression for the bubble radius,

$$r_b = \frac{2\sigma}{p_{sat} - p_l}$$

where p_{sat} is the pressure of the saturated vapor and p_l is the pressure of the superheated liquid outside the bubble.

(b) On a pressure–volume diagram, represent the bubble and liquid states. Discuss what changes in these conditions will cause the bubble to grow or collapse.

(c) Calculate the bubble size under equilibrium conditions for which the vapor is saturated at 101°C and the liquid pressure corresponds to a saturation temperature of 100°C.

10.4 A long, 1-mm-diameter wire passes an electrical current dissipating 3150 W/m and reaches a surface temperature of 126°C when submerged in water at 1 atm. What is the boiling heat transfer coefficient? Estimate the value of the correlation coefficient $C_{s,f}$.

10.5 Estimate the nucleate pool boiling heat transfer coefficient for water boiling at atmospheric pressure on the outer surface of a platinum-plated, 10-mm-diameter tube maintained 10°C above the saturation temperature.

10.6 Estimate the nucleate pool boiling heat transfer coefficient for water under atmospheric pressure in contact with mechanically polished stainless steel when the excess temperature is 15°C.

10.7 A simple expression to account for the effect of pressure on the nucleate boiling convection coefficient in water is [50]

$$h = C(\Delta T_e)^n \left(\frac{p}{p_a}\right)^{0.4}$$

where p and p_a are the system pressure and standard atmospheric pressure, respectively. For a horizontal plate and the range $15 < q_s'' < 235\ \text{kW/m}^2$, $C = 5.56$ and $n = 3$. Units of ΔT_e and h are K and $\text{W/m}^2 \cdot \text{K}$, respectively. Compare predictions from this expression with the Rohsenow correlation ($C_{s,f} = 0.013$, $n = 1$) for pressures of 2 and 5 bar with $\Delta T_e = 10°\text{C}$.

10.8 Calculate the critical heat flux for the following fluids at 1 atm: mercury, ethanol, and refrigerant R-12. Compare these results to the critical heat flux for water at 1 atm.

10.9 The bottom of a copper pan, 150 mm in diameter, is maintained at 115°C by the heating element of an electric range. Estimate the power required to boil the water in this pan. Determine the evaporation rate. What is the ratio of the surface heat flux to the critical heat flux? What pan temperature is required to achieve the critical heat flux?

10.10 Advances in large scale integration (LSI) of electronic devices, particularly for larger computers, is often restricted by the ability to cool the individual chips. Typical construction of a device package involves an array of several hundred chips, each of area 25 mm × 25 mm, mounted on a ceramic substrate. A method of cooling the array is by immersion of the device in a low boiling point fluid such as refrigerant R-113. Estimate the permissible power level of a chip if the operation point is 50% of the critical heat flux. What is the temperature rise of the chip for this condition? Assume $C_{s,f} = 0.004$ and $n = 1.7$.

10.11 Saturated ethylene glycol at 1 atm is heated by a chromium-plated surface of diameter 200 mm maintained at 480 K. Estimate the heating power requirement and the rate of evaporation. What fraction is the power requirement of the maximum power associated with the critical heat flux?

10.12 Copper tubes 25 mm in diameter and 0.75 m long are to be used for boiling saturated water at 1 atm. If the tubes are to be operated at 75% of the critical heat flux, how many tubes are required to provide an evaporation rate of 750 kg/h? What is the surface temperature of the tubes for this condition?

10.13 Estimate the current at which a 1-mm-diameter nickel wire will burn out when submerged in water at atmospheric pressure. The electrical resistance of the wire is 0.129 Ω/m.

10.14 Estimate the power (W/m^2) required to maintain a brass plate at 115°C while boiling saturated water at 1 atm. What is the power requirement if the water is pressurized to 10 atm? At what fraction of the critical heat flux is the plate operating?

10.15 It has been demonstrated experimentally that the critical heat flux is highly dependent on pressure, primarily through the pressure dependence of the fluid surface tension and latent heat of vaporization. Using Equation 10.7, calculate values of q''_{max} for water as a function of pressure. Demonstrate that the peak critical heat flux occurs at approximately one-third the critical pressure ($p_c = 221$ bars). Since all common fluids have this characteristic, suggest what coordinates should be used to plot critical heat flux–pressure values to obtain a universal curve.

10.16 In applying dimensional analysis, Kutateladze [8] postulated that the critical heat flux varies with the heat of vaporization, vapor density, surface tension, and a bubble diameter parameter

$$D_b = \left[\frac{\sigma}{g(\rho_l - \rho_v)}\right]^{1/2}$$

(a) Verify that dimensional analysis would yield the following expression for the critical heat flux

$$q''_{max} - C h_{fg} \rho_v^{1/2} D_b^{-1/2} \sigma^{1/2}$$

(b) The ratio of the characteristic length L of a heater surface (its width or diameter) to the bubble diameter parameter D_b is referred to as the Bond number (Bo) and compares buoyant and capillary forces. The correlation of Equation 10.7 strictly applies when $Bo \geq 3$. Estimate the heater size that will satisfy this requirement for water at 1 atm.

10.17 When the heater surface is not infinite, as strictly required for use of the critical heat flux correlation of Equation 10.7, more detailed correlations are available in the literature. For the small horizontal cylinder of radius r, Lienhard and Dhir [51] recommend the following expression

$$q''_{max} = q''_{max,Z}[0.94(Bo)^{-1/4}] \qquad 0.15 \leq Bo \leq 1.2$$

where $q''_{max,Z}$ is the critical heat flux as predicted by the Zuber-Kutateladze relation, Equation 10.6, and for Bo, the Bond number, the characteristic length is the radius of the cylinder. Estimate the critical heat flux for horizontal cylinders of 1 and 3 mm diameter in saturated water at 1 atm and compare the result with that for a large horizontal cylinder.

10.18 What is the critical heat flux for boiling water at 1 atm on the surface of the moon, where the gravitational acceleration is one-sixth that of the earth?

10.19 A steel bar, 20 mm in diameter and 200 mm long, with an emissivity of 0.9, is removed from a furnace at 455°C and suddenly submerged in a water bath under atmospheric pressure. Estimate the initial heat transfer rate from the bar.

10.20 The passage of electrical current through a horizontal, 2-mm-diameter conductor with emissivity 0.5 causes its surface temperature to reach 555°C when immersed in water under atmospheric pressure. Estimate the power dissipation per unit length of the conductor.

10.21 A polished stainless steel bar of 50 mm diameter having an emissivity 0.10 is maintained at a surface temperature of 250°C while horizontally submerged in water at 25°C under atmospheric pressure. Estimate the heat rate per unit length of the bar.

10.22 A cylinder of 120 mm diameter at 1000 K is quenched in saturated water at 1 atm.

Describe the quenching process and estimate the maximum heat removal rate per unit length during the process.

10.23 A copper sphere of 10 mm diameter is initially at 500°C and is quenched in a saturated (1 atm) water bath. Using the lumped capacitance method and selecting typical heat fluxes from the boiling curve, estimate the time for the sphere to reach: 220, 130, and 105°C.

10.24 A tube of 2 mm diameter is used to heat saturated water at 1 atm which is in cross flow over the tube. Calculate and plot the critical heat flux as a function of water velocity over the range 0–2 m/s. On your plot identify the pool boiling region and the transition region between the low and high velocity ranges.

10.25 Saturated water at 1 atm and velocity 2 m/s flows over a circular heating element of diameter 5 mm. What is the maximum heating rate (W/m) for the element?

10.26 For forced-convection local boiling of water inside vertical tubes, the heat transfer coefficient can be estimated by the correlation

$$h = 2.54(\Delta T_e)^3 \exp\left(\frac{p}{15.3}\right)$$

with h in units of $W/m^2 \cdot K$, ΔT_e in units of K, and p in units of bar. Consider water at 4 bars flowing through a vertical tube of 50 mm inner diameter. Local boiling occurs when the tube wall is 15°C above the saturation temperature. Estimate the boiling heat transfer rate per unit length of the tube.

10.27 For forced-convection boiling in smooth tubes, the heat flux can be estimated by combining the separate effects of boiling and forced convection. The Rohsenow and Dittus-Boelter correlations may be used to predict nucleate boiling and forced–convection effects, with 0.019 replacing 0.023 in the latter expression. Consider water at 1 atm with a mean velocity of 1.5 m/s and a mean temperature of 95°C flowing through a 15-mm-diameter brass tube whose surface is maintained at 110°C. Estimate the heat transfer rate per unit length of the tube.

10.28 Saturated steam at 0.1 bar condenses with a convection coefficient of $6800\ W/m^2 \cdot K$ on the outside of a brass tube having inner and outer diameters of 16.5 and 19 mm, respectively. The convection coefficient for water flowing inside the tube is $5200\ W/m^2 \cdot K$. Estimate the steam condensation rate per unit length of the tube when the mean water temperature is 30°C.

10.29 If advection effects in laminar film condensation on a vertical plate are not neglected, the energy balance, Equation 10.20, on the control volume of Figure 10.13 has the form

$$q_s'' = \frac{d\Gamma}{dx}\left[h_{fg} + \frac{1}{\Gamma}\int_0^\delta \rho_l u c_{p,l}(T_{sat} - T)\, dy\right]$$

Verify this expression and show that the influence of advection for a linear temperature distribution across the condensate film is to replace h_{fg} with $h'_{fg} = h_{fg}(1 + \frac{3}{8}Ja)$ in the Nusselt analysis. The more appropriate expression for h'_{fg} is Equation 10.26, which results from use of a nonlinear temperature distribution in the film.

10.30 Consider a container exposed to a saturated vapor T_{sat} having a cold bottom surface, $T_s < T_{sat}$, and with insulated side walls.

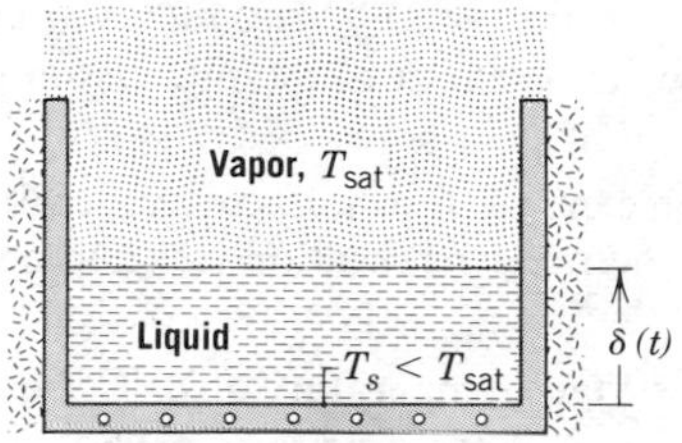

Assuming a linear temperature distribution for the liquid, perform a surface energy balance on the liquid–vapor interface to obtain the following expression for the growth rate of the liquid layer.

$$\delta(t) = \left[\frac{2k_l (T_{sat} - T_s)}{\rho_l h_{fg}} t \right]^{1/2}$$

Calculate the thickness of the liquid layer formed in 1 h for a 200 mm × 200 mm bottom surface maintained at 80°C and exposed to saturated steam at 1 atm. Compare this result with the condensate formed by a *vertical* plate of the same dimensions for the same period of time.

10.31 Saturated steam at 1 atm is exposed to a vertical plate 1 m high and 0.5 m wide having a uniform surface temperature of 70°C. Estimate the heat transfer rate to the plate and the steam condensation rate.

10.32 Saturated steam at 1 atm condenses on the outer surface of a vertical, 100-mm-diameter pipe 1 m long at a uniform surface temperature of 94°C. Estimate the total condensation rate and the heat transfer rate to the pipe.

10.33 Determine the total condensation rate and the heat transfer rate for Problem 10.32 when the steam is saturated at 1.5 bars.

10.34 Consider the condensation process of Problem 10.32. To maintain the pipe wall at the uniform surface temperature of 94°C, cooling water passes through the steel pipe of inner diameter 92 mm. What water flow rate will result in a 4°C temperature rise of the water between the outlet and inlet of the pipe?

10.35 A vertical plate 500 mm high and 200 mm wide is to be used to condense saturated steam at 1 atm. At what surface temperature must the plate be maintained to achieve a condensation rate of 25 kg/h?

10.36 Saturated ethylene glycol vapor at 1 atm is exposed to a vertical plate 300 mm high and 100 mm wide having a uniform temperature of 420 K. Estimate the heat transfer rate to the plate and the condensation rate.

10.37 A vertical plate 2.5 m high, maintained at a uniform temperature of 54°C, is exposed to saturated steam at atmospheric pressure. Estimate the condensation rate and the heat transfer rate per unit width of the plate. If the plate height were halved, would turbulent flow still exist?

10.38 Determine the total condensation rate and heat transfer rate for the condensation process of Problem 10.32 when the pipe is horizontal.

10.39 Consider laminar film condensation on the outer surface of a tube of length L and diameter D. Obtain an expression for the ratio of the condensation heat transfer rate in the horizontal orientation relative to the vertical orientation.

10.40 An uninsulated, 25-mm-diameter pipe with a surface temperature of 15°C passes through a room having an air temperature of 37°C and relative humidity of 75%. Estimate the condensation rate per unit length of the pipe assuming film, rather than dropwise, condensation.

10.41 A horizontal tube 50 mm in diameter, with a surface temperature of 34°C, is exposed to steam at 0.2 bar. Estimate the condensation rate and heat transfer rate per unit length of the tube.

10.42 A horizontal tube 1 m long with a surface temperature of 70°C is used to condense saturated steam at 1 atm. What diameter is required to achieve a condensation rate of 125 kg/h?

10.43 Refrigerant 12 at 1 atm condenses on the outside of a horizontal tube of 10 mm diameter and 1 m length. What surface temperature is required for a condensate rate of 50 kg/h?

10.44 Saturated steam at a pressure of 0.1 bar is condensed over a square array of 100 tubes each of diameter 8 mm. If the tube surfaces are maintained at 27°C, estimate the condensation rate per unit length of the tubes.

10.45 Determine the rate of condensation on a 100-mm-diameter sphere with a surface temperature of 150°C in saturated ethylene glycol vapor at 1 atm.

10.46 A 10-mm-diameter copper sphere, initially at a uniform temperature of 50°C, is placed in a large container filled with saturated steam at 1 atm. Using the lumped capacitance method, estimate the time required for the sphere to reach an equilibrium condition. How much condensate (kg) was formed during this period?

10.47 Water at a mean temperature and velocity of 17°C and 2 m/s, respectively, flows through a horizontal, brass tube with inner and outer diameters of 28.0 and 30.5 mm, respectively. Estimate the condensation rate per unit tube length of saturated steam at a pressure of 0.15 bar surrounding the tube.

10.48 Saturated steam at 1.5 bars condenses inside a horizontal 75-mm-diameter pipe whose surface is maintained at 100°C. Assuming low vapor velocities and film condensation, estimate the heat transfer coefficient and the condensation rate per unit length of the pipe.

10.49 Consider the conditions of Problem 10.40. Estimate the condensation rate for dropwise condensation.

Eleven

HEAT EXCHANGERS

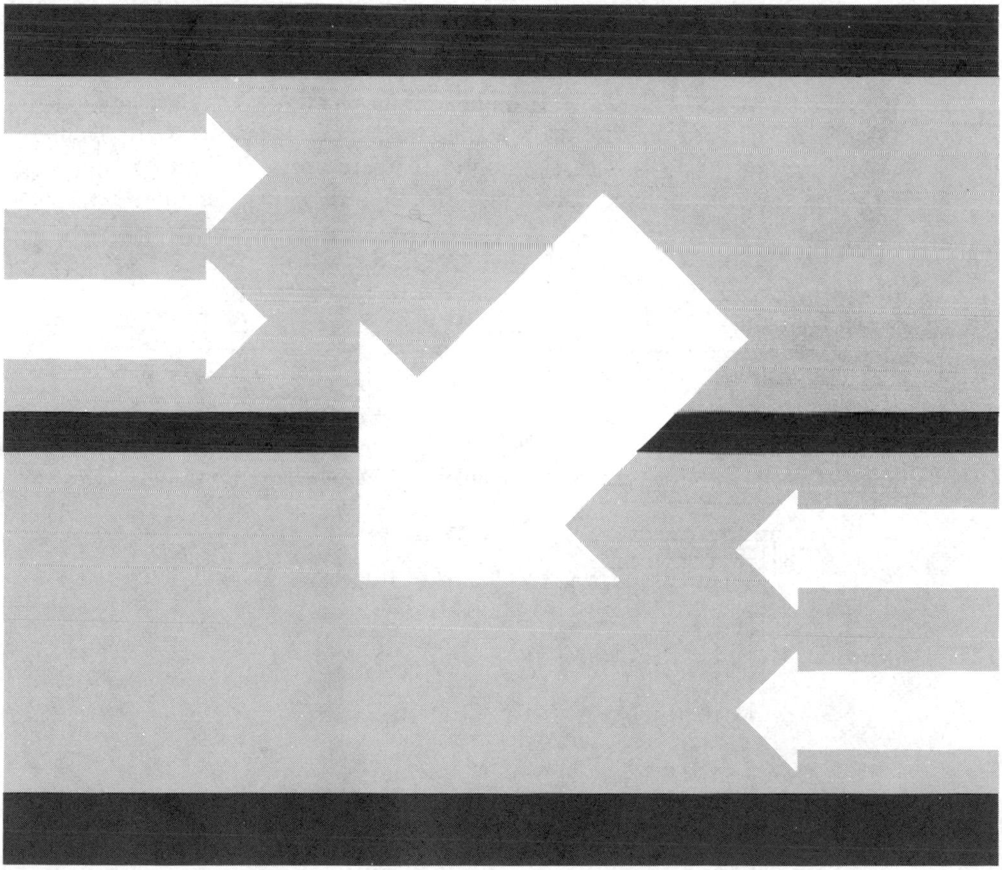

The process of heat exchange between two fluids that are at different temperatures and separated by a solid wall occurs in many engineering applications. The device used to implement this exchange is termed a *heat exchanger*, and specific applications may be found in space heating and air-conditioning, power production, waste heat recovery, and chemical processing. In this chapter we consider the principles of heat transfer needed to design and/or to evaluate the performance of a heat exchanger.

11.1 HEAT EXCHANGER TYPES

Heat exchangers are typically classified according to *flow arrangement* and *type of construction*. The simplest heat exchanger is one for which the hot and cold fluids move in the same or opposite directions in a *concentric tube* (or *double-pipe*) construction. In the *parallel-flow* arrangement of Figure 11.1*a*, the hot and cold fluids enter at the same end, flow in the same direction, and leave at the same end. In the *counterflow* arrangement of Figure 11.1*b*, the fluids enter at opposite ends, flow in opposite directions, and leave at opposite ends.

Alternatively, the fluids may move in *cross flow* (perpendicular to each other), as shown by the *finned* and *unfinned* tubular heat exchangers of Figure 11.2. The two configurations differ according to whether the fluid moving over the tubes is *unmixed* or *mixed*. In Figure 11.2*a*, the fluid is said to be unmixed because the fins prevent motion in a direction (y) that is transverse to the main-flow direction (x). In this case the fluid temperature varies with x and y. In contrast, for the unfinned tube bundle of Figure 11.2*b*, fluid motion, hence

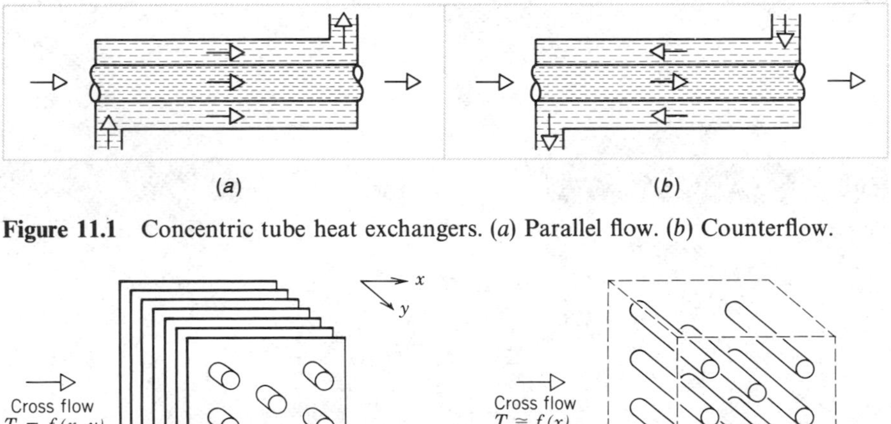

Figure 11.1 Concentric tube heat exchangers. (*a*) Parallel flow. (*b*) Counterflow.

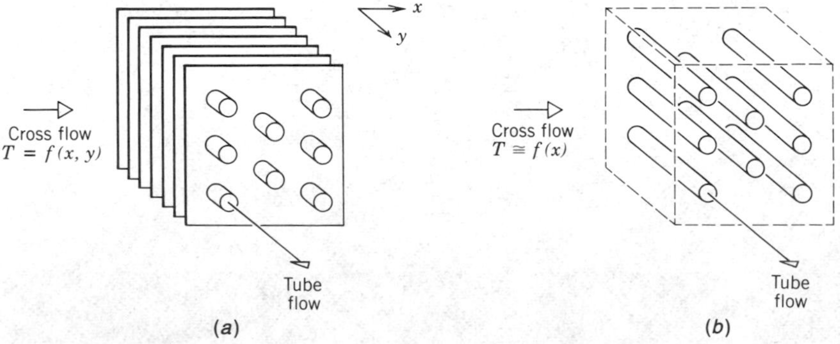

Figure 11.2 Cross-flow heat exchangers. (*a*) Finned with both fluids unmixed. (*b*) Unfinned with one fluid mixed and the other unmixed.

mixing, in the transverse direction is possible, and temperature variations are primarily in the main-flow direction. Since the tube flow is unmixed, both fluids are unmixed in the finned exchanger, while one fluid is mixed and the other unmixed in the unfinned exchanger. The nature of the mixing condition can significantly influence heat exchanger performance.

Another common configuration is the *shell-and-tube* heat exchanger [1]. Specific forms differ according to the number of shell and tube passes, and the simplest form, which involves single tube and shell *passes*, is shown in Figure 11.3. Baffles are usually installed to increase the convection coefficient of the shell-side fluid by inducing turbulence and a cross-flow velocity component. Baffled heat exchangers with one shell pass and two tube passes and with two shell passes and four tube passes are shown in Figures 11.4*a* and 11.4*b*, respectively.

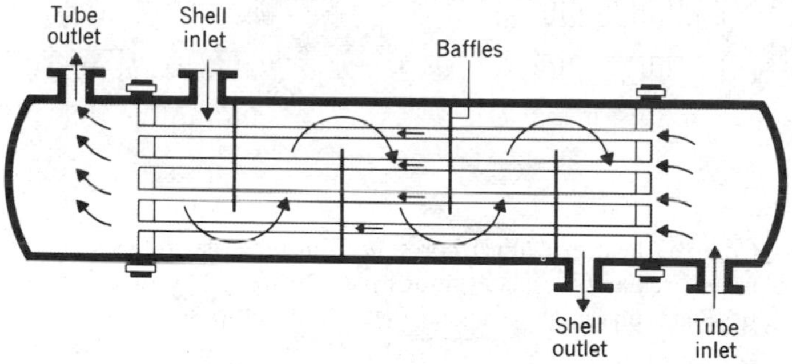

Figure 11.3 Shell-and-tube heat exchanger with one shell pass and one tube pass (cross-counterflow mode of operation).

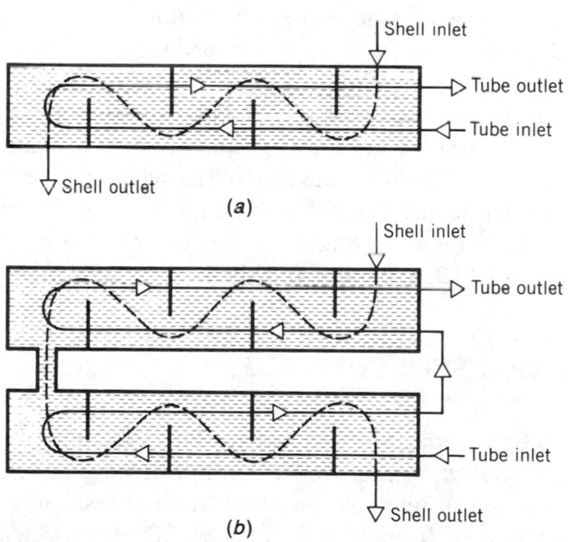

Figure 11.4 Shell-and-tube heat exchangers. (*a*) One shell pass and two tube passes. (*b*) Two shell passes and four tube passes.

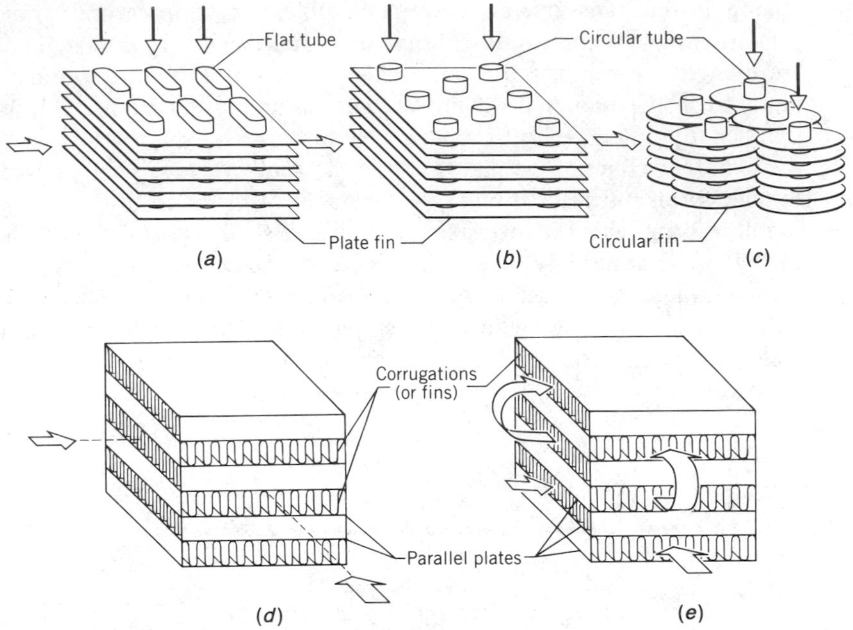

Figure 11.5 Compact heat exchanger cores. (*a*) Fin–tube (flat tubes, continuous plate fins). (*b*) Fin–tube (circular tubes, continuous plate fins). (*c*) Fin–tube (circular tubes, circular fins). (*d*) Plate–fin (single pass). (*e*) Plate–fin (multipass).

A special and important class of heat exchangers is used to achieve a very large ($\gtrsim 700\ m^2/m^3$) heat transfer surface area per unit volume. Termed *compact heat exchangers*, these devices have dense arrays of finned tubes or plates and are typically used when at least one of the fluids is a gas, and is hence characterized by a small convection coefficient. The tubes may be *flat* or *circular*, as in Figures 11.5*a* and 11.5*b*, respectively, and the fins may be *plate* or *circular*, as in Figures 11.5*a, b* and 11.5*c*, respectively. Parallel plate heat exchangers may be finned or corrugated and may be used in single-pass (Figure 11.5*d*) or multipass (Figure 11.5*e*) modes of operation. Flow passages associated with compact heat exchangers are typically small ($D_h \lesssim 5$ mm), and the flow is usually laminar.

11.2 THE OVERALL HEAT TRANSFER COEFFICIENT

An essential, and often the most uncertain, part of any heat exchanger analysis is determination of the *overall heat transfer coefficient*. Recall from Equation 3.18 that this coefficient is defined in terms of the total thermal resistance to heat transfer between two fluids. In Equations 3.17 and 3.30, the coefficient was determined by accounting for conduction and convection resistances between fluids separated by composite plane and cylindrical walls, respectively. It is important to acknowledge, however, that such results apply only to *clean, unfinned* surfaces.

During normal heat exchanger operation, surfaces are often subject to fouling by fluid impurities, rust formation, or other reactions between the fluid and the wall material. The subsequent deposition of a film or scale on the surface can greatly increase the resistance to heat transfer between the fluids. This effect can be treated by introducing an additional thermal resistance, termed the *fouling factor*, R_f. Its value depends on the operating temperature, fluid velocity, and length of service of the heat exchanger.

In addition, we know that fins are often added to surfaces exposed to either or both fluids and that, by increasing the surface area, they reduce the resistance to convection heat transfer. Accordingly, with inclusion of surface fouling and fin (extended surface) effects, the overall heat transfer coefficient may be expressed as

$$\frac{1}{UA} = \frac{1}{U_c A_c} = \frac{1}{U_h A_h}$$

$$= \frac{1}{(\eta_o hA)_c} + \frac{R''_{f,c}}{(\eta_o A)_c} + R_w + \frac{R''_{f,h}}{(\eta_o A)_h} + \frac{1}{(\eta_o hA)_h} \tag{11.1}$$

where c and h refer to the cold and hot fluids, respectively. Note that calculation of the UA product does not require designation of the hot or cold side ($U_c A_c = U_h A_h$). However, calculation of an overall coefficient depends on whether it is based on the cold or hot side surface area, since $U_c \neq U_h$ if $A_c \neq A_h$. The conduction resistance R_w is obtained from Equation 3.6 for a plane wall or Equation 3.27 for a cylindrical wall. Although representative fouling factors are listed in Table 11.1, the factor is a variable during heat exchanger operation (increasing from zero for a clean surface, as deposits accumulate on the surface). Comprehensive discussions of fouling are provided in References 2 through 4.

Table 11.1 Representative fouling factors [1]

FLUID	R''_f ($m^2 \cdot K/W$)
Seawater and treated boiler feedwater (below 50°C)	0.0001
Seawater and treated boiler feedwater (above 50°C)	0.0002
River water (below 50°C)	0.0002–0.0001
Fuel oil	0.0009
Refrigerating liquids	0.0002
Steam (nonoil bearing)	0.0001

The quantity η_o in Equation 11.1 is termed the *temperature effectiveness* of a fin and is defined such that, for the hot or cold surface, the heat transfer rate is

$$q = \eta_o hA(T_s - T_\infty) \tag{11.2}$$

where T_s is the surface temperature (Figure 3.12*b*). Decomposing the total surface area into contributions due to the fin and the prime (unfinned) surface $(A = A_f + A_p)$, the heat rate may also be expressed as

$$q = hA_p(T_s - T_\infty) + hA_f\eta_f(T_s - T_\infty)$$

where η_f is the fin efficiency defined by Equation 3.82. Rearranging the right-hand side of this expression

$$q = h(A_p + A_f\eta_f)(T_s - T_\infty) = hA\left[1 - \frac{A_f}{A}(1 - \eta_f)\right](T_s - T_\infty)$$

and combining with Equation 11.2, the fin effectiveness is

$$\eta_o = 1 - \frac{A_f}{A}(1 - \eta_f) \tag{11.3}$$

If a straight or pin fin of length L (Figure 3.16) is used and an adiabatic tip is assumed, Equations 3.75 and 3.82 yield

$$\eta_f = \frac{\tanh(mL)}{mL} \tag{11.4}$$

where $m = (2h/kt)^{1/2}$ and t is the fin thickness. For an annular fin the efficiency may be obtained from Figure 3.19.

Note that the wall conduction term in Equation 11.1 may often be neglected, as a thin wall of high thermal conductivity is generally used. Also, one of the convection coefficients is often much smaller than the other, and hence dominates determination of the overall coefficient. For example, if one of the fluids is a gas and the other is a liquid–vapor mixture experiencing boiling or condensation, the gas-side convection coefficient is much smaller. It is in such situations that fins are used to enhance gas-side convection. Representative values of the overall coefficient are summarized in Table 11.2.

For the unfinned, tubular heat exchangers of Figures 11.1 to 11.4, Equation 11.1 reduces to

$$\frac{1}{UA} = \frac{1}{U_iA_i} = \frac{1}{U_oA_o}$$

$$= \frac{1}{h_iA_i} + \frac{R''_{f,i}}{A_i} + \frac{\ln(D_o/D_i)}{2\pi kL} + \frac{R''_{f,o}}{A_o} + \frac{1}{h_oA_o} \tag{11.5}$$

where subscripts i and o refer to inner and outer tube surfaces ($A_i = \pi D_i L$, $A_o = \pi D_o L$), which may be exposed to either the hot or the cold fluid.

The overall heat transfer coefficient may be determined from knowledge of the hot and cold fluid convection coefficients and fouling factors and from appropriate geometric parameters. For unfinned surfaces, the convection

Table 11.2 Representative values of the overall heat transfer coefficient

FLUID COMBINATION	U ($W/m^2 \cdot K$)
Water to water	850–1700
Water to oil	110–350
Steam condenser (water in tubes)	1000–6000
Ammonia condenser (water in tubes)	800–1400
Alcohol condenser (water in tubes)	250–700
Finned-tube heat exchanger (water in tubes, air in cross flow)	25–50

coefficients may be estimated from correlations presented in Chapters 7 and 8. For standard fin configurations, the coefficients may be obtained from results compiled by Kays and London [5].

11.3 HEAT EXCHANGER ANALYSIS: USE OF THE LOG MEAN TEMPERATURE DIFFERENCE

To design or to predict the performance of a heat exchanger, it is essential to relate the total heat transfer rate to quantities such as the inlet and outlet fluid temperatures, the overall heat transfer coefficient, and the total surface area for heat transfer. Two such relations may readily be obtained by applying overall energy balances to the hot and cold fluids, as shown in Figure 11.6. In particular, if q is the total rate of heat transfer between the hot and cold fluids and there is negligible heat transfer between the exchanger and its surroundings, as well as negligible potential and kinetic energy changes, application of an energy balance, Equation 1.10a, gives

$$q = \dot{m}_h(i_{h,i} - i_{h,o}) \tag{11.6a}$$

and

$$q = \dot{m}_c(i_{c,o} - i_{c,i}) \tag{11.7a}$$

where i is the fluid enthalpy. The subscripts h and c refer to the hot and cold fluids, whereas i and o designate the fluid inlet and outlet conditions. If the fluids

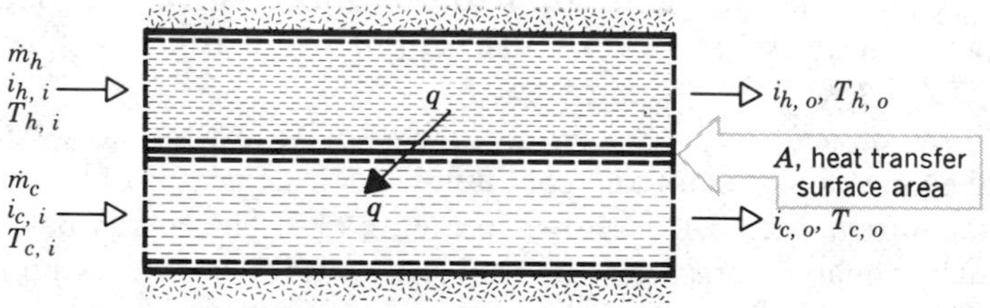

Figure 11.6 Overall energy balances for the hot and cold fluids of a two-fluid heat exchanger.

are not undergoing a phase change and constant specific heats are assumed, these expressions reduce to

$$q = \dot{m}_h c_{p,h}(T_{h,i} - T_{h,o}) \tag{11.6b}$$

and

$$q = \dot{m}_c c_{p,c}(T_{c,o} - T_{c,i}) \tag{11.7b}$$

where the temperatures appearing in the expressions refer to the *mean* fluid temperatures at the designated locations. Note that Equations 11.6 and 11.7 are independent of the flow arrangement and heat exchanger type.

Another useful expression may be obtained by relating the total heat transfer rate q to the temperature difference ΔT between the hot and cold fluids, where

$$\Delta T \equiv T_h - T_c \tag{11.8}$$

Such an expression would be an extension of Newton's law of cooling, with the overall heat transfer coefficient U used in place of the single convection coefficient h. However, since ΔT varies with position in the heat exchanger, it is necessary to work with a rate equation of the form

$$q = UA\Delta T_m \tag{11.9}$$

where ΔT_m is an appropriate *mean* value of the temperature difference. Equation 11.9 may be used with Equations 11.6 and 11.7 to perform a heat exchanger analysis, but before this can be done, the specific form of ΔT_m must be established. Consider first the parallel-flow heat exchanger.

11.3.1 The Parallel-Flow Heat Exchanger

The hot and cold fluid temperature distributions associated with a parallel-flow heat exchanger are shown in Figure 11.7. The temperature difference ΔT is initially large but decays rapidly with increasing x, approaching zero asymptotically. It is important to note that, for such an exchanger, the outlet temperature of the cold fluid never exceeds that of the hot fluid. In Figure 11.7 the subscripts 1 and 2 designate opposite ends of the heat exchanger. This convention is used for all types of heat exchangers considered. For parallel flow, it follows that $T_{h,i} = T_{h,1}$, $T_{h,o} = T_{h,2}$, $T_{c,i} = T_{c,1}$ and $T_{c,o} = T_{c,2}$.

The form of ΔT_m may be determined by applying an energy balance to differential elements in the hot and cold fluids. Each element is of length dx and heat transfer surface area dA, as shown in Figure 11.7. The energy balances and the subsequent analysis are performed subject to the following assumptions.

1. The heat exchanger is insulated from its surroundings, in which case the only heat exchange is between the hot and cold fluids.

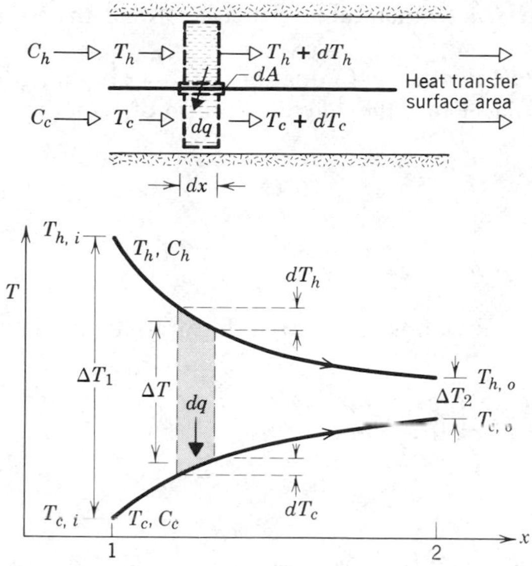

Figure 11.7 Temperature distributions for a parallel-flow heat exchanger.

2. Axial conduction along the tubes is negligible.
3. Potential and kinetic energy changes are negligible.
4. The fluid specific heats are constant.
5. The overall heat transfer coefficient is constant.

The specific heats may of course change as a result of temperature variations, and the overall heat transfer coefficient may change because of variations in fluid properties and flow conditions. However, in many applications such variations are not significant, and it is reasonable to work with average values of $c_{p,c}$, $c_{p,h}$, and U for the heat exchanger.

Applying an energy balance to each of the differential elements of Figure 11.7, it follows that

$$dq = -\dot{m}_h c_{p,h}\, dT_h \equiv -C_h\, dT_h \tag{11.10}$$

and

$$dq = \dot{m}_c c_{p,c}\, dT_c \equiv C_c\, dT_c \tag{11.11}$$

where C_h and C_c are the hot and cold fluid *heat capacity rates*, respectively. These expressions may be integrated across the heat exchanger to obtain the overall energy balances given by Equations 11.6b and 11.7b. The heat transfer across the surface area dA may also be expressed as

$$dq = U\Delta T\, dA \tag{11.12}$$

where $\Delta T = T_h - T_c$ is the *local* temperature difference between the hot and cold fluids.

To determine the integrated form of Equation 11.12, we begin by substituting Equations 11.10 and 11.11 into the differential form of Equation 11.8

$$d(\Delta T) = dT_h - dT_c$$

to obtain

$$d(\Delta T) = -dq\left(\frac{1}{C_h} + \frac{1}{C_c}\right)$$

Substituting for dq from Equation 11.12 and integrating across the heat exchanger, we obtain

$$\int_1^2 \frac{d(\Delta T)}{\Delta T} = -U\left(\frac{1}{C_h} + \frac{1}{C_c}\right)\int_1^2 dA$$

or

$$\ln\left(\frac{\Delta T_2}{\Delta T_1}\right) = -UA\left(\frac{1}{C_h} + \frac{1}{C_c}\right) \tag{11.13}$$

Substituting for C_h and C_c from Equations 11.6b and 11.7b, respectively, it follows that

$$\ln\left(\frac{\Delta T_2}{\Delta T_1}\right) = -UA\left(\frac{T_{h,i} - T_{h,o}}{q} + \frac{T_{c,o} - T_{c,i}}{q}\right)$$

$$= -\frac{UA}{q}[(T_{h,i} - T_{c,i}) - (T_{h,o} - T_{c,o})]$$

Recognizing that, for the parallel-flow heat exchanger of Figure 11.7, $\Delta T_1 = (T_{h,i} - T_{c,i})$ and $\Delta T_2 = (T_{h,o} - T_{c,o})$, we then obtain

$$q = UA\frac{\Delta T_2 - \Delta T_1}{\ln(\Delta T_2/\Delta T_1)}$$

Comparing the above expression with Equation 11.9, we conclude that the appropriate average temperature difference is a *log mean temperature difference*, ΔT_{lm}. Accordingly, we may write

$$q = UA\Delta T_{\text{lm}} \tag{11.14}$$

where

$$\Delta T_{\text{lm}} = \frac{\Delta T_2 - \Delta T_1}{\ln(\Delta T_2/\Delta T_1)} = \frac{\Delta T_1 - \Delta T_2}{\ln(\Delta T_1/\Delta T_2)} \tag{11.15}$$

Remember that, for the *parallel-flow exchanger*,

$$\begin{bmatrix} \Delta T_1 \equiv T_{h,1} - T_{c,1} = T_{h,i} - T_{c,i} \\ \Delta T_2 \equiv T_{h,2} - T_{c,2} = T_{h,o} - T_{c,o} \end{bmatrix} \tag{11.16}$$

11.3.2 The Counterflow Heat Exchanger

The hot and cold fluid temperature distributions associated with a counterflow heat exchanger are shown in Figure 11.8. In contrast to the parallel-flow exchanger, this configuration provides for heat transfer between the hotter portions of the two fluids near the entrance, as well as between the colder portions near the exit. For this reason, the change in the temperature difference, $\Delta T = T_h - T_c$, with respect to x is nowhere as large as it is for the inlet region of the parallel-flow exchanger. Note that the outlet temperature of the cold fluid may now exceed the outlet temperature of the hot fluid.

Equations 11.6b and 11.7b apply to any heat exchanger, and hence may be used for the counterflow arrangement. Moreover, from an analysis like that performed in Section 11.3.1, it may be shown that Equations 11.14 and 11.15 also apply. However, for the *counterflow exchanger* the endpoint temperature differences must now be defined as

$$\begin{bmatrix} \Delta T_1 \equiv T_{h,1} - T_{c,1} = T_{h,i} - T_{c,o} \\ \Delta T_2 \equiv T_{h,2} - T_{c,2} = T_{h,o} - T_{c,i} \end{bmatrix} \tag{11.17}$$

Note that, for the same inlet and outlet temperatures, the log mean temperature difference for counterflow exceeds that for parallel flow, $\Delta T_{\text{lm,CF}} > \Delta T_{\text{lm,PF}}$. Hence the surface area required to effect a prescribed heat transfer rate q is smaller for the counterflow than for the parallel-flow arrangement, assuming the same value of U. Also note that $T_{c,o}$ can exceed $T_{h,o}$ for counterflow but not for parallel flow.

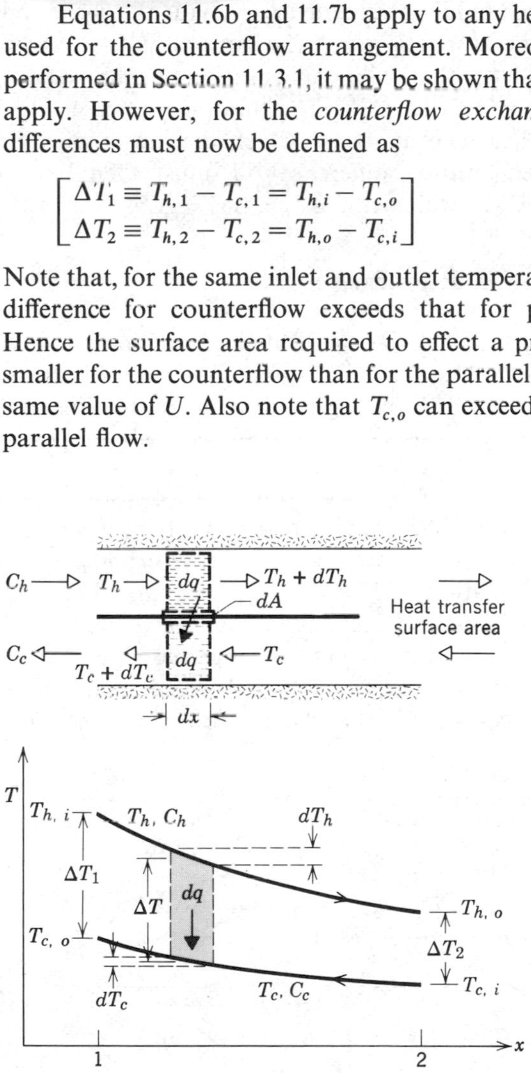

Figure 11.8 Temperature distributions for a counterflow heat exchanger.

11.3.3 Special Operating Conditions

It is useful to note certain special conditions under which heat exchangers may be operated. Figure 11.9*a* shows temperature distributions for a heat exchanger in which the hot fluid has a heat capacity rate, $C_h \equiv \dot{m}_h c_{p,h}$, which is much larger than that of the cold fluid, $C_c \equiv \dot{m}_c c_{p,c}$. For this case the temperature of the hot fluid remains approximately constant throughout the heat exchanger, while the temperature of the cold fluid increases. The same condition is achieved if the hot fluid is a condensing vapor. Condensation occurs at constant temperature, and, for all practical purposes, $C_h \rightarrow \infty$. Conversely, in an evaporator or a boiler (Figure 11.9*b*), it is the cold fluid that experiences a change in phase and remains at a nearly uniform temperature ($C_c \rightarrow \infty$). The same effect is achieved without phase change if $C_h \ll C_c$. Note that, with condensation or evaporation, the heat rate is given by Equation 11.6a or 11.7a. The third special case (Figure 11.9*c*) involves a counterflow heat exchanger for which the heat capacity rates are equal ($C_h = C_c$). The temperature difference ΔT must then be a constant throughout the exchanger, in which case $\Delta T_1 = \Delta T_2 = \Delta T_{\text{lm}}$.

11.3.4 Multipass and Cross-Flow Heat Exchangers

Although flow conditions are more complicated in multipass and cross-flow heat exchangers, Equations 11.6, 11.7, 11.14 and 11.15 may still be used if the following modification is made to the log mean temperature difference [6].

$$\Delta T_{\text{lm}} = F \Delta T_{\text{lm,CF}} \tag{11.18}$$

That is, the appropriate form of ΔT_{lm} is obtained by applying a correction factor to the value of ΔT_{lm} that would be computed *under the assumption of counterflow conditions.* Hence from Equation 11.17 $\Delta T_1 = T_{h,i} - T_{c,o}$ and $\Delta T_2 = T_{h,o} - T_{c,i}$.

Algebraic expressions for the correction factor F have been developed for various shell-and-tube and cross-flow heat exchanger configurations [1, 6, 7], and the results may be represented graphically. Selected results are shown in Figures 11.10 to 11.13 for common heat exchanger configurations. The notation

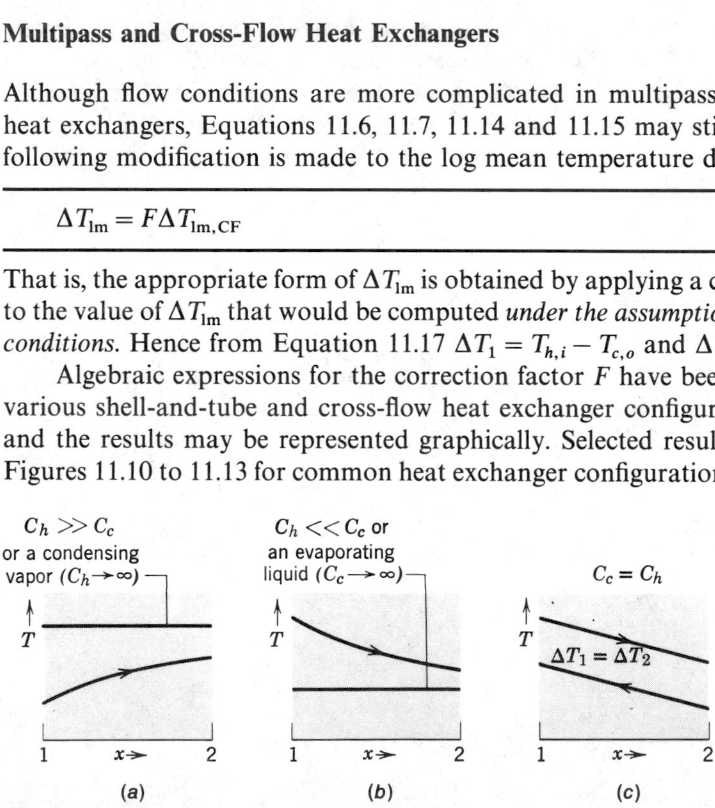

Figure 11.9 Special heat exchanger conditions. (*a*) $C_h \gg C_c$ or a condensing vapor. (*b*) An evaporating liquid or $C_h \ll C_c$. (*c*) A counterflow heat exchanger with equivalent fluid heat capacities ($C_h = C_c$).

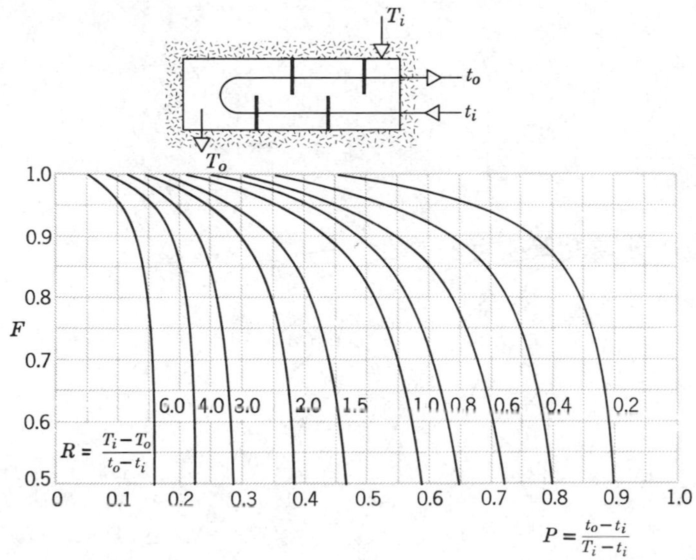

Figure 11.10 Correction factor for a shell-and-tube heat exchanger with one shell and any multiple of two tube passes (two, four, etc. tube passes).

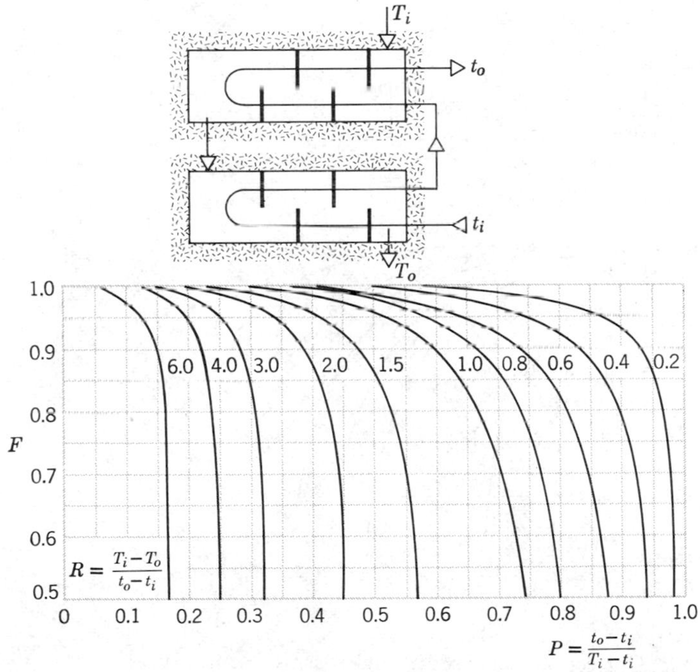

Figure 11.11 Correction factor for a shell-and-tube heat exchanger with two shell passes and any multiple of four tube passes (four, eight, etc. tube passes).

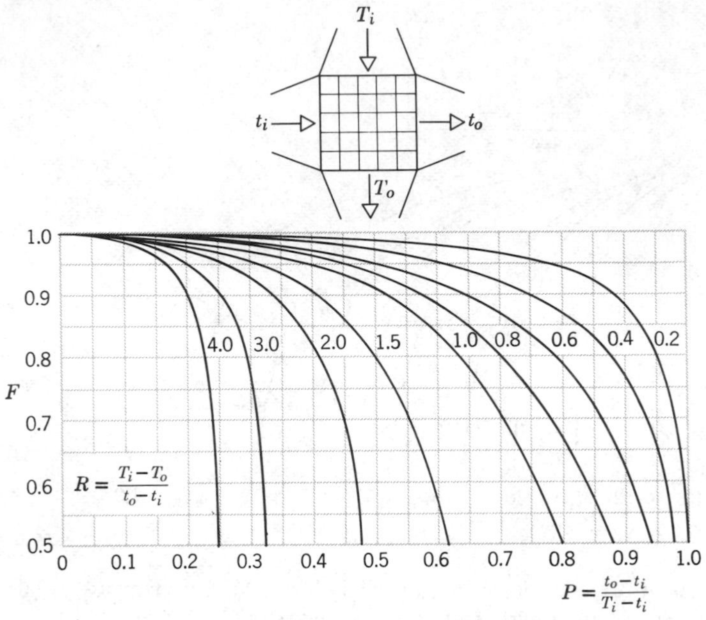

Figure 11.12 Correction factor for a single-pass, cross-flow heat exchanger with both fluids unmixed.

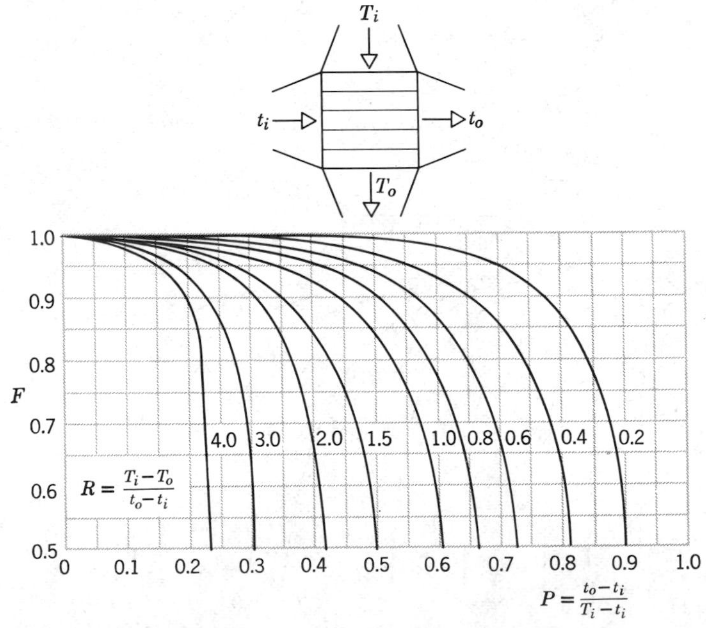

Figure 11.13 Correction factor for a single-pass, cross-flow heat exchanger with one fluid mixed and the other unmixed.

(T, t) is used to specify the fluid temperatures, with the variable t always assigned to the tube-side fluid. With this convention it does not matter whether the hot fluid or the cold fluid flows through the shell or the tubes. An important implication of Figures 11.10 to 11.13 is that, *if the temperature change of one fluid is negligible*, either P or R is zero and F is 1. *Hence heat exchanger behavior is independent of the specific configuration.* Such would be the case if one of the fluids underwent a phase change.

EXAMPLE 11.1

A counterflow, concentric tube heat exchanger is used to cool the lubricating oil for a large industrial gas turbine engine. The flow rate of cooling water through the inner tube ($D_i = 25$ mm) is 0.2 kg/s, while the flow rate of oil through the outer annulus ($D_o = 45$ mm) is 0.1 kg/s. The oil and water enter at temperatures of 100 and 30°C, respectively. How long must the tube be made if the outlet temperature of the oil is to be 60°C?

SOLUTION

Known:

Fluid flow rates and inlet temperatures for a counterflow, concentric tube heat exchanger of prescribed inner and outer diameter.

Find: Tube length to achieve a desired hot fluid outlet temperature.

Schematic:

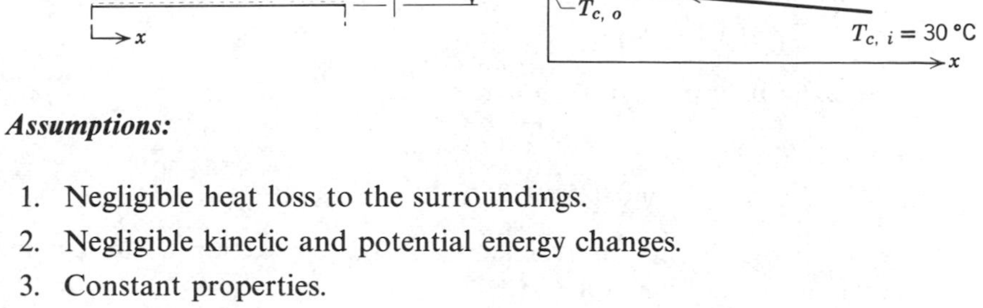

Assumptions:

1. Negligible heat loss to the surroundings.
2. Negligible kinetic and potential energy changes.
3. Constant properties.

4. Negligible tube wall thermal resistance and fouling factors.
5. Fully developed conditions for the water and oil (U independent of x).

Properties:

Table A.5, unused engine oil ($\bar{T}_h = 80°\text{C} = 353$ K): $c_p = 2131$ J/kg·K, $\mu = 3.25 \times 10^{-2}$ N·s/m², $k = 0.138$ W/m·K.
Table A.6, water ($\bar{T}_c \approx 35°\text{C}$): $c_p = 4178$ J/kg·K, $\mu = 725 \times 10^{-6}$ N·s/m², $k = 0.625$ W/m·K, $Pr = 4.85$.

Analysis:

The required heat transfer rate may be obtained from the overall energy balance for the hot fluid, Equation 11.6b.

$$q = \dot{m}_h c_{p,h}(T_{h,i} - T_{h,o})$$

$$q = 0.1 \text{ kg/s} \times 2131 \text{ J/kg}\cdot\text{K } (100 - 60)°\text{C} = 8524 \text{ W}$$

Applying Equation 11.7b, the water outlet temperature is

$$T_{c,o} = \frac{q}{\dot{m}_c c_{p,c}} + T_{c,i}$$

$$T_{c,o} = \frac{8524 \text{ W}}{0.2 \text{ kg/s} \times 4178 \text{ J/kg}\cdot\text{K}} + 30°\text{C} = 40.2°\text{C}$$

Accordingly, use of $\bar{T}_c = 35°\text{C}$ to evaluate the water properties was a good choice. The required heat exchanger length may now be obtained from Equation 11.14

$$q = UA\Delta T_{\text{lm}}$$

where $A = \pi D_i L$ and from Equations 11.15 and 11.17

$$\Delta T_{\text{lm}} = \frac{(T_{h,i} - T_{c,o}) - (T_{h,o} - T_{c,i})}{\ln\left[(T_{h,i} - T_{c,o})/(T_{h,o} - T_{c,i})\right]} = \frac{59.8 - 30}{\ln (59.8/30)} = 43.2°\text{C}$$

From Equation 11.5 the overall heat transfer coefficient is

$$U = \frac{1}{(1/h_i) + (1/h_o)}$$

For water flow through the tube,

$$Re_D = \frac{4\dot{m}_c}{\pi D_i \mu} = \frac{4 \times 0.2 \text{ kg/s}}{\pi(0.025 \text{ m})725 \times 10^{-6} \text{ N}\cdot\text{s/m}^2} = 14{,}050$$

Accordingly, the flow is turbulent and the convection coefficient may be computed from Equation 8.58

$$Nu_D = 0.023 Re_D^{4/5} Pr^{0.4}$$

$$Nu_D = 0.023(14{,}050)^{4/5}(4.85)^{0.4} = 90$$

Hence

$$h_i = Nu_D \frac{k}{D_i} = \frac{90 \times 0.625 \text{ W/m}\cdot\text{K}}{0.025 \text{ m}} = 2250 \text{ W/m}^2\cdot\text{K}$$

For the flow of oil through the annulus, the hydraulic diameter is, from Equation 8.67, $D_h = D_o - D_i = 0.02$ m, and the Reynolds number is

$$Re_D = \frac{\rho u_m D_h}{\mu} = \frac{\rho(D_o - D_i)}{\mu} \times \frac{\dot{m}_h}{\rho\pi(D_o^2 - D_i^2)/4}$$

$$Re_D = \frac{4\dot{m}_h}{\pi(D_o + D_i)\mu} = \frac{4 \times 0.1 \text{ kg/s}}{\pi(0.045 + 0.025) \text{ m} \times 3.25 \times 10^{-2} \text{ kg/s}\cdot\text{m}} = 56.0$$

The annular flow is therefore laminar. Assuming uniform temperature along the inner surface of the annulus and a perfectly insulated outer surface, the convection coefficient at the inner surface may be obtained from Table 8.2. With $(D_i/D_o) = 0.56$, linear interpolation provides

$$Nu_i = \frac{h_o D_h}{k} = 5.56$$

and

$$h_o = 5.56 \frac{0.138 \text{ W/m}\cdot\text{K}}{0.020 \text{ m}} = 38.4 \text{ W/m}^2\cdot\text{K}$$

The overall convection coefficient is then

$$U = \frac{1}{(1/2250 \text{ W/m}^2\cdot\text{K}) + (1/38.4 \text{ W/m}^2\cdot\text{K})} = 37.8 \text{ W/m}^2\cdot\text{K}$$

and from the rate equation it follows that

$$L = \frac{q}{U\pi D_i \Delta T_{\text{lm}}} = \frac{8524 \text{ W}}{37.8 \text{ W/m}^2\cdot\text{K}\ \pi(0.025 \text{ m})(43.2°\text{C})} = 66.5 \text{ m} \qquad \triangleleft$$

Comments:

1. The hot side convection coefficient controls the rate of heat transfer between the two fluids, and the low value of h_o is responsible for the large value of L. A spiral tube arrangement would be needed.
2. Because $h_i \gg h_o$, the tube wall temperature will follow closely that of the coolant water. Accordingly, the assumption of uniform wall temperature used to obtain h_o is reasonable.

EXAMPLE 11.2

A shell-and-tube heat exchanger must be designed to heat 2.5 kg/s of water from 15 to 85°C. The heating is to be accomplished by passing hot engine oil, which is available at 160°C, through the shell side of the exchanger. The oil is known to provide an average convection coefficient of $h_o = 400\ \mathrm{W/m^2 \cdot K}$ on the outside of the tubes. Ten tubes pass the water through the shell. Each tube is thin walled, of diameter $D = 25$ mm, and makes eight passes through the shell. If the oil leaves the exchanger at 100°C, what is its flow rate? How long must the tubes be to accomplish the desired heating?

SOLUTION

Known:

Fluid inlet and outlet temperatures for a shell-and-tube heat exchanger with 10 tubes making eight passes.

Find:

1. Oil flow rate required to achieve specified outlet temperature.
2. Tube length required to achieve specified water heating.

Schematic:

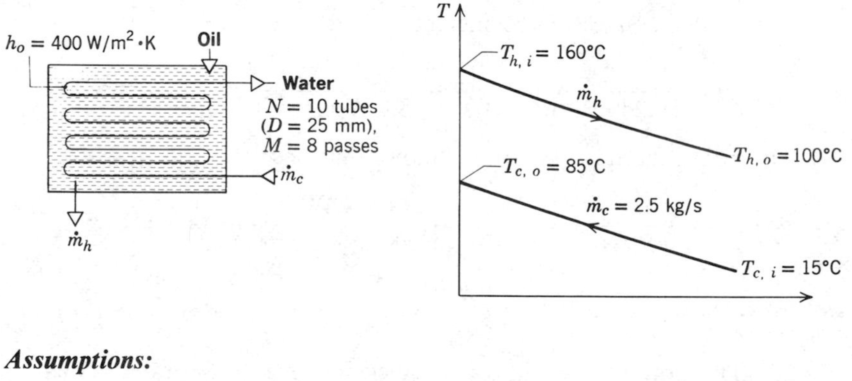

Assumptions:

1. Negligible heat loss to the surroundings and kinetic and potential energy changes.
2. Constant properties.
3. Negligible tube wall thermal resistance and fouling effects.
4. Fully developed water flow in tubes.

Properties:

Table A.5, unused engine oil ($\bar{T}_h = 130°C$): $c_p = 2350$ J/kg · K.
Table A.6, water ($\bar{T}_c = 50°C$): $c_p = 4181$ J/kg · K, $\mu = 548 \times 10^{-6}$ N · s/m^2, $k = 0.643$ W/m · K, $Pr = 3.56$.

Analysis:

1. From the overall energy balance, Equation 11.7b, the heat transfer required of the exchanger is

$$q = \dot{m}_c c_{p,c}(T_{c,o} - T_{c,i}) = 2.5 \text{ kg/s} \times 4181 \text{ J/kg} \cdot \text{K } (85 - 15)°\text{C}$$

$$q = 7.317 \times 10^5 \text{ W}$$

Hence, from Equation 11.6b,

$$\dot{m}_h = \frac{q}{c_{p,h}(T_{h,i} - T_{h,o})} = \frac{7.317 \times 10^5 \text{ W}}{2350 \text{ J/kg} \cdot \text{K} \times (160 - 100)°\text{C}} = 5.19 \text{ kg/s} \qquad \triangleleft$$

2. The required tube length may be obtained from Equations 11.14 and 11.18, where

$$q = UAF\Delta T_{\text{lm,CF}}$$

From Equation 11.5,

$$U = \frac{1}{(1/h_i) + (1/h_o)}$$

where h_i may be obtained by first calculating Re_D. With $\dot{m}_1 \equiv \dot{m}_c/N = 0.25$ kg/s defined as the water flow rate per tube, Equation 8.6 yields

$$Re_D = \frac{4\dot{m}_1}{\pi D \mu} = \frac{4 \times 0.25 \text{ kg/s}}{\pi(0.025 \text{ m})548 \times 10^{-6} \text{ kg/s} \cdot \text{m}} = 23{,}234$$

Hence the water flow is turbulent, and from Equation 8.58

$$Nu_D = 0.023 Re_D^{4/5} Pr^{0.4} = 0.023(23{,}234)^{4/5}(3.56)^{0.4} = 119$$

$$h_i = \frac{k}{D} Nu_D = \frac{0.643 \text{ W/m} \cdot \text{K}}{0.025 \text{ m}} 119 = 3061 \text{ W/m}^2 \cdot \text{K}$$

Hence

$$U = \frac{1}{(1/400) + (1/3061)} = 354 \text{ W/m}^2 \cdot \text{K}$$

The correction factor F may be obtained from Figure 11.10, where

$$R = \frac{160 - 100}{85 - 15} = 0.86 \qquad P = \frac{85 - 15}{160 - 15} = 0.48$$

Hence $F \approx 0.87$. From Equations 11.15 and 11.17, it follows that

$$\Delta T_{\mathrm{lm,CF}} = \frac{(T_{h,i} - T_{c,o}) - (T_{h,o} - T_{c,i})}{\ln\,[(T_{h,i} - T_{c,o})/(T_{h,o} - T_{c,i})]} = \frac{75 - 85}{\ln\,(75/85)} = 79.9^\circ\mathrm{C}$$

Hence, since $A = N\pi DL$, where $N = 10$ is the number of tubes,

$$L = \frac{q}{UN\pi DF\Delta T_{\mathrm{lm,CF}}} = \frac{7.317 \times 10^5\ \mathrm{W}}{354\ \mathrm{W/m^2 \cdot K} \times 10\ \pi(0.025\ \mathrm{m})0.87(79.9^\circ\mathrm{C})}$$

$$L = 37.9\ \mathrm{m}$$ ◁

Comments:

1. With $(L/D) = 37.9\ \mathrm{m}/0.025\ \mathrm{m} = 1516$, the assumption of fully developed conditions throughout the tube is justified.
2. With eight passes, the shell length is approximately $L/M = 4.7$ m.

11.4 HEAT EXCHANGER ANALYSIS: THE EFFECTIVENESS–NTU METHOD

It is a simple matter to use the log mean temperature difference (LMTD) method of heat exchanger analysis when the fluid inlet temperatures are known and the outlet temperatures are specified or readily determined from the energy balance expressions, Equations 11.6b and 11.7b. The value of ΔT_{lm} for the exchanger may then be determined. However, if only the inlet temperatures are known, use of the LMTD method requires an iterative procedure. In such cases it is preferable to use an alternative approach, termed the *effectiveness-NTU* method.

11.4.1 Definitions

To define the *effectiveness of a heat exchanger*, we must first determine the *maximum possible heat transfer rate*, q_{max}, for the exchanger. This heat transfer rate could, in principle, be achieved in a counterflow heat exchanger (Figure 11.8) of infinite length. In such an exchanger, one of the fluids would experience the maximum possible temperature difference, $T_{h,i} - T_{c,i}$. To illustrate this point, consider a situation for which $C_c < C_h$, in which case, from Equations 11.10 and 11.11, $|dT_c| > |dT_h|$. The cold fluid would then experience the larger temperature change, and since $L \rightarrow \infty$, it would be heated to the inlet temperature of the hot fluid ($T_{c,o} = T_{h,i}$). Accordingly, from Equation 11.7b

$$C_c < C_h: \qquad q_{\mathrm{max}} = C_c(T_{h,i} - T_{c,i})$$

Similarly, if $C_h < C_c$, the hot fluid would experience the larger temperature change and would be cooled to the inlet temperature of the cold fluid ($T_{h,o} = T_{c,i}$). From Equation 11.6b we then obtain

$$C_h < C_c: \qquad q_{max} = C_h(T_{h,i} - T_{c,i})$$

From the foregoing results we are then prompted to write the general expression

$$q_{max} = C_{min}(T_{h,i} - T_{c,i}) \tag{11.19}$$

where C_{min} is equal to C_c or C_h, whichever is smaller. For prescribed hot and cold fluid inlet temperatures, Equation 11.19 provides the maximum heat transfer rate that could possibly be delivered by the exchanger.

It is now logical to define the *effectiveness*, ε, as the ratio of the actual heat transfer rate for a heat exchanger to the maximum possible heat transfer rate.

$$\varepsilon \equiv \frac{q_{act}}{q_{max}} \qquad 0 \le \varepsilon \le 1 \tag{11.20}$$

From Equations 11.6b, 11.7b, and 11.19, it follows that

$$\varepsilon = \frac{C_h(T_{h,i} - T_{h,o})}{C_{min}(T_{h,i} - T_{c,i})} \tag{11.21}$$

or

$$\varepsilon = \frac{C_c(T_{c,o} - T_{c,i})}{C_{min}(T_{h,i} - T_{c,i})} \tag{11.22}$$

By definition the effectiveness, which is dimensionless, must be in the range $0 \le \varepsilon \le 1$. It is useful because, if ε, $T_{h,i}$, and $T_{c,i}$ are known, the actual heat transfer rate may readily be determined from the expression

$$q = \varepsilon C_{min}(T_{h,i} - T_{c,i}) \tag{11.23}$$

For any heat exchanger it can be shown that [5]

$$\varepsilon = f\left(NTU, \frac{C_{min}}{C_{max}}\right) \tag{11.24}$$

where C_{min}/C_{max} is equal to C_c/C_h or C_h/C_c, depending on the relative magnitudes of the hot and cold fluid heat capacity rates. The *number of transfer units* (NTU) is a dimensionless parameter that is widely used for heat exchanger analysis and is defined as

$$NTU \equiv \frac{UA}{C_{min}} \tag{11.25}$$

11.4.2 Effectiveness–NTU Relations

To determine a specific form of the effectiveness–NTU relation, Equation 11.24, consider a parallel-flow heat exchanger for which $C_{\min} = C_h$. From Equation 11.21 we then obtain

$$\varepsilon = \frac{T_{h,i} - T_{h,o}}{T_{h,i} - T_{c,i}} \tag{11.26}$$

and from Equations 11.6b and 11.7b it follows that

$$\frac{C_{\min}}{C_{\max}} = \frac{\dot{m}_h c_{p,h}}{\dot{m}_c c_{p,c}} = \frac{T_{c,o} - T_{c,i}}{T_{h,i} - T_{h,o}} \tag{11.27}$$

Now consider Equation 11.13, which may be expressed as

$$\ln\left(\frac{T_{h,o} - T_{c,o}}{T_{h,i} - T_{c,i}}\right) = -\frac{UA}{C_{\min}}\left(1 + \frac{C_{\min}}{C_{\max}}\right)$$

or from Equation 11.25

$$\frac{T_{h,o} - T_{c,o}}{T_{h,i} - T_{c,i}} = \exp\left[-\text{NTU}\left(1 + \frac{C_{\min}}{C_{\max}}\right)\right] \tag{11.28}$$

Rearranging the left-hand side of this expression as

$$\frac{T_{h,o} - T_{c,o}}{T_{h,i} - T_{c,i}} = \frac{T_{h,o} - T_{h,i} + T_{h,i} - T_{c,o}}{T_{h,i} - T_{c,i}}$$

and substituting for $T_{c,o}$ from Equation 11.27, it follows that

$$\frac{T_{h,o} - T_{c,o}}{T_{h,i} - T_{c,i}} = \frac{(T_{h,o} - T_{h,i}) + (T_{h,i} - T_{c,i}) - (C_{\min}/C_{\max})(T_{h,i} - T_{h,o})}{T_{h,i} - T_{c,i}}$$

or from Equation 11.26

$$\frac{T_{h,o} - T_{c,o}}{T_{h,i} - T_{c,i}} = -\varepsilon + 1 - \left(\frac{C_{\min}}{C_{\max}}\right)\varepsilon = 1 - \varepsilon\left(1 + \frac{C_{\min}}{C_{\max}}\right)$$

Substituting the above expression into Equation 11.28 and solving for ε, we obtain for the *parallel-flow heat exchanger*

$$\varepsilon = \frac{1 - \exp\{-\text{NTU}[1 + (C_{\min}/C_{\max})]\}}{1 + (C_{\min}/C_{\max})} \tag{11.29a}$$

Since precisely the same result may be obtained for $C_{\min} = C_c$, Equation 11.29a applies for any parallel-flow heat exchanger, irrespective of whether the minimum heat capacity rate is associated with the hot or cold fluid.

Similar expressions have been developed for a variety of heat exchangers [5], and representative results are summarized in Table 11.3, where C_r is the *heat capacity ratio* $C_r \equiv C_{\min}/C_{\max}$. In deriving Equation 11.32a, it is assumed that the

Table 11.3 Heat exchanger effectiveness relations [5]

FLOW ARRANGEMENT	RELATION	
Concentric tube		
Parallel flow	$\varepsilon = \dfrac{1-\exp[-\text{NTU}(1+C_r)]}{1+C_r}$	(11.29a)
Counterflow	$\varepsilon = \dfrac{1-\exp[-\text{NTU}(1-C_r)]}{1-C_r\exp[-\text{NTU}(1-C_r)]}$	(11.30a)
Shell and tube		
One shell pass (2, 4, ... tube passes)	$\varepsilon_1 = 2\left\{1+C_r+(1+C_r^2)^{1/2}\dfrac{1+\exp[-\text{NTU}(1+C_r^2)^{1/2}]}{1-\exp[-\text{NTU}(1+C_r^2)^{1/2}]}\right\}^{-1}$	(11.31a)
n Shell passes (2n, 4n, ... tube passes)	$\varepsilon = \left[\left(\dfrac{1-\varepsilon_1 C_r}{1-\varepsilon_1}\right)^n - 1\right]\left[\left(\dfrac{1-\varepsilon_1 C_r}{1-\varepsilon_1}\right)^n - C_r\right]^{-1}$	(11.32a)
Cross flow (single pass)		
Both fluids unmixed	$\varepsilon = 1-\exp[(1/C_r)(\text{NTU})^{0.22}\{\exp[-C_r(\text{NTU})^{0.78}]-1\}]$	(11.33)
$C_{\max}$ (mixed), $C_{\min}$ (unmixed)	$\varepsilon = (1/C_r)(1-\exp\{-C_r[1-\exp(-\text{NTU})]\})$	(11.34a)
$C_{\min}$ (mixed), $C_{\max}$ (unmixed)	$\varepsilon = 1-\exp(-C_r^{-1}\{1-\exp[-C_r(\text{NTU})]\})$	(11.35a)
All exchangers ($C_r = 0$)	$\varepsilon = 1-\exp(-\text{NTU})$	(11.36a)

Table 11.4 Heat exchanger NTU relations

FLOW ARRANGEMENT	RELATION	
Concentric tube		
Parallel flow	$\text{NTU} = -\dfrac{\ln[1-\varepsilon(1+C_r)]}{1+C_r}$	(11.29b)
Counterflow	$\text{NTU} = \dfrac{1}{C_r-1}\ln\left(\dfrac{\varepsilon-1}{\varepsilon C_r-1}\right)$	(11.30b)
Shell and tube		
One shell pass	$\text{NTU} = -(1+C_r^2)^{-1/2}\ln\left(\dfrac{E-1}{E+1}\right)$	(11.31b)
(2, 4, ... tube passes)	$E = \dfrac{2/\varepsilon_1-(1+C_r)}{(1+C_r^2)^{1/2}}$	(11.31c)
n Shell passes	Use Equations 11.31b and 11.31c with	
(2n, 4n, ... tube passes)	$\varepsilon_1 = \dfrac{F-1}{F-C_r}, \quad F = \left(\dfrac{\varepsilon C_r-1}{\varepsilon-1}\right)^{1/n}$	(11.32b, c)
Cross flow (single pass)		
$C_{\max}$ (mixed), $C_{\min}$ (unmixed)	$\text{NTU} = -\ln[1+(1/C_r)\ln(1-\varepsilon C_r)]$	(11.34b)
$C_{\min}$ (mixed), $C_{\max}$ (unmixed)	$\text{NTU} = -(1/C_r)\ln[C_r\ln(1-\varepsilon)+1]$	(11.35b)
All exchangers ($C_r = 0$)	$\text{NTU} = -\ln(1-\varepsilon)$	(11.36b)

total NTU is equally distributed between shell passes of the same arrangement, $NTU = n(NTU)_1$. Hence, when using ε_1 with this expression, NTU is replaced by (NTU)/n in Equation 11.31a. Note that for $C_r = 0$, as in a boiler or condenser, ε is given by Equation 11.36a for *all flow arrangements*. Hence, for this special case, it again follows that heat exchanger behavior is independent of flow arrangement. For the cross-flow heat exchanger with both fluids unmixed, Equation 11.33 is exact only for $C_r = 1$. However, it may be used to an excellent approximation for all $0 < C_r \leq 1$. For $C_r = 0$, Equation 11.36a must be used.

In heat exchanger design calculations, it is more convenient to work with ε–NTU relations of the form

$$NTU = f\left(\varepsilon, \frac{C_{min}}{C_{max}}\right)$$

Explicit relations for NTU as a function of ε and C_r are provided in Table 11.4. Note that Equation 11.33 may not be manipulated to yield a direct relationship for NTU as a function of ε and C_r. Note also, that in using Equations 11.32b, c with 11.31b, c, it is the NTU *per shell pass* that is computed from Equation 11.31b. This result is multiplied by n to obtain the NTU for the entire exchanger.

The foregoing expressions are represented graphically in Figures 11.14 to 11.19. For Figure 11.19 the solid curves correspond to C_{min} mixed and C_{max} unmixed, while the dashed curves correspond to C_{min} unmixed and C_{max} mixed. Note that for $C_r = 0$, all heat exchangers have the same effectiveness, which may be computed from Equation 11.36a. Moreover, if $NTU \lesssim 0.25$, all heat exchangers have the same effectiveness regardless of the value of C_r, and ε may again be computed from Equation 11.36a. More generally, for $C_r > 0$ and $NTU \gtrsim 0.25$, the counterflow exchanger is the most effective. For any exchanger, maximum and minimum values of the effectiveness are associated with $C_r = 0$ and $C_r = 1$, respectively.

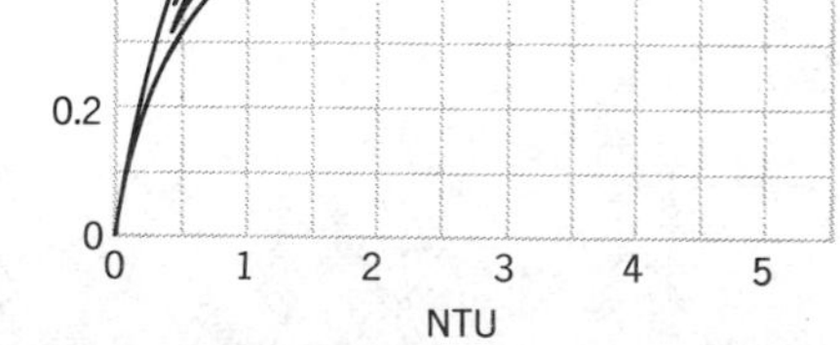

Figure 11.14 Effectiveness of a parallel-flow heat exchanger (Equation 11.29).

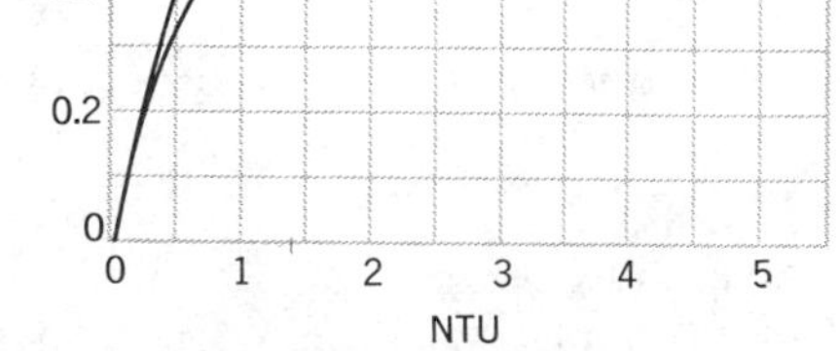

Figure 11.15 Effectiveness of a counterflow heat exchanger (Equation 11.30).

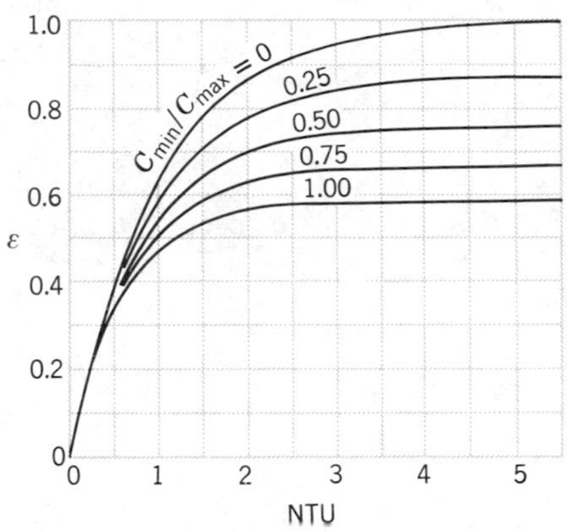

Figure 11.16 Effectiveness of a shell-and-tube heat exchanger with one shell and any multiple of two tube passes (two, four, etc. tube passes) (Equation 11.31).

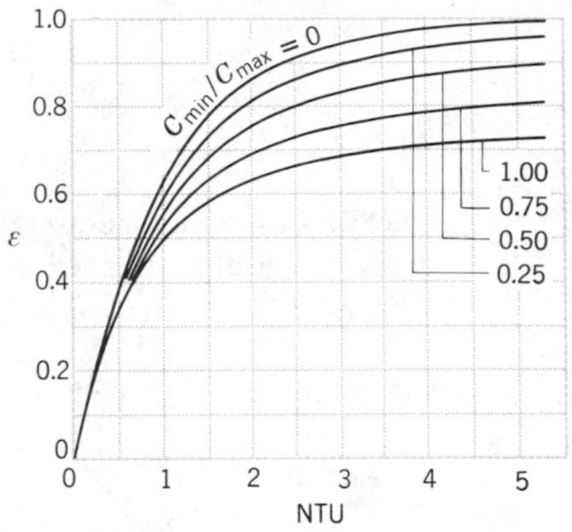

Figure 11.17 Effectiveness of a shell-and-tube heat exchanger with two shell passes and any multiple of four tube passes (four, eight, etc. tube passes) (Equation 11.32 with $n = 2$).

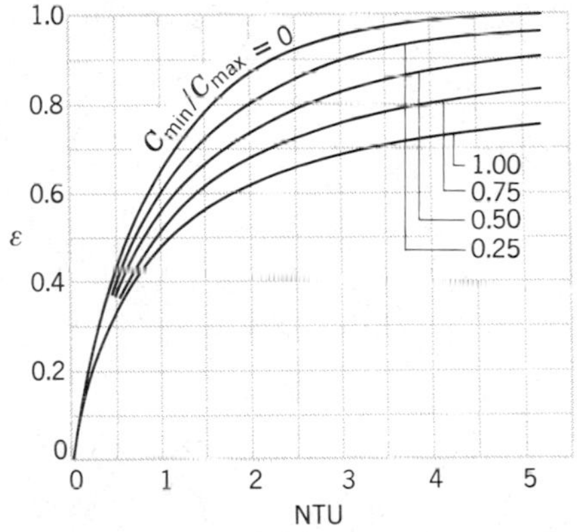

Figure 11.18 Effectiveness of a single-pass, cross-flow heat exchanger with both fluids unmixed (Equation 11.33).

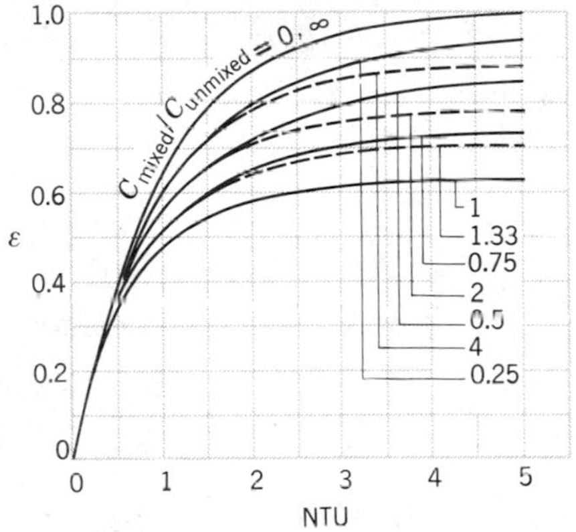

Figure 11.19 Effectiveness of a single-pass, cross-flow heat exchanger with one fluid mixed and the other unmixed (Equations 11.34, 11.35).

EXAMPLE 11.3

Hot exhaust gases, which enter a finned-tube, cross-flow heat exchanger at 300°C and leave at 100°C, are used to heat pressurized water at a flow rate of 1 kg/s from 35 to 125°C. The exhaust gas specific heat is approximately 1000 J/kg · K, and the overall heat transfer coefficient based on the gas-side surface area is $U_h = 100$ W/m^2 · K. Determine the required gas-side surface area A_h using the NTU method.

SOLUTION

Known:

Inlet and outlet temperatures of hot gases and water used in a finned-tube, cross-flow heat exchanger. Water flow rate and gas-side overall heat transfer coefficient.

Find: Required gas-side surface area.

Schematic:

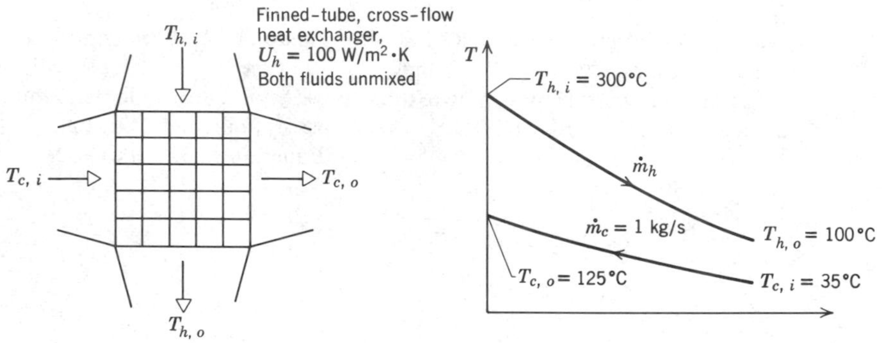

Assumptions:

1. Negligible heat loss to the surroundings and kinetic and potential energy changes.
2. Constant properties.

Properties:

Table A.6, water ($\bar{T}_c = 80°C$): $c_{p,c} = 4197$ J/kg·K.
Exhaust gases: $c_{p,h} = 1000$ J/kg·K.

Analysis:

The required surface area may be obtained from knowledge of the number of transfer units, which, in turn, may be obtained from knowledge of the ratio of heat capacity rates and the effectiveness. To determine the minimum heat capacity rate, we begin by computing

$$C_c = \dot{m}_c c_{p,c} = 1 \text{ kg/s} \times 4197 \text{ J/kg}\cdot\text{K} = 4197 \text{ W/K}$$

Since $\dot{m}_h$ is not specified, C_h is obtained by combining the overall energy balances, Equations 11.6b and 11.7b.

$$C_h = \dot{m}_h c_{p,h} = C_c \frac{T_{c,o} - T_{c,i}}{T_{h,i} - T_{h,o}} = 4197 \frac{125 - 35}{300 - 100} = 1889 \text{ W/K} = C_{\min}$$

From Equation 11.19

$$q_{\max} = C_{\min}(T_{h,i} - T_{c,i}) = 1889 \text{ W/K } (300 - 35)°\text{C} = 5.01 \times 10^5 \text{ W}$$

From Equation 11.7b the actual heat transfer rate is

$$q = \dot{m}_c c_{p,c}(T_{c,o} - T_{c,i}) = 1 \text{ kg/s} \times 4197 \text{ J/kg} \cdot \text{K}(125 - 35)°\text{C}$$

$$q = 3.77 \times 10^5 \text{ W}$$

Hence from Equation 11.20 the effectiveness is

$$\varepsilon = \frac{q}{q_{\max}} = \frac{3.77 \times 10^5 \text{ W}}{5.01 \times 10^5 \text{ W}} = 0.75$$

With

$$\frac{C_{\min}}{C_{\max}} = \frac{1889}{4197} = 0.45$$

it follows from Figure 11.18 that

$$\text{NTU} = \frac{U_h A_h}{C_{\min}} \approx 2.1$$

or

$$A_h = \frac{2.1(1889 \text{ W/K})}{100 \text{ W/m}^2 \cdot \text{K}} = 39.7 \text{ m}^2$$

Comments:

The desired heat transfer area may also be determined by using the LMTD method. From Equations 11.14 and 11.18,

$$A_h = \frac{q}{U_h F \Delta T_{\text{lm,CF}}}$$

With

$$P = \frac{t_o - t_i}{T_i - t_i} = \frac{125 - 35}{300 - 35} = 0.34 \qquad R = \frac{T_i - T_o}{t_o - t_i} = \frac{300 - 100}{125 - 35} = 2.22$$

it follows from Figure 11.12 that $F \approx 0.87$. From Equations 11.15 and 11.17,

$$\Delta T_{\text{lm,CF}} = \frac{(T_{h,i} - T_{c,o}) - (T_{h,o} - T_{c,i})}{\ln [(T_{h,i} - T_{c,o})/(T_{h,o} - T_{c,i})]} = 111°\text{C}$$

in which case

$$A_h = \frac{3.77 \times 10^5 \text{ W}}{100 \text{ W/m}^2 \cdot \text{K} \times 0.87 \times 111°\text{C}} = 39.1 \text{ m}^2 \qquad \triangleleft$$

Agreement of the two results is well within the accuracy associated with reading the F and ε, NTU, $C_{\min}/C_{\max}$ charts.

11.5 METHODOLOGY OF A HEAT EXCHANGER CALCULATION

We have developed two procedures for performing a heat exchanger analysis, the LMTD method and the NTU approach. For any problem both methods may be used to obtain equivalent results. However, depending on the nature of the problem, the NTU approach may be easier to implement.

Clearly, use of the LMTD method, Equations 11.14 and 11.15, is facilitated by knowledge of the hot and cold fluid inlet and outlet temperatures, since ΔT_{lm} may then be readily computed. Problems for which these temperatures are known may be classified as *heat exchanger design problems.* Typically, the fluid inlet temperatures and flow rates, as well as a desired hot or cold fluid outlet temperature, are prescribed. The design problem is then one of selecting an appropriate heat exchanger type and determining the size, that is, the heat transfer surface area A required to achieve the desired outlet temperature. For example, consider an application for which $\dot{m}_c$, $\dot{m}_h$, $T_{c,i}$, and $T_{h,i}$ are known, and the objective is to specify a heat exchanger that will provide a desired value of $T_{c,o}$. The corresponding values of q and $T_{h,o}$ may be computed from the energy balances, Equations 11.7b and 11.6b, respectively, and the value of ΔT_{lm} may be found from its definition, Equation 11.15. Using the rate equation (11.14), it is then a simple matter to determine the required value of A. Of course, the NTU method may also be used to obtain A by first calculating ε and $(C_{\min}/C_{\max})$. The appropriate chart (or equation) may then be used to obtain the NTU value, which in turn may be used to determine A.

Alternatively, the heat exchanger type and size may be known and the objective is to determine the heat transfer rate and the fluid outlet temperatures for prescribed fluid flowrates and inlet temperatures. Although the LMTD method may be used for such a heat exchanger *performance calculation,* the computations would be tedious, requiring iteration. For example, a guess could be made for the value of $T_{c,o}$, and Equations 11.7b and 11.6b could be used to determine q and $T_{h,o}$, respectively. Knowing all fluid temperatures, ΔT_{lm} could be determined and Equation 11.14 could then be used to again compute the value of q. The original guess for $T_{c,o}$ would be correct, if the values of q obtained from Equations 11.7b and 11.14 were in agreement. Such agreement would be fortuitous, however, and it is likely that some iteration on the value of $T_{c,o}$ would be needed.

The iterative nature of the above solution could be eliminated by using the

NTU method. From knowledge of the heat exchanger type and size and the fluid flow rates, the NTU and (C_{min}/C_{max}) values may be computed and ε may then be determined from the appropriate chart (or equation). Since q_{max} may also be computed from Equation 11.19, it is a simple matter to determine the actual heat transfer rate from the requirement that $q = \varepsilon q_{max}$. Both fluid outlet temperatures may then be determined from Equations 11.6b and 11.7b.

EXAMPLE 11.4

Consider the heat exchanger design of Example 11.3, that is, a finned-tube, cross-flow heat exchanger with a gas-side heat transfer coefficient and area of $100\ W/m^2 \cdot K$ and $40\ m^2$, respectively. The water flow rate and inlet temperature remain at 1 kg/s and 35°C. However, a change in operating conditions for the hot gas generator causes the gases to now enter the heat exchanger with a flow rate of 1.5 kg/s and a temperature of 250°C. What is the rate of heat transfer by the exchanger, and what are the gas and water outlet temperatures?

SOLUTION

Known:

Hot and cold fluid inlet conditions for a finned-tube, cross-flow heat exchanger of known surface area and overall heat transfer coefficient.

Find: Heat transfer rate and fluid outlet temperatures.

Schematic:

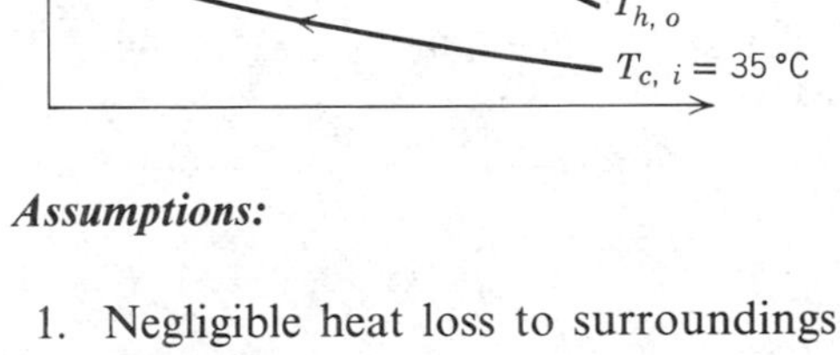

Assumptions:

1. Negligible heat loss to surroundings and kinetic and potential energy changes.
2. Constant properties (unchanged from Example 11.3).

Analysis:

The problem may be classified as one requiring a heat exchanger *performance calculation.* Accordingly, it is expedient to base the calculations on the NTU method. The heat capacity rates are

$$C_c = \dot{m}_c c_{p,c} = 1 \text{ kg/s} \times 4197 \text{ J/kg} \cdot \text{K} = 4197 \text{ W/K}$$

$$C_h = \dot{m}_h c_{p,h} = 1.5 \text{ kg/s} \times 1000 \text{ J/kg} \cdot \text{K} = 1500 \text{ W/K} = C_{\min}$$

in which case

$$\frac{C_{\min}}{C_{\max}} = \frac{1500}{4197} = 0.357$$

The number of transfer units is

$$\text{NTU} = \frac{U_h A_h}{C_{\min}} = \frac{100 \text{ W/m}^2 \cdot \text{K} \times 40 \text{ m}^2}{1500 \text{ W/K}} = 2.67$$

From Figure 11.18 the heat exchanger effectiveness is then $\varepsilon \approx 0.82$, and from Equation 11.19 the maximum possible heat transfer rate is

$$q_{\max} = C_{\min}(T_{h,i} - T_{c,i}) = 1500 \text{ W/K } (250 - 35)^\circ\text{C} = 3.23 \times 10^5 \text{ W}$$

Accordingly, from the definition of ε, Equation 11.20, the actual heat transfer rate is

$$q = \varepsilon q_{\max} = 0.82 \times 3.23 \times 10^5 \text{ W} = 2.65 \times 10^5 \text{ W} \qquad \triangleleft$$

It is now a simple matter to determine the outlet temperatures from the overall energy balances. From Equation 11.6b

$$T_{h,o} = T_{h,i} - \frac{q}{\dot{m}_h c_{p,h}} = 250^\circ\text{C} - \frac{2.65 \times 10^5 \text{ W}}{1500 \text{ W/K}} = 73.3^\circ\text{C} \qquad \triangleleft$$

and from Equation 11.7b

$$T_{c,o} = T_{c,i} + \frac{q}{\dot{m}_c c_{p,c}} = 35^\circ\text{C} + \frac{2.65 \times 10^5 \text{ W}}{4197 \text{ W/K}} = 98.1^\circ\text{C} \qquad \triangleleft$$

Comments:

From Equation 11.33, $\varepsilon = 0.845$, which is in good agreement with the estimate obtained from the charts.

EXAMPLE 11.5

The condenser of a 1 MW steam power plant is a heat exchanger in which steam is condensed to liquid water. Assume the condenser of a large power plant to be a *shell-and-tube* heat exchanger consisting of a single shell and 30,000 tubes, each executing two passes. The tubes are of thin wall construction with $D = 25$ mm, and steam condenses on their outer surface with an associated convection coefficient of $h_o = 11{,}000$ W/m² · K. The heat transfer rate that must be effected by the exchanger is $q = 2 \times 10^9$ W, and this is accomplished by passing cooling water through the tubes at a rate of 3×10^4 kg/s (the flow rate per tube is therefore 1 kg/s). The water enters at 20°C, while the steam condenses at 50°C. What is the temperature of the cooling water emerging from the condenser? What is the required tube length L per pass?

SOLUTION

Known:

Heat exchanger consisting of single shell and 30,000 tubes with two passes each.

Find:

1. Outlet temperature of the cooling water.
2. Tube length per pass to achieve required heat transfer.

Schematic:

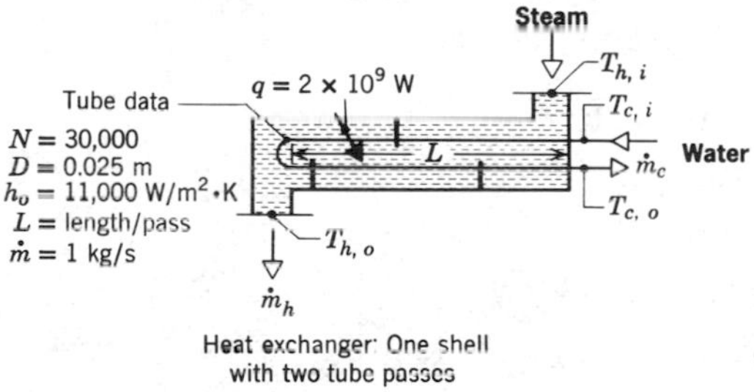

Heat exchanger: One shell with two tube passes

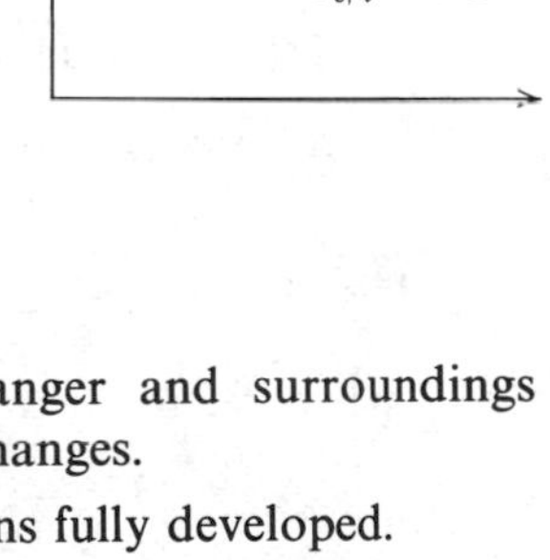

Assumptions:

1. Negligible heat transfer between exchanger and surroundings and negligible kinetic and potential energy changes.
2. Tube internal flow and thermal conditions fully developed.
3. Negligible thermal resistance of tube material and fouling effects.
4. Constant properties.

Properties:

Table A.6, water (assume $\bar{T}_c \approx 27°\text{C} = 300$ K): $c_p = 4179$ J/kg · K, $\mu = 855 \times 10^{-6}$ N · s/m², $k = 0.613$ W/m · K, $Pr = 5.83$.

Analysis:

1. The cooling water outlet temperature may be obtained from the overall energy balance, Equation 11.7b. Accordingly,

$$T_{c,o} = T_{c,i} + \frac{q}{\dot{m}_c c_{p,c}} = 20°\text{C} + \frac{2 \times 10^9 \text{ W}}{3 \times 10^4 \text{ kg/s} \times 4179 \text{ J/kg} \cdot \text{K}}$$

$$T_{c,o} = 36.0°\text{C} \qquad \triangleleft$$

2. The problem may be classified as one requiring a heat exchanger *design calculation.* Accordingly, either the LMTD or NTU method may be conveniently applied. Using the LMTD method, it follows from Equations 11.14 and 11.18 that

$$q = UAF\Delta T_{\text{lm,CF}}$$

where $A = N \times 2L \times \pi D$. From Equation 11.5

$$U = \frac{1}{(1/h_i) + (1/h_o)}$$

where h_i may be estimated from an internal flow correlation. With

$$Re_D = \frac{4\dot{m}}{\pi D \mu} = \frac{4 \times 1 \text{ kg/s}}{\pi(0.025 \text{ m})855 \times 10^{-6} \text{ N} \cdot \text{s/m}^2} = 59{,}567$$

the flow is turbulent and from Equation 8.58

$$Nu_D = 0.023 Re_D^{4/5} Pr^{0.4} = 0.023(59{,}567)^{0.8}(5.83)^{0.4} = 308$$

Hence

$$h_i = Nu_D \frac{k}{D} = 308 \frac{0.613 \text{ W/m} \cdot \text{K}}{0.025 \text{ m}} = 7552 \text{ W/m}^2 \cdot \text{K}$$

$$U = \frac{1}{[(1/7552) + (1/11{,}000)]\text{m}^2 \cdot \text{K/W}} = 4478 \text{ W/m}^2 \cdot \text{K}$$

From Equations 11.15 and 11.17 the LMTD is

$$\Delta T_{\text{lm,CF}} = \frac{(T_{h,i} - T_{c,o}) - (T_{h,o} - T_{c,i})}{\ln [(T_{h,i} - T_{c,o})/(T_{h,o} - T_{c,i})]} = \frac{(50 - 36) - (50 - 20)}{\ln (14/30)} = 21°\text{C}$$

The correction factor F may be obtained from Figure 11.10, where

$$P = \frac{t_o - t_i}{T_i - t_i} = \frac{(36 - 20)}{(50 - 20)} = 0.53 \qquad R = \frac{T_i - T_o}{t_o - t_i} = \frac{50 - 50}{36 - 20} = 0$$

Accordingly, $F = 1$, and it follows that the tube length per pass is

$$L = \frac{q}{U(N2\pi D)F\Delta T_{\text{lm,CF}}}$$

$$L = \frac{2 \times 10^9 \text{ W}}{4478 \text{ W/m}^2 \cdot \text{K} \, (30{,}000 \times 2\pi \times 0.025 \text{ m}) \times 1 \times 21°\text{C}} = 4.51 \text{ m} \qquad \triangleleft$$

Comments:

1. Recognize that L is the tube length per pass, in which case the total tube length is 9.02 m.
2. Using the NTU method, $C_h = C_{max} = \infty$ and $C_{min} = \dot{m}_c c_{p,c} = 3 \times 10^4$ (kg/s) $\times$ 4179 (J/kg $\cdot$ K) $= 1.25 \times 10^8$ W/K. It follows that $q_{max} = C_{min}(T_{h,i} - T_{c,i}) = 3.76 \times 10^9$ W, and hence that $\varepsilon = 0.53$. From Figure 11.16 it also follows that NTU ≈ 0.75, from which it may then be shown that $L = 4.46$ m. From Equation 11.36b, note that NTU $= 0.755$.

11.6 COMPACT HEAT EXCHANGERS

As discussed in Section 11.1, *compact heat exchangers* are typically used when a large heat transfer surface area per unit volume is desired and at least one of the fluids is a gas. Many different tubular and plate configurations have been considered, where differences are due primarily to fin design and arrangement. Heat transfer and flow characteristics have been determined for specific configurations and are typically presented in the format of Figures 11.20 and

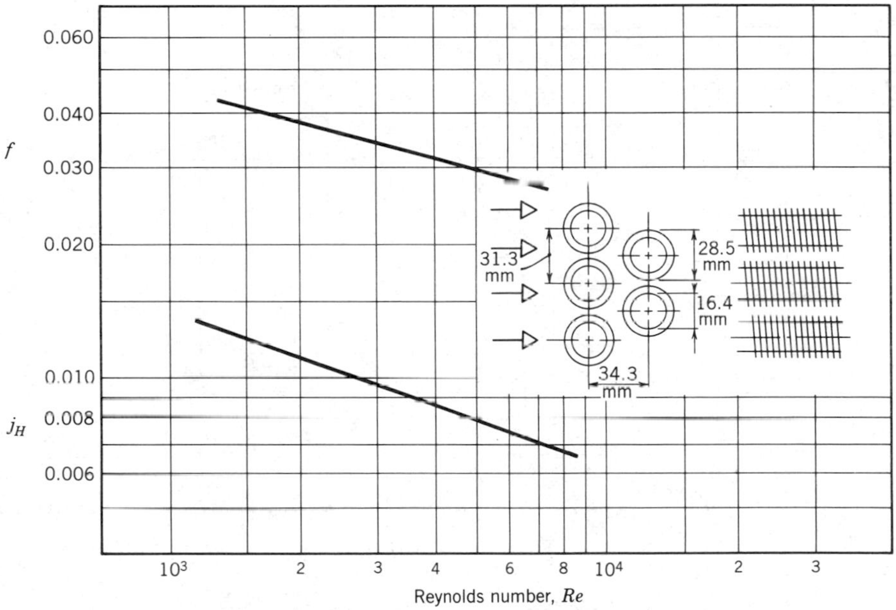

Tube outside diameter, D_o = 16.4 mm
Fin pitch = 275 per meter
Flow passage hydraulic diameter, D_h = 6.68 mm
Fin thickness, t = 0.254 mm
Free-flow area/frontal area, σ = 0.449
Heat transfer area/total volume, α = 269 m^2/m^3
Fin area/total area, A_f/A = 0.830
Note: Minimum free-flow area is in spaces transverse to flow.

Figure 11.20 Heat transfer and friction factor for a circular tube–circular fin heat exchanger, surface CF-7.0-5/8J from Kays and London [5].

11.21. Heat transfer results are correlated in terms of the Colburn j factor $j_H = StPr^{2/3}$ and the Reynolds number, where both the Stanton ($St = h/Gc_p$) and Reynolds ($Re = GD_h/\mu$) numbers are based on the maximum mass velocity

$$G \equiv \rho V_{\max} = \frac{\rho V A_{\text{fr}}}{A_{\text{ff}}} = \frac{\dot{m}}{A_{\text{ff}}} = \frac{\dot{m}}{\sigma A_{\text{fr}}} \tag{11.37}$$

The quantity σ is the ratio of the minimum free-flow area (cross-sectional area perpendicular to flow direction), A_{ff}, to the frontal area A_{fr} of the exchanger. Values of σ, D_h (the hydraulic diameter of the flow passage), α (the heat transfer surface area per total heat exchanger volume), A_f/A (the ratio of fin to total heat transfer surface area), and other geometrical parameters are listed for each configuration. The ratio A_f/A is used in Equation 11.3 to evaluate the temperature effectiveness η_o. In a design calculation, α would be used to determine the required heat exchanger volume, after the total heat transfer

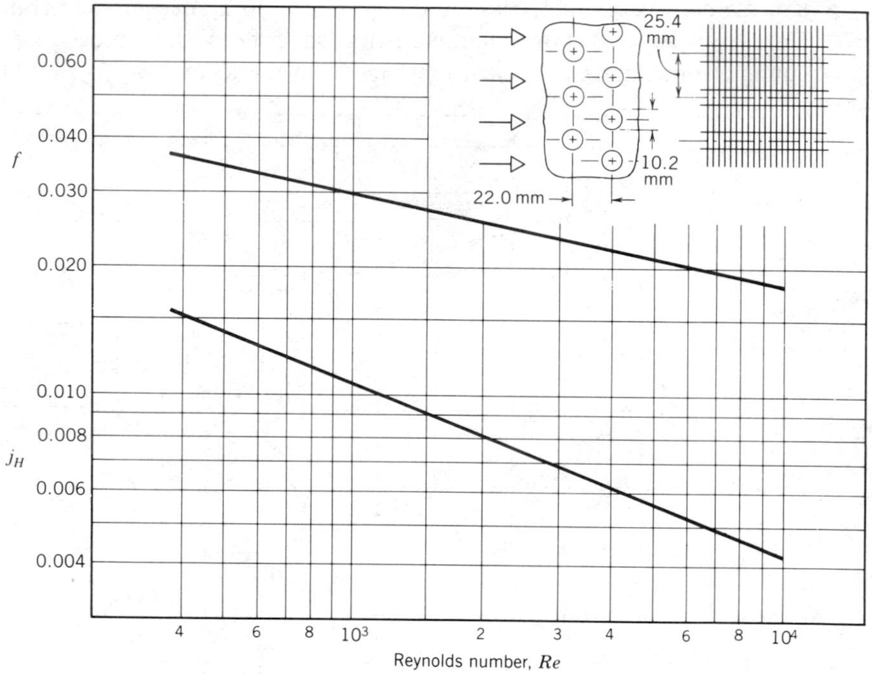

Tube outside diameter, D_o = 10.2 mm
Fin pitch = 315 per meter
Flow passage hydraulic diameter, D_h = 3.63 mm
Fin thickness = 0.330 mm
Free-flow area/frontal area, σ = 0.534
Heat transfer area/total volume, α = 587 m^2/m^3
Fin area/total area = 0.913
Note: Minimum free-flow area in spaces transverse to flow.

Figure 11.21 Heat transfer and friction factor for a circular tube–continuous fin heat exchanger, surface 8.0-3/8T from Kays and London [5].

surface area has been found; in a performance calculation it would be used to determine the surface area from knowledge of the heat exchanger volume.

The pressure drop associated with flow through a compact heat exchanger may be computed from [5]

$$\Delta p = \frac{G^2 v_i}{2}\left[(1+\sigma^2)\left(\frac{v_o}{v_i}-1\right)+f\frac{A}{A_{\text{ff}}}\frac{v_m}{v_i}\right] \tag{11.38}$$

where v_i and v_o are the fluid inlet and outlet specific volumes and $v_m = (v_i + v_o)/2$. For a prescribed core configuration, the friction factor is known as a function of Reynolds number, as, for example, from Figures 11.20 and 11.21; and for a prescribed heat exchanger size, the area ratio may be evaluated from the relation $(A/A_{\text{ff}}) = (\alpha V/\sigma A_{\text{fr}})$, where V is the total heat exchanger volume.

Detailed information on the design and performance of compact heat exchangers may be found in the classic work of Kays and London [5], as well as in the recent publications [3, 4, 8, 9].

EXAMPLE 11.6

Consider a finned-tube, compact heat exchanger having the core configuration of Figure 11.20. The core is fabricated from aluminum, and the tubes have an inside diameter of 13.8 mm. In a waste heat recovery application, water flow through the tubes provides an inside convection coefficient of $h_i = 1500$ W/m$^2\cdot$K, while combustion gases at 1 atm and 700 K are in cross flow over the tubes. If the gas flow rate is 1.25 kg/s and the frontal area is 0.20 m^2, what is the gas-side overall heat transfer coefficient?

SOLUTION

Known:

Compact heat exchanger geometry, gas-side flow rate and temperature, and water-side convection coefficient.

Find: Gas-side overall heat transfer coefficient.

Schematic:

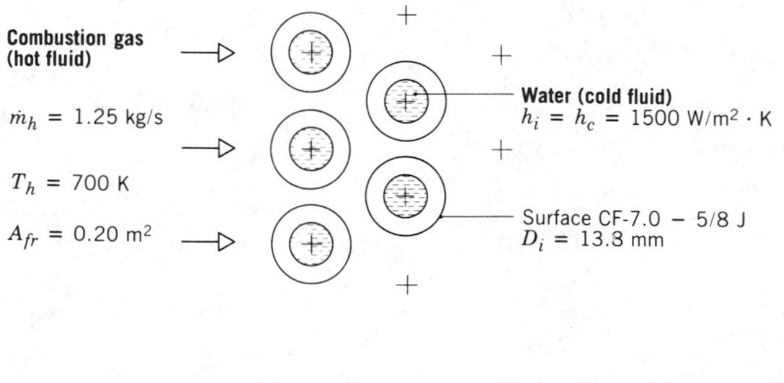

Assumptions:

1. Gas has properties of atmospheric air.
2. Fouling is negligible.

Properties:

Table A.1, aluminum ($T \approx 300$ K): $k = 237$ W/m · K.
Table A.4, air ($p = 1$ atm, $T = 700$ K): $c_p = 1075$ J/kg · K, $\mu = 338.8 \times 10^{-7}$ N · s/m^2, $Pr = 0.695$.

Analysis:

Referring to Equation 11.1, the combustion gas and the water are the hot and cold fluids, respectively. Hence, neglecting fouling effects and acknowledging that the tube inner surface is not finned ($\eta_{o,c} = 1$), the overall heat transfer coefficient based on the gas- (hot-) side surface area is given by

$$\frac{1}{U_h} = \frac{1}{h_c(A_c/A_h)} + A_h R_w + \frac{1}{\eta_{o,h} h_h}$$

where A_h and A_c are the total gas-side (hot) and water-side (cold) surface areas, respectively. If the fin thickness is assumed to be negligible, it is readily shown that

$$\frac{A_c}{A_h} \approx \frac{D_i}{D_o}\left(1 - \frac{A_{f,h}}{A_h}\right)$$

where $A_{f,h}$ is that portion of the total gas-side area associated with the fins. The approximation is valid to within 10%, and for the heat exchanger core conditions (Figure 11.20)

$$\frac{A_c}{A_h} \approx \frac{13.8}{16.4}(1 - 0.830) = 0.143$$

Obtaining the wall conduction resistance from Equation 3.27, it follows that

$$A_h R_w = \frac{\ln(D_o/D_i)}{2\pi Lk/A_h} = \frac{D_i \ln(D_o/D_i)}{2k(A_c/A_h)}$$

Hence

$$A_h R_w = \frac{(0.0138 \text{ m}) \ln(16.4/13.8)}{2(237 \text{ W/m} \cdot \text{K})(0.143)} = 3.51 \times 10^{-5} \text{ m}^2 \cdot \text{K/W}$$

The gas-side convection coefficient may be obtained by first using Equation 11.37 to evaluate the mass velocity.

$$G = \frac{\dot{m}}{\sigma A_{fr}} = \frac{1.25 \text{ kg/s}}{0.449 \times 0.20 \text{ m}^2} = 13.9 \text{ kg/s} \cdot \text{m}^2$$

Hence

$$Re = \frac{13.9\ \text{kg/s}\cdot\text{m}^2 \times 6.68 \times 10^{-3}\ \text{m}}{338.8 \times 10^{-7}\ \text{kg/s}\cdot\text{m}} = 2740$$

and from Figure 11.20, $j_H \approx 0.0096$. Hence

$$h_h \approx 0.0096\frac{Gc_p}{Pr^{2/3}} = 0.0096\frac{(13.9\ \text{kg/s}\cdot\text{m}^2)(1075\ \text{J/kg}\cdot\text{K})}{(0.695)^{2/3}}$$

$$= 183\ \text{W/m}^2\cdot\text{K}$$

To obtain the hot-side temperature effectiveness from Equation 11.3, the fin efficiency must first be determined from Figure 3.19. With $r_{2c} = 14.38$ mm, $r_{2c}/r_1 = 1.75$, $L_c = 6.18$ mm, $A_p = 1.57 \times 10^{-6}\ \text{m}^2$, and $L_c^{3/2}(h_h/kA_p)^{1/2} = 0.34$, it follows that $\eta_f \approx 0.89$. Hence

$$\eta_{o,h} = 1 - \frac{A_f}{A}(1 - \eta_f) = 1 - 0.830(1 - 0.89) = 0.91$$

We then obtain

$$\frac{1}{U_h} = \left(\frac{1}{1500 \times 0.143} + 3.51 \times 10^{-5}\frac{1}{0.91 \times 183}\right)\ \text{m}^2\cdot\text{K/W}$$

$$\frac{1}{U_h} = (4.66 \times 10^{-3} + 3.51 \times 10^{-5} + 6.00 \times 10^{-3}) = 0.0107\ \text{m}^2\cdot\text{K/W}$$

or

$$U_h = 93.4\ \text{W/m}^2\cdot\text{K} \qquad \triangleleft$$

Comments:

The effect of the tube wall thermal conduction resistance is negligible, while contributions due to the cold- and hot-side convection resistances are comparable.

11.7 SUMMARY

Because there are many important applications, heat exchanger research and development has had a long history. Such activity is by no means complete, however, as many talented workers continue to seek ways of improving design and performance. In fact, with heightened concern for energy conservation, there has been a steady and substantial increase in activity. A focal point for this work has been *heat transfer enhancement*, which includes the search for special heat exchanger surfaces through which enhancement may be achieved. In this chapter we have attempted to develop tools that will allow you to perform approximate heat exchanger calculations. More detailed considerations of the subject may be found in several recent references [3, 4, 8, 10, 11].

Although we have restricted ourselves to heat exchangers involving separation of hot and cold fluids by a stationary wall, there are other important options. For example, *evaporative* heat exchangers enable *direct contact* between a liquid and a gas (there is no separating wall); because of latent energy effects, large heat transfer rates per unit volume are possible. Also, for gas-to-gas heat exchange, use is often made of *regenerators* in which the same space is alternately occupied by the hot and cold gases. In a fixed regenerator such as a packed bed, the hot and cold gases alternately enter a stationary, porous solid. In a rotary regenerator, the porous solid is a rotating wheel, which alternately exposes its surfaces to the continuously flowing hot and cold gases. Detailed descriptions of such heat exchangers are available in the literature [3, 4, 8, 11–15].

REFERENCES

1. *Standards of the Tubular Exchange Manufacturers Association*, 6th ed., Tubular Exchanger Manufacturers Association, New York, 1978.
2. Chenoweth, J. M., and M. Impagliazzo, Eds., *Fouling in Heat Exchange Equipment*, American Society of Mechanical Engineers Symposium Volume HTD-17, ASME, New York, 1981.
3. Kakac, S., A. E. Bergles, and F. Mayinger, Eds., *Heat Exchangers*, Hemisphere Publishing Corp., New York, 1981.
4. Kakac, S., R. K. Shah, and A. E. Bergles, Eds., *Low Reynolds Number Flow Heat Exchangers*, Hemisphere Publishing Corp., New York, 1983.
5. Kays, W. M., and A. L. London, *Compact Heat Exchangers*, 3rd ed., McGraw-Hill, New York, 1984.
6. Bowman, R. A., A. C. Mueller, and W. M. Nagle, "Mean Temperature Difference in Design," *Trans. ASME*, *62*, 283, 1940.
7. Jakob, M., *Heat Transfer*, Vol. 2, Wiley, New York, 1957.
8. Shah, R. K., C. F. McDonald, and C. P. Howard, Eds., *Compact Heat Exchangers*, American Society of Mechanical Engineers Symposium Volume HTD-10, ASME, New York, 1980.
9. Webb, R. L., "Compact Heat Exchangers," in E. U. Schlünder, Ed., *Heat Exchanger Design Handbook*, Section 3.9, Hemisphere Publishing Corp., New York, 1983.
10. Marner, W. J., A. E. Bergles, and J. M. Chenoweth, "On the Presentation of Performance Data for Enhanced Tubes Used in Shell-and-Tube Heat Exchangers," *Trans. ASME, J. Heat Transfer*, *105*, 358, 1983.
11. Schlünder, E. U., Ed.-in-Chief, *Heat Exchanger Design Handbook*, Vols. 1–5, Hemisphere Publishing Corp., New York, 1983.
12. Coppage, J. E., and A. L. London, "The Periodic Flow Regenerator: A Summary of Design Theory," *Trans. ASME*, *75*, 779, 1953.
13. Treybal, R. E., *Mass Transfer Operations*, 2nd ed., McGraw-Hill, New York, 1968.

14. Sherwood, T. K., R. L. Pigford, and C. R. Wilkie, *Mass Transfer*, McGraw-Hill, New York, 1975.
15. Schmidt, F. W., and A. J. Willmott, *Thermal Energy Storage and Regeneration*, Hemisphere Publishing Corp., New York, 1981.

PROBLEMS

11.1 A shell-and-tube exchanger (two shells, four tube passes) is used to heat 10,000 kg/h of pressurized water from 35 to 120°C with 5000 kg/h water entering the exchanger at 300°C. If the overall heat transfer coefficient is 1500 $W/m^2 \cdot K$, determine the required heat exchanger area.

11.2 Consider the heat exchanger of Problem 11.1. After several years of operation, it is observed that the outlet temperature of the cold water reaches only 95°C rather than the desired 120°C for the same flow rates and inlet temperatures of the fluids. Determine the cumulative (inner and outer surface) fouling factor that is the cause of the poorer performance.

11.3 A finned-tube, cross-flow heat exchanger is to use the exhaust of a gas turbine to heat pressurized water. Laboratory measurements are performed on a prototype version of the exchanger, which has a surface area of 10 m^2, to determine the overall heat transfer coefficient as a function of operating conditions. Measurements made under particular conditions, for which $\dot{m}_h = 2$ kg/s, $T_{h,i} = 325°C$, $\dot{m}_c = 0.5$ kg/s, and $T_{c,i} = 25°C$, reveal a water outlet temperature of $T_{c,o} = 150°C$. What is the overall heat transfer coefficient of the exchanger?

11.4 Water at a rate of 45,500 kg/h is heated from 80 to 150°C in a heat exchanger having two shell passes and eight tube passes with a total surface area of 925 m^2. Hot exhaust gases having approximately the same thermophysical properties as air enter at 350°C and exit at 175°C. Determine the overall heat transfer coefficient.

11.5 A thin-walled concentric tube heat exchanger is to be used to cool engine oil from 160 to 60°C, and water, which is available at 25°C, is to be used as the coolant. The oil and water flow rates are each 2 kg/s, and the diameter of the inner tube is 0.5 m. The corresponding value of the overall heat transfer coefficient is 250 $W/m^2 \cdot K$. How long must the heat exchanger be to accomplish the desired cooling?

11.6 A single-pass, cross-flow heat exchanger uses hot exhaust gases (mixed) to heat water (unmixed) from 30 to 80°C at a rate of 3 kg/s. The exhaust gases, having thermophysical properties similar to air, enter and exit the exchanger at 225 and 100°C, respectively. If the overall heat transfer coefficient is 200 $W/m^2 \cdot K$, estimate the required surface area.

11.7 Consider the fluid conditions and overall heat transfer coefficient of Problem 11.6 for a concentric tube heat exchanger operating in parallel flow. The thin-walled separator tube has a diameter of 100 mm. (a) Determine the required length for the exchanger. (b) Assuming water flow inside the separator tube to be fully developed, estimate the convection heat transfer coefficient. (c) If the length of the exchanger is twice the value found in part a, determine the heat transfer rate and exit temperatures of the fluids, assuming the overall heat transfer coefficient and the inlet temperatures remain the same. (d) If the exchanger is operated in counter flow, what would be the reduction in required length compared to that found in part a?

11.8 Steam at 0.14 bar is condensed in a shell-and-tube heat exchanger with one shell pass and two tube passes consisting of 130 brass tubes, each with a length per pass of 2 m. The tubes have inner and outer diameters of 13.4 and 15.9 mm, respectively. Cooling water enters the tubes at 20°C with a mean velocity of 1.25 m/s. The heat transfer coefficient for condensation on the outer surfaces of the tubes is 13,500 $W/m^2 \cdot K$. Determine the overall heat transfer coefficient, the cooling water outlet temperature, and the steam condensation rate.

11.9 A two-fluid heat exchanger has inlet and outlet temperatures of 65 and 40°C for the hot fluid and 15 and 30°C for the cold fluid. Can you tell whether this exchanger is operating under counterflow or parallel-flow conditions? What is the effectiveness of the exchanger if the cold fluid has the minimum capacity rate?

11.10 Water at 225 kg/h is to be heated from 35 to 95°C by means of a concentric tube heat exchanger. Oil at 225 kg/h and 210°C, with a specific heat of 2095 $J/kg \cdot K$, is to be used as the hot fluid. If the overall heat transfer coefficient based on the outer diameter of the inner tube is 550 $W/m^2 \cdot K$, determine the length of the exchanger if the outer diameter is 100 mm.

11.11 A hot fluid enters a concentric pipe heat exchanger at 150°C and is to be cooled to 100°C by a cold fluid entering at 35°C and heated to 65°C. Would you use parallel-flow or counterflow for the most effective design?

11.12 Consider a *very long*, concentric tube heat exchanger having hot and cold water inlet temperatures of 85 and 15°C. The flow rate of the hot water is twice that of the cold water. Assuming equivalent hot and cold water specific heats, determine the hot water outlet temperature for the following modes of operation: (a) counterflow, (b) parallel flow.

11.13 A shell-and-tube heat exchanger is to heat 10,000 kg/h of water from 16 to 84°C by hot engine oil flowing through the shell. The oil makes a single shell pass, entering at 160°C and leaving at 94°C, with an average heat transfer coefficient of 400 $W/m^2 \cdot K$. The water flows through 11 brass tubes of 22.9 mm inside diameter and 25.4 mm outside diameter, with each tube making four passes through the shell. Assuming fully developed flow for the water, determine the required tube length per pass.

11.14 A shell-and-tube heat exchanger consists of 135 thin-walled tubes in a double-pass arrangement, each of 12.5 mm diameter with a total surface area 47.5 m^2. Water (the tube-side fluid) enters the heat exchanger at 15°C and 6.5 kg/s and is heated by exhaust gas entering at 200°C and 5 kg/s. The gas may be assumed to have the properties of atmospheric air, and the overall heat transfer coefficient is approximately 200 $W/m^2 \cdot K$. What are the gas and water outlet temperatures? Assuming fully developed flow, what is the tube-side convection coefficient?

11.15 An ocean thermal energy conversion system is being proposed for electric power generation. Such a system is based on the standard power cycle for which the working fluid is evaporated, passed through a turbine, and subsequently condensed. The system is to be used in very special locations for which the oceanic water temperature near the surface is approximately 300 K, while the temperature at reasonable depths is approximately 280 K. The warmer water is used as a heat source to evaporate the working fluid, while the colder water is used as a heat sink for condensation of the fluid. Consider a power plant that is to generate 2 MW of electricity at an efficiency (electric power output per heat input) of 3%. The evaporator is a heat exchanger consisting of a single shell with many tubes executing two passes. If the working fluid is evaporated at its phase change temperature of 290 K, with ocean water entering at 300 K and leaving at 292 K, what is the heat exchanger area required for the evaporator? What flow rate must be maintained for the water passing through the evaporator? The overall heat transfer coefficient may be approximated as 1200 $W/m^2 \cdot K$.

11.16 The feedwater heater for a boiler supplies 10,000 kg/h of water at 65°C. The feedwater has an inlet temperature of 20°C and is to be heated in a single-shell, two-tube pass heat exchanger by condensing steam at 1.30 bars. The overall heat transfer coefficient is 2000 $W/m^2 \cdot K$. Using both the LMTD and NTU methods, determine the required heat transfer area. What is the steam condensation rate?

11.17 A shell-and-tube heat exchanger with single shell and tube passes (Figure 11.3) is used to cool the oil of a large marine engine. Lake water (the shell-side fluid) enters the heat exchanger at 2 kg/s and 15°C, while the oil enters at 1 kg/s and 140°C. The oil flows through 100 copper tubes, each 500 mm long and having inner and outer diameters of 6 and 8 mm. The shell-side convection coefficient is approximately 500 $W/m^2 \cdot K$. What is the oil outlet temperature?

11.18 A shell-and-tube heat exchanger with one shell pass and 20 tube passes uses hot water on the tube side to heat oil on the shell side. The single copper tube has inner and outer diameters of 20 and 24 mm and a length per pass of 3 m. The water enters at 87°C and 0.2 kg/s and leaves at 27°C. Inlet and outlet temperatures of the oil are 7 and 37°C. What is the average convection coefficient for the tube outer surface?

11.19 The oil in an engine is cooled by air in a cross-flow heat exchanger where both fluids are unmixed. Atmospheric air enters at 30°C and 0.53 kg/s. Oil at 0.026 kg/s enters at 75°C and flows through a tube of 10 mm diameter. Assuming fully developed flow and constant wall heat flux, estimate the oil-side heat transfer coefficient. If the overall convection coefficient is 53 $W/m^2 \cdot K$ and the total heat transfer area is 1 m^2, determine the effectiveness. What is the exit temperature of the oil?

11.20 A concentric tube heat exchanger uses water, which is available at 15°C, to cool ethylene glycol from 100 to 60°C. The water and glycol flow rates are each 0.5 kg/s. What are the maximum possible heat transfer rate and effectiveness of the exchanger? Which is preferred, a parallel-flow or counterflow mode of operation?

11.21 Water is used for both fluids (unmixed) flowing through a single-pass, cross-flow heat exchanger. The hot water enters at 90°C and 10,000 kg/h, while the cold water enters at 10°C and 20,000 kg/h. If the effectiveness of the exchanger is 60%, determine the cold water exit temperature.

11.22 A cross-flow heat exchanger consists of a bundle of 32 tubes in a 0.6-m^2 duct. Hot water at 150°C and a mean velocity of 0.5 m/s enters the tubes having inner and outer diameters of 10.2 and 12.5 mm. Atmospheric air at 10°C enters the exchanger with a volumetric flow rate of 1.0 m^3/s. The convection heat transfer coefficient on the tube outer surfaces is 400 $W/m^2 \cdot K$. Estimate the fluid outlet temperatures.

11.23 Exhaust gas from a furnace is used to preheat the combustion air supplied to the furnace burners. The gas, which has a flow rate of 15 kg/s and an inlet temperature of 1100 K, passes through a bundle of tubes, while the air, which has a flow rate of 10 kg/s and an inlet temperature of 300 K, is in cross flow over the tubes. The tubes are unfinned, and the overall heat transfer coefficient is 100 $W/m^2 \cdot K$. Determine the total tube surface area required to achieve an air outlet temperature of

850 K. The exhaust gas and the air may each be assumed to have a specific heat of 1075 J/kg·K.

11.24 Saturated steam at 100°C condenses in a shell-and-tube exchanger (single shell, two tube passes) with a surface area of 0.5 m^2 and overall heat transfer coefficient of 2000 $W/m^2 \cdot K$. Water enters at 0.5 kg/s and 15°C. Determine the outlet temperature of the water and the rate of steam condensation.

11.25 Consider a concentric tube heat exchanger characterized by a uniform overall heat transfer coefficient and operating under the following conditions.

	$\dot{m}$ (kg/s)	c_p (J/kg·K)	T_i (°C)	T_o (°C)
Cold fluid	0.125	4200	40	95
Hot fluid	0.125	2100	210	

What is the maximum possible heat transfer rate? What is the heat exchanger effectiveness? Should the heat exchanger be operated in parallel flow or in counterflow? What is the ratio of the required areas for these two flow conditions?

11.26 Consider a concentric tube heat exchanger with hot and cold water inlet temperatures of 200 and 35°C, respectively. The flow rates of the hot and cold fluids are 42 and 84 kg/h, respectively. Assume the overall heat transfer coefficient is 180 $W/m^2 \cdot K$.

(a) What is the maximum heat transfer rate that could be achieved for the prescribed inlet conditions?

(b) If the exchanger is operated in counterflow with a heat transfer area of 0.33 m^2, determine the outlet fluid temperatures.

(c) What is the largest possible heat transfer rate that could be achieved for the prescribed inlet conditions if the exchanger is operated in parallel flow and is very long? What is the effectiveness of the exchanger in this configuration?

11.27 Hot water at 2.5 kg/s and 100°C enters a concentric tube counterflow heat exchanger having a total area of 23 m^2. Cold water at 20°C enters at 5.0 kg/s, and the overall heat transfer coefficient is 1000 $W/m^2 \cdot K$. Determine the total heat transfer rate and the outlet temperatures of the hot and cold fluids.

11.28 Hot exhaust gases are used in a shell-and-tube exchanger to heat 2.5 kg/s of water from 35 to 85°C. The gases, assumed to have the properties of air, enter at 200°C and leave at 93°C. The overall heat transfer coefficient is 180 $W/m^2 \cdot K$. Using the NTU-effectiveness method, calculate the area of the heat exchanger.

11.29 In open heart surgery under hypothermic conditions, the patient's blood is cooled before the surgery and rewarmed afterwards. It is proposed that a concentric tube, counterflow heat exchanger of length 0.5 m be used for this purpose with the thin-walled inner tube having a diameter of 55 mm. If water at 60°C and 0.10 kg/s is used to heat blood entering the exchanger at 18°C and 0.05 kg/s, what is the temperature of the blood leaving the exchanger? The overall heat transfer coefficient is 500 $W/m^2 \cdot K$, and the specific heat of the blood is 3500 J/kg·K.

11.30 Ethylene glycol and water, at 60 and 10°C, respectively, enter a shell-and-tube heat exchanger for which the total heat transfer area is 15 m^2. With ethylene glycol and water flow rates of 2 and 5 kg/s, respectively, the overall heat transfer coefficient is

800 $W/m^2 \cdot K$. Determine the rate of heat transfer and the fluid outlet temperatures.

11.31 A boiler used to generate saturated steam is in the form of an unfinned, cross-flow heat exchanger, with water flowing through the tubes and a high temperature gas in cross flow over the tubes. The gas, which has a specific heat of 1120 $J/kg \cdot K$ and a mass flow rate of 10 kg/s, enters the heat exchanger at 1400 K. The water, which has a flow rate of 3 kg/s, enters as saturated liquid at 450 K and leaves as saturated vapor at the same temperature. If the overall heat transfer coefficient is 50 $W/m^2 \cdot K$ and there are 500 tubes, each of 0.025 m diameter, what is the required tube length?

11.32 Waste heat from the exhaust gas of an industrial furnace is recovered by mounting a bank of unfinned tubes in the furnace stack. Pressurized water at a flow rate of 0.025 kg/s makes a single pass through *each* of the tubes, while the exhaust gas, which has an upstream velocity of 5.0 m/s, moves in cross flow over the tubes at 2.25 kg/s. The tube bank consists of a square array of 100 thin-walled tubes (10×10), each 25 mm in diameter and 4 m long. The tubes are aligned with a transverse pitch of 50 mm. The inlet temperatures of the water and the exhaust gas are 300 and 800 K, respectively. What is the overall heat transfer coefficient between the water and the gas? The water flow is fully developed, and the gas properties may be assumed to be those of atmospheric air. What are the fluid outlet temperatures?

11.33 A heat exchanger consists of a bank of 1200 thin-walled tubes with air in cross flow over the tubes. The tubes are arranged in-line, with 40 longitudinal rows (along the direction of air flow) and 30 transverse rows. The tubes are 0.07 m in diameter and 2 m long, with transverse and longitudinal pitches of 0.14 m. The hot fluid flowing through the tubes consists of saturated steam condensing at 400 K. The convection coefficient of the condensing steam is much larger than that of the air. If air enters the heat exchanger at 12 kg/s, 300 K, and 1 atm, what is its outlet temperature?

11.34 Consider the compact heat exchanger conditions of Example 11.6. After extended use, fouling factors of 0.0005 and 0.001 $m^2 \cdot K/W$ are associated with the water- and gas-side conditions, respectively. What is gas-side overall heat transfer coefficient?

11.35 Consider the heat exchanger core geometry and frontal area prescribed in Example 11.6. The exchanger must heat 2 kg/s of water from 300 to 350 K, using 1.25 kg/s of combustion gases entering at 700 K. Using the overall heat transfer coefficient determined in the example, find the required heat exchanger volume, assuming single-pass operation. What is the number of tube rows N_L in the longitudinal (gas-flow) direction? If the velocity of water flowing through the tubes is 100 mm/s, what is the number of tube rows N_T in the transverse direction? What is the required tube length?

Twelve

RADIATION: PROCESSES AND PROPERTIES

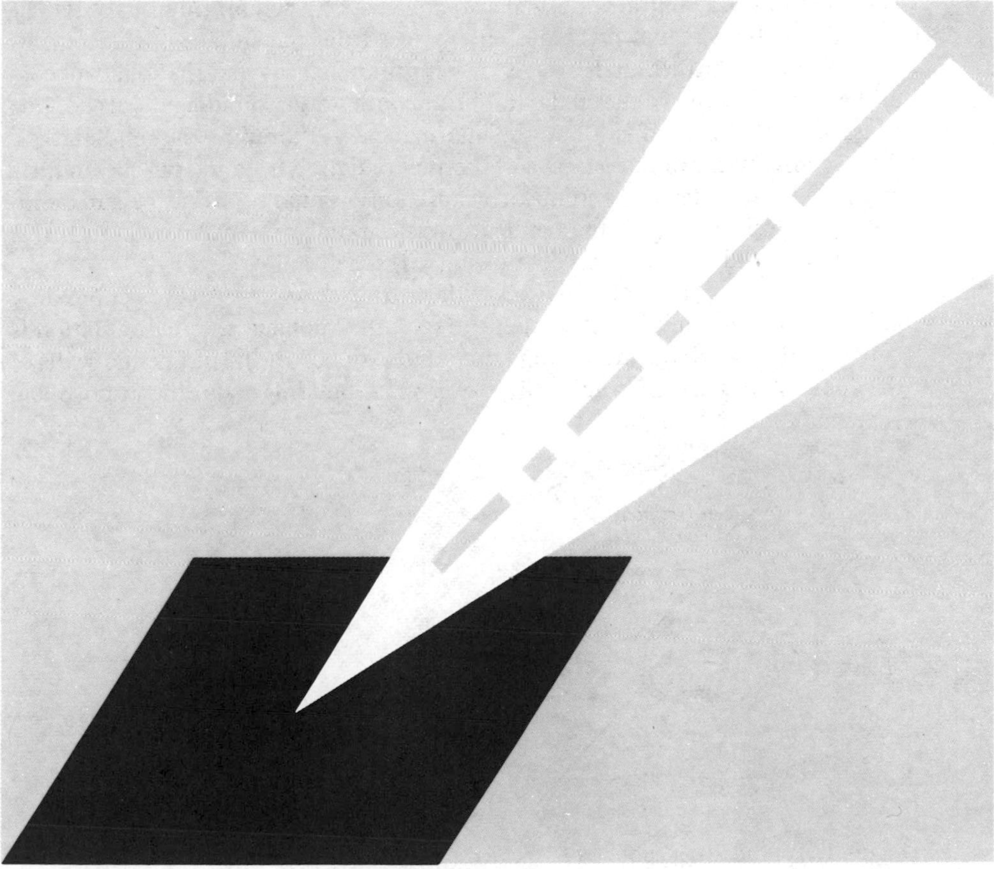

We have come to recognize that heat transfer by conduction and convection requires the presence of a temperature gradient in some form of matter. In contrast, heat transfer by *thermal radiation* requires no matter. It is an extremely important process, and in the physical sense it is the most interesting of the heat transfer modes. It is relevant to many industrial heating, cooling, and drying processes, as well as to energy conversion methods that involve fossil fuel combustion and solar radiation.

In this chapter we consider the means by which thermal radiation is generated, the specific nature of the radiation, and the manner in which the radiation interacts with matter. We give particular attention to radiative interactions at a surface and to the properties that must be introduced to describe these interactions. In Chapter 13 we focus on means for computing radiative exchange between two or more surfaces.

12.1 FUNDAMENTAL CONCEPTS

Consider a solid that is initially at a higher temperature T_s than that of its surroundings T_{sur}, but around which there exists a vacuum (Figure 12.1). The presence of the vacuum precludes energy loss from the surface of the solid by conduction or convection. However, our intuition tells us that the solid will cool and eventually achieve thermal equilibrium with its surroundings. This cooling is associated with a reduction in the internal energy stored by the solid and is a direct consequence of the *emission* of thermal radiation from the surface. In turn, the surface will intercept and absorb radiation originating from the surroundings. However, if $T_s > T_{sur}$ the *net* heat transfer rate by radiation $q_{rad,net}$ is *from* the surface, and the surface will cool until T_s reaches T_{sur}.

We associate thermal radiation with the rate at which energy is emitted by matter as a result of its finite temperature. At this moment thermal radiation is being emitted by all the matter that surrounds you: by the furniture and walls of the room, if you are indoors, or by the ground, buildings, and the atmosphere

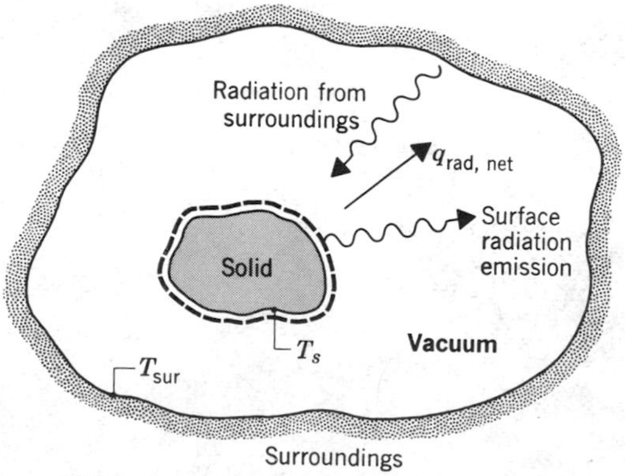

Figure 12.1 Radiation cooling of a heated solid.

and sun, if you are outdoors. The mechanism of emission is related to energy released as a result of oscillations or transitions of the many electrons that constitute matter. These oscillations are, in turn, sustained by the internal energy, and therefore the temperature, of the matter. Hence we associate the emission of thermal radiation with thermally excited conditions within the matter.

All forms of matter emit radiation. For gases and for semitransparent solids, such as glass and salt crystals at elevated temperatures, emission is a *volumetric phenomenon*, as illustrated in Figure 12.2. That is, radiation emerging from a finite volume of matter is the integrated effect of local emission throughout the volume. However, in this text we concentrate on situations for which radiation is a *surface phenomenon.* In most solids and liquids, radiation emitted from interior molecules is strongly absorbed by adjoining molecules. Accordingly, radiation that is emitted from a solid or a liquid originates from molecules that are within a distance of approximately 1 μm from the exposed surface. It is for this reason that emission from a solid or a liquid into an adjoining gas or a vacuum is viewed as a surface phenomenon.

We know that radiation originates due to emission by matter and that its subsequent transport does not require the presence of any matter. But what is the nature of this transport? One theory views radiation as the propagation of a collection of particles termed *photons* or *quanta.* Alternatively, radiation may be viewed as the propagation of *electromagnetic waves.* In any case we wish to attribute to radiation the standard wave properties of frequency ν and wavelength λ. For radiation propagating in a particular medium, the two properties are related by

$$\lambda = \frac{c}{\nu} \tag{12.1}$$

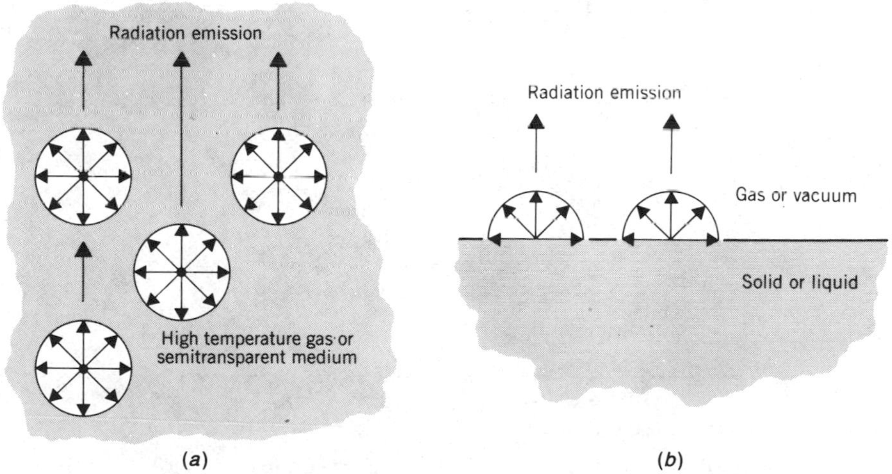

Figure 12.2 The emission process. (*a*) As a volumetric phenomenon. (*b*) Surface phenomenon.

where c is the speed of light in the medium. For propagation in a vacuum, $c_o = 2.998 \times 10^8$ m/s. The unit of wavelength is commonly the micrometer (μm), where 1 $\mu\text{m} = 10^{-6}$ m.

The complete electromagnetic spectrum is delineated in Figure 12.3. The short wavelength gamma rays, X rays, and ultraviolet (UV) radiation are primarily of interest to the high energy physicist and the nuclear engineer, while the long wavelength microwaves and radio waves are of concern to the electrical engineer. It is the intermediate portion of the spectrum, which extends from approximately 0.1 to 100 μm and includes a portion of the UV and all of the visible and infrared (IR), that is termed *thermal radiation* and is pertinent to heat transfer.

Thermal radiation emitted by a surface encompasses a range of wavelengths. As shown in Figure 12.4*a*, the magnitude of the radiation varies with wavelength, and the term *spectral* is used to refer to the nature of this dependence. Emitted radiation consists of a continuous, nonuniform distribution of *monochromatic* (single-wavelength) components. As we will find, both

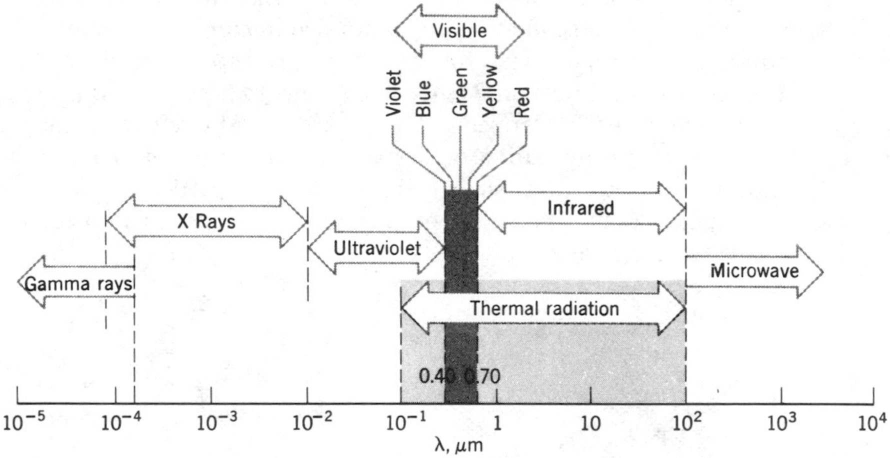

Figure 12.3 Spectrum of electromagnetic radiation.

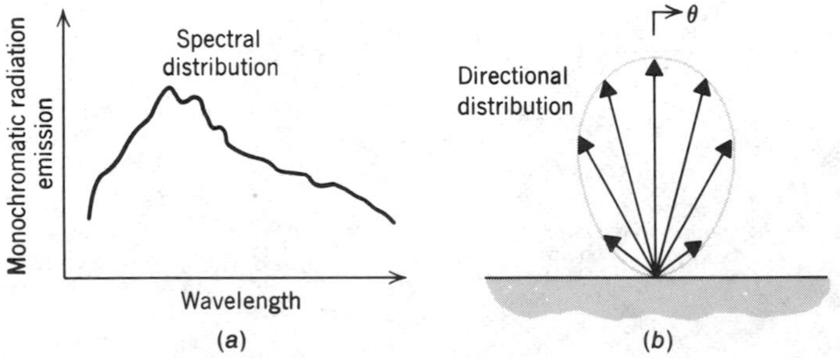

Figure 12.4 Radiation emitted by a surface. (*a*) Spectral distribution. (*b*) Directional distribution.

the magnitude of the radiation at any wavelength and the *spectral distribution* vary with the nature and temperature of the emitting surface.

The spectral nature of thermal radiation is one of two features that complicate its description. The second feature relates to its *directionality*. As shown in Figure 12.4*b*, a surface may emit preferentially in certain directions, creating a *directional distribution* of the emitted radiation. To properly quantify radiation heat transfer, we must be able to treat both spectral and directional effects.

12.2 RADIATION INTENSITY

Radiation emitted by a surface propagates in all possible directions (Figure 12.4*b*), and we are often interested in knowing its directional distribution. Also, radiation incident on a surface may come from different directions, and the manner in which the surface responds to this radiation depends on the direction. Such directional effects may be treated by introducing the concept of *radiation intensity*.

12.2.1 Definitions

Consider emission in a particular direction from an element of area dA_1, as shown in Figure 12.5*a*. This direction may be specified in terms of the zenith and azimuthal angles, θ and ϕ, respectively, of a spherical coordinate system (Figure 12.5*b*). A differentially small surface in space dA_n, through which this radiation passes, subtends a solid angle $d\omega$ when viewed from a point on dA_1. From Figure 12.6*a* we see that the differential plane angle $d\alpha$ is defined by a region between

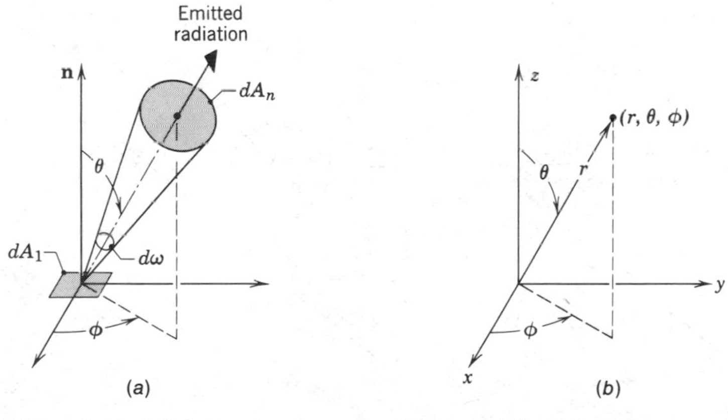

Figure 12.5 Directional nature of radiation. (*a*) Emission of radiation from a differential area dA_1 into a solid angle $d\omega$ subtended by dA_n at a point on dA_1. (*b*) The spherical coordinate system.

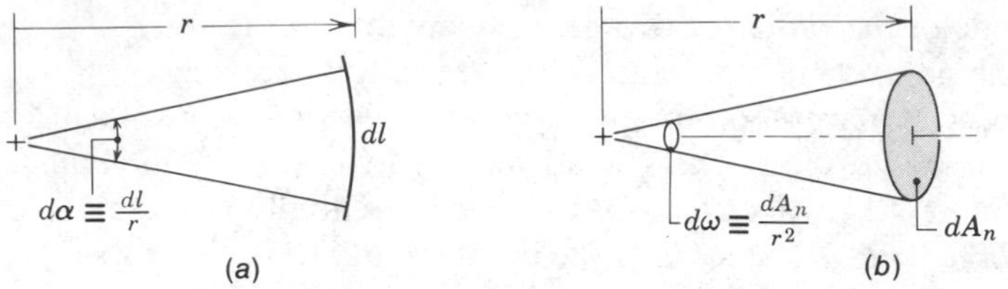

Figure 12.6 Definition of (*a*) plane and (*b*) solid angles.

the rays of a circle and is measured as the ratio of the element of arc length dl on the circle to the radius r of the circle. Similarly, from Figure 12.6*b* the differential solid angle $d\omega$ is defined by a region between the rays of a sphere and is measured as the ratio of the element of area dA_n on the sphere to the square of the sphere's radius. Accordingly,

$$d\omega \equiv \frac{dA_n}{r^2} \tag{12.2}$$

The area dA_n is normal to the (θ, ϕ) direction, and as shown in Figure 12.7, it may be represented as $dA_n = r^2 \sin\theta \, d\theta \, d\phi$ for a spherical surface. Accordingly,

$$d\omega = \sin\theta \, d\theta \, d\phi \tag{12.3}$$

Whereas the plane angle $d\alpha$ has the unit of radians (rad), the unit of the solid angle is the steradian (sr).

Returning to Figure 12.5*a*, we now consider the rate at which emission from dA_1 passes through dA_n. This quantity may be expressed in terms of the *spectral intensity* $I_{\lambda,e}$ of the emitted radiation. We formally define $I_{\lambda,e}$ as the *rate at which radiant energy is emitted at the wavelength* λ *in the* (θ, ϕ) *direction, per unit area of*

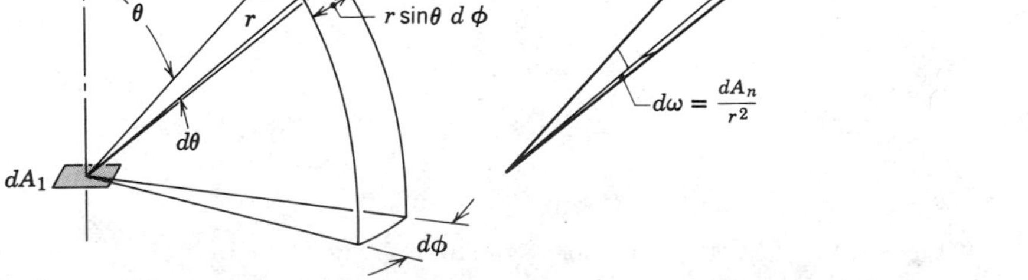

Figure 12.7 The solid angle subtended by dA_n at a point on dA_1 in the spherical coordinate system.

the emitting surface normal to this direction, per unit solid angle about this direction, and per unit wavelength interval $d\lambda$ about λ. Note that the area used to define the intensity is the component of dA_1 perpendicular to the direction of the radiation. From Figure 12.8 we see that this projected area is equal to $dA_1 \cos\theta$. In effect it is how dA_1 would appear to an observer situated on dA_n. The spectral intensity, which has units of $W/m^2 \cdot sr \cdot \mu m$, is then

$$I_{\lambda,e}(\lambda,\theta,\phi) \equiv \frac{dq}{dA_1 \cos\theta \cdot d\omega \cdot d\lambda} \tag{12.4}$$

where $(dq/d\lambda) \equiv dq_\lambda$ is the rate at which radiation of wavelength λ leaves dA_1 and passes through dA_n. Rearranging Equation 12.4, it follows that

$$dq_\lambda = I_{\lambda,e}(\lambda,\theta,\phi)\, dA_1 \cos\theta\, d\omega \tag{12.5}$$

where dq_λ has the units of $W/\mu m$. This important expression allows us to compute the rate at which radiation emitted by a surface propagates into the region of space defined by the solid angle $d\omega$ about the (θ,ϕ) direction. However, to compute this rate, the spectral intensity $I_{\lambda,e}$ of the emitted radiation must be known. The manner in which this quantity may be determined is discussed in subsequent sections. Expressing Equation 12.5 per unit area of the emitting surface and substituting from Equation 12.3, the spectral radiation *flux* associated with dA_1 is

$$dq''_\lambda = I_{\lambda,e}(\lambda,\theta,\phi) \cos\theta \sin\theta\, d\theta\, d\phi \tag{12.6}$$

If the spectral and directional distributions of $I_{\lambda,e}$ are known, that is, $I_{\lambda,e}(\lambda,\theta,\phi)$, the heat flux associated with emission into any finite solid angle or over any finite wavelength interval may be determined by integrating Equation 12.6. For example, the spectral heat flux associated with emission into a hypothetical hemisphere above dA_1, as shown in Figure 12.9, is

$$q''_\lambda(\lambda) = \int_0^{2\pi} \int_0^{\pi/2} I_{\lambda,e}(\lambda,\theta,\phi) \cos\theta \sin\theta\, d\theta\, d\phi \tag{12.7}$$

The solid angle associated with the entire hemisphere may be obtained by

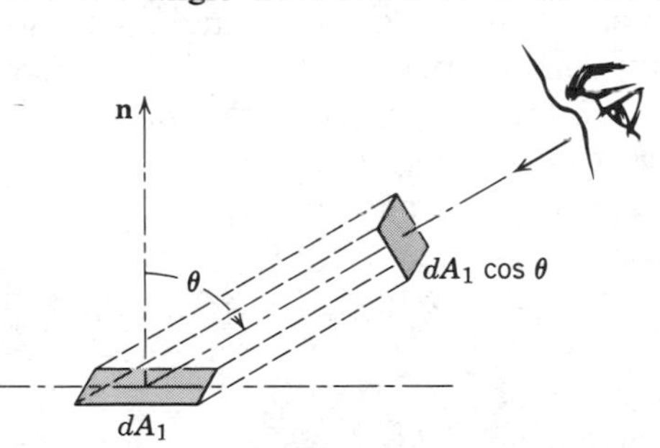

Figure 12.8 The projection of dA_1 normal to the direction of radiation.

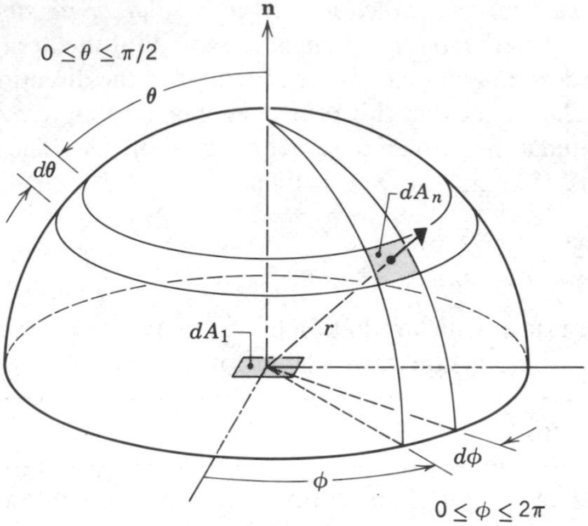

Figure 12.9 Emission from a differential element of area dA_1 into a hypothetical hemisphere centered at a point on dA_1.

integrating Equation 12.3 over the limits $\phi = 0$ to $\phi = 2\pi$ and $\theta = 0$ to $\theta = \pi/2$. Hence

$$\int_h d\omega = \int_0^{2\pi} \int_0^{\pi/2} \sin\theta \, d\theta \, d\phi = 2\pi \int_0^{\pi/2} \sin\theta \, d\theta = 2\pi \tag{12.8}$$

where the subscript h refers to integration over the hemisphere. The total heat flux associated with emission in all directions and at all wavelengths is then

$$q'' = \int_0^{\infty} q''_\lambda(\lambda) \, d\lambda \tag{12.9}$$

12.2.2 Relation to Emission

The radiation intensity is related to several important radiation fluxes. Recall that emission occurs from any surface that is at a finite temperature. The concept of *emissive power* is introduced to quantify the amount of radiation emitted per unit surface area. The *spectral, hemispherical emissive power* E_λ ($\mathrm{W/m^2 \cdot \mu m}$) is defined as the rate at which radiation of wavelength λ is emitted in *all directions* from a surface per unit wavelength $d\lambda$ about λ and per unit surface area. Accordingly, it is related to the spectral intensity of the emitted radiation by an expression of the form

$$E_\lambda(\lambda) = \int_0^{2\pi} \int_0^{\pi/2} I_{\lambda,e}(\lambda, \theta, \phi) \cos\theta \sin\theta \, d\theta \, d\phi \tag{12.10}$$

in which case it is equivalent to the heat flux given by Equation 12.7. Note that E_λ is a flux based on the *actual* surface area, whereas $I_{\lambda,e}$ is based on the *projected* area. The cos θ term appearing in the integrand is a consequence of this difference.

The *total, hemispherical emissive power*, E (W/m^2), is the rate at which radiation is emitted per unit area at all possible wavelengths and in all possible directions. Accordingly,

$$E = \int_0^\infty E_\lambda(\lambda)\, d\lambda \tag{12.11}$$

or from Equation 12.10

$$E = \int_0^\infty \int_0^{2\pi} \int_0^{\pi/2} I_{\lambda,e}(\lambda,\theta,\phi) \cos\theta \sin\theta \, d\theta \, d\phi \, d\lambda \tag{12.12}$$

Since the term "emissive power" implies emission in all directions, the adjective "hemispherical" is redundant and is often dropped. One then speaks of the *spectral emissive power* E_λ, or the *total emissive power* E.

Although the directional distribution of surface emission varies according to the nature of the surface, there is a special case that provides a reasonable approximation for many surfaces. We speak of a *diffuse emitter* as a surface for which the intensity of the emitted radiation is independent of direction, in which case $I_{\lambda,e}(\lambda,\theta,\phi) = I_{\lambda,e}(\lambda)$. Removing $I_{\lambda,e}$ from the integrand of Equation 12.10 and performing the integration, it follows that

$$E_\lambda(\lambda) = \pi I_{\lambda,e}(\lambda) \tag{12.13}$$

Similarly, from Equation 12.12

$$E = \pi I_e \tag{12.14}$$

where I_e is the *total intensity* of the emitted radiation. Note that the constant appearing in the above expressions is π, not 2π, and has the unit steradians.

EXAMPLE 12.1

A small surface of area $A_1 = 10^{-3}$ m^2 is known to emit diffusely, and from measurements the total intensity associated with emission in the normal direction is $I_n = 7000$ W/m$^2 \cdot$ sr.

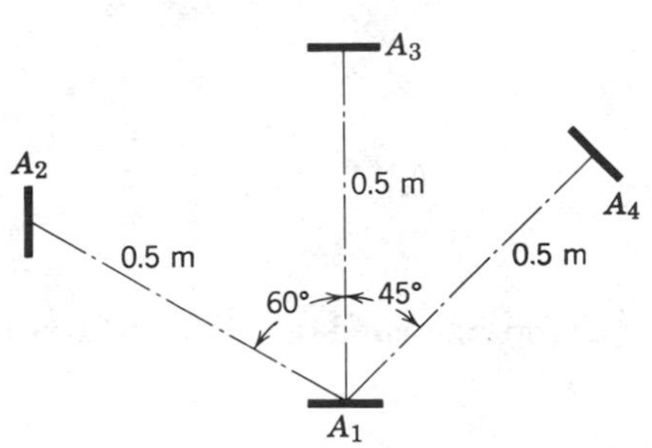

Radiation emitted from the surface is intercepted by three other surfaces of area $A_2 = A_3 = A_4 = 10^{-3}\ m^2$, which are 0.5 m from A_1 and are oriented as shown. What is the intensity associated with emission in each of the three directions? What are the solid angles subtended by the three surfaces when viewed from A_1? What is the rate at which radiation emitted by A_1 is intercepted by the three surfaces?

SOLUTION

Known:

Normal intensity of diffuse emitter of area A_1 and orientation of three surfaces relative to A_1.

Find:

1. Intensity of emission in each of the three directions.
2. Solid angles subtended by the three surfaces.
3. Rate at which radiation is intercepted by the three surfaces.

Schematic:

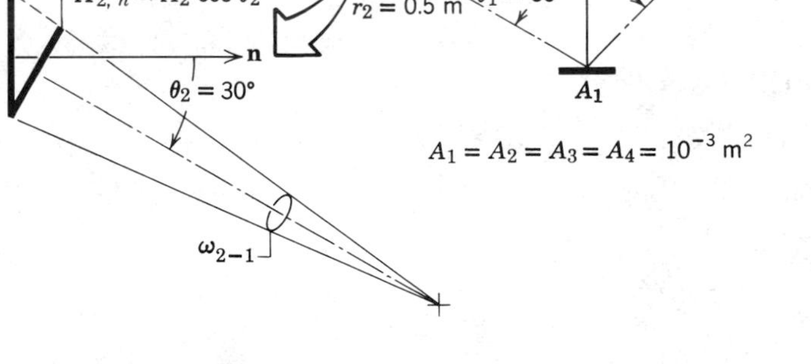

Assumptions:

1. Surface A_1 emits diffusely.
2. A_1, A_2, A_3, and A_4 may be approximated as differential surfaces, $(A_j/r_j^2) \ll 1$.

Analysis:

1. From the definition of a diffuse emitter, we know that the intensity of the emitted radiation is independent of direction. Hence

   $$I = 7000\ \text{W/m}^2 \cdot \text{sr} \qquad \triangleleft$$

   for each of the three directions.

2. Treating A_2, A_3, and A_4 as differential surface areas, the solid angles may be computed from Equation 12.2

   $$d\omega \equiv \frac{dA_n}{r^2}$$

   where dA_n is the projection of the surface normal to the direction of the radiation. Accordingly,

   $$dA_{n,j} = dA_j \times \cos\theta_j$$

   where θ_j is the angle between the surface normal and the direction of the radiation. The solid angle subtended by surface A_2 with respect to A_1 is then

   $$\omega_{2-1} \approx \frac{A_2 \times \cos\theta_2}{r^2} = \frac{10^{-3}\ \text{m}^2 \times \cos 30^\circ}{(0.5\ \text{m})^2} = 3.46 \times 10^{-3}\ \text{sr} \qquad \triangleleft$$

   Similarly,

   $$\omega_{3-1} \approx \frac{A_3 \times \cos\theta_3}{r^2} = \frac{10^{-3}\ \text{m}^2 \times \cos 0^\circ}{(0.5\ \text{m})^2} = 4.00 \times 10^{-3}\ \text{sr} \qquad \triangleleft$$

   $$\omega_{4-1} \approx \frac{A_4 \times \cos\theta_4}{r^2} = \frac{10^{-3}\ \text{m}^2 \times \cos 0^\circ}{(0.5\ \text{m})^2} = 4.00 \times 10^{-3}\ \text{sr} \qquad \triangleleft$$

3. Approximating A_1 as a differential surface, the rate at which radiation is intercepted by each of the three surfaces may be estimated from Equation 12.5, which, for the total radiation, may be expressed as

   $$q_{1-j} \approx I \times A_1 \cos\theta_1 \times \omega_{j-1}$$

   where θ_1 is the angle between the normal to surface 1 and the direction of the radiation. Hence

   $$q_{1-2} \approx 7000\ \text{W/m}^2 \cdot \text{sr}\ (10^{-3}\ \text{m}^2 \times \cos 60^\circ)\ 3.46 \times 10^{-3}\ \text{sr}$$
   $$= 12.1 \times 10^{-3}\ \text{W} \qquad \triangleleft$$

   $$q_{1-3} \approx 7000\ \text{W/m}^2 \cdot \text{sr}\ (10^{-3}\ \text{m}^2 \times \cos 0^\circ)\ 4.00 \times 10^{-3}\ \text{sr}$$
   $$= 28.0 \times 10^{-3}\ \text{W} \qquad \triangleleft$$

   $$q_{1-4} \approx 7000\ \text{W/m}^2 \cdot \text{sr}\ (10^{-3}\ \text{m}^2 \times \cos 45^\circ)\ 4.00 \times 10^{-3}\ \text{sr}$$
   $$= 19.8 \times 10^{-3}\ \text{W} \qquad \triangleleft$$

Comments:

1. Note the different values of θ_1 for the emitting surface and the values of θ_2, θ_3, and θ_4 for the receiving surfaces.
2. If the surfaces were not small relative to the square of the separation distance, the solid angles and radiation heat transfer rates would have to be obtained by integrating Equations 12.3 and 12.5, respectively.
3. Any spectral component of the radiation rate could also be obtained using these procedures, if the spectral intensity I_λ is known.
4. Even though the intensity of the emitted radiation is independent of direction, the rate at which radiation is intercepted by the three surfaces differs significantly due to differences in the solid angles and projected areas.

12.2.3 Relation to Irradiation

Although we have focused on radiation emitted by a surface, the foregoing concepts may be extended to *incident* radiation (Figure 12.10). Such radiation may originate from emission and reflection occurring at other surfaces and will have spectral and directional distributions determined by the spectral intensity $I_{\lambda,i}(\lambda,\theta,\phi)$. This quantity is defined as the rate at which radiant energy of wavelength λ is incident from the (θ,ϕ) direction, per unit area of the *intercepting surface* normal to this direction, per unit solid angle about this direction, and per unit wavelength interval $d\lambda$ about λ.

The intensity of the incident radiation may be related to an important radiative flux, termed the *irradiation*, which encompasses radiation incident *from all directions*. The *spectral irradiation* G_λ (W/m$^2\cdot\mu$m) is defined as the rate at which radiation of wavelength λ is incident on a surface, per unit area of the

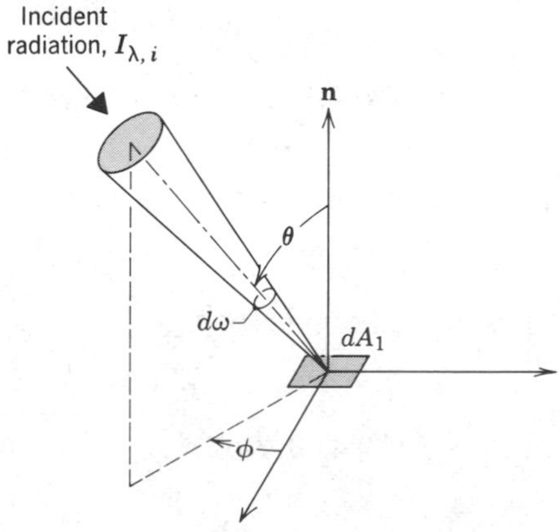

Figure 12.10 Directional nature of incident radiation.

surface and per unit wavelength interval $d\lambda$ about λ. Accordingly,

$$G_\lambda(\lambda) = \int_0^{2\pi} \int_0^{\pi/2} I_{\lambda,i}(\lambda,\theta,\phi) \cos\theta \sin\theta \, d\theta \, d\phi \tag{12.15}$$

where $\sin\theta \, d\theta \, d\phi$ is the unit solid angle. The $\cos\theta$ factor originates because G_λ is a flux based on the actual surface area, whereas $I_{\lambda,i}$ is defined in terms of the projected area. If the *total irradiation* G (W/m^2) represents the rate at which radiation is incident per unit area from all directions and at all wavelengths, it follows that

$$G = \int_0^\infty G_\lambda(\lambda) \, d\lambda \tag{12.16}$$

or from Equation 12.15

$$G = \int_0^\infty \int_0^{2\pi} \int_0^{\pi/2} I_{\lambda,i}(\lambda,\theta,\phi) \cos\theta \sin\theta \, d\theta \, d\phi \, d\lambda \tag{12.17}$$

If the incident radiation is *diffuse*, that is, the intercepting area is *irradiated diffusely*, $I_{\lambda,i}$ is independent of θ and ϕ and it follows that

$$G_\lambda(\lambda) = \pi I_{\lambda,i}(\lambda) \tag{12.18}$$

and

$$G = \pi I_i \tag{12.19}$$

EXAMPLE 12.2

The spectral distribution of surface irradiation is as follows.

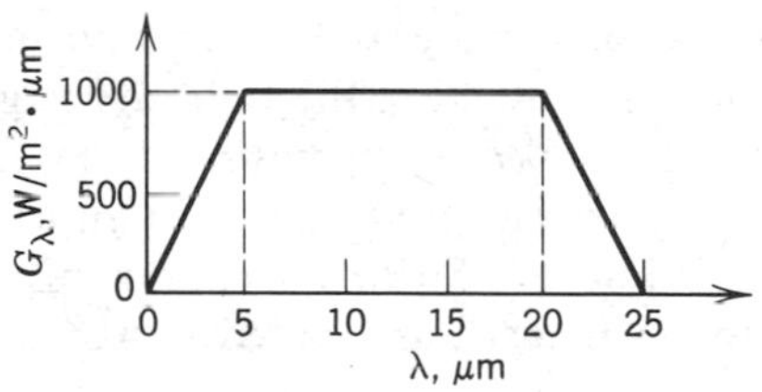

What is the total irradiation?

SOLUTION

Known: Spectral distribution of surface irradiation.

Find: Total irradiation.

Analysis:

The total irradiation may be obtained from Equation 12.16

$$G = \int_0^\infty G_\lambda \, d\lambda$$

The integral is readily evaluated by breaking it into parts. That is,

$$G = \int_0^{5\,\mu m} G_\lambda \, d\lambda + \int_{5\,\mu m}^{20\,\mu m} G_\lambda \, d\lambda + \int_{20\,\mu m}^{25\,\mu m} G_\lambda \, d\lambda + \int_{25\,\mu m}^{\infty} G_\lambda \, d\lambda$$

or

$$G = \tfrac{1}{2}(1000 \text{ W/m}^2 \cdot \mu\text{m})(5 - 0)\ \mu\text{m} + (1000 \text{ W/m}^2 \cdot \mu\text{m})(20 - 5)\ \mu\text{m}$$
$$+ \tfrac{1}{2}(1000 \text{ W/m}^2 \cdot \mu\text{m})(25 - 20)\ \mu\text{m} + 0 = (2500 + 15000 + 2500) \text{ W/m}^2$$

or

$$G = 20{,}000 \text{ W/m}^2 \qquad \triangleleft$$

Comments:

Generally, radiation sources do not provide such a regular spectral distribution for the irradiation. However, the procedure of computing the total irradiation from knowledge of the spectral distribution remains the same, although evaluation of the integral is likely to involve more detail.

12.2.4 Relation to Radiosity

The last radiative flux of interest, termed *radiosity*, accounts for *all* of the radiant energy leaving a surface. Since this radiation includes the *reflected* portion of the irradiation, as well as direct emission (Figure 12.11), the radiosity is generally different from the emissive power. The *spectral radiosity* J_λ (W/m$^2 \cdot \mu$m) represents the rate at which radiation of wavelength λ leaves a unit area of the surface, per unit wavelength interval $d\lambda$ about λ. Since it accounts for radiation leaving in all directions, it is related to the intensity associated with emission and

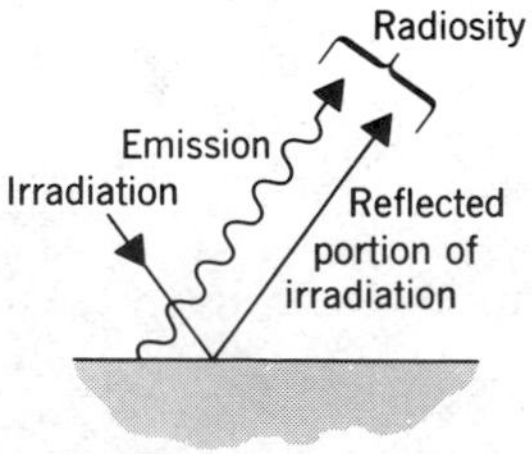

Figure 12.11 Surface radiosity.

reflection, $I_{\lambda,e+r}(\lambda,\theta,\phi)$, by an expression of the form

$$J_\lambda(\lambda) = \int_0^{2\pi} \int_0^{\pi/2} I_{\lambda,e+r}(\lambda,\theta,\phi) \cos\theta \sin\theta \, d\theta \, d\phi \tag{12.20}$$

Hence the *total radiosity* J (W/m^2) associated with the entire spectrum is

$$J = \int_0^\infty J_\lambda(\lambda) \, d\lambda \tag{12.21}$$

or

$$J = \int_0^\infty \int_0^{2\pi} \int_0^{\pi/2} I_{\lambda,e+r}(\lambda,\theta,\phi) \cos\theta \sin\theta \, d\theta \, d\phi \, d\lambda \tag{12.22}$$

If the surface is both a *diffuse reflector* and a *diffuse emitter*, $I_{\lambda,e+r}$ is independent of θ and ϕ, and it follows that

$$J_\lambda(\lambda) = \pi I_{\lambda,e+r} \tag{12.23}$$

and

$$J = \pi I_{e+r} \tag{12.24}$$

Again, note that the radiation flux, in this case the radiosity, is based on the actual surface area, while the intensity is based on the projected area.

12.3 BLACKBODY RADIATION

When describing the radiation characteristics of real surfaces, it is useful to introduce the concept of a *blackbody*. The blackbody is an *ideal* surface having the following properties.

1. *A blackbody absorbs all incident radiation, regardless of wavelength and direction.*
2. *For a prescribed temperature and wavelength, no surface can emit more energy than a blackbody.*
3. *Although the radiation emitted by a blackbody is a function of wavelength and temperature, it is independent of direction. That is, the blackbody is a diffuse emitter.*

As the perfect absorber and emitter, the blackbody serves as a *standard* against which the radiative properties of actual surfaces may be compared.

Although closely approximated by some surfaces, it is important to note that no surfaces have precisely the properties of a blackbody. The closest approximation is achieved by a *cavity* whose inner surface is at a uniform temperature. If radiation enters the cavity through a small aperture (Figure 12.12*a*), it is likely to experience many reflections before reemergence. Hence it is

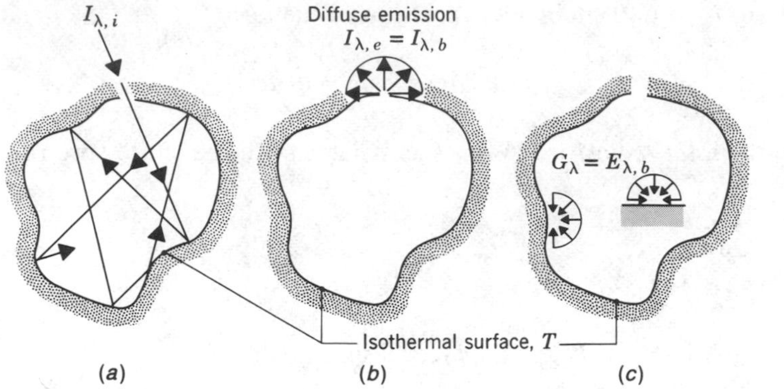

Figure 12.12 Characteristics of an isothermal blackbody cavity. (*a*) Complete absorption. (*b*) Diffuse emission from an aperture. (*c*) Diffuse irradiation of interior surfaces.

almost entirely absorbed by the cavity, and blackbody behavior is approximated. From thermodynamic principles it may then be argued that radiation leaving the aperture depends only on the surface temperature and corresponds to blackbody emission (Figure 12.12*b*). Since blackbody emission is diffuse, the spectral intensity $I_{\lambda,b}$ of radiation leaving the cavity is independent of direction. Moreover, since the radiation field in the cavity, which is the cumulative effect of emission and reflection from the cavity surface, must be of the same form as the radiation emerging from the aperture, it also follows that a blackbody radiation field exists within the cavity. Accordingly, any small surface in the cavity (Figure 12.12*c*), experiences irradiation for which $G_\lambda = E_{\lambda,b}(\lambda, T)$. This surface is diffusely irradiated, regardless of its orientation. *Blackbody radiation exists within the cavity irrespective of whether the cavity surface is highly reflecting or absorbing.*

12.3.1 The Planck Distribution

The spectral distribution of blackbody emission is well known, having first been determined by Planck [1]. It is of the form

$$I_{\lambda,b}(\lambda, T) = \frac{2hc_o^2}{\lambda^5[\exp(hc_o/\lambda kT) - 1]} \tag{12.25}$$

where $h = 6.6256 \times 10^{-34}$ J·s and $k = 1.3805 \times 10^{-23}$ J/K are the universal Planck and Boltzmann constants, respectively, $c_o = 2.998 \times 10^8$ m/s is the speed of light in vacuum, and T is the *absolute* temperature of the blackbody (K). Since the blackbody is a diffuse emitter, it follows from Equation 12.13 that its spectral emissive power is of the form

$$E_{\lambda,b}(\lambda, T) = \pi I_{\lambda,b}(\lambda, T) = \frac{C_1}{\lambda^5[\exp(C_2/\lambda T) - 1]} \tag{12.26}$$

where the first and second radiation constants are $C_1 = 2\pi hc_o^2 = 3.742 \times 10^8$ $W \cdot \mu m^4/m^2$ and $C_2 = (hc_o/k) = 1.439 \times 10^4$ $\mu m \cdot K$.

Equation 12.26, known as the *Planck distribution*, is plotted in Figure 12.13 for selected temperatures. Several important features should be noted.

1. The emitted radiation varies *continuously* with wavelength.
2. At any wavelength the magnitude of the emitted radiation increases with increasing temperature.
3. The spectral region in which the radiation is concentrated depends on temperature, with *comparatively* more radiation appearing at shorter wavelengths as the temperature increases.
4. A significant fraction of the radiation emitted by the sun, which may be approximated as a blackbody at 5800 K, is in the visible region of the spectrum. In contrast for $T \lesssim 800$ K, emission is predominantly in the infrared region of the spectrum and is not visible to the eye.

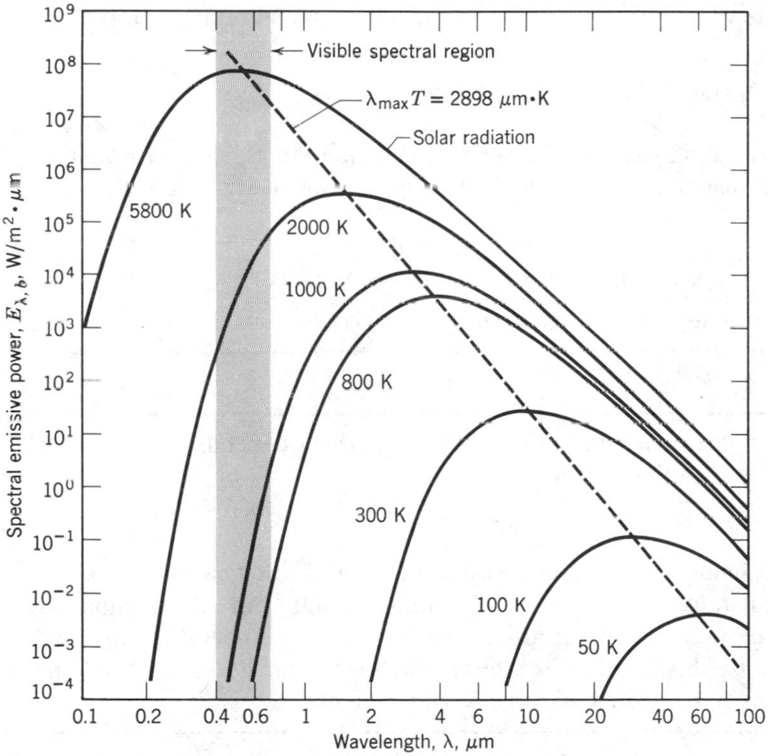

Figure 12.13 Spectral blackbody emissive power.

12.3.2 Wien's Displacement Law

From Figure 12.13 we see that the blackbody spectral distribution has a maximum and that the corresponding wavelength λ_{max} depends on temperature. The nature of this dependence may be obtained by differentiating Equation 12.26 with respect to λ and setting the result equal to zero. In so doing, we obtain

$$\lambda_{max} T = C_3 \tag{12.27}$$

where the third radiation constant is $C_3 = 2897.6\ \mu m \cdot K$.

Equation 12.27 is known as *Wien's displacement law*, and the locus of points described by the law is plotted as the dashed line of Figure 12.13. According to this result, the maximum spectral emissive power is displaced to shorter wavelengths with increasing temperature. This emission is in the middle of the visible spectrum ($\lambda \approx 0.50\ \mu m$) for solar radiation, since the sun emits as a blackbody at approximately 5800 K. For a blackbody at 1000 K, peak emission occurs at 2.90 μm, with some of the emitted radiation appearing visible as red light. With increasing temperature, shorter wavelengths become more prominent, until eventually significant emission occurs over the entire visible spectrum. For example, a tungsten filament lamp operating at 2900 K ($\lambda_{max} = 1\ \mu m$) emits white light, although most of the emission remains in the IR region.

12.3.3 The Stefan-Boltzmann Law

Substituting the Planck distribution, Equation 12.26, into Equation 12.11, the total emissive power of a blackbody E_b may be expressed as

$$E_b = \int_0^\infty \frac{C_1}{\lambda^5[\exp(C_2/\lambda T) - 1]} d\lambda$$

Performing the integration, it may be shown that

$$E_b = \sigma T^4 \tag{12.28}$$

where the *Stefan-Boltzmann* constant, which depends on C_1 and C_2, has the numerical value

$$\sigma = 5.670 \times 10^{-8}\ W/m^2 \cdot K^4$$

This simple, yet important, result is termed the *Stefan-Boltzmann law*. It enables calculation of the amount of radiation emitted in all directions and over all wavelengths simply from knowledge of the temperature of the blackbody. Because this emission is diffuse, it follows from Equation 12.14 that the total intensity associated with blackbody emission is

$$I_b = \frac{E_b}{\pi} \tag{12.29}$$

12.3.4 Band Emission

It is often necessary to know the fraction of the total emission from a blackbody that is in a certain wavelength interval or *band*. For a prescribed temperature and the interval from 0 to λ, this fraction is determined by the ratio of the shaded section to the total area under the curve of Figure 12.14. Hence

$$F_{(0\to\lambda)} \equiv \frac{\int_0^\lambda E_{\lambda,b}\, d\lambda}{\int_0^\infty E_{\lambda,b}\, d\lambda} = \frac{\int_0^\lambda E_{\lambda,b}\, d\lambda}{\sigma T^4} = \int_0^{\lambda T} \frac{E_{\lambda,b}}{\sigma T^5}\, d(\lambda T) = f(\lambda T) \qquad (12.30)$$

Since the integrand $(E_{\lambda,b}/\sigma T^5)$ is exclusively a function of the wavelength–temperature product λT, the integral of Equation 12.30 may be evaluated to obtain $F_{(0\to\lambda)}$ as a function of only λT. The results are presented in Table 12.1 and Figure 12.15. They may also be used to obtain the fraction of the radiation between any two wavelengths λ_1 and λ_2, since

$$F_{(\lambda_1\to\lambda_2)} = \frac{\int_0^{\lambda_2} E_{\lambda,b}\, d\lambda - \int_0^{\lambda_1} E_{\lambda,b}\, d\lambda}{\sigma T^4} = F_{(0\to\lambda_2)} - F_{(0\to\lambda_1)} \qquad (12.31)$$

Additional blackbody functions are listed in the third and fourth columns of Table 12.1. The third column facilitates calculation of the spectral intensity for a prescribed wavelength and temperature. In lieu of computing this quantity from Equation 12.25, it may be obtained by simply multiplying the tabulated value of $I_{\lambda,b}/\sigma T^5$ by σT^5. The fourth column is used to obtain a quick comparison of the ratio of the spectral intensity at any wavelength to that at λ_{max}.

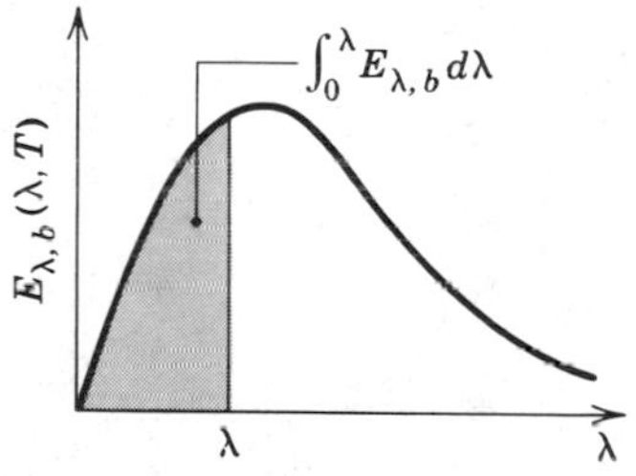

Figure 12.14 Radiation emission from a blackbody in the spectral band 0 to λ.

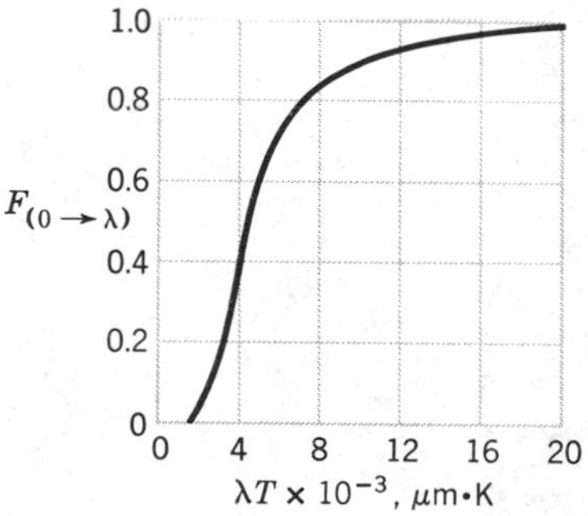

Figure 12.15 Fraction of the total blackbody emission in the spectral band from 0 to λ as a function of λT.

Table 12.1 Blackbody radiation functions[a]

λT (μm · K)	$F_{(0\to\lambda)}$	$I_{\lambda,b}(\lambda,T)/\sigma T^5$ (μm · K · sr)$^{-1}$	$\dfrac{I_{\lambda,b}(\lambda,T)}{I_{\lambda,b}(\lambda_{max},T)}$
200	0.000000	0.375034×10^{-27}	0.000000
400	0.000000	0.490335×10^{-13}	0.000000
600	0.000000	0.104046×10^{-8}	0.000014
800	0.000016	0.991126×10^{-7}	0.001372
1,000	0.000321	0.118505×10^{-5}	0.016406
1,200	0.002134	0.523927×10^{-5}	0.072534
1,400	0.007790	0.134411×10^{-4}	0.186082
1,600	0.019718	0.249130	0.344904
1,800	0.039341	0.375568	0.519949
2,000	0.066728	0.493432	0.683123
2,200	0.100888	0.589649×10^{-4}	0.816329
2,400	0.140256	0.658866	0.912155
2,600	0.183120	0.701292	0.970891
2,800	0.227897	0.720239	0.997123
2,898	0.250108	0.722318×10^{-4}	1.000000
3,000	0.273232	0.720254×10^{-4}	0.997143
3,200	0.318102	0.705974	0.977373
3,400	0.361735	0.681544	0.943551
3,600	0.403607	0.650396	0.900429
3,800	0.443382	0.615225	0.851737
4,000	0.480877	0.578064	0.800291
4,200	0.516014	0.540394×10^{-4}	0.748139
4,400	0.548796	0.503253	0.696720
4,600	0.579280	0.467343	0.647004
4,800	0.607559	0.433109	0.599610
5,000	0.633747	0.400813	0.554898
5,200	0.658970	0.370580×10^{-4}	0.513043
5,400	0.680360	0.342445	0.474092
5,600	0.701046	0.316376	0.438002
5,800	0.720158	0.292301	0.404671
6,000	0.737818	0.270121	0.373965
6,200	0.754140	0.249723×10^{-4}	0.345724
6,400	0.769234	0.230985	0.319783
6,600	0.783199	0.213786	0.295973
6,800	0.796129	0.198008	0.274128
7,000	0.808109	0.183534	0.254090
7,200	0.819217	0.170256×10^{-4}	0.235708
7,400	0.829527	0.158073	0.218842
7,600	0.839102	0.146891	0.203360
7,800	0.848005	0.136621	0.189143
8,000	0.856288	0.127185	0.176079
8,500	0.874608	0.106772×10^{-4}	0.147819
9,000	0.890029	0.901463×10^{-5}	0.124801
9,500	0.903085	0.765338	0.105956
10,000	0.914199	0.653279	0.090442
10,500	0.923710	0.560522	0.077600

Table 12.1 Continued

λT (μm · K)	$F_{(0\rightarrow\lambda)}$	$I_{\lambda,b}(\lambda, T)/\sigma T^5$ (μm · K · sr)$^{-1}$	$\dfrac{I_{\lambda,b}(\lambda, T)}{I_{\lambda,b}(\lambda_{max}, T)}$
11,000	0.931890	0.483321×10^{-5}	0.066913
11,500	0.939959	0.418725	0.057970
12,000	0.945098	0.364394	0.050448
13,000	0.955139	0.279457	0.038689
14,000	0.962898	0.217641	0.030131
15,000	0.969981	0.171866×10^{-5}	0.023794
16,000	0.973814	0.137429	0.019026
18,000	0.980860	0.908240×10^{-6}	0.012574
20,000	0.985602	0.623310	0.008629
25,000	0.992215	0.276171	0.003828
30,000	0.995340	0.140469×10^{-6}	0.001945
40,000	0.997967	0.473891×10^{-7}	0.000656
50,000	0.998953	0.201605	0.000279
75,000	0.999713	0.418597×10^{-8}	0.000058
100,000	0.999905	0.135752	0.000019

[a] The radiation constants used to generate these blackbody functions are
$C_1 = 3.7420 \times 10^8$ μm^4/m^2
$C_2 = 1.4388 \times 10^4$ μm · K
$\sigma = 5.670 \times 10^{-8}$ W/m^2 · K^4.

EXAMPLE 12.3

Consider a large isothermal enclosure that is maintained at a uniform temperature of 2000 K. Calculate the emissive power of the radiation that emerges from a small aperture on the enclosure surface. What is the wavelength λ_1 below which 10% of the emission is concentrated? What is the wavelength λ_2 above which 10% of the emission is concentrated? Determine the maximum spectral emissive power and the wavelength at which this emission occurs. What is the irradiation incident on a small object placed inside the enclosure?

SOLUTION

Known: Large isothermal enclosure at uniform temperature.

Find:

1. Emissive power of a small aperture on the enclosure.
2. Wavelengths below which and above which 10% of the radiation is concentrated.

3. Maximum spectral emissive power and wavelength associated with maximum emission.
4. Irradiation on a small object inside the enclosure.

Schematic:

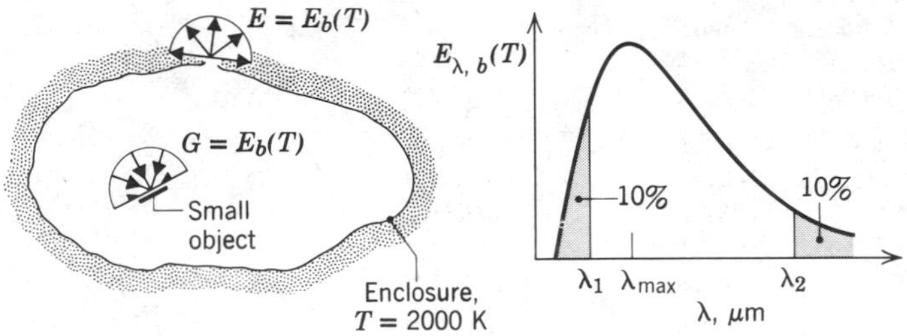

Assumptions:

Areas of aperture and object are very small relative to enclosure surface.

Analysis:

1. Emission from the aperture of any isothermal enclosure will have the characteristics of blackbody radiation. Hence, from Equation 12.28

   $E = E_b(T) = \sigma T^4 = 5.670 \times 10^{-8}\ \text{W/m}^2 \cdot \text{K}^4\ (2000\ \text{K})^4$

   $E = 9.07 \times 10^5\ \text{W/m}^2$ ◁

2. The wavelength λ_1 corresponds to the upper limit of the spectral band $(0 \rightarrow \lambda_1)$ containing 10% of the emitted radiation. With $F_{(0\rightarrow\lambda_1)} = 0.10$ it follows from Table 12.1 that $\lambda_1 T \approx 2200\ \mu\text{m} \cdot \text{K}$. Hence

   $\lambda_1 = 1.1\ \mu\text{m}$ ◁

   The wavelength λ_2 corresponds to the lower limit of the spectral band $(\lambda_2 \rightarrow \infty)$ containing 10% of the emitted radiation. With

   $F_{(\lambda_2\rightarrow\infty)} = 1 - F_{(0\rightarrow\lambda_2)} = 0.1$

   $F_{(0\rightarrow\lambda_2)} = 0.9$

   it follows from Table 12.1 that $\lambda_2 T = 9382\ \mu\text{m} \cdot \text{K}$. Hence

   $\lambda_2 = 4.69\ \mu\text{m}$ ◁

3. From Wien's law, Equation 12.27, $\lambda_{\text{max}} T = 2898\ \mu\text{m} \cdot \text{K}$. Hence

   $\lambda_{\text{max}} = 1.45\ \mu\text{m}$ ◁

   The spectral emissive power associated with this wavelength may be computed from Equation 12.26 or from the third column of Table 12.1.

For $\lambda_{max}T = 2898\ \mu m \cdot K$ it follows from Table 12.1 that

$I_{\lambda,b}(1.45\ \mu m, T) = 0.722 \times 10^{-4}\sigma T^5$

Hence

$$I_{\lambda,b}(1.45\ \mu m, 2000\ K) = 0.722 \times 10^{-4}(1/\mu m \cdot K \cdot sr)5.67 \times 10^{-8}\ W/m^2 \cdot K^4\ (2000\ K)^5$$

$$I_{\lambda,b}(1.45\ \mu m, 2000\ K) = 1.31 \times 10^5\ W/m^2 \cdot sr \cdot \mu m$$

Since the emission is diffuse, it follows from Equation 12.13 that

$$E_{\lambda,b} = \pi I_{\lambda,b} = 4.12 \times 10^5\ W/m^2 \cdot \mu m \quad \triangleleft$$

4 Irradiation of any small object inside the enclosure may be approximated as being equal to emission from a blackbody at the enclosure surface temperature. Hence $G = E_b(T)$, in which case

$$G = 9.07 \times 10^5\ W/m^2 \quad \triangleleft$$

EXAMPLE 12.4

A surface emits as a blackbody at 1500 K. What is the rate per unit area (W/m^2) at which it emits radiation in directions corresponding to $0° \le \theta \le 60°$ and in the wavelength interval $2\ \mu m \le \lambda \le 4\ \mu m$?

SOLUTION

Known: Temperature of a surface that emits as a blackbody.

Find:

Rate of emission per unit area in directions between $\theta = 0°$ and $60°$ and at wavelengths between $\lambda = 2$ and $4\ \mu m$.

Schematic:

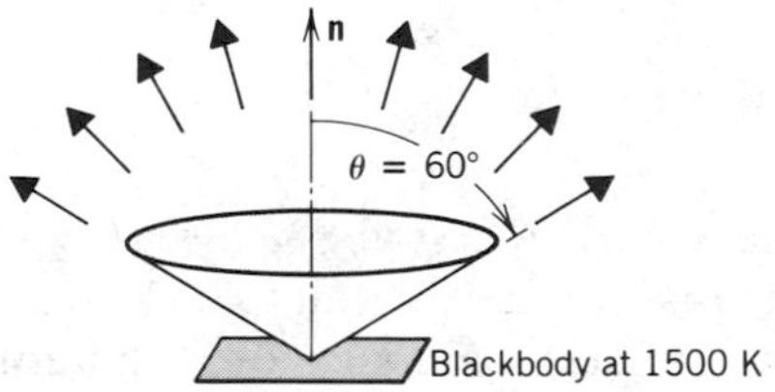

Assumptions: Surface emits as a blackbody.

Analysis:

The desired emission may be inferred from Equation 12.12, with the limits of integration restricted as follows

$$\Delta E = \int_2^4 \int_0^{2\pi} \int_0^{\pi/3} I_{\lambda,b} \cos\theta \sin\theta \, d\theta \, d\phi \, d\lambda$$

or, since a blackbody emits diffusely,

$$\Delta E = \int_2^4 I_{\lambda,b} \left(\int_0^{2\pi} \int_0^{\pi/3} \cos\theta \sin\theta \, d\theta \, d\phi \right) d\lambda$$

$$\Delta E = \int_2^4 I_{\lambda,b} \left(2\pi \frac{\sin^2\theta}{2} \bigg|_0^{\pi/3} \right) d\lambda = 0.75 \int_2^4 \pi I_{\lambda,b} \, d\lambda$$

Substituting from Equation 12.13 and multiplying and dividing by E_b, this result may be put in a form that allows for use of Table 12.1 in evaluating the spectral integration. In particular

$$\Delta E = 0.75 E_b \int_2^4 \frac{E_{\lambda,b}}{E_b} d\lambda = 0.75 E_b [F_{(0\to 4)} - F_{(0\to 2)}]$$

where from Table 12.1

$$\lambda_1 T = 2 \ \mu\text{m} \times 1500 \ \text{K} = 3000 \ \mu\text{m} \cdot \text{K}, \ F_{(0\to 2)} = 0.273$$

$$\lambda_2 T = 4 \ \mu\text{m} \times 1500 \ \text{K} = 6000 \ \mu\text{m} \cdot \text{K}, \ F_{(0\to 4)} = 0.738$$

Hence

$$\Delta E = 0.75(0.738 - 0.273) E_b = 0.75(0.465) E_b$$

From Equation 12.28, it then follows that

$$\Delta E = 0.75(0.465) 5.67 \times 10^{-8} \ \text{W/m}^2 \cdot \text{K}^4 \ (1500 \ \text{K})^4 = 10^5 \ \text{W/m}^2$$ ◁

Comments:

The total, hemispherical emissive power is reduced by 25% and 53.5% due to the directional and spectral restrictions, respectively.

12.4 SURFACE EMISSION

Having developed the notion of a blackbody to describe *ideal* surface behavior, we may now consider the behavior of *real* surfaces. Recall that the blackbody is an ideal emitter in the sense that no surface can emit more radiation than a blackbody at the same temperature. It is therefore convenient to choose the blackbody as a reference in describing emission from a real surface. A surface radiative property known as the *emissivity*[1] may then be defined as the *ratio* of the radiation emitted by the surface to the radiation emitted by a blackbody at the same temperature.

It is important to acknowledge that, in general, the spectral radiation emitted by a real surface differs from the Planck distribution (Figure 12.16*a*). Moreover, the directional distribution (Figure 12.16*b*) may be other than diffuse. Hence the emissivity may assume different values according to whether one is interested in emission at a given wavelength or in a given direction, or in integrated averages over wavelength and direction.

We define the *spectral, directional emissivity* $\varepsilon_{\lambda,\theta}(\lambda,\theta,\phi,T)$ of a surface at the temperature T as the ratio of the intensity of the radiation emitted at the wavelength λ and in the direction of θ and ϕ to the intensity of the radiation emitted by a blackbody at the same values of T and λ. Hence

$$\varepsilon_{\lambda,\theta}(\lambda,\theta,\phi,T) \equiv \frac{I_{\lambda,e}(\lambda,\theta,\phi,T)}{I_{\lambda,b}(\lambda,T)} \tag{12.32}$$

Note how the subscripts λ and θ designate interest in a specific wavelength and direction for the emissivity. In contrast, terms appearing within parentheses designate functional dependence on wavelength, direction, and/or temperature. The absence of directional variables in the parentheses of the denominator in Equation 12.32 implies that the intensity is independent of direction, which is, of course, a characteristic of blackbody emission. In like manner a *total, directional emissivity* ε_θ, which represents a spectral average of $\varepsilon_{\lambda,\theta}$, may be defined as

$$\varepsilon_\theta(\theta,\phi,T) \equiv \frac{I_e(\theta,\phi,T)}{I_b(T)} \tag{12.33}$$

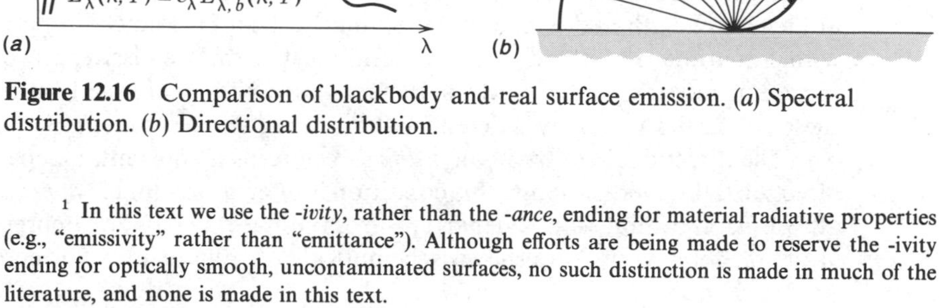

Figure 12.16 Comparison of blackbody and real surface emission. (*a*) Spectral distribution. (*b*) Directional distribution.

[1] In this text we use the *-ivity*, rather than the *-ance*, ending for material radiative properties (e.g., "emissivity" rather than "emittance"). Although efforts are being made to reserve the -ivity ending for optically smooth, uncontaminated surfaces, no such distinction is made in much of the literature, and none is made in this text.

For most engineering calculations, it is desirable to work with surface properties that represent directional averages. A *spectral, hemispherical emissivity* is therefore defined as

$$\varepsilon_\lambda(\lambda, T) \equiv \frac{E_\lambda(\lambda, T)}{E_{\lambda,b}(\lambda, T)} \tag{12.34}$$

It may be related to the directional emissivity $\varepsilon_{\lambda,\theta}$ by substituting the expression for the spectral emissive power, Equation 12.10, to obtain

$$\varepsilon_\lambda(\lambda, T) = \frac{\int_0^{2\pi}\int_0^{\pi/2} I_{\lambda,e}(\lambda, \theta, \phi, T) \cos\theta \sin\theta \, d\theta \, d\phi}{\int_0^{2\pi}\int_0^{\pi/2} I_{\lambda,b}(\lambda, T) \cos\theta \sin\theta \, d\theta \, d\phi}$$

In contrast to Equation 12.10, the temperature dependence of emission is now acknowledged. From Equation 12.32 and the fact that $I_{\lambda,b}$ is independent of θ and ϕ, it follows that

$$\varepsilon_\lambda(\lambda, T) = \frac{\int_0^{2\pi}\int_0^{\pi/2} \varepsilon_{\lambda,\theta}(\lambda, \theta, \phi, T) \cos\theta \sin\theta \, d\theta \, d\phi}{\int_0^{2\pi}\int_0^{\pi/2} \cos\theta \sin\theta \, d\theta \, d\phi} \tag{12.35}$$

Assuming $\varepsilon_{\lambda,\theta}$ to be independent of ϕ, which is a reasonable assumption for most surfaces, and evaluating the denominator, we obtain

$$\varepsilon_\lambda(\lambda, T) = 2 \int_0^{\pi/2} \varepsilon_{\lambda,\theta}(\lambda, \theta, T) \cos\theta \sin\theta \, d\theta \tag{12.36}$$

The *total, hemispherical emissivity*, which represents an average over all possible directions and wavelengths, is defined as

$$\varepsilon(T) \equiv \frac{E(T)}{E_b(T)} \tag{12.37}$$

Substituting from Equations 12.11 and 12.34, it follows that

$$\varepsilon(T) = \frac{\int_0^\infty \varepsilon_\lambda(\lambda, T) E_{\lambda,b}(\lambda, T) \, d\lambda}{E_b(T)} \tag{12.38}$$

If the emissivities of a surface are known, it is a simple matter to compute its emission characteristics. For example, if $\varepsilon_\lambda(\lambda, T)$ is known, it may be used with Equations 12.26 and 12.34 to compute the spectral emissive power of the surface at any wavelength and temperature. Similarly, if $\varepsilon(T)$ is known, it may be used with Equations 12.28 and 12.37 to compute the total emissive power of the surface at any temperature. Measurements have been performed to determine these properties for many different materials and surface coatings.

The directional emissivity of a *diffuse emitter* is a constant, independent of direction. However, although this condition is often a reasonable *approximation*, all surfaces exhibit some departure from diffuse behavior. Representative variations of ε_θ with θ are shown schematically in Figure 12.17 for conducting

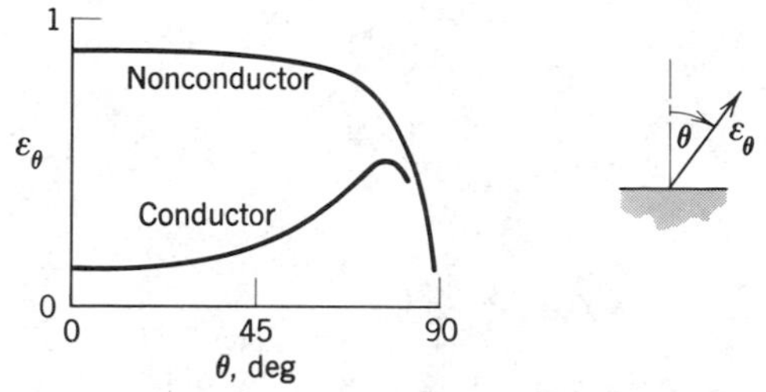

Figure 12.17 Representative directional distributions of the total, directional emissivity.

and nonconducting materials. For conductors ε_θ is approximately constant over the range $\theta \lesssim 40°$, after which it increases with increasing θ but ultimately decays to zero. In contrast, for nonconductors ε_θ is approximately constant for $\theta \lesssim 70°$, beyond which it decreases sharply with increasing θ. One implication of these variations is that, although there are preferential directions for emission, the hemispherical emissivity ε will not differ markedly from the normal emissivity ε_n, corresponding to $\theta = 0$. In fact the ratio rarely falls outside the range $1.0 \leq (\varepsilon/\varepsilon_n) \leq 1.3$ for conductors and the range of $0.95 \leq (\varepsilon/\varepsilon_n) \leq 1.0$ for nonconductors. Hence to a reasonable approximation,

$$\varepsilon \approx \varepsilon_n \tag{12.39}$$

Note that, although the foregoing statements were made for the total emissivity, they also apply to spectral components.

Since the spectral distribution of emission from real surfaces departs from the Planck distribution (Figure 12.16*a*), we do not expect the value of the spectral emissivity ε_λ to be independent of wavelength. Representative spectral distributions of ε_λ are shown in Figure 12.18. The manner in which ε_λ varies with λ depends on whether the solid is a conductor or nonconductor, as well as the nature of the surface coating.

Representative values of the total, normal emissivity ε_n are plotted in Figures 12.19 and 12.20 and listed in Table A.11. Several generalizations may be made.

1. The emissivity of metallic surfaces is generally small, achieving values as low as 0.02 for highly polished gold and silver.
2. The presence of oxide layers may significantly increase the emissivity of metallic surfaces. Contrast the value of 0.10 for lightly oxidized stainless steel with the value of approximately 0.50 for the heavily oxidized form.
3. The emissivity of nonconductors is comparatively large, generally exceeding 0.6.
4. The emissivity of conductors increases with increasing temperature; however, depending on the specific material, the emissivity of nonconductors

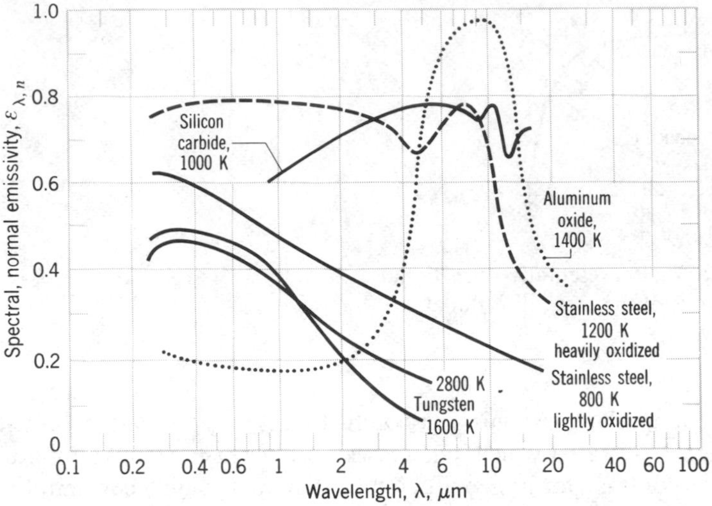

Figure 12.18 Spectral dependence of the spectral, normal emissivity $\varepsilon_{\lambda,n}$ of selected materials.

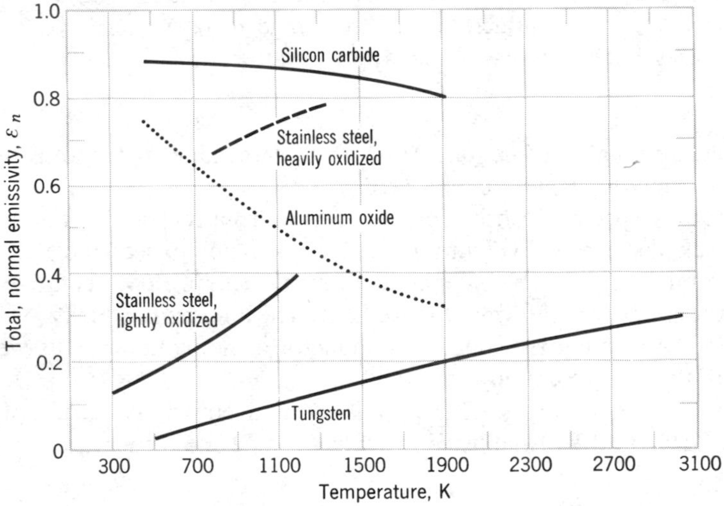

Figure 12.19 Temperature dependence of the total, normal emissivity ε_n of selected materials.

may either increase or decrease with increasing temperature. Note that the variations of ε_n with T shown in Figure 12.19 are consistent with the spectral distributions of $\varepsilon_{\lambda,n}$ shown in Figure 12.18. These trends follow from Equation 12.38. Although the spectral distribution of $\varepsilon_{\lambda,n}$ is approximately independent of temperature, there is proportionately more emission at lower wavelengths with increasing temperature. Hence, if $\varepsilon_{\lambda,n}$

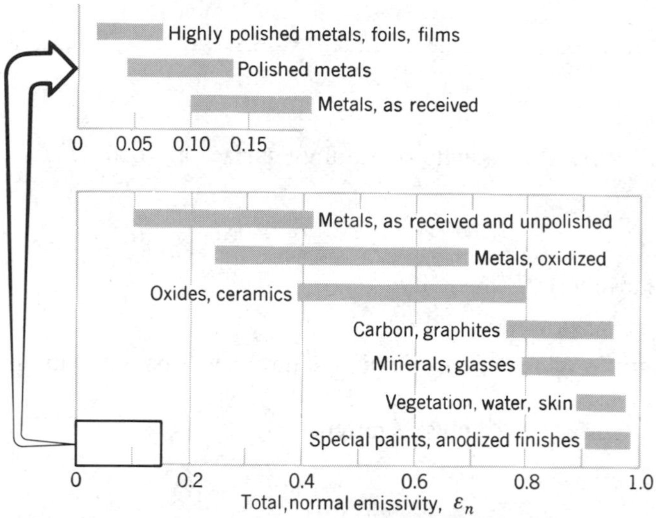

Figure 12.20 Representative values of the total, normal emissivity ε_n.

increases with decreasing wavelength for a particular material, ε_n will increase with increasing temperature for that material.

It should be recognized that the emissivity depends strongly on the nature of the surface, which can be influenced by the method of fabrication, thermal cycling, and chemical reaction with its environment. More comprehensive surface emissivity compilations are available in the literature [2–5].

EXAMPLE 12.5

A diffuse surface at 1600 K has the spectral, hemispherical emissivity shown below.

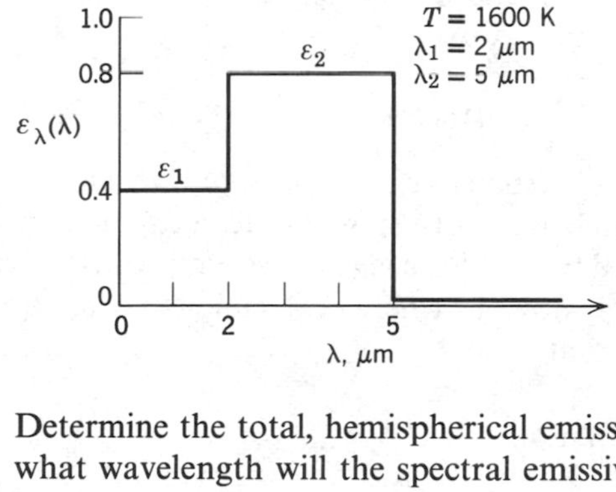

Determine the total, hemispherical emissivity and the total emissive power. At what wavelength will the spectral emissive power be a maximum?

SOLUTION

Known:

Spectral, hemispherical emissivity of a diffuse surface at 1600 K.

Find:

1. Total, hemispherical emissivity.
2. Total emissive power.
3. Wavelength at which spectral emissive power will be a maximum.

Assumptions: Surface is a diffuse emitter.

Analysis:

1. The total, hemispherical emissivity is given by Equation 12.38, where the integration may be performed in parts as follows.

$$\varepsilon = \frac{\int_0^\infty \varepsilon_\lambda E_{\lambda,b}\, d\lambda}{E_b} = \frac{\varepsilon_1 \int_0^2 E_{\lambda,b}\, d\lambda}{E_b} + \frac{\varepsilon_2 \int_2^5 E_{\lambda,b}\, d\lambda}{E_b}$$

or

$$\varepsilon = \varepsilon_1 F_{(0\to 2\ \mu m)} + \varepsilon_2 [F_{(0\to 5\ \mu m)} - F_{(0\to 2\ \mu m)}]$$

From Table 12.1 we obtain

$$\lambda_1 T = 2\ \mu m \times 1600\ K = 3200\ \mu m \cdot K{:}\ F_{(0\to 2\ \mu m)} = 0.318$$

$$\lambda_2 T = 5\ \mu m \times 1600\ K = 8000\ \mu m \cdot K{:}\ F_{(0\to 5\ \mu m)} = 0.856$$

Hence

$$\varepsilon = 0.4 \times 0.318 + 0.8[0.856 - 0.318] = 0.558 \qquad \triangleleft$$

2. From Equation 12.37 the total emissive power is

$$E = \varepsilon E_b = \varepsilon \sigma T^4$$

$$E = 0.558(5.67 \times 10^{-8}\ W/m^2 \cdot K^4)(1600\ K)^4 = 207\ kW/m^2 \qquad \triangleleft$$

3. If the surface emitted as a blackbody or if its emissivity were a constant, independent of λ, the wavelength corresponding to maximum spectral emission could be obtained from Wien's law. However, because ε_λ varies with λ, it is not immediately obvious where peak emission occurs. From Equation 12.27 we know that

$$\lambda_{max} = \frac{2898\ \mu m \cdot K}{1600\ K} = 1.81\ \mu m$$

The spectral emissive power at this wavelength may be obtained by using Equation 12.34 with Table 12.1. That is,

$$E_\lambda(\lambda_{max}, T) = \varepsilon_\lambda(\lambda_{max}) E_{\lambda,b}(\lambda_{max}, T)$$

or, since the surface is a diffuse emitter,

$$E_\lambda(\lambda_{max}, T) = \pi \varepsilon_\lambda(\lambda_{max}) I_{\lambda,b}(\lambda_{max}, T) = \pi \varepsilon_\lambda(\lambda_{max}) \frac{I_{\lambda,b}(\lambda_{max}, T)}{\sigma T^5} \times \sigma T^5$$

$$E_\lambda(1.81\ \mu\text{m}, 1600\ \text{K}) = \pi \times 0.4 \times 0.722 \times 10^{-4}\ (1/\mu\text{m}\cdot\text{K}\cdot\text{sr})5.67 \times 10^{-8}\ \text{W/m}^2\cdot\text{K}^4 \times (1600\ \text{K})^5 = 54\ \text{kW/m}^2\cdot\mu\text{m}$$

Since $\varepsilon_\lambda = 0.4$ from $\lambda = 0$ to $\lambda = 2\ \mu$m, the foregoing result provides the maximum spectral emissive power for the region $\lambda < 2\ \mu$m. However, with the change in ε_λ that occurs at $\lambda = 2\ \mu$m, the value of E_λ at $\lambda = 2\ \mu$m may be larger than that for $\lambda = 1.81\ \mu$m. To determine whether this is, in fact, the case, we compute

$$E_\lambda(\lambda_1, T) = \pi \varepsilon_\lambda(\lambda_1) \frac{I_{\lambda,b}(\lambda_1, T)}{\sigma T^5} \times \sigma T^5$$

where, for $\lambda_1 T = 3200\ \mu\text{m}\cdot\text{K}$, $[I_{\lambda,b}(\lambda_1, T)/\sigma T^5] = 0.706 \times 10^{-4}$ $(\mu\text{m}\cdot\text{K}\cdot\text{sr})^{-1}$. Hence

$$E_\lambda(2\ \mu\text{m}, 1600\ \text{K}) = \pi \times 0.80 \times 0.706 \times 10^{-4}\ (1/\mu\text{m}\cdot\text{K}\cdot\text{sr})5.67 \times 10^{-8}\ \text{W/m}^2\cdot\text{K}^4\ (1600\ \text{K})^5$$

$$E_\lambda(2\ \mu\text{m}, 1600\ \text{K}) = 105.5\ \text{kW/m}^2\cdot\mu\text{m} > E_\lambda(1.81\ \mu\text{m}, 1600\ \text{K})$$

and peak emission occurs at

$$\lambda = \lambda_1 = 2\ \mu\text{m}$$ ◁

Comments:

For the prescribed spectral distribution of ε_λ, the spectral emissive power will vary with wavelength as shown.

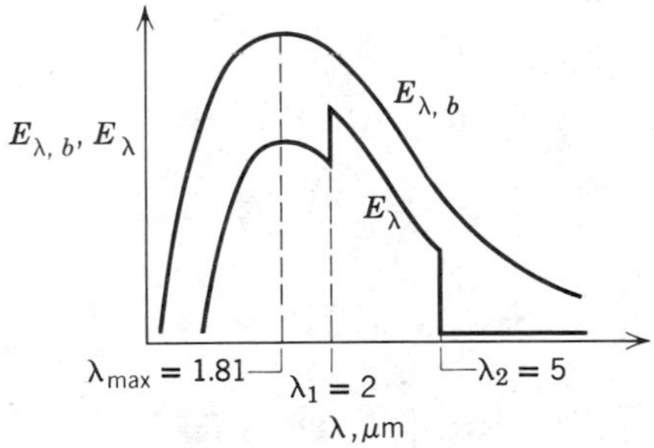

EXAMPLE 12.6

Measurements of the spectral, directional emissivity of a metallic surface at $T = 2000$ K and $\lambda = 1.0\ \mu$m yield a directional distribution that may be approximated as follows.

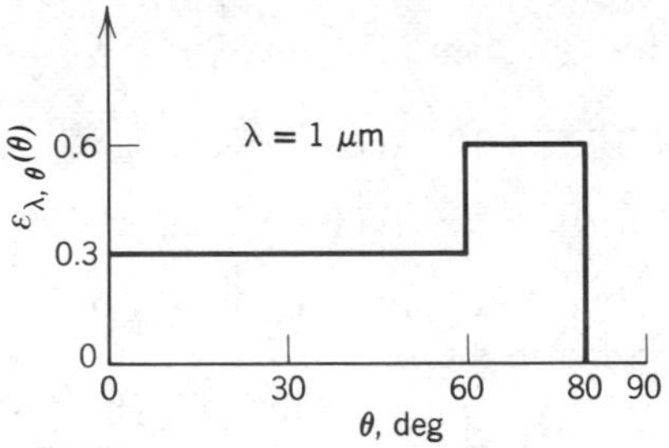

Determine corresponding values of the spectral, normal emissivity; the spectral, hemispherical emissivity; the spectral intensity of radiation emitted in the normal direction; and the spectral emissive power.

SOLUTION

Known:

Directional distribution of $\varepsilon_{\lambda,\theta}$ at $\lambda = 1\ \mu$m for a metallic surface at 2000 K.

Find:

1. Spectral, normal emissivity $\varepsilon_{\lambda,n}$ and spectral, hemispherical emissivity ε_λ.
2. Spectral, normal intensity $I_{\lambda,n}$ and spectral emissive power E_λ.

Analysis:

1. From the measurement of $\varepsilon_{\lambda,\theta}$ at $\lambda = 1\ \mu$m, we see that

$$\varepsilon_{\lambda,n} = \varepsilon_{\lambda,\theta}(1\ \mu\text{m}, 0^\circ) = 0.3 \qquad \triangleleft$$

From Equation 12.36, the spectral, hemispherical emissivity is

$$\varepsilon_\lambda(1\ \mu\text{m}) = 2\int_0^{\pi/2} \varepsilon_{\lambda,\theta} \cos\theta \sin\theta\, d\theta$$

or

$$\varepsilon_\lambda(1\ \mu\text{m}) = 2\left[0.3\int_0^{\pi/3} \cos\theta \sin\theta\, d\theta + 0.6\int_{\pi/3}^{4\pi/9} \cos\theta \sin\theta\, d\theta\right]$$

$$\varepsilon_\lambda(1\ \mu\text{m}) = 2\left[0.3\frac{\sin^2\theta}{2}\bigg|_0^{\pi/3} + 0.6\frac{\sin^2\theta}{2}\bigg|_{\pi/3}^{4\pi/9}\right]$$

$$= 2\left[\frac{0.3}{2}(0.75) + \frac{0.6}{2}(0.97 - 0.75)\right]$$

$$\varepsilon_\lambda(1\ \mu\text{m}) = 0.36 \qquad \triangleleft$$

2. From Equation 12.32 the spectral intensity of radiation emitted at $\lambda = 1\ \mu$m in the normal direction is

$$I_{\lambda,n}(1\ \mu\text{m}, 0°, 2000\text{ K}) = \varepsilon_{\lambda,\theta}(1\ \mu\text{m}, 0°)I_{\lambda,b}(1\ \mu\text{m}, 2000\text{ K})$$

where $\varepsilon_{\lambda,\theta}(1\ \mu\text{m}, 0°) = 0.3$ and $I_{\lambda,b}(1\ \mu\text{m}, 2000\text{ K})$ may be obtained from Table 12.1. For $\lambda T = 2000\ \mu\text{m}\cdot\text{K}$, $(I_{\lambda,b}/\sigma T^5) = 0.493 \times 10^{-4}$ $(\mu\text{m}\cdot\text{K}\cdot\text{sr})^{-1}$ and

$$I_{\lambda,b} = 0.493 \times 10^{-4}(\mu\text{m}\cdot\text{K}\cdot\text{sr})^{-1} \times 5.67 \times 10^{-8}\text{ W/m}^2\cdot\text{K}^4(2000\text{ K})^5$$

$$I_{\lambda,b} = 8.95 \times 10^4\text{ W/m}^2\cdot\mu\text{m}\cdot\text{sr}$$

Hence

$$I_{\lambda,n}(1\ \mu\text{m}, 0°, 2000\text{ K}) = 0.3 \times 8.95 \times 10^4\text{ W/m}^2\cdot\mu\text{m}\cdot\text{sr}$$

$$I_{\lambda,n}(1\ \mu\text{m}, 0°, 2000\text{ K}) = 2.69 \times 10^4\text{ W/m}^2\cdot\mu\text{m}\cdot\text{sr} \qquad \triangleleft$$

From Equation 12.34 the spectral emissive power for $\lambda = 1\ \mu$m and $T = 2000$ K is

$$E_\lambda(1\ \mu\text{m}, 2000\text{ K}) = \varepsilon_\lambda(1\ \mu\text{m})E_{\lambda,b}(1\ \mu\text{m}, 2000\text{ K})$$

where

$$E_{\lambda,b}(1\ \mu\text{m}, 2000\text{ K}) = \pi I_{\lambda,b}(1\ \mu\text{m}, 2000\text{ K})$$

$$E_{\lambda,b}(1\ \mu\text{m}, 2000\text{ K}) = \pi\text{ sr} \times 8.95 \times 10^4\text{ W/m}^2\cdot\mu\text{m}\cdot\text{sr}$$

$$= 2.81 \times 10^5\text{ W/m}^2\cdot\mu\text{m}$$

Hence

$$E_\lambda(1\ \mu\text{m}, 2000\text{ K}) = 0.36 \times 2.81 \times 10^5\text{ W/m}^2\cdot\mu\text{m}$$

or

$$E_\lambda(1\ \mu\text{m}, 2000\text{ K}) = 1.01 \times 10^5\text{ W/m}^2\cdot\mu\text{m} \qquad \triangleleft$$

12.5 SURFACE ABSORPTION, REFLECTION, AND TRANSMISSION

In Section 12.2.3 we defined the *spectral irradiation* G_λ (W/m$^2\cdot\mu$m) as the rate at which radiation of wavelength λ is incident on a surface per unit area of the surface and per unit wavelength interval $d\lambda$ about λ. It may be incident from all possible directions, and it may originate from several different sources. The *total irradiation* G (W/m^2) encompasses all spectral contributions and may be evaluated from Equation 12.16. In this section we consider the *processes* resulting from interception of this radiation by a solid (or liquid) medium.

In the most general situation the irradiation interacts with a *semitransparent medium*, such as a layer of water or a glass plate. As shown in Figure 12.21 for a spectral component of the irradiation, portions of this radiation may be *reflected*, *absorbed*, and *transmitted*. From a radiation balance on the medium, it follows that

$$G_\lambda = G_{\lambda,\mathrm{ref}} + G_{\lambda,\mathrm{abs}} + G_{\lambda,\mathrm{tr}} \tag{12.40}$$

In general, determination of these components is complex, depending on the upper and lower surface conditions, the wavelength of the radiation, and the composition and thickness of the medium. Moreover, conditions may be strongly influenced by *volumetric* effects occurring within the medium.

In a simpler situation, which pertains to most engineering applications, the medium is *opaque* to the incident radiation. In this case, $G_{\lambda,\mathrm{tr}} = 0$ and the remaining absorption and reflection processes may be treated as *surface phenomena*. That is, they are controlled by processes occurring within a fraction of a micrometer from the irradiated surface. It is therefore appropriate to speak of irradiation being absorbed and reflected *by the surface*, with the relative magnitudes of $G_{\lambda,\mathrm{abs}}$ and $G_{\lambda,\mathrm{ref}}$ depending on λ and the nature of the surface material. There is no net effect of the reflection process on the medium, while absorption has the effect of increasing the internal thermal energy of the medium.

It is interesting to note that surface absorption and reflection are responsible for our perception of *color*. Unless the surface is at a high temperature ($T_s \gtrsim 1000$ K), such that it is *incandescent*, color is in no way due to emission, which is concentrated in the IR region, and is hence imperceptible to the eye. Color is instead due to selective reflection and absorption of the visible portion of irradiation that is incident from the sun or an artificial source of light. A shirt is "red" because it contains a pigment that preferentially absorbs the blue, green,

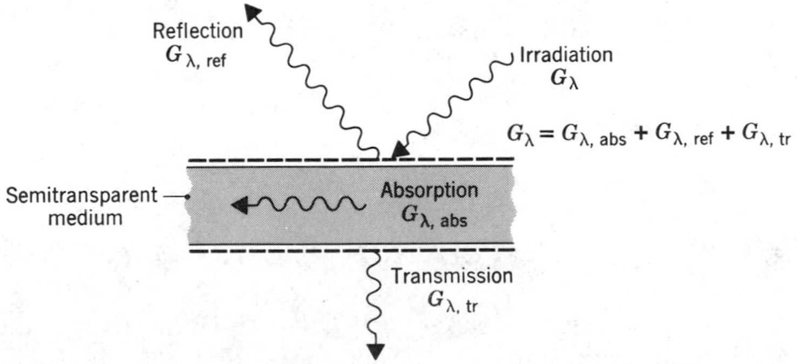

Figure 12.21 Absorption, reflection, and transmission processes associated with a semitransparent medium.

and yellow components of the incident light. Hence the relative contributions of these components to the reflected light, which is seen, is diminished, and the red component is dominant. Similarly, a leaf is "green" because its cells contain chlorophyll, a pigment that shows strong absorption in the blue and the red and preferential reflection in the green. A surface appears "black" if it absorbs all incident visible radiation, and it is "white" if it reflects this radiation. However, we must be careful how we interpret such *visual* effects. For a prescribed irradiation, the "color" of a surface may not indicate its overall capacity as an absorber or reflector, since much of the irradiation may be in the IR region. A "white" surface such as snow, for example, is highly reflective to visible radiation but strongly absorbs IR radiation, thereby approximating blackbody behavior at long wavelengths.

In Section 12.4 we introduced a property, which we called the emissivity, to characterize the process of surface emission. In the subsections that follow we shall introduce properties to characterize the absorption, reflection, and transmission processes. In general these properties depend on surface material and finish, surface temperature, and the wavelength and direction of the incident radiation.

12.5.1 Absorptivity

The absorptivity is a property that determines the fraction of the irradiation absorbed by a surface. Determination of the property is complicated by the fact that, like emission, it may be characterized by both a directional and spectral dependence. The *spectral, directional absorptivity,* $\alpha_{\lambda,\theta}(\lambda,\theta,\phi)$, of a surface is defined as the fraction of the spectral intensity incident in the direction of θ and ϕ that is absorbed by the surface. Hence,

$$\alpha_{\lambda,\theta}(\lambda,\theta,\phi) \equiv \frac{I_{\lambda,i,\text{abs}}(\lambda,\theta,\phi)}{I_{\lambda,i}(\lambda,\theta,\phi)} \tag{12.41}$$

In this expression, we have neglected any dependence of the absorptivity on the surface temperature. Such a dependence is small for most spectral radiative properties.

It is implicit in the foregoing result that surfaces may exhibit selective absorption with respect to the wavelength and direction of the incident radiation. For most engineering calculations, however, it is desirable to work with surface properties that represent directional averages. We therefore define a *spectral, hemispherical absorptivity* $\alpha_\lambda(\lambda)$ as

$$\alpha_\lambda(\lambda) \equiv \frac{G_{\lambda,\text{abs}}(\lambda)}{G_\lambda(\lambda)} \tag{12.42}$$

which, from Equations 12.15 and 12.41, may be expressed as

$$\alpha_\lambda(\lambda) = \frac{\int_0^{2\pi}\int_0^{\pi/2} \alpha_{\lambda,\theta}(\lambda,\theta,\phi) I_{\lambda,i}(\lambda,\theta,\phi)\cos\theta\sin\theta\, d\theta\, d\phi}{\int_0^{2\pi}\int_0^{\pi/2} I_{\lambda,i}(\lambda,\theta,\phi)\cos\theta\sin\theta\, d\theta\, d\phi} \tag{12.43}$$

Hence α_λ depends on the directional distribution of the incident radiation, as well as on the wavelength of the radiation and the nature of the absorbing surface. Note that, if the incident radiation is diffusely distributed and $\alpha_{\lambda,\theta}$ is independent of ϕ, Equation 12.43 reduces to

$$\alpha_\lambda(\lambda) = \int_0^{\pi/2} \alpha_{\lambda,\theta}(\lambda,\theta) \cos\theta \sin\theta \, d\theta \tag{12.44}$$

The *total, hemispherical absorptivity*, α, represents an integrated average over both direction and wavelength. It is defined as the fraction of the total irradiation absorbed by a surface

$$\alpha \equiv \frac{G_{\text{abs}}}{G} \tag{12.45}$$

and, from Equations 12.16 and 12.42, may be expressed as

$$\alpha = \frac{\int_0^\infty \alpha_\lambda(\lambda) G_\lambda(\lambda) \, d\lambda}{\int_0^\infty G_\lambda(\lambda) \, d\lambda} \tag{12.46}$$

Accordingly, α depends on the spectral distribution of the incident radiation, as well as on its directional distribution and the nature of the absorbing surface. Note that, although α is approximately independent of temperature, the same may not be said for the total, hemispherical emissivity, ε. From Equation 12.38 it is evident that this property is strongly temperature dependent.

Because α depends on the spectral distribution of the irradiation, its value for a surface exposed to solar radiation may differ appreciably from its value for the same surface exposed to longer wavelength radiation originating from a source of lower temperature. Since the spectral distribution of solar radiation is nearly proportional to that of emission from a blackbody at 5800 K, it follows from Equation 12.46 that the total absorptivity to solar radiation α_S may be approximated as

$$\alpha_S \approx \frac{\int_0^\infty \alpha_\lambda(\lambda) E_{\lambda,b}(\lambda, 5800 \text{ K}) \, d\lambda}{\int_0^\infty E_{\lambda,b}(\lambda, 5800 \text{ K}) \, d\lambda} \tag{12.47}$$

The integrals appearing in this equation may be evaluated by using the blackbody radiation function $F_{(0\rightarrow\lambda)}$ of Table 12.1.

12.5.2 Reflectivity

The reflectivity is a property that determines the fraction of the incident radiation reflected by a surface. However, its specific definition may take several different forms, because the property is inherently *bidirectional* [6]. That is, in addition to depending on the direction of the incident radiation, it also depends on the direction of the reflected radiation. We shall avoid this complication by working exclusively with a reflectivity that represents an integrated average over the hemisphere associated with the reflected radiation and therefore provides no information concerning the directional distribution of this radiation. Accordingly, the *spectral, directional reflectivity*, $\rho_{\lambda,\theta}(\lambda,\theta,\phi)$, of a surface is defined as the fraction of the spectral intensity incident in the direction of θ and ϕ, which is reflected by the surface. Hence,

$$\rho_{\lambda,\theta}(\lambda,\theta,\phi) \equiv \frac{I_{\lambda,i,\mathrm{ref}}(\lambda,\theta,\phi)}{I_{\lambda,i}(\lambda,\theta,\phi)} \tag{12.48}$$

The *spectral, hemispherical reflectivity* $\rho_\lambda(\lambda)$ is then defined as the fraction of the spectral irradiation that is reflected by the surface. Accordingly,

$$\rho_\lambda(\lambda) \equiv \frac{G_{\lambda,\mathrm{ref}}(\lambda)}{G_\lambda(\lambda)} \tag{12.49}$$

which is equivalent to

$$\rho_\lambda(\lambda) = \frac{\int_0^{2\pi}\int_0^{\pi/2} \rho_{\lambda,\theta}(\lambda,\theta,\phi) I_{\lambda,i}(\lambda,\theta,\phi)\cos\theta\sin\theta\, d\theta\, d\phi}{\int_0^{2\pi}\int_0^{\pi/2} I_{\lambda,i}(\lambda,\theta,\phi)\cos\theta\sin\theta\, d\theta\, d\phi} \tag{12.50}$$

The *total, hemispherical reflectivity* ρ is then defined as

$$\rho \equiv \frac{G_{\mathrm{ref}}}{G} \tag{12.51}$$

in which case

$$\rho = \frac{\int_0^\infty \rho_\lambda(\lambda) G_\lambda(\lambda)\, d\lambda}{\int_0^\infty G_\lambda(\lambda)\, d\lambda} \tag{12.52}$$

Surfaces may be idealized as *diffuse* or *specular*, according to the manner in which they reflect radiation (Figure 12.22). Diffuse reflection occurs if, regardless of the direction of the incident radiation, the intensity of the reflected radiation is independent of the reflection angle. In contrast, if all the reflection is in the direction of θ_2, which equals the incident angle θ_1, specular reflection is said to occur. Although no surface is perfectly diffuse or specular, the latter condition is more closely approximated by polished, mirrorlike surfaces and the former

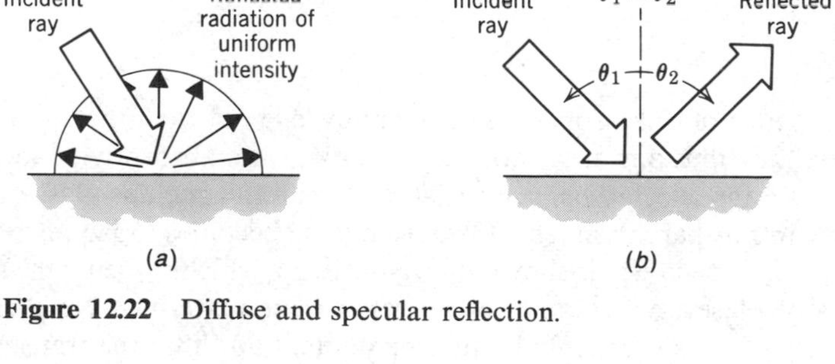

Figure 12.22 Diffuse and specular reflection.

condition by rough surfaces. The assumption of diffuse reflection is reasonable for most engineering applications.

12.5.3 Transmissivity

Although treatment of the response of a semitransparent material to incident radiation is a complicated problem [6], reasonable results may often be obtained through the use of hemispherical transmissivities defined as

$$\tau_\lambda = \frac{G_{\lambda,\text{tr}}(\lambda)}{G_\lambda(\lambda)} \tag{12.53}$$

and

$$\tau = \frac{G_{\text{tr}}}{G} \tag{12.54}$$

The total transmissivity τ is related to the spectral component τ_λ by

$$\tau = \frac{\int_0^\infty G_{\lambda,\text{tr}}(\lambda)\,d\lambda}{\int_0^\infty G_\lambda(\lambda)\,d\lambda} = \frac{\int_0^\infty \tau_\lambda(\lambda)G_\lambda(\lambda)\,d\lambda}{\int_0^\infty G_\lambda(\lambda)\,d\lambda} \tag{12.55}$$

12.5.4 Special Considerations

We conclude this section by noting that, from the radiation balance of Equation 12.40 and the foregoing definitions,

$$\rho_\lambda + \alpha_\lambda + \tau_\lambda = 1 \tag{12.56}$$

for a semitransparent medium. For properties that are averaged over the entire spectrum, it also follows that

$$\rho + \alpha + \tau = 1 \tag{12.57}$$

Of course, if the medium is opaque, there is no transmission, and absorption and reflection are surface processes for which

$$\alpha_\lambda + \rho_\lambda = 1 \tag{12.58}$$

and

$$\alpha + \rho = 1 \tag{12.59}$$

Hence knowledge of one property implies knowledge of the other.

Spectral distributions of the normal reflectivity and absorptivity are plotted in Figure 12.23 for selected *opaque* surfaces. A material such as glass or water, which is semitransparent at short wavelengths, becomes opaque at longer wavelengths. This behavior is shown in Figure 12.24, which presents the spectral transmissivity of several common semitransparent materials. Note that the transmissivity of glass is affected by its iron content and that the transmissivity of plastics, such as Tedlar, is greater than that of glass in the IR region. These

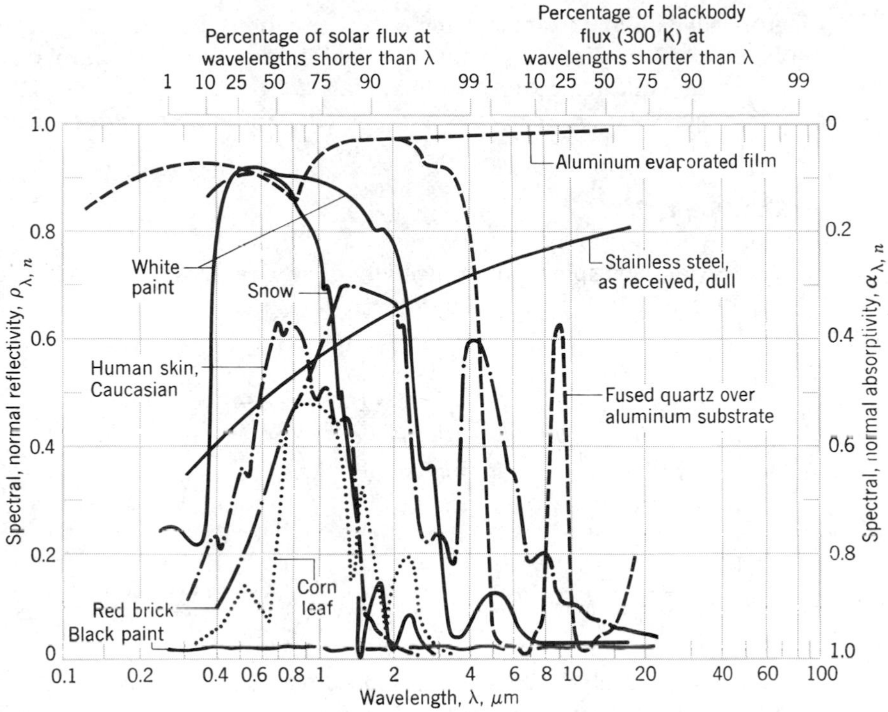

Figure 12.23 Spectral dependence of the spectral, normal absorptivity $\alpha_{\lambda,n}$ and reflectivity $\rho_{\lambda,n}$ of selected materials.

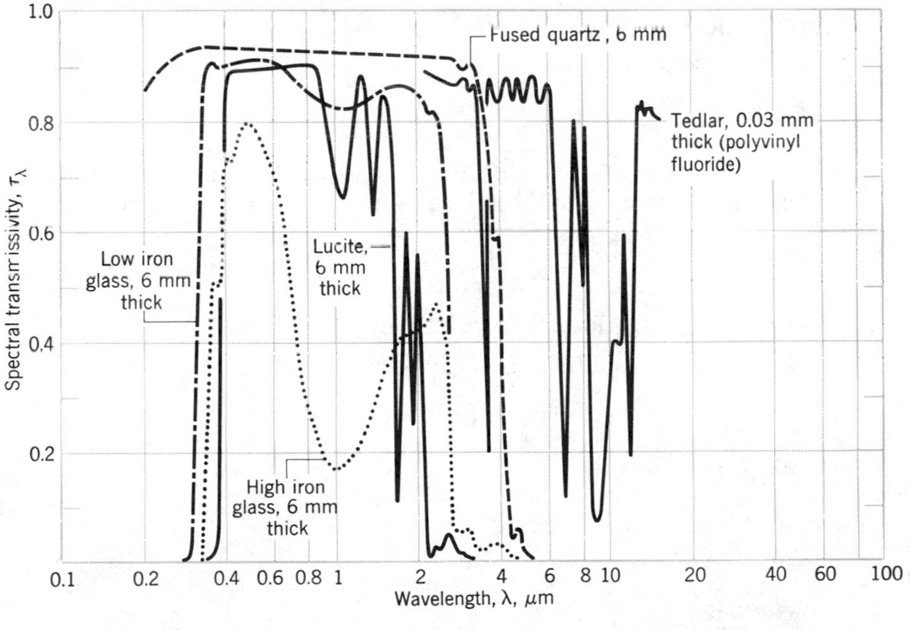

Figure 12.24 Spectral dependence of the spectral transmissivities τ_λ of selected materials.

factors have an important bearing on the selection of cover plate materials for solar collector applications. Values for the total transmissivity to solar radiation of common collector cover plate materials are presented in Table A.12, along with surface solar absorptivities and low temperature emissivities.

EXAMPLE 12.7

The spectral, hemispherical absorptivity of an opaque surface and the spectral irradiation at the surface are as shown.

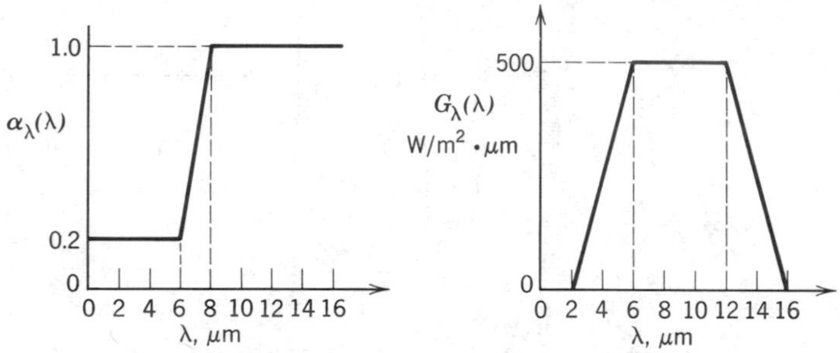

How does the spectral, hemispherical reflectivity vary with wavelength? What is the total, hemispherical absorptivity of the surface? If the surface is initially at 500 K and has a total, hemispherical emissivity of 0.8, how will its temperature change upon exposure to the irradiation?

SOLUTION

Known:

Spectral, hemispherical absorptivity and irradiation of a surface. Surface temperature (500 K) and total, hemispherical emissivity (0.8).

Find:

1. Spectral distribution of reflectivity.
2. Total, hemispherical absorptivity.
3. Time rate of surface temperature change.

Schematic:

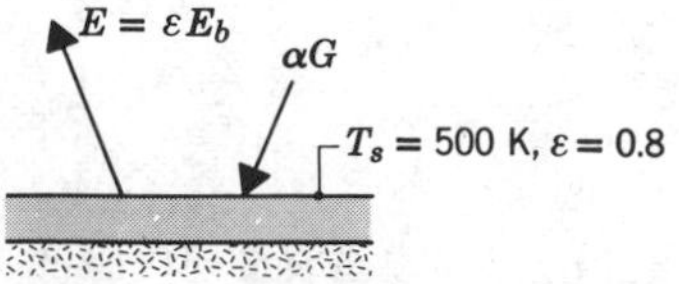

Assumptions:

1. Surface is opaque.
2. Surface convection effects are negligible.
3. Back surface is insulated.

Analysis:

1. From Equation 12.58, $\rho_\lambda = 1 - \alpha_\lambda$. Hence from knowledge of $\alpha_\lambda(\lambda)$, the spectral distribution of ρ_λ is as shown.

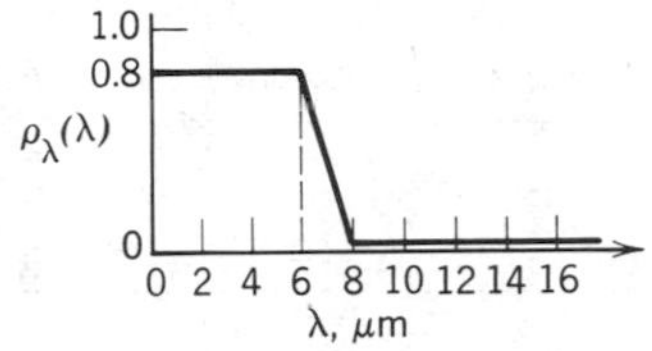

◁

2. From Equations 12.45 and 12.46

$$\alpha = \frac{G_{abs}}{G} = \frac{\int_0^\infty \alpha_\lambda G_\lambda \, d\lambda}{\int_0^\infty G_\lambda \, d\lambda}$$

or, subdividing the integral into parts,

$$\alpha = \frac{0.2\int_2^6 G_\lambda \, d\lambda + 500\int_6^8 \alpha_\lambda \, d\lambda + 1.0\int_8^{16} G_\lambda \, d\lambda}{\int_2^6 G_\lambda \, d\lambda + \int_6^{12} G_\lambda \, d\lambda + \int_{12}^{16} G_\lambda \, d\lambda}$$

$$\alpha = \{0.2(\tfrac{1}{2})500 \text{ W/m}^2\cdot\mu\text{m } (6-2)\ \mu\text{m} + 500 \text{ W/m}^2\cdot\mu\text{m } [0.2(8-6)\ \mu\text{m}$$
$$+ (1-0.2)(\tfrac{1}{2})(8-6)\ \mu\text{m}] + [1 \times 500 \text{ W/m}^2\cdot\mu\text{m } (12-8)\ \mu\text{m}$$
$$+ 1(\tfrac{1}{2})500 \text{ W/m}^2\cdot\mu\text{m } (16-12)\ \mu\text{m}]\} \div [(\tfrac{1}{2})500 \text{ W/m}^2\cdot\mu\text{m } (6-2)\ \mu\text{m}$$
$$+ 500 \text{ W/m}^2\cdot\mu\text{m } (12-6)\ \mu\text{m} + (\tfrac{1}{2})500 \text{ W/m}^2\cdot\mu\text{m } (16-12)\ \mu\text{m}]$$

Hence

$$\alpha = \frac{G_{abs}}{G} = \frac{(200+600+3000)\text{ W/m}^2}{(1000+3000+1000)\text{ W/m}^2} = \frac{3800 \text{ W/m}^2}{5000 \text{ W/m}^2} = 0.76$$ ◁

3. Neglecting convection effects, the net heat flux *to* the surface is

$$q''_{net} = \alpha G - E = \alpha G - \varepsilon\sigma T^4$$

Hence

$$q''_{net} = 0.76(5000 \text{ W/m}^2) - 0.8 \times 5.67 \times 10^{-8} \text{ W/m}^2\cdot\text{K}^4(500 \text{ K})^4$$

$$q''_{net} = 3800 - 2835 = 965 \text{ W/m}^2$$

Since $q''_{net} > 0$, the surface temperature will *increase* with time. ◁

EXAMPLE 12.8

The cover glass on a flat plate solar collector has a low iron content, and its spectral transmissivity may be approximated by the following distribution.

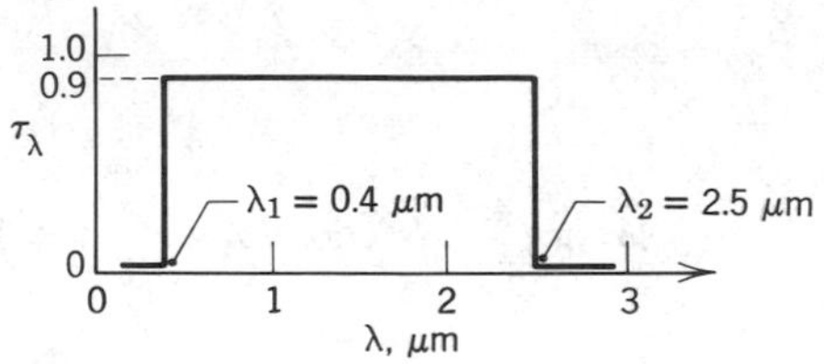

What is the total transmissivity of the cover glass to solar radiation?

SOLUTION

Known: Spectral transmissivity of solar collector cover glass.

Find: Total transmissivity of cover glass to solar radiation.

Assumptions:

Spectral distribution of solar irradiation is proportional to that of the blackbody emission at 5800 K.

Analysis:

From Equation 12.55 the total transmissivity of the cover is

$$\tau = \frac{\int_0^\infty \tau_\lambda G_\lambda \, d\lambda}{\int_0^\infty G_\lambda \, d\lambda}$$

where the irradiation G_λ is due to solar emission. Having assumed that the sun emits as a blackbody at 5800 K, it follows that

$$G_\lambda(\lambda) \propto E_{\lambda,b}(5800 \text{ K})$$

With the proportionality constant canceling from the numerator and denominator of the expression for τ, we obtain

$$\tau = \frac{\int_0^\infty \tau_\lambda E_{\lambda,b}(5800 \text{ K}) \, d\lambda}{\int_0^\infty E_{\lambda,b}(5800 \text{ K}) \, d\lambda}$$

or, for the prescribed spectral distribution of $\tau_\lambda(\lambda)$,

$$\tau = 0.90 \frac{\int_{0.4}^{2.5} E_{\lambda,b}(5800 \text{ K}) \, d\lambda}{E_b(5800 \text{ K})}$$

From Table 12.1

$$\lambda_1 = 0.4\ \mu\text{m},\ T = 5800\ \text{K}: \lambda_1 T = 2320\ \mu\text{m}\cdot\text{K},\ F_{(0\to\lambda_1)} = 0.1245$$

$$\lambda_2 = 2.5\ \mu\text{m},\ T = 5800\ \text{K}: \lambda_2 T = 14{,}500\ \mu\text{m}\cdot\text{K},\ F_{(0\to\lambda_2)} = 0.9660$$

Hence from Equation 12.31

$$\tau = 0.90[F_{(0\to\lambda_2)} - F_{(0\to\lambda_1)}] = 0.90(0.9660 - 0.1245) = 0.76 \qquad \triangleleft$$

Comments:

It is important to recognize that the irradiation at the cover plate is not equal to the emissive power of a blackbody at 5800 K, $G_\lambda \neq E_{\lambda,b}$ (5800 K). It is simply assumed to be proportional to this emissive power, in which case it is assumed to have a spectral distribution of the same form. With G_λ appearing in both the numerator and denominator of the expression for τ, it is then possible to replace G_λ by $E_{\lambda,b}$.

12.6 KIRCHHOFF'S LAW

In the foregoing sections we separately considered surface properties associated with emission and absorption. In Sections 12.6 and 12.7 we consider conditions for which these properties are equal.

Consider a large, isothermal enclosure of surface temperature T_s, within which several small bodies are confined (Figure 12.25). Since these bodies are small relative to the enclosure, they have a negligible influence on the radiation field, which is due to the cumulative effect of emission and reflection by the enclosure surface. Recall that, regardless of its radiative properties, such a surface forms a *blackbody cavity*. Accordingly, regardless of its orientation, the

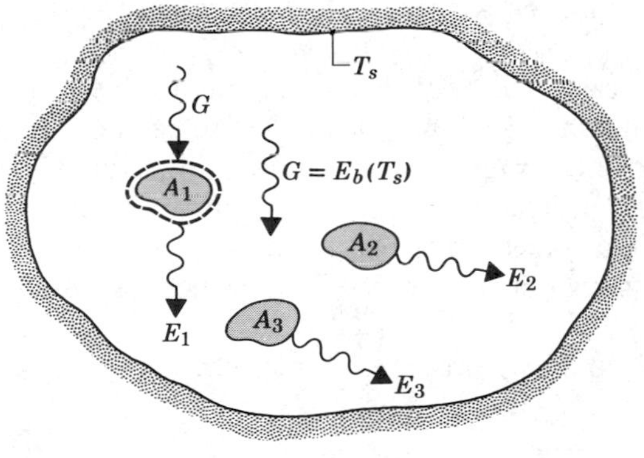

Figure 12.25 Radiative exchange in an isothermal enclosure.

irradiation experienced by any body in the cavity is diffuse and equal to emission from a blackbody at T_s.

$$G = E_b(T_s) \tag{12.60}$$

Under steady-state conditions, *thermal equilibrium* must exist between the bodies and the enclosure. Hence $T_1 = T_2 = \cdots = T_s$, and the net rate of energy transfer to each surface must be zero. Applying an energy balance to a control surface about body 1, it follows that

$$\alpha_1 G A_1 - E_1(T_s) A_1 = 0$$

or, from Equation 12.60

$$\frac{E_1(T_s)}{\alpha_1} = E_b(T_s)$$

Since this result must apply to each of the confined bodies, we then obtain

$$\frac{E_1(T_s)}{\alpha_1} = \frac{E_2(T_s)}{\alpha_2} = \cdots = E_b(T_s) \tag{12.61}$$

This relation is known as *Kirchhoff's law*. A major implication is that, since $\alpha \leq 1$, $E(T_s) \leq E_b(T_s)$. Hence *no real surface can have an emissive power exceeding that of a black surface at the same temperature*, and the notion of the blackbody as an ideal emitter is confirmed.

From the definition of the total, hemispherical emissivity, Equation 12.37, an alternative form of Kirchhoff's law is

$$\frac{\varepsilon_1}{\alpha_1} = \frac{\varepsilon_2}{\alpha_2} = \cdots = 1$$

Hence, for any surface in the enclosure

$$\varepsilon = \alpha \tag{12.62}$$

That is, the total, hemispherical emissivity of the surface is equal to its total, hemispherical absorptivity.

We will later find that calculations of radiative exchange between surfaces are greatly simplified if Equation 12.62 may be applied to each of the surfaces. However, the restrictive conditions inherent in its derivation should be remembered. In particular the surface irradiation has been assumed to correspond to emission from a blackbody at the same temperature as the surface. In Section 12.7 we consider other, less restrictive, conditions for which Equation 12.62 is applicable.

The preceding derivation may be repeated for spectral conditions, and for any surface in the enclosure it follows that

$$\varepsilon_\lambda = \alpha_\lambda \tag{12.63}$$

Conditions associated with the use of Equation 12.63 are less restrictive than those associated with Equation 12.62. In particular it will be shown that Equation 12.63 is applicable if the irradiation is diffuse *or* if the surface is diffuse. A form of Kirchhoff's law for which there are *no restrictions* involves the spectral, directional properties.

$$\varepsilon_{\lambda,\theta} = \alpha_{\lambda,\theta} \tag{12.64}$$

This equality is always applicable because $\varepsilon_{\lambda,\theta}$ and $\alpha_{\lambda,\theta}$ are *inherent* surface properties. That is, respectively, they are independent of the spectral and directional distributions of the emitted and incident radiation.

More detailed developments of Kirchhoff's law are provided by Planck [1] and Siegel and Howell [6].

12.7 THE GRAY SURFACE

In Chapter 13 we will find that the problem of predicting radiant energy exchange between surfaces is greatly simplified if Equation 12.62 may be assumed to apply for each surface. It is therefore important to examine whether this equality may be applied to conditions other than those for which it was derived, namely, irradiation due to emission from a blackbody at the same temperature as the surface.

Accepting the fact that the spectral, directional emissivity and absorptivity are equal under any conditions, Equation 12.64, we begin by considering the conditions associated with using Equation 12.63. According to the definitions of the spectral, hemispherical properties, Equations 12.35 and 12.43, we are really asking under what conditions, if any, the following equality will hold.

$$\varepsilon_\lambda = \frac{\int_0^{2\pi}\int_0^{\pi/2} \varepsilon_{\lambda,\theta} \cos\theta \sin\theta \, d\theta \, d\phi}{\int_0^{2\pi}\int_0^{\pi/2} \cos\theta \sin\theta \, d\theta \, d\phi} \overset{?}{=} \frac{\int_0^{2\pi}\int_0^{\pi/2} \alpha_{\lambda,\theta} I_{\lambda,i} \cos\theta \sin\theta \, d\theta \, d\phi}{\int_0^{2\pi}\int_0^{\pi/2} I_{\lambda,i} \cos\theta \sin\theta \, d\theta \, d\phi} = \alpha_\lambda \tag{12.65}$$

Since $\varepsilon_{\lambda,\theta} = \alpha_{\lambda,\theta}$, it follows by inspection that Equation 12.63 is applicable if *either* of the following conditions is satisfied

1. The *irradiation* is *diffuse* ($I_{\lambda,i}$ is independent of θ and ϕ).
2. The *surface* is *diffuse* ($\varepsilon_{\lambda,\theta}$ and $\alpha_{\lambda,\theta}$ are independent of θ and ϕ).

The first condition is a reasonable approximation for many engineering calculations; the second condition is reasonable for many surfaces, particularly for electrically nonconducting materials (Figure 12.17).

Assuming the existence of either diffuse irradiation or a diffuse surface, we now consider what *additional* conditions must be satisfied for Equation 12.62 to be valid. From Equations 12.38 and 12.46, the equality applies if

$$\varepsilon = \frac{\int_0^\infty \varepsilon_\lambda E_{\lambda,b}(\lambda, T) \, d\lambda}{E_b(T)} \overset{?}{=} \frac{\int_0^\infty \alpha_\lambda G_\lambda(\lambda) \, d\lambda}{G} = \alpha \tag{12.66}$$

Since $\varepsilon_\lambda = \alpha_\lambda$, it follows by inspection that Equation 12.62 applies if *either* of the following conditions is satisfied.

1. The irradiation corresponds to emission from a blackbody at the surface temperature T, in which case $G_\lambda(\lambda) = E_{\lambda,b}(\lambda, T)$ and $G = E_b(T)$.
2. The *surface* is *gray* (α_λ and ε_λ are independent of λ).

Note that the first condition corresponds to the major assumption required for the derivation of Kirchhoff's law (Section 12.6).

Because the total absorptivity of a surface depends on the spectral distribution of the irradiation, it cannot be stated unequivocally that $\alpha = \varepsilon$. For example, a particular surface may be highly absorbing to radiation in one spectral region and virtually nonabsorbing in another region (Figure 12.26*a*). Accordingly, for the two possible irradiation fields $G_{\lambda,1}(\lambda)$ and $G_{\lambda,2}(\lambda)$ of Figure 12.26*b*, the values of α will differ drastically. In contrast the value of ε is independent of the irradiation. Hence, there is *no* basis for stating that α is always equal to ε.

To assume gray surface behavior, hence the validity of Equation 12.62, it is not necessary for α_λ and ε_λ to be independent of λ over the entire spectrum. Practically speaking, a *gray surface* may be defined as *one for which α_λ and ε_λ are independent of λ over the spectral regions of the irradiation and the surface emission.* From Equation 12.66 it is readily shown that gray surface behavior may be assumed for the conditions of Figure 12.27. That is, the irradiation and surface emission are concentrated in a region for which the spectral properties of the surface are approximately constant. Accordingly,

$$\varepsilon = \frac{\varepsilon_{\lambda,o} \int_{\lambda_1}^{\lambda_2} E_{\lambda,b}(\lambda, T)\, d\lambda}{E_b(T)} = \varepsilon_{\lambda,o} \qquad \text{and} \qquad \alpha = \frac{\alpha_{\lambda,o} \int_{\lambda_3}^{\lambda_4} G_\lambda(\lambda)\, d\lambda}{G} = \alpha_{\lambda,o}$$

in which case $\alpha = \varepsilon = \varepsilon_{\lambda,o}$. However, if the irradiation were in a spectral region corresponding to $\lambda < \lambda_1$ or $\lambda > \lambda_4$, gray surface behavior could not be assumed.

A surface for which $\alpha_{\lambda,\theta}$ and $\varepsilon_{\lambda,\theta}$ are independent of θ and λ is termed a *diffuse, gray surface* (diffuse because of the directional independence and gray because of the wavelength independence). It is a surface for which both Equations 12.62 and 12.63 are satisfied. We assume such surface conditions in many of our subsequent considerations. However, although the assumption of a

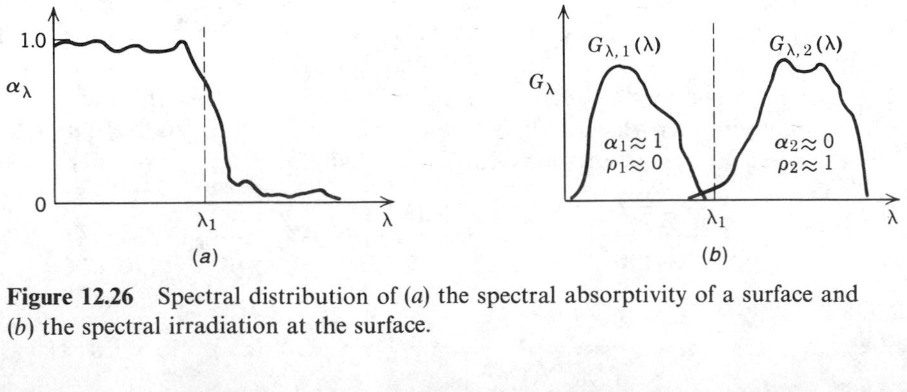

Figure 12.26 Spectral distribution of (*a*) the spectral absorptivity of a surface and (*b*) the spectral irradiation at the surface.

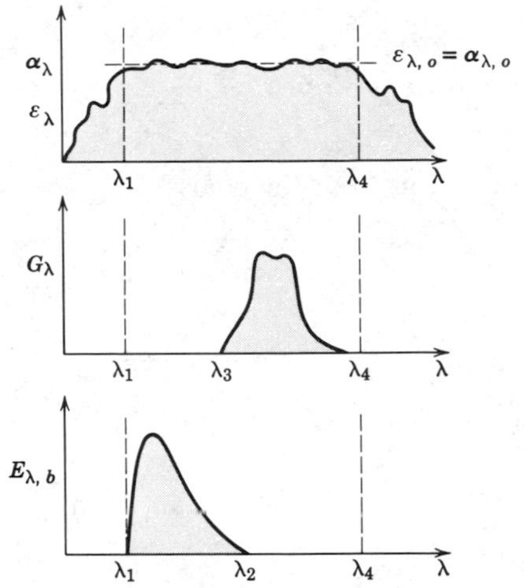

Figure 12.27 A set of conditions for which gray surface behavior may be assumed.

gray surface is reasonable for many practical applications, some caution should be exercised in its use, particularly if the spectral regions of the irradiation and emission are widely separated.

EXAMPLE 12.9

A diffuse, fire brick wall of temperature $T_s = 500$ K has the spectral emissivity shown and is exposed to a bed of coals at 2000 K.

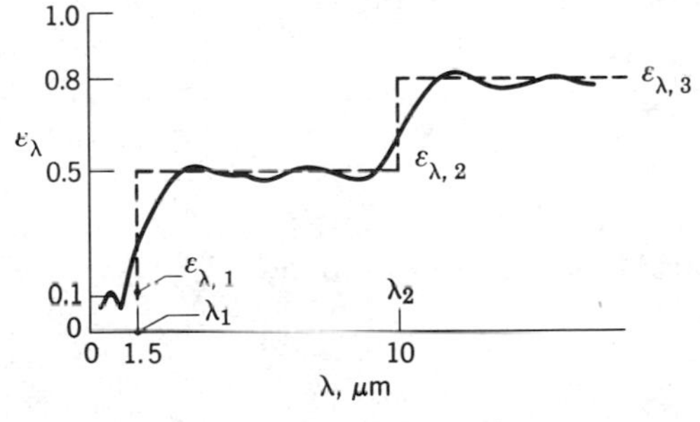

Determine the total, hemispherical emissivity and emissive power of the fire brick wall. What is the total absorptivity of the wall to irradiation resulting from emission by the coals?

SOLUTION

Known:

Brick wall of surface temperature $T_s = 500$ K and prescribed $\varepsilon_\lambda(\lambda)$ is exposed to coals at $T_c = 2000$ K.

Find:

1. Total, hemispherical emissivity of the fire brick wall.
2. Total emissive power of the brick wall.
3. Absorptivity of the wall to irradiation from the coals.

Schematic:

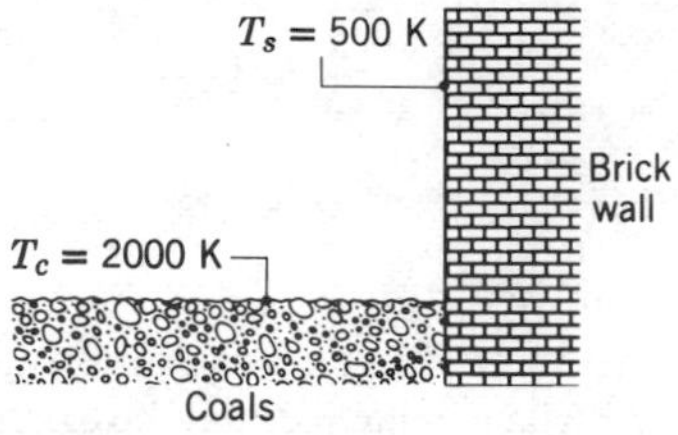

Assumptions:

1. Brick wall is opaque and diffuse.
2. Spectral distribution of irradiation at the brick wall approximates that due to emission from a blackbody at 2000 K.

Analysis:

1. The total, hemispherical emissivity follows from Equation 12.38.

$$\varepsilon(T_s) = \frac{\int_0^\infty \varepsilon_\lambda(\lambda) E_{\lambda,b}(\lambda, T_s)\, d\lambda}{E_b(T_s)}$$

Breaking the integral into parts,

$$\varepsilon(T_s) = \varepsilon_{\lambda,1}\frac{\int_0^{\lambda_1} E_{\lambda,b}\, d\lambda}{E_b} + \varepsilon_{\lambda,2}\frac{\int_{\lambda_1}^{\lambda_2} E_{\lambda,b}\, d\lambda}{E_b} + \varepsilon_{\lambda,3}\frac{\int_{\lambda_2}^{\infty} E_{\lambda,b} d\lambda}{E_b}$$

and introducing the blackbody functions, it follows that

$$\varepsilon(T_s) = \varepsilon_{\lambda,1} F_{(0\to\lambda_1)} + \varepsilon_{\lambda,2}[F_{(0\to\lambda_2)} - F_{(0\to\lambda_1)}] + \varepsilon_{\lambda,3}[1 - F_{(0\to\lambda_2)}]$$

From Table 12.1

$$\lambda_1 T_s = 1.5\ \mu\text{m} \times 500\ \text{K} = 750\ \mu\text{m}\cdot\text{K}:\ F_{(0\to\lambda_1)} = 0.000$$

$$\lambda_2 T_s = 10\ \mu\text{m} \times 500\ \text{K} = 5000\ \mu\text{m}\cdot\text{K}:\ F_{(0\to\lambda_2)} = 0.634$$

Hence

$$\varepsilon(T_s) = 0.1 \times 0 + 0.5 \times 0.634 + 0.8(1 - 0.634) = 0.610 \qquad \triangleleft$$

2. From Equations 12.28 and 12.37, the total emissive power is

$$E(T_s) = \varepsilon(T_s)E_b(T_s) = \varepsilon(T_s)\sigma T_s^4$$

$$E(T_s) = 0.61 \times 5.67 \times 10^{-8}\ \text{W/m}^2\cdot\text{K}^4\ (500\ \text{K})^4 = 2161\ \text{W/m}^2 \qquad \triangleleft$$

3. From Equation 12.46, the total absorptivity of the wall to radiation from the coals is

$$\alpha = \frac{\int_0^\infty \alpha_\lambda(\lambda) G_\lambda(\lambda)\, d\lambda}{\int_0^\infty G_\lambda(\lambda)\, d\lambda}$$

Since the surface is diffuse, $\alpha_\lambda(\lambda) = \varepsilon_\lambda(\lambda)$. Moreover, since the spectral distribution of the irradiation approximates that due to emission from a blackbody at 2000 K, $G_\lambda(\lambda) \propto E_{\lambda,b}(\lambda, T_c)$. It follows that

$$\alpha = \frac{\int_0^\infty \varepsilon_\lambda(\lambda) E_{\lambda,b}(\lambda, T_c)\, d\lambda}{\int_0^\infty E_{\lambda,b}(\lambda, T_c)\, d\lambda}$$

Breaking the integral into parts and introducing the blackbody functions, we then obtain

$$\alpha = \varepsilon_{\lambda,1} F_{(0\rightarrow\lambda_1)} + \varepsilon_{\lambda,2}[F_{(0\rightarrow\lambda_2)} - F_{(0\rightarrow\lambda_1)}] + \varepsilon_{\lambda,3}[1 - F_{(0\rightarrow\lambda_3)}]$$

From Table 12.1

$$\lambda_1 T_c = 1.5\ \mu\text{m} \times 2000\ \text{K} = 3000\ \mu\text{m} \cdot \text{K}:\ F_{(0\rightarrow\lambda_1)} = 0.273$$

$$\lambda_2 T_c = 10\ \mu\text{m} \times 2000\ \text{K} = 20{,}000\ \mu\text{m} \cdot \text{K}:\ F_{(0\rightarrow\lambda_2)} = 0.986$$

Hence

$$\alpha = 0.1 \times 0.273 + 0.5(0.986 - 0.273) + 0.8(1 - 0.986) = 0.395 \qquad \triangleleft$$

Comments:

1. The emissivity depends on the surface temperature T_s, while the absorptivity depends on the spectral distribution of the irradiation, which depends on the temperature of the source T_c.
2. The surface is not gray, $\alpha \neq \varepsilon$. This result is to be expected. Since emission is associated with $T_s = 500$ K, its spectral maximum occurs at $\lambda_{max} \approx 6\ \mu$m. In contrast, since irradiation is associated with emission from a source at $T_c = 2000$ K, its spectral maximum occurs at $\lambda_{max} \approx 1.5\ \mu$m. Because $\varepsilon_\lambda = \alpha_\lambda$ is not constant over the spectral ranges of both emission and irradiation, $\alpha \neq \varepsilon$.

EXAMPLE 12.10

A *small*, solid metallic sphere has an opaque, diffuse coating for which $\alpha_\lambda = 0.8$ in the spectral region for $\lambda \leq 5\ \mu$m and $\alpha_\lambda = 0.1$ for $\lambda > 5\ \mu$m. The sphere, which is initially at a uniform temperature of 300 K, is inserted into a *large* furnace whose walls are at 1200 K. Determine the total, hemispherical absorptivity and emissivity of the coating for the initial condition and for the final, steady-state condition.

SOLUTION

Known:

Small metallic sphere with spectrally selective absorptivity, initially at $T_s = 300$ K, is inserted into a large furnace at $T_f = 1200$ K.

Find:

1. Total, hemispherical absorptivity and emissivity of sphere coating for the initial condition.
2. Values of α and ε after sphere has been in furnace a long time.

Schematic:

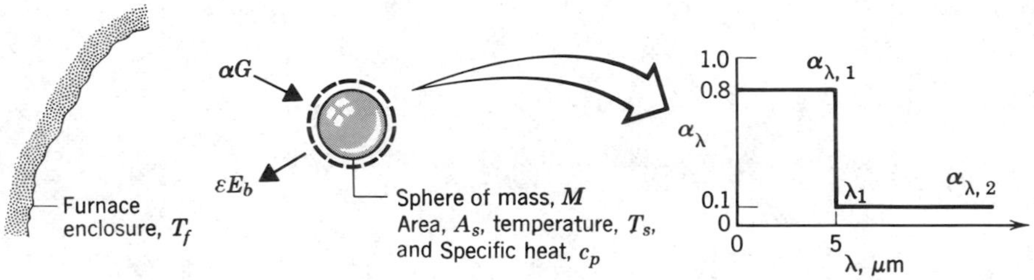

Assumptions:

1. Coating is opaque and diffuse.
2. Since furnace surface is much larger than that of sphere, irradiation approximates emission from a blackbody at T_f.

Analysis:

1. From Equation 12.46 the total, hemispherical absorptivity is

$$\alpha = \frac{\int_0^\infty \alpha_\lambda(\lambda) G_\lambda(\lambda)\, d\lambda}{\int_0^\infty G_\lambda(\lambda)\, d\lambda}$$

or, with $G_\lambda = E_{\lambda,b}(T_f) = E_{\lambda,b}(\lambda, 1200 \text{ K})$

$$\alpha = \frac{\int_0^\infty \alpha_\lambda(\lambda) E_{\lambda,b}(\lambda, 1200 \text{ K})\, d\lambda}{E_b(1200 \text{ K})}$$

Hence

$$\alpha = \alpha_{\lambda,1} \frac{\int_0^{\lambda_1} E_{\lambda,b}(\lambda, 1200 \text{ K})\, d\lambda}{E_b(1200 \text{ K})} + \alpha_{\lambda,2} \frac{\int_{\lambda_1}^\infty E_{\lambda,b}(\lambda, 1200 \text{ K})\, d\lambda}{E_b(1200 \text{ K})}$$

or

$$\alpha = \alpha_{\lambda,1} F_{(0\to\lambda_1)} + \alpha_{\lambda,2}[1 - F_{(0\to\lambda_1)}]$$

From Table 12.1,

$\lambda_1 T_f = 5\ \mu\text{m} \times 1200\ \text{K} = 6000\ \mu\text{m} \cdot \text{K}$: $F_{(0 \to \lambda_1)} = 0.738$

Hence

$\alpha = 0.8 \times 0.738 + 0.1(1 - 0.738) = 0.62$ ◁

The total, hemispherical emissivity follows from Equation 12.38.

$$\varepsilon(T_s) = \frac{\int_0^\infty \varepsilon_\lambda E_{\lambda,b}(\lambda, T_s)\, d\lambda}{E_b(T_s)}$$

Since the surface is diffuse, $\varepsilon_\lambda = \alpha_\lambda$ and it follows that

$$\varepsilon = \alpha_{\lambda,1} \frac{\int_0^{\lambda_1} E_{\lambda,b}(\lambda, 300\ \text{K})\, d\lambda}{E_b(300\ \text{K})} + \alpha_{\lambda,2} \frac{\int_{\lambda_1}^\infty E_{\lambda,b}(\lambda, 300\ \text{K})\, d\lambda}{E_b(300\ \text{K})}$$

or,

$$\varepsilon = \alpha_{\lambda,1} F_{(0 \to \lambda_1)} + \alpha_{\lambda,2}[1 - F_{(0 \to \lambda_1)}]$$

From Table 12.1,

$\lambda_1 T_s = 5\ \mu\text{m} \times 300\ \text{K} = 1500\ \mu\text{m} \cdot \text{K}$: $F_{(0 \to \lambda_1)} = 0.014$

Hence

$\varepsilon = 0.8 \times 0.014 + 0.1(1 - 0.014) = 0.11$ ◁

2. Because the spectral characteristics of the coating and the furnace temperature remain fixed, there is no change in the value of α with increasing time. However, as T_s increases with time, the value of ε will change. After a sufficiently long time, $T_s = T_f$, and $\varepsilon = \alpha (\varepsilon = 0.62)$.

Comments:

1. The equilibrium condition that eventually exists ($T_s = T_f$) corresponds precisely to the condition for which Kirchhoff's law was derived. Hence α must equal ε.
2. Approximating the sphere as a lumped capacitance and neglecting convection, an energy balance for a control volume about the sphere yields

$$\dot{E}_{\text{in}} - \dot{E}_{\text{out}} = \dot{E}_{\text{st}}$$

$$(\alpha G)A_s - (\varepsilon \sigma T_s^4)A_s = Mc_p \frac{dT_s}{dt}$$

The differential equation could be solved to determine $T(t)$ for $t > 0$, and the variation in ε that occurs with increasing time would have to be included in the solution.

12.8 ENVIRONMENTAL RADIATION

It would be inappropriate to conclude this chapter without commenting on the radiation that comprises our natural environment. Solar radiation is, of course, essential to all life on earth. Through the process of photosynthesis, it satisfies our need for food, fiber, and fuel. Moreover, through thermal and photovoltaic processes, it has the potential to satisfy much of our demand for space heat, process heat, and electricity.

The sun is a nearly spherical radiation source that is 1.39×10^9 m in diameter and is located 1.50×10^{11} m from the earth. With respect to the magnitude and the spectral and directional dependence of the incident solar radiation, it is necessary to distinguish between conditions at the earth's surface and at the outer edge of the earth's atmosphere. For a horizontal surface outside the earth's atmosphere, solar radiation appears as a beam of nearly *parallel rays* that form an angle θ, the *zenith angle*, relative to the surface normal (Figure 12.28). The *extraterrestrial* solar irradiation $G_{S,o}$ depends on the geographic latitude, as well as the time of day and year. It may be determined from an expression of the form

$$G_{S,o} = S_c \cdot f \cdot \cos\theta \tag{12.67}$$

where S_c, the *solar constant*, is the flux of solar energy incident on a surface oriented *normal* to the sun's rays, when the earth is at its mean distance from the sun. It is known to have a value of $S_c = 1353$ W/m^2. The quantity f is a small correction factor to account for the eccentricity of the earth's orbit about the sun ($0.97 \lesssim f \lesssim 1.03$).

The spectral distribution of solar radiation is significantly different from that associated with emission by engineering surfaces. As shown in Figure 12.29, this distribution approximates that of a blackbody at 5800 K. The radiation is concentrated in the low wavelength region ($0.2 \lesssim \lambda \lesssim 3$ μm) of the thermal spectrum with the peak occurring at approximately 0.50 μm. It is this short wavelength concentration that frequently precludes the assumption of graybody behavior for solar irradiated surfaces, since emission is generally in the spectral region beyond 4 μm and it is unlikely that surface spectral properties would be constant over such a wide spectral range.

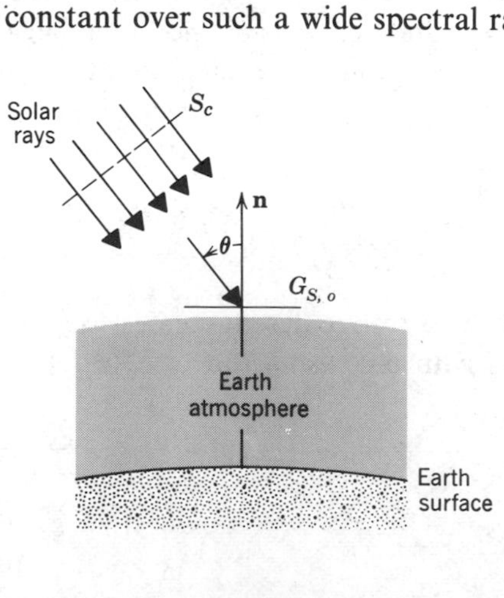

Figure 12.28 Directional nature of solar radiation outside the earth's atmosphere.

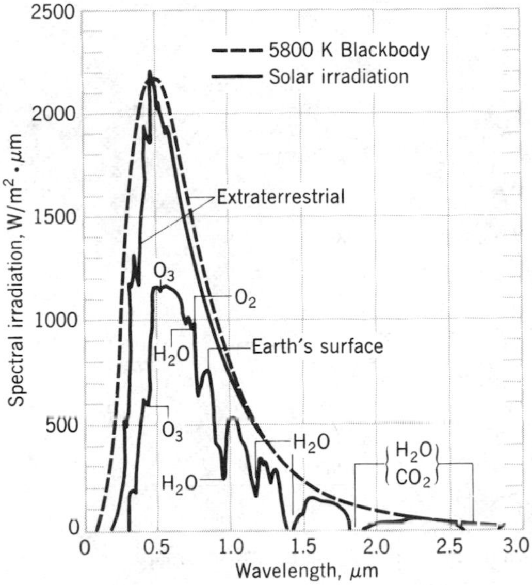

Figure 12.29 Spectral distribution of solar radiation.

As solar radiation passes through the earth's atmosphere, its magnitude and its spectral and directional distributions experience significant change. The change is due to *absorption* and *scattering* of the radiation by the atmospheric constituents. The effect of absorption by the atmospheric gases O_3 (ozone), H_2O, O_2, and CO_2 is shown by the lower curve of Figure 12.29. Absorption by ozone is strong in the UV region, providing considerable attenuation below 0.4 μm and complete attenuation below 0.3 μm. In the visible region there is some absorption by O_3 and O_2; and in the near and far IR regions, absorption is dominated by water vapor. Throughout the solar spectrum, there is also continuous absorption of radiation by the dust and aerosol content of the atmosphere.

Atmospheric scattering provides for *redirection* of the sun's rays and consists of two kinds (Figure 12.30). *Rayleigh* (or *molecular) scattering* by the gas molecules provides for nearly uniform scattering of radiation in all directions. Hence approximately half of the scattered radiation is redirected to space, while the remaining portion strikes the earth's surface. At any point on this surface, the scattered radiation is incident from all directions. In contrast, *Mie scattering* by the dust and aerosol particles of the atmosphere is concentrated in directions that are close to that of the incident rays. Hence, virtually all of this radiation strikes the earth's surface in directions close to that of the sun's rays.

The cumulative effect of the scattering processes on the directional distribution of solar radiation striking the earth's surface is shown in Figure 12.31*a*. That portion of the radiation that has penetrated the atmosphere without having been scattered (or absorbed) is in the direction of the zenith angle and is termed the *direct radiation.* The scattered radiation is incident from

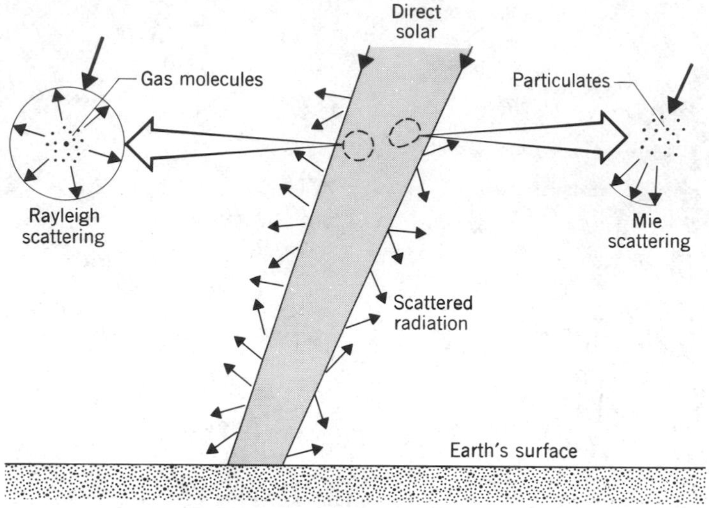

Figure 12.30 Scattering of solar radiation in the earth's atmosphere.

all directions, although its intensity is largest for directions close to that of the direct radiation. However, because the radiation intensity is often *assumed* to be independent of direction (Figure 12.31*b*), the radiation is termed *diffuse*. The total solar radiation reaching the earth's surface is therefore the sum of direct and diffuse contributions. The diffuse contribution may vary from approximately 10% of the total solar radiation on a clear day to nearly 100% on a totally overcast day.

The foregoing discussion has focused on the *nature* of solar radiation. Heat transfer analyses related to its use are considered in many of the examples and problems of the text. Detailed treatment of solar energy technologies is left to the literature [7–11].

Long wavelength forms of environmental radiation include emission from

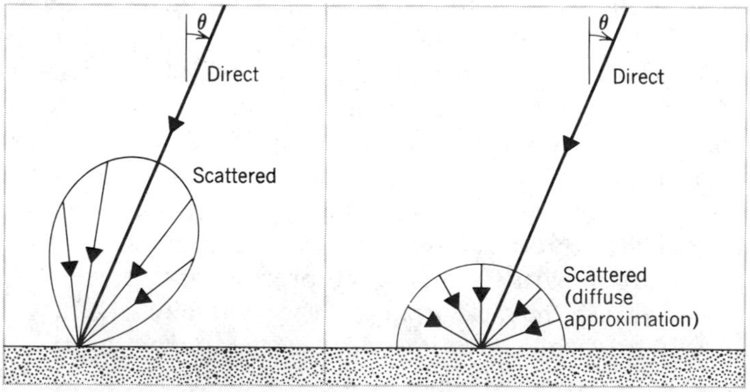

Figure 12.31 Directional distribution of solar radiation at the earth's surface. (*a*) Actual distribution. (*b*) Diffuse approximation.

the earth's surface, as well as emission from certain atmospheric constituents. The emissive power associated with the earth's surface may be computed in the conventional manner. That is,

$$E = \varepsilon\sigma T^4 \tag{12.68}$$

where ε and T are the surface emissivity and temperature, respectively. Emissivities are generally close to unity, and that of water, for example, is approximately 0.97. Since temperatures typically range from 250 to 320 K, emission is concentrated in the spectral region from approximately 4 to 40 μm, with the peak occurring at approximately 10 μm.

Atmospheric emission is largely from the CO_2 and H_2O molecules and is concentrated in the spectral regions from 5 to 8 μm and above 13 μm. Although the spectral distribution of atmospheric emission does not correspond to that of a blackbody, its contribution to irradiation of the earth's surface can be estimated using Equation 12.28. In particular, earth irradiation due to atmospheric emission may be expressed in the form

$$G_{\text{atm}} = \sigma T_{\text{sky}}^4 \tag{12.69}$$

where T_{sky} is termed the effective sky temperature. Its value depends on atmospheric conditions, ranging from a low of 230 K under cold, clear sky conditions to a high of approximately 285 K under warm, cloudy conditions. At night, atmospheric emission is the only source of earth irradiation. When its value is small, as on a cool, clear night, water may freeze even though the air temperature exceeds 273 K.

We close by recalling that values of the spectral properties of a surface at short wavelengths may be appreciably different from values at long wavelengths (Figures 12.18 and 12.23). Since solar radiation is concentrated in the short wavelength region of the spectrum and surface emission is at much longer wavelengths, it follows that many surfaces may not be approximated as gray in their response to solar irradiation. In other words, the solar absorptivity of a surface α_S may differ from its emissivity ε. Values of α_S and the emissivity at moderate temperatures are presented in Table 12.2 for representative surfaces.

Table 12.2 Solar absorptivity α_S and emissivity ε of surfaces having spectral absorptivity given in Figure 12.23

SURFACE	α_S	ε (300 K)	α_S/ε
Evaporated aluminum film	0.09	0.03	3.0
Fused quartz on aluminum film	0.19	0.81	0.24
White paint on metallic substrate	0.21	0.96	0.22
Black paint on metallic substrate	0.97	0.97	1.0
Stainless steel, as received, dull	0.50	0.21	2.4
Red brick	0.63	0.93	0.68
Human skin (Caucasian)	0.62	0.97	0.64
Snow	0.28	0.97	0.29
Corn leaf	0.76	0.97	0.78

Note that the ratio α_S/ε is an important engineering parameter. Small values are desired if the surface is intended to reject heat; large values are required if the surface is intended to collect solar energy.

EXAMPLE 12.11

A flat plate solar collector with no cover plate has a selective absorber surface of emissivity 0.1 and solar absorptivity 0.95. At a given time of day the absorber surface temperature T_s is 120°C when the solar irradiation is 750 W/m^2, the effective sky temperature is -10°C, and the ambient air temperature T_∞ is 30°C. Assume that the heat transfer convection coefficient for the calm day conditions can be estimated from

$$\bar{h} = 0.22(T_s - T_\infty)^{1/3} \text{ W/m}^2 \cdot \text{K}$$

Calculate the useful heat removal rate (W/m^2) from the collector for these conditions. What is the corresponding efficiency of the collector?

SOLUTION

Known: Operating conditions for a flat plate solar collector.

Find:

1. Useful heat removal rate per unit area, q''_u (W/m^2).
2. Efficiency η of the collector.

Schematic:

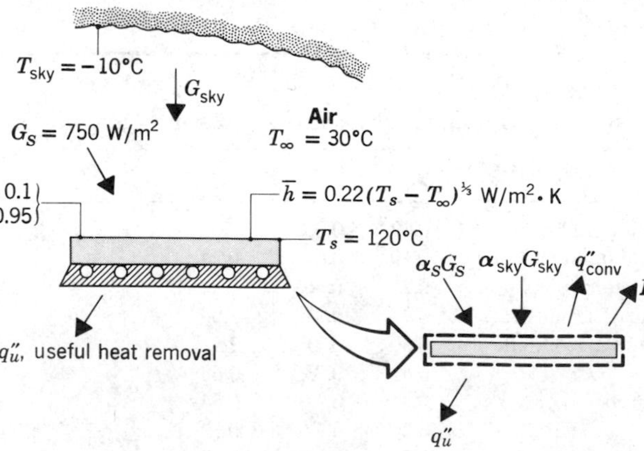

Assumptions:

1. Steady-state conditions.
2. Bottom of collector well insulated.
3. Absorber surface diffuse.

Analysis:

1. Performing an energy balance on the absorber,

   $$\dot{E}_{in} - \dot{E}_{out} = 0$$

   or, per unit surface area,

   $$\alpha_S G_S + \alpha_{sky} G_{sky} - q''_{conv} - E - q''_u = 0$$

   From Equation 12.69

   $$G_{sky} = \sigma T_{sky}^4$$

   Since the sky radiation is concentrated in approximately the same spectral region as that of surface emission, it is reasonable to assume that

   $$\alpha_{sky} \approx \varepsilon = 0.1$$

   With

   $$q''_{conv} = \bar{h}(T_s - T_\infty) = 0.22(T_s - T_\infty)^{4/3}$$

   and

   $$E = \varepsilon\sigma T_s^4$$

   it follows that

   $$q''_u = \alpha_S G_S + \varepsilon\sigma T_{sky}^4 - 0.22(T_s - T_\infty)^{4/3} - \varepsilon\sigma T_s^4$$

   $$q''_u = \alpha_S G_S - 0.22(T_s - T_\infty)^{4/3} - \varepsilon\sigma(T_s^4 - T_{sky}^4)$$

   $$q''_u = 0.95 \times 750\ \text{W/m}^2 - 0.22(120 - 30)^{4/3}\ \text{W/m}^2 - 0.1 \times 5.67 \times 10^{-8}\ \text{W/m}^2 \cdot \text{K}^4 (393^4 - 263^4)\ \text{K}^4$$

   $$q''_u = (712.5 - 88.7 - 108.1)\ \text{W/m}^2 = 516\ \text{W/m}^2 \quad \triangleleft$$

2. The collector efficiency, defined as the fraction of the solar irradiation extracted as useful energy, is then

   $$\eta = \frac{q''_u}{G_S} = \frac{516\ \text{W/m}^2}{750\ \text{W/m}^2} = 0.69 \quad \triangleleft$$

Comments:

1. Since the spectral range of G_{sky} is entirely different from that of G_S, it would be *incorrect* to assume that $\alpha_{sky} = \alpha_S$.

2. The convection heat transfer coefficient is extremely low ($\bar{h} \approx 1 \text{ W/m}^2 \cdot \text{K}$). With a modest increase to $\bar{h} = 5 \text{ W/m}^2 \cdot \text{K}$, the useful heat flux and the efficiency are reduced to $q''_u = 161 \text{ W/m}^2$ and $\eta = 0.21$. A cover plate can contribute significantly to reducing convection (and radiation) heat loss from the absorber plate.

12.9 SUMMARY

Many new and important ideas have been introduced in this chapter, and at this stage you may well be confused, particularly by the terminology. However, the subject matter has been developed in a systematic fashion, and a careful rereading of the material should make you more comfortable with its application. A glossary has been provided in Table 12.3 to assist you in assimilating the terminology.

You should be able to answer the following questions. What is radiation, and what position does *thermal radiation* occupy in the *electromagnetic spectrum*? What are the physical origins of thermal radiation? What are meant by the terms *irradiation, emissive power*, and *radiosity*? What material property characterizes the ability of a surface to emit thermal radiation? What processes and

Table 12.3 Glossary of radiative terms

TERM	DEFINITION
Absorption	The process of converting radiation intercepted by matter to internal thermal energy.
Absorptivity	Fraction of the incident radiation absorbed by matter. Equations 12.41, 12.42, and 12.45. Modifiers: *directional, hemispherical, spectral, total.*
Blackbody	The ideal emitter and absorber. Modifier referring to ideal behavior. Denoted by the subscript *b*.
Diffuse	Modifier referring to the directional independence of the intensity associated with emitted, reflected, or incident radiation.
Directional	Modifier referring to a particular direction. Denoted by the subscript θ.
Directional distribution	Refers to variation with direction.
Emission	The process of radiation production by matter at a finite temperature. Modifiers: *diffuse, blackbody, spectral.*
Emissive power	Rate of radiant energy emitted by a surface in all directions per unit area of the surface, E (W/m^2). Modifiers: *spectral, total, blackbody.*
Emissivity	Ratio of the radiation emitted by a surface to the radiation emitted by a blackbody at the same temperature. Equations 12.32, 12.33, 12.34, and 12.37. Modifiers: *directional, hemispherical, spectral, total.*

Table 12.3 Continued

TERM	DEFINITION
Gray surface	A surface for which the spectral absorptivity and the emissivity are independent of wavelength over the spectral regions of surface irradiation and emission.
Hemispherical	Modifier referring to all directions in the space above a surface.
Intensity	Rate of radiant energy propagation in a particular direction, per unit area normal to the direction, per unit solid angle about the direction, I ($W/m^2 \cdot sr$). Modifier: *spectral*.
Irradiation	Rate at which radiation is incident on a surface from all directions per unit area of the surface, G (W/m^2). Modifiers: *spectral, total, diffuse*.
Kirchhoff's law	Relation between emission and absorption properties for surfaces irradiated by a blackbody at the same temperature. Equations 12.61, 12.62, 12.63, and 12.64.
Planck's law	Spectral distribution of emission from a blackbody. Equations 12.25 and 12.26.
Radiosity	Rate at which radiation leaves a surface due to emission and reflection in all directions per unit area of the surface, J (W/m^2). Modifiers: *spectral, total*.
Reflection	The process of redirection of radiation incident on a surface. Modifiers: *diffuse, specular*.
Reflectivity	Fraction of the incident radiation reflected by matter. Equations 12.48, 12.49, and 12.51. Modifiers: *directional, hemispherical, spectral, total*.
Semitransparent	Refers to a medium in which radiation absorption is a volumetric process.
Solid angle	Region subtended by an element of area on the surface of a sphere with respect to the origin of the sphere, ω (sr). Equations 12.2 and 12.3.
Spectral	Modifier referring to a single-wavelength (monochromatic) component. Denoted by the subscript λ.
Spectral distribution	Refers to variation with wavelength.
Specular	Refers to a surface for which the angle of reflected radiation is equal to the angle of incident radiation.
Stefan-Boltzmann law	Emissive power of a blackbody. Equation 12.28.
Thermal radiation	Electromagnetic energy emitted by matter at a finite temperature and concentrated in the spectral region from approximately 0.1 to 100 μm.
Total	Modifier referring to all wavelengths.
Transmission	The process of thermal radiation passing through matter.
Transmissivity	Fraction of the incident radiation transmitted by matter. Equations 12.53 and 12.54. Modifiers: *hemispherical, spectral, total*.
Wien's law	Locus of the wavelength corresponding to peak emission by a blackbody. Equation 12.27.

associated material properties characterize the manner in which a surface responds to irradiation? What is an *opaque* surface? A *semitransparent* surface?

You should recognize the difference between *directional* and *spectral* radiation properties on the one hand and *hemispherical* and *total* properties on the other. Moreover, you should be able to proceed from knowledge of the former to determination of the latter. You should also appreciate the unique role of the *blackbody* in the description of thermal radiation. In what sense is the blackbody *perfect*? Why is it an idealization, and how may it be approximated in practice? What is the *Planck distribution*? How is it altered with increasing surface temperature? What are the *Wien* and *Stefan-Boltzmann* laws? For what purposes may the blackbody emission functions of Table 12.1 be used?

Relations between emissivity and absorptivity are often extremely important in radiative exchange calculations. What is *Kirchhoff's law*, and what restrictive conditions are inherent in its derivation? What is a *gray surface*, and under what conditions might the assumption of a gray surface be particularly suspect? Finally, what are the characteristics of *solar radiation*? In what region of the spectrum is this radiation concentrated, and how is it altered due to passage through the earth's atmosphere? What is the nature of its directional distribution at the earth's surface?

REFERENCES

1. Planck, M., *The Theory of Heat Radiation*, Dover Publications, New York, 1959.
2. Gubareff, G. G., J. E. Janssen, and R. H. Torberg, *Thermal Radiation Properties Survey*, 2nd ed., Honeywell Research Center, Minneapolis, 1960.
3. Wood, W. D., H. W. Deem, and C. F. Lucks, *Thermal Radiative Properties*, Plenum Press, New York, 1964.
4. Touloukian, Y. S., *Thermophysical Properties of High Temperature Solid Materials*, Macmillan, New York, 1967.
5. Touloukian, Y. S., and D. P. DeWitt, *Thermal Radiative Properties*, Vols. 7, 8, and 9, from *Thermophysical Properties of Matter*, TPRC Data Series, Y. S. Touloukian and C. Y. Ho, Eds., IFI Plenum, New York, 1970–1972.
6. Siegel, R., and J. R. Howell, *Thermal Radiation Heat Transfer*, 2nd ed., McGraw-Hill, New York, 1981.
7. Duffie, J. A., and W. A. Beckman, *Solar Energy Thermal Processes*, Wiley, New York, 1974.
8. Meinel, A. B., and M. P. Meinel, *Applied Solar Energy: An Introduction*, Addison-Wesley, Reading, MA, 1976.
9. Sayigh, A. A. M., Ed., *Solar Energy Engineering*, Academic Press, New York, 1977.
10. Anderson, E. E., *Solar Energy Fundamentals for Designers and Engineers*, Addison-Wesley, Reading, MA, 1982.
11. Howell, J. R., R. B. Bannerot, and G. C. Vliet, *Solar–Thermal Energy Systems, Analysis and Design*, McGraw-Hill, New York, 1982.

PROBLEMS

12.1 Consider a small surface of area $A_1 = 10^{-4}\ \text{m}^2$, which emits diffusely with a total, hemispherical emissive power of $E_1 = 5 \times 10^4\ \text{W/m}^2$. At what rate is this emission intercepted by a small surface of area $A_2 = 5 \times 10^{-4}\ \text{m}^2$, which is oriented as shown.

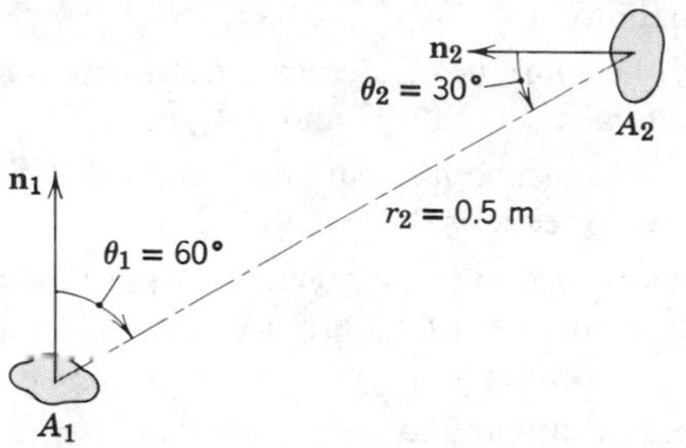

12.2 What is the irradiation at surfaces A_2, A_3, and A_4 of Example 12.1 due to emission from A_1?

12.3 According to its directional distribution, solar radiation incident on the earth's surface may be divided into two components. The *direct* component consists of parallel rays incident at a fixed zenith angle θ, while the *diffuse* component consists of radiation that may be approximated as being diffusely distributed with θ.

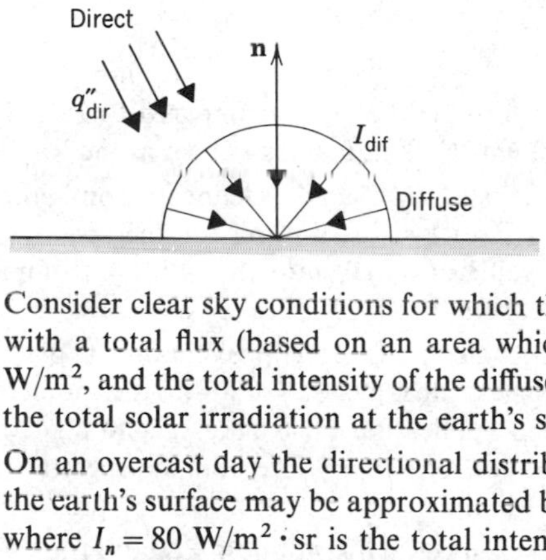

Consider clear sky conditions for which the direct radiation is incident at $\theta = 30°$, with a total flux (based on an area which is normal to the rays) of $q''_{\text{dir}} = 1000$ W/m², and the total intensity of the diffuse radiation is $I_{\text{dif}} = 70\ \text{W/m}^2 \cdot \text{sr}$. What is the total solar irradiation at the earth's surface?

12.4 On an overcast day the directional distribution of the solar radiation incident on the earth's surface may be approximated by an expression of the form $I_i = I_n \cos\theta$, where $I_n = 80\ \text{W/m}^2 \cdot \text{sr}$ is the total intensity of radiation directed normal to the surface and θ is the zenith angle. What is the solar irradiation at the earth's surface?

12.5 The spectral distribution of the radiation emitted by a diffuse surface may be approximated as follows.

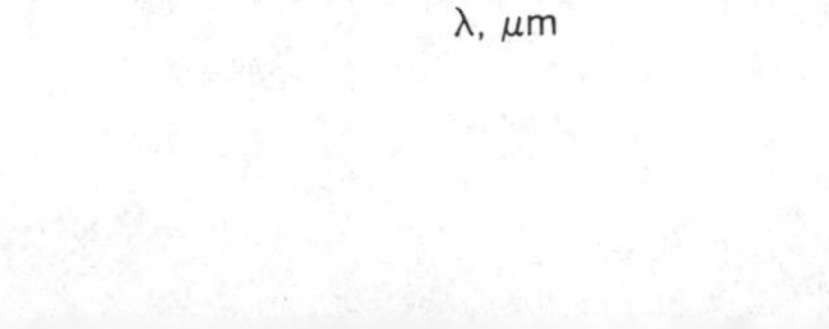

What is the total emissive power? What is the total intensity of the radiation emitted in the normal direction and at an angle of 30° from the normal?

12.6 A furnace with an aperture of 20 mm diameter and emissive power of 3.72×10^5 W/m^2 is used to calibrate a heat flux gage having a sensitive area of 1.6×10^{-5} m^2. At what distance, measured along a normal from the aperture, should the gage be positioned to receive irradiation of 1000 W/m^2? If the gage is tilted off normal by 20°, what will be its irradiation?

12.7 Determine the fraction of the total, hemispherical emissive power that leaves a diffuse surface in the directions $\pi/4 \leq \theta \leq \pi/2$ and $0 \leq \phi \leq \pi$.

12.8 What fraction of the total, hemispherical emissive power E (W/m^2) leaving a diffuse radiator will be in the directions $0 \leq \theta \leq 30°$?

12.9 Approximations to Planck's law for the spectral emissive power are the Wien and Rayleigh-Jeans spectral distributions, which are useful for the extreme low and high limits of the product λT, respectively.

(a) Show that Planck's spectral distribution will have the form

$$E_{\lambda,b}(\lambda, T) \approx \frac{C_1}{\lambda^5} \exp\left(-\frac{C_2}{\lambda T}\right)$$

when $C_2/\lambda T \gg 1$ and determine the error (compared to Planck's law) for the condition $\lambda T = 2898\ \mu\text{m} \cdot \text{K}$. This form is known as Wien's law.

(b) Show that the Planck distribution will have the form

$$E_{\lambda,b}(\lambda, T) \approx \frac{C_1}{C_2} \frac{T}{\lambda^4}$$

when $C_2/\lambda T \ll 1$ and determine the error (compared to Planck's law) for the condition $\lambda T = 100{,}000\ \mu\text{m} \cdot \text{K}$. This form is known as the Rayleigh-Jeans law.

12.10 The fourth radiation constant, C_4, is defined from the relationship $I_{\lambda,b}(\lambda_{max}, T) = C_4 T^5$. Using Planck's distribution for the spectral intensity of a blackbody, determine the value of C_4 (W/m$^2 \cdot \mu$m $\cdot$ sr $\cdot$ K^5). Compare your result with the appropriate value from Table 12.1.

12.11 Isothermal furnaces with small apertures approximating a blackbody are frequently used to calibrate heat flux gages, radiation thermometers, and other radiometric devices. In such applications, it is necessary to control power to the furnace such that the variation of temperature and the spectral intensity of the aperture are within desired limits.

(a) By considering the Planck spectral distribution, Equation 12.26, show that the ratio of the fractional change in the spectral intensity to the fractional change in the temperature of the furnace has the form

$$\frac{dI_\lambda/I_\lambda}{dT/T} = \frac{C_2}{\lambda T} \frac{1}{1 - \exp(-C_2/\lambda T)}$$

(b) Using this relation, determine the allowable variation in temperature of the furnace operating at 2000 K to ensure that the spectral intensity at 0.65 μm will not vary by more than 0.5%. What is the allowable variation at 10 μm?

12.12 Substitute the Planck distribution, Equation 12.26, into Equation 12.11 and perform the spectral integration to obtain Equation 12.28 for the total emissive power of a blackbody. Show that

$$\sigma = \left(\frac{\pi}{C_2}\right)^4 \frac{C_1}{15}$$

and calculate the value of the Stefan-Boltzmann constant using values of the radiation constants C_1 and C_2.

12.13 A spherical aluminum shell of inside diameter $D = 2$ m is evacuated and is used as a radiation test chamber. If the inner surface is coated with carbon black and maintained at 600 K, what is the irradiation on a small test surface placed in the chamber? If the inner surface were not coated and maintained at 600 K, what would the irradiation be?

12.14 An enclosure has an inside area of 100 m^2, and its inside surface is black and is maintained at a constant temperature. A small opening in the enclosure has an area of 0.02 m^2. The radiant power emitted from this opening is 70 W. What is the temperature of the interior enclosure wall? If the interior surface is maintained at this temperature, but is now polished, what will be the value of the radiant power emitted from the opening?

12.15 The energy flux associated with solar radiation incident on the outer surface of the earth's atmosphere has been accurately measured and is known to be 1353 W/m^2. The diameters of the sun and earth are 1.39×10^9 and 1.29×10^7 m, respectively, and the distance between the sun and the earth is 1.5×10^{11} m.

(a) What is the emissive power of the sun?

(b) Approximating the sun's surface as black, what is its temperature?

(c) At what wavelength is the spectral emissive power of the sun a maximum?

(d) Assuming the earth's surface to be black and the sun to be the only source of energy for the earth, estimate the earth surface temperature.

12.16 Estimate the wavelength corresponding to maximum emission from each of the following surfaces: the sun, a tungsten filament at 2500 K, a heated metal at 1500 K, human skin at 305 K, and a cryogenically cooled metal surface at 60 K. Estimate the fraction of the solar emission that is in the following spectral regions: the ultraviolet, the visible, and the infrared.

12.17 What is the efficiency of a 100-W light source consisting of a filament that radiates as a blackbody at 2900 K? You may assume that the glass bulb transmits all the incident visible radiation and that the filament is a thin rectangular strip 5 mm long by 2 mm wide.

12.18 Consider the radiation emerging from a small aperture of a furnace operating at 1000 K. Calculate the emissive power of the aperture. Determine the spectral intensity at 2 μm. What is the ratio of the spectral intensity at 2 μm to the spectral intensity at 6 μm? What fraction of the emissive power is in the spectral range 2 to 6 μm?

12.19 Determine the fraction of the radiation emitted by the sun in the visible region of the spectrum.

12.20 The spectral, hemispherical emissivity of tungsten may be approximated by the distribution on the following page. Consider a cylindrical tungsten filament that is of diameter $D = 0.8$ mm and length $L = 20$ mm. The filament is enclosed in an evacuated bulb and is heated by an electrical current to a steady-state temperature of 2900 K.

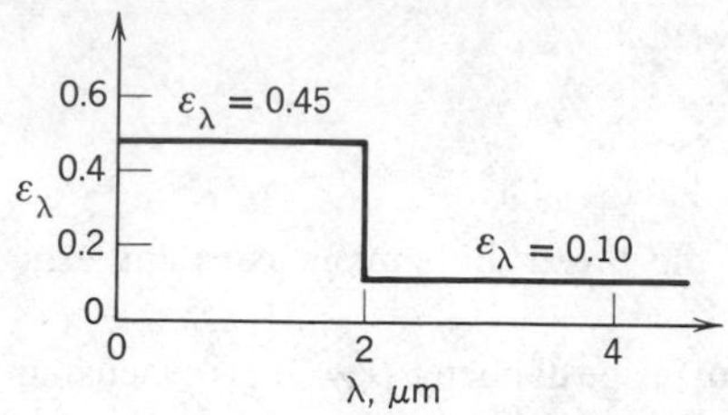

What is the initial rate of cooling (K/s) of the filament following discontinuation of the current? Estimate the time required for the filament to cool to a temperature of 1300 K, below which the light is barely visible.

12.21 For materials A and B, whose spectral, hemispherical emissivities vary with wavelength as shown, how does the total, hemispherical emissivity vary with temperature? Explain briefly.

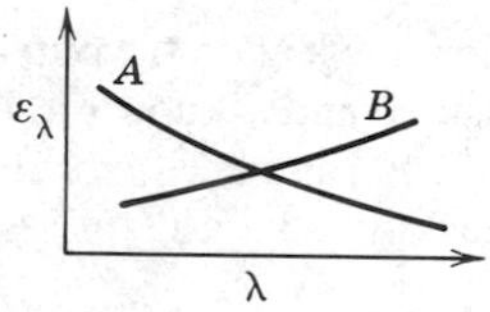

12.22 Various materials are to be evaluated for design of more energy-efficient household cooking pans. Compare the emissive powers of the following materials at 200°C and explain what influence this quantity would have on energy consumption: polished copper, Teflon-coated copper, cleaned stainless steel, Pyrex, and pyroceram.

12.23 The spectral, directional emissivity of a diffuse material at 2000 K has the following distribution.

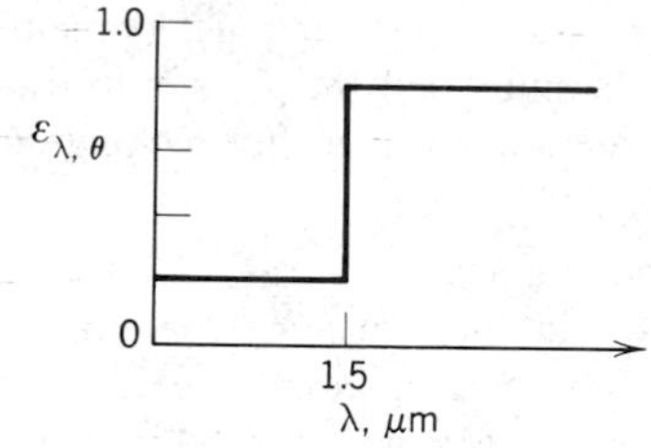

Determine the total, hemispherical emissivity at 2000 K. Determine the emissive power over the spectral range 0.8 to 2.5 μm *and* for the directions $0 \le \theta \le 30°$.

12.24 Consider the directionally selective surface having the directional emissivity ε_θ, as shown. Assuming that the surface is isotropic in the ϕ direction, calculate the ratio of the normal emissivity ε_n to the hemispherical emissivity ε_h.

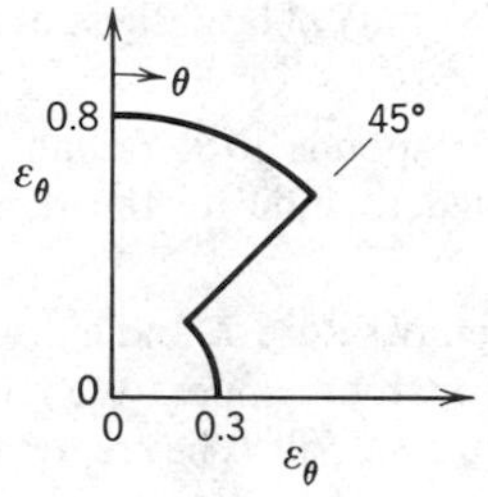

12.25 Consider the metallic surface of Example 12.6. Additional measurements of the spectral, hemispherical emissivity yield a spectral distribution which may be approximated as follows.

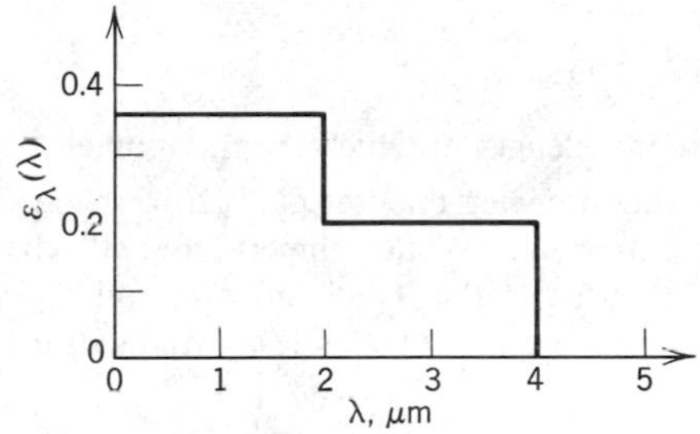

Determine corresponding values of the total, hemispherical emissivity and the total emissive power at 2000 K.

12.26 A detector of area $A_2 = 4 \times 10^{-6}\ m^2$ is used to measure the total radiation emitted by a surface of area $A_1 = 5 \times 10^{-6}\ m^2$ and temperature $T_1 = 1000$ K.

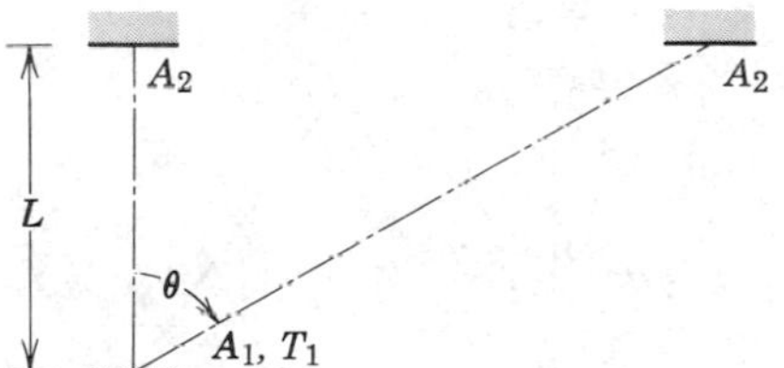

When the detector surface views radiation emitted by A_1 in the normal direction ($\theta = 0°$) at a distance of $L = 0.5$ m, it measures a radiant power of 1.155×10^{-6} W. What is the total, normal emissivity of surface 1? When the detector is displaced horizontally, such that $\theta = 60°$, it measures a radiant power of 5.415×10^{-8} W. Is surface 1 a diffuse emitter?

12.27 A sphere is suspended in air in a dark room and maintained at a uniform incandescent temperature. When first viewed with the naked eye, the sphere appears to be brighter around the rim. After several hours, however, it appears to be brighter in the center. Of what type material would you reason the sphere is made? Give plausible reasons for the nonuniformity of brightness of the sphere and for the changing appearance with time.

12.28 A radiation thermometer is a device that responds to a radiant flux within a prescribed spectral interval and is calibrated to indicate the temperature of a blackbody that produces the same flux.

(a) When viewing a surface at an elevated temperature T_s and emissivity less than unity, the thermometer will indicate an apparent temperature referred to as the brightness or spectral radiance temperature T_λ. Will T_λ be greater than, less than, or equal to T_s?

(b) Write an expression for the spectral emissive power of the surface in terms of Wien's spectral distribution (see Problem 12.9) and the spectral emissivity of the surface. Write the equivalent expression using the spectral radiance temperature of the surface and show that

$$\frac{1}{T_s} = \frac{1}{T_\lambda} + \frac{\lambda}{C_2} \ln \varepsilon_\lambda$$

where λ represents the wavelength at which the thermometer operates.

(c) Consider a radiation thermometer that responds to a spectral flux centered about the wavelength 0.65 μm. What temperature will the thermometer indicate when viewing a surface with ε_λ (0.65 μm) = 0.9 and T_s = 1000 K? Verify that Wien's spectral distribution is a reasonable approximation to Planck's law for this situation.

12.29 A diffusely emitting surface is exposed to a radiant source causing the irradiation on the surface to be 1000 W/m^2. The intensity for emission is 143 W/m$^2 \cdot$ sr and the reflectivity of the surface is 0.8. Determine the emissive power, E (W/m^2), and radiosity, J (W/m^2), for the surface. What is the net heat flux to the surface by the radiation mode?

12.30 An opaque surface with the prescribed spectral, hemispherical reflectivity distribution is subjected to the spectral irradiation shown.

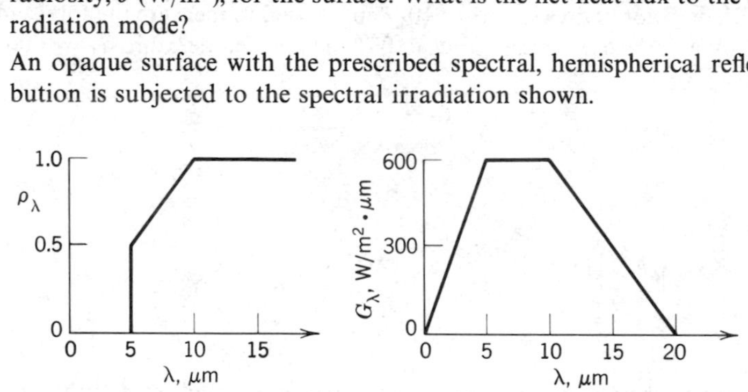

(a) Sketch the spectral, hemispherical absorptivity distribution.

(b) Determine the total irradiation on the surface.

(c) Determine the radiant flux that is absorbed by the surface.

(d) What is the total, hemispherical absorptivity of this surface?

12.31 Radiation leaves a furnace of inside surface temperature 1500 K through an aperture 20 mm in diameter. A portion of the radiation is intercepted by a detector that is 1 m from the aperture, has a surface area of 10^{-5} m^2, and is oriented as shown.

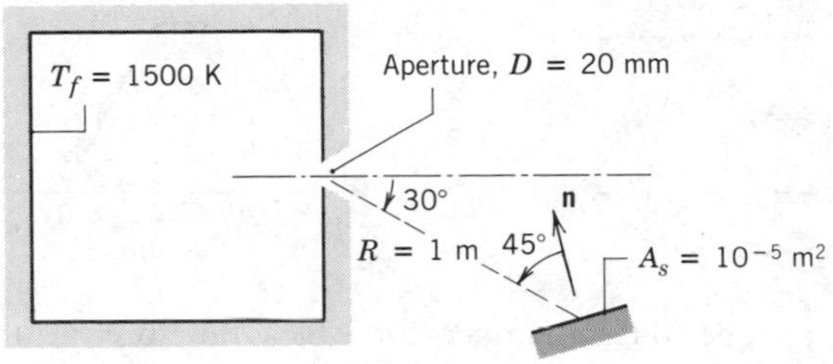

If the aperture is open, what is the rate at which radiation leaving the furnace is intercepted by the detector? If the aperture is covered with a diffuse, semi-transparent material of spectral transmissivity $\tau_\lambda = 0.8$ for $\lambda \leq 2\ \mu$m and $\tau_\lambda = 0$ for $\lambda > 2\ \mu$m, what is the rate at which radiation leaving the furnace is intercepted by the detector?

12.32 A 20-mm-diameter peephole is maintained in the wall of a large furnace whose interior surfaces are at a temperature of 1700 K. What is the heat loss by radiation through an uncovered hole? What is the heat loss by radiation through a fused quartz cover for which $\tau_\lambda = 0.9$ for $\lambda \leq 3.5\ \mu$m and $\tau_\lambda = 0$ for $\lambda > 3.5\ \mu$m?

12.33 The spectral transmissivity of plain and tinted glass can be approximated as follows:

Plain glass:	$\tau_\lambda = 0.9$	$0.3 \leq \lambda \leq 2.5\ \mu$m
Tinted glass:	$\tau_\lambda = 0.9$	$0.5 \leq \lambda \leq 1.5\ \mu$m

Outside the specified wavelength ranges, the spectral transmissivity is zero for both glasses. Compare the solar energy that could be transmitted through the glasses. With solar irradiation on the glasses, compare the visible radiant energy that could be transmitted.

12.34 Referring to the distribution of the spectral transmissivity of low iron glass (Figure 12.24), describe briefly what is meant by the "greenhouse effect." That is, how does the glass influence energy transfer to and from the contents of a greenhouse?

12.35 The spectral, hemispherical absorptivity of an opaque surface is as shown.

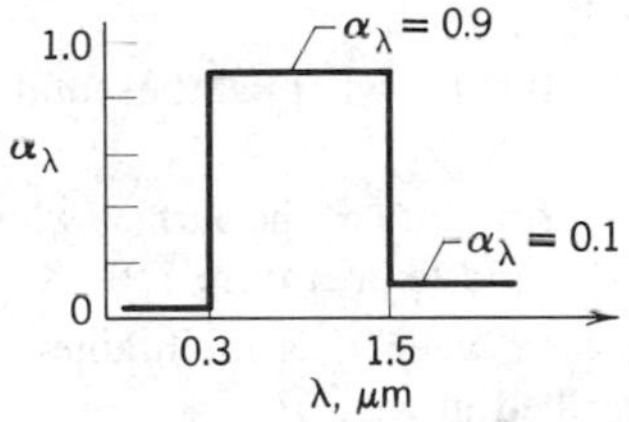

What is the solar absorptivity, α_S? If it is assumed that $\varepsilon_\lambda = \alpha_\lambda$ and that the surface is at a temperature of 340 K, what is its total, hemispherical emissivity?

12.36 The spectral, hemispherical absorptivity of an opaque surface and the spectral distribution of radiation incident on the surface are shown on the following page. What is the total, hemispherical absorptivity of the surface? If it is assumed that $\varepsilon_\lambda = \alpha_\lambda$ and that the surface is at 1000 K, what is its total, hemispherical emissivity? What is the net radiant heat flux to the surface?

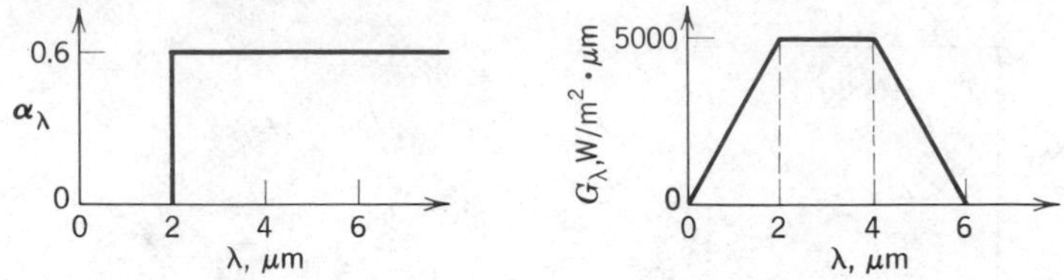

12.37 Consider an opaque, diffuse surface for which the spectral absorptivity and irradiation are as follows.

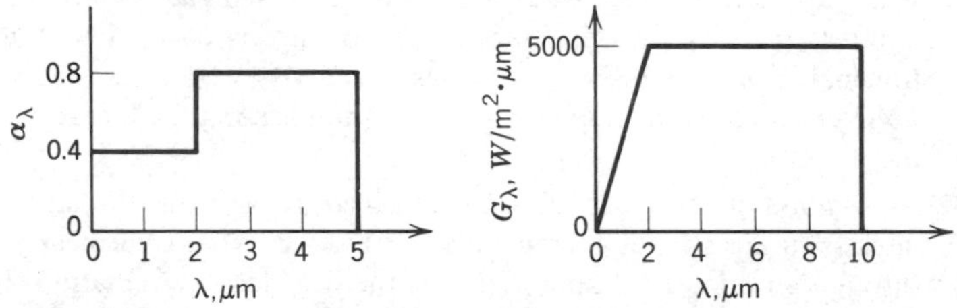

What is the total absorptivity of the surface for the prescribed irradiation? If the surface is at a temperature of 1250 K, what is its emissive power? How will the surface temperature vary with time, for the prescribed conditions?

12.38 The spectral emissivity of an opaque, diffuse surface is as shown.

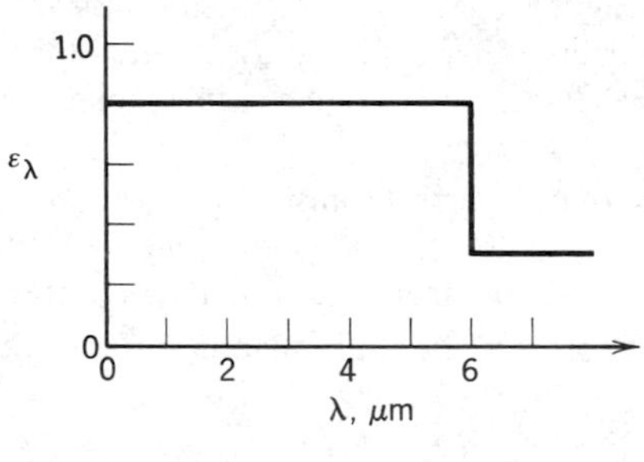

(a) If the surface is maintained at 1000 K, what is the total, hemispherical emissivity?

(b) What is the total, hemispherical absorptivity of the surface when irradiated by large surroundings of emissivity 0.8 and temperature 1500 K.

(c) What is the radiosity of the surface when it is maintained at 1000 K and subjected to the irradiation prescribed in part b?

(d) Determine the net radiation flux into the surface for the conditions of part c.

12.39 Two small surfaces, A and B, are placed inside an isothermal enclosure at a uniform temperature. The enclosure provides an irradiation of 6300 W/m^2 to each of the surfaces, and surfaces A and B absorb incident radiation at rates of 5600 and 630 W/m^2, respectively. Consider conditions after a long time has elapsed.

(a) What are the net heat fluxes for each surface? What are their temperatures?

(b) Determine the absorptivity of each surface.

(c) What are the emissive powers of each surface?

(d) Determine the emissivity of each surface.

12.40 Solar irradiation of 1100 W/m^2 is incident on a large, flat, horizontal metal roof on a day when the wind blowing over the roof causes a convection heat transfer coefficient of 25 $W/m^2 \cdot K$. The outside air temperature is 27°C, the metal surface absorptivity for incident solar radiation is 0.60, the metal surface emissivity is 0.20, and the roof is well insulated from below. Estimate the roof temperature under steady-state conditions.

12.41 The 50-mm peephole of a large furnace operating at 450°C is covered with a material having $\tau = 0.8$ and $\rho = 0$ for irradiation originating from the furnace. The material has an emissivity of 0.8 and is opaque to irradiation from a source at room temperature. The outer surface of the cover is exposed to surroundings and ambient air at 27°C with a convection heat transfer coefficient of 50 $W/m^2 \cdot K$. Assuming that convection effects on the inner surface of the cover are negligible, calculate the heat loss by the furnace and the temperature of the cover.

12.42 A deep cavity of 50 mm diameter approximates a blackbody and is maintained at 250°C while exposed to solar irradiation of 800 W/m^2 and surroundings and ambient air at 25°C. A thin window of spectral transmissivity and reflectivity 0.9 and 0, respectively, for the spectral range 0.2 to 4 μm is placed over the cavity opening. In the spectral range beyond 4 μm, the window behaves as an opaque, diffuse-gray body of emissivity 0.95.

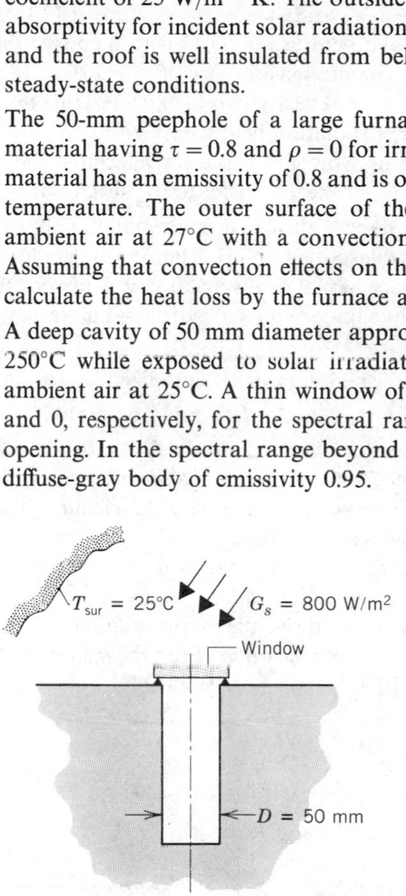

Assuming that the convection coefficient on the upper surface of the window is 10 $W/m^2 \cdot K$, determine the temperature of the window and the power required to maintain the cavity at 250°C.

12.43 Consider the evacuated tube solar collector described in Problem 1.26g of Chapter 1. In the interest of maximizing collector efficiency, what spectral radiative characteristics are desired for the outer tube and for the inner tube?

12.44 Solar flux of 900 W/m^2 is incident on the top side of a plate whose surface has a solar absorptivity of 0.9 and an emissivity of 0.1. The air and surroundings are at 17°C and the convection heat transfer coefficient between the plate and air is 20 $W/m^2 \cdot K$. Assuming that the bottom side of the plate is insulated, determine the steady-state temperature of the plate.

12.45 Consider an opaque horizontal plate that is well insulated on its back side. The irradiation on the plate is 2500 W/m^2, of which 500 W/m^2 is reflected. The plate is at 227°C and has an emissive power of 1200 W/m^2. Air at 127°C flows over the plate with a heat transfer convection coefficient of 15 W/m$^2 \cdot$K. Determine the emissivity, absorptivity, and radiosity of the plate. What is the net heat transfer rate per unit area?

12.46 A horizontal, opaque surface at a steady-state temperature of 77°C is exposed to an airflow having a freestream temperature of 27°C with a convection heat transfer coefficient of 28 W/m$^2 \cdot$K. The emissive power of the surface is 628 W/m^2, the irradiation is 1380 W/m^2, and the reflectivity is 0.40. Determine the absorptivity of the surface. Determine the net radiation heat transfer rate for this surface. Is this heat transfer to the surface or from the surface? Determine the combined heat transfer rate for the surface. Is this heat transfer to the surface or from the surface?

12.47 Consider small cylindrical shapes fabricated from materials of different composition and surface finishes as described below. These cylinders, initially at 300 K, are inserted into a large furnace at 600 K. Using tabular values for the emissivity of the materials, indicate which cylinder you expect to heat up faster. Briefly explain your reasoning.

(a) Aluminum: (A) highly polished, film or (B) anodized.

(b) Stainless steel: (A) typical, polished or (B) stably oxidized.

12.48 An apparatus commonly used for measuring the reflectivity of materials is shown. A water-cooled sample, of 30 mm diameter and temperature $T_s = 300$ K, is mounted flush with the inner surface of a large enclosure. The walls of the enclosure are gray and diffuse with an emissivity of 0.8 and a uniform temperature $T_f = 1000$ K. A small aperture is located at the bottom of the enclosure to permit sighting of the sample or the enclosure wall.

The spectral reflectivity ρ_λ of an opaque, diffuse sample material is as shown. The heat transfer coefficient for convection between the sample and the air within the cavity, which is also at 1000 K, is $h = 10$ W/m$^2 \cdot$K.

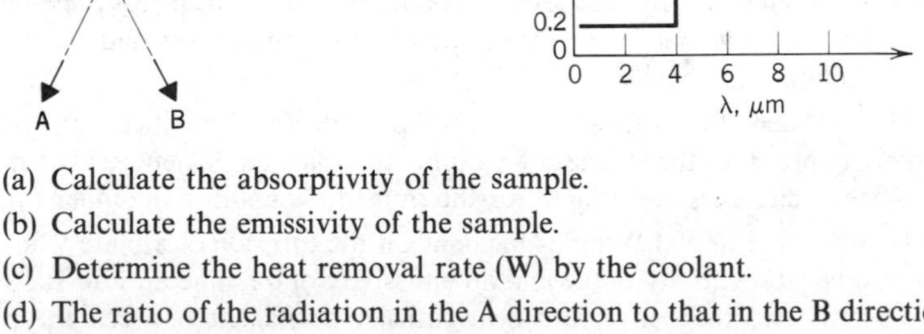

(a) Calculate the absorptivity of the sample.

(b) Calculate the emissivity of the sample.

(c) Determine the heat removal rate (W) by the coolant.

(d) The ratio of the radiation in the A direction to that in the B direction will give the reflectivity of the sample. Briefly explain why this is so.

12.49 The spectral absorptivity α_λ and spectral reflectivity ρ_λ for a spectrally selective, diffuse material are as shown.

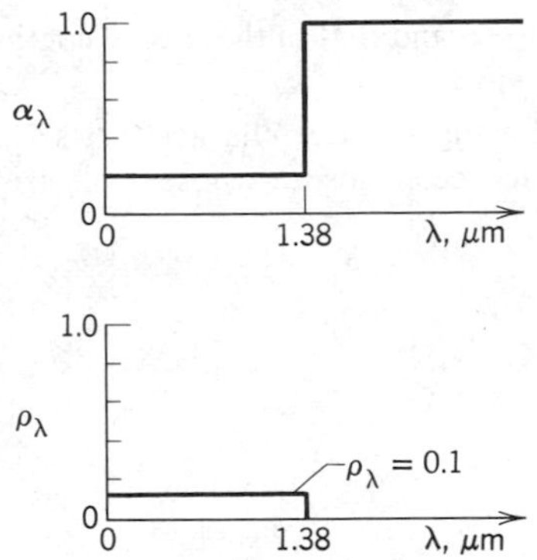

(a) Sketch the spectral transmissivity τ_λ

(b) If solar irradiation with $G_S = 750\ W/m^2$, having the spectral distribution of a blackbody at 5800 K, is incident on this material, determine the fractions of the irradiation that are transmitted, reflected, and absorbed by the material.

(c) If the temperature of this material is 350 K, determine the emissivity ε.

(d) Determine the net heat flux by radiation to the material.

12.50 The window of a large vacuum chamber is fabricated from a material having the spectral characteristics shown.

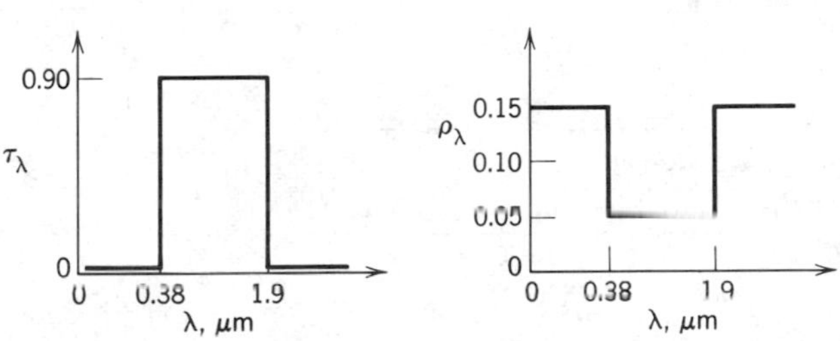

A collimated beam of radiant energy from a solar simulator is incident on the window and has a flux of 3000 W/m^2. The inside walls of the chamber, which are large compared to the window area, are maintained at 77 K. The outer surface of the window is subjected to surroundings and room air at 25°C, with a convection heat transfer coefficient of 15 W/m$^2 \cdot$ K.

(a) Determine the transmissivity of the window material to radiation from the solar simulator, which approximates the solar spectral distribution.

(b) Assuming that the window is insulated from its chamber mounting arrangement, what steady-state temperature does the window reach?

(c) Calculate the net radiation transfer per unit area of the window to the vacuum chamber wall, excluding the transmitted simulated solar flux.

12.51 Two plates, one with a black painted surface and the other with a special coating (chemically oxidized copper) are in earth orbit and are exposed to solar radiation. The solar rays make an angle of 30° with the normal to the plate. Estimate the equilibrium temperature of each plate assuming they are diffuse and that the solar flux is 1353 W/m^2. The spectral absorptivity of the black painted surface can be approximated by $\alpha_\lambda = 0.95$, $0 \leq \lambda \leq \infty$ and that of the special coating by $\alpha_\lambda = 0.95$, $0 \leq \lambda < 3\ \mu m$ and $\alpha_\lambda = 0.05$, $\lambda \geq 3\ \mu m$.

12.52 A diffuse surface having the following spectral characteristics is maintained at 500 K when situated in a large furnace enclosure whose walls are maintained at 1500 K.

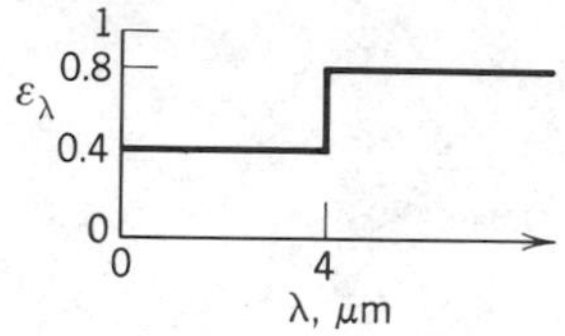

(a) Sketch the spectral distribution of the surface emissive power E_λ and the emissive power $E_{\lambda,b}$ that the surface would have if it were a blackbody.

(b) Neglecting convection effects, what is the net heat flux to the surface for the prescribed conditions?

12.53 The spectral, hemispherical emissivity distributions for two diffuse panels to be used in a spacecraft are as shown.

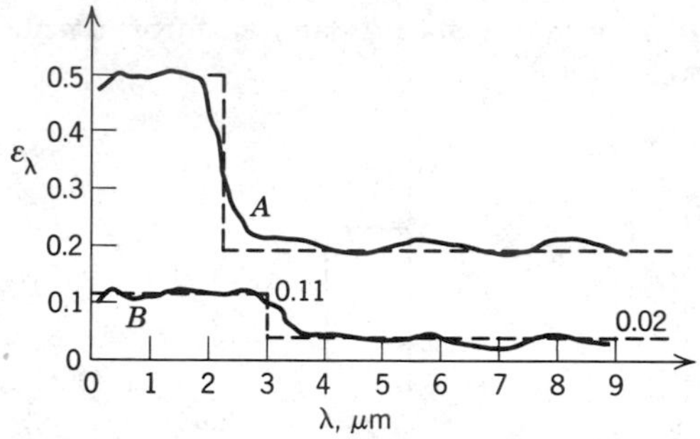

Assuming that the back sides of the panels are insulated and that the panels are oriented normal to the solar flux at 1353 W/m^2, determine which panel has the higher steady-state temperature.

12.54 Consider an opaque, diffuse surface whose spectral reflectivity varies with wavelength as shown. The surface is at 750 K, and irradiation on one side varies with wavelength as shown. The other side of the surface is insulated.

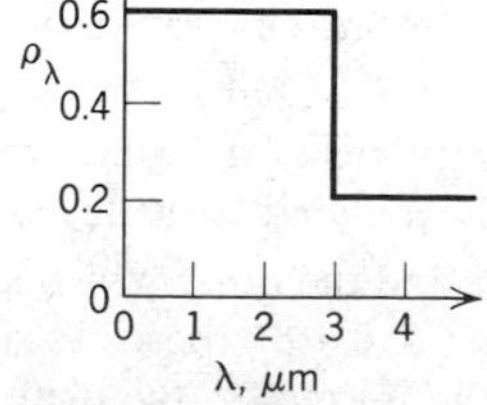

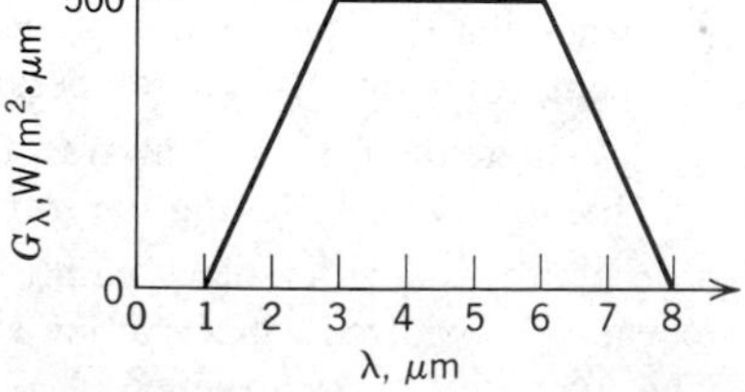

What are the total absorptivity and emissivity of the surface? What is the net radiative heat flux to the surface?

12.55 A very small sample of an opaque surface is initially at 1200 K and has the spectral, hemispherical absorptivity shown.

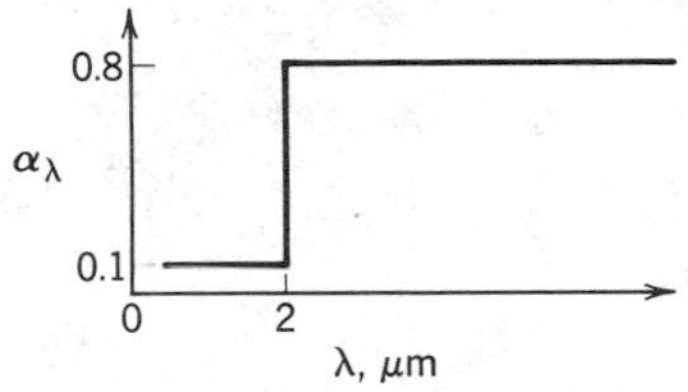

The sample is placed inside a very large enclosure whose walls have an emissivity of 0.2 and are maintained at 2400 K. Determine the total, hemispherical absorptivity and emissivity of the sample surface. What are the values of the absorptivity and emissivity after the sample has been in the enclosure a long time?

12.56 An opaque, gray surface at 27°C is exposed to irradiation of 1000 W/m^2, and 800 W/m^2 is reflected. Air at 17°C flows over the surface and the heat transfer convection coefficient is 15 W/m^2·K. Determine the net heat flux from the surface.

12.57 A thermocouple whose surface is diffuse and gray with an emissivity of 0.6 indicates a temperature of 180°C when used to measure the temperature of a gas flowing through a large duct whose walls have an emissivity of 0.85 and a uniform temperature of 450°C. If the convection heat transfer coefficient between the thermocouple and the gas stream is 125 W/m^2·K and there are negligible conduction losses from the thermocouple, determine the temperature of the gas.

12.58 A thermocouple inserted in a 4-mm-diameter stainless steel tube having a diffuse, gray surface with an emissivity of 0.4 is positioned horizontally in a large air-conditioned room whose walls and air temperature are 30 and 20°C, respectively. What temperature will the thermocouple indicate if the air is quiescent?

12.59 A horizontal, uninsulated pipe of 125 mm diameter passes through a large room having walls and quiescent air at 37 and 25°C, respectively. The pipe is maintained at 125°C; its outer surface is diffuse and gray with an emissivity of 0.8. Calculate the heat loss per unit length from the pipe by convection and radiation.

12.60 A sphere of 30 mm diameter whose surface is diffuse and gray with emissivity of 0.8 is placed in a large oven whose walls are of uniform temperature at 600 K. The temperature of the air in the oven is 400 K, and the convection heat transfer coefficient between the sphere and the oven air is 15 W/m^2·K. Determine the net heat transfer rate to the sphere when its temperature is 300 K.

12.61 A procedure for measuring the thermal conductivity of solids at elevated temperatures involves placement of a sample at the bottom of a large furnace. The sample is of thickness L and is placed in a square container of width W on a side. The sides are well insulated. (See figure on next page). The walls of the cavity are maintained at T_w, while the bottom surface of the sample is maintained at a much lower temperature T_c by circulating coolant through the sample container. The sample surface is diffuse and gray with an emissivity ε_s. Its temperature T_s is measured optically.

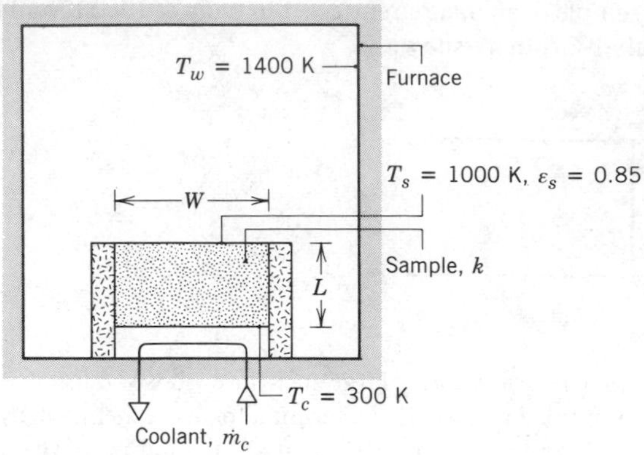

(a) Neglecting convection effects, obtain an expression from which the sample thermal conductivity may be evaluated in terms of measured and known quantities ($T_w, T_s, T_c, \varepsilon_s, L$). The measurements are made under steady-state conditions. If $T_w = 1400$ K, $T_s = 1000$ K, $\varepsilon_s = 0.85$, $L = 0.015$ m, and $T_c = 300$ K, what is the sample thermal conductivity?

(b) If $W = 0.10$ m and the coolant is water with a flow rate of $\dot{m}_c = 0.1$ kg/s, is it reasonable to assume a uniform bottom surface temperature T_c?

12.62 A thermograph is a device responding to the radiant power from the scene, which reaches its radiation detector within the spectral region 9–12 μm. The thermograph provides an image of the scene, such as the side of a furnace, from which the surface temperature can be determined.

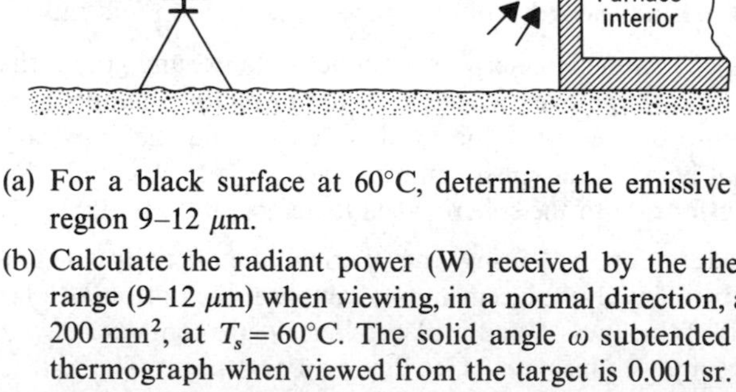

(a) For a black surface at 60°C, determine the emissive power for the spectral region 9–12 μm.

(b) Calculate the radiant power (W) received by the thermograph in the same range (9–12 μm) when viewing, in a normal direction, a small black wall area, 200 mm^2, at $T_s = 60$°C. The solid angle ω subtended by the aperture of the thermograph when viewed from the target is 0.001 sr.

(c) Determine the radiant power (W) received by the thermograph for the same wall area (200 mm^2) and solid angle (0.001 sr) when the wall is a gray, opaque, diffuse material at $T_s = 60$°C with emissivity 0.7 and the surroundings are black at $T_{sur} = 23$°C.

12.63 A small disk 5 mm in diameter is positioned at the center of an isothermal, hemispherical enclosure. The disk is diffuse and gray with an emissivity of 0.7 and is maintained at 900 K. The hemispherical enclosure, maintained at 300 K, has a radius of 100 mm and an emissivity of 0.85.

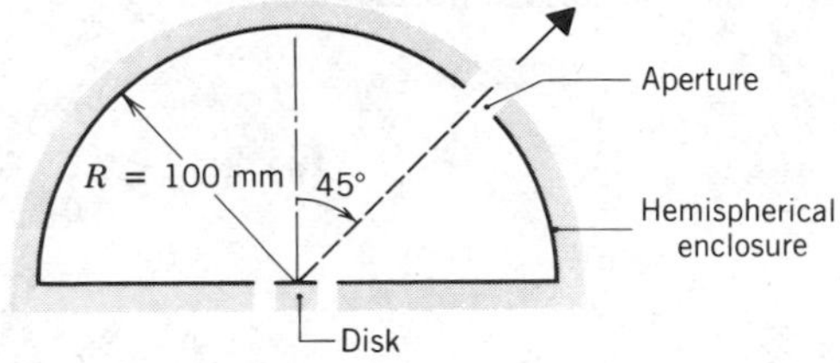

Calculate the radiant power leaving an aperture of diameter 2 mm located on the enclosure as shown.

12.64 A radiation thermometer is a radiometer calibrated to indicate the temperature of a blackbody. A steel billet having a diffuse, gray surface of emissivity 0.8 is heated in a furnace whose walls are at 1500 K. Estimate the temperature of the billet when the radiation thermometer viewing the billet through a small hole in the furnace indicates 1160 K.

12.65 The directional absorptivity of a gray surface varies with θ as follows.

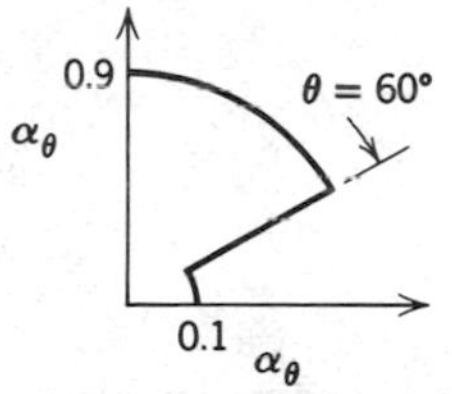

(a) What is the ratio of the normal absorptivity α_n to the hemispherical emissivity of the surface?

(b) Consider a plate with these surface characteristics on both sides in earth orbit. If the solar flux incident on one side of the plate is $q''_S = 1353$ W/m^2, what equilibrium temperature will the plate assume if it is oriented normal to the sun's rays? What temperature will it assume if it is oriented at 75° to the sun's rays?

12.66 Four diffuse surfaces having the spectral characteristics shown are at 300 K and are exposed to solar radiation.

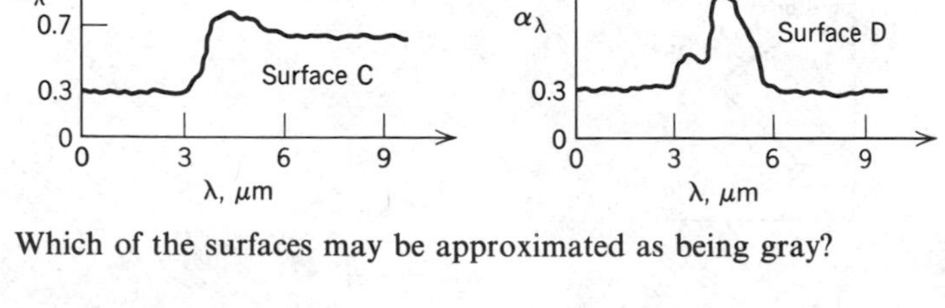

Which of the surfaces may be approximated as being gray?

12.67 A radiation detector has an aperture of area $A_d = 10^{-6}\ m^2$ and is positioned at a distance of $r = 1$ m from a surface of area $A_s = 10^{-4}\ m^2$. The angle formed by the normal to the detector and the surface normal is $\theta = 30°$.

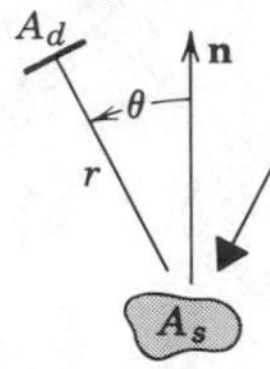

The surface is at 500 K, is opaque, and is diffuse/gray with an emissivity of 0.7. If the surface irradiation is 1500 W/m^2, what is the rate at which the detector intercepts radiation from the surface?

12.68 Two special coatings are available for application to an absorber plate installed below the cover glass described in Example 12.8. Each coating is diffuse and is characterized by the spectral distributions shown.

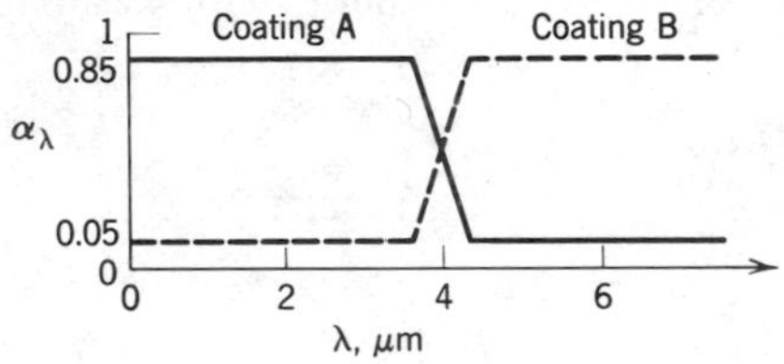

Which coating would you select for the absorber plate? Explain briefly. For the selected coating, what is the rate at which radiation is absorbed per unit area of the absorber plate if the total solar irradiation at the cover glass is $G_S = 1000$ W/m^2?

12.69 A small anodized aluminum block at 35°C is heated in a large oven whose walls are diffuse and gray with $\varepsilon = 0.85$ and maintained at a uniform temperature 175°C. The anodized coating is also diffuse and gray with $\varepsilon = 0.92$. A radiation detector views the block through a small opening in the oven and receives the radiant energy from a small area, referred to as the target, A_t, on the block. The target has a diameter of 3 mm, and the detector receives radiation within a solid angle 0.001 sr centered about the normal from the block.

(a) If the radiation detector views a small, but deep, hole drilled into the block, what is the total power (W) received by the detector?

(b) If the radiation detector now views an area on the block surface, what is the total power (W) received by the detector?

12.70 A flat plate whose surface is gray but has the total, directional emissivity ε_θ shown, is in earth orbit, where the solar flux is 1353 W/m^2.

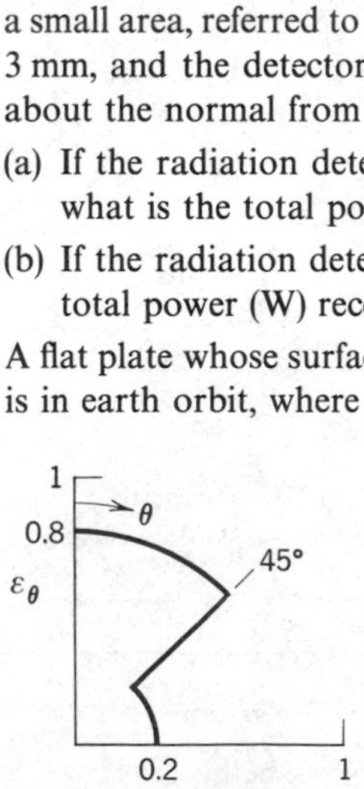

Determine the equilibrium temperatures for the plate when it is oriented normal to the solar flux and at 60° to the solar flux.

12.71 Consider an opaque, gray surface whose directional absorptivity is 0.8 for $0° \leq \theta \leq 60°$ and 0.1 for $\theta > 60°$. The surface is horizontal and exposed to solar irradiation comprised of direct and diffuse components.

(a) What is the surface absorptivity to direct solar radiation that is incident at an angle of 45° from the normal? What is the absorptivity to diffuse irradiation?

(b) Neglecting convection heat transfer between the surface and the surrounding air, what would be the equilibrium temperature of the surface if the direct and diffuse components of the irradiation were 600 and 100 W/m^2, respectively? The back side of the surface is insulated.

12.72 A contractor must select a roof covering material from the two diffuse, opaque coatings with $\alpha_\lambda(\lambda)$ as shown. Which of the two coatings would result in a lower roof temperature? Which is preferred for summer use? For winter use? Sketch the spectral distribution of α_λ that would be ideal for summer use. For winter use.

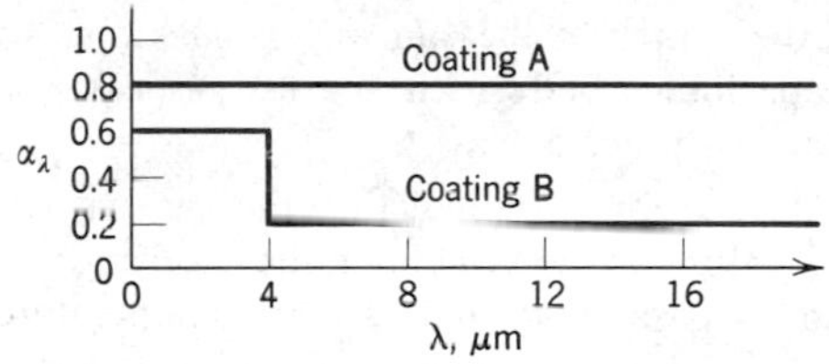

12.73 A radiator on a proposed satellite solar power station must dissipate heat being generated within the satellite by radiating it into space. The radiator surface has a solar absorptivity of 0.5 and an emissivity of 0.95. What is the equilibrium surface temperature when the solar irradiation is 1000 W/m^2 and the required heat dissipation is 1500 W/m^2?

12.74 A diffuse, gray plate of emissivity 0.6 insulated from the ground is exposed to solar irradiation of 1000 W/m^2. The effective sky temperature is $-40°C$ and the heat transfer coefficient between the plate and the ambient air at 25°C is 6 $W/m^2 \cdot K$. The temperature of the plate is 65°C at this particular time.

(a) What is the radiosity of the plate?

(b) What is the net heat flux between the plate and the environment?

(c) In what manner will the temperature of the plate change with respect to time?

12.75 It is not uncommon for the night sky temperature in desert regions to drop to $-40°C$. If the ambient air temperature is 20°C and the convection coefficient for still air conditions is approximately 5 $W/m^2 \cdot K$, can a shallow pan of water freeze?

12.76 A spherical satellite of diameter D is in orbit about the earth and is coated with a diffuse material for which the spectral absorptivity is $\alpha_\lambda = 0.6$ for $\lambda \leq 3$ μm and $\alpha_\lambda = 0.3$ for $\lambda > 3$ μm. When it is on the "dark" side of the earth, the satellite sees irradiation from the earth's surface only. The irradiation may be assumed to be incident as parallel rays, and its magnitude is $G_E = 340$ W/m^2. On the "bright" side of the earth the satellite sees the earth irradiation G_E plus the solar irradiation $G_S = 1353$ W/m^2. The spectral distribution of radiation from the earth may be approximated as that of a blackbody at 280 K, and the temperature of the satellite may be assumed to remain below 500 K.

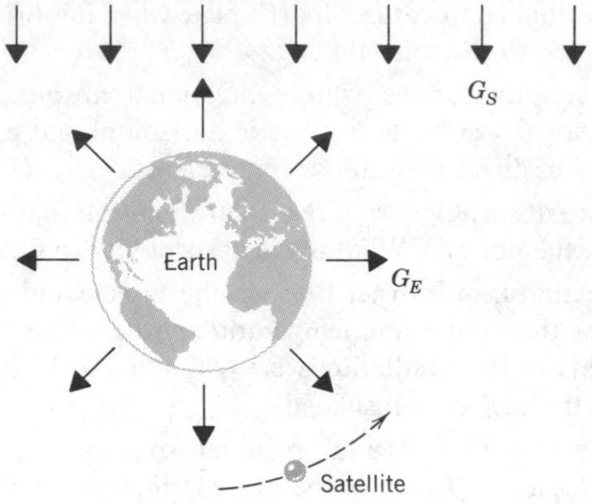

What is the steady-state temperature of the satellite when it is on the dark side of the earth and when it is on the bright side?

12.77 A spherical capsule of 3 m radius is fired from a space platform in earth orbit, such that it travels toward the center of the sun at 16,000 km/s. Assume that the capsule is a lumped capacitance body with a density–specific heat product of 4×10^6 J/m$^3 \cdot$K and that its surface is black.

(a) Derive a differential equation for predicting the capsule temperature as a function of time. Solve this equation to obtain the temperature as a function of time in terms of capsule parameters and its initial temperature T_i.

(b) If the capsule begins its journey at 20°C, predict the position of the capsule relative to the sun at which its destruction temperature, 500°C, is reached.

12.78 A thin sheet of glass is used on the roof of a greenhouse and is irradiated as shown.

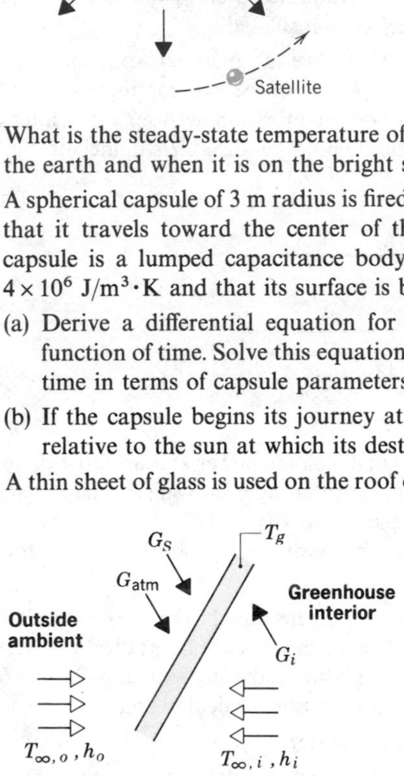

The irradiation comprises the total solar flux G_S, the flux G_{atm} due to atmospheric emission (sky radiation), and the flux G_i due to emission from interior surfaces. The fluxes G_{atm} and G_i are concentrated in the far IR region ($\lambda \gtrsim 8\ \mu$m). The glass may also exchange energy by convection with the outside and inside atmospheres. The glass may be assumed to be totally transparent for $\lambda < 1\ \mu$m ($\tau_\lambda = 1.0$ for $\lambda < 1\ \mu$m) and opaque, with $\alpha_\lambda = 1.0$ for $\lambda \geq 1\ \mu$m.

(a) Assuming steady-state conditions, with all radiative fluxes uniformly distributed over the surfaces and the glass characterized by a uniform temperature T_g, write an appropriate energy balance for a unit area of the glass.

(b) For the following conditions, calculate the temperature of the greenhouse ambient air $T_{\infty,i}$.

$T_g = 27°$C $\qquad h_i = 10$ W/m$^2 \cdot$K $\qquad G_S = 1100$ W/m^2

$T_{\infty,o} = 24°$C $\qquad h_o = 55$ W/m$^2 \cdot$K $\qquad G_{atm} = 250$ W/m^2

$G_i = 440$ W/m^2

Thirteen

RADIATION EXCHANGE BETWEEN SURFACES

Having thus far restricted our attention to radiative processes that occur at a *single surface*, we now consider the problem of radiative exchange between two or more surfaces. This exchange depends strongly on the surface geometries and orientations, as well as on their radiative properties and temperatures. We assume that the surfaces are separated by a *nonparticipating medium.* Since such a medium neither emits, absorbs, nor scatters, it has no effect on the transfer of radiation between surfaces. A vacuum meets these requirements exactly, and most gases meet them to an excellent approximation.

We initially focus on the geometrical features of the radiation exchange problem by developing the notion of a *view factor.* We then consider radiation exchange between *black surfaces* and follow with a treatment of exchange between *diffuse, gray surfaces.* We also consider the problem of radiation transfer in an *enclosure*, a term that describes the region enveloped by a collection of surfaces.

13.1 THE VIEW FACTOR

To compute radiation exchange between any two surfaces, we must first introduce the concept of a *view factor* (also called a *configuration* or *shape factor*).

13.1.1 The View Factor Integral

The view factor F_{ij} is defined as the *fraction of the radiation leaving surface i, which is intercepted by surface j.* To develop a general expression for F_{ij}, we consider the arbitrarily oriented surfaces A_i and A_j of Figure 13.1. Elemental areas on each surface, dA_i and dA_j, are connected by a line of length R, which forms the polar angles θ_i and θ_j, respectively, with the surface normals $\mathbf{n}_i$ and $\mathbf{n}_j$.

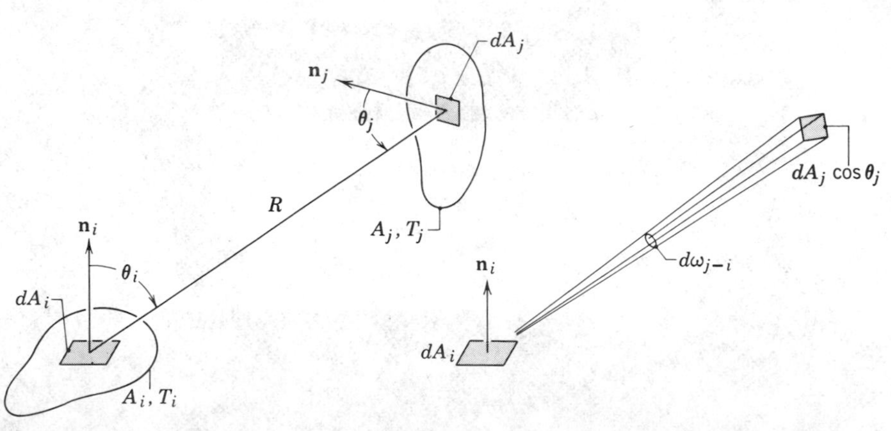

Figure 13.1 View factor associated with radiation exchange between elemental surfaces of area dA_i and dA_j.

The values of R, θ_i, and θ_j vary with the position of the elemental areas on A_i and A_j.

From the definition of the radiation intensity, Section 12.2.1, and Equation 12.5, the rate at which radiation *leaves* dA_i and is *intercepted* by dA_j may be expressed as

$$dq_{i \to j} = I_i \cos\theta_i \, dA_i \, d\omega_{j-i}$$

where I_i is the intensity of the radiation leaving surface i and $d\omega_{j-i}$ is the solid angle subtended by dA_j when viewed from dA_i. With $d\omega_{j-i} = (\cos\theta_j \, dA_j)/R^2$ from Equation 12.2, it follows that

$$dq_{i \to j} = I_i \frac{\cos\theta_i \cos\theta_j}{R^2} \, dA_i \, dA_j$$

Assuming that surface i *emits* and *reflects diffusely* and substituting from Equation 12.24, we then obtain

$$dq_{i \to j} = J_i \frac{\cos\theta_i \cos\theta_j}{\pi R^2} \, dA_i \, dA_j$$

The total rate at which radiation leaves surface i and is intercepted by j may then be obtained by integrating over the two surfaces. That is,

$$q_{i \to j} = J_i \int_{A_i} \int_{A_j} \frac{\cos\theta_i \cos\theta_j}{\pi R^2} \, dA_i \, dA_j$$

where it is assumed that the radiosity J_i is uniform over the surface A_i. From the definition of the view factor as the fraction of the radiation that leaves A_i and is intercepted by A_j,

$$F_{ij} = \frac{q_{i \to j}}{A_i J_i}$$

it follows that

$$F_{ij} = \frac{1}{A_i} \int_{A_i} \int_{A_j} \frac{\cos\theta_i \cos\theta_j}{\pi R^2} \, dA_i \, dA_j \qquad (13.1)$$

Similarly, the view factor F_{ji} is defined as the fraction of the radiation that leaves A_j and is intercepted by A_i. The same development then yields

$$F_{ji} = \frac{1}{A_j} \int_{A_i} \int_{A_j} \frac{\cos\theta_i \cos\theta_j}{\pi R^2} \, dA_i \, dA_j \qquad (13.2)$$

Either Equation 13.1 or 13.2 may be used to determine the view factor associated with any two surfaces that are *diffuse emitters* and *reflectors* and have *uniform radiosity*.

13.1.2 View Factor Relations

An important view factor relation is suggested by Equations 13.1 and 13.2. In particular, equating the integrals appearing in these equations, it follows that

$$A_i F_{ij} = A_j F_{ji} \tag{13.3}$$

This expression, termed the *reciprocity relation,* is useful in determining one view factor from knowledge of the other.

Another important view factor relation pertains to the surfaces of an *enclosure* (Figure 13.2). From the definition of the view factor, the *summation rule*

$$\sum_{j=1}^{N} F_{ij} = 1 \qquad \text{Enclosure} \tag{13.4}$$

may be applied to each of the N surfaces in the enclosure. This rule follows from the conservation requirement that all the radiation leaving surface i be intercepted by the enclosure surfaces. The term F_{ii} appearing in this summation represents the fraction of the radiation that leaves surface i and is directly intercepted by i. If the surface is concave, it *sees itself* and F_{ii} is nonzero. However, for a plane or convex surface, $F_{ii} = 0$.

To calculate radiation exchange in an enclosure of N surfaces, a total of N^2 view factors is needed. This requirement becomes evident when the view factors are arranged in the matrix form.

$$\begin{bmatrix} F_{11} & F_{12} \cdots F_{1N} \\ F_{21} & F_{22} \cdots F_{2N} \\ \vdots & \vdots \qquad \vdots \\ F_{N1} & F_{N2} \quad F_{NN} \end{bmatrix}$$

However, all the view factors need not be calculated *directly.* A total of N view factors may be obtained from the N equations associated with application of the summation rule, Equation 13.4, to each of the surfaces in the enclosure. In addition, $N(N-1)/2$ view factors may be obtained from the $N(N-1)/2$ applications of the reciprocity relation, Equation 13.3, which are possible for the enclosure. Accordingly, only $[N^2 - N - N(N-1)/2] = N(N-1)/2$ view factors

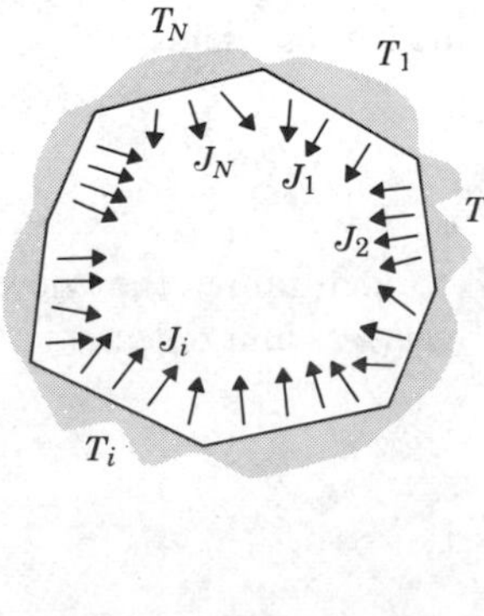

Figure 13.2 Radiation exchange in an enclosure.

need be determined directly. For example, in a three-surface enclosure this requirement corresponds to only $3(3-1)/2 = 3$ view factors. The remaining six view factors may be obtained by solving the six equations that result from use of Equations 13.3 and 13.4.

To illustrate the foregoing procedure, consider a simple, two-surface enclosure involving the spherical surfaces of Figure 13.3. Although the enclosure is characterized by $N^2 = 4$ view factors ($F_{11}, F_{12}, F_{21}, F_{22}$), only $N(N-1)/2 = 1$ view factor need be determined directly. In this case such a determination may be made by *inspection*. In particular, since all radiation leaving the inner surface must reach the outer surface, it follows that $F_{12} = 1$. The same may not be said of radiation leaving the outer surface, since this surface sees itself. However, from the reciprocity relation, Equation 13.3, we obtain

$$F_{21} = \left(\frac{A_1}{A_2}\right) F_{12} = \left(\frac{A_1}{A_2}\right)$$

From the summation rule, we also obtain

$$F_{11} + F_{12} = 1$$

in which case $F_{11} = 0$, and

$$F_{21} + F_{22} = 1$$

in which case

$$F_{22} = 1 - \left(\frac{A_1}{A_2}\right)$$

For more complicated geometries, the view factor must be determined by solving the double integral of Equation 13.1. Such solutions have been obtained for many different surface arrangements and are available in equation, graphical, or tabular form [1–4]. Results for several common geometries are presented in Tables 13.1 and 13.2 and Figures 13.4 to 13.6. The configurations of Table 13.1 are assumed to be infinitely long (in a direction perpendicular to the page) and are hence two-dimensional. The configurations of Table 13.2 and Figures 13.4 to 13.6 are three-dimensional.

It is useful to note that the results of Figures 13.4 to 13.6 may be used to determine other view factors. For example, the view factor for an end surface of a

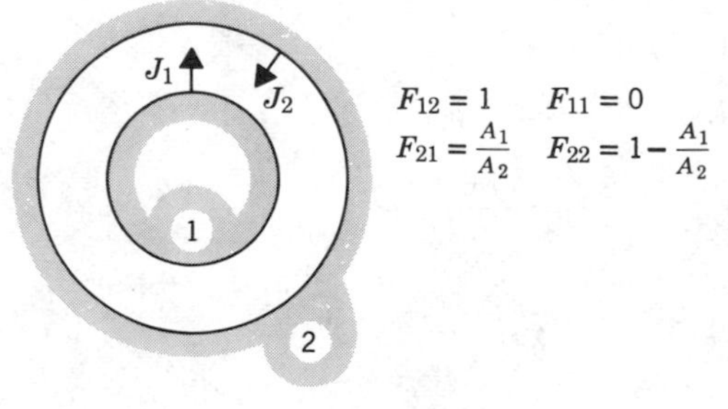

Figure 13.3 View factors for the enclosure formed by two spheres.

Table 13.1 View factors for two-dimensional geometries [4]

GEOMETRY	RELATION

Parallel Plates with Midlines Connected by Perpendicular

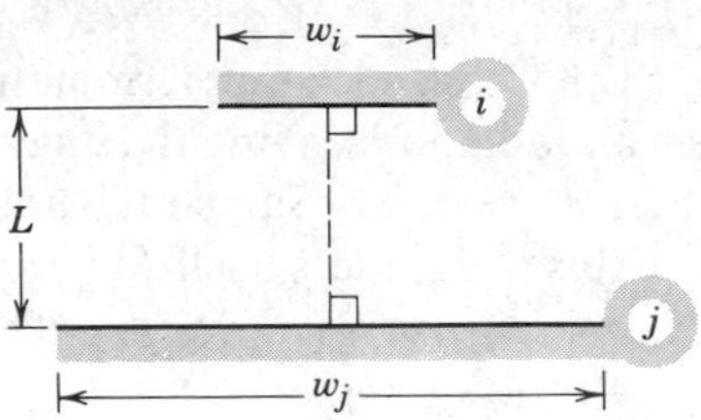

$$F_{ij} = \frac{[(W_i + W_j)^2 + 4]^{1/2} - [(W_j - W_i)^2 + 4]^{1/2}}{2W_i}$$

$$W_i = w_i/L,\ W_j = w_j/L$$

Inclined Parallel Plates of Equal Width and a Common Edge

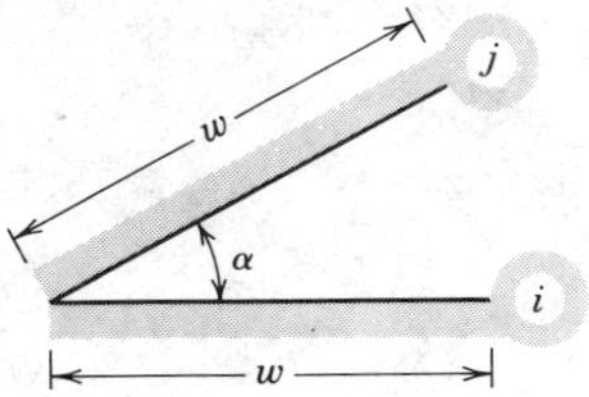

$$F_{ij} = 1 - \sin\left(\frac{\alpha}{2}\right)$$

Perpendicular Plates with a Common Edge

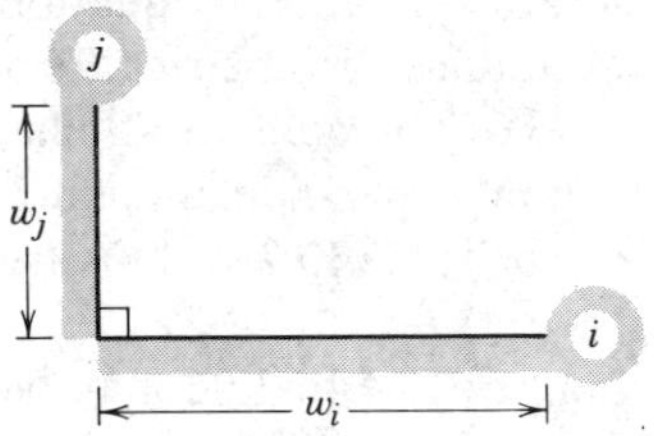

$$F_{ij} = \frac{1 + (w_j/w_i) - [1 + (w_j/w_i)^2]^{1/2}}{2}$$

Three-Sided Enclosure

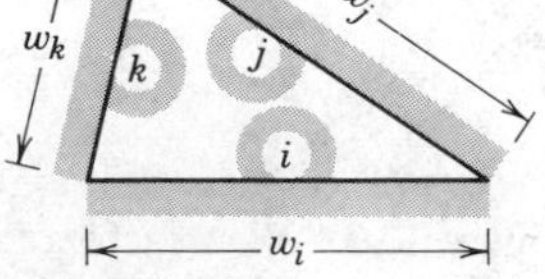

$$F_{ij} = \frac{w_i + w_j - w_k}{2w_i}$$

Table 13.1 Continued

GEOMETRY	RELATION

Parallel Cylinders of Different Radius

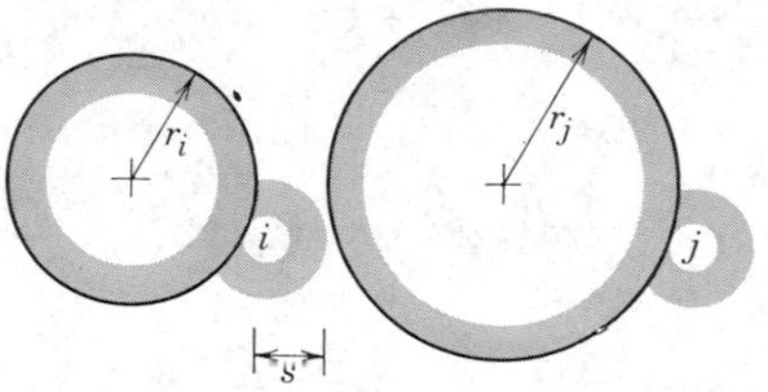

$$F_{ij} = \frac{1}{2\pi}\left\{\pi + [C^2 - (R+1)^2]^{1/2} - [C^2 - (R-1)^2]^{1/2} + (R-1)\cos^{-1}\left[\left(\frac{R}{C}\right) - \left(\frac{1}{C}\right)\right] - (R+1)\cos^{-1}\left[\left(\frac{R}{C}\right) + \left(\frac{1}{C}\right)\right]\right\}$$

$R = r_j/r_i,\ S = s/r_i$

$C = 1 + R + S$

Cylinder and Parallel Rectangle

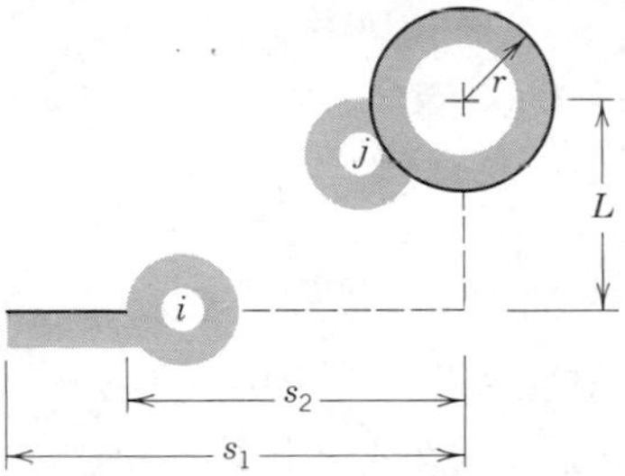

$$F_{ij} = \frac{r}{s_1 - s_2}\left[\tan^{-1}\frac{s_1}{L} - \tan^{-1}\frac{s_2}{L}\right]$$

Infinite Plane and Row of Cylinders

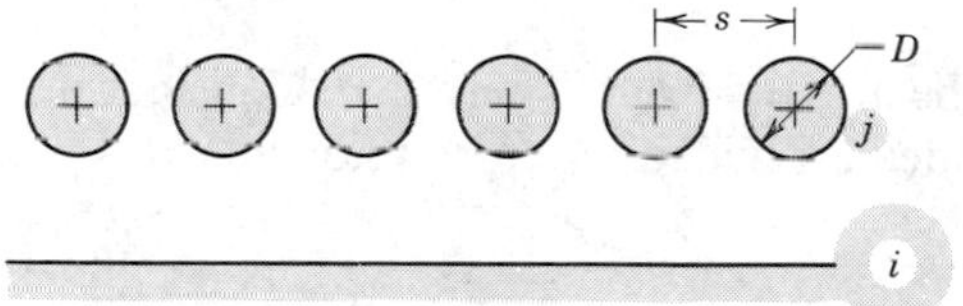

$$F_{ij} = 1 - \left[1 - \left(\frac{D}{s}\right)^2\right]^{1/2} + \left(\frac{D}{s}\right)\tan^{-1}\left(\frac{s^2 - D^2}{D^2}\right)^{1/2}$$

cylinder (or a truncated cone) relative to the lateral surface may be obtained by using the results of Figure 13.5 with the summation rule, Equation 13.4. Moreover, Figures 13.4 and 13.6 may be used to obtain other useful results if two additional view factor relations are developed.

The first relation concerns the additive nature of the view factor for a

Table 13.2 View factors for three-dimensional geometries [4]

GEOMETRY	RELATION
Aligned Parallel Rectangles (Figure 13.4) 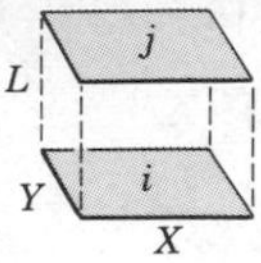	$\bar{X} = X/L, \bar{Y} = Y/L$ $F_{ij} = \frac{2}{\pi \bar{X}\bar{Y}} \left\{ \ln\left[\frac{(1+\bar{X}^2)(1+\bar{Y}^2)}{1+\bar{X}^2+\bar{Y}^2}\right]^{1/2} + \bar{X}(1+\bar{Y}^2)^{1/2} \tan^{-1}\frac{\bar{X}}{(1+\bar{Y}^2)^{1/2}} + \bar{Y}(1+\bar{X}^2)^{1/2} \tan^{-1}\frac{\bar{Y}}{(1+\bar{X}^2)^{1/2}} - \bar{X}\tan^{-1}\bar{X} - \bar{Y}\tan^{-1}\bar{Y} \right\}$
Coaxial Parallel Disks (Figure 13.5) 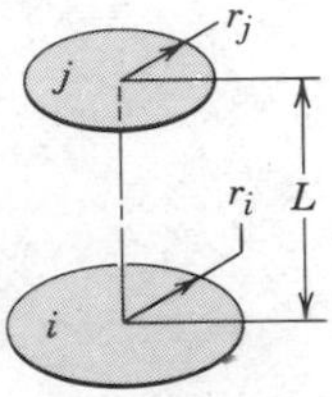	$R_i = r_i/L, R_j = r_j/L$ $S = 1 + \frac{1+R_j^2}{R_i^2}$ $F_{ij} = \frac{1}{2}\{S - [S^2 - 4(r_j/r_i)^2]^{1/2}\}$
Perpendicular Rectangles with a Common Edge (Figure 13.6) 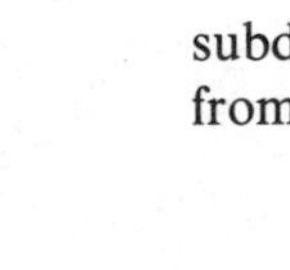	$H = Z/X, W = Y/X$ $F_{ij} = \frac{1}{\pi W}\left(W\tan^{-1}\frac{1}{W} + H\tan^{-1}\frac{1}{H} - (H^2+W^2)^{1/2}\tan^{-1}\frac{1}{(H^2+W^2)^{1/2}} + \frac{1}{4}\ln\left\{\frac{(1+W^2)(1+H^2)}{1+W^2+H^2}\left[\frac{W^2(1+W^2+H^2)}{(1+W^2)(W^2+H^2)}\right]^{W^2} \times \left[\frac{H^2(1+H^2+W^2)}{(1+H^2)(H^2+W^2)}\right]^{H^2}\right\}\right)$

subdivided surface and may be inferred from Figure 13.7. Considering radiation from surface *i* to surface *j*, which is divided into *n* components, it is evident that

$$F_{i(j)} = \sum_{k=1}^{n} F_{ik} \tag{13.5}$$

where the parentheses around a subscript indicate that it is a composite surface, in which case (j) is equivalent to $(1, 2, \ldots, k, \ldots, n)$. This expression simply states that radiation reaching a composite surface is the sum of the radiation reaching its parts. Although it pertains to subdivision of the receiving surface, it may also be used to obtain the second view factor relation, which pertains to subdivision of the originating surface. Multiplying Equation 13.5 by A_i and applying the reciprocity relation, Equation 13.3, to each of the resulting terms, it follows that

$$A_j F_{(j)i} = \sum_{k=1}^{n} A_k F_{ki} \tag{13.6}$$

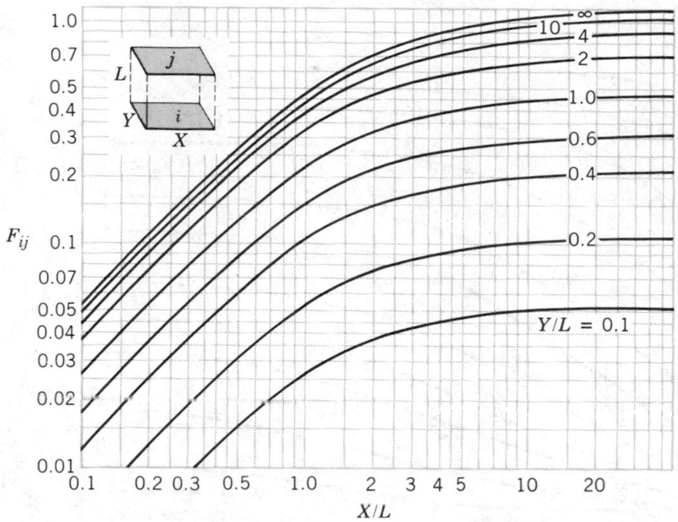

Figure 13.4 View factor for aligned parallel rectangles.

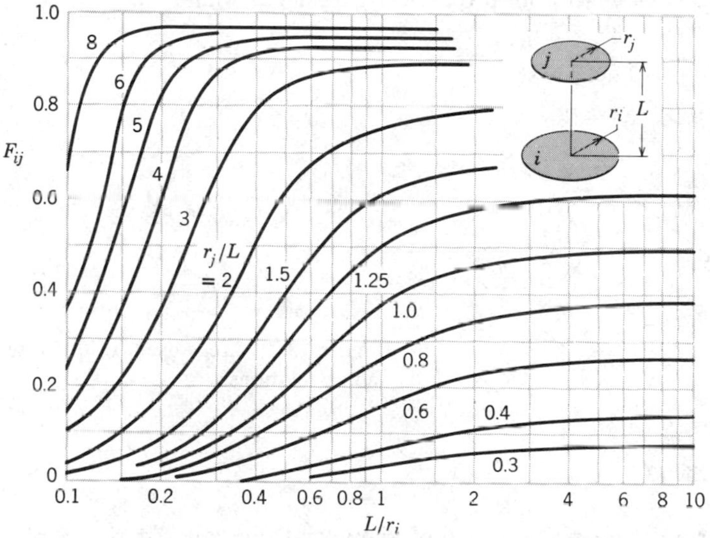

Figure 13.5 View factor for coaxial parallel disks.

or

$$F_{(j)i} = \frac{\sum_{k=1}^{n} A_k F_{ki}}{\sum_{k=1}^{n} A_k} \tag{13.7}$$

Equations 13.6 and 13.7 may be applied when the originating surface is composed of several parts.

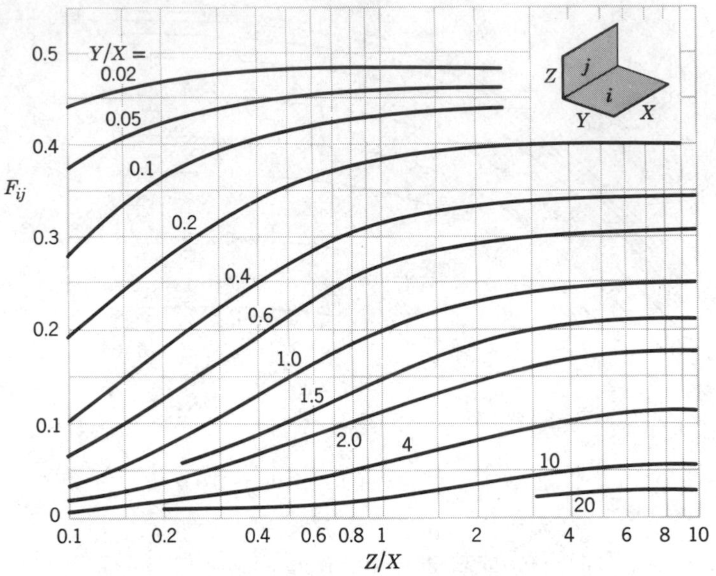

Figure 13.6 View factor for perpendicular rectangles with a common edge.

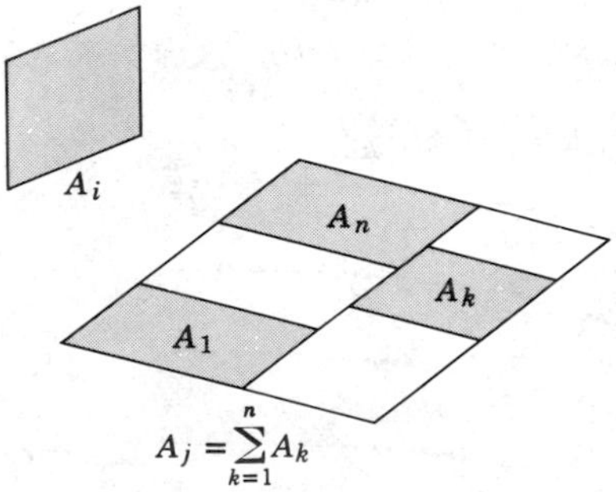

Figure 13.7 Areas used to illustrate view factor relations.

EXAMPLE 13.1

Consider a diffuse circular disk of diameter D and area A_j and a plane diffuse surface of area $A_i \ll A_j$. The surfaces are parallel, and A_i is located at a distance L from the center of A_j. Obtain an expression for the view factor F_{ij}.

SOLUTION

Known: Orientation of small surface relative to large circular disk.

Find: View factor of small surface with respect to disk, F_{ij}.

Schematic:

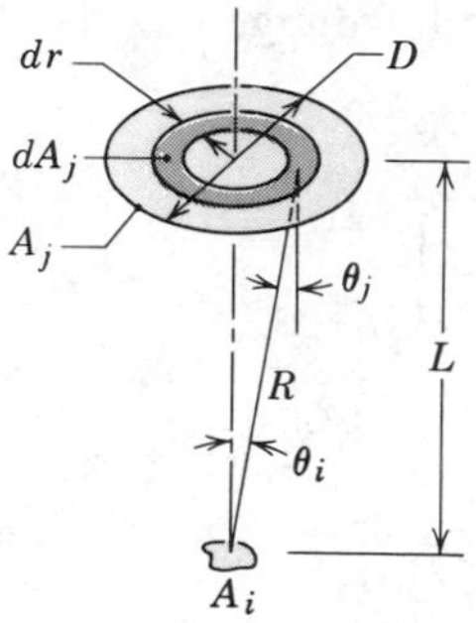

Assumptions:

1. Diffuse surfaces.
2. $A_i \ll A_j$.

Analysis:

The desired view factor may be obtained from Equation 13.1.

$$F_{ij} = \frac{1}{A_i} \int_{A_i} \int_{A_j} \frac{\cos \theta_i \cos \theta_j}{\pi R^2} \, dA_i \, dA_j$$

Recognizing that θ_i, θ_j, and R are approximately independent of position on A_i, this expression reduces to

$$F_{ij} = \int_{A_j} \frac{\cos \theta_i \cos \theta_j}{\pi R^2} \, dA_j$$

or, with $\theta_i = \theta_j \equiv \theta$,

$$F_{ij} = \int_{A_j} \frac{\cos^2 \theta}{\pi R^2} \, dA_j$$

With $R^2 = r^2 + L^2$, $\cos \theta = (L/R)$ and $dA_j = 2\pi r \, dr$, it follows that

$$F_{ij} = 2L^2 \int_0^{D/2} \frac{r \, dr}{(r^2 + L^2)^2} = \frac{D^2}{D^2 + 4L^2} \quad \triangleleft \qquad (13.8)$$

Comments:

The preceding geometry is one of the simplest cases for which the view factor may be obtained from Equation 13.1. Geometries involving more detailed integrations are considered in the literature [1, 3].

EXAMPLE 13.2

Determine the view factors F_{12} and F_{21} for the following geometries.

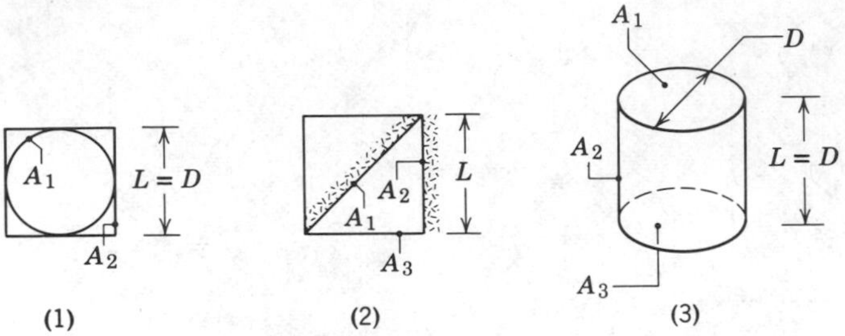

1. Sphere of diameter D inside a cubical box of length $L = D$.
2. Diagonal partition within a long square duct.
3. End and side of a circular tube of equal length and diameter.

SOLUTION

Known: Surface geometries.

Find: View factors.

Assumptions: Diffuse surfaces.

Analysis:

The desired view factors may be obtained from inspection, the reciprocity rule, the summation rule, and/or use of the charts.

1. Sphere within a cube:

   By inspection, $F_{12} = 1$ ◁

   By reciprocity, $F_{21} = \dfrac{A_1}{A_2} F_{12} = \dfrac{\pi D^2}{6L^2} \times 1 = \dfrac{\pi}{6}$ ◁

2. Partition within a square duct:

   From summation rule, $F_{11} + F_{12} + F_{13} = 1$

   where $F_{11} = 0$

   By symmetry, $F_{12} = F_{13}$

   Hence $F_{12} = 0.50$ ◁

   By reciprocity, $F_{21} = \dfrac{A_1}{A_2} F_{12} = \dfrac{\sqrt{2}\,L}{L} \times 0.5 = 0.71$ ◁

3. Circular tube:

From Figure 13.5, with $(r_3/L) = 0.5$ and $(L/r_1) = 2$, $F_{13} \approx 0.17$

From summation rule, $F_{11} + F_{12} + F_{13} = 1$

or, with $F_{11} = 0$, $F_{12} = 1 - F_{13} = 0.83$ ◁

From reciprocity, $F_{21} = \frac{A_1}{A_2} F_{12} = \frac{\pi D^2/4}{\pi D L} \times 0.83 = 0.21$ ◁

13.2 BLACKBODY RADIATION EXCHANGE

In general, radiation may leave a surface due to both reflection and emission, and on reaching a second surface, experience reflection as well as absorption. However, matters are simplified for surfaces that may be approximated as blackbodies, since there is no reflection. Hence energy only leaves as a result of emission, and all incident radiation is absorbed.

Consider radiation exchange between two black surfaces of arbitrary shape (Figure 13.8). Defining $q_{i \to j}$ as the rate at which radiation *leaves* surface i and is *intercepted* by surface j, it follows that

$$q_{i \to j} = (A_i J_i) F_{ij} \tag{13.9}$$

or, since radiosity equals emissive power for a black surface ($J_i = E_{bi}$),

$$q_{i \to j} = A_i F_{ij} E_{bi} \tag{13.10}$$

Similarly,

$$q_{j \to i} = A_j F_{ji} E_{bj} \tag{13.11}$$

The *net radiative exchange* between the two surfaces may then be defined as

$$q_{ij} = q_{i \to j} - q_{j \to i} \tag{13.12}$$

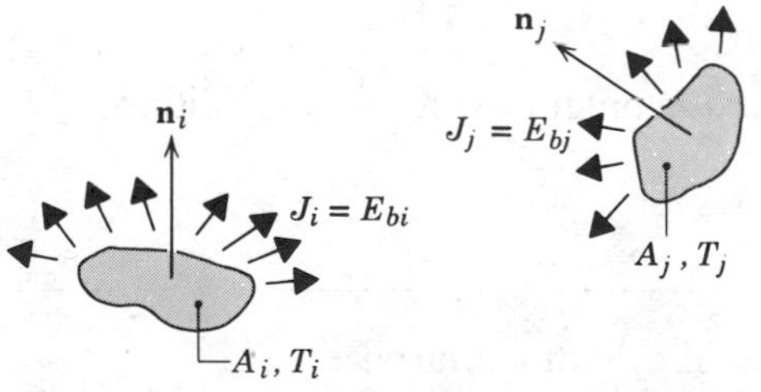

Figure 13.8 Radiation transfer between two surfaces that may be approximated as blackbodies.

from which it follows that

$$q_{ij} = A_i F_{ij} E_{bi} - A_j F_{ji} E_{bj}$$

or, from Equations 12.28 and 13.3

$$q_{ij} = A_i F_{ij} \sigma (T_i^4 - T_j^4) \tag{13.13}$$

Equation 13.13 provides the *net* rate at which radiation *leaves* surface i as a result of its interaction with j, which is equal to the *net* rate at which j *gains* radiation due to its interaction with i.

The foregoing result may also be used to evaluate the net radiation transfer from any surface in an *enclosure* of black surfaces. With N surfaces maintained at different temperatures, the net transfer of radiation from surface i is due to exchange with the remaining surfaces and may be expressed as

$$q_i = \sum_{j=1}^{N} A_i F_{ij} \sigma (T_i^4 - T_j^4) \tag{13.14}$$

EXAMPLE 13.3

A furnace cavity, which is in the form of a cylinder of 75 mm diameter and 150 mm length, is open at one end to surroundings that are at 27°C. The sides and bottom may be approximated as blackbodies, are heated electrically, are well insulated, and are maintained at temperatures of 1350 and 1650°C, respectively.

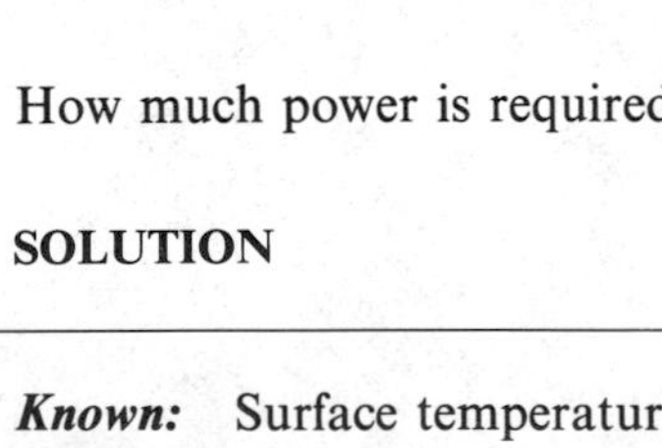

How much power is required to maintain the furnace conditions?

SOLUTION

Known: Surface temperatures of cylindrical furnace.

Find: Power required to maintain prescribed temperatures.

Schematic:

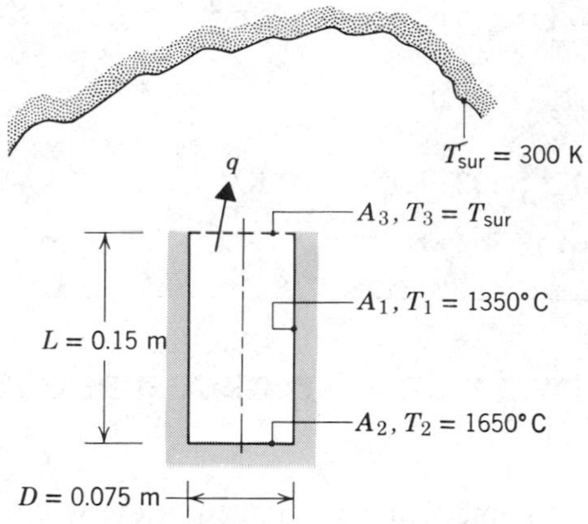

Assumptions:

1. Interior surfaces behave as blackbodies.
2. Heat transfer by convection is negligible.
3. Outer surface of furnace is adiabatic.

Analysis:

The power needed to operate the furnace at the prescribed conditions must balance heat losses from the furnace. Subject to the foregoing assumptions, the only heat loss is by radiation through the opening, which may be treated as a hypothetical surface of area A_3. Because the surroundings are large, radiation exchange between the furnace and the surroundings may be treated by approximating the surface as a blackbody at $T_3 = T_{sur}$. The heat loss may then be expressed as

$$q = q_{13} + q_{23}$$

or, from Equation 13.13,

$$q = A_1 F_{13} \sigma (T_1^4 - T_3^4) + A_2 F_{23} \sigma (T_2^4 - T_3^4)$$

From Figure 13.5, it follows that, with $(r_j/L) = (0.0375 \text{ m}/0.15 \text{ m}) = 0.25$ and $(L/r_i) = (0.15 \text{ m}/0.0375 \text{ m}) = 4$, $F_{23} = 0.06$. From the summation rule

$$F_{21} = 1 - F_{23} = 1 - 0.06 = 0.94$$

and from reciprocity

$$F_{12} = \frac{A_2}{A_1} F_{21} = \frac{\pi (0.075 \text{ m})^2/4}{\pi (0.075 \text{ m})(0.15 \text{ m})} \times 0.94 = 0.118$$

Hence, since $F_{13} = F_{12}$ from symmetry,

$$q = (\pi \times 0.075 \text{ m} \times 0.15 \text{ m})0.118 \times 5.67 \times 10^{-8} \text{ W/m}^2 \cdot \text{K}^4 \, [(1623 \text{ K})^4 - (300 \text{ K})^4] + \left(\frac{\pi}{4}\right)(0.075 \text{ m})^2 \times 0.06 \times 5.67 \times 10^{-8} \text{ W/m}^2 \cdot \text{K}^4 \, [(1923 \text{ K})^4 - (300 \text{ K})^4]$$

$$q = 1639 \text{ W} + 205 \text{ W} = 1844 \text{ W}$$ ◁

13.3 RADIATION EXCHANGE BETWEEN DIFFUSE, GRAY SURFACES IN AN ENCLOSURE

Although useful to a point, the foregoing results are limited by the assumption of blackbody behavior. The blackbody is, of course, an idealization, which, although closely approximated by some surfaces, is never precisely achieved. A major complication associated with radiation exchange between nonblack surfaces is due to surface reflection. In an enclosure, such as that of Figure 13.9*a*,

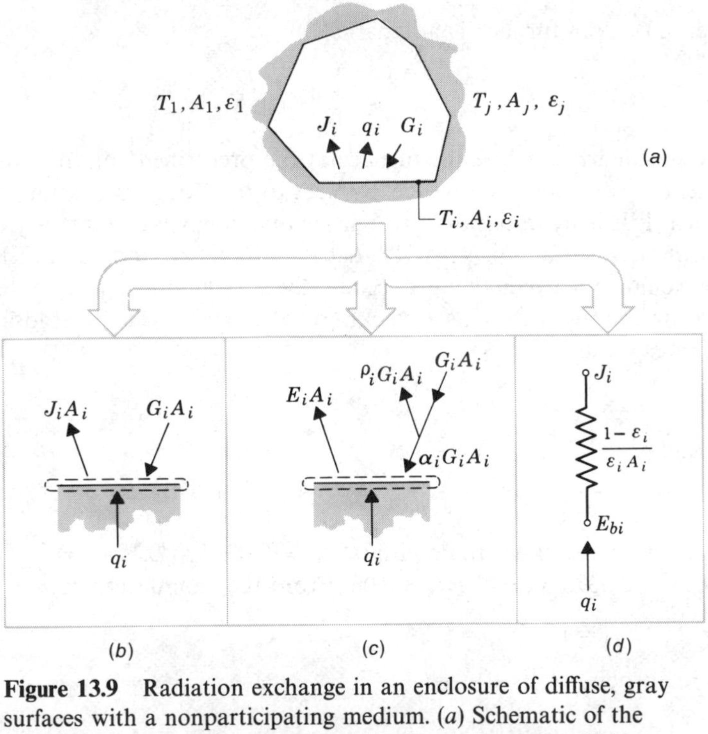

Figure 13.9 Radiation exchange in an enclosure of diffuse, gray surfaces with a nonparticipating medium. (*a*) Schematic of the enclosure. (*b*) Radiative balance according to Equation 13.15. (*c*) Radiative balance according to Equation 13.17. (*d*) Network element representing the net radiation transfer from a surface.

radiation may experience multiple reflections off all surfaces, with partial absorption occurring at each.

Analyzing radiation exchange in an enclosure may be simplified by making certain assumptions. Each surface of the enclosure is assumed to be *isothermal* and to be characterized by a *uniform radiosity* and *irradiation. Opaque, diffuse, gray* surface behavior is also assumed, and the medium within the enclosure is taken to be *nonparticipating.* The problem is generally one in which the temperature T_i associated with each of the surfaces is known, and the objective is to determine the *net radiative heat flux* q_i'' from each surface.

13.3.1 Net Radiation Exchange at a Surface

The term q_i, which is the *net* rate at which radiation *leaves* surface *i*, represents the net effect of radiative interactions occurring at the surface (Figure 13.9*b*). It is the rate at which energy would have to be transferred to the surface by other means to maintain it at a constant temperature. It is equal to the difference between the surface radiosity and irradiation and may be expressed as

$$q_i = A_i(J_i - G_i) \tag{13.15}$$

From Figure 13.9*c* and the definition of the radiosity J_i,

$$J_i \equiv E_i + \rho_i G_i \tag{13.16}$$

it is evident that the net radiative transfer from the surface may also be expressed in terms of the surface emissive power and the absorbed irradiation.

$$q_i = A_i(E_i - \alpha_i G_i) \tag{13.17}$$

Substituting from Equation 12.37 and recognizing that $\rho_i = 1 - \alpha_i = 1 - \varepsilon_i$ for an opaque, diffuse, gray surface, the radiosity may also be expressed as

$$J_i = \varepsilon_i E_{bi} + (1 - \varepsilon_i)G_i \tag{13.18}$$

Solving for G_i and substituting into Equation 13.15, it follows that

$$q_i = A_i\left(J_i - \frac{J_i - \varepsilon_i E_{bi}}{1 - \varepsilon_i}\right)$$

or

$$q_i = \frac{E_{bi} - J_i}{(1 - \varepsilon_i)/\varepsilon_i A_i} \tag{13.19}$$

Equation 13.19 provides a convenient representation for the net radiative heat transfer rate from a surface. This transfer, which may be represented by the network element of Figure 13.9*d*, is associated with the driving potential $(E_{bi} - J_i)$ and a *surface radiative resistance* of the form $(1 - \varepsilon_i)/\varepsilon_i A_i$. Hence if the emissive power that the surface would have if it were black exceeds its radiosity, there is net radiation heat transfer from the surface; if the inverse is true, the net transfer is to the surface.

13.3.2 Radiation Exchange Between Surfaces

To use Equation 13.19, the surface radiosity J_i must be known. To determine this quantity, it is necessary to consider radiation exchange between the surfaces of the enclosure.

The irradiation of surface i can be evaluated from the radiosities of all the surfaces in the enclosure. In particular from the definition of the view factor, it follows that the total rate at which radiation reaches surface i from all surfaces, including i, is

$$A_i G_i = \sum_{j=1}^{N} F_{ji} A_j J_j$$

or from the reciprocity relation, Equation 13.3,

$$A_i G_i = \sum_{j=1}^{N} A_i F_{ij} J_j$$

Canceling the area A_i and substituting into Equation 13.15 for G_i,

$$q_i = A_i \left(J_i - \sum_{j=1}^{N} F_{ij} J_j \right)$$

or, from the summation rule, Equation 13.4,

$$q_i = A_i \left(\sum_{j=1}^{N} F_{ij} J_i - \sum_{j=1}^{N} F_{ij} J_j \right)$$

Hence

$$q_i = \sum_{j=1}^{N} A_i F_{ij} (J_i - J_j) = \sum_{j=1}^{N} q_{ij} \tag{13.20}$$

This result equates the net rate of radiation transfer from surface i, q_i, to the sum of components q_{ij} related to radiative exchange with the other surfaces. Each component may be represented by a network element for which $(J_i - J_j)$ is the driving potential and $(A_i F_{ij})^{-1}$ is a *space* or *geometrical resistance* (Figure 13.10).

Combining Equations 13.19 and 13.20, we then obtain

$$\frac{E_{bi} - J_i}{(1 - \varepsilon_i)/\varepsilon_i A_i} = \sum_{j=1}^{N} \frac{J_i - J_j}{(A_i F_{ij})^{-1}} \tag{13.21}$$

As shown in Figure 13.10 this expression represents a radiation balance for the *node* associated with surface i. The rate of radiation transfer (current flow) to i

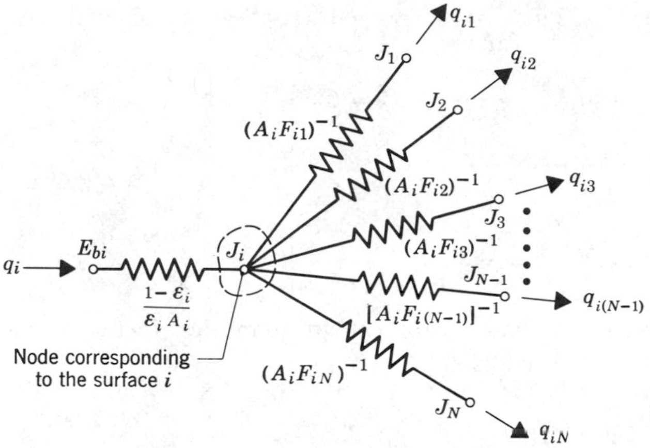

Figure 13.10 Network representation of radiative exchange between surface i and the remaining surfaces of an enclosure.

through its surface resistance must equal the rate of radiation transfer (current flows) from i to all other surfaces through the corresponding geometrical resistances.

Note that Equation 13.21 is especially useful when the surface temperature T_i (hence E_{bi}) is known. Although this situation is typical, it does not always apply. In particular, situations may arise for which the net radiation transfer rate at the surface q_i, rather than the temperature T_i, is known. In such cases the preferred form of the radiation balance is Equation 13.20, rearranged as

$$q_i = \sum_{j=1}^{N} \frac{J_i - J_j}{(A_i F_{ij})^{-1}} \tag{13.22}$$

Use of network representations to solve enclosure radiation problems was first suggested by Oppenheim [5]. The method provides a useful tool for visualizing radiation exchange in the enclosure and, at least for simple enclosures, may be used as the basis for predicting this exchange. However, a more direct approach, which lends itself to use with a computer, simply involves working with Equations 13.21 and 13.22. Equation 13.21 is written for each surface at which T_i is known, and Equation 13.22 is written for each surface at which q_i is known. The resulting set of N linear, algebraic equations is solved for the N unknowns, $J_1, J_2, \ldots, J_N$. With knowledge of the J_i, Equation 13.19 may be used to determine the net radiation heat transfer rate q_i at each surface of known T_i or the value of T_i at each surface of known q_i.

For any number N of surfaces in the enclosure, the foregoing problem may be readily solved by iteration or matrix inversion. For each of the N surfaces

Equation 13.21 or 13.22 may be rearranged to obtain the following system of N equations

$$\begin{bmatrix} a_{11}J_1 + \cdots + a_{1i}J_i + \cdots + a_{1N}J_N = C_1 \\ \vdots \\ a_{i1}J_1 + \cdots + a_{ii}J_i + \cdots + a_{iN}J_N = C_i \\ \vdots \\ a_{N1}J_1 + \cdots + a_{Ni}J_i + \cdots + a_{NN}J_N = C_N \end{bmatrix}$$

where the coefficients a_{ij} and C_i are known quantities. In matrix form these equations may be expressed as

$$[A][J] = [C]$$

where

$$[A] = \begin{bmatrix} a_{11} \cdots a_{1i} \cdots a_{1N} \\ \vdots \\ a_{i1} \cdots a_{ii} \cdots a_{iN} \\ \vdots \\ a_{N1} \cdots a_{Ni} \cdots a_{NN} \end{bmatrix} \qquad [J] = \begin{bmatrix} J_1 \\ \vdots \\ J_i \\ \vdots \\ J_N \end{bmatrix} \qquad [C] = \begin{bmatrix} C_1 \\ \vdots \\ C_i \\ \vdots \\ C_N \end{bmatrix}$$

Expressing the unknown radiosities as,

$$\begin{bmatrix} J_1 = b_{11}C_1 + \cdots + b_{1i}C_i + \cdots + b_{1N}C_N \\ \vdots \\ J_i = b_{i1}C_1 + \cdots + b_{ii}C_i + \cdots + b_{iN}C_N \\ \vdots \\ J_N = b_{N1}C_1 + \cdots + b_{Ni}C_i + \cdots + b_{NN}C_N \end{bmatrix}$$

they may be found by obtaining the inverse of $[A]$, $[A]^{-1}$, such that

$$[J] = [A]^{-1}[C]$$

where

$$[A]^{-1} = \begin{bmatrix} b_{11} \cdots b_{1i} \cdots b_{1N} \\ \vdots \\ b_{i1} \cdots b_{ii} \cdots b_{iN} \\ \vdots \\ b_{N1} \cdots b_{Ni} \cdots b_{NN} \end{bmatrix}$$

The foregoing matrix inversion may be readily obtained by using any of numerous computer routines available for this purpose or, for simple problems, from a programmable hand calculator. The system of equations may also be solved by the Gauss-Seidel iteration method described in Chapter 4.

EXAMPLE 13.4

In manufacturing, the special coating on a curved solar absorber surface of area $A_2 = 15\ m^2$ is cured by exposing it to an infrared heater of width $W = 1$ m. The absorber and heater are each of length $L = 10$ m and are separated by a distance of $H = 1$ m.

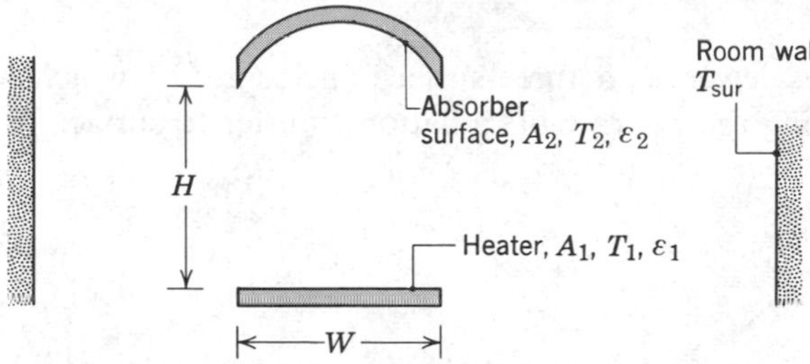

The heater is at $T_1 = 1000$ K and has an emissivity of $\varepsilon_1 = 0.9$, while the absorber is at $T_2 = 600$ K and has an emissivity of $\varepsilon_2 = 0.5$. The system is in a large room whose walls are at 300 K. What is the net rate of heat transfer to the absorber surface?

SOLUTION

Known:

A curved, solar absorber surface with a special coating is being cured by use of an infrared heater in a large room.

Find: Net rate of heat transfer to the absorber surface.

Schematic:

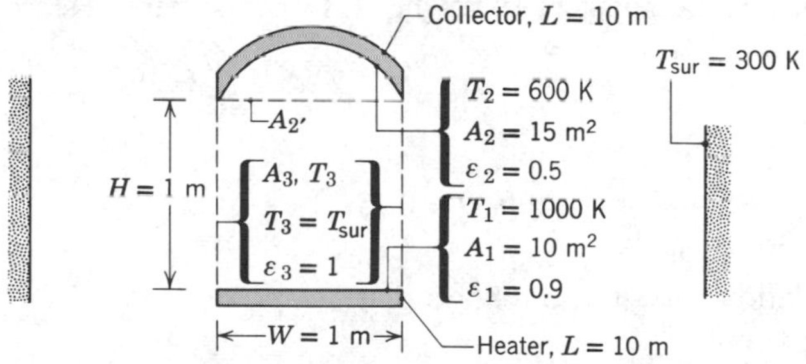

Assumptions:

1. Steady-state conditions exist.
2. Convection effects are negligible.

3. Absorber and heater surfaces are diffuse and gray.
4. The surroundings may be represented by a hypothetical surface A_3, which completes the enclosure and is approximated as a blackbody of temperature $T_3 = 300$ K.

Analysis:

The system may be viewed as a three-surface enclosure for which we are interested in obtaining the net rate of radiation transfer to surface 2. From Equation 13.19

$$q_2 = \frac{E_{b2} - J_2}{(1 - \varepsilon_2)/\varepsilon_2 A_2} \quad (1)$$

where all quantities are known except J_2. In any enclosure problem all unknown radiosities must be determined simultaneously, and the matrix inversion procedure is well suited for this purpose. In the present problem, however, matters are simplified by approximation of the hypothetical surface as a blackbody of known temperature. Hence $J_3 = E_{b3}$ is known, and the only unknowns are J_1 and J_2. Since T_1 and T_2 are known, J_1 and J_2 may be obtained by expressing Equation 13.21 for the heater and absorber surfaces. For the absorber surface, it follows that

$$\frac{E_{b2} - J_2}{(1 - \varepsilon_2)/\varepsilon_2 A_2} = \frac{J_2 - J_1}{1/A_2 F_{21}} + \frac{J_2 - J_3}{1/A_2 F_{23}}$$

where $J_3 = E_{b3} = \sigma T_3^4 = 459$ W/m^2 and $E_{b2} = \sigma T_2^4 = 7348$ W/m^2. The quantity $A_2 F_{21}$ may be obtained by recognizing that, from reciprocity,

$$A_2 F_{21} = A_1 F_{12}$$

where F_{12} may be obtained from Figure 13.4. That is, $F_{12} = F_{12'}$, where $A_{2'}$ is simply the rectangular base of the absorber surface. Hence with $Y/L = 10/1 = 10$ and $X/L = 1/1 = 1$,

$$F_{12} = 0.39$$

$$F_{21} = \frac{A_1}{A_2} F_{12} = \frac{1 \text{ m} \times 10 \text{ m}}{15 \text{ m}^2} \times 0.39 = 0.26$$

From the summation rule, it also follows that

$$F_{13} = 1 - F_{12} = 1 - 0.39 = 0.61$$

and from reciprocity

$$F_{31} = \frac{A_1}{A_3} F_{13} = \frac{1 \text{ m} \times 10 \text{ m}}{(10 + 10 + 1 + 1) \text{ m}^2} \times 0.61 = 0.28$$

But from symmetry, $F_{31} = F_{32'} = F_{32}$. Hence

$$F_{23} = \frac{A_3}{A_2} F_{32} = \frac{22 \text{ m}^2}{15 \text{ m}^2} \times 0.28 = 0.41$$

Canceling the area A_2, the radiation balance for surface 2 is then

$$\frac{7348 - J_2}{(1 - 0.5)/0.5} = \frac{J_2 - J_1}{1/0.26} + \frac{J_2 - 459}{1/0.41}$$

or

$$7348 - J_2 = 0.26J_2 - 0.26J_1 + 0.41J_2 - 188$$

$$0.26J_1 - 1.67J_2 = -7536 \quad (2)$$

Expressing Equation 13.21 for the heater surface, it follows that

$$\frac{E_{b1} - J_1}{(1 - \varepsilon_1)/\varepsilon_1 A_1} = \frac{J_1 - J_2}{1/A_1F_{12}} + \frac{J_1 - J_3}{1/A_1F_{13}}$$

where $E_{b1} = \sigma T_1^4 = 56{,}700 \text{ W/m}^2$. Hence canceling the area A_1,

$$\frac{56{,}700 - J_1}{(1 - 0.9)/0.9} = \frac{J_1 - J_2}{1/0.39} + \frac{J_1 - 459}{1/0.61}$$

or

$$-10J_1 + 0.39J_2 = -510{,}002 \quad (3)$$

From Equation 3, we then obtain

$$J_1 = 51{,}000 + 0.039J_2$$

Substituting into Equation 2

$$0.26(51{,}000 + 0.039J_2) - 1.67J_2 = -7536$$

in which case

$$J_2 = 12{,}528 \text{ W/m}^2$$

Substituting into Equation 1, the net heat rate to the absorber is

$$q_2 = \frac{(7348 - 12{,}528) \text{ W/m}^2}{(1 - 0.5)/0.5 \times 15 \text{ m}^2} = -77.7 \text{ kW} \quad \triangleleft$$

Comments:

Recognize the utility of using the hypothetical surfaces, in one case to complete the enclosure with A_3 and in the other case to simplify the evaluation of the shape factor with $A_{2'}$.

13.3.3 The Two-Surface Enclosure

The simplest example of an enclosure is one involving two surfaces that exchange radiation only with each other. Such a two-surface enclosure is shown schematically in Figure 13.11*a*. Since there are only two surfaces, the net rate of radiation transfer *from* surface 1, q_1, must equal the net rate of radiation transfer *to* surface 2, $-q_2$, and both quantities must equal the net rate at which radiation is exchanged between 1 and 2. Accordingly,

$$q_1 = -q_2 = q_{12}$$

The radiation transfer rate may be determined by applying Equation 13.21 to surfaces 1 and 2 and solving the resulting two equations for J_1 and J_2. The results could then be used with Equation 13.19 to determine q_1 (or q_2). However, in this case the desired result is more readily obtained by working with the network representation of the enclosure shown in Figure 13.11*b*.

From Figure 13.11*b* we see that the total resistance to radiation exchange between surfaces 1 and 2 is comprised of the two surface resistances and the geometrical resistance. Hence, substituting from Equation 12.28, the net radiation exchange between surfaces may be expressed as

$$q_{12} = q_1 = -q_2 = \frac{\sigma(T_1^4 - T_2^4)}{\dfrac{1-\varepsilon_1}{\varepsilon_1 A_1} + \dfrac{1}{A_1 F_{12}} + \dfrac{1-\varepsilon_2}{\varepsilon_2 A_2}} \tag{13.23}$$

The foregoing result may be used for any two diffuse, gray surfaces *that form an enclosure*. Important special cases are summarized in Table 13.3.

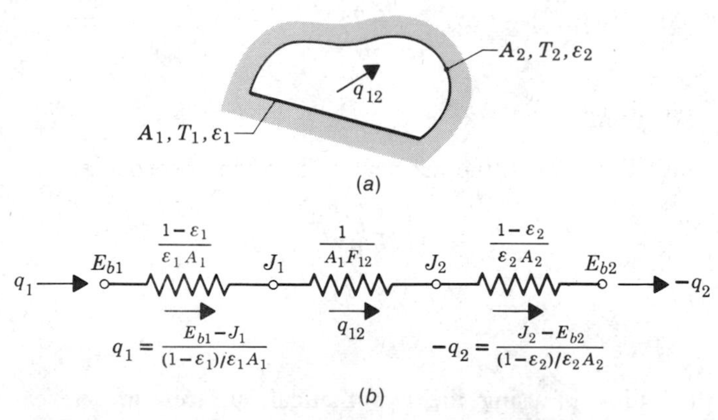

Figure 13.11 The two-surface enclosure. (*a*) Schematic. (*b*) Network representation.

Table 13.3 Special diffuse, gray, two-surface enclosures

Large (Infinite) Parallel Planes

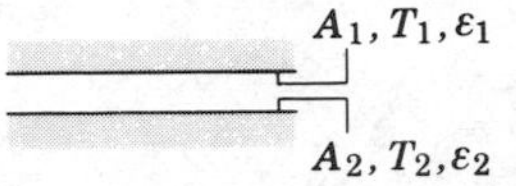

$A_1 = A_2 = A$

$F_{12} = 1$

$$q_{12} = \frac{A\sigma(T_1^4 - T_2^4)}{\dfrac{1}{\varepsilon_1} + \dfrac{1}{\varepsilon_2} - 1} \qquad (13.24)$$

Long (Infinite) Concentric Cylinders

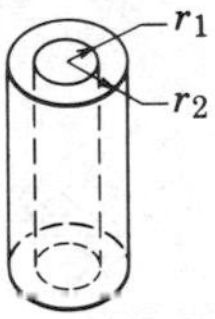

$\dfrac{A_1}{A_2} = \dfrac{r_1}{r_2}$

$F_{12} = 1$

$$q_{12} = \frac{\sigma A_1(T_1^4 - T_2^4)}{\dfrac{1}{\varepsilon_1} + \dfrac{1 - \varepsilon_2}{\varepsilon_2}\left(\dfrac{r_1}{r_2}\right)} \qquad (13.25)$$

Concentric Spheres

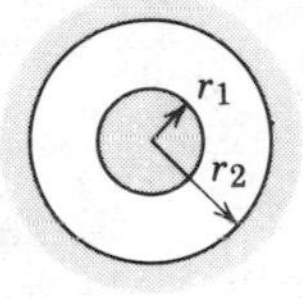

$\dfrac{A_1}{A_2} = \dfrac{r_1^2}{r_2^2}$

$F_{12} = 1$

$$q_{12} = \frac{\sigma A_1(T_1^4 - T_2^4)}{\dfrac{1}{\varepsilon_1} + \dfrac{1 - \varepsilon_2}{\varepsilon_2}\left(\dfrac{r_1}{r_2}\right)^2} \qquad (13.26)$$

Small Convex Object in a Large Cavity

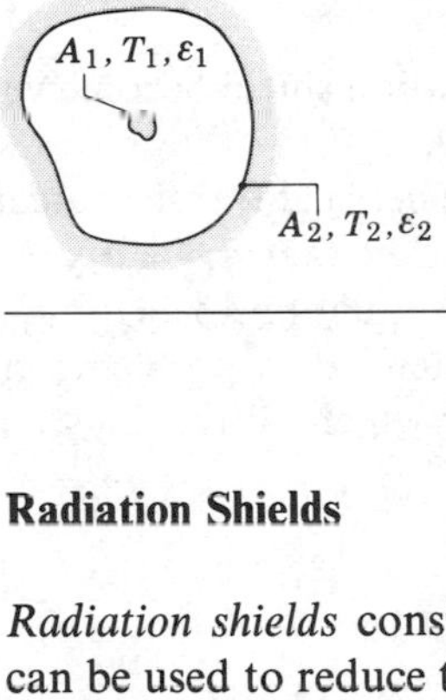

$\dfrac{A_1}{A_2} \approx 0$

$F_{12} = 1$

$$q_{12} = \sigma A_1 \varepsilon_1 (T_1^4 - T_2^4) \qquad (13.27)$$

13.3.4 Radiation Shields

Radiation shields constructed from low emissivity (high reflectivity) materials can be used to reduce the net radiation transfer between two surfaces. Consider placing a radiation shield, surface 3, between the two large, parallel planes of Figure 13.12*a*. Without the radiation shield, the net rate of radiation transfer between surfaces 1 and 2 is given by Equation 13.24. However, with the radiation shield, additional resistances are present, as shown in Figure 13.12*b*, and the heat transfer rate is reduced. Note that the emissivity associated with one side of the shield ($\varepsilon_{3,1}$) may differ from that associated with the opposite side ($\varepsilon_{3,2}$) and the radiosities will always differ. Summing the resistances and

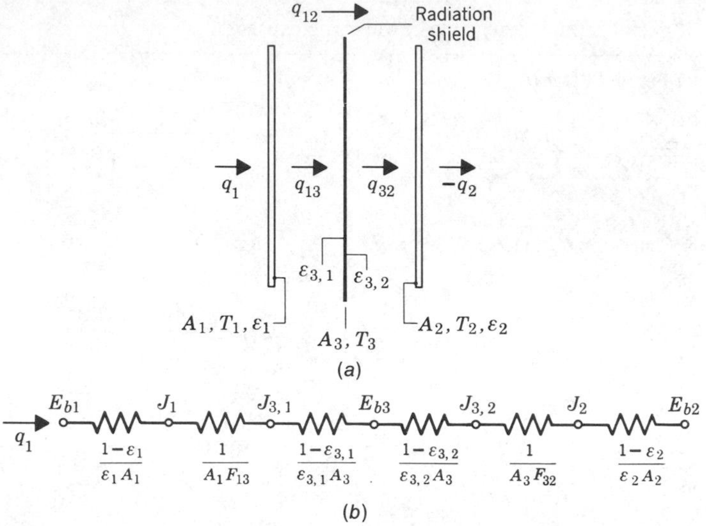

Figure 13.12 Radiation exchange between large parallel planes with a radiation shield. (*a*) Schematic. (*b*) Network representation.

recognizing that $F_{13} = F_{32} = 1$, it follows that

$$q_{12} = \frac{A_1\sigma(T_1^4 - T_2^4)}{\dfrac{1}{\varepsilon_1} + \dfrac{1}{\varepsilon_2} + \dfrac{1-\varepsilon_{3,1}}{\varepsilon_{3,1}} + \dfrac{1-\varepsilon_{3,2}}{\varepsilon_{3,2}}} \tag{13.28}$$

Note that the resistances associated with the radiation shield become very large when the emissivities $\varepsilon_{3,1}$ and $\varepsilon_{3,2}$ are very small.

Equation 13.28 may be used to determine the net heat transfer rate if T_1 and T_2 are known. From knowledge of q_{12} and the fact that $q_{12} = q_{13} = q_{32}$, the value of T_3 may then be determined by expressing Equation 13.24 for q_{13} or q_{32}.

The foregoing procedure may readily be extended to problems involving multiple radiation shields. In the special case for which all the emissivities are equal, it may be shown that, with N shields,

$$(q_{12})_N = \frac{1}{N+1}(q_{12})_0 \tag{13.29}$$

where $(q_{12})_0$ is the radiation transfer rate with no shields ($N = 0$).

EXAMPLE 13.5

A cryogenic fluid flows through a long tube of 20 mm diameter, the outer surface of which is diffuse and gray with $\varepsilon_1 = 0.02$ and $T_1 = 77$ K. This tube is concentric with a larger tube of 50 mm diameter, the inner surface of which is diffuse and gray with $\varepsilon_2 = 0.05$ and $T_2 = 300$ K. The space between the surfaces is evacuated.

Calculate the heat gain by the cryogenic fluid per unit length of tubes. If a thin radiation shield of 35 mm diameter and $\varepsilon_3 = 0.02$ (both sides) is inserted midway between the inner and outer surfaces, calculate the change (percentage) in heat gain per unit length of the tubes.

SOLUTION

Known:

Concentric tube arrangement with diffuse, gray surfaces of different emissivities and temperatures.

Find:

1. Heat gain by the cryogenic fluid passing through the inner tube.
2. Percentage change in heat gain with radiation shield inserted midway between inner and outer tubes.

Schematic:

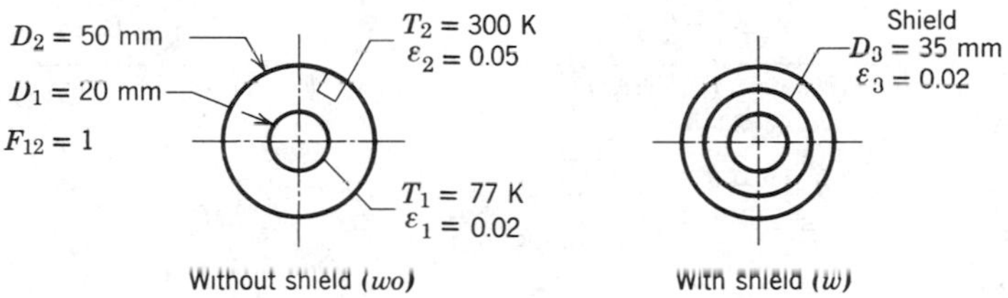

Assumptions:

1. Surfaces are diffuse gray.
2. Space between tubes is evacuated.
3. Conduction resistance for radiation shield is negligible.
4. Concentric tubes form a two-surface enclosure (end effects are negligible).

Analysis:

1. The network representation of the system without the shield is shown in Figure 13.11, and the desired heat rate may be obtained from Equation 13.25, where

$$q = \frac{\sigma(\pi D_1 L)(T_1^4 - T_2^4)}{\dfrac{1}{\varepsilon_1} + \dfrac{1-\varepsilon_2}{\varepsilon_2}\left(\dfrac{D_1}{D_2}\right)}$$

Hence

$$q' = \frac{q}{L} = \frac{5.67 \times 10^{-8}\ \text{W/m}^2 \cdot \text{K}^4\ (\pi \times 0.02\ \text{m})[(77\ \text{K})^4 - (300\ \text{K})^4]}{\dfrac{1}{0.02} + \dfrac{1 - 0.05}{0.05}\left(\dfrac{0.02\ \text{m}}{0.05\ \text{m}}\right)}$$

$$q' = -0.50\ \text{W/m} \quad \triangleleft$$

2. The network representation of the system with the shield is shown in Figure 13.12, and the desired heat rate is now

$$q = \frac{E_{b1} - E_{b2}}{R_{\text{tot}}} = \frac{\sigma(T_1^4 - T_2^4)}{R_{\text{tot}}}$$

where

$$R_{\text{tot}} = \frac{1 - \varepsilon_1}{\varepsilon_1(\pi D_1 L)} + \frac{1}{(\pi D_1 L)F_{13}} + 2\left[\frac{1 - \varepsilon_3}{\varepsilon_3(\pi D_3 L)}\right] + \frac{1}{(\pi D_3 L)F_{32}} + \frac{1 - \varepsilon_2}{\varepsilon_2(\pi D_2 L)}$$

or

$$R_{\text{tot}} = \frac{1}{L}\left\{\frac{1 - 0.02}{0.02(\pi \times 0.02\ \text{m})} + \frac{1}{(\pi \times 0.02\ \text{m})1} + 2\left[\frac{1 - 0.02}{0.02(\pi \times 0.035\ \text{m})}\right] + \frac{1}{(\pi \times 0.035\ \text{m})1} + \frac{1 - 0.05}{0.05(\pi \times 0.05\ \text{m})}\right\}$$

$$R_{\text{tot}} = \frac{1}{L}(779.9 + 15.9 + 891.3 + 9.1 + 121.0) = \frac{1817}{L}\left(\frac{1}{\text{m}^2}\right)$$

Hence

$$q' = \frac{q}{L} = \frac{5.67 \times 10^{-8}\ \text{W/m}^2 \cdot \text{K}^4\ [(77\ \text{K})^4 - (300\ \text{K})^4]}{1817\ (1/\text{m})} = -0.25\ \text{W/m} \quad \triangleleft$$

The percentage change in the heat gain is then

$$\frac{q'_w - q'_{wo}}{q'_{wo}} \times 100 = \frac{(-0.25\ \text{W/m}) - (-0.50\ \text{W/m})}{-0.50\ \text{W/m}} \times 100 = -50\%$$

13.3.5 The Reradiating Surface

The assumption of a *reradiating surface* is common to many industrial applications. This idealized surface is characterized by *zero* net radiation transfer ($q_i = 0$). It is closely approached by real surfaces that are well insulated on one side and for which convection effects may be neglected on the opposite (radiating) side. With $q_i = 0$, it follows from Equations 13.15 and 13.19 that $G_i = J_i = E_{bi}$. Hence, if the radiosity of a reradiating surface is known, its temperature is readily determined. In an enclosure, the equilibrium temperature of a reradiating surface is determined by its interaction with the other surfaces, and it is *independent of the emissivity of the reradiating surface.*

A three-surface enclosure, for which the third surface, surface R, is reradiating, is shown in Figure 13.13*a*, and the corresponding network is shown in Figure 13.13*b*. Surface R is presumed to be well insulated, and convection effects are assumed to be negligible. Hence, with $q_R = 0$, the net radiation *transfer* from surface 1 must equal the net radiation transfer to surface 2. The network is a simple series-parallel arrangement, and from its analysis it is readily shown that

$$q_1 = -q_2 = \frac{E_{b1} - E_{b2}}{\dfrac{1-\varepsilon_1}{\varepsilon_1 A_1} + \dfrac{1}{A_1 F_{12} + [(1/A_1 F_{1R}) + (1/A_2 F_{2R})]^{-1}} + \dfrac{1-\varepsilon_2}{\varepsilon_2 A_2}} \tag{13.30}$$

Knowing $q_1 = -q_2$, Equation 13.19 may be applied to surfaces 1 and 2 to determine their radiosities J_1 and J_2. Knowing J_1, J_2, and the geometrical resistances, the radiosity of the reradiating surface J_R may be determined from the radiation balance

$$\frac{J_1 - J_R}{(1/A_1 F_{1R})} - \frac{J_R - J_2}{(1/A_2 F_{2R})} = 0 \tag{13.31}$$

The temperature of the reradiating surface may then be determined from the requirement that $\sigma T_R^4 = J_R$. $= E_{bR}$

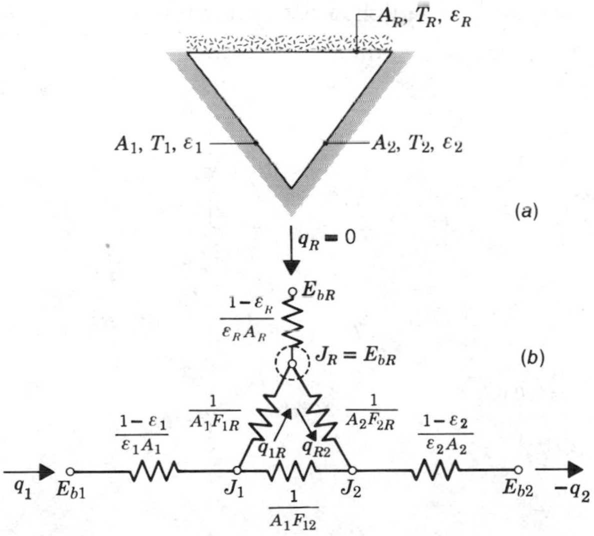

Figure 13.13 A three-surface enclosure with one surface reradiating. (*a*) Schematic. (*b*) Network representation.

Note that the general procedure described in Section 13.3.2 may be applied to enclosures with reradiating surfaces. For each such surface, it is appropriate to use Equation 13.22 with $q_i = 0$.

EXAMPLE 13.6

A paint baking oven consists of a long, triangular duct in which a heated surface is maintained at 1200 K and another surface is insulated. Painted panels, which are maintained at 500 K, occupy the third surface. The triangle is of width $W = 1$ m on a side, and the heated and insulated surfaces have an emissivity of 0.8. The emissivity of the panels is 0.4. During steady-state operation, at what rate must energy be supplied to the heated side per unit length of the duct to maintain its temperature at 1200 K? What is the temperature of the insulated surface?

SOLUTION

Known:

Surface properties of a long triangular duct that is insulated on one side and heated and cooled on the other sides.

Find:

1. Rate at which heat must be supplied per unit length of duct.
2. Temperature of the insulated surface.

Schematic:

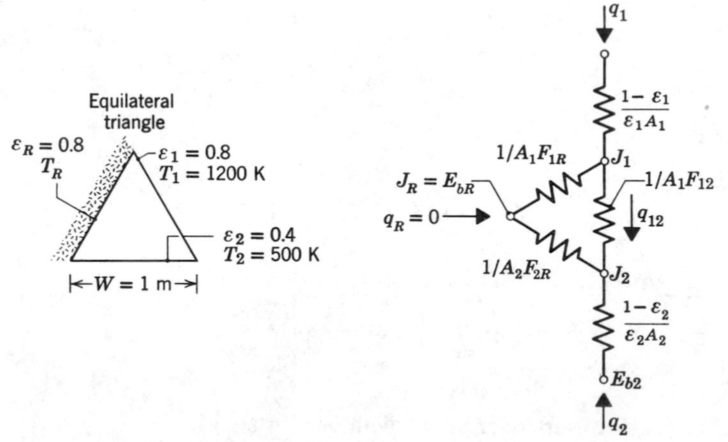

Assumptions:

1. Steady-state conditions exist.
2. All surfaces are opaque, diffuse, gray, and of uniform radiosity.
3. Convection effects are negligible.
4. Surface R is reradiating.
5. End effects are negligible.

Analysis:

1. The system may be modeled as a three-surface enclosure with one surface reradiating. The rate at which energy must be supplied to the heated surface may then be obtained from Equation 13.30

$$q_1 = \frac{E_{b1} - E_{b2}}{\dfrac{1-\varepsilon_1}{\varepsilon_1 A_1} + \dfrac{1}{A_1 F_{12} + [(1/A_1 F_{1R}) + (1/A_2 F_{2R})]^{-1}} + \dfrac{1-\varepsilon_2}{\varepsilon_2 A_2}}$$

From symmetry, $F_{12} = F_{1R} = F_{2R} = 0.5$. Also, $A_1 = A_2 = W \cdot L$, where L is the duct length. Hence

$$q_1' = \frac{q_1}{L} = \frac{5.67 \times 10^{-8}\ \mathrm{W/m^2 \cdot K^4}\ (1200^4 - 500^4)\ \mathrm{K^4}}{\dfrac{1-0.8}{0.8 \times 1\ \mathrm{m}} + \dfrac{1}{1\ \mathrm{m} \times 0.5 + (2+2)^{-1}\ \mathrm{m}} + \dfrac{1-0.4}{0.4 \times 1\ \mathrm{m}}}$$

or

$$q_1' = 37\ \mathrm{kW/m} = -q_2' \qquad \triangleleft$$

2. The temperature of the insulated surface may be obtained from the requirement that $J_R = E_{bR}$, where J_R may be obtained from Equation 13.31. However, to use this expression J_1 and J_2 must be known. Applying the surface energy balance, Equation 13.19, to surfaces 1 and 2, it follows that

$$J_1 = E_{b1} - \frac{1-\varepsilon_1}{\varepsilon_1 W} q_1' = 5.67 \times 10^{-8}\ \mathrm{W/m^2 \cdot K^4}\ (1200\ \mathrm{K})^4 - \frac{1-0.8}{0.8 \times 1\ \mathrm{m}}$$

$$\times\ 37{,}000\ \mathrm{W/m} = 108{,}323\ \mathrm{W/m^2}$$

$$J_2 = E_{b2} - \frac{1-\varepsilon_2}{\varepsilon_2 W} q_2' = 5.67 \times 10^{-8}\ \mathrm{W/m^2 \cdot K^4}\ (500\ \mathrm{K})^4$$

$$-\ \frac{1-0.4}{0.4 \times 1\ \mathrm{m}}(-37{,}000\ \mathrm{W/m}) = 59{,}043\ \mathrm{W/m^2}$$

From the energy balance for the reradiating surface, Equation 13.31, it follows that

$$\frac{108{,}323 - J_R}{\dfrac{1}{W \times L \times 0.5}} - \frac{J_R - 59{,}043}{\dfrac{1}{W \times L \times 0.5}} = 0$$

Hence

$$J_R = 83{,}683 \text{ W/m}^2 = E_{bR} = \sigma T_R^4$$

$$T_R = \left(\frac{83{,}683 \text{ W/m}^2}{5.67 \times 10^{-8} \text{ W/m}^2 \cdot \text{K}^4}\right)^{1/4} = 1102 \text{ K} \qquad \triangleleft$$

Comments:

1. Note that temperature and radiosity discontinuities cannot exist at the corners, and the assumptions of uniform temperature and radiosity are weakest in these regions.
2. The results are independent of the value of ε_R.
3. This problem may also be solved using the matrix inversion method. The solution involves first determining the three unknown radiosities J_1, J_2, and J_R. The governing equations are obtained by writing Equation 13.21 for the two surfaces of known temperature, 1 and 2, and Equation 13.22 for surface R. The three equations are

$$\frac{E_{b1} - J_1}{(1 - \varepsilon_1)/\varepsilon_1 A_1} = \frac{J_1 - J_2}{(A_1 F_{12})^{-1}} + \frac{J_1 - J_R}{(A_1 F_{1R})^{-1}}$$

$$\frac{E_{b2} - J_2}{(1 - \varepsilon_2)/\varepsilon_2 A_2} = \frac{J_2 - J_1}{(A_2 F_{21})^{-1}} + \frac{J_2 - J_R}{(A_2 F_{2R})^{-1}}$$

$$0 = \frac{J_R - J_1}{(A_R F_{R1})^{-1}} + \frac{J_R - J_2}{(A_R F_{R2})^{-1}}$$

Canceling the area A_1, the first equation reduces to

$$\frac{117{,}573 - J_1}{0.25} = \frac{J_1 - J_2}{2} + \frac{J_1 - J_R}{2}$$

or

$$10J_1 - J_2 - J_R = 940{,}584 \qquad (1)$$

Similarly, for surface 2,

$$\frac{3544 - J_2}{1.50} = \frac{J_2 - J_1}{2} + \frac{J_2 - J_R}{2}$$

or

$$-J_1 + 3.33J_2 - J_R = 4725 \tag{2}$$

and for the reradiating surface

$$0 = \frac{J_R - J_1}{2} + \frac{J_R - J_2}{2}$$

or

$$-J_1 - J_2 + 2J_R = 0 \tag{3}$$

From Equations 1, 2, and 3 the coefficient and right-hand side matrices are

$$\begin{vmatrix} a_{11} & a_{12} & a_{1R} \\ a_{21} & a_{22} & a_{2R} \\ a_{R1} & a_{R2} & a_{RR} \end{vmatrix} = \begin{vmatrix} 10 & -1 & -1 \\ -1 & 3.33 & -1 \\ -1 & -1 & +2 \end{vmatrix}, \quad \begin{vmatrix} C_1 \\ C_2 \\ C_R \end{vmatrix} = \begin{vmatrix} 940{,}584 \\ 4725 \\ 0 \end{vmatrix}$$

Inverting the coefficient matrix, it follows that

$J_1 = 108{,}328\ \text{W/m}^2 \quad J_2 = 59{,}018\ \text{W/m}^2$ and $J_R = 83{,}673\ \text{W/m}^2$

Recognizing that $J_R = \sigma T_R^4$, it follows that

$$T_R = \left(\frac{J_R}{\sigma}\right)^{1/4} = \left(\frac{83{,}673\ \text{W/m}^2}{5.67 \times 10^{-8}\ \text{W/m}^2 \cdot \text{K}^4}\right)^{1/4} = 1102\ \text{K}$$

13.4 ADDITIONAL EFFECTS

Although we have developed means for predicting radiation exchange between surfaces, it is important to be cognizant of the inherent limitations. Recall that we have considered *isothermal*, *opaque*, *gray* surfaces that *emit* and *reflect diffusely* and that are characterized by *uniform* surface *radiosity* and *irradiation.* For enclosures we have also considered the medium that separates the surfaces to be *nonparticipating*; that is, it neither absorbs nor scatters the surface radiation, and it emits no radiation.

The foregoing conditions and the related equations may often be used to obtain reliable first estimates and, in some cases, highly accurate results for radiation transfer in an enclosure. Sometimes, however, the assumptions are grossly inappropriate and more refined prediction methods are needed. Although beyond the scope of this text, the methods are discussed in more advanced treatments of radiation transfer [3, 6–11].

We have said little about gaseous radiation, having confined our attention to radiation exchange at the surface of an opaque solid or liquid. For *nonpolar* gases, such as O_2 or N_2, such neglect is justified, since the gases do not emit radiation and are essentially transparent to incident thermal radiation. However, the same may not be said for polar molecules, such as CO_2, H_2O (vapor), NH_3, and hydrocarbon gases, which emit and absorb over a wide temperature range. For such gases matters are complicated by the fact that, unlike radiation from a solid or a liquid, which is distributed continuously with wavelength, gaseous radiation is concentrated in specific *wavelength intervals* (called bands). Moreover, gaseous radiation is not a surface phenomenon, but is instead a *volumetric* phenomenon.

13.4.1 Volumetric Absorption

Spectral radiation absorption in a gas (or in a semitransparent liquid or solid) is a function of the absorption coefficient κ_λ (1/m) and the thickness L of the medium (Figure 13.14). If a monochromatic beam of intensity $I_{\lambda,0}$ is incident on the medium, the intensity is reduced due to absorption, and the reduction occurring in an infinitesimal layer of thickness dx may be expressed as

$$dI_\lambda(x) = -\kappa_\lambda I_\lambda(x)\, dx \tag{13.32}$$

Separating variables and integrating over the entire layer, we obtain

$$\int_{I_{\lambda,0}}^{I_{\lambda,L}} \frac{dI_\lambda(x)}{I_\lambda(x)} = -\kappa_\lambda \int_0^L dx$$

where κ_λ is assumed to be independent of x. It follows that

$$\frac{I_{\lambda,L}}{I_{\lambda,0}} = e^{-\kappa_\lambda L} \tag{13.33}$$

This exponential decay, termed *Beer's law*, is a useful tool in approximate radiation analysis. It may, for example, be used to infer the overall spectral absorptivity of the medium. In particular with the transmissivity defined as

$$\tau_\lambda = \frac{I_{\lambda,L}}{I_{\lambda,0}} = e^{-\kappa_\lambda L} \tag{13.34}$$

the absorptivity is

$$\alpha_\lambda = 1 - \tau_\lambda = 1 - e^{-\kappa_\lambda L} \tag{13.35}$$

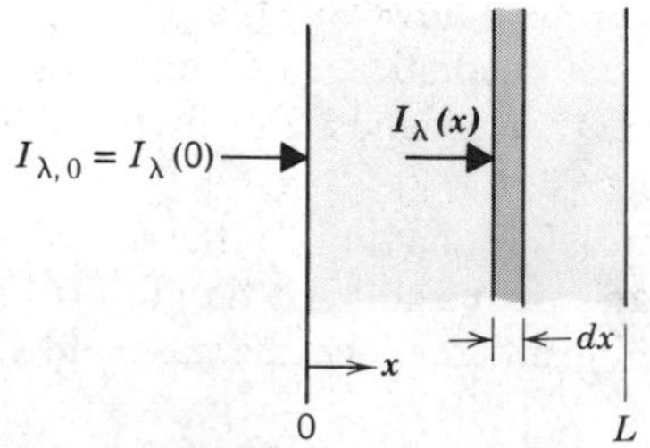

Figure 13.14 Absorption in a gas or liquid layer.

If Kirchhoff's law is assumed to be valid, $\alpha_\lambda = \varepsilon_\lambda$, Equation 13.35 also provides the spectral emissivity of the medium.

13.4.2 Gaseous Emission and Absorption

A common engineering calculation is one that requires determination of the radiant heat flux from a gas to an adjoining surface. Despite the complicated spectral and directional effects inherent in such calculations, a simplified procedure may be used. The method was developed by Hottel [12] and involves determining radiation emission from a hemispherical gas mass of temperature T_g to a surface element dA_1, which is located at the center of the hemisphere's base. Emission from the gas per unit area of the surface is expressed as

$$E_g = \varepsilon_g \sigma T_g^4 \tag{13.36}$$

where the gas emissivity ε_g was determined by correlating available data. In particular ε_g was correlated in terms of the temperature T_g and total pressure p of the gas, the partial pressure p_g of the radiating species, and the radius L of the hemisphere.

Results for the emissivity of water vapor are plotted in Figure 13.15 as a

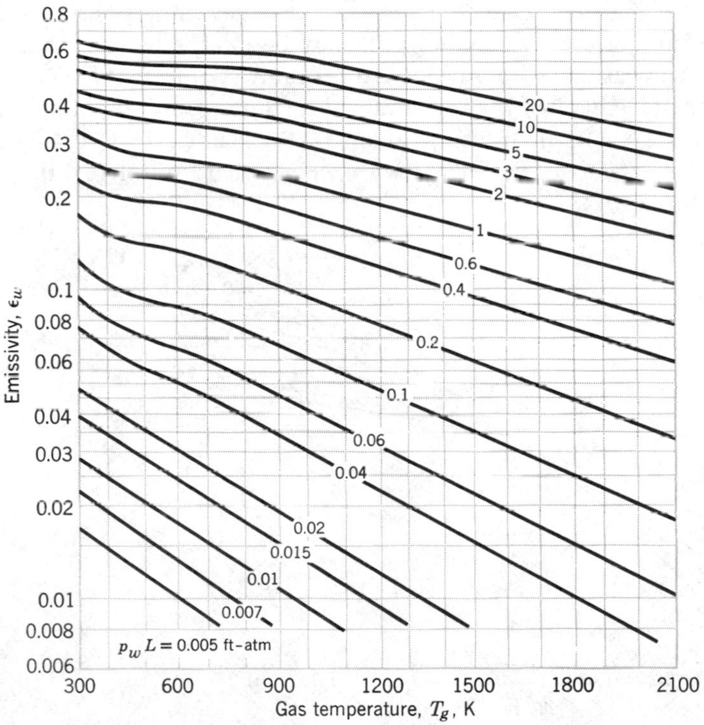

Figure 13.15 Emissivity of water vapor in a mixture with nonradiating gases at 1 atm total pressure and of hemispherical shape [12]. Used with permission.

function of the gas temperature, for a total pressure of 1 atm, and for different values of the product of the vapor partial pressure and the hemisphere radius. To evaluate the emissivity for total pressures other than one atmosphere, the emissivity from Figure 13.15 must be multiplied by the correction factor C_w from Figure 13.16. Similar results were obtained for carbon dioxide and are presented in Figures 13.17 and 13.18.

The foregoing results apply when water vapor or carbon dioxide appear *separately* in a mixture with other species that are nonradiating. However, the results may readily be extended to situations in which water vapor and carbon dioxide appear *together* in a mixture with other nonradiating gases. In particular the total gas emissivity may be expressed as

$$\varepsilon_g = \varepsilon_w + \varepsilon_c - \Delta\varepsilon \tag{13.37}$$

where the correction factor $\Delta\varepsilon$ is presented in Figure 13.19 for different values of the gas temperature. This factor accounts for the reduction in emission associated with the mutual absorption of radiation between the two species.

Recall that the foregoing results provide the emissivity of a hemispherical gas mass of radius L radiating to an element of area at the center of its base. However, the results may be extended to other gas geometries by introducing the concept of a *mean beam length*, L_e. The quantity was introduced to correlate, in terms of a single parameter, the dependence of gas emissivity on both the size and the shape of the gas geometry. It may be interpreted as the radius of a hemispherical gas mass whose emissivity is equivalent to that for the geometry of interest. Its value has been determined for numerous gas shapes [12], and representative results are listed in Table 13.4. Replacing L by L_e in Figures 13.15 to 13.19, the emissivity associated with the geometry of interest may then be determined.

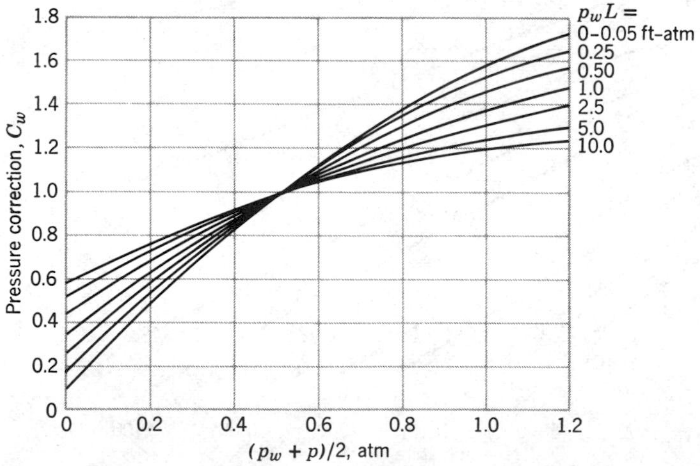

Figure 13.16 Correction factor for obtaining water vapor emissivities at pressures other than 1 atm ($\varepsilon_{w,p\neq 1\text{ atm}} = C_w \varepsilon_{w,p=1\text{ atm}}$) [12]. Used with permission.

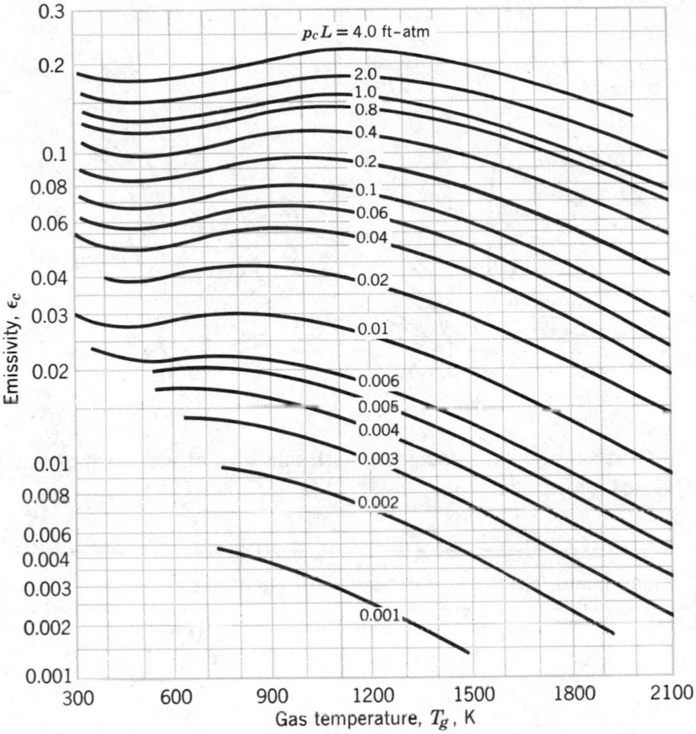

Figure 13.17 Emissivity of carbon dioxide in a mixture with non-radiating gases at 1 atm total pressure and of hemispherical shape [12]. Used with permission.

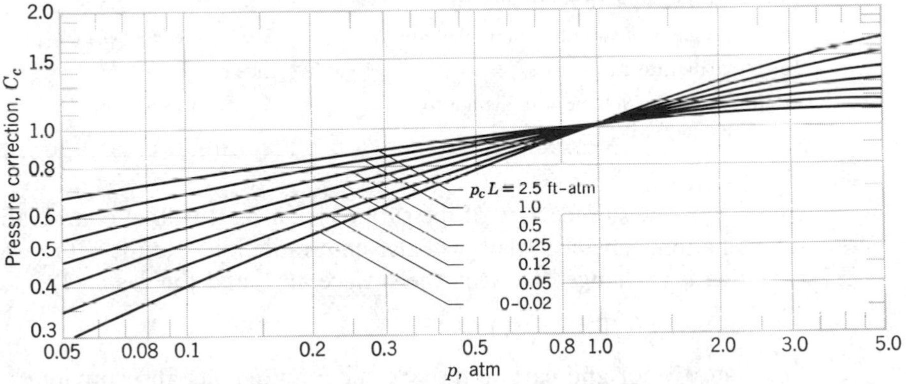

Figure 13.18 Correction factor for obtaining carbon dioxide emissivities at pressures other than 1 atm ($\varepsilon_{c,p\neq 1\text{ atm}} = C_c\varepsilon_{c,p=1\text{ atm}}$) [12]. Used with permission.

Using the results of Table 13.4 with Figures 13.15 to 13.19, it is possible to determine the rate of radiant heat transfer to a surface due to emission from an adjoining gas. This heat rate may be expressed as

$$q = \varepsilon_g A_s \sigma T_g^4 \qquad (13.38)$$

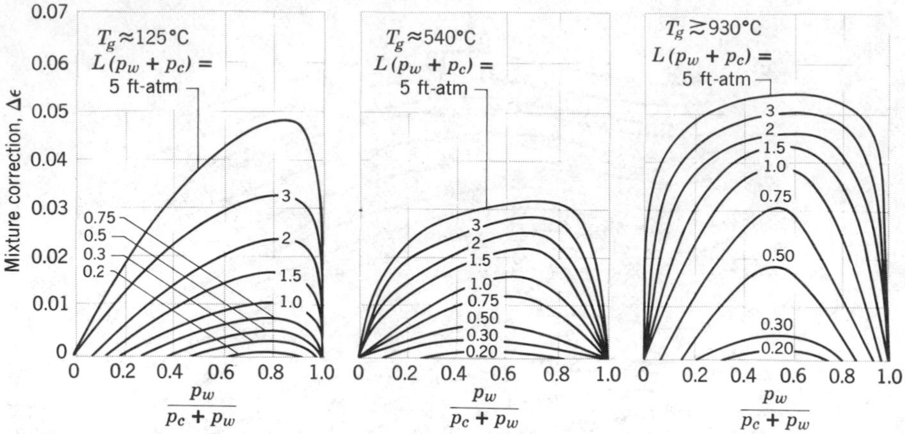

Figure 13.19 Correction factor associated with mixtures of water vapor and carbon dioxide [12]. Used with permission.

Table 13.4 Mean beam lengths L_e for various gas geometries

GEOMETRY	CHARACTERISTIC LENGTH	L_e
Sphere (radiation to surface)	Diameter (D)	$0.65D$
Infinite circular cylinder (radiation to curved surface)	Diameter (D)	$0.95D$
Semiinfinite circular cylinder (radiation to base)	Diameter (D)	$0.65D$
Circular cylinder of equal weight and diameter (radiation to entire surface)	Diameter (D)	$0.60D$
Infinite parallel planes (radiation to planes)	Spacing between planes (L)	$1.80L$
Cube (radiation to any surface)	Side (L)	$0.66L$
Arbitrary shape of volume V (radiation to surface of area A)	Volume to area ratio (V/A)	$3.6V/A$

where A_s is the surface area. If the surface is black, it will, of course, absorb all this radiation. A black surface will also emit radiation, and the net rate at which radiation is exchanged between the surface at T_s and the gas at T_g is

$$q_{net} = A_s \sigma(\varepsilon_g T_g^4 - \alpha_g T_s^4) \tag{13.39}$$

For water vapor and carbon dioxide the required gas absorptivity α_g may be evaluated from the emissivity by expressions of the form [12]

$$\text{Water:} \quad \alpha_w = C_w \left(\frac{T_g}{T_s}\right)^{0.45} \times \varepsilon_w\left(T_s, p_w L_e \frac{T_s}{T_g}\right) \tag{13.40}$$

$$\text{Carbon Dioxide:} \quad \alpha_c = C_c \left(\frac{T_g}{T_s}\right)^{0.65} \times \varepsilon_c\left(T_s, p_c L_e \frac{T_s}{T_g}\right) \tag{13.41}$$

where ε_w and ε_c are evaluated from Figures 13.15 and 13.17, respectively, and C_w and C_c are evaluated from Figures 13.16 and 13.18, respectively. Note, however, that in using Figures 13.15 and 13.17, T_g is replaced by T_s and $p_w L_e$ or $p_c L_e$ is replaced by $p_w L_e(T_s/T_g)$ or $p_c L_e(T_s/T_g)$, respectively. Note also that, in the presence of both water vapor and carbon dioxide, the total gas absorptivity may be expressed as

$$\alpha_g = \alpha_w + \alpha_c - \Delta\alpha \tag{13.42}$$

where $\Delta\alpha = \Delta\varepsilon$ is obtained from Figure 13.19.

13.5 SUMMARY

In this chapter we focused on the analysis of radiation exchange between the surfaces of an enclosure, and in treating this exchange we introduced the *view factor* concept. Because knowing this geometrical quantity is essential to determining radiation exchange between any two diffuse surfaces, you should be familiar with the means by which it may be determined. You should also be adept at performing radiation calculations for an enclosure of *isothermal, opaque, diffuse*, and *gray* surfaces of *uniform radiosity* and *irradiation*. Moreover, you should be familiar with the results that apply to simple cases such as the two-surface enclosure or the three-surface enclosure with a reradiating surface.

REFERENCES

1. Hamilton, D. C., and W. R. Morgan, "Radiant Interchange Configuration Factors," National Advisory Committee for Aeronautics, Technical Note 2836, 1952.
2. Eckert, E. R. G., "Radiation: Relations and Properties," in W. M. Rohsenow and J. P. Hartnett, Eds., *Handbook of Heat Transfer*, 2nd ed., McGraw-Hill, New York, 1973.
3. Siegel, R., and J. R. Howell, *Thermal Radiation Heat Transfer*, McGraw-Hill, New York, 1981.
4. Howell, J. R., *A Catalog of Radiation Configuration Factors*, McGraw-Hill, New York, 1982.
5. Oppenheim, A. K., *Trans. ASME*, *65*, 725, 1956.
6. Hottel, H. C., and A. F. Sarofim, *Radiative Transfer*, McGraw-Hill, New York, 1967.
7. Tien, C. L., "Thermal Radiation Properties of Gases," in J. P. Hartnett and T. F. Irvine, Eds., *Advances in Heat Transfer*, Vol. 5, Academic Press, New York, 1968.
8. Sparrow, E. M., "Radiant Interchange Between Surfaces Separated by Nonabsorbing and Nonemitting Media," in W. M. Rohsenow and J. P. Hartnett, Eds., *Handbook of Heat Transfer*, McGraw-Hill, New York, 1973.
9. Dunkle, R. V., "Radiation Exchange in an Enclosure with a Participating Gas," in W. M. Rohsenow and J. P. Hartnett, Eds., *Handbook of Heat Transfer*, McGraw-Hill, New York, 1973.

10. Sparrow, E. M., and R. D. Cess, *Radiation Heat Transfer*, Hemisphere Publishing Corp., New York, 1978.
11. Edwards, D. K., *Radiation Heat Transfer Notes*, Hemisphere Publishing Corp., New York, 1981.
12. Hottel, H. C., "Radiant-Heat Transmission," in W. H. McAdams, Ed., *Heat Transmission*, 3rd ed., McGraw-Hill, New York, 1954.

PROBLEMS

13.1 Determine F_{12} and F_{21} for the following configurations using the reciprocity theorem and other basic shape factor relations. Do not use tables or charts.

(a)
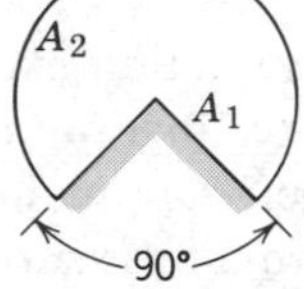

Long duct

(b)
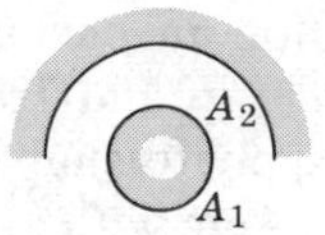

Small sphere of area A_1 under a concentric hemisphere of area $A_2 = 2A_1$

(c)
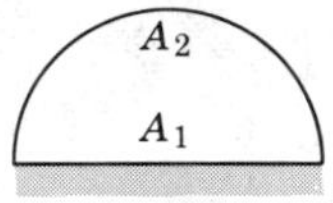

Long duct. What is F_{22} for this case?

(d)
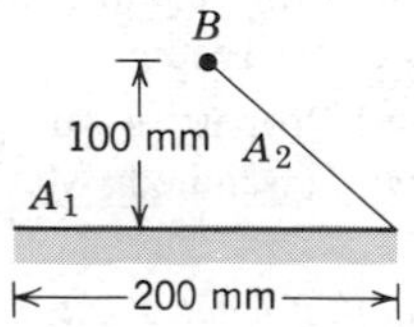

Long inclined plates (point B is directly above the center of A_1)

(e)
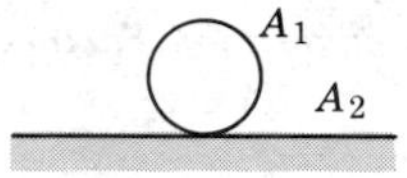

Sphere lying on infinite plane

(f)
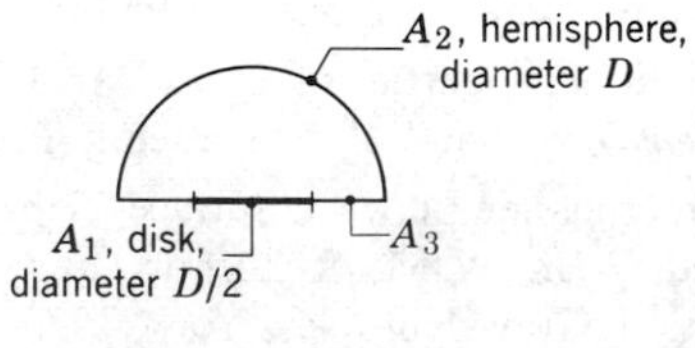

Hemisphere–disk arrangement

(g)
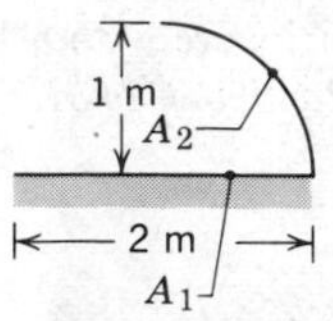

Long open channel

13.2 Consider the following grooves, each of width W, that have been machined from a solid block of material.

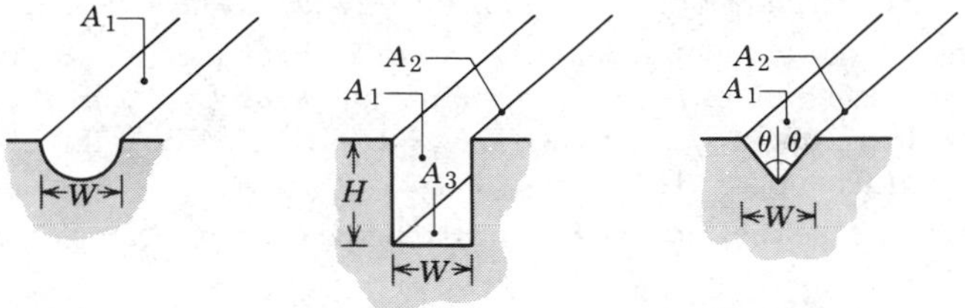

Semicircular groove Rectangular groove V groove

(a) For each case obtain an expression for the view factor of the groove with respect to the surroundings outside the groove.

(b) For the V groove, obtain an expression for the view factor F_{12}, where A_1 and A_2 are opposite surfaces.

(c) If $H = 2W$ in the rectangular groove, what is the view factor F_{12}?

13.3 Consider a long cylindrical tube having an inner surface area A_2. Surfaces represented by dashed lines (A_1) in the sketches are inserted into the tube. The curved surfaces (b,c) are formed by cylindrical segments tangent to one another at their tips.

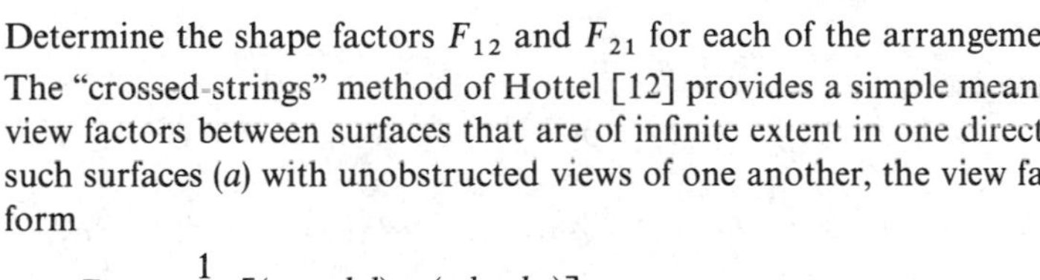

(a) (b) (c)

Determine the shape factors F_{12} and F_{21} for each of the arrangements.

13.4 The "crossed-strings" method of Hottel [12] provides a simple means to calculate view factors between surfaces that are of infinite extent in one direction. For two such surfaces (a) with unobstructed views of one another, the view factor is of the form

$$F_{12} = \frac{1}{2w_1}[(ac + bd) - (ad + bc)]$$

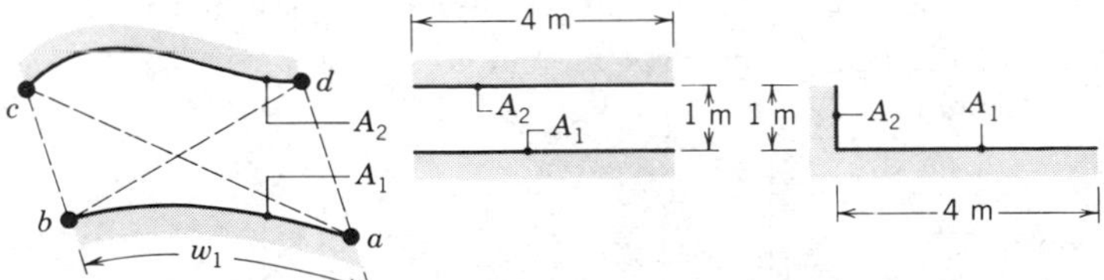

(a) (b) (c)

Use this method to evaluate the view factors F_{12} for sketches (b) and (c). Compare your results with those from the appropriate graphs and analytical expressions.

13.5 Consider a rectangular parallelepiped enclosure.

(a) How many view factors are involved and how many of these view factors must be determined directly?

(b) Acknowledging symmetry and the fact that all surfaces are planar, how many of the view factors must be determined directly?

(c) If the enclosure is cubical, how many view factors must be determined directly?

13.6 Consider the two diffuse surfaces shown in Example 13.1. Beginning with the general definition of $F_{ij} = q''_{i \to j}/J_i$, show that when $L \gg D$, the shape factor for a small area element (i) to a disk (j) is $F_{ij} = D^2/4L^2$. How does this compare with the result of Example 13.1?

13.7 Determine the shape factor F_{12} for the perpendicular rectangles shown.

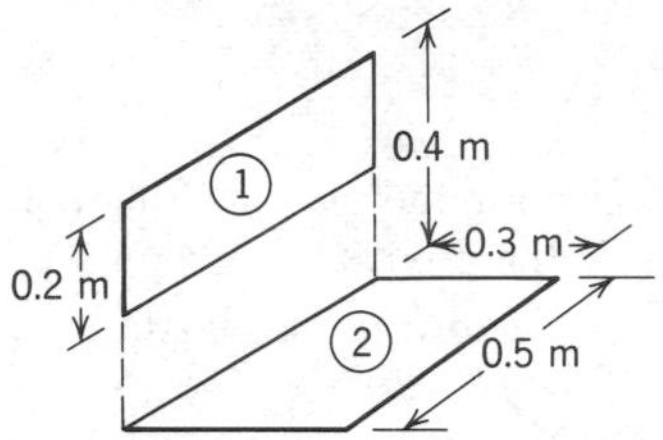

13.8 The reciprocity relation, the summation rule, and Equations 13.5 to 13.7 can be used to develop view factor relations that allow for applications of Figure 13.4 and/or 13.6 to more complex configurations. Consider the view factor F_{14} for surfaces 1 and 4 of the following geometry. These surfaces are perpendicular but do not share a common edge.

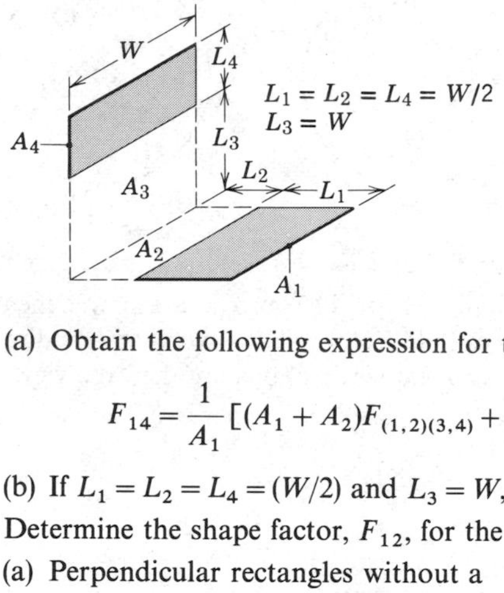

(a) Obtain the following expression for the view factor F_{14}.

$$F_{14} = \frac{1}{A_1}[(A_1 + A_2)F_{(1,2)(3,4)} + A_2F_{23} - (A_1 + A_2)F_{(1,2)3} - A_2F_{2(3,4)}]$$

(b) If $L_1 = L_2 = L_4 = (W/2)$ and $L_3 = W$, what is the value of F_{14}?

13.9 Determine the shape factor, F_{12}, for the rectangles shown.

(a) Perpendicular rectangles without a common edge.

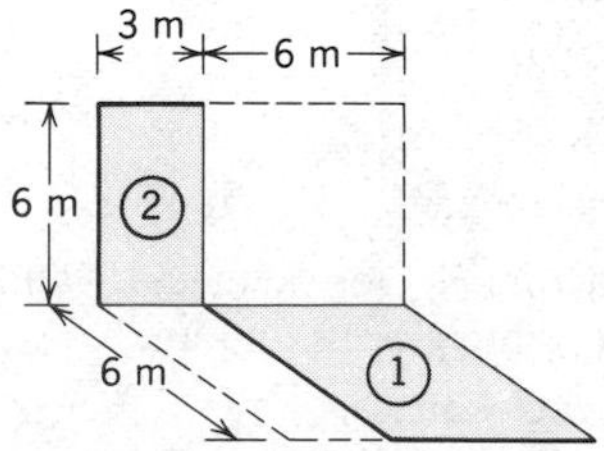

(b) Parallel rectangles of unequal areas.

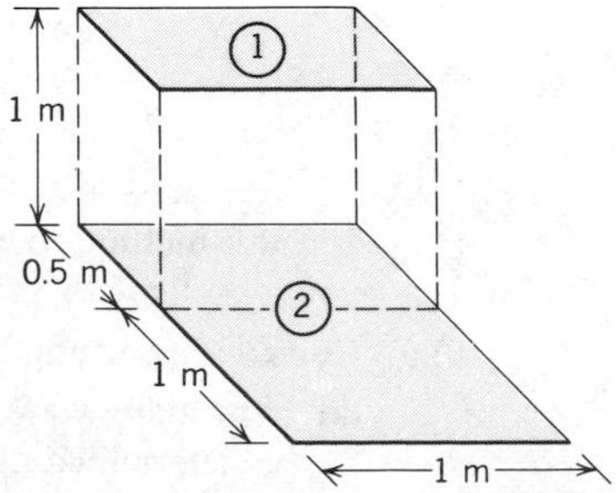

13.10 Determine the radiant power from the tubular heater that is incident on the disk. The inner surface of the heater is black and at a uniform temperature of 1000 K.

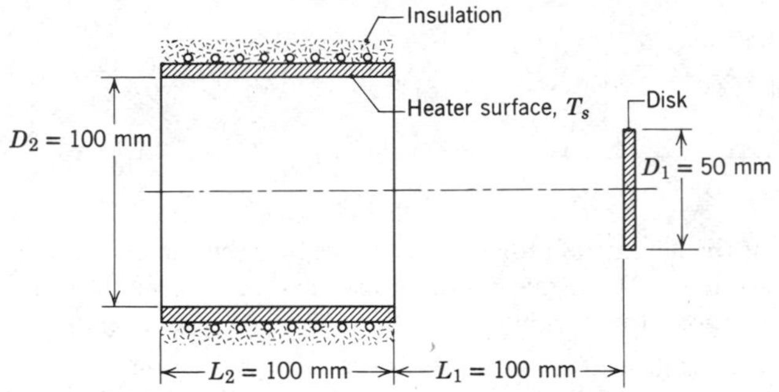

13.11 Consider the arrangement of the three black surfaces shown, where A_1 is small compared to A_2 or A_3.

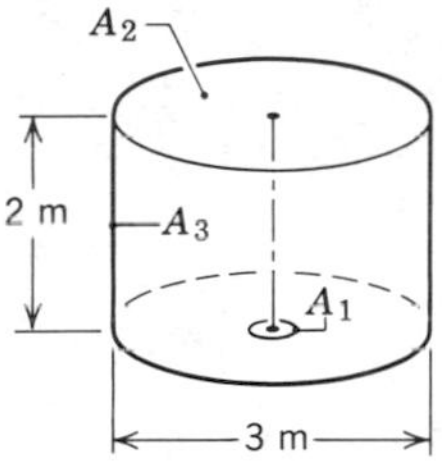

Determine the value of F_{13}. Calculate the net radiation heat transfer from A_1 to A_3 if $A_1 = 0.05\ m^2$, $T_1 = 1000$ K, and $T_3 = 500$ K.

13.12 Two perfectly black, parallel disks 1 m in diameter are separated a distance of 0.25 m. One disk is maintained at 60°C and the other at 20°C. The disks are located in a large room whose walls are 40°C. Assume that the exterior surfaces of the disks (the surfaces that do not face each other) are very well insulated. Determine the net radiation exchange between the disks. Determine the net radiation exchange between the disks and the room.

13.13 A drying oven consists of a long semicircular duct of diameter $D = 1$ m.

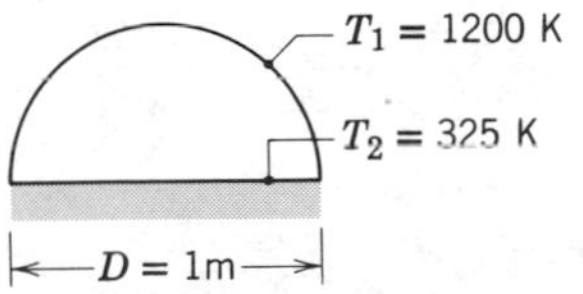

Materials to be dried cover the base of the oven, while the wall is maintained at 1200 K. What is the drying rate per unit length of the oven (kg/s · m), if a water coated layer of material is maintained at 325 K during the drying process? Blackbody behavior may be assumed for the water surface and for the oven wall.

13.14 A circular disk of diameter $D_1 = 20$ mm is located at the base of an enclosure that has a cylindrical sidewall and a hemispherical dome. The enclosure is of diameter $D = 0.5$ m, and the height of the cylindrical section is $L = 0.3$ m. The disk and the enclosure surface are black and at temperatures of 1000 and 300 K, respectively.

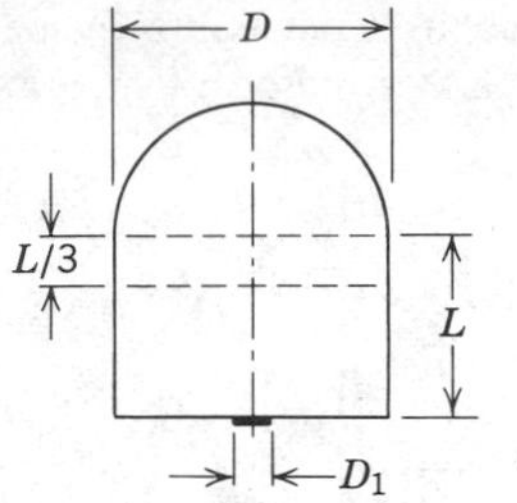

What is the net rate of radiation exchange between the disk and the hemispherical dome? What is the net rate of radiation exchange between the disk and the top one-third portion of the cylindrical section?

13.15 In the arrangement shown, the lower disk has a diameter of 30 mm and a temperature of 500 K. The upper surface, which is at 1000 K, is a ring-shaped disk with inner and outer diameters of 0.15 m and 0.2 m. This upper surface is aligned with and parallel to the lower disk and is separated by a distance of 1 m.

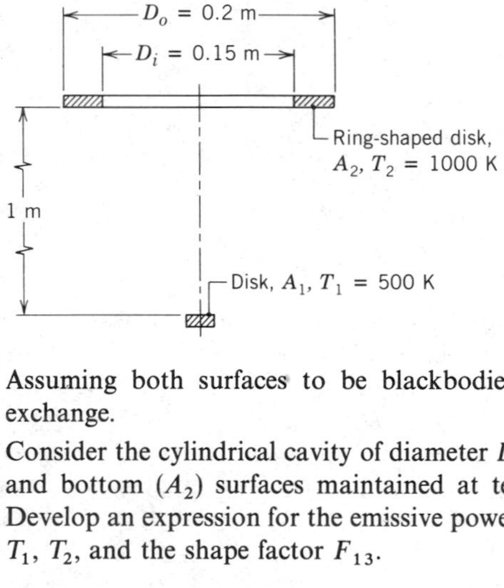

Assuming both surfaces to be blackbodies, calculate their net radiative heat exchange.

13.16 Consider the cylindrical cavity of diameter D and length L having the lateral (A_1) and bottom (A_2) surfaces maintained at temperatures T_1 and T_2, respectively. Develop an expression for the emissive power of the cavity opening A_3 in terms of T_1, T_2, and the shape factor F_{13}.

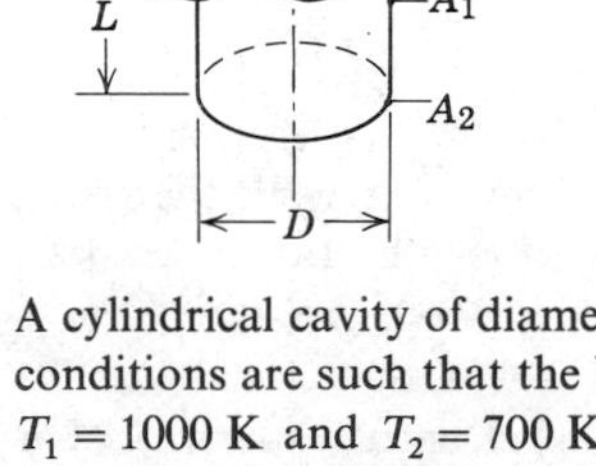

13.17 A cylindrical cavity of diameter D and depth L is machined in a metal block, and conditions are such that the base and side surfaces of the cavity are maintained at $T_1 = 1000$ K and $T_2 = 700$ K, respectively. Approximating the surfaces as black, determine the emissive power of the cavity if $L = 20$ mm and $D = 10$ mm.

13.18 Two parallel plates 1 m × 1 m, insulated on their back sides and separated by 1 m, may be approximated as blackbodies at 500 and 750 K. The plates are located in a large room whose walls are maintained at 300 K. Determine the net radiative heat transfer from each plate and the net radiative heat transfer to the room walls.

13.19 The arrangement shown is to be used to calibrate a heat flux gage. The gage has a black surface that is 10 mm in diameter and is maintained at 17°C by means of a water-cooled backing plate. The heater, 200 mm in diameter, has a black surface that is maintained at 800 K and is located 0.5 m from the gage. The surroundings and the air are at 27°C and the convection heat transfer coefficient between the gage and the air is 15 $W/m^2 \cdot K$.

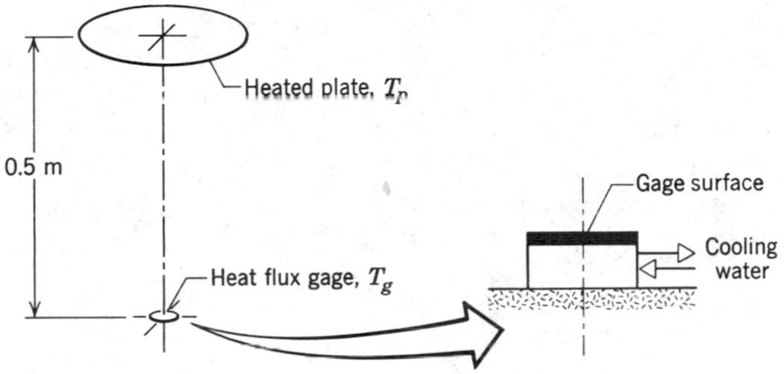

(a) Determine the net radiation exchange between the heater and the gage.

(b) Determine the net transfer of radiation to the gage per unit area of the gage.

(c) What is the net heat transfer rate to the gage per unit area of the gage?

(d) If the gage is constructed according to the description of Problem 3.28, what heat flux will it indicate?

13.20 A long, cylindrical heating element of 20 mm diameter operating at 700 K in vacuum is located 40 mm from an insulated wall.

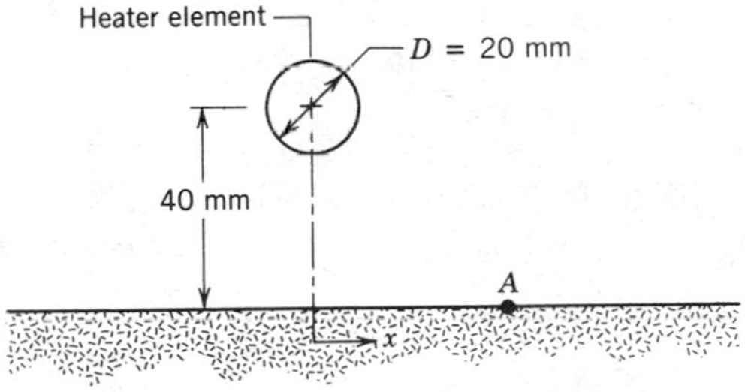

Assuming that the wall thermal conductivity is very low and that both element and wall are black, estimate the maximum temperature reached by the wall when the surroundings are at 300 K. Estimate the wall temperature at point A, where $x = 40$ mm.

13.21 A row of regularly spaced, cylindrical heating elements is used to maintain an insulated furnace wall at 500 K. The opposite wall is at a uniform temperature of 300 K.

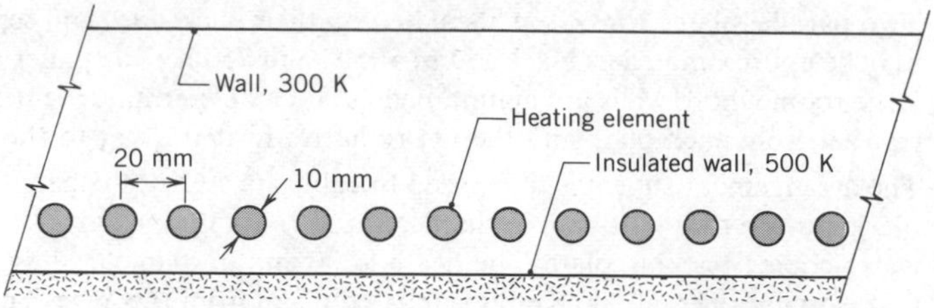

The insulated wall experiences convection with air at 450 K and a convection coefficient of 200 $W/m^2 \cdot K$. Assuming the walls and elements are black, estimate the required operating temperature for the elements.

13.22 A manufacturing process calls for heating long copper rods, which are coated with a thin film having $\varepsilon = 1$, by placing them in a large evacuated oven whose surface is maintained at 1650 K. If the rods are of 10 mm diameter and are placed in the oven with an initial temperature of 300 K, what is the initial rate of change of the rod temperature?

13.23 Consider the very long, inclined black surfaces (A_1, A_2) maintained at uniform temperatures of $T_1 = 1000$ K and $T_2 = 800$ K.

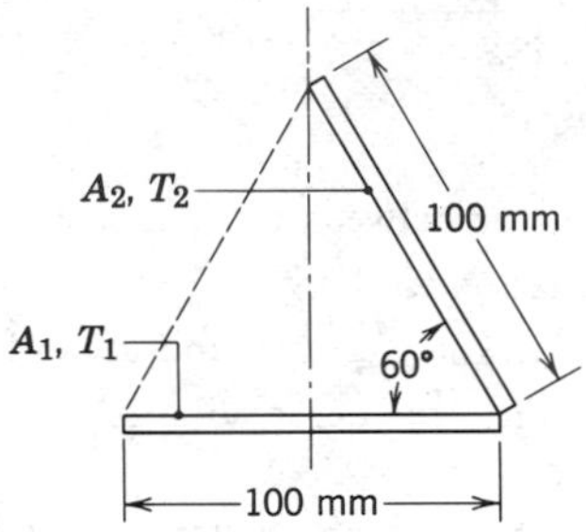

Determine the net radiation exchange between the surfaces per unit length of the surfaces. Consider the configuration when a black surface (A_3), whose back side is insulated, is positioned along the dashed line shown. Calculate the net radiation transfer to surface A_2 per unit length of the surface and determine the temperature of the insulated surface A_3.

13.24 Two plane coaxial disks are separated by a distance $L = 0.20$ m. The lower disk (A_1) is solid with a diameter $D_o = 0.80$ m and a temperature $T_1 = 300$ K. The upper disk (A_2), at temperature $T_2 = 1000$ K, has the same outer diameter but is ring

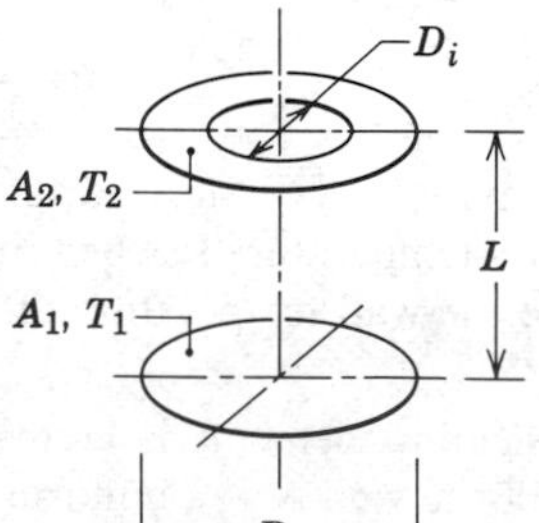

shaped with an inner diameter $D_i = 0.40$ m. Assuming the disks to be blackbodies, calculate the net radiative heat exchange between them.

13.25 A circular ice rink 25 m in diameter is enclosed by a hemispherical dome 35 m in diameter. If the ice and dome surfaces may be approximated as blackbodies and are at 0° and 15°C, respectively, what is the net rate of radiative transfer from the dome to the rink?

13.26 A round tube with a diameter of 0.75 m and a length of 0.33 m has an electrical heater wrapped around the outside, and a heavy layer of insulation is wrapped over the tube–heater combination. The tube is open at both ends and is suspended in a very large vacuum chamber whose walls are at 27°C. The inside surface of the tube is black and is maintained at a steady uniform temperature of 127°C. Determine the electrical power that is supplied to the heater.

13.27 Consider two very large parallel plates with diffuse, gray surfaces.

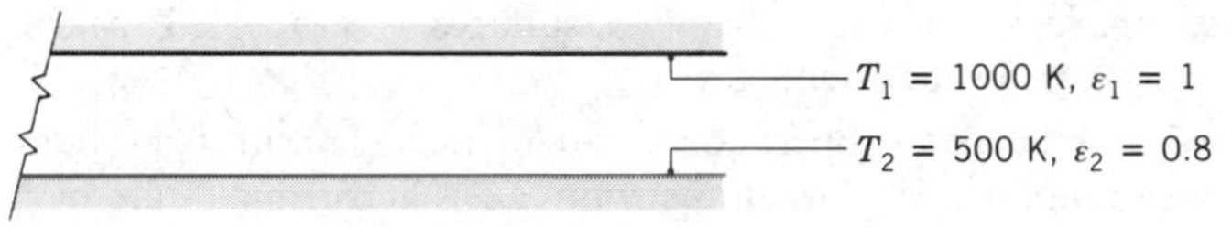

Determine the irradiation and radiosity for the upper plate. What is the radiosity for the lower plate? What is the net radiation exchange between the plates per unit area of the plates?

13.28 A flat-bottomed hole 6 mm in diameter is drilled to a depth of 24 mm in a diffuse, gray material having an emissivity of 0.8 and a uniform temperature of 1000 K.

(a) Determine the radiant power leaving the opening of the cavity.

(b) The effective emissivity ε_e of a cavity is defined as the ratio of the radiant power leaving the cavity to that from a blackbody having the area of the cavity opening and a temperature of the inner surfaces of the cavity. Calculate the effective emissivity of the cavity described above.

(c) If the depth of the hole were increased, would ε_e increase or decrease? What is the limit of ε_e as the depth increases?

13.29 Consider a long V groove 10 mm deep machined on a block that is maintained at 1000 K.

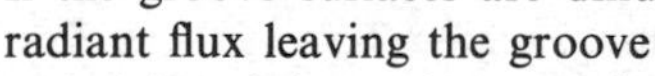

If the groove surfaces are diffuse gray with an emissivity of 0.6, determine the radiant flux leaving the groove to its surroundings. Also determine the effective emissivity of the cavity, as defined in Problem 13.28.

13.30 Consider a flat-bottomed hole of 10 mm diameter which is drilled to a depth of 40 mm in a metallic block of emissivity $\varepsilon = 0.7$. If the block is maintained at 1000 K, what is the rate at which radiation leaves the hole? What is the effective, or apparent, emissivity of the hole, as defined in Problem 13.28.

13.31 Consider a cylindrical cavity 100 mm in diameter and 50 mm in depth whose walls are diffuse gray with an emissivity of 0.6 and have a uniform temperature of 1500 K. Assuming the surroundings of the cavity opening are very large and at 300 K, calculate the net radiation transfer from the cavity.

13.32 A long, thin-walled horizontal tube 100 mm in diameter is maintained at 120°C by the passage of steam through its interior. A radiation shield is installed around the tube, providing an air gap of 10 mm between the tube and the shield, and reaches a surface temperature of 35°C. The tube and shield are diffuse, gray surfaces with emissivities of 0.80 and 0.10, respectively. What is the radiant heat transfer from the tube per unit length?

13.33 A very long electrical conductor 10 mm in diameter is concentric with a cooled cylindrical tube 50 mm in diameter whose surface is diffuse gray with an emissivity of 0.9 and temperature of 27°C. The electrical conductor has a diffuse, gray surface with an emissivity of 0.6 and is dissipating 6.0 W per meter of length. Assuming that the space between the two surfaces is evacuated, calculate the surface temperature of the conductor.

13.34 Under steady-state operation a 50-W incandescent light bulb has a surface temperature of 135°C when the room air is at a temperature of 25°C. If the bulb may be approximated as a 60-mm diameter sphere with a diffuse, gray surface of emissivity 0.8, what is the radiant heat transfer from the bulb surface to its surroundings?

13.35 An arrangement for direct conversion of thermal energy to electrical power is shown. The inner cylinder of diameter $D_i = 25$ mm is heated internally by a combustion process that brings the ceramic cylinder ($\varepsilon_i = 0.9$) to a surface temperature of $T_i = 1675$°C. The outer cylinder of diameter $D_o = 0.38$ m consists of a semiconductor material ($\varepsilon_o = 0.5$) that converts absorbed incident irradiation to electrical current; the backing material for the semiconductor is a highly conducting metal that is water cooled to 20°C. The converter is assumed to be very long compared to the outer diameter. The space between the two concentric cylinders is evacuated.

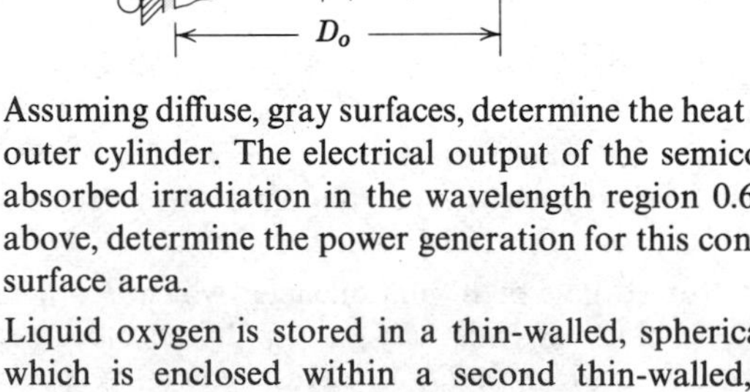

Assuming diffuse, gray surfaces, determine the heat transfer rate per unit area of the outer cylinder. The electrical output of the semiconductor surface is 10% of the absorbed irradiation in the wavelength region 0.6 to 2.0 μm. For the conditions above, determine the power generation for this converter in watts per unit of outer surface area.

13.36 Liquid oxygen is stored in a thin-walled, spherical container 0.8 m in diameter, which is enclosed within a second thin-walled, spherical container 1.2 m in diameter. The opaque, diffuse, gray container surfaces have an emissivity of 0.05

and are separated by an evacuated space. If the outer surface is at 280 K and the inner surface is at 95 K, what is the mass rate of oxygen lost due to evaporation? (The latent heat of vaporization of oxygen is 2.13×10^5 J/kg.)

13.37 The end of a cylindrical liquid cryogenic propellant tank in free space is to be protected from external (solar) radiation by placing a thin metallic shield in front of the tank. Assume the view factor F_{ts} between the tank and the shield is unity; all surfaces are diffuse gray, and the surroundings are at 0 K.

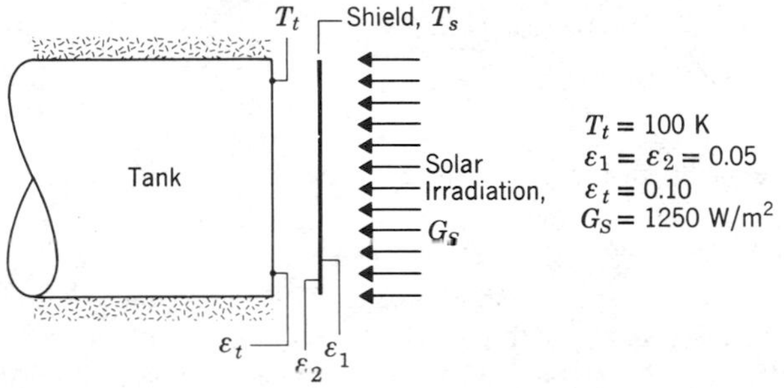

Find the temperature of the shield T_s and the heat flux (W/m^2) to the end of the tank.

13.38 Consider two large, diffuse, gray, parallel surfaces separated by a small distance. If the surface emissivities are 0.8, what emissivity should a thin radiation shield have to reduce the radiation heat transfer rate between the two surfaces by a factor of 10?

13.39 Two concentric spheres of diameters $D_1 = 0.8$ m and $D_2 = 1.2$ m are separated by an air space and have surface temperatures of $T_1 = 400$ K and $T_2 = 300$ K.

(a) If the surfaces are black, what is the net rate of radiation exchange between the spheres?

(b) What is the net rate of radiation exchange between the surfaces if they are diffuse gray with $\varepsilon_1 = 0.5$ and $\varepsilon_2 = 0.05$?

(c) What is the net rate of radiation exchange if D_2 is increased to 20 m, with $\varepsilon_2 = 0.05$, $\varepsilon_1 = 0.5$ and $D_1 = 0.8$ m? What error would be introduced by assuming blackbody behavior for the outer surface ($\varepsilon_2 = 1$), with all other conditions remaining the same?

13.40 A cryogenic fluid flows through a tube 20 mm in diameter, the outer surface of which is diffuse gray with an emissivity of 0.02 and temperature of 77 K. This tube is concentric with a larger tube of 50 mm diameter, the inner surface of which is diffuse gray with an emissivity of 0.05 and temperature of 300 K. The space between the surfaces is evacuated. Determine the heat gain by the cryogenic fluid per unit length of the inner tube. If a thin-walled radiation shield that is diffuse gray with an emissivity of 0.02 (both sides) is inserted midway between the inner and outer surfaces, calculate the change (percentage) in heat gain per unit length of the inner tube.

13.41 A diffuse, gray radiation shield of 60 mm diameter and emissivities of $\varepsilon_{2,i} = 0.01$ and $\varepsilon_{2,o} = 0.1$ on the inner and outer surfaces, respectively, is concentric with a long tube transporting a hot process fluid. The tube surface is black with a diameter of 20 mm. The region interior to the shield is evacuated. The exterior surface of the shield is exposed to a large room whose walls are at 17°C and experiences convection with air at 27°C and a convection heat transfer coefficient of 10 $W/m^2 \cdot K$.

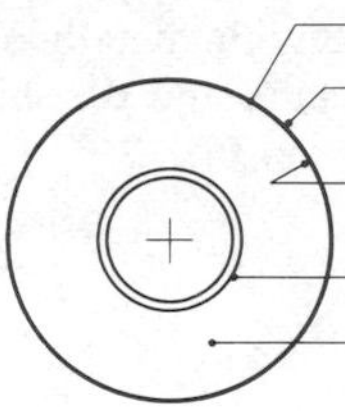

Determine the operating temperature for the inner tube if the shield temperature is maintained at 42°C.

13.42 At the bottom of a very large vacuum chamber whose walls are at 300 K, a black panel 0.1 m in diameter is maintained at 77 K. To reduce the heat gain to this panel, a radiation shield, of the same diameter D, and with an emissivity of 0.05, is placed very close to the panel. Calculate the net heat gain to the panel.

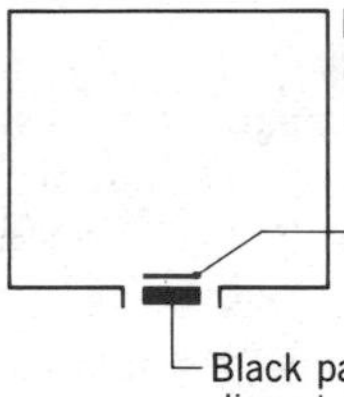

13.43 Heat transfer by radiation occurs between two large parallel plates, which are maintained at temperatures $T_1 > T_2$. To reduce the rate of heat transfer between the plates, it is proposed that they be separated by a thin shield that has different emissivities on opposite surfaces. In particular one surface has the emissivity $\varepsilon_s < 0.5$, while the opposite surface has an emmissivity of $2\varepsilon_s$.

(a) How should the shield be oriented to provide the largest reduction in heat transfer between the plates? That is, should the surface of emissivity ε_s or that of emissivity $2\varepsilon_s$ be oriented towards the plate at T_1?

(b) What orientation will result in the larger value of the shield temperature T_s?

13.44 A large slab of meat 0.15 m × 0.30 m in area is taken from a refrigerator at 4°C and is placed on a wire grill 0.15 m above and parallel to a bed of hot coals at 850°C with approximately the same area. Assume that the meat and the coals are essentially black and neglect convection.

(a) What is the initial heat transfer rate between the coals and the meat?

(b) By what percentage would this rate increase if well-insulated sidewalls were placed around the perimeter of the system?

(c) What is the average temperature of the insulated wall?

13.45 The bottom of a steam-producing still of 200 mm diameter is heated by radiation. The heater, maintained at 1000°C and separated 100 mm from the still, has the same diameter as the still bottom. The still bottom and heater surfaces are black.

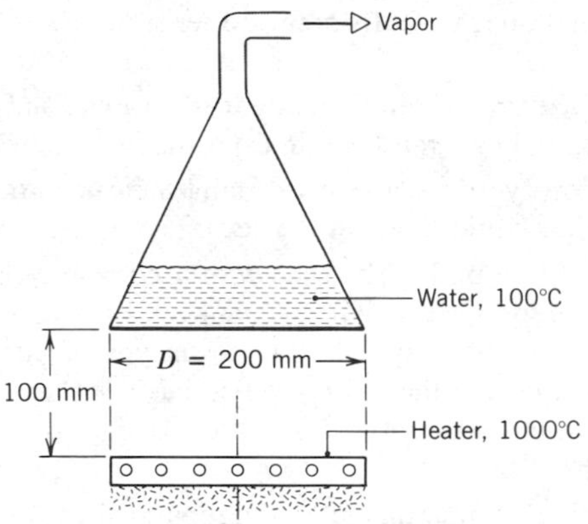

By what factor could the vapor production rate be increased if the cylindrical sides (dashed surface) were insulated rather than open to the surroundings maintained at 27°C?

13.46 A long cylindrical heater element of diameter $D = 10$ mm, temperature $T_1 = 1500$ K, and emissivity $\varepsilon_1 = 1$ is used in the furnace shown as follows. The bottom area A_2 is a diffuse, gray surface with $\varepsilon_2 = 0.6$ and is maintained at $T_2 = 500$ K. The side and top walls are fabricated from an insulating, refractory brick that is diffuse gray with $\varepsilon = 0.9$. The length of the furnace normal to the page, l, is very large compared to the width w and height h.

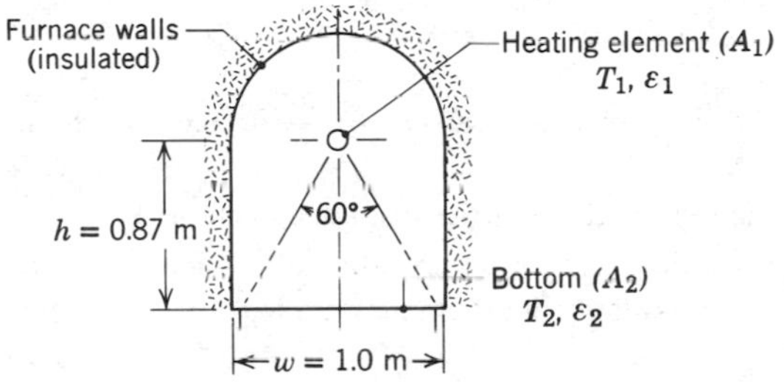

Neglecting convection and treating the furnace walls as isothermal, determine the power per unit length that must be provided to the heating element to maintain steady-state conditions. Calculate the temperature of the furnace wall area.

13.47 Consider two aligned, parallel, square planes (0.4 m × 0.4 m) spaced 0.8 m apart and maintained at $T_1 = 500$ K and $T_2 = 800$ K. Calculate the net radiative heat transfer *from surface 1* for the following special conditions.

(a) Both planes are black and the surroundings are at 0 K.

(b) Both planes are black with connecting, reradiating walls.

(c) Both planes are diffuse gray with $\varepsilon_1 = 0.6$, $\varepsilon_2 = 0.8$, and the surroundings at 0 K.

(d) Both planes are diffuse gray ($\varepsilon_1 = 0.6$ and $\varepsilon_2 = 0.8$) with connecting, reradiating walls.

13.48 Two parallel, diffuse gray surfaces, each 1 m × 2 m, face each other and are separated by 1 m. Each surface has an emissivity of 0.2. One surface is at 27°C and the other is at 277°C.

(a) Calculate the net radiation transfer from the hotter surface, if the surroundings are at 0 K.

(b) If well-insulated, diffuse gray sidewalls are added to this configuration, what will be the net radiation heat transfer rate from the hotter surface?

(c) Review the assumptions you made in performing the calculations and briefly explain what effect they could have on the accuracy of your results.

13.49 Consider a room that is 4 m long by 3 m wide with a floor-to-ceiling distance of 2.5 m. The four walls of the room are well insulated, while the surface of the floor is maintained at a uniform temperature of 30°C by means of electric resistance heaters. Heat loss occurs through the ceiling, which has a surface temperature of 12°C. If all surfaces have an emissivity of 0.9, what is the rate of heat loss by radiation from the room?

13.50 Two parallel, aligned disks, 0.4 m in diameter and separated by 0.1 m, are located in a large room whose walls are maintained at 300 K. One of the disks is maintained at a uniform temperature of 500 K with an emissivity of 0.6, while the back side of the second disk is well insulated. If the disks are diffuse, gray surfaces, determine the temperature of the insulated disk.

13.51 Consider a long duct constructed with diffuse, gray walls 1 m wide.

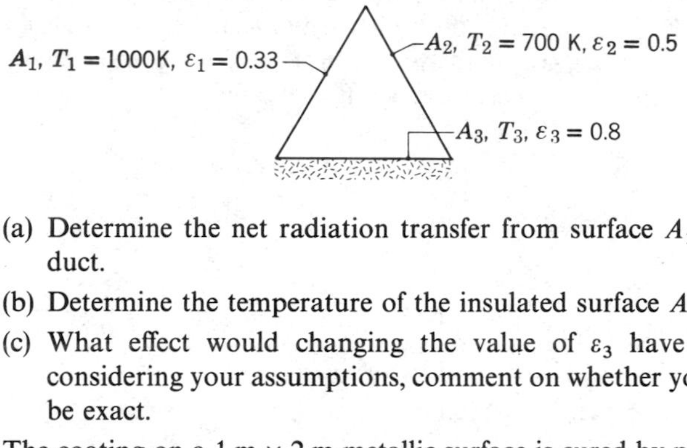

(a) Determine the net radiation transfer from surface A_1 per unit length of the duct.

(b) Determine the temperature of the insulated surface A_3.

(c) What effect would changing the value of ε_3 have on your result? After considering your assumptions, comment on whether you expect your results to be exact.

13.52 The coating on a 1 m × 2 m metallic surface is cured by placing it 0.5 m below an electrically heated surface of equivalent dimensions.

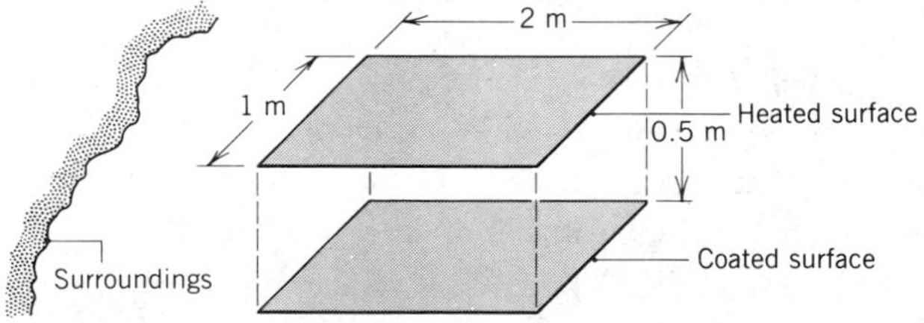

The heater is well insulated on its top side and is aligned with the coated surface. Both the heater and coated surfaces may be approximated as blackbodies. During the curing procedure, the heater and coated surfaces are maintained at 700 and 400 K, respectively. Convection effects may be neglected. If the assembly is exposed to large surroundings at 300 K, what is the electric power required to operate the heater? If the surfaces are connected with reradiating sidewalls, what is the electric power required to operate the heater?

13.53 A cubical furnace 2 m on a side is used for heat treating steel plate. The top surface of the furnace consists of electrical radiant heaters that have an emissivity of 0.8 and a power input of 1.5×10^5 W. The sidewalls consist of a well insulated refractory material, while the bottom consists of the steel plate, which has an emissivity of 0.4. Assume diffuse, gray surface behavior for the heater and the plate, and consider conditions for which the plate is at 300 K. What are the corresponding temperatures of the heater surface and the sidewalls?

13.54 An opaque, diffuse, gray (200 mm × 200 mm) plate with an emissivity of 0.8 is placed over the opening of a furnace and is known to be at 400 K at a certain instant. The bottom of the furnace, having the same dimensions as the plate, is black and operates at 1000 K. The sidewalls of the furnace are well insulated. The top of the plate is exposed to ambient air with a convection coefficient of $25\ W/m^2 \cdot K$ and to large surroundings. The air and surroundings are each at 300 K.

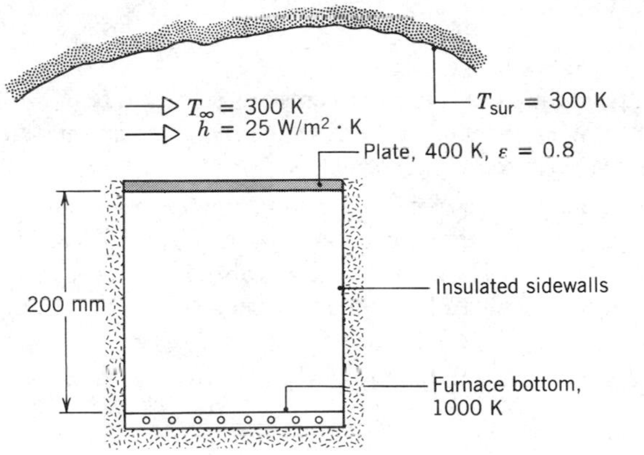

(a) Evaluate the net radiative heat transfer to the bottom surface of the plate.

(b) If the plate has mass and specific heat of 2 kg and 900 J/kg·K, respectively, what will be the change in temperature of the plate with time? Assume convection to the bottom surface of the plate to be negligible.

13.55 An electric furnace consisting of two heater sections, top and bottom, is used to heat treat a coating that is applied to both surfaces of a thin metal plate inserted midway between the heaters.

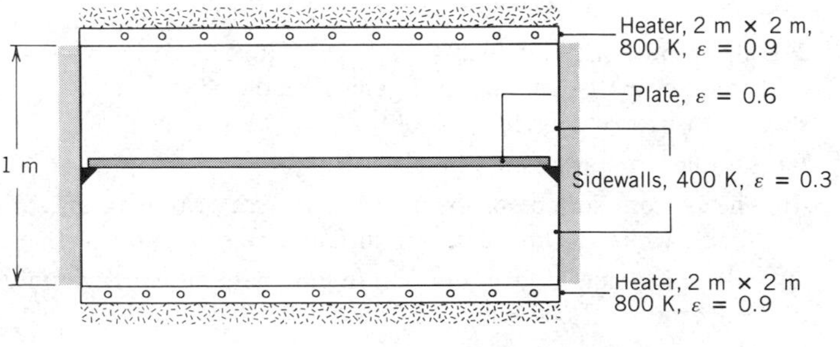

The heaters and the plate are 2 m × 2 m on a side, and each heater is separated from the plate by a distance of 0.5 m. Each heater is well insulated on its back side and has an emissivity of 0.9 at its exposed surface. The plate and the sidewalls have emissivities of 0.6 and 0.3, respectively. Sketch the equivalent radiation network for the system and label all pertinent resistances and potentials. For the prescribed conditions, obtain the required electrical power and the plate temperature.

13.56 A solar collector consists of a long duct through which air is blown; its cross section forms an equilateral triangle 1 m on a side. One side consists of a glass cover of emissivity $\varepsilon_1 = 0.9$, while the other two sides are absorber plates, $\varepsilon_2 = \varepsilon_3 = 1.0$.

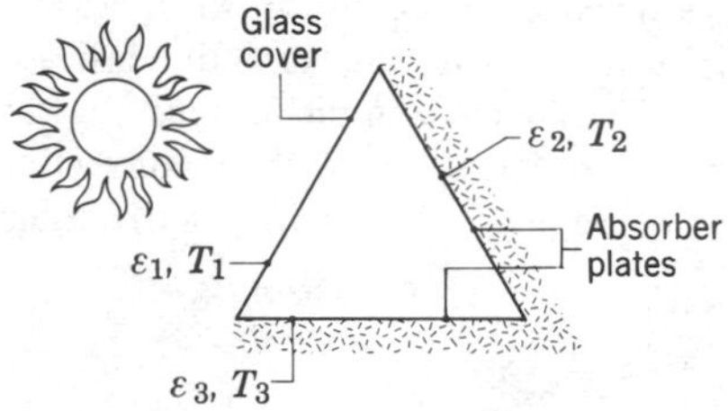

During operation the surface temperatures are known to be $T_1 = 25°C$, $T_2 = 60°C$, and $T_3 = 70°C$. What is the net rate at which radiation is transferred to the cover due to exchange with the absorber plates?

13.57 Consider a circular furnace that is 0.3 m long and 0.3 m in diameter. The two ends have diffuse, gray surfaces that are maintained at 400 and 500 K with emissivities of 0.4 and 0.5, respectively. The lateral surface is also diffuse gray with an emissivity of 0.8 and temperature of 800 K. Determine the net radiative heat transfer from each of the surfaces.

13.58 Consider the cylindrical cavity closed at the bottom with an opening in the top surface as shown in cross section.

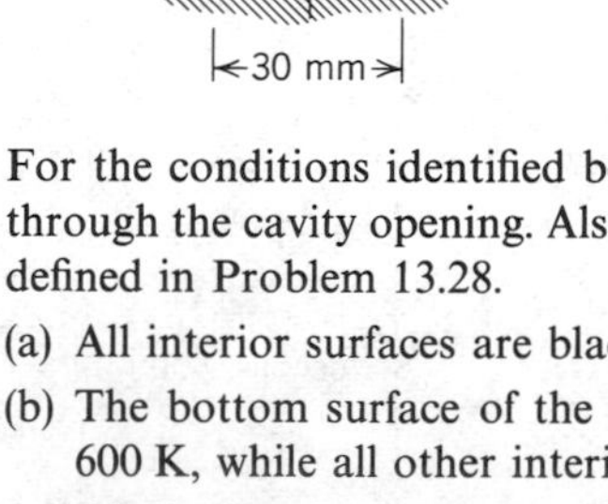

For the conditions identified below, calculate the rate at which radiation passes through the cavity opening. Also determine the effective emissivity of the cavity as defined in Problem 13.28.

(a) All interior surfaces are black at 600 K.

(b) The bottom surface of the cavity is diffuse gray with an emissivity of 0.6 at 600 K, while all other interior surfaces are reradiating.

(c) All the interior surfaces are diffuse gray with an emissivity of 0.6 and uniform temperature of 600 K.

13.59 Treat the system of Problem 12.63 as a three-surface enclosure to evaluate the radiant power leaving the 2-mm-diameter hole.

13.60 The enclosure analysis of diffuse, gray surfaces presented in Section 13.3 requires that the irradiation over any surface be uniform. When the geometry of the enclosure precludes this requirement, it is necessary to divide the isothermal surface into zones over which the irradiation is approximately uniform. In the cylindrical cavity shown the irradiations along the lateral surface and on the base are likely to be different.

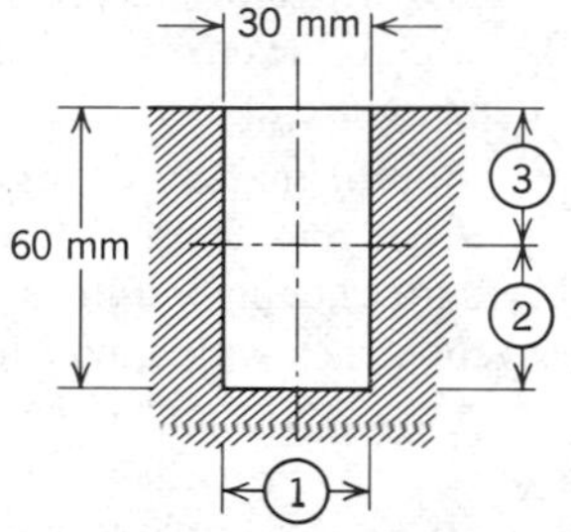

Dividing the cavity into three zones, each having the same surface temperature, perform an enclosure analysis to determine the radiative heat transfer leaving the cavity opening. The surfaces of the cavity are diffuse gray with an emissivity of 0.2 and with a uniform temperature of 1000 K. Compare your result to the calculation treating the cavity walls (base plus lateral side) as a single surface.

13.61 Consider the diffuse, gray, four-surface enclosure with all sides equal as shown. The temperatures of three surfaces are specified, while the fourth surface is well insulated and can be treated as a reradiating surface. Determine the temperature of the reradiating surface (4).

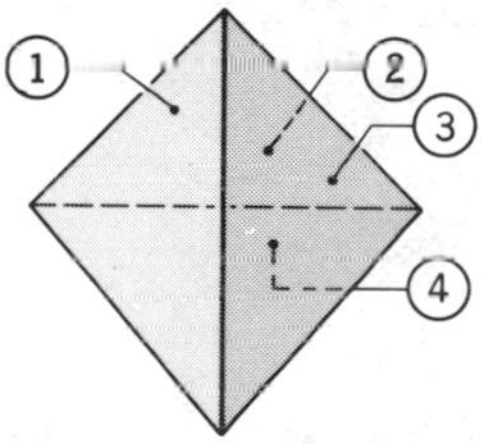

$T_1 = 700$ K $\quad \varepsilon_1 = 0.7$
$T_2 = 500$ K $\quad \varepsilon_2 = 0.5$
$T_3 = 300$ K $\quad \varepsilon_3 = 0.3$

13.62 A room is represented by the enclosure shown as follows, where the ceiling (1) has an emissivity of 0.8 and is maintained at 40°C by embedded electrical heating

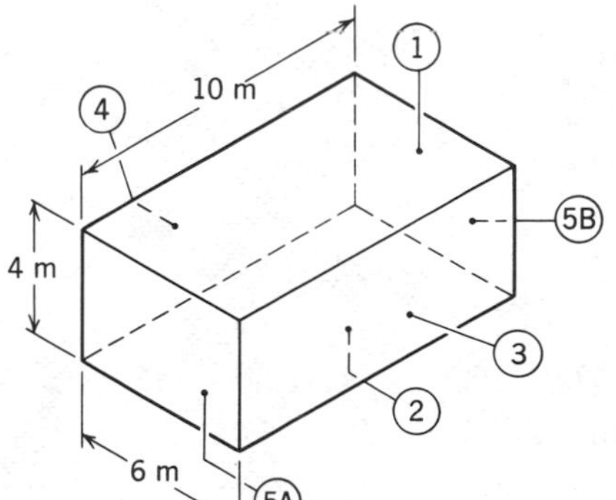

elements. Heaters are also used to maintain the floor (2) with an emissivity of 0.9 at 50°C. The right wall (3) with an emissivity of 0.7 reaches a temperature of 15°C on a cold, winter day. The left wall (4) and end walls (5A, 5B) are very well insulated. To simplify the analysis, treat the two end walls as a single surface (5). Assuming the surfaces are diffuse gray, find the net radiation heat transfer from each surface.

13.63 A furnace having a spherical cavity of 0.5 m diameter contains a gas mixture at 1 atm and 1400 K. The mixture consists of CO_2 with a partial pressure of 0.25 atm and nitrogen with a partial pressure of 0.75 atm. If the cavity wall is black, what is the cooling rate needed to maintain its temperature at 500 K?

13.64 A gas turbine combustion chamber may be approximated as a long tube of 0.4 m diameter. The combustion gas is at a pressure and temperature of 1 atm and 1000°C, respectively, while the chamber surface temperature is 500°C. If the combustion gas contains CO_2 and water vapor, each with a mole fraction of 0.15, what is the net radiative heat flux between the gas and the chamber surface, which may be approximated as a blackbody?

13.65 A flue gas at 1 atm total pressure and a temperature of 1400 K contains CO_2 and water vapor at partial pressures of 0.05 and 0.10 atm, respectively. If the gas flows through a long flue of 1 m diameter and 400 K surface temperature, determine the net radiative heat flux from the gas to the surface. Blackbody behavior may be assumed for the surface.

13.66 A furnace consists of two large parallel plates separated by 0.75 m. A gas mixture comprised of O_2, N_2, CO_2, and water vapor, with mole fractions of 0.20, 0.50, 0.15, and 0.15, respectively, flows between the plates at a total pressure of 2 atm and a temperature 1300 K. If the plates may be approximated as blackbodies and are maintained at 500 K, what is the net radiative heat flux to the plates?

Fourteen

MULTIMODE HEAT TRANSFER

In Chapter 1 we introduced the different modes of heat transfer and stressed the importance of being able to identify which modes are important in any given problem. In subsequent chapters, however, we isolated each mode to develop the computational tools needed for their respective treatments. In Chapters 2 through 5, for example, we focused on means for determining temperatures and heat transfer rates in solids experiencing heat transfer by conduction. Although we recognized that convection heat transfer occurring at the surface of a solid strongly influences conditions within the solid, we were forced to presume that the convection coefficient was known. In Chapters 6 through 10 we considered means of determining the convection coefficient for a variety of flow conditions. However, in most cases we presumed the surface temperature of the adjoining solid to be known, when, in fact, this temperature could be strongly influenced by conduction within the solid. In Chapters 2 through 11 we gave little attention to the important effect that radiation exchange at a surface or between surfaces can have on thermal conditions. The situation was rectified in Chapters 12 and 13, but little attention was given to conduction and convection effects.

The time has come to recognize that most technically important problems involve *multimode heat transfer* effects. For example, conduction may occur within a solid, while convection and radiation occur at its boundaries. Also, radiation exchange may occur between two or more solid surfaces, while heat is transferred by convection at the surfaces and by conduction within the solids. To solve such problems, it may be necessary to work simultaneously with the heat equation, surface energy balances, overall energy balances, and appropriate rate equations for conduction, convection and radiation heat transfer. Moreover, the problems may involve several unknowns and require the use of iterative solution methods.

In this chapter we consider several multimode problems that illustrate the manner in which previously developed tools may be integrated to obtain solutions. We hope that you will begin to develop a state of mind that allows you to identify relevant transport processes and to implement appropriate rate equations and conservation principles.

EXAMPLE 14.1

The rear window of an automobile is of thickness $L = 8$ mm and height $H = 0.5$ m and contains fine-meshed heating wires that can induce nearly volumetric heating, $\dot{q}$ (W/m^3). Consider steady-state conditions for which the interior surface of the window is exposed to air at 10°C, while the exterior surface is exposed to air at −10°C and a very cold sky with an effective temperature of −33°C. The air moves in parallel flow over the surface with a velocity of 20 m/s. Determine the volumetric heating rate needed to maintain the interior window surface at $T_{s,i} = 15$°C.

SOLUTION

Known:

Boundary conditions associated with a rear window experiencing uniform volumetric heating.

Find:

Volumetric heating rate $\dot{q}$ needed to maintain inner surface temperature at $T_{s,i} = 15°C$.

Schematic:

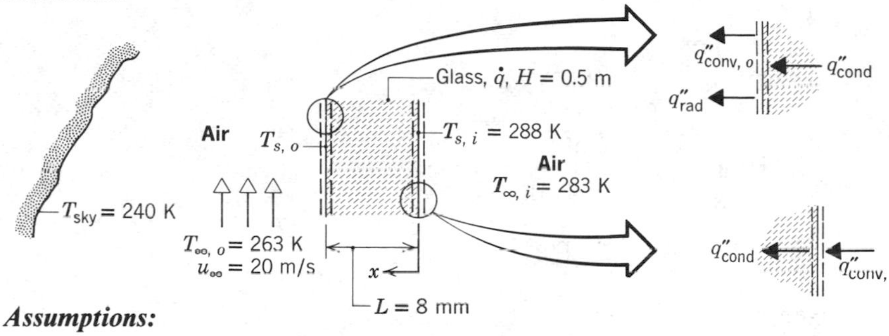

Assumptions:

1. Steady-state, one-dimensional conditions.
2. Constant properties.
3. Uniform volumetric heating in window.
4. Convection heat transfer from interior surface of window to interior air may be approximated as free convection from a vertical plate. Radiation exchange with automobile interior is negligible.
5. Heat transfer from outer surface is due to forced convection over a flat plate and radiation exchange with the sky.
6. Negligible boundary layer separation on the outer surface.
7. Outer surface is diffuse and gray.

Properties:

Table A.3, glass (300 K): $k = 1.4$ W/m·K.
Table A.8, glass (300 K): $\varepsilon \approx 0.90$.
Table A.4, air ($T_{f,i} = 12.5°C$): $\nu = 14.6 \times 10^{-6}$ m^2/s, $k = 0.0251$ W/m·K, $\alpha = 20.59 \times 10^{-6}$ m^2/s, $\beta = (1/285.5) = 3.503 \times 10^{-3}$ K^{-1}, $Pr = 0.711$. Air ($T_{f,o} \approx 0°C$): $\nu = 13.49 \times 10^{-6}$ m^2/s, $k = 0.0241$ W/m·K, $Pr = 0.714$.

Analysis:

The temperature distribution in the glass is governed by the appropriate form of the heat equation, Equation 3.38, whose general solution is given by Equation 3.39.

$$T(x) = -\frac{\dot{q}}{2k}x^2 + C_1 x + C_2$$

The constants of integration may be evaluated by applying appropriate boundary conditions at $x=0$. In particular, with $T(0)=T_{s,i}$, $C_2=T_{s,i}$. Applying an energy balance to the inner surface, it also follows that

$$q''_{\text{cond}} = q''_{\text{conv},i}$$

or

$$-k\frac{dT}{dx}\bigg|_{x=0} = \bar{h}_i(T_{\infty,i} - T_{s,i})$$

$$-k\left(-\frac{\dot{q}}{k}x + C_1\right)\bigg|_{x=0} = \bar{h}_i(T_{\infty,i} - T_{s,i})$$

$$C_1 = -\frac{\bar{h}_i}{k}(T_{\infty,i} - T_{s,i})$$

Hence

$$T(x) = -\frac{\dot{q}}{2k}x^2 - \frac{\bar{h}_i(T_{\infty,i} - T_{s,i})}{k}x + T_{s,i} \tag{1}$$

The required generation may then be obtained by formulating an energy balance at the outer surface. That is, at $x=L$

$$q''_{\text{cond}} = q''_{\text{conv},o} + q''_{\text{rad}}$$

or

$$-k\frac{dT}{dx}\bigg|_{x=L} = \bar{h}_o(T_{s,o} - T_{\infty,o}) + \varepsilon\sigma(T_{s,o}^4 - T_{\text{sky}}^4) \tag{2}$$

From Equation 1 it follows that

$$-k\frac{dT}{dx}\bigg|_{x=L} = -k\left(-\frac{\dot{q}L}{k}\right) + \bar{h}_i(T_{\infty,i} - T_{s,i}) = \dot{q}L + \bar{h}_i(T_{\infty,i} - T_{s,i}) \tag{3}$$

Substituting Equation 3 into Equation 2, the energy balance becomes

$$\dot{q}L = \bar{h}_o(T_{s,o} - T_{\infty,o}) + \bar{h}_i(T_{s,i} - T_{\infty,i}) + \varepsilon\sigma(T_{s,o}^4 - T_{\text{sky}}^4) \tag{4}$$

where $T_{s,o}$ may be evaluated by applying Equation 1 at $x=L$.

$$T_{s,o} = -\frac{\dot{q}L^2}{2k} - \frac{\bar{h}_i(T_{\infty,i} - T_{s,i})}{k}L + T_{s,i} \tag{5}$$

The required generation rate may now be obtained by substituting appropriate numerical values into Equations 4 and 5 and solving for $\dot{q}$ by trial and error.

The inside convection coefficient may be obtained from Equation 9.26

$$\overline{Nu_H} = \left\{0.825 + \frac{0.387 Ra_H^{1/6}}{[1+(0.492/Pr)^{9/16}]^{8/27}}\right\}^2$$

where

$$Ra_H = \frac{g\beta(T_{s,i} - T_{\infty,i})H^3}{\nu\alpha}$$

$$Ra_H = \frac{9.8\ \text{m/s}^2\ (3.503\times10^{-3}\ \text{K}^{-1})(15-10)\ \text{K}\ (0.5\ \text{m})^3}{14.60\times10^{-6}\ \text{m}^2/\text{s}\times 20.59\times10^{-6}\ \text{m}^2/\text{s}} = 7.137\times10^7$$

Hence

$$\overline{Nu_H} = \left\{0.825 + \frac{0.387(7.137\times10^7)^{1/6}}{[1+(0.492/0.711)^{9/16}]^{8/27}}\right\}^2 = 56$$

and

$$\bar{h}_i = \overline{Nu_H}\frac{k}{H} = \frac{56\times0.0251\ \text{W/m}\cdot\text{K}}{0.5\ \text{m}} = 2.81\ \text{W/m}^2\cdot\text{K}$$

The outside convection coefficient may be obtained by first evaluating the Reynolds number. With

$$Re_H = \frac{u_\infty H}{\nu} = \frac{20\ \text{m/s}\times0.5\ \text{m}}{13.49\times10^{-6}\ \text{m}^2/\text{s}} = 7.413\times10^5$$

and $Re_{x,c}$ assumed to equal 5×10^5, mixed boundary layer conditions exist on the outer surface. Hence, from Equation 7.32,

$$\overline{Nu_H} = (0.037 Re_H^{4/5} - 871)Pr^{1/3}$$

$$\overline{Nu_H} = [0.037(7.413\times10^5)^{4/5} - 871](0.714)^{1/3} = 864$$

and

$$\bar{h}_o = \overline{Nu_H}\frac{k}{H} = \frac{864\times0.0241\ \text{W/m}\cdot\text{K}}{0.5\ \text{m}} = 41.6\ \text{W/m}^2\cdot\text{K}$$

Equation 5 may now be expressed as

$$T_{s,o} = -\frac{\dot{q}(0.008\ \text{m})^2}{2(1.4\ \text{W/m}\cdot\text{K})} - \frac{2.81\ \text{W/m}^2\cdot\text{K}\ (10-15)\ \text{K}}{1.4\ \text{W/m}\cdot\text{K}}\times0.008\ \text{m} + 288\ \text{K}$$

$$T_{s,o} = -2.286\times10^{-5}\dot{q} + 288.1\ \text{K}$$

or, solving for $\dot{q}$,

$$\dot{q} = -43745(T_{s,o} - 288.1) \tag{6}$$

and substituting into Equation 4,

$$-43745\,(T_{s,o}-288.1)\,(0.008\text{ m})=41.6\text{ W/m}^2\cdot\text{K}\,(T_{s,o}-263\text{ K})$$
$$+2.81\text{ W/m}^2\cdot\text{K}\,(288-283)\text{ K}+0.90\times 5.67$$
$$\times 10^{-8}\text{ W/m}^2\cdot\text{K}^4\,(T_{s,o}^4-240^4)\text{ K}^4$$

From trial-and-error calculations, it follows that

$$T_{s,o}\approx 285\text{ K}$$

in which case, from Equation 6

$$\dot{q}\approx 136\text{ kW/m}^3 \qquad \triangleleft$$

Comments:

In lieu of performing a surface energy balance at $x=L$, Equation 4 may also be obtained by applying an energy balance to a control volume about the entire window.

EXAMPLE 14.2

The composite insulation shown below and described in Problem 1.26d is being considered as a roofing material.

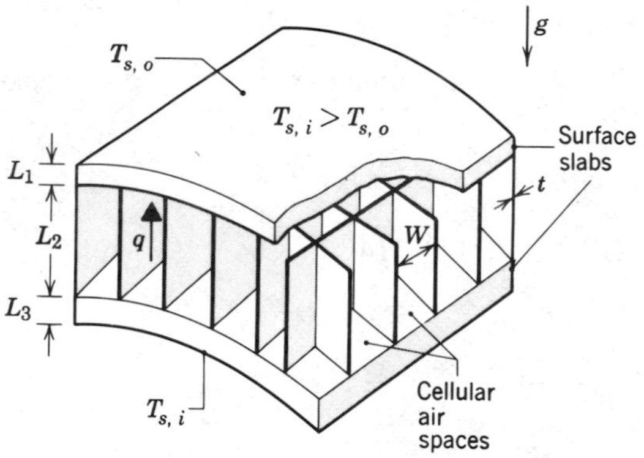

It is proposed that the outer and inner slabs be made from low density particle board of thicknesses $L_1=L_3=12.5$ mm and that the honeycomb core be constructed from a high density particle board. The square cells of the core are to be of length $L_2=50$ mm, width $W=10$ mm, and wall thickness $t=2$ mm. The emissivity of both particle boards is approximately 0.85, and the honeycomb cells are filled with air at 1 atm pressure. To assess the effectiveness of the insulation, its total thermal resistance must be evaluated under representative operating conditions for which the bottom (inner) surface temperature is $T_{s,i} = 25°C$ and the top (outer) surface temperature is $T_{s,o} = -10°C$.

SOLUTION

Known:

Dimensions of a composite insulation consisting of a honeycomb core sandwiched between solid slabs.

Find: Total thermal resistance.

Schematic:

Because of the repetitive nature of the honeycomb core, the cell sidewalls will be adiabatic. That is, there is no lateral heat transfer from cell to cell, and it suffices to consider the heat transfer across a single cell.

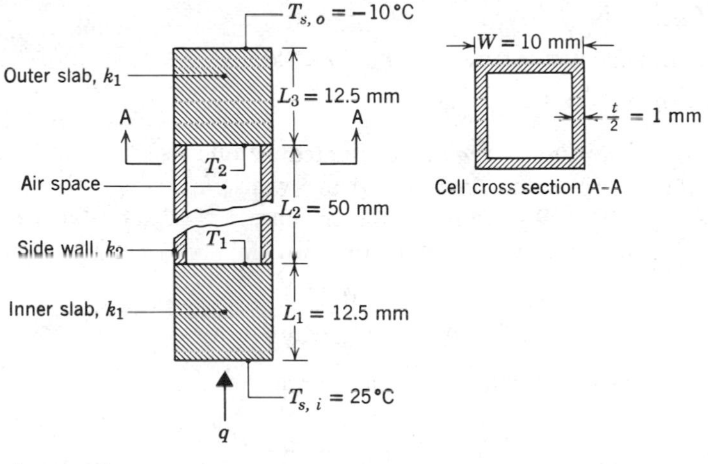

Assumptions:

1. One-dimensional, steady-state conditions.
2. Equivalent conditions for each cell.
3. Constant properties.
4. Diffuse, gray surface behavior.

Properties:

Table A.3, particle board (low density): $k_1 = 0.078$ W/m·K. Particle board (high density): $k_2 = 0.170$ W/m·K.
For both board materials, $\varepsilon = 0.85$.

Table A.4, air ($\bar{T} \approx 7.5°C$): $\nu = 14.15 \times 10^{-6}$ m^2/s, $k = 0.0247$ W/m·K, $\alpha = 19.9 \times 10^{-6}$ m^2/s, $Pr = 0.71$, $\beta = 3.57 \times 10^{-3}$ K^{-1}.

Analysis:

The total resistance of the composite is determined by conduction, convection, and radiation processes occurring within the honeycomb and by conduction across the inner and outer slabs. The corresponding thermal circuit follows.

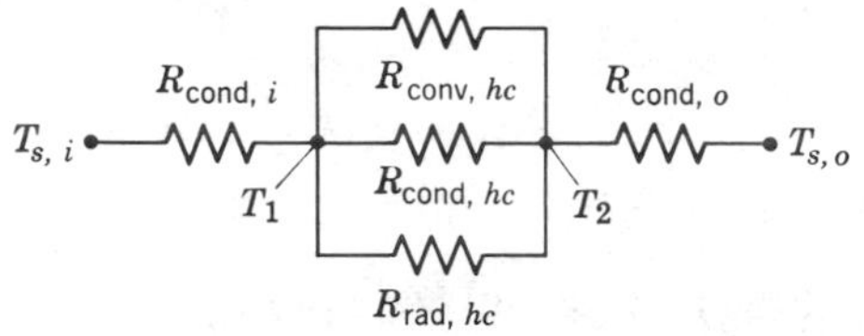

The total resistance of the composite is

$$R = R_{\text{cond},i} + R_{\text{eq}} + R_{\text{cond},o}$$

where the equivalent resistance for the honeycomb is

$$R_{\text{eq}}^{-1} = (R_{\text{cond}}^{-1} + R_{\text{conv}}^{-1} + R_{\text{rad}}^{-1})_{hc}$$

The component resistances may be evaluated as follows.

The inner and outer slabs are plane walls, for which the thermal resistance is given by Equation 3.6. Hence, since $L_1 = L_3$ and the slabs are constructed from low density particle board,

$$R_{\text{cond},i} = R_{\text{cond},o} = \frac{L_1}{k_1 W^2} = \frac{0.0125 \text{ m}}{0.078 \text{ W/m·K } (0.01 \text{ m})^2} = 1603 \text{ K/W}$$

Similarly, applying Equation 3.6 to the sidewalls of the cell,

$$R_{\text{cond},hc} = \frac{L_2}{k_2[W^2 - (W-t)^2]} = \frac{L_2}{k_2(2Wt - t^2)}$$

$$R_{\text{cond},hc} = \frac{0.050 \text{ m}}{0.170 \text{ W/m·K } [2 \times 0.01 \text{ m} \times 0.002 \text{ m} - (0.002 \text{ m})^2]}$$

$$= 8170 \text{ K/W}$$

From Equation 3.9 the convection resistance associated with the cellular air space may be expressed as

$$R_{\text{conv},hc} = \frac{1}{h(W-t)^2}$$

The cell forms an enclosure that may be classified as a horizontal cavity heated from below, and the appropriate form of the Rayleigh number is

$$Ra_L = \frac{g\beta(T_1 - T_2)L_2^3}{\alpha\nu}$$

To evaluate this parameter, however, it is necessary to *assume* a value of the cell temperature difference. As a first approximation we choose

$$(T_1 - T_2) = [15°\text{C} - (-5°\text{C})] = 20°\text{C}$$

in which case

$$Ra_L = \frac{9.8\ \text{m/s}^2\ (3.57\times10^{-3}\ \text{K}^{-1})(20\ \text{K})(0.05\ \text{m})^3}{19.9\times10^{-6}\ \text{m}^2/\text{s}\times14.15\times10^{-6}\ \text{m}^2/\text{s}} = 3.11\times10^5$$

Applying Equation 9.37 as a first approximation, it follows that

$$h = \frac{k}{L_2}[0.069 Ra_L^{1/3} Pr^{0.074}]$$

$$h = \frac{0.0247\ \text{W/m}\cdot\text{K}}{0.05\ \text{m}}[0.069(3.11\times10^5)^{1/3}(0.71)^{0.074}] = 2.25\ \text{W/m}^2\cdot\text{K}$$

The convection resistance is then

$$R_{\text{conv},hc} = \frac{1}{2.25\ \text{W/m}^2\cdot\text{K}\ (0.01\ \text{m} - 0.002\ \text{m})^2} = 6944\ \text{K/W}$$

The resistance to heat transfer by radiation may be obtained by first noting that the cell forms a three-surface enclosure for which the sidewalls are reradiating. The net radiation heat transfer between the end surfaces of the cell is then given by Equation 13.30. With $\varepsilon_1 = \varepsilon_2 = \varepsilon$ and $A_1 = A_2 = (W - t)^2$, the equation reduces to

$$q_{\text{rad}} = \frac{(W-t)^2\sigma(T_1^4 - T_2^4)}{2\left(\frac{1}{\varepsilon} - 1\right) + \frac{1}{F_{12} + [(F_{1R} + F_{2R})/F_{1R}F_{2R}]^{-1}}}$$

However, with $F_{1R} = F_{2R} = (1 - F_{12})$ it follows that

$$q_{\text{rad}} = \frac{(W-t)^2\sigma(T_1^4 - T_2^4)}{2\left(\frac{1}{\varepsilon} - 1\right) + \left[F_{12} + \frac{(1-F_{12})^2}{2(1-F_{12})}\right]^{-1}} = \frac{(W-t)^2\sigma(T_1^4 - T_2^4)}{2\left(\frac{1}{\varepsilon} - 1\right) + \frac{2}{1 + F_{12}}}$$

The view factor F_{12} may be obtained from Figure 13.4, where

$$\frac{X}{L} = \frac{Y}{L} = \frac{(W-t)}{L_2} = \frac{(10-2)\ \text{mm}}{50\ \text{mm}} = 0.16$$

Hence $F_{12} \approx 0.01$. Defining the radiation resistance as

$$R_{\text{rad},hc} = \frac{T_1 - T_2}{q_{\text{rad}}}$$

it follows that

$$R_{\mathrm{rad},hc}=\frac{2\left(\frac{1}{\varepsilon}-1\right)+\frac{2}{1+F_{12}}}{(W-t)^2\sigma(T_1^2+T_2^2)(T_1+T_2)}$$

where

$$(T_1^4-T_2^4)=(T_1^2+T_2^2)(T_1+T_2)(T_1-T_2).$$

Accordingly,

$$R_{\mathrm{rad},hc}=\left[2\left(\frac{1}{0.85}-1\right)+\frac{2}{1+0.01}\right]\div$$

$$\{(0.01\ \mathrm{m}-0.002\ \mathrm{m})^2\times 5.67\times 10^{-8}\ \mathrm{W/m^2\cdot K^4}\ [(288\ \mathrm{K})^2$$
$$+(268\ \mathrm{K})^2](288+268)\ \mathrm{K}\}$$

where, again, it is *assumed* that $T_1=15°\mathrm{C}$ and $T_2=-5°\mathrm{C}$. From the above expression, it follows that

$$R_{\mathrm{rad},hc}=\frac{0.353+1.980}{3.123\times 10^{-4}}=7471\ \mathrm{K/W}$$

In summary the component resistances are then

$$R_{\mathrm{cond},i}=R_{\mathrm{cond},o}=1603\ \mathrm{K/W}$$

$$R_{\mathrm{cond},hc}=8170\ \mathrm{K/W}\qquad R_{\mathrm{conv},hc}=6944\ \mathrm{K/W}\qquad R_{\mathrm{rad},hc}=7471\ \mathrm{K/W}$$

The equivalent resistance is then

$$R_{\mathrm{eq}}=\left(\frac{1}{8170}+\frac{1}{6944}+\frac{1}{7471}\right)^{-1}=2498\ \mathrm{K/W}$$

and the total resistance is

$$R=1603+2498+1603=5704\ \mathrm{K/W} \qquad \triangleleft$$

Comments:

1. The problem is iterative, since values of T_1 and T_2 were assumed to calculate $R_{\mathrm{conv},hc}$ and $R_{\mathrm{rad},hc}$. To check the validity of the assumed values, we first obtain the heat transfer rate q from the expression

$$q=\frac{T_{s,1}-T_{s,2}}{R}=\frac{25°\mathrm{C}-(-10°\mathrm{C})}{5704\ \mathrm{K/W}}=6.14\times 10^{-3}\ \mathrm{W}$$

Hence

$$T_1=T_{s,i}-qR_{\mathrm{cond},i}=25°\mathrm{C}-6.14\times 10^{-3}\ \mathrm{W}\times 1603\ \mathrm{K/W}=15.2°\mathrm{C}$$

$$T_2=T_{s,o}+qR_{\mathrm{cond},o}=-10°\mathrm{C}+6.14\times 10^{-3}\ \mathrm{W}\times 1603\ \mathrm{K/W}$$
$$=-0.2°\mathrm{C}$$

Using these values of T_1 and T_2, $R_{conv,hc}$ and $R_{rad,hc}$ should be recomputed and the process repeated until satisfactory agreement is obtained between the initial and computed values of T_1 and T_2.

2. The resistance of a section of low density particle board 75 mm thick $(L_1+L_2+L_3)$ of area W^2 is 9615 K/W, which exceeds the total resistance of the composite by approximately 70%. Accordingly, use of the honeycomb structure offers no advantages as an insulating material. Its effectiveness as an insulator could be improved (R_{eq} increased) by reducing the wall thickness t to increase R_{cond}, evacuating the cell to increase R_{conv}, and/or by decreasing ε to increase R_{rad}. A significant increase in $R_{rad,hc}$ could be achieved by aluminizing the top and bottom surfaces of the cell.

EXAMPLE 14.3

In Chapter 13 radiation exchange in an enclosure was considered under conditions for which conduction and convection could be neglected. However, in many cases, radiation, convection, and/or conduction rates are comparable, and the enclosure analysis must be modified to account for multimode effects. The most general condition for an enclosure surface would involve external heat addition $q_{i,ext}$ as, for example, by electric heating, and heat transfer by radiation, convection, and conduction.

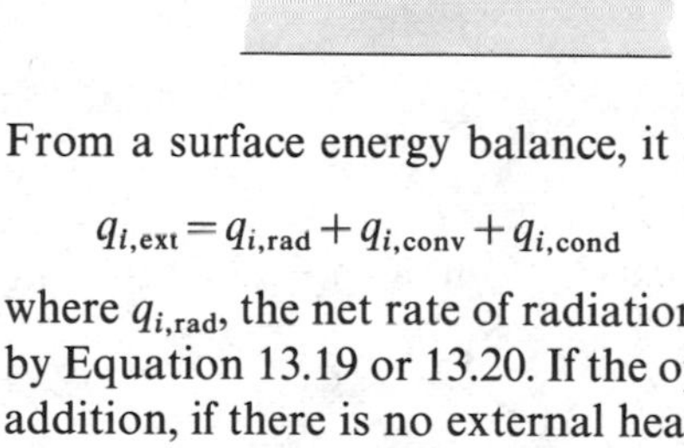

From a surface energy balance, it follows that

$$q_{i,ext} = q_{i,rad} + q_{i,conv} + q_{i,cond}$$

where $q_{i,rad}$, the net rate of radiation transfer from the surface, is still determined by Equation 13.19 or 13.20. If the opposite surface is well insulated, $q_{i,cond}=0$. In addition, if there is no external heating and convection is negligible, the surface is *reradiating*.

As a special case, consider an air heater consisting of a semicircular tube for which the plane surface is maintained at 1000 K and the other surface is well insulated. The tube radius is 20 mm, and both surfaces have an emissivity of 0.8. If atmospheric air flows through the tube at 0.01 kg/s and $T_m = 400$ K, what is the rate at which heat must be supplied per unit length to maintain the plane surface at 1000 K? What is the temperature of the insulated surface?

SOLUTION

Known: Airflow conditions in tubular heater and heater surface conditions.

Find:

Rate at which heat must be supplied and temperature of insulated surface.

Schematic:

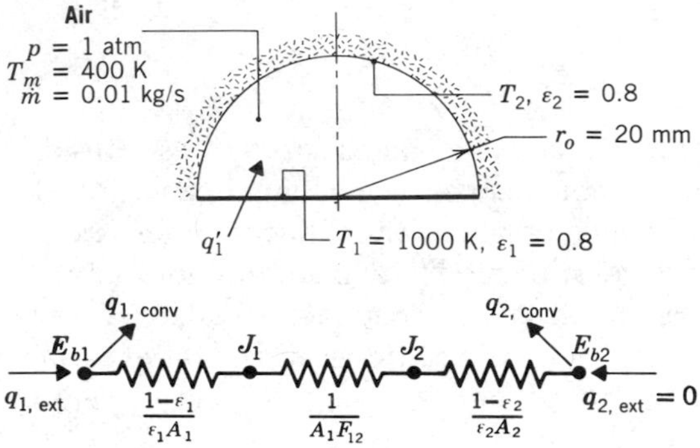

Assumptions:

1. Steady-state conditions.
2. Diffuse, gray surfaces.
3. Negligible tube end effects and axial variations in gas temperature.
4. Fully developed flow.

Properties:

Table A.4, air (1 atm, 400 K): $k = 0.0338$ W/m·K, $\mu = 230 \times 10^{-7}$ kg/s·m, $c_p = 1014$ J/kg·K, $Pr = 0.69$.

Analysis:

Because the semicircular surface is well insulated and there is no external heat addition, a surface energy balance yields

$$q_{2,\text{rad}} = q_{2,\text{conv}}$$

Since the tube constitutes a two-surface enclosure, the net radiation transfer to surface 2 may be evaluated from Equation 13.23. Hence

$$\frac{\sigma(T_1^4 - T_2^4)}{\dfrac{1-\varepsilon_1}{\varepsilon_1 A_1} + \dfrac{1}{A_1 F_{12}} + \dfrac{1-\varepsilon_2}{\varepsilon_2 A_2}} = hA_2(T_2 - T_m)$$

where the view factor is $F_{12}=1$ and, per unit length, the surface areas are $A_1 = 2r_o$ and $A_2 = \pi r_o$. With

$$Re_D = \frac{\rho u_m D_h}{\mu} = \frac{\dot{m} D_h}{A_c \mu} = \frac{\dot{m} D_h}{(\pi r_o^2/2)\mu}$$

the hydraulic diameter is

$$D_h = \frac{4A_c}{P} = \frac{2\pi r_o}{\pi + 2} = \frac{0.04\pi m}{\pi + 2} = 0.0244 \text{ m}$$

Hence

$$Re_D = \frac{0.01 \text{ kg/s} \times 0.0244 \text{ m}}{(\pi/2)(0.02 \text{ m})^2 \times 230 \times 10^{-7} \text{ kg/s} \cdot \text{m}} = 16{,}900$$

From the Dittus-Boelter equation,

$$Nu_D = 0.023 Re_D^{4/5} Pr^{0.4}$$

$$Nu_D = 0.023(16{,}900)^{4/5}(0.69)^{0.4} = 47.8$$

$$h = \frac{k}{D_h} Nu_D = \frac{0.0338 \text{ W/m} \cdot \text{K}}{0.0244 \text{ m}} 47.8 = 66.2 \text{ W/m}^2 \text{ K}$$

Dividing both sides of the energy balance by A_1, it follows that

$$\frac{5.67 \times 10^{-8} \text{ W/m}^2 \cdot \text{K}^4 \, [(1000)^4 - T_2^4] \text{ K}^4}{\dfrac{1-0.8}{0.8} + 1 + \dfrac{1-0.8}{0.8}\dfrac{2}{\pi}} = 66.2 \frac{\pi}{2} (T_2 - 400) \text{ W/m}^2$$

or

$$5.67 \times 10^{-8} T_2^4 + 146.5 T_2 - 115{,}313 = 0$$

A trial-and-error solution yields

$$T_2 = 696 \text{ K}$$ ◁

From an energy balance at the heated surface,

$$q_{1,\text{ext}} = q_{1,\text{rad}} + q_{1,\text{conv}} = q_{2,\text{conv}} + q_{1,\text{conv}}$$

Hence, on a unit length basis,

$$q'_{1,\text{ext}} = h\pi r_o(T_2 - T_m) + h2r_o(T_1 - T_m)$$

$$q'_{1,\text{ext}} = 66.2 \times 0.02[\pi(696 - 400) + 2(1000 - 400)] \text{ W/m}$$

$$q'_{1,\text{ext}} = (1231 + 1589) \text{ W/m} = 2820 \text{ W/m} \qquad \triangleleft$$

Comments:

Applying an energy balance to a differential control volume about the air, it follows that

$$\frac{dT_m}{dx} = \frac{q'_1}{\dot{m}c_p} = \frac{2820 \text{ W/m}}{0.01 \text{ kg/s } (1014 \text{ J/kg}\cdot\text{K})} = 278 \text{ K/m}$$

Hence the air temperature change is significant, and a more representative analysis would subdivide the tube into axial zones and allow for variations in air and insulated surface temperatures between zones. Moreover, a two-surface analysis of radiation exchange would no longer be appropriate.

PROBLEMS

14.1 A square plate of pure aluminum, 0.5 m on a side and 16 mm thick, is initially at 300°C and is suspended in a large chamber. The walls of the chamber and the enclosed air are each maintained at 27°C. If the surface emissivity of the plate is 0.25, what is the initial cooling rate? Is it reasonable to assume a uniform plate temperature during the cooling process?

14.2 A horizontal pipe with an outside diameter of 0.17 m is used to transport steam through a building in which the air and walls are at a temperature of 25°C. If the pipe is insulated with a layer of calcium silicate insulation 25 mm thick and the inner surface of the insulation is at 650 K, what is the heat loss per unit length of pipe?

14.3 The 4 m × 4 m horizontal roof of an uninsulated aluminum melting furnace is composed of a 0.08-m-thick fireclay brick refractory, which is covered by a 5-mm-thick steel (AISI 1010) plate. The refractory surface exposed to the furnace gases is maintained at 1700 K during operation, while the outer surface of the steel is exposed to the air and walls of a large room at 25°C. The total hemispherical emissivity of the steel is $\varepsilon = 0.3$. What is the rate of heat loss from the roof? If a 20-mm-thick layer of alumina-silica insulation (64 kg/m^3) is placed between the refractory and the steel, what is the new rate of heat loss from the roof? What is the temperature at the inner surface of the insulation?

14.4 A square (0.25 m × 0.25 m) channel is used to transport preheated atmospheric air for an industrial process application. The channel is constructed from steel, except for a special glass viewing panel, which is of width $w = 80$ mm and extends from $x = L_1 = 0.8$ m to $x = L_2 = 1$ m from the channel entrance. The glass is mounted flush with the steel on the inside surface of the channel, and its sides are separated from the steel by insulating strips.

The glass has a thickness of $t_g = 50$ mm, an emissivity of $\varepsilon_g = 0.8$, and a thermal conductivity of $k_g = 2$ W/m·K. It may be assumed to be opaque to radiation of

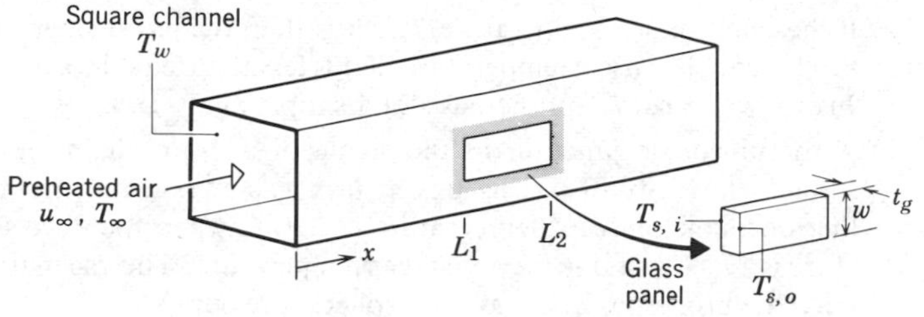

wavelengths greater than 1 μm. During operation air at a velocity of $u_\infty = 15$ m/s and a temperature of $T_\infty = 800$ K passes through the channel, while the outer surface of the glass is uniformly cooled to a temperature of $T_{s,o} = 300$ K. Steady-state conditions exist, and it may be assumed that the inner surface of the glass exchanges radiation only with the steel walls, which have an emissivity of $\varepsilon_w = 1.0$ and a temperature of $T_w = 800$ K. Estimate the heat loss through the glass panel and the glass inner surface temperature, $T_{s,i}$.

14.5 Hot air flows from a furnace through a 0.15-m-diameter, thin-walled steel duct with a velocity of 3 m/s. The duct passes through the crawlspace of a house, and its uninsulated exterior surface is exposed to quiescent air and surroundings at 0°C. At a location in the duct for which the mean air temperature is 70°C, determine the heat loss per unit duct length and the duct wall temperature.

14.6 A radiant heating oven is one whose walls are maintained at an elevated surface temperature T_s, while a gas at a lower temperature T_∞ flows through the oven. A product placed in the oven is then heated to an equilibrium temperature T_p, which is less than T_s and larger than T_∞. Consider the product to be in the form of a solid cylinder whose length is much larger than its diameter, $D = 50$ mm. The surface emissivity of the solid is known to be $\varepsilon = 0.8$. If the solid is placed in an oven whose surfaces are maintained at $T_s = 650$ K and atmospheric nitrogen is in cross flow over the cylinder with $T_\infty = 350$ K and $V = 3$ m/s, what is the equilibrium temperature T_p of the product?

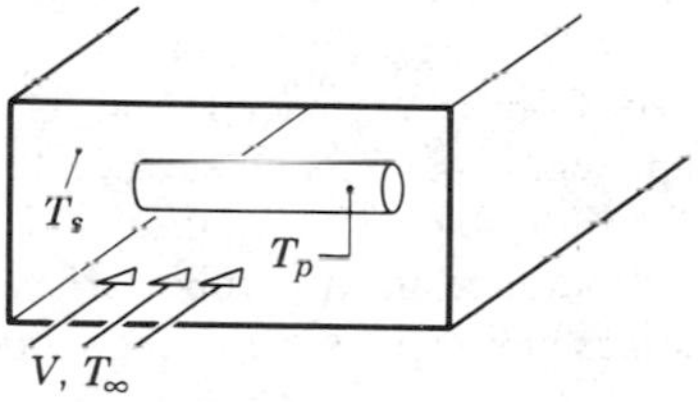

14.7 A thermocouple junction is inserted in a large duct to measure the temperature of hot gases flowing through the duct.

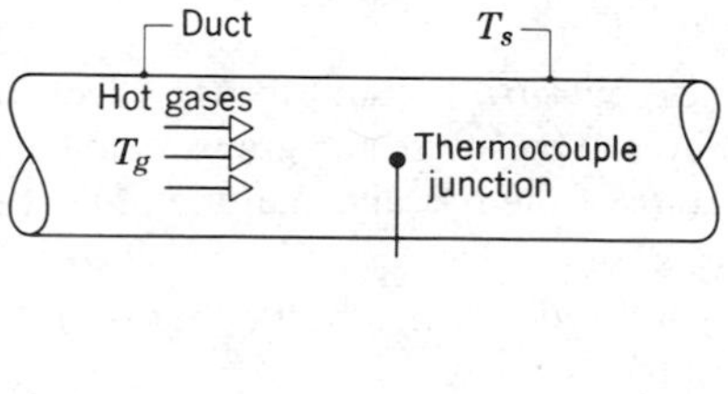

(a) If the duct surface temperature T_s is less than the gas temperature T_g, will the thermocouple sense a temperature that is less than, equal to, or greater than T_g? Justify your answer on the basis of a simple analysis.

(b) A thermocouple junction in the shape of a 2-mm-diameter sphere with a surface emissivity of 0.60 is placed in a gas stream moving at 3 m/s. If the thermocouple senses a temperature of 320°C when the duct surface temperature is 175°C, what is the actual gas temperature? The gas may be assumed to have the properties of air at atmospheric pressure.

14.8 Consider temperature measurement in a gas stream using the thermocouple junction described in Problem 14.7 ($D=2$ mm, $\varepsilon=0.60$). If the gas velocity and temperature are 3 m/s and 500°C, respectively, what temperature will be indicated by the thermocouple if the duct surface temperature is 200°C? The gas may be assumed to have the properties of atmospheric air. What temperature will be indicated by the thermocouple if the gas pressure is doubled and all other conditions remain the same?

14.9 The temperature of flue gases flowing through the stack of a boiler is measured by means of a thermocouple enclosed within a cylindrical tube as shown.

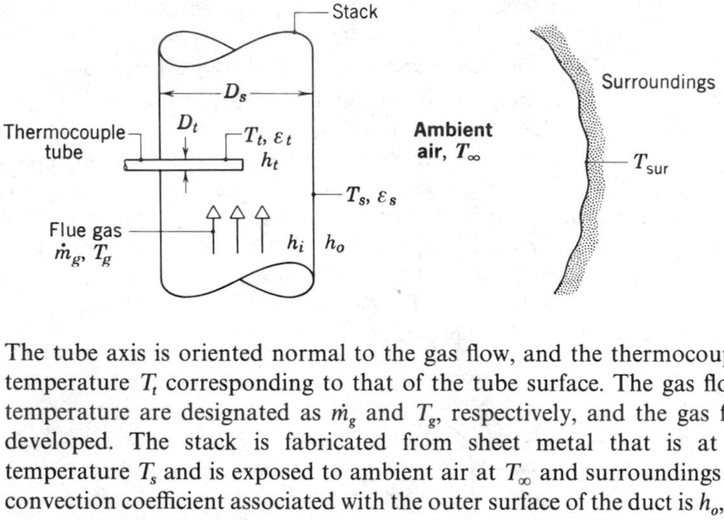

The tube axis is oriented normal to the gas flow, and the thermocouple senses a temperature T_t corresponding to that of the tube surface. The gas flow rate and temperature are designated as $\dot{m}_g$ and T_g, respectively, and the gas flow is fully developed. The stack is fabricated from sheet metal that is at a uniform temperature T_s and is exposed to ambient air at T_∞ and surroundings at T_{sur}. The convection coefficient associated with the outer surface of the duct is h_o, while those associated with the inner surface of the duct and the tube surface are h_i and h_t, respectively. The tube and duct surface emissivities are ε_t and ε_s, respectively.

(a) Neglecting conduction losses along the thermocouple tube, develop an analysis that could be used to predict the error $(T_g - T_t)$ in the temperature measurement.

(b) Assuming the flue gas to have the properties of atmospheric air, evaluate the error for $T_t=300$°C, $D_s=0.6$ m, $D_t=10$ mm, $\dot{m}_g=1$ kg/s, $T_\infty=T_{sur}=27$°C, $\varepsilon_t=\varepsilon_s=0.8$, and $h_o=25\ \text{W/m}^2\cdot\text{K}$.

14.10 Consider part b of Problem 8.27, but now account for the effects of radiation by assuming the rod and tube surfaces to be diffuse and gray and to have an emissivity of 0.8. Determine the temperature of the rod and the rate of heat transfer to the water per unit length of tube.

14.11 Hot coffee is contained in a cylindrical thermos bottle that is of length $L=0.3$ m and is lying on its side (horizontally).

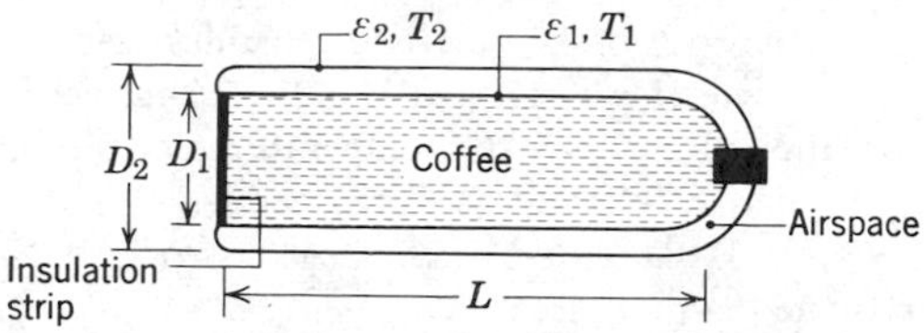

The coffee container consists of a glass flask of diameter $D_1 = 0.07$ m, separated from an aluminum housing of diameter $D_2 = 0.08$ m by an air at atmospheric pressure. The outer surface of the flask and the inner surface of the housing are silver coated to provide emissivities of $\varepsilon_1 = \varepsilon_2 = 0.25$. If these surfaces are at temperatures of $T_1 = 75°C$ and $T_2 = 35°C$, what is the heat loss from the coffee?

14.12 A room heater panel consists of a vertical array of N copper pipes of diameter D embedded in a solid panel of thermal conductivity k_s. Hot water flows through the pipes with mean velocity and temperature of u_m and T_m, respectively, maintaining the pipe wall at a temperature $T_{s,1} < T_m$. The surface of the panel is at a uniform temperature $T_{s,2}$, which exceeds the mean temperature T_a of the air in the room. The outside wall of the room, which is of thermal conductivity k_w and without windows, has uniform inner and outer surface temperatures of $T_{s,3}$ and $T_{s,4}$, respectively. The outer surface is exposed to a cold ambient airflow of temperature T_∞ and velocity u_∞, to solar irradiation of magnitude G_S, and to surroundings characterized by an effective temperature T_e. The room is long in the direction normal to the sketch; and the floor, ceiling, and rear wall, which houses the panel, are well insulated.

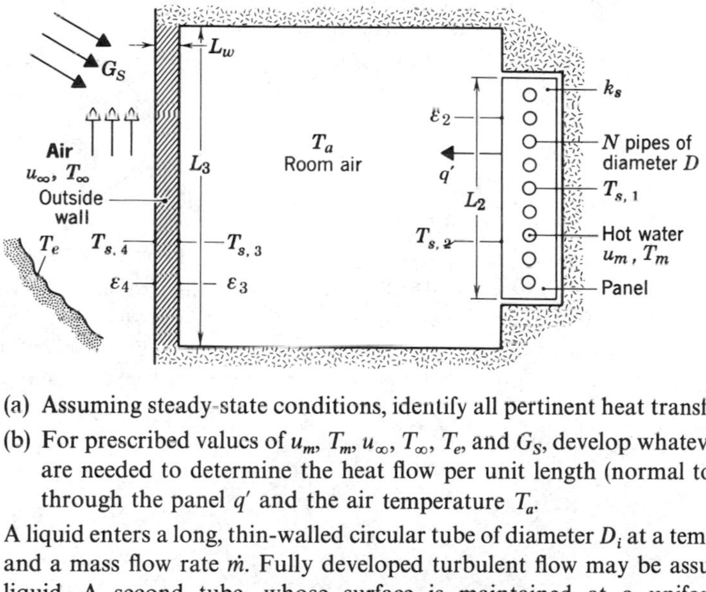

(a) Assuming steady-state conditions, identify all pertinent heat transfer processes.

(b) For prescribed values of u_m, T_m, u_∞, T_∞, T_e, and G_S, develop whatever equations are needed to determine the heat flow per unit length (normal to the sketch) through the panel q' and the air temperature T_a.

14.13 A liquid enters a long, thin-walled circular tube of diameter D_i at a temperature $T_{m,i}$ and a mass flow rate $\dot{m}$. Fully developed turbulent flow may be assumed for the liquid. A second tube, whose surface is maintained at a uniform, elevated temperature T_h, surrounds the first tube, and the space between the two tubes is evacuated. Both tube surfaces may be approximated as blackbodies. Obtain expressions that could be used to determine the variation of the liquid mean temperature T_m and the inner tube surface temperature T_s as a function of distance x along the tube. Express your results in terms of the foregoing variables and appropriate liquid thermophysical properties.

14.14 A vertical air space in the wall of a home is 0.1 m thick and 3 m high. The air separates a brick exterior from a plaster board wall; each surface has an emissivity of 0.9. Consider conditions for which the temperatures of the brick and plaster surfaces exposed to the air are -10 and 18°C, respectively. What is the rate of heat loss per unit surface area? What would be the heat loss per unit area if the space were filled with urethane foam?

14.15 To estimate the energy savings associated with using double-paned (thermopane) versus single-paned window construction, consider a window that is oriented vertically and is 0.6 m × 0.6 m on a side. The inner surface of the window is exposed to the air and walls of a room that are maintained at a temperature of T_i, while the outer surface is exposed to ambient air at T_∞ and to surroundings (the sky, ground, etc.) characterized by an effective temperature T_{sur}. The convection coefficient associated with the outer surface and the ambient air is h_o. The window is constructed of glass that is opaque to long wave radiation and of known emissivity ε_g. The single-paned window is of thickness t_1, while each pane of the thermopane window is of thickness t_2 with an intervening air space of thickness t_a.

(a) Assuming steady-state, nighttime conditions and negligible temperature gradients within the glass, develop a model that could be used to determine the rate of heat transfer through the window and the glass temperature(s) for both single pane and thermopane constructions.

(b) Determine the heat transfer rate and the glass temperatures when $T_i = 18°C$, $T_\infty = -20°C$, $T_{sur} = -30°C$, $h_o = 25\ W/m^2 \cdot K$, $\varepsilon_g = 0.9$, $t_1 = 10$ mm, $t_2 = 5$ mm, and $t_a = 15$ mm.

14.16 The radiation generated in an evacuated 60-W light bulb may be approximated as having the spectral distribution of radiation emitted by a blackbody at 2900 K.

(a) If the radiative properties of the glass envelope are as follows,

$$0 < \lambda < 1\ \mu m \qquad \tau_\lambda = 1,\ \alpha_\lambda = 0$$

$$1\ \mu m < \lambda \qquad \tau_\lambda = 0,\ \alpha_\lambda = 1$$

what fraction of the radiation emitted by the filament is transmitted to the surroundings of the bulb?

(b) Neglecting conduction losses through the socket, what will be the temperature of the glass bulb, which may be approximated as a sphere of diameter $D = 50$ mm, when it is situated in a room whose walls and air are at 20°C?

14.17 Growers use giant fans to prevent grapes from freezing when the effective sky temperature is too low. The grape, which may be viewed as a thin skin of negligible

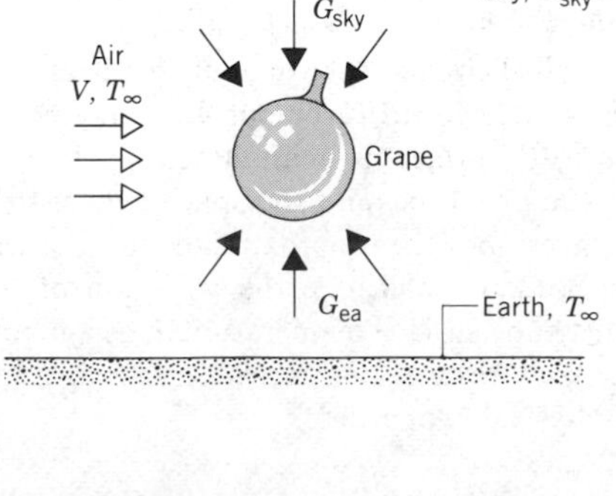

thermal resistance enclosing a volume of sugar water, is exposed to ambient air and is irradiated from the sky and from the ground. Assume the grape to be an isothermal sphere of 15 mm diameter, and assume uniform blackbody irradiation over its top and bottom hemispheres due to emission from the sky and the earth, respectively.

(a) Derive an expression for the rate of change of the grape temperature. Express your result in terms of a convection coefficient and appropriate temperatures and radiative quantities.

(b) Under conditions for which $T_{sky} = 235$ K, $T_{\infty} = 273$ K, and the fan is off ($V = 0$), determine whether the grape will freeze. To a good approximation, the skin emissivity is 1 and the grape's thermophysical properties are those of sugarless water. However, because of its sugar content, a grape freezes at -5°C.

(c) With all conditions remaining the same, except that the fans are now operating with $V = 1$ m/s, will the grape freeze?

14.18 A thin coating, which is applied to long, cylindrical copper rods of 10 mm diameter, is cured by placing the rods horizontally in a large furnace whose walls are maintained at 1300 K. The furnace is filled with nitrogen gas, which is also at 1300 K and a pressure of 1 atm. The coating is diffuse, and its spectral emissivity has the following distribution.

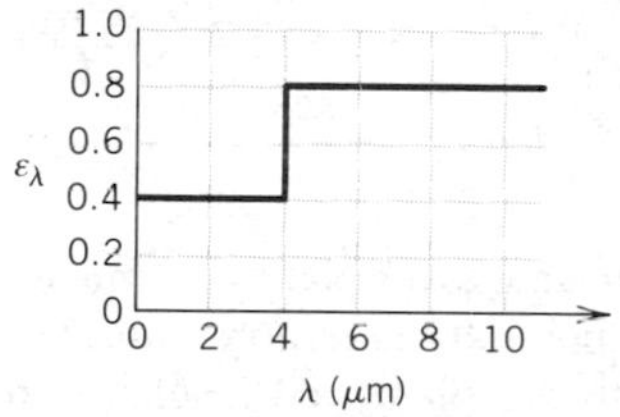

If the rods are inserted at 300 K, what is the initial rate of change of their temperature?

14.19 A large combination convection-radiation oven is used to heat treat a small product in the form of a cylinder of diameter 25 mm and length 0.2 m. The oven walls are diffuse and gray, with an emissivity of 0.85 and operate at a uniform temperature of 1000 K. The circulating hot air at 750 K is in cross flow over the cylinder with a velocity of 5 m/s. The cylinder surface is opaque-diffuse with the following spectral emissivity distribution.

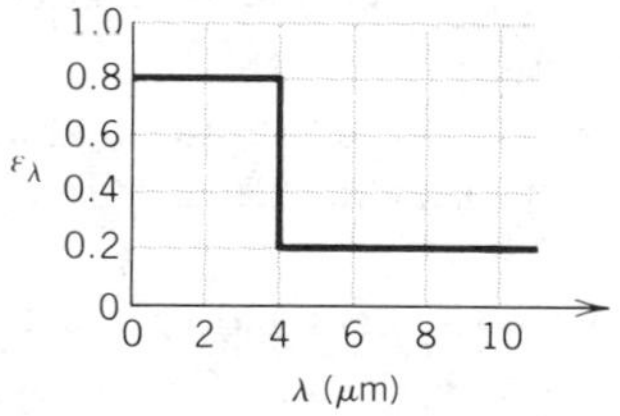

Determine the heat transfer rate to the cylinder when first placed in the oven at 300 K. Obtain the steady-state temperature of the cylinder.

14.20 A flat plate solar collector, consisting of an absorber plate and single cover plate, is inclined at an angle of $\tau = 60°$ relative to the horizontal.

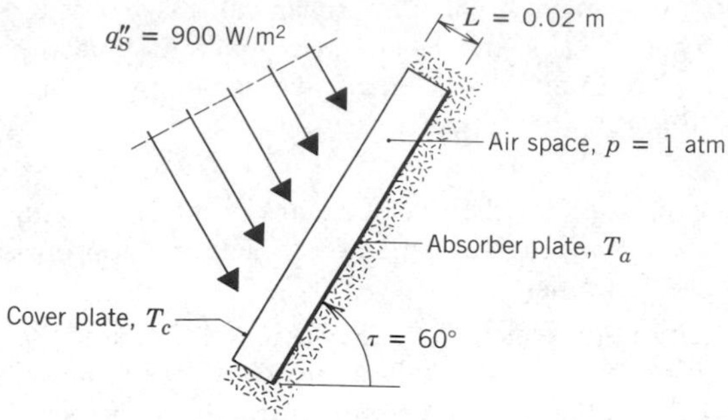

Consider conditions for which the incident solar radiation is collimated at an angle of 60° relative to the horizontal and the solar flux is 900 W/m^2. The cover plate is perfectly transparent to solar radiation ($\lambda \lesssim 3$ μm) and is opaque to radiation of larger wavelengths. The cover and absorber plates are diffuse surfaces having the following spectral absorptivities.

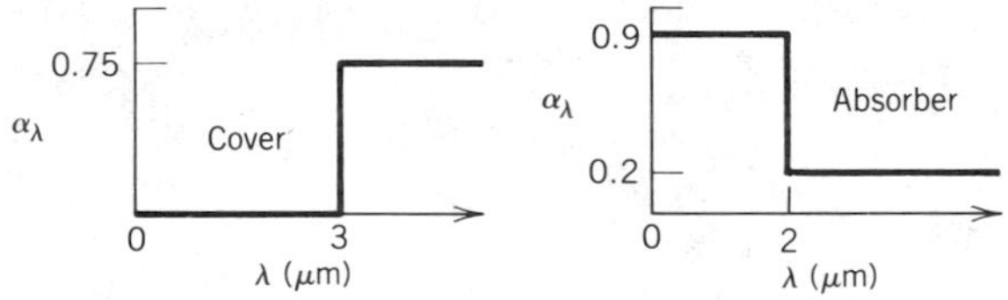

The length and width of the absorber and cover plates are much larger than the plate spacing L. What is the rate at which solar radiation is absorbed per unit area of the absorber plate? With the absorber plate well insulated from below and absorber and cover plate temperatures of $T_a = 70$°C and $T_c = 27$°C, what is the heat loss per unit area of the absorber plate?

14.21 A circular metal disk having a diameter of 0.4 m is placed firmly against the ground in a barren horizontal region where the earth is at a temperature of 280 K. The effective sky temperature is also 280 K. The disk is exposed to quiescent ambient air at 300 K and direct solar irradiation of 745 W/m^2. The surface of the disk is diffuse with $\varepsilon_\lambda = 0.9$ for $0 < \lambda < 1$ μm and $\varepsilon_\lambda = 0.2$ for $\lambda > 1$ μm. After some time has elapsed, the disk achieves a uniform, steady-state temperature. The thermal conductivity of the soil is 0.52 W/m·K.

(a) Determine the fraction of the incident solar irradiation that is absorbed.

(b) What is the emissivity of the disk surface?

(c) For a steady-state disk temperature of 340 K, employ a suitable correlation to determine the average free convection heat transfer coefficient at the upper surface of the disk.

(d) Show that a disk temperature of 340 K does indeed yield a steady-state condition for the disk.

14.22 The lower side of a 400-mm-diameter disk is heated by an electric furnace, whereas the upper side is exposed to quiescent ambient air and surroundings at 300 K. The radiant furnace (negligible convection) is of circular construction with the bottom surface ($\varepsilon_1 = 0.6$) and cylindrical side surface ($\varepsilon_2 = 1.0$) maintained at

$T_1 = T_2 = 500$ K. The surface of the disk toward the radiant furnace is black ($\varepsilon_{d,1} = 1.0$), while the upper surface has an emissivity of $\varepsilon_{d,2} = 0.8$. Assuming the plate and furnace surfaces to be diffuse and gray, determine whether net heat transfer is to or from the disk when $T_d = 400$ K.

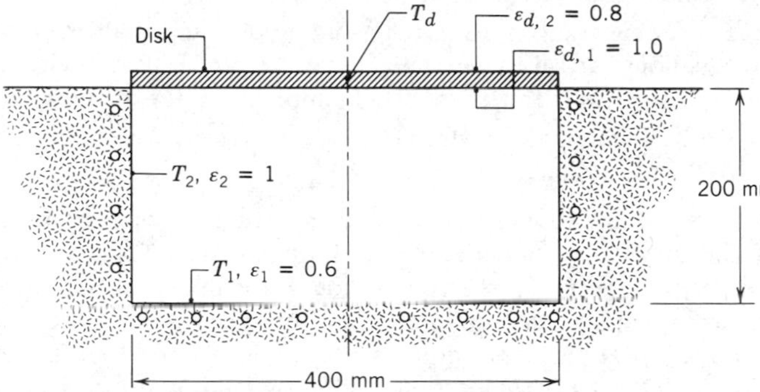

14.23 A special surface coating on a square panel that is 5 m × 5 m on a side is cured by placing the panel directly under a radiant heat source that has the same dimensions, is diffuse and gray, and operates with a power input of 75 kW. The top surface of the heater, as well as the bottom surface of the panel, are well insulated, and the arrangement exists in a large room with air and wall temperatures of 25°C. The surface coating is diffuse and gray, with an emissivity of $\varepsilon = 0.30$ and an upper temperature limit of 400 K. Neglecting convection effects, what is the minimum spacing that may be maintained between the heater and the panel to ensure that the panel temperature will not exceed 400 K? Allowing for convection effects at the coated surface of the panel, what is the minimum spacing?

14.24 A long rod heater of diameter $D_1 = 10$ mm and emissivity $\varepsilon_1 = 1.0$ is coaxial with a well-insulated, semicylindrical reflector of diameter $D_2 = 1$ m. A long panel of width $W = 1$ m is aligned with the reflector and is separated from the heater by a distance of $H = 1$ m. The panel is coated with a special paint ($\varepsilon_3 = 0.7$), which is cured by maintaining it at a temperature of 400 K. The panel is well insulated on its

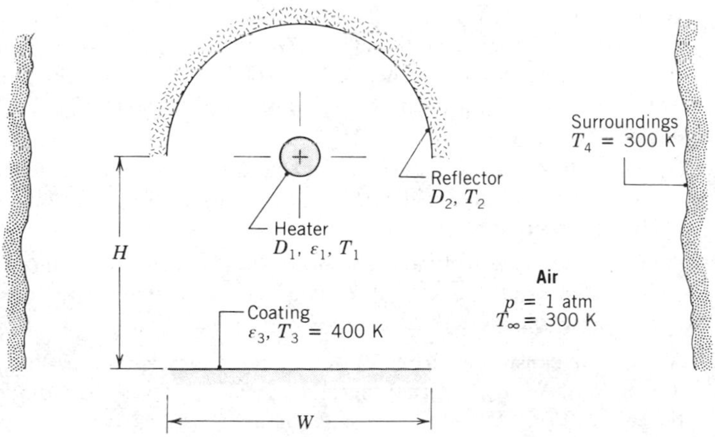

back side, and the entire system is located in a large room for which the walls and the atmospheric quiescent air are at 300 K. Heat transfer by convection may be neglected for the reflector surface.

(a) Sketch the equivalent thermal circuit for the system and label all pertinent resistances and potentials.

(b) Expressing your results in terms of appropriate variables, write the system of equations needed to determine the heater and reflector temperatures, T_1 and T_2, respectively. Determine these temperatures for the prescribed conditions.

(c) Determine the rate at which electrical power must be supplied per unit length of the rod heater.

14.25 In an industrial process, products of combustion at a temperature and pressure of 2000 K and 1 atm, respectively, flow through a long, 0.25-m-diameter pipe whose inner surface is black. The combustion gas contains carbon dioxide and water vapor, each at a partial pressure of 0.10 atm. The gas may be assumed to have the thermophysical properties of atmospheric air and to be in fully developed flow with $\dot{m} = 0.25$ kg/s. The pipe is cooled by passing water in cross flow over its outer surface. The upstream velocity and temperature of the water are 0.30 m/s and 300 K, respectively. Determine the pipe wall temperature and heat flux. *Hint:* Emission from the pipe wall may be neglected.

14.26 It is desired to estimate the effectiveness of a horizontal, straight fin of rectangular cross section when applied to a surface operating at 45°C in an environment for which the surroundings and ambient air are at 25°C. The fin is to be fabricated from aluminum alloy (2024-T6) with an anodized finish, 2 mm thick and 100 mm long.

(a) Considering only free convection from the fin surface and estimating an average heat transfer coefficient, determine the effectiveness of the fin.

(b) Estimate the effectiveness of the fin including the influence of radiation exchange with the surroundings.

(c) Using a numerical method, develop the finite-difference equations and outline the procedure for obtaining the fin effectiveness. The free convection and radiation exchange modes should be based on local, rather than average, values for the fin.

14.27 In the analytical treatment of the fin with uniform cross-sectional area, it was assumed that the convection heat transfer coefficient is constant along the length of the fin. Consider a stainless steel (AISI 316) fin 6 mm in diameter and 50 mm long (with insulated tip) operating under conditions for which $T_b = 125°C$, $T_\infty = 27°C$, $T_{sur} = 27°C$, and $\varepsilon = 0.6$.

(a) Estimate average values of the fin heat transfer coefficients for free convection (h_c) and radiation exchange (h_r). Use these values to predict the tip temperature and fin effectiveness.

(b) Use a numerical method of solution to estimate the foregoing parameters when the convection and radiation coefficients are based on local, rather than average, values for the fin.

14.28 A heat sink constructed from 2024 aluminum alloy is used to cool a power diode dissipating 5 W. The internal thermal resistance between the diode junction and the case is 0.8°C/W, and the thermal contact resistance between the case and heat sink is 10^{-5} m$^2\cdot$°C/W. Assuming that all of the diode power will be transferred to the ambient air through the rectangular fins, estimate the operating temperature of the diode.

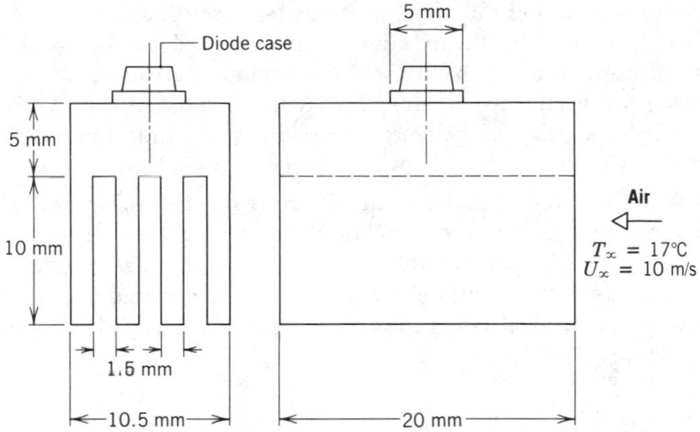

14.29 Four high power silicon-controlled rectifiers (SCR), each dissipating 150 W, are mounted on a water-cooled heat sink.

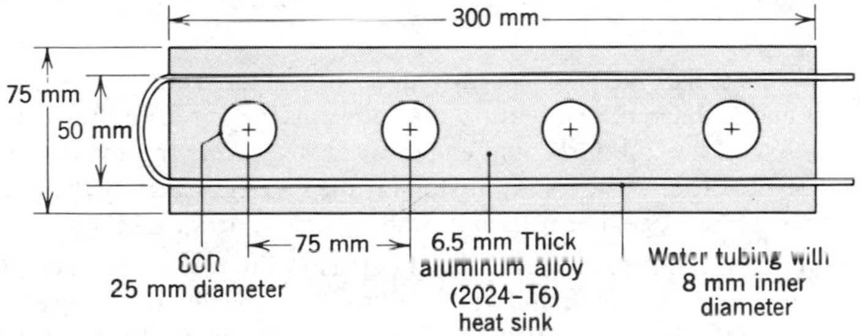

Water at 15°C is supplied at a rate of 4 liters/min. The thermal contact resistance between an SCR and the heat sink is estimated to be 0.1°C/W. Estimate the operating temperature of the SCRs by performing the following series of calculations.

(a) Determine the temperature rise of the water coolant. What is the average mean temperature of the coolant?

(b) Using the proper correlation for forced convection inside a tube, estimate the heat transfer coefficient between the water and the tube wall. What is the approximate temperature rise between the coolant and the tube wall?

(c) Determine the shape factor for the heat sink material by using the two-dimensional graphical flux plotting method. Be sure to recognize symmetry of the system. What is the temperature rise between the tube wall and the location where an SCR contacts the heat sink?

(d) Draw the thermal resistance circuit and identify each of the elements present. Label the temperatures at each node and indicate the operating temperature of the SCRs.

(e) Summarize briefly the assumptions that were made in arriving at the SCR operating temperatures.

14.30 A wire manufacturing process involves extrusion of wire of diameter D through platinum dies. The wire is extruded at a constant speed V and is cooled by forced-air convection and radiation, as it passes some distance between fixed points at $x=0$ and $x=L$. The convection is characterized by the coefficient h, and the temperature of the ambient air is T_∞. Radiation is exchanged with surroundings having an effective temperature T_{sur}. After it is cooled, the wire is rolled onto large spools for subsequent shipment. Develop an equation that governs the temperature distribution along the wire between $x=0$ and $x=L$.

14.31 A thin plastic rod of density ρ and diameter D, which is extruded from a die at T_0, moves with a velocity V to a rolling station where the rod temperature is T_L. Cool air at T_∞ is blown across the rod, and the local heat transfer coefficient h_x is constant over the length of the rod. It can be assumed that radial temperature gradients and axial conduction in the rod are negligible and that steady-state conditions exist.

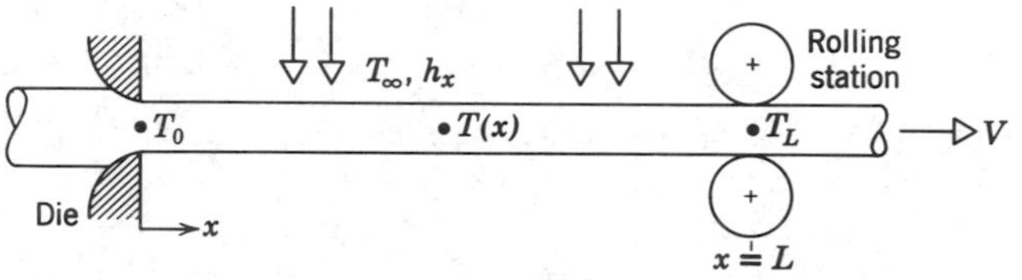

Use the conservation of energy requirement on a properly defined control volume to find a differential equation that governs the variation of the rod temperature with x. Solve the foregoing equation to determine the temperature distribution along the rod. Write an expression for the total heat loss by the rod in terms of the heat transfer coefficient h_x and other appropriate variables.

14.32 A process for thermal curing of a thin plastic film of thickness l employs a radiation source that provides a uniform heat flux q_o'' on the film traveling at a lineal speed V. The film is at a temperature T_o as it enters under the radiant heater section, and it is desired to have the film reach the curing temperature T_c before it exits the section. Energy exchange by convection occurs between the film and the ambient air at T_∞, but radiative exchange between the film and its surroundings may be neglected.

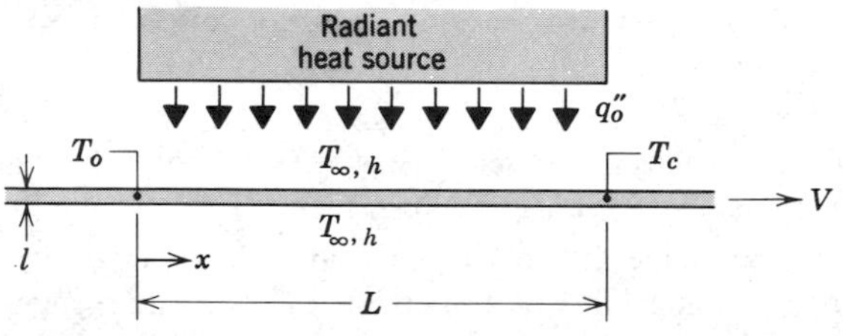

Beginning with a properly defined control volume or control mass, perform an energy balance to obtain a differential equation for the temperature distribution in the film. Indicate how you would solve this equation. Be specific but do not attempt to solve it. To *estimate* the length required to reach T_c, neglect internal conduction and surface convection effects to obtain a simplified model. Calculate the length of the heating section L using these parameters: $c_p = 1$ kJ/kg·K, $l = 0.10$ mm, $q_o'' = 2500$ W/m^2, $\rho = 1000$ kg/m^3, $V = 0.5$ m/s, $T_o = 30$°C, and $T_c = 175$°C.

14.33 The process for production of plastic tubing is illustrated below. Plastic powder is fed into one end of the extruder and the plastic melt is forced through the die forming the product, which in this case is a tube. The product is cooled by passing through a water tank causing its dimensions to stabilize (referred to as sizing). To verify proper sizing conditions and to determine coolant requirements, it is necessary to perform a heat transfer analysis for predicting the temperature distribution in the product for prescribed process parameters.

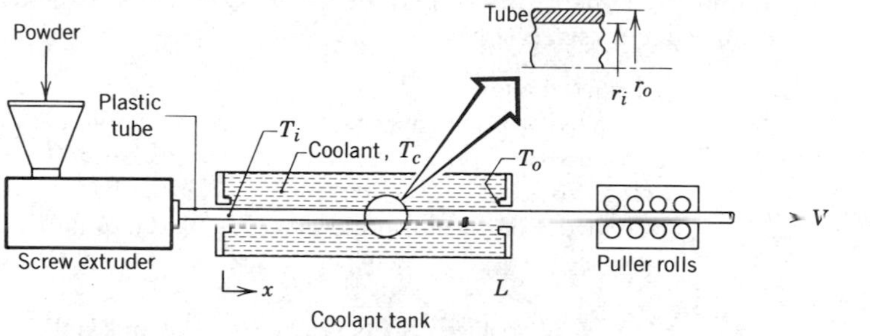

(a) Beginning with a differential control volume, derive an energy equation for the tube, accounting for radial and axial conduction.

(b) Write the energy equation, assuming that radial temperature gradients are negligible and that the convection heat transfer coefficient between the tube and coolant is constant along the length of the tube. Solve the resulting differential equation for the coolant tank exit temperature T_o in terms of product parameters (dimensions, properties, and velocity), coolant tank inlet temperature T_i, and length L.

(c) Using a numerical method that considers both radial and axial conduction, estimate inner and outer surface temperatures of the tube $T(r_i,x)$ and $T(r_o,x)$ as a function of distance in the tank for $r_i = 7.3$ mm, $r_o = 11.0$ mm, $T_i = 175°$C, $V = 0.65$ m/s, $L = 6$ m, and $T_c = 35°$C. The thermophysical properties of the polypropylene tube material are $k = 0.180$ W/m·K, $\rho = 905$ kg/m^3, and $c_p = 2100$ J/kg·K.

14.34 In recent years there has been considerable interest in developing flat plate solar collectors for home heating needs. The typical collector configuration shown

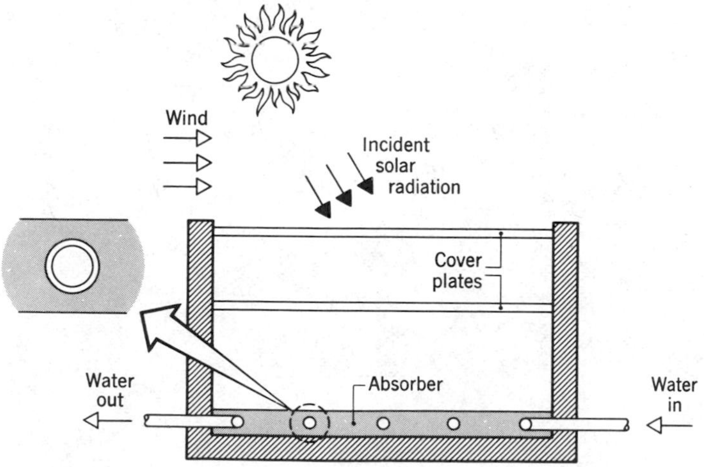

consists of an absorbing surface that is thermally insulated on the edges and on the back side. Solar radiation is transmitted through one or more glass *cover plates* before reaching the absorber. The collected energy is removed by circulating water through tubes that are in good thermal contact with the absorber.

Identify all the energy exchange processes that influence the performance of the collector. Suggest design features that would increase collection efficiency.

14.35 Consider a flat plate solar collector, such as that shown schematically in Problem 14.34, but with one, rather than two, cover plates. The following quantities are presumed to be known.

G_S	solar irradiation
$\alpha_{cp,S}$	absorptivity of the cover plate to solar radiation
$\tau_{cp,S}$	transmissivity of the cover plate to solar radiation
$\alpha_{ap,S}$	absorptivity of the absorber plate to solar radiation
$\varepsilon_{cp}, \varepsilon_{ap}$	emissivities of the cover and absorber plates, respectively
h_i	air space convection coefficient based on the temperature difference, $T_{ap} - T_{cp}$
h_o	convection coefficient between cover plate and ambient air
T_{ap}	absorber plate temperature
T_∞	ambient air temperature
T_{sky}	effective sky temperature

Assuming the absorber–plate tube assembly to be perfectly insulated from below, develop expressions that could be used to obtain the rate at which useful energy is collected per unit area of the absorber plate surface q''_u under steady-state conditions.

Appendix A

THERMOPHYSICAL PROPERTIES OF MATTER[1]

[1] The convention used to present numerical values of the properties is illustrated by this example:

T (K)	$\nu \cdot 10^7$ (m^2/s)	$k \cdot 10^3$ ($W/m \cdot K$)
300	0.349	521

where $\nu = 0.349 \times 10^{-7}$ m^2/s and $k = 521 \times 10^{-3} = 0.521$ $W/m \cdot K$ at 300 K.

Table A.1 Thermophysical properties of selected metallic solids

COMPOSITION	MELTING POINT (K)	PROPERTIES AT 300 K: ρ (kg/m^3)	c_p (J/kg · K)	k (W/m · K)	$\alpha \cdot 10^6$ (m^2/s)	PROPERTIES AT VARIOUS TEMPERATURES (K), k (W/m · K) / c_p (J/kg · K): 100	200	400	600	800	1000	1200	1500	2000	2500
Aluminum															
Pure	933	2702	903	237	97.1	302	237	240	231	218					
						482	798	949	1033	1146					
Alloy 2024-T6 (4.5% Cu, 1.5% Mg, 0.6% Mn)	775	2770	875	177	73.0	65	163	186	186						
						473	787	925	1042						
Alloy 195, Cast (4.5% Cu)		2790	883	168	68.2			174	185						
								—	—						
Beryllium	1550	1850	1825	200	59.2	990	301	161	126	106	90.8	78.7			
						203	1114	2191	2604	2823	3018	3227	3519		
Bismuth	545	9780	122	7.86	6.59	16.5	9.69	7.04							
						112	120	127							
Boron	2573	2500	1107	27.0	9.76	190	55.5	16.8	10.6	9.60	9.85				
						128	600	1463	1892	2160	2338				
Cadmium	594	8650	231	96.8	48.4	203	99.3	94.7							
						198	222	242							
Chromium	2118	7160	449	93.7	29.1	159	111	90.9	80.7	71.3	65.4	61.9	57.2	49.4	
						192	384	484	542	581	616	682	779	937	
Cobalt	1769	8862	421	99.2	26.6	167	122	85.4	67.4	58.2	52.1	49.3	42.5		
						236	379	450	503	550	628	733	674		
Copper															
Pure	1358	8933	385	401	117	482	413	393	379	366	352	339			
						252	356	397	417	433	451	480			
Commercial bronze (90% Cu, 10% Al)	1293	8800	420	52	14		42	52	59						
							785	460	545						
Phosphor gear bronze (89% Cu, 11% Sn)	1104	8780	355	54	17		41	65	74						
							—	—	—						
Cartridge brass (70% Cu, 30% Zn)	1188	8530	380	110	33.9	75	95	137	149						
							360	395	425						
Constantan (55% Cu, 45% Ni)	1493	8920	384	23	6.71	17	19								
						237	362								

Table A.1 Continued

COMPOSITION	MELTING POINT (K)	PROPERTIES AT 300 K: ρ (kg/m^3)	c_p (J/kg·K)	k (W/m·K)	$\alpha \cdot 10^6$ (m^2/s)	PROPERTIES AT VARIOUS TEMPERATURES (K): k (W/m·K) / c_p (J/kg·K): 100	200	400	600	800	1000	1200	1500	2000	2500
Germanium	1211	5360	322	59.9	34.7	232 190	96.8 290	43.2 337	27.3 348	19.8 357	17.4 375	17.4 395			
Gold	1336	19300	129	317	127	327 109	323 124	311 131	298 135	284 140	270 145	255 155			
Iridium	2720	22500	130	147	50.3	172 90	153 122	144 133	138 138	132 144	126 153	120 161	111 172		
Iron															
Pure	1810	7870	447	80.2	23.1	134 216	94.0 384	69.5 490	54.7 574	43.3 680	32.8 975	28.3 609	32.1 654		
Armco (99.75% pure)		7870	447	72.7	20.7	95.6 215	80.6 384	65.7 490	53.1 574	42.2 680	32.3 975	28.7 609	31.4 654		
Carbon steels															
Plain carbon ($Mn \leq 1\%$, $Si \leq 0.1\%$)		7854	434	60.5	17.7			56.7 487	48.0 559	39.2 685	30.0 1169				
AISI 1010		7832	434	63.9	18.8			58.7 487	48.8 559	39.2 685	31.3 1168				
Carbon-silicon ($Mn \leq 1\%$, $0.1\% < Si \leq 0.6\%$)		7817	446	51.9	14.9			49.8 501	44.0 582	37.4 699	29.3 971				
Carbon-manganese-silicon ($1\% < Mn \leq 1.65\%$, $0.1\% < Si \leq 0.6\%$)		8131	434	41.0	11.6			42.2 487	39.7 559	35.0 685	27.6 1090				
Chromium (low) steels															
$\frac{1}{2}$ Cr-$\frac{1}{4}$ Mo-Si (0.18% C, 0.65% Cr, 0.23% Mo, 0.6% Si)		7822	444	37.7	10.9			38.2 492	36.7 575	33.3 688	26.9 969				
1 Cr-$\frac{1}{2}$ Mo (0.16% C, 1% Cr, 0.54% Mo, 0.39% Si)		7858	442	42.3	12.2			42.0 492	39.1 575	34.5 688	27.4 969				
1 Cr-V (0.2% C, 1.02% Cr, 0.15% V)		7836	443	48.9	14.1			46.8 492	42.1 575	36.3 688	28.2 969				

Stainless steels															
AISI 302		8055	480	15.1	3.91			17.3 512	20.0 559	22.8 585	25.4 606				
AISI 304	1670	7900	477	14.9	3.95	9.2 272	12.6 402	16.6 515	19.8 557	22.6 582	25.4 611	28.0 640	31.7 682		
AISI 316		8238	468	13.4	3.48			15.2 504	18.3 550	21.3 576	24.2 602				
AISI 347		7978	480	14.2	3.71			15.8 513	18.9 559	21.9 585	24.7 606				
Lead	601	11340	129	35.3	24.1	39.7 118	36.7 125	34.0 132	31.4 142						
Magnesium	923	1740	1024	156	87.6	169 649	159 934	153 1074	149 1170	146 1267					
Molybdenum	2894	10240	251	138	53.7	179 141	143 224	134 261	126 275	118 285	112 295	105 308	98 330	90 380	86 459
Nickel															
Pure	1728	8900	444	90.7	23.0	164 232	107 383	80.2 485	65.6 592	67.6 530	71.8 562	76.2 594	82.6 616		
Nichrome (80% Ni, 20% Cr)	1672	8400	420	12	3.4			14 480	16 525	21 545					
Inconel X-750 (73% Ni, 15% Cr, 6.7% Fe)	1665	8510	439	11.7	3.1	8.7 —	10.3 372	13.5 473	17.0 510	20.5 546	24.0 626	27.6 —	33.0 —		
Niobium	2741	8570	265	53.7	23.6	55.2 188	52.6 249	55.2 274	58.2 283	61.3 292	64.4 301	67.5 310	72.1 324	79.1 347	
Palladium	1827	12020	244	71.8	24.5	76.5 168	71.6 227	73.6 251	79.7 261	86.9 271	94.2 281	102 291	110 307		
Platinum															
Pure	2045	21450	133	71.6	25.1	77.5 100	72.6 125	71.8 136	73.2 141	75.6 145	78.7 152	82.6 157	89.5 165	99.4 179	
Alloy 60Pt-40Rh (60% Pt, 40% Rh)	1800	16630	162	47	17.4			52 —	59 —	65 —	69 —	73 —	76 —		
Rhenium	3453	21100	136	47.9	16.7	58.9 97	51.0 127	46.1 139	44.2 145	44.1 151	44.6 156	45.7 162	47.8 171	51.9 186	
Rhodium	2236	12450	243	150	49.6	186 147	154 220	146 253	136 274	127 293	121 311	116 327	110 349	112 376	
Silicon	1685	2330	712	148	89.2	884 259	264 556	98.9 790	61.9 867	42.2 913	31.2 946	25.7 967	22.7 992		

Table A.1 Continued

COMPOSITION	MELTING POINT (K)	PROPERTIES AT 300 K				PROPERTIES AT VARIOUS TEMPERATURES (K) k (W/m · K) / c_p (J/kg · K)									
		ρ (kg/m³)	c_p (J/kg · K)	k (W/m · K)	$\alpha \cdot 10^6$ (m²/s)	100	200	400	600	800	1000	1200	1500	2000	2500
Silver	1235	10500	235	429	174	444	430	425	412	396	379	361			
						187	225	239	250	262	277	292			
Tantalum	3269	16600	140	57.5	24.7	59.2	57.5	57.8	58.6	59.4	60.2	61.0	62.2	64.1	65.6
						110	133	144	146	149	152	155	160	172	189
Thorium	2023	11700	118	54.0	39.1	59.8	54.6	54.5	55.8	56.9	56.9	58.7			
						99	112	124	134	145	156	167			
Tin	505	7310	227	66.6	40.1	85.2	73.3	62.2							
						188	215	243							
Titanium	1953	4500	522	21.9	9.32	30.5	24.5	20.4	19.4	19.7	20.7	22.0	24.5		
						300	465	551	591	633	675	620	686		
Tungsten	3660	19300	132	174	68.3	208	186	159	137	125	118	113	107	100	95
						87	122	137	142	145	148	152	157	167	176
Uranium	1406	19070	116	27.6	12.5	21.7	25.1	29.6	34.0	38.8	43.9	49.0			
						94	108	125	146	176	180	161			
Vanadium	2192	6100	489	30.7	10.3	35.8	31.3	31.3	33.3	35.7	38.2	40.8	44.6	50.9	
						258	430	515	540	563	597	645	714	867	
Zinc	693	7140	389	116	41.8	117	118	111	103						
						297	367	402	436						
Zirconium	2125	6570	278	22.7	12.4	33.2	25.2	21.6	20.7	21.6	23.7	26.0	28.8	33.0	
						205	264	300	322	342	362	344	344	344	

Adapted from References 1–7.

Table A.2 Thermophysical properties of selected nonmetallic solids

		PROPERTIES AT 300 K				PROPERTIES AT VARIOUS TEMPERATURES (K) k (W/m·K) / c_p (J/kg·K)									
COMPOSITION	MELTING POINT (K)	ρ (kg/m³)	c_p (J/kg·K)	k (W/m·K)	$\alpha \cdot 10^6$ (m²/s)	100	200	400	600	800	1000	1200	1500	2000	2500
Aluminum oxide, sapphire	2323	3970	765	46	15.1	450 —	82 —	32.4 940	18.9 1110	13.0 1180	10.5 1225				
Aluminum oxide, polycrystalline	2323	3970	765	36.0	11.9	133 —	55 —	26.4 940	15.8 1110	10.4 1180	7.85 1225	6.55 —	5.66 —	6.00 —	
Beryllium oxide	2725	3000	1030	272	88.0			196 1350	111 1690	70 1865	47 1975	33 2055	21.5 2145	15 2750	
Boron	2573	2500	1105	27.6	9.99	190 —	52.5 —	18.7 1490	11.3 1880	8.1 2135	6.3 2350	5.2 2555			
Boron fiber epoxy (30% vol) composite	590	2080													
k, ∥ to fibers				2.29		2.10	2.23	2.28							
k, ⊥ to fibers				0.59		0.37	0.49	0.60							
c_p			1122			364	757	1431							
Carbon															
Amorphous	1500	1950	—	1.50	—	0.67 —	1.18 —	1.89 —	2.19 —	2.37 —	2.53 —	2.84 —	3.48 —		
Diamond, type IIa insulator	—	3500	509	2300		10000 21	4000 194	1540 853							
Graphite, pyrolytic	2273	2210													
k, ∥ to layers				1950		4970	3230	1390	892	667	534	448	357	262	
k, ⊥ to layers				5.70		16.8	9.23	4.09	2.68	2.01	1.60	1.34	1.08	0.81	
c_p			709			136	411	992	1406	1650	1793	1890	1974	2043	
Graphite fiber epoxy (25% vol) composite	450	1400													
k, heat flow ∥ to fibers				11.1		5.7	8.7	13.0							
k, heat flow ⊥ to fibers				0.87		0.46	0.68	1.1							
c_p			935			337	642	1216							
Pyroceram, Corning 9606	1623	2600	808	3.98	1.89	5.25 —	4.78 —	3.64 908	3.28 1038	3.08 1122	2.96 1197	2.87 1264	2.79 1498		
Silicon carbide	3100	3160	675	490	230			— 880	— 1050	— 1135	87 1195	58 1243	30 1310		

Table A.2 Continued

COMPOSITION	MELTING POINT (K)	PROPERTIES AT 300 K ρ (kg/m^3)	c_p (J/kg · K)	k (W/m · K)	$\alpha \cdot 10^6$ (m^2/s)	PROPERTIES AT VARIOUS TEMPERATURES (K) k (W/m · K) / c_p (J/kg · K) 100	200	400	600	800	1000	1200	1500	2000	2500
Silicon dioxide, crystalline (quartz)	1883	2650													
k, ∥ to c axis				10.4		39	16.4	7.6	5.0	4.2					
k, ⊥ to c axis				6.21		20.8	9.5	4.70	3.4	3.1					
c_p			745			—	—	885	1075	1250					
Silicon dioxide, polycrystalline (fused silica)	1883	2220	745	1.38	0.834	0.69	1.14	1.51	1.75	2.17	2.87	4.00			
						—	—	905	1040	1105	1155	1195			
Silicon nitride	2173	2400	691	16.0	9.65	—	—	13.9	11.3	9.88	8.76	8.00	7.16	6.20	
						—	578	778	937	1063	1155	1226	1306	1377	
Sulfur	392	2070	708	0.206	0.141	0.165	0.185								
						403	606								
Thorium dioxide	3573	9110	235	13	6.1			10.2	6.6	4.7	3.68	3.12	2.73	2.5	
								255	274	285	295	303	315	330	
Titanium dioxide, polycrystalline	2133	4157	710	8.4	2.8			7.01	5.02	3.94	3.46	3.28			
								805	880	910	930	945			

Adapted from References 1, 2, 3, and 6.

Table A.3 Thermophysical properties of common materials

Structural building materials

	TYPICAL PROPERTIES AT 300 K		
DESCRIPTION/COMPOSITION	DENSITY, ρ (kg/m^3)	THERMAL CONDUCTIVITY, k (W/m · K)	SPECIFIC HEAT, c_p (J/kg · K)
Building boards			
Asbestos-cement board	1920	0.58	—
Gypsum or plaster board	800	0.17	—
Plywood	545	0.12	1215
Sheathing, regular density	290	0.055	1300
Acoustic tile	290	0.058	1340
Hardboard, siding	640	0.094	1170
Hardboard, high density	1010	0.15	1380
Particle board, low density	590	0.078	1300
Particle board, high density	1000	0.170	1300
Woods			
Hardwoods (oak, maple)	720	0.16	1255
Softwoods (fir, pine)	510	0.12	1380
Masonry materials			
Cement mortar	1860	0.72	780
Brick, common	1920	0.72	835
Brick, face	2083	1.3	—
Clay tile, hollow			
1 cell deep, 10 cm thick	—	0.52	—
3 cells deep, 30 cm thick	—	0.69	—
Concrete block, 3 oval cores			
sand/gravel, 20 cm thick	—	1.0	—
cinder aggregate, 20 cm thick	—	0.67	—
Concrete block, rectangular core			
2 cores, 20 cm thick, 16 kg	—	1.1	—
same with filled cores	—	0.60	—
Plastering materials			
Cement plaster, sand aggregate	1860	0.72	—
Gypsum plaster, sand aggregate	1680	0.22	1085
Gypsum plaster, vermiculite aggregate	720	0.25	—

Table A.3 Continued

Insulating materials and systems

DESCRIPTION/COMPOSITION	TYPICAL PROPERTIES AT 300 K: DENSITY, ρ (kg/m^3)	THERMAL CONDUCTIVITY, k (W/m · K)	SPECIFIC HEAT, c_p (J/kg · K)
Blanket and batt			
Glass fiber, paper faced	16	0.046	—
	28	0.038	—
	40	0.035	—
Glass fiber, coated; duct liner	32	0.038	835
Board and slab			
Cellular glass	145	0.058	1000
Glass fiber, organic bonded	105	0.036	795
Polystyrene, expanded			
extruded (R-12)	55	0.027	1210
molded beads	16	0.040	1210
Mineral fiberboard; roofing material	265	0.049	—
Wood, shredded/cemented	350	0.087	1590
Cork	120	0.039	1800
Loose fill			
Cork, granulated	160	0.045	—
Diatomaceous silica, coarse powder	350	0.069	—
	400	0.091	—
Diatomaceous silica, fine powder	200	0.052	—
	275	0.061	—
Glass fiber, poured or blown	16	0.043	835
Vermiculite, flakes	80	0.068	835
	160	0.063	1000
Formed or foamed in place			
Mineral wool granules with asbestos/inorganic binders, sprayed	190	0.046	—
Polyvinyl acetate cork mastic; sprayed or troweled	—	0.100	—
Urethane, two-part mixture; rigid foam	70	0.026	1045
Reflective			
Aluminum foil separating fluffy glass mats; 10–12 layers; evacuated; for cryogenic application (150 K)	40	0.00016	—
Aluminum foil and glass paper laminate; 75–150 layers; evacuated; for cryogenic application (150 K)	120	0.000017	—
Typical silica powder, evacuated	160	0.0017	—

Table A.3 Thermophysical properties of common materials Continued

Industrial insulation

DESCRIPTION/COMPOSITION	MAX SERVICE TEMP (K)	TYPICAL DENSITY (kg/m^3)	TYPICAL THERMAL CONDUCTIVITY, k (W/m · K), AT VARIOUS TEMPERATURES (K)													
			200	215	230	240	255	270	285	300	310	365	420	530	645	750
Blankets																
Blanket, mineral fiber, metal	920	96–192									0.038	0.046	0.056	0.078		
reinforced	815	40–96									0.035	0.045	0.058	0.088		
Blanket, mineral fiber, glass;	450	10				0.036	0.038	0.040	0.043	0.048	0.052	0.076				
fine fiber, organic bonded		12				0.035	0.036	0.039	0.042	0.046	0.049	0.069				
		16				0.033	0.035	0.036	0.039	0.042	0.046	0.062				
		24				0.030	0.032	0.033	0.036	0.039	0.040	0.053				
		32				0.029	0.030	0.032	0.033	0.036	0.038	0.048				
		48				0.027	0.029	0.030	0.032	0.033	0.035	0.045				
Blanket, alumina-silica fiber	1530	48												0.071	0.105	0.150
		64												0.059	0.087	0.125
		96												0.052	0.076	0.100
		128												0.049	0.068	0.091
Felt, semirigid; organic bonded	480	50–125						0.035	0.036	0.038	0.039	0.051	0.063			
	730	50	0.023	0.025	0.026	0.027	0.029	0.030	0.032	0.033	0.035	0.051	0.079			
Felt, laminated; no binder	920	120											0.051	0.065	0.087	
Blocks, boards, and pipe insulations																
Asbestos paper, laminated and corrugated																
4-ply	420	190								0.078	0.082	0.098				
6-ply	420	255								0.071	0.074	0.085				
8-ply	420	300								0.068	0.071	0.082				
Magnesia, 85%	590	185									0.051	0.055	0.061			
Calcium silicate	920	190									0.055	0.059	0.063	0.075	0.089	0.104
Cellular glass	700	145			0.046	0.048	0.051	0.052	0.055	0.058	0.062	0.069	0.079			
Diatomaceous silica	1145	345												0.092	0.098	0.104
	1310	385												0.101	0.100	0.115
Polystyrene, rigid																
Extruded (R-12)	350	56	0.023	0.023	0.022	0.023	0.023	0.025	0.026	0.027	0.029					
Extruded (R-12)	350	35	0.023	0.023	0.023	0.025	0.025	0.026	0.027	0.029						
Molded beads	350	16	0.026	0.029	0.030	0.033	0.035	0.036	0.038	0.040						
Rubber, rigid foamed	340	70						0.029	0.030	0.032	0.033					

Table A.3 Continued

Industrial insulation Continued

DESCRIPTION/COMPOSITION	MAX SERVICE TEMP (K)	TYPICAL DENSITY (kg/m³)	TYPICAL THERMAL CONDUCTIVITY, k (W/m · K), AT VARIOUS TEMPERATURES (K) 200	215	230	240	255	270	285	300	310	365	420	530	645	750
Insulating cement																
Mineral fiber (rock, slag, or glass)																
with clay binder	1255	430									0.071	0.079	0.088	0.105	0.123	
with hydraulic setting binder	922	560									0.108	0.115	0.123	0.137		
Loose fill																
Cellulose, wood, or paper pulp	—	45							0.038	0.039	0.042					
Perlite, expanded	—	105	0.036	0.039	0.042	0.043	0.046	0.049	0.051	0.053	0.056					
Vermiculite, expanded	—	122			0.056	0.058	0.061	0.063	0.065	0.068	0.071					
		80			0.049	0.051	0.055	0.058	0.061	0.063	0.066					

Table A.3 Thermophysical properties of common materials Continued

Other materials

DESCRIPTION/COMPOSITION	TEMPERATURE (K)	DENSITY, ρ (kg/m^3)	THERMAL CON-DUCTIVITY, k (W/m · K)	SPECIFIC HEAT, c_p (J/kg · K)
Asphalt	300	2115	0.062	920
Bakelite	300	1300	1.4	1465
Brick, refractory				
Carborundum	872	—	18.5	—
	1672	—	11.0	—
Chrome brick	473	3010	2.3	835
	823		2.5	
	1173		2.0	
Diatomaceous silica, fired	478	—	0.25	—
	1145	—	0.30	
Fireclay, burnt 1600 K	773	2050	1.0	960
	1073	—	1.1	
	1373	—	1.1	
Fireclay, burnt 1725 K	773	2325	1.3	960
	1073		1.4	
	1373		1.4	
Fireclay brick	478	2645	1.0	960
	922		1.5	
	1478		1.8	
Magnesite	478	—	3.8	1130
	922		2.8	
	1478		1.9	
Clay	300	1460	1.3	880
Coal, anthracite	300	1350	0.26	1260
Concrete (stone mix)	300	2300	1.4	880
Cotton	300	80	0.06	1300
Foodstuffs				
Banana (75.7% water content)	300	980	0.481	3350
Apple, red (75% water content)	300	840	0.513	3600
Cake, batter	300	720	0.223	—
Cake, fully baked	300	280	0.121	—
Chicken meat, white	198	—	1.60	—
(74.4% water content)	233	—	1.49	
	253		1.35	
	263		1.20	
	273		0.476	
	283		0.480	
	293		0.489	
Glass				
Plate (soda lime)	300	2500	1.4	750
Pyrex	300	2225	1.4	835

Table A.3 Continued

Other materials Continued

DESCRIPTION/COMPOSITION	TEMPERATURE (K)	DENSITY ρ (kg/m^3)	THERMAL CONDUCTIVITY, k ($W/m \cdot K$)	SPECIFIC HEAT, c_p ($J/kg \cdot K$)
Ice	273	920	1.88	2040
	253	—	2.03	1945
Leather (sole)	300	998	0.159	—
Paper	300	930	0.180	1340
Paraffin	300	900	0.240	2890
Rock				
Granite, Barre	300	2630	2.79	775
Limestone, Salem	300	2320	2.15	810
Marble, Halston	300	2680	2.80	830
Quartzite, Sioux	300	2640	5.38	1105
Sandstone, Berea	300	2150	2.90	745
Rubber, vulcanized				
Soft	300	1100	0.13	2010
Hard	300	1190	0.16	—
Sand	300	1515	0.27	800
Soil	300	2050	0.52	1840
Snow	273	110	0.049	—
		500	0.190	—
Teflon	300	2200	0.35	—
	400		0.45	—
Tissue, human				
Skin	300	—	0.37	—
Fat layer (adipose)	300	—	0.2	—
Muscle	300	—	0.41	—
Wood, cross grain				
Balsa	300	140	0.055	—
Cypress	300	465	0.097	—
Fir	300	415	0.11	2720
Oak	300	545	0.17	2385
Yellow pine	300	640	0.15	2805
White pine	300	435	0.11	—
Wood, radial				
Oak	300	545	0.19	2385
Fir	300	420	0.14	2720

Adapted from References 1, 8–13.

Table A.4 Thermophysical properties of gases at atmospheric pressure

T (K)	ρ (kg/m^3)	c_p (kJ/kg·K)	$\mu \cdot 10^7$ (N·s/m^2)	$\nu \cdot 10^6$ (m^2/s)	$k \cdot 10^3$ (W/m·K)	$\alpha \cdot 10^6$ (m^2/s)	Pr
Air							
100	3.5562	1.032	71.1	2.00	9.34	2.54	0.786
150	2.3364	1.012	103.4	4.426	13.8	5.84	0.758
200	1.7458	1.007	132.5	7.590	18.1	10.3	0.737
250	1.3947	1.006	159.6	11.44	22.3	15.9	0.720
300	1.1614	1.007	184.6	15.89	26.3	22.5	0.707
350	0.9950	1.009	208.2	20.92	30.0	29.9	0.700
400	0.8711	1.014	230.1	26.41	33.8	38.3	0.690
450	0.7740	1.021	250.7	32.39	37.3	47.2	0.686
500	0.6964	1.030	270.1	38.79	40.7	56.7	0.684
550	0.6329	1.040	288.4	45.57	43.9	66.7	0.683
600	0.5804	1.051	305.8	52.69	46.9	76.9	0.685
650	0.5356	1.063	322.5	60.21	49.7	87.3	0.690
700	0.4975	1.075	338.8	68.10	52.4	98.0	0.695
750	0.4643	1.087	354.6	76.37	54.9	109	0.702
800	0.4354	1.099	369.8	84.93	57.3	120	0.709
850	0.4097	1.110	384.3	93.80	59.6	131	0.716
900	0.3868	1.121	398.1	102.9	62.0	143	0.720
950	0.3666	1.131	411.3	112.2	64.3	155	0.723
1000	0.3482	1.141	424.4	121.9	66.7	168	0.726
1100	0.3166	1.159	449.0	141.8	71.5	195	0.728
1200	0.2902	1.175	473.0	162.9	76.3	224	0.728
1300	0.2679	1.189	496.0	185.1	82	238	0.719
1400	0.2488	1.207	530	213	91	303	0.703
1500	0.2322	1.230	557	240	100	350	0.685
1600	0.2177	1.248	584	268	106	390	0.688
1700	0.2049	1.267	611	298	113	435	0.685
1800	0.1935	1.286	637	329	120	482	0.683
1900	0.1833	1.307	663	362	128	534	0.677
2000	0.1741	1.337	689	396	137	589	0.672
2100	0.1658	1.372	715	431	147	646	0.667
2200	0.1582	1.417	740	468	160	714	0.655
2300	0.1513	1.478	766	506	175	783	0.647
2400	0.1448	1.558	792	547	196	869	0.630
2500	0.1389	1.665	818	589	222	960	0.613
3000	0.1135	2.726	955	841	486	1570	0.536
Ammonia (NH_3)							
300	0.6894	2.158	101.5	14.7	24.7	16.6	0.887
320	0.6448	2.170	109	16.9	27.2	19.4	0.870
340	0.6059	2.192	116.5	19.2	29.3	22.1	0.872
360	0.5716	2.221	124	21.7	31.6	24.9	0.872
380	0.5410	2.254	131	24.2	34.0	27.9	0.869
400	0.5136	2.287	138	26.9	37.0	31.5	0.853
420	0.4888	2.322	145	29.7	40.4	35.6	0.833
440	0.4664	2.357	152.5	32.7	43.5	39.6	0.826

Table A.4 Continued

T (K)	ρ (kg/m^3)	c_p (kJ/kg·K)	$\mu \cdot 10^7$ (N·s/m^2)	$\nu \cdot 10^6$ (m^2/s)	$k \cdot 10^3$ (W/m·K)	$\alpha \cdot 10^6$ (m^2/s)	Pr
Ammonia (NH_3) Continued							
460	0.4460	2.393	159	35.7	46.3	43.4	0.822
480	0.4273	2.430	166.5	39.0	49.2	47.4	0.822
500	0.4101	2.467	173	42.2	52.5	51.9	0.813
520	0.3942	2.504	180	45.7	54.5	55.2	0.827
540	0.3795	2.540	186.5	49.1	57.5	59.7	0.824
560	0.3708	2.577	193	52.0	60.6	63.4	0.827
580	0.3533	2.613	199.5	56.5	63.8	69.1	0.817
Carbon Dioxide (CO_2)							
280	1.9022	0.830	140	7.36	15.20	9.63	0.765
300	1.7730	0.851	149	8.40	16.55	11.0	0.766
320	1.6609	0.872	156	9.39	18.05	12.5	0.754
340	1.5618	0.891	165	10.6	19.70	14.2	0.746
360	1.4743	0.908	173	11.7	21.2	15.8	0.741
380	1.3961	0.926	181	13.0	22.75	17.6	0.737
400	1.3257	0.942	190	14.3	24.3	19.5	0.737
450	1.1782	0.981	210	17.8	28.3	24.5	0.728
500	1.0594	1.02	231	21.8	32.5	30.1	0.725
550	0.9625	1.05	251	26.1	36.6	36.2	0.721
600	0.8826	1.08	270	30.6	40.7	42.7	0.717
650	0.8143	1.10	288	35.4	44.5	49.7	0.712
700	0.7564	1.13	305	40.3	48.1	56.3	0.717
750	0.7057	1.15	321	45.5	51.7	63.7	0.714
800	0.6614	1.17	337	51.0	55.1	71.2	0.716
Carbon Monoxide (CO)							
200	1.6888	1.045	127	7.52	17.0	9.63	0.781
220	1.5341	1.044	137	8.93	19.0	11.9	0.753
240	1.4055	1.043	147	10.5	20.6	14.1	0.744
260	1.2967	1.043	157	12.1	22.1	16.3	0.741
280	1.2038	1.042	166	13.8	23.6	18.8	0.733
300	1.1233	1.043	175	15.6	25.0	21.3	0.730
320	1.0529	1.043	184	17.5	26.3	23.9	0.730
340	0.9909	1.044	193	19.5	27.8	26.9	0.725
360	0.9357	1.045	202	21.6	29.1	29.8	0.725
380	0.8864	1.047	210	23.7	30.5	32.9	0.729
400	0.8421	1.049	218	25.9	31.8	36.0	0.719
450	0.7483	1.055	237	31.7	35.0	44.3	0.714
500	0.67352	1.065	254	37.7	38.1	53.1	0.710
550	0.61226	1.076	271	44.3	41.1	62.4	0.710
600	0.56126	1.088	286	51.0	44.0	72.1	0.707
650	0.51806	1.101	301	58.1	47.0	82.4	0.705
700	0.48102	1.114	315	65.5	50.0	93.3	0.702
750	0.44899	1.127	329	73.3	52.8	104	0.702
800	0.42095	1.140	343	81.5	55.5	116	0.705

Table A.4 Continued

T (K)	ρ (kg/m^3)	c_p ($kJ/kg \cdot K$)	$\mu \cdot 10^7$ ($N \cdot s/m^2$)	$\nu \cdot 10^6$ (m^2/s)	$k \cdot 10^3$ ($W/m \cdot K$)	$\alpha \cdot 10^6$ (m^2/s)	Pr
Helium (He)							
100	0.4871	5.193	96.3	19.8	73.0	28.9	0.686
120	0.4060	5.193	107	26.4	81.9	38.8	0.679
140	0.3481	5.193	118	33.9	90.7	50.2	0.676
160	—	5.193	129	—	99.2	—	—
180	0.2708	5.193	139	51.3	107.2	76.2	0.673
200	—	5.193	150	—	115.1	—	—
220	0.2216	5.193	160	72.2	123.1	107	0.675
240	—	5.193	170	—	130	—	—
260	0.1875	5.193	180	96.0	137	141	0.682
280	—	5.193	190	—	145	—	—
300	0.1625	5.193	199	122	152	180	0.680
350	—	5.193	221	—	170	—	
400	0.1219	5.193	243	199	187	295	0.675
450		5.193	263	—	204	—	—
500	0.09754	5.193	283	290	220	434	0.668
550	—	5.193	—	—	—	—	—
600	—	5.193	320	—	252	—	—
650	—	5.193	332	—	264	—	—
700	0.06969	5.193	350	502	278	768	0.654
750	—	5.193	364	—	291	—	—
800	—	5.193	382	—	304	—	—
900	—	5.193	414	—	330	—	—
1000	0.04879	5.193	446	914	354	1400	0.654
Hydrogen (H_2)							
100	0.24255	11.23	42.1	17.4	67.0	24.6	0.707
150	0.16156	12.60	56.0	34.7	101	49.6	0.699
200	0.12115	13.54	68.1	56.2	131	79.9	0.704
250	0.09693	14.06	78.9	81.4	157	115	0.707
300	0.08078	14.31	89.6	111	183	158	0.701
350	0.06924	14.43	98.8	143	204	204	0.700
400	0.06059	14.48	108.2	179	226	258	0.695
450	0.05386	14.50	117.2	218	247	316	0.689
500	0.04848	14.52	126.4	261	266	378	0.691
550	0.04407	14.53	134.3	305	285	445	0.685
600	0.04040	14.55	142.4	352	305	519	0.678
700	0.03463	14.61	157.8	456	342	676	0.675
800	0.03030	14.70	172.4	569	378	849	0.670
900	0.02694	14.83	186.5	692	412	1030	0.671
1000	0.02424	14.99	201.3	830	448	1230	0.673
1100	0.02204	15.17	213.0	966	488	1460	0.662
1200	0.02020	15.37	226.2	1120	528	1700	0.659
1300	0.01865	15.59	238.5	1279	568	1955	0.655
1400	0.01732	15.81	250.7	1447	610	2230	0.650
1500	0.01616	16.02	262.7	1626	655	2530	0.643

Table A.4 Continued

T (K)	ρ (kg/m^3)	c_p ($kJ/kg \cdot K$)	$\mu \cdot 10^7$ ($N \cdot s/m^2$)	$\nu \cdot 10^6$ (m^2/s)	$k \cdot 10^3$ ($W/m \cdot K$)	$\alpha \cdot 10^6$ (m^2/s)	Pr
Hydrogen (H_2) Continued							
1600	0.0152	16.28	273.7	1801	697	2815	0.639
1700	0.0143	16.58	284.9	1992	742	3130	0.637
1800	0.0135	16.96	296.1	2193	786	3435	0.639
1900	0.0128	17.49	307.2	2400	835	3730	0.643
2000	0.0121	18.25	318.2	2630	878	3975	0.661
Nitrogen (N_2)							
100	3.4388	1.070	68.8	2.00	9.58	2.60	0.768
150	2.2594	1.050	100.6	4.45	13.9	5.86	0.759
200	1.6883	1.043	129.2	7.65	18.3	10.4	0.736
250	1.3488	1.042	154.9	11.48	22.2	15.8	0.727
300	1.1233	1.041	178.2	15.86	25.9	22.1	0.716
350	0.9625	1.042	200.0	20.78	29.3	29.2	0.711
400	0.8425	1.045	220.4	26.16	32.7	37.1	0.704
450	0.7485	1.050	239.6	32.01	35.8	45.6	0.703
500	0.6739	1.056	257.7	38.24	38.9	54.7	0.700
550	0.6124	1.065	274.7	44.86	41.7	63.9	0.702
600	0.5615	1.075	290.8	51.79	44.6	73.9	0.701
700	0.4812	1.098	321.0	66.71	49.9	94.4	0.706
800	0.4211	1.22	349.1	82.90	54.8	116	0.715
900	0.3743	1.146	375.3	100.3	59.7	139	0.721
1000	0.3368	1.167	399.9	118.7	64.7	165	0.721
1100	0.3062	1.187	423.2	138.2	70.0	193	0.718
1200	0.2807	1.204	445.3	158.6	75.8	224	0.707
1300	0.2591	1.219	466.2	179.9	81.0	256	0.701
Oxygen (O_2)							
100	3.945	0.962	76.4	1.94	9.25	2.44	0.796
150	2.585	0.921	114.8	4.44	13.8	5.80	0.766
200	1.930	0.915	147.5	7.64	18.3	10.4	0.737
250	1.542	0.915	178.6	11.58	22.6	16.0	0.723
300	1.284	0.920	207.2	16.14	26.8	22.7	0.711
350	1.100	0.929	233.5	21.23	29.6	29.0	0.733
400	0.9620	0.942	258.2	26.84	33.0	36.4	0.737
450	0.8554	0.956	281.4	32.90	36.3	44.4	0.741
500	0.7698	0.972	303.3	39.40	41.2	55.1	0.716
550	0.6998	0.988	324.0	46.30	44.1	63.8	0.726
600	0.6414	1.003	343.7	53.59	47.3	73.5	0.729
700	0.5498	1.031	380.8	69.26	52.8	93.1	0.744
800	0.4810	1.054	415.2	86.32	58.9	116	0.743
900	0.4275	1.074	447.2	104.6	64.9	141	0.740
1000	0.3848	1.090	477.0	124.0	71.0	169	0.733
1100	0.3498	1.103	505.5	144.5	75.8	196	0.736
1200	0.3206	1.115	532.5	166.1	81.9	229	0.725
1300	0.2960	1.125	588.4	188.6	87.1	262	0.721

Table A.4 Continued

T (K)	ρ (kg/m^3)	c_p (kJ/kg · K)	$\mu \cdot 10^7$ (N · s/m^2)	$\nu \cdot 10^6$ (m^2/s)	$k \cdot 10^3$ (W/m · K)	$\alpha \cdot 10^6$ (m^2/s)	Pr
Water vapor (steam)							
380	0.5863	2.060	127.1	21.68	24.6	20.4	1.06
400	0.5542	2.014	134.4	24.25	26.1	23.4	1.04
450	0.4902	1.980	152.5	31.11	29.9	30.8	1.01
500	0.4405	1.985	170.4	38.68	33.9	38.8	0.998
550	0.4005	1.997	188.4	47.04	37.9	47.4	0.993
600	0.3652	2.026	206.7	56.60	42.2	57.0	0.993
650	0.3380	2.056	224.7	66.48	46.4	66.8	0.996
700	0.3140	2.085	242.6	77.26	50.5	77.1	1.00
750	0.2931	2.119	260.4	88.84	54.9	88.4	1.00
800	0.2739	2.152	278.6	101.7	59.2	100	1.01
850	0.2579	2.186	296.9	115.1	63.7	113	1.02

Adapted from References 8, 14, and 15.

Table A.5 Thermophysical properties of saturated liquids

Saturated liquids[a]

T (K)	ρ (kg/m^3)	c_p (kJ/kg·K)	$\mu \cdot 10^2$ (N·s/m^2)	$\nu \cdot 10^6$ (m^2/s)	$k \cdot 10^3$ (W/m·K)	$\alpha \cdot 10^7$ (m^2/s)	Pr	$\beta \cdot 10^3$ (K^{-1})
Engine oil (unused)								
273	899.1	1.796	385	4,280	147	0.910	47,000	0.70
280	895.3	1.827	217	2,430	144	0.880	27,500	0.70
290	890.0	1.868	99.9	1,120	145	0.872	12,900	0.70
300	884.1	1.909	48.6	550	145	0.859	6,400	0.70
310	877.9	1.951	25.3	288	145	0.847	3,400	0.70
320	871.8	1.993	14.1	161	143	0.823	1,965	0.70
330	865.8	2.035	8.36	96.6	141	0.800	1,205	0.70
340	859.9	2.076	5.31	61.7	139	0.779	793	0.70
350	853.9	2.118	3.56	41.7	138	0.763	546	0.70
360	847.8	2.161	2.52	29.7	138	0.753	395	0.70
370	841.8	2.206	1.86	22.0	137	0.738	300	0.70
380	836.0	2.250	1.41	16.9	136	0.723	233	0.70
390	830.6	2.294	1.10	13.3	135	0.709	187	0.70
400	825.1	2.337	0.874	10.6	134	0.695	152	0.70
410	818.9	2.381	0.698	8.52	133	0.682	125	0.70
420	812.1	2.427	0.564	6.94	133	0.675	103	0.70
430	806.5	2.471	0.470	5.83	132	0.662	88	0.70
Ethylene glycol [$C_2H_4(OH)_2$]								
273	1,130.8	2.294	6.51	57.6	242	0.933	617	0.65
280	1,125.8	2.323	4.20	37.3	244	0.933	400	0.65
290	1,118.8	2.368	2.47	22.1	248	0.936	236	0.65
300	1,111.4	2.415	1.57	14.1	252	0.939	151	0.65
310	1,103.7	2.460	1.07	9.65	255	0.939	103	0.65
320	1,096.2	2.505	0.757	6.91	258	0.940	73.5	0.65
330	1,089.5	2.549	0.561	5.15	260	0.936	55.0	0.65
340	1,083.8	2.592	0.431	3.98	261	0.929	42.8	0.65
350	1,079.0	2.637	0.342	3.17	261	0.917	34.6	0.65
360	1,074.0	2.682	0.278	2.59	261	0.906	28.6	0.65
370	1,066.7	2.728	0.228	2.14	262	0.900	23.7	0.65
373	1,058.5	2.742	0.215	2.03	263	0.906	22.4	0.65
Glycerin [$C_3H_5(OH)_3$]								
273	1,276.0	2.261	1,060	8,310	282	0.977	85,000	0.47
280	1,271.9	2.298	534	4,200	284	0.972	43,200	0.47
290	1,265.8	2.367	185	1,460	286	0.955	15,300	0.48
300	1,259.9	2.427	79.9	634	286	0.935	6,780	0.48
310	1,253.9	2.490	35.2	281	286	0.916	3,060	0.49
320	1,247.2	2.564	21.0	168	287	0.897	1,870	0.50
Freon (refrigerant-12) (CCl_2F_2)								
230	1,528.4	0.8816	0.0457	0.299	68	0.505	5.9	1.85
240	1,498.0	0.8923	0.0385	0.257	69	0.516	5.0	1.90
250	1,469.5	0.9037	0.0354	0.241	70	0.527	4.6	2.00
260	1,439.0	0.9163	0.0322	0.224	73	0.554	4.0	2.10

Table A.5 Continued

Saturated liquids Continued

T (K)	ρ (kg/m^3)	c_p (kJ/kg · K)	$\mu \cdot 10^2$ (N · s/m^2)	$\nu \cdot 10^6$ (m^2/s)	$k \cdot 10^3$ (W/m · K)	$\alpha \cdot 10^7$ (m^2/s)	Pr	$\beta \cdot 10^3$ (K^{-1})
Freon (refrigerant-12) (CCl_2F_2) Continued								
270	1,407.2	0.9301	0.0304	0.216	73	0.558	3.9	2.25
280	1,374.4	0.9450	0.0283	0.206	73	0.562	3.7	2.35
290	1,340.5	0.9609	0.0265	0.198	73	0.567	3.5	2.55
300	1,305.8	0.9781	0.0254	0.195	72	0.564	3.5	2.75
310	1,268.9	0.9963	0.0244	0.192	69	0.546	3.4	3.05
320	1,228.6	1.0155	0.0233	0.190	68	0.545	3.5	3.5
Mercury (Hg)								
273	13,595	0.1404	0.1688	0.1240	8,180	42.85	0.0290	0.181
300	13,529	0.1393	0.1523	0.1125	8,540	45.30	0.0248	0.181
350	13,407	0.1377	0.1309	0.0976	9,180	49.75	0.0196	0.181
400	13,287	0.1365	0.1171	0.0882	9,800	54.05	0.0163	0.181
450	13,167	0.1357	0.1075	0.0816	10,400	58.10	0.0140	0.181
500	13,048	0.1353	0.1007	0.0771	10,950	61.90	0.0125	0.182
550	12,929	0.1352	0.0953	0.0737	11,450	65.55	0.0112	0.184
600	12,809	0.1355	0.0911	0.0711	11,950	68.80	0.0103	0.187

Saturated liquid–vapor, 1 atm[b]

FLUID	T_{sat} (K)	h_{fg} (kJ/kg)	ρ_f (kg/m^3)	ρ_g (kg/m^3)	$\sigma \cdot 10^3$ (N/m)
Ethanol	351	846	757	1.44	17.7
Ethylene glycol	470	812	1,111[c]	—	32.7
Glycerin	563	974	1,260[c]	—	63.0[c]
Mercury	630	301	12,740	3.90	417
Refrigerant R-12	243	165	1,488	6.32	15.8
Refrigerant R-113	321	147	1,511	7.38	15.9

[a] Adapted from references 15 and 16.
[b] Adapted from references 8, 17, and 18.
[c] Property value corresponding to 300 K.

Table A.6 Thermophysical properties of saturated water

TEMPERATURE, T (K)	PRESSURE, P (bars)[a]	SPECIFIC VOLUME (m³/kg) $v_f \cdot 10^3$	v_g	HEAT OF VAPORIZATION, h_{fg} (kJ/kg)	SPECIFIC HEAT (kJ/kg·K) $c_{p,f}$	$c_{p,g}$	VISCOSITY (N·s/m²) $\mu_f \cdot 10^6$	$\mu_g \cdot 10^6$	THERMAL CONDUCTIVITY (W/m·K) $k_f \cdot 10^3$	$k_g \cdot 10^3$	PRANDTL NUMBER Pr_f	Pr_g	SURFACE TENSION, $\sigma_f \cdot 10^3$ (N/m)	EXPANSION COEFFICIENT, $\beta_f \cdot 10^6$ (K⁻¹)	TEMPERATURE, T (K)
273.15	0.00611	1.000	206.3	2502	4.217	1.854	1750	8.02	569	18.2	12.99	0.815	75.5	−68.05	273.15
275	0.00697	1.000	181.7	2497	4.211	1.855	1652	8.09	574	18.3	12.22	0.817	75.3	−32.74	275
280	0.00990	1.000	130.4	2485	4.198	1.858	1422	8.29	582	18.6	10.26	0.825	74.8	46.04	280
285	0.01387	1.000	99.4	2473	4.189	1.861	1225	8.49	590	18.9	8.81	0.833	74.3	114.1	285
290	0.01917	1.001	69.7	2461	4.184	1.864	1080	8.69	598	19.3	7.56	0.841	73.7	174.0	290
295	0.02617	1.002	51.94	2449	4.181	1.868	959	8.89	606	19.5	6.62	0.849	72.7	227.5	295
300	0.03531	1.003	39.13	2438	4.179	1.872	855	9.09	613	19.6	5.83	0.857	71.7	276.1	300
305	0.04712	1.005	27.90	2426	4.178	1.877	769	9.29	620	20.1	5.20	0.865	70.9	320.6	305
310	0.06221	1.007	22.93	2414	4.178	1.882	695	9.49	628	20.4	4.62	0.873	70.0	361.9	310
315	0.08132	1.009	17.82	2402	4.179	1.888	631	9.69	634	20.7	4.16	0.883	69.2	400.4	315
320	0.1053	1.011	13.98	2390	4.180	1.895	577	9.89	640	21.0	3.77	0.894	68.3	436.7	320
325	0.1351	1.013	11.06	2378	4.182	1.903	528	10.09	645	21.3	3.42	0.901	67.5	471.2	325
330	0.1719	1.016	8.82	2366	4.184	1.911	489	10.29	650	21.7	3.15	0.908	66.6	504.0	330
335	0.2167	1.018	7.09	2354	4.186	1.920	453	10.49	656	22.0	2.88	0.916	65.8	535.5	335
340	0.2713	1.021	5.74	2342	4.188	1.930	420	10.69	660	22.3	2.66	0.925	64.9	566.0	340
345	0.3372	1.024	4.683	2329	4.191	1.941	389	10.89	668	22.6	2.45	0.933	64.1	595.4	345
350	0.4163	1.027	3.846	2317	4.195	1.954	365	11.09	668	23.0	2.29	0.942	63.2	624.2	350
355	0.5100	1.030	3.180	2304	4.199	1.968	343	11.29	671	23.3	2.14	0.951	62.3	652.3	355
360	0.6209	1.034	2.645	2291	4.203	1.983	324	11.49	674	23.7	2.02	0.960	61.4	697.9	360
365	0.7514	1.038	2.212	2278	4.209	1.999	306	11.69	677	24.1	1.91	0.969	60.5	707.1	365
370	0.9040	1.041	1.861	2265	4.214	2.017	289	11.89	679	24.5	1.80	0.978	59.5	728.7	370
373.15	1.0133	1.044	1.679	2257	4.217	2.029	279	12.02	680	24.8	1.76	0.984	58.9	750.1	373.15
375	1.0815	1.045	1.574	2252	4.220	2.036	274	12.09	681	24.9	1.70	0.987	58.6	761	375
380	1.2869	1.049	1.337	2239	4.226	2.057	260	12.29	683	25.4	1.61	0.999	57.6	788	380
385	1.5233	1.053	1.142	2225	4.232	2.080	248	12.49	685	25.8	1.53	1.004	56.6	814	385
390	1.794	1.058	0.980	2212	4.239	2.104	237	12.69	686	26.3	1.47	1.013	55.6	841	390
400	2.455	1.067	0.731	2183	4.256	2.158	217	13.05	688	27.2	1.34	1.033	53.6	896	400
410	3.302	1.077	0.553	2153	4.278	2.221	200	13.42	688	28.2	1.24	1.054	51.5	952	410
420	4.370	1.088	0.425	2123	4.302	2.291	185	13.79	688	29.8	1.16	1.075	49.4	1010	420
430	5.699	1.099	0.331	2091	4.331	2.369	173	14.14	685	30.4	1.09	1.10	47.2		430
440	7.333	1.110	0.261	2059	4.36	2.46	162	14.50	682	31.7	1.04	1.12	45.1		440
450	9.319	1.123	0.208	2024	4.40	2.56	152	14.85	678	33.1	0.99	1.14	42.9		450
460	11.71	1.137	0.167	1989	4.44	2.68	143	15.19	673	34.6	0.95	1.17	40.7		460
470	14.55	1.152	0.136	1951	4.48	2.79	136	15.54	667	36.3	0.92	1.20	38.5		470
480	17.90	1.167	0.111	1912	4.53	2.94	129	15.88	660	38.1	0.89	1.23	36.2		480

490	21.83	1.184	0.0922	1870	4.59	3.10	124	16.23	651	40.1	0.87	1.25	33.9	—	490
500	26.40	1.203	0.0766	1825	4.66	3.27	118	16.59	642	42.3	0.86	1.28	31.6	—	500
510	31.66	1.222	0.0631	1779	4.74	3.47	113	16.95	631	44.7	0.85	1.31	29.3	—	510
520	37.70	1.244	0.0525	1730	4.84	3.70	108	17.33	621	47.5	0.84	1.35	26.9	—	520
530	44.58	1.268	0.0445	1679	4.95	3.96	104	17.72	608	50.6	0.85	1.39	24.5	—	530
540	52.38	1.294	0.0375	1622	5.08	4.27	101	18.1	594	54.0	0.86	1.43	22.1	—	540
550	61.19	1.323	0.0317	1564	5.24	4.64	97	18.6	580	58.3	0.87	1.47	19.7	—	550
560	71.08	1.355	0.0269	1499	5.43	5.09	94	19.1	563	63.7	0.90	1.52	17.3	—	560
570	82.16	1.392	0.0228	1429	5.68	5.67	91	19.7	548	76.7	0.94	1.59	15.0	—	570
580	94.51	1.433	0.0193	1353	6.00	6.40	88	20.4	528	76.7	0.99	1.68	12.8	—	580
590	108.3	1.482	0.0163	1274	6.41	7.35	84	21.5	513	84.1	1.05	1.84	10.5	—	590
600	123.5	1.541	0.0137	1176	7.00	8.75	81	22.7	497	92.9	1.14	2.15	8.4	—	600
610	137.3	1.612	0.0115	1068	7.85	11.1	77	24.1	467	103	1.30	2.60	6.3	—	610
620	159.1	1.705	0.0094	941	9.35	15.4	72	25.9	444	114	1.52	3.46	4.5	—	620
625	169.1	1.778	0.0085	858	10.6	18.3	70	27.0	430	121	1.65	4.20	3.5	—	625
630	179.7	1.856	0.0075	781	12.6	22.1	67	28.0	412	130	2.0	4.8	2.6	—	630
635	190.9	1.935	0.0066	683	16.4	27.6	64	30.0	392	141	2.7	6.0	1.5	—	635
640	202.7	2.075	0.0057	560	26	42	59	32.0	367	155	4.2	9.6	0.8	—	640
645	215.2	2.351	0.0045	361	90	—	54	37.0	331	178	12	26	0.1	—	645
647.3[b]	221.2	3.170	0.0032	0	∞	∞	45	45.0	238	238	∞	∞	0.0	—	647.3[b]

[a] 1 bar = 10^5 N/m^2.
[b] Critical temperature.
Adapted from Reference 19.

Table A.7 Thermophysical properties of liquid metals

COMPOSITION	MELTING POINT (K)	T (K)	ρ (kg/m^3)	c_p (kJ/kg·K)	$\nu \cdot 10^7$ (m^2/s)	k (W/m·K)	$\alpha \cdot 10^5$ (m^2/s)	Pr
Bismuth	544	589	10,011	0.1444	1.617	16.4	0.138	0.0142
		811	9,739	0.1545	1.133	15.6	1.035	0.0110
		1033	9,467	0.1645	0.8343	15.6	1.001	0.0083
Lead	600	644	10,540	0.159	2.276	16.1	1.084	0.024
		755	10,412	0.155	1.849	15.6	1.223	0.017
		977	10,140	—	1.347	14.9	—	—
Potassium	337	422	807.3	0.80	4.608	45.0	6.99	0.0066
		700	741.7	0.75	2.397	39.5	7.07	0.0034
		977	674.4	0.75	1.905	33.1	6.55	0.0029
Sodium	371	366	929.1	1.38	7.516	86.2	6.71	0.011
		644	860.2	1.30	3.270	72.3	6.48	0.0051
		977	778.5	1.26	2.285	59.7	6.12	0.0037
NaK (45%/55%)	292	366	887.4	1.130	6.522	25.6	2.552	0.026
		644	821.7	1.055	2.871	27.5	3.17	0.0091
		977	740.1	1.043	2.174	28.9	3.74	0.0058
NaK (22%/78%)	262	366	849.0	0.946	5.797	24.4	3.05	0.019
		672	775.3	0.879	2.666	26.7	3.92	0.0068
		1033	690.4	0.883	2.118	—	—	—
PbBi (44.5%/55.5%)	398	422	10,524	0.147	—	9.05	0.586	—
		644	10,236	0.147	1.496	11.86	0.790	0.189
		922	9,835	—	1.171	—	—	—
Mercury	234	See Table A.5						

Adapted from *Liquid Materials Handbook*, 23rd ed., the Atomic Energy Commission, Department of the Navy, Washington, D.C., 1952.

Metallic solids and their oxides

DESCRIPTION/COMPOSITION		ε_n OR ε_h AT VARIOUS TEMPERATURES (K) 100	200	300	400	600	800	1000	1200	1500	2000	2500
Aluminum												
Highly polished, film	(*h*)	0.02	0.03	0.04	0.05	0.06						
Foil, bright	(*h*)	0.06	0.06	0.07								
Anodized	(*h*)			0.82	0.76							
Chromium												
Polished or plated	(*n*)	0.05	0.07	0.10	0.12	0.14						
Copper												
Highly polished	(*h*)			0.03	0.03	0.04	0.04	0.04				
Stably oxidized	(*h*)					0.50	0.58	0.80				
Gold												
Highly polished or film	(*h*)	0.01	0.02	0.03	0.03	0.04	0.05	0.06				
Foil, bright	(*h*)	0.06	0.07	0.07								
Molybdenum												
Polished	(*h*)					0.06	0.08	0.10	0.12	0.15	0.21	0.26
Shot-blasted, rough	(*h*)					0.25	0.28	0.31	0.35	0.42		
Stably oxidized	(*h*)					0.80	0.82					
Nickel												
Polished	(*h*)					0.09	0.11	0.14	0.17			
Stably oxidized	(*h*)					0.40	0.49	0.57				
Platinum												
Polished	(*h*)						0.10	0.13	0.15	0.18		
Silver												
Polished	(*h*)			0.02	0.02	0.03	0.05	0.08				
Stainless steels												
Typical, polished	(*n*)			0.17	0.17	0.19	0.23	0.30				
Typical, cleaned	(*n*)			0.22	0.22	0.24	0.28	0.35				
Typical, lightly oxidized	(*n*)						0.33	0.40				
Typical, highly oxidized	(*n*)						0.67	0.70	0.76			
AISI 347, stably oxidized	(*n*)					0.87	0.88	0.89	0.90			
Tantalum												
Polished	(*h*)								0.11	0.17	0.23	0.28
Tungsten												
Polished	(*h*)							0.10	0.13	0.18	0.25	0.29

Table A.8 Total normal (*n*) or hemispherical (*h*) emissivity of selected surfaces Continued

Nonmetallic substances[b]

DESCRIPTION/COMPOSITION		TEMPERATURE (K)	EMISSIVITY ε
Aluminum oxide	(*n*)	600	0.69
		1000	0.55
		1500	0.41
Asphalt pavement	(*h*)	300	0.85–0.93
Building materials			
Asbestos sheet	(*h*)	300	0.93–0.96
Brick, red	(*h*)	300	0.93–0.96
Gypsum or plaster board	(*h*)	300	0.90–0.92
Wood	(*h*)	300	0.82–0.92
Cloth	(*h*)	300	0.75–0.90
Concrete	(*h*)	300	0.88–0.93
Glass, window	(*h*)	300	0.90–0.95
Ice	(*h*)	273	0.95–0.98
Paints			
Black (Parsons)	(*h*)	300	0.98
White, acrylic	(*h*)	300	0.90
White, zinc oxide	(*h*)	300	0.92
Paper, white	(*h*)	300	0.92–0.97
Pyrex	(*n*)	300	0.82
		600	0.80
		1000	0.71
		1200	0.62
Pyroceram	(*n*)	300	0.85
		600	0.78
		1000	0.69
		1500	0.57
Refractories (furnace liners)			
Alumina brick	(*n*)	800	0.40
		1000	0.33
		1400	0.28
		1600	0.33
Magnesia brick	(*n*)	800	0.45
		1000	0.36
		1400	0.31
		1600	0.40
Kaolin insulating brick	(*n*)	800	0.70
		1200	0.57
		1400	0.47
		1600	0.53
Sand	(*h*)	300	0.90

Table A.8 Continued

Nonmetallic substances Continued

DESCRIPTION/COMPOSITION		TEMPERATURE (K)	EMISSIVITY ε
Silicon carbide	(*n*)	600	0.87
		1000	0.87
		1500	0.85
Skin	(*h*)	300	0.95
Snow	(*h*)	273	0.82–0.90
Soil	(*h*)	300	0.93–0.96
Rocks	(*h*)	300	0.88–0.95
Teflon	(*h*)	300	0.85
		400	0.87
		500	0.92
Vegetation	(*h*)	300	0.92–0.96
Water	(*h*)	300	0.96

[a]Adapted from Reference 1.

[b]Adapted from References 1, 9, 20, and 21.

Table A.9 Solar radiative properties for selected materials

DESCRIPTION/COMPOSITION	α_S	ε^a	α_S/ε	τ_S
Aluminum				
Polished	0.09	0.03	3.0	
Anodized	0.14	0.84	0.17	
Quartz overcoated	0.11	0.37	0.30	
Foil	0.15	0.05	3.0	
Brick, red (Purdue)	0.63	0.93	0.68	
Concrete	0.60	0.88	0.68	
Galvanized sheet metal				
Clean, new	0.65	0.13	5.0	
Oxidized, weathered	0.80	0.28	2.9	
Glass, 3.2 mm thickness				
Float or tempered				0.79
Low iron oxide type				0.88
Metals, plated				
Black sulfide	0.92	0.10	9.2	
Black cobalt oxide	0.93	0.30	3.1	
Black nickel oxide	0.92	0.08	11	
Black chrome	0.87	0.09	9.7	
Mylar, 0.13 mm thickness				0.87
Paints				
Black (Parsons)	0.98	0.98	1.0	
White, acrylic	0.26	0.90	0.29	
White, zinc oxide	0.16	0.93	0.17	
Plexiglas, 3.2 mm thickness				0.90
Snow				
Fine particles, fresh	0.13	0.82	0.16	
Ice granules	0.33	0.89	0.37	
Tedlar, 0.10 mm thickness				0.92
Teflon, 0.13 mm thickness				0.92

[a]The emissivity values in this table correspond to a surface temperature of approximately 300 K.
Adapted with permission from Reference 21.

REFERENCES

1. Touloukian, Y. S., and C. Y. Ho, Eds., *Thermophysical Properties of Matter*, Plenum Press, New York, 1972: Vol. 1, *Thermal Conductivity of Metallic Solids*; Vol. 2, *Thermal Conductivity of Nonmetallic Solids*; Vol. 4, *Specific Heat of Metallic Solids*; Vol. 5, *Specific Heat of Nonmetallic Solids*; Vol. 7, *Thermal Radiative Properties of Metallic Solids*; Vol. 8, *Thermal Radiative Properties of Nonmetallic Solids*; Vol. 9, *Thermal Radiative Properties of Coatings*.
2. Touloukian, Y. S., and C. Y. Ho, Eds., *Thermophysical Properties of Selected Aerospace Materials*, Thermophysical and Electronic Properties Information Analysis Center, CINDAS, Purdue University, West Lafayette, Ind., 1976: Part I, Thermal Radiative Properties; Part II, Thermophysical Properties of Seven Materials.
3. Ho, C. Y., R. W. Powell, and P. E. Liley, "Thermal Conductivity of the Elements: A Comprehensive Review," *J. Phys. Chem. Ref. Data*, 3, Suppl. 1, 1974.
4. Desai, P. D., T. K. Chu, R. H. Bogaard, M. W. Ackermann, and C. Y. Ho, Part I: Thermophysical Properties of Carbon Steels, Part II: Thermophysical Properties of Low Chromium Steels, Part III: Thermophysical Properties of Nickel Steels, Part IV: Thermophysical Properties of Stainless Steels, CINDAS Special Report, Purdue University, West Lafayette, Ind., September 1976.
5. American Society for Metals, *Metals Handbook*, Vol. 1, *Properties and Selection of Metals*, 8th ed., ASM, Metals Park, Ohio, 1961.
6. Hultgren, R., P. D. Desai, D. T. Hawkins, M. Gleiser, K. K. Kelley, and D. D. Wagman, *Selected Values of the Thermodynamic Properties of the Elements*, American Society for Metals, Metals Park, Ohio, 1973.
7. Hultgren, R., P. D. Desai, D. T. Hawkins, M. Gleiser, and K. K. Kelley, *Selected Values of the Thermodynamic Properties of Binary Alloys*, American Society for Metals, Metals Park, Ohio, 1973.
8. American Society of Heating, Refrigerating and Air Conditioning Engineers, *ASHRAE Handbook of Fundamentals*, ASHRAE, New York, 1981.
9. Mallory, J. F., *Thermal Insulation*, Van Nostrand Reinhold, New York, 1969.
10. Hanley, E. J., D. P. DeWitt, and R. E. Taylor, "The Thermal Transport Properties at Normal and Elevated Temperature of Eight Representative Rocks," *Proceedings of the Seventh Symposium on Thermophysical Properties*, American Society of Mechanical Engineers, New York,1977.
11. Sweat, V. E., "A Miniature Thermal Conductivity Probe for Foods," American Society of Mechanical Engineers, Paper 76-HT-60, August 1976.
12. Kothandaraman, C. P., and S. Subramanyan, *Heat and Mass Transfer Data Book*, Halsted Press/Wiley, New York, 1975.
13. Chapman, A. J., *Heat Transfer*, 4th ed., Macmillan, New York, 1984.
14. Vargaftik, N. B., *Tables of Thermophysical Properties of Liquids and Gases*, 2nd ed., Hemisphere Publishing Corp., New York, 1975.
15. Eckert, E. R. G., and R. M. Drake, *Analysis of Heat and Mass Transfer*, McGraw-Hill, New York, 1972.
16. Vukalovich, M. P., A. I. Ivanov, L. R. Fokin, and A. T. Yakovelev, *Thermophysical Properties of Mercury*, State Committee on Standards, State Service for Standards and Handbook Data, Monograph Series No. 9, Izd. Standartov, Moscow, 1971.

17. Bolz, R. E., and G. L. Tuve, Eds., *CRC Handbook of Tables for Applied Engineering Science*, 2nd ed., CRC Press, Boca Raton, Fla., 1979.
18. Liley, P. E., Private communication, School of Mechanical Engineering, Purdue University, West Lafayette, Ind., May 1984.
19. Liley, P. E., Steam Tables in SI Units, Private communication, School of Mechanical Engineering, Purdue University, West Lafayette, Ind., December 1978.
20. Gubareff, G. G., J. E. Janssen, and R. H. Torborg, *Thermal Radiation Properties Survey*, Minneapolis-Honeywell Regulator Co., Minneapolis, Minn., 1960.
21. Kreith, F., and J. F. Kreider, *Principles of Solar Energy*, Hemisphere Publishing Corp., New York, 1978.

Appendix B

MATHEMATICAL RELATIONS AND FUNCTIONS

B.1 HYPERBOLIC FUNCTIONS[a]

x	sinh x	cosh x	tanh x	x	sinh x	cosh x	tanh x
0.00	0.0000	1.0000	0.00000	2.00	3.6269	3.7622	0.96403
0.10	0.1002	1.0050	0.09967	2.10	4.0219	4.1443	0.97045
0.20	0.2013	1.0201	0.19738	2.20	4.4571	4.5679	0.97574
0.30	0.3045	1.0453	0.29131	2.30	4.9370	5.0372	0.98010
0.40	0.4108	1.0811	0.37995	2.40	5.4662	5.5569	0.98367
0.50	0.5211	1.1276	0.46212	2.50	6.0502	6.1323	0.98661
0.60	0.6367	1.1855	0.53705	2.60	6.6947	6.7690	0.98903
0.70	0.7586	1.2552	0.60437	2.70	7.4063	7.4735	0.99101
0.80	0.8881	1.3374	0.66404	2.80	8.1919	8.2527	0.99263
0.90	1.0265	1.4331	0.71630	2.90	9.0596	9.1146	0.99396
1.00	1.1752	1.5431	0.76159	3.00	10.018	10.068	0.99505
1.10	1.3356	1.6685	0.80050	3.50	16.543	16.573	0.99818
1.20	1.5095	1.8107	0.83365	4.00	27.290	27.308	0.99933
1.30	1.6984	1.9709	0.86172	4.50	45.003	45.014	0.99975
1.40	1.9043	2.1509	0.88535	5.00	74.203	74.210	0.99991
1.50	2.1293	2.3524	0.90515	6.00	201.71	201.72	0.99999
1.60	2.3756	2.5775	0.92167	7.00	548.32	548.32	1.0000
1.70	2.6456	2.8283	0.93541	8.00	1490.5	1490.5	1.0000
1.80	2.9422	3.1075	0.94681	9.00	4051.5	4051.5	1.0000
1.90	3.2682	3.4177	0.95624	10.00	11013	11013	1.0000

[a]The hyperbolic functions are defined as

$$\sinh x = \tfrac{1}{2}(e^x - e^{-x}) \qquad \cosh x = \tfrac{1}{2}(e^x + e^{-x}) \qquad \tanh x = \frac{e^x - e^{-x}}{e^x + e^{-x}} = \frac{\sinh x}{\cosh x}$$

The derivatives of the hyperbolic functions of the variable u are given as

$$\frac{d}{dx}(\sinh u) = (\cosh u)\frac{du}{dx} \qquad \frac{d}{dx}(\cosh u) = (\sinh u)\frac{du}{dx} \qquad \frac{d}{dx}(\tanh u) = \left(\frac{1}{\cosh^2 u}\right)\frac{du}{dx}$$

B.2 GAUSSIAN ERROR FUNCTION[a]

w	erf w	w	erf w	w	erf w
0.00	0.00000	0.36	0.38933	1.04	0.85865
0.02	0.02256	0.38	0.40901	1.08	0.87333
0.04	0.04511	0.40	0.42839	1.12	0.88679
0.06	0.06762	0.44	0.46622	1.16	0.89910
0.08	0.09008	0.48	0.50275	1.20	0.91031
0.10	0.11246	0.52	0.53790	1.30	0.93401
0.12	0.13476	0.56	0.57162	1.40	0.95228
0.14	0.15695	0.60	0.60386	1.50	0.96611
0.16	0.17901	0.64	0.63459	1.60	0.97635
0.18	0.20094	0.68	0.66378	1.70	0.98379
0.20	0.22270	0.72	0.69143	1.80	0.98909
0.22	0.24430	0.76	0.71754	1.90	0.99279
0.24	0.26570	0.80	0.74210	2.00	0.99532
0.26	0.28690	0.84	0.76514	2.20	0.99814
0.28	0.30788	0.88	0.78669	2.40	0.99931
0.30	0.32863	0.92	0.80677	2.60	0.99976
0.32	0.34913	0.96	0.82542	2.80	0.99992
0.34	0.36936	1.00	0.84270	3.00	0.99998

[a]The Gaussian error function is defined as

$$\text{erf } w \equiv \frac{2}{\sqrt{\pi}} \int_0^w e^{-v^2}\, dv$$

The complementary error function is defined as

$$\text{erfc } w \equiv 1 - \text{erf } w$$

B.3 THE FIRST FOUR ROOTS OF THE TRANSCENDENTAL EQUATION, $\xi_n \tan \xi_n = Bi$, FOR TRANSIENT CONDUCTION IN A PLANE WALL

$Bi = \frac{hL}{k}$	ξ_1	ξ_2	ξ_3	ξ_4
0	0	3.1416	6.2832	9.4248
0.001	0.0316	3.1419	6.2833	9.4249
0.002	0.0447	3.1422	6.2835	9.4250
0.004	0.0632	3.1429	6.2838	9.4252
0.006	0.0774	3.1435	6.2841	9.4254
0.008	0.0893	3.1441	6.2845	9.4256
0.01	0.0998	3.1448	6.2848	9.4258
0.02	0.1410	3.1479	6.2864	9.4269
0.04	0.1987	3.1543	6.2895	9.4290
0.06	0.2425	3.1606	6.2927	9.4311
0.08	0.2791	3.1668	6.2959	9.4333
0.1	0.3111	3.1731	6.2991	9.4354
0.2	0.4328	3.2039	6.3148	9.4459
0.3	0.5218	3.2341	6.3305	9.4565
0.4	0.5932	3.2636	6.3461	9.4670
0.5	0.6533	3.2923	6.3616	9.4775
0.6	0.7051	3.3204	6.3770	9.4879
0.7	0.7506	3.3477	6.3923	9.4983
0.8	0.7910	3.3744	6.4074	9.5087
0.9	0.8274	3.4003	6.4224	9.5190
1.0	0.8603	3.4256	6.4373	9.5293
1.5	0.9882	3.5422	6.5097	9.5801
2.0	1.0769	3.6436	6.5783	9.6296
3.0	1.1925	3.8088	6.7040	9.7240
4.0	1.2646	3.9352	6.8140	9.8119
5.0	1.3138	4.0336	6.9096	9.8928
6.0	1.3496	4.1116	6.9924	9.9667
7.0	1.3766	4.1746	7.0640	10.0339
8.0	1.3978	4.2264	7.1263	10.0949
9.0	1.4149	4.2694	7.1806	10.1502
10.0	1.4289	4.3058	7.2281	10.2003
15.0	1.4729	4.4255	7.3959	10.3898
20.0	1.4961	4.4915	7.4954	10.5117
30.0	1.5202	4.5615	7.6057	10.6543
40.0	1.5325	4.5979	7.6647	10.7334
50.0	1.5400	4.6202	7.7012	10.7832
60.0	1.5451	4.6353	7.7259	10.8172
80.0	1.5514	4.6543	7.7573	10.8606
100.0	1.5552	4.6658	7.7764	10.8871
∞	1.5708	4.7124	7.8540	10.9956

B.4 BESSEL FUNCTIONS OF THE FIRST KIND

x	$J_0(x)$	$J_1(x)$
0.0	1.0000	0.0000
0.1	0.9975	0.0499
0.2	0.9900	0.0995
0.3	0.9776	0.1483
0.4	0.9604	0.1960
0.5	0.9385	0.2423
0.6	0.9120	0.2867
0.7	0.8812	0.3290
0.8	0.8463	0.3688
0.9	0.8075	0.4059
1.0	0.7652	0.4400
1.1	0.7196	0.4709
1.2	0.6711	0.4983
1.3	0.6201	0.5220
1.4	0.5669	0.5419
1.5	0.5118	0.5579
1.6	0.4554	0.5699
1.7	0.3980	0.5778
1.8	0.3400	0.5815
1.9	0.2818	0.5812
2.0	0.2239	0.5767
2.1	0.1666	0.5683
2.2	0.1104	0.5560
2.3	0.0555	0.5399
2.4	0.0025	0.5202

Appendix C

AN INTEGRAL LAMINAR BOUNDARY LAYER SOLUTION FOR PARALLEL FLOW OVER A FLAT PLATE

An alternative approach to solving the boundary layer equations involves the use of an approximate *integral* method. The approach was originally proposed by von Kármán [1] in 1921 and first applied by Pohlhausen [2]. It is without the mathematical complications inherent in the *exact* (*similarity*) method of Section 7.2.1; yet it can be used to obtain reasonably accurate results for the key boundary layer parameters (δ, δ_t, C_f, and h). Although the method has been used with some success for a variety of flow conditions, we restrict our attention to parallel flow over a flat plate, subject to the same restrictions enumerated in Section 7.2.1; that is, *incompressible laminar flow* with *constant fluid properties* and *negligible viscous dissipation.*

To use the method, the boundary layer equations, (7.3) to (7.5), must be cast in integral form. These forms are obtained by integrating the equations in the y direction across the boundary layer. For example, integrating Equation 7.3, we obtain

$$\int_0^\delta \frac{\partial u}{\partial x} dy + \int_0^\delta \frac{\partial v}{\partial y} dy = 0 \tag{C.1}$$

or, since $v = 0$ at $y = 0$,

$$v(y = \delta) = -\int_0^\delta \frac{\partial u}{\partial x} dy \tag{C.2}$$

Similarly, from Equation 7.4, we obtain

$$\int_0^\delta u \frac{\partial u}{\partial x} dy + \int_0^\delta v \frac{\partial u}{\partial y} dy = \nu \int_0^\delta \frac{\partial}{\partial y}\left(\frac{\partial u}{\partial y}\right) dy$$

or, integrating the second term on the left-hand side by parts,

$$\int_0^\delta u\frac{\partial u}{\partial x}\,dy + uv\Big|_0^\delta - \int_0^\delta u\frac{\partial v}{\partial y}\,dy = \nu\frac{\partial u}{\partial y}\Big|_0^\delta$$

Substituting from Equations 7.3 and C.2 we obtain

$$\int_0^\delta u\frac{\partial u}{\partial x}\,dy - u_\infty\int_0^\delta \frac{\partial u}{\partial x}\,dy + \int_0^\delta u\frac{\partial u}{\partial x}\,dy = -\nu\frac{\partial u}{\partial y}\Big|_{y=0}$$

or

$$u_\infty\int_0^\delta \frac{\partial u}{\partial x}\,dy - \int_0^\delta 2u\frac{\partial u}{\partial x}\,dy = \nu\frac{\partial u}{\partial y}\Big|_{y=0}$$

Therefore

$$\int_0^\delta \frac{\partial}{\partial x}(u_\infty\cdot u - u\cdot u)\,dy = \nu\frac{\partial u}{\partial y}\Big|_{y=0}$$

Rearranging, we then obtain

$$\frac{d}{dx}\left[\int_0^\delta (u_\infty - u)u\,dy\right] = \nu\frac{\partial u}{\partial y}\Big|_{y=0} \tag{C.3}$$

Equation C.3 is the integral form of the boundary layer momentum equation. In a similar fashion the following integral form of the boundary layer energy equation may be obtained.

$$\frac{d}{dx}\left[\int_0^{\delta_t} (T_\infty - T)u\,dy\right] = \alpha\frac{\partial T}{\partial y}\Big|_{y=0} \tag{C.4}$$

Equations C.3 and C.4 satisfy the x-momentum and energy conservation requirements in an *integral* (or *average*) fashion over the entire boundary layer. In contrast, the original conservation equations, (7.4) and (7.5), satisfy the conservation requirements *locally*, that is, at each point in the boundary layer.

The integral equations can be used to obtain *approximate* boundary layer solutions. The procedure involves first *assuming* reasonable functional forms for the unknowns u and T in terms of the corresponding (*unknown*) boundary layer thicknesses. The assumed forms must satisfy appropriate boundary conditions. Substituting these forms into the integral equations, expressions for the boundary layer thicknesses may be determined and the assumed functional forms may then be completely specified. Although this method is approximate, it frequently leads to accurate results for the surface parameters.

Consider the hydrodynamic boundary layer, for which appropriate boundary conditions are

$$u(y=0) = \frac{\partial u}{\partial y}\Big|_{y=\delta} = 0 \qquad \text{and} \qquad u(y=\delta) = u_\infty$$

From Equation 7.4 it also follows that, since $u = v = 0$ at $y = 0$,

$$\left.\frac{\partial^2 u}{\partial y^2}\right|_{y=0} = 0$$

With the foregoing conditions, we can approximate the velocity profile as a third-degree polynomial of the form

$$\frac{u}{u_\infty} = a_1 + a_2\left(\frac{y}{\delta}\right) + a_3\left(\frac{y}{\delta}\right)^2 + a_4\left(\frac{y}{\delta}\right)^3$$

and apply the conditions to determine the coefficients a_1 to a_4. It is easily verified that $a_1 = a_3 = 0$, $a_2 = \frac{3}{2}$ and $a_4 = -\frac{1}{2}$, in which case

$$\frac{u}{u_\infty} = \frac{3}{2}\frac{y}{\delta} - \frac{1}{2}\left(\frac{y}{\delta}\right)^3 \tag{C.5}$$

The velocity profile is then specified in terms of the unknown boundary layer thickness δ. This unknown may be determined by substituting Equation C.5 into C.3 and integrating over y to obtain

$$\frac{d}{dx}\left(\frac{39}{280}u_\infty^2\,\delta\right) = \frac{3}{2}\frac{\nu u_\infty}{\delta}$$

Separating variables and integrating over x, we obtain

$$\frac{\delta^2}{2} = \frac{140}{13}\frac{\nu x}{u_\infty} + \text{constant}$$

However, since $\delta = 0$ at the leading edge of the plate ($x = 0$), the integration constant must be zero and

$$\delta = 4.64\left(\frac{\nu x}{u_\infty}\right)^{1/2} = \frac{4.64x}{Re_x^{1/2}} \tag{C.6}$$

Substituting Equation C.6 into Equation C.5 and evaluating $\tau_s = \mu(\partial u/\partial y)_s$, we also obtain

$$C_{f,x} = \frac{\tau_s}{\rho u_\infty^2/2} = \frac{0.646}{Re_x^{1/2}} \tag{C.7}$$

Despite the approximate nature of the foregoing procedure, Equations C.6 and C.7 compare quite well with results obtained from the exact solution, Equations 7.17 and 7.18.

In a similar fashion one could assume a temperature profile of the form

$$T^* = \frac{T - T_s}{T_\infty - T_s} = b_1 + b_2\left(\frac{y}{\delta_t}\right) + b_3\left(\frac{y}{\delta_t}\right)^2 + b_4\left(\frac{y}{\delta_t}\right)^3$$

and determine the coefficients from the conditions

$$T^*(y=0) = \left.\frac{\partial T^*}{\partial y}\right|_{y=\delta_t} = 0$$

$$T^*(y=\delta_t) = 1$$

as well as

$$\left.\frac{\partial^2 T^*}{\partial y^2}\right|_{y=0} = 0$$

which is inferred from the energy equation (7.5). We then obtain

$$T^* = \frac{3}{2}\frac{y}{\delta_t} - \frac{1}{2}\left(\frac{y}{\delta_t}\right)^3 \tag{C.8}$$

Substituting Equations C.5 and C.8 into Equation C.4, we obtain, after some manipulation and assuming $Pr \gtrsim 1$,

$$\frac{\delta_t}{\delta} = \frac{Pr^{-1/3}}{1.026} \tag{C.9}$$

This result is in good agreement with that obtained from the exact solution, Equation 7.22. Moreover, the heat transfer coefficient may be then computed from

$$h = \frac{-k\partial T/\partial y|_{y=0}}{T_s - T_\infty} = \frac{3}{2}\frac{k}{\delta_t}$$

Substituting from Equations C.6 and C.9, we obtain

$$Nu_x = \frac{hx}{k} = 0.332 Re_x^{1/2} Pr^{1/3} \tag{C.10}$$

This result agrees precisely with that obtained from the exact solution, Equation 7.21.

REFERENCES

1. von Kárman, T., *Z. Angew. Math. Mech.*, *1*, 232, 1921.
2. Pohlhausen, K., *Z. Angew. Math. Mech.*, *1*, 252, 1921.

INDEX

CONVERSION FACTORS

Acceleration	1 m/s^2	$= 4.2520 \times 10^7$ ft/h^2
Area	1 m^2	$= 1550.0$ in.^{2} $= 10.764$ ft^2
Energy	1 J	$= 9.4787 \times 10^{-4}$ Btu
Force	1 N	$= 0.22481$ lb$_f$
Heat transfer rate	1 W	$= 3.4123$ Btu/h
Heat flux	1 W/m^2	$= 0.3171$ Btu/h·ft^2
Heat generation rate	1 W/m^3	$= 0.09665$ Btu/h·ft^3
Heat transfer coefficient	1 W/m^2·K	$= 0.17612$ Btu/h·ft^2·°F
Kinematic viscosity and diffusivities	1 m^2/s	$= 3.875 \times 10^4$ ft^2/h
Latent heat	1 J/kg	$= 4.2995 \times 10^{-4}$ Btu/lb$_m$
Length	1 m	$= 39.370$ in. $= 3.2808$ ft
	1 km	$= 0.62137$ mile
Mass	1 kg	$= 2.2046$ lb$_m$
Mass density	1 kg/m^3	$= 0.062428$ lb$_m$/ft^3
Mass flow rate	1 kg/s	$= 7936.6$ lb$_m$/h
Mass transfer coefficient	1 m/s	$= 1.1811 \times 10^4$ ft/h
Pressure and stress[1]	1 N/m^2	$= 0.020886$ lb$_f$/ft^2 $= 1.4504 \times 10^{-4}$ lb$_f$/in.^{2} $= 4.015 \times 10^{-3}$ in. water $= 2.953 \times 10^{-4}$ in. Hg
	1.0133×10^5 N/m^2	$= 1$ standard atmosphere
	1×10^5 N/m^2	$= 1$ bar
Specific heat	1 J/kg·K	$= 2.3886 \times 10^{-4}$ Btu/lb$_m$·°F
Temperature	K	$= (5/9)$°R $= (5/9)$(°F $+ 459.67$) $=$ °C $+ 273.15$
Temperature difference	1 K	$= 1$°C $= (9/5)$°R $= (9/5)$°F
Thermal conductivity	1 W/m·K	$= 0.57782$ Btu/h·ft·°F
Thermal resistance	1 K/W	$= 0.52750$ °F/h·Btu
Viscosity (dynamic)[2]	1 N·s/m^2	$= 2419.1$ lb$_m$/ft·h $= 5.8016 \times 10^{-6}$ lb$_f$·h/ft^2
Volume	1 m^3	$= 6.1023 \times 10^4$ in.^{3} $= 35.314$ ft^3 $= 264.17$ gal
Volume flow rate	1 m^3/s	$= 1.2713 \times 10^5$ ft^3/h $= 2.1189 \times 10^3$ ft^3/min $= 1.5850 \times 10^4$ gal/min

[1] The SI name for the quantity pressure is pascal (Pa) having units N/m^2 or kg/m·s^2.

[2] Also expressed in equivalent units of kg/s·m.